Quantum Mechanics

Classical Results, Modern Systems, and
Visualized Examples

Quantum Mechanics

Classical Results, Modern Systems, and Visualized Examples

Second Edition

Richard W. Robinett

Pennsylvania State University

OXFORD
UNIVERSITY PRESS

OXFORD
UNIVERSITY PRESS

Great Clarendon Street, Oxford OX2 6DP

Oxford University Press is a department of the University of Oxford.
It furthers the University's objective of excellence in research, scholarship,
and education by publishing worldwide in

Oxford New York

Auckland Cape Town Dar es Salaam Hong Kong Karachi
Kuala Lumpur Madrid Melbourne Mexico City Nairobi
New Delhi Shanghai Taipei Toronto

With offices in

Argentina Austria Brazil Chile Czech Republic France Greece
Guatemala Hungary Italy Japan Poland Portugal Singapore
South Korea Switzerland Thailand Turkey Ukraine Vietnam

Oxford is a registered trade mark of Oxford University Press
in the UK and in certain other countries

Published in the United States
by Oxford University Press Inc., New York

© Oxford University Press 2006

British Library Cataloguing in Publication Data

Data available

Library of Congress Cataloging in Publication Data

Robinett, Richard W. (Richard Wallace)
　　Quantum mechanics : classical results, modern systems, and
　　　visualized examples / Richard W. Robinett.—2nd ed.
　　　　p. cm.
　　ISBN-13: 978–0–19–853097–8 (alk. paper)
　　ISBN-10: 0–19–853097–8 (alk. paper)
　1. Quantum theory.　I. Title.
　QC174.12.R6 2006
　　530.12—dc22　　　　　　　　　　　　　2006000424

Typeset by Newgen Imaging Systems (P) Ltd., Chennai, India

ISBN 978–0–19–853097–8

10 9

Preface to the Second Edition

One of the hallmarks of science is the continual quest to refine and expand one's understanding and vision of the universe, seeking not only new answers to old questions, but also proactively searching out new avenues of inquiry based on past experience. In much the same way, teachers of science (including textbook authors) can and should explore the pedagogy of their disciplines in a scientific way, maintaining and streamlining what has been documented to work, but also improving, updating, and expanding their educational materials in response to new knowledge in their fields, in basic, applied, and educational research. For that reason, I am very pleased to have been given the opportunity to produce a Second Edition of this textbook on quantum mechanics at the advanced undergraduate level.

The First Edition of *Quantum Mechanics* had a number of novel features, so it may be useful to first review some aspects of that work, in the context of this Second Edition. The descriptive subtitle of the text, *Classical Results, Modern Systems, and Visualized Examples*, was, and still is, intended to suggest a number of the inter-related approaches to the teaching and learning of quantum mechanics which have been adopted here.

- Many of the expected familiar topics and examples (the *Classical Results*) found in standard quantum texts are indeed present in both editions, but we also continue to focus extensively on the classical–quantum connection as one of the best ways to help students learn the subject. Topics such as momentum-space probability distributions, time-dependent wave packet solutions, and the correspondence principle limit of large quantum numbers can all help students use their existing intuition to make contact with new quantum ideas; classical wave physics continues to be emphasized as well, with its own separate chapter, for the same reason. Additional examples of quantum wave packet solutions have been included in this new Edition, as well as a self-contained discussion of the Wigner quasi-probability (phase-space) distribution, designed to help make contact with related ideas in statistical mechanics, classical mechanics, and even quantum optics.

- An even larger number of examples of the application of quantum mechanics to *Modern Systems* is provided, including discussions of experimental realizations of quantum phenomena which have only appeared since the First Edition. Advances in such areas as materials science and laser trapping/cooling have meant a large number of quantum systems which have historically been only considered as "textbook" examples have become physically realizable. For example, the "quantum bouncer", once discussed only in pedagogical journals, has been explored experimentally in the *Quantum states of neutrons in the Earth's gravitational field.*[1] The production of atomic wave packets which exhibit the classical periodicity of Keplerian orbits[2] is another example of a *Classical Result* which has become a *Modern System.*

The ability to manipulate nature at the extremes of small distance (nano- and even atomic-level) and low temperatures (as with Bose–Einstein condensates) implies that a knowledge of quantum mechanics is increasingly important in modern physical science, and a number of new discussions of applications have been added to both the text and to the *Problems*, including ones on such topics as expanding/interfering Bose–Einstein condensates, the quantum Hall effect, and quantum wave packet revivals, all in the context of familiar textbook level examples.

- We continue to emphasize the use of *Visualized Examples* (with 200 figures included) to reinforce students' conceptual understanding of the basic ideas and to enhance their mathematical facility in solving problems. This includes not only pictorial representations of stationary state wavefunctions and time-dependent wave packets, but also real data. The graphical representation of such information often provides the map of the meeting ground of the sometimes arcane formalism of a theorist, the observations of an experimentalist, and the rest of the scientific community; the ability to "follow such maps" is an important part of a physics education.

Motivated in this Edition (even more than before) by results appearing from Physics Education Research (PER), we still stress concepts which PER studies have indicated can pose difficulties for many students, such as notions of probability, reading potential energy diagrams, and the time-development of eigenstates and wave packets.

As with any textbook revision, the opportunity to streamline the presentation and pedagogy, based on feedback from actual classroom use, is one of the most important aspects of a new Edition, and I have taken this opportunity to remove some topics (moving them, however, to an accompanying Web site) and adding new ones. New sections on *The Wigner Quasi-Probability Distribution* (and many related problems), an *Infinite Array of δ-functions: Periodic Potentials and the Dirac Comb, Time-Dependent Perturbation Theory,* and *Timescales in Bound State*

---

[1] The title of a paper by V. V. Nesvizhevsky *et al.* (2002). *Nature* 415, 297.
[2] See Yeazell *et al.* (1989).

Systems: Classical Period and Quantum Revival Times reflect suggestions from various sources on (hopefully) useful new additions. A number of new in-text *Examples* and end-of-chapter *Problems* have been added for similar reasons, as well as an expanded set of *Appendices*, on dimensions and mathematical methods.

An exciting new feature of the Second Edition is the development of a Web site[3] in support of the textbook, for use by both students and instructors, linked from the Oxford University Press[4] web page for this text. Students will find many additional (extended) homework problems in the form of *Worksheets* on both formal and applied topics, such as "slow light", femtosecond chemistry, and quantum wave packet revivals. Additional material in the form of *Supplementary Chapters* on such topics as neutrino oscillations, quantum Monte Carlo approximation methods, supersymmetry in quantum mechanics, periodic orbit theory of quantum billiards, and quantum chaos are available.

For instructors, copies of a complete *Solutions Manual* for the textbook, as well as *Worksheet Solutions*, will be provided on a more secure portion of the site, in addition to copies of the *Transparencies* for the figures in the text. An 85-page *Guide to the Pedagogical Literature on Quantum Mechanics* is also available there, surveying articles from The American Journal of Physics, The European Journal of Physics, and The Journal of Chemical Education from their earliest issues, through the publication date of this text (with periodic updates planned.) In addition, a quantum mechanics assessment test (the so-called *Quantum Mechanics Visualization Instrument* or QMVI) is available at the Instructors site, along with detailed information on its development and sample results from earlier educational studies. Given my long-term interest in the science, as well as the pedagogy, of quantum mechanics, I trust that this site will continually grow in both size and coverage as new and updated materials are added. Information on accessing the Instructors area of the Web site is available through the publisher at the Oxford University Press Web site describing this text.

I am very grateful to all those from whom I have had help in learning quantum mechanics over the years, including faculty and fellow students in my undergraduate, graduate, and postdoctoral days, current faculty colleagues (here at Penn State and elsewhere), my own undergraduate students over the years, and numerous authors of textbooks and both research and pedagogical articles, many of whom I have never met, but to whom I owe much. I would like to thank all those who helped very directly in the production of the Second Edition of this text, specifically including those who provided useful suggestions for improvement or who found corrections, namely, J. Banavar, A. Bernacchi, B. Chasan,

---

[3] See robinett.phys.psu.edu/qm
[4] See www.oup.co.uk

J. Edmonds, M. Cole, C. Patton, and J. Yeazell. I have truly enjoyed recent collaborations with both M. Belloni and M. A. Doncheski on pedagogical issues related to quantum theory, and some of our recent work has found its way into the Second Edition (including the cover) and I thank them for their insights, and patience.

No work done in a professional context can be separated from one's personal life (nor should it be) and so I want to thank my family for all of their help and understanding over my entire career, including during the production of this new Edition. The First Edition of this text was thoroughly proof-read by my mother-in-law (Nancy Malone) who graciously tried to teach me the proper use of the English language; her recent passing has saddened us all. My own mother (Betty Robinett) has been, and continues to be, the single most important role model in my life—both personal and professional—and I am deeply indebted to her far more than I can ever convey. Finally, to my wife (Sarah) and children (James and Katherine), I give thanks everyday for the richness and joy they bring to my life.

Richard Robinett
December, 2005
State College, PA

Contents

PART I

The Quantum Paradigm

ONE
A First Look at Quantum Physics

1.1 How This Book Approaches Quantum Mechanics

It can easily be argued that a fully mature and complete knowledge of quantum mechanics should include historical, axiomatic, formal mathematical, and even philosophical background to the subject. However, for a student approaching quantum theory for the first time in a serious way, it can be the case that an approach utilizing his or her existing knowledge of, and intuition for, classical physics (including mechanics, wave physics, and electricity and magnetism) as well as emphasizing connections to experimental results can be the most productive. That, at least, is the point of view adopted in this text and can be illustrated by a focus on the following general topics:

(1) The incorporation of a wave property description of matter into a consistent wave equation, via the Schrödinger equation;

(2) The statistical interpretation of the Schrödinger wavefunction in terms of a probability density (in both position- and momentum-space);

(3) The study of single-particle solutions of the Schrödinger equation, for both time-independent energy eigenstates as well as time-dependent systems, for many model systems, in a variety of spatial dimensions, and finally;

(4) The influence of both quantum mechanical effects and the constraints arising from the indistinguishability of particles (and how that depends on their spin) on the properties of multiparticle systems, and the resulting implications for the structure of different forms of matter.

By way of example of our approach, we first note that Fig. 1.1 illustrates an example of a precision measurement of the wave properties of ultracold neutrons, exhibiting a Fresnel diffraction pattern arising from scattering from

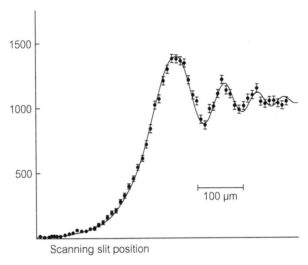

Figure 1.1. Fresnel diffraction pattern obtained from scattering at a sharp edge, obtained using ultracold neutrons by Gähler and Zeilinger (1991).

a sharp edge, nicely explained by classical optical analogies. We devote Chapter 2 to a discussion of classical wave physics and Chapter 3 to the description of such wave effects for material particles, via the Schrödinger equation. Figure 1.2 demonstrates an interference pattern using electron beams, built up "electron by electron," with the obvious fringes resulting only from a large number of individual measurements. The important statistical aspect of quantum mechanics, simply illustrated by this experiment, is discussed in Chapter 4 and beyond.

It can be argued that much of the early success of quantum theory can be traced to the fact that many exactly soluble quantum models are surprisingly coincident with naturally occurring physical systems, such as the hydrogen atom and the rotational/vibrational states of molecules and such systems are, of course, discussed here. The standing wave patterns obtained from scanning tunneling microscopy of "electron waves" in a circular corral geometry constructed from arrays of iron atoms on a copper surface, seen in Fig. 1.3, reminds us of the continuing progress in such areas as materials science and atom trapping in developing artificial systems (and devices) for which quantum mechanics is applicable. In that context, many exemplary quantum mechanical models, which have historically been considered as only textbook idealizations, have also recently found experimental realizations. Examples include "designer" potential wells approximating square and parabolic shapes made using molecular beam techniques, as well as magnetic or optical traps. The solution of the Schrödinger equation, in a wide variety of standard (and not-so-standard) one-, two-, and three-dimensional applications, is therefore emphasized here, in Chapters 5, 8, 9,

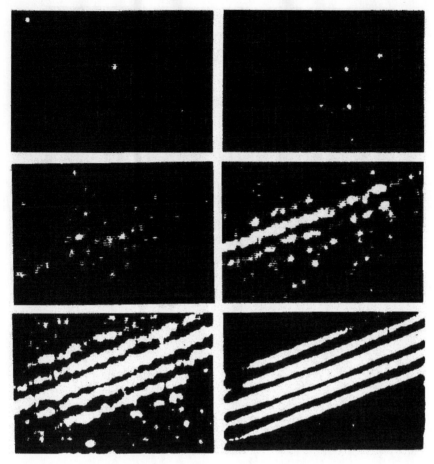

Figure 1.2. Interference patterns obtained by using an electron microscope showing the fringes being "built up" from an increasingly large number of measurements of individual events. From Merli, Missiroli, and Pozzi (1976). (Photo reproduced by permission of the American Institute of Physics.)

and 15–17. In parallel to these examples, more formal aspects of quantum theory are outlined in Chapters 7, 10, 12, 13, and 14.

The *quantum* in quantum mechanics is often associated with the discrete energy levels observed in bound-state systems, most famously for atomic systems such as the hydrogen atom, which we discuss in Chapter 17, emphasizing that this is the quantum version of the classical Kepler problem. We also show, in Fig. 1.4, experimental measurements leading to a map of the momentum-space probability density for the $1S$ state of hydrogen and the emphasis on momentum-space methods suggested by this result is stressed throughout the text. The influence of additional "real-life" effects, such as gravity and electromagnetism, on atomic and other systems are then discussed in Chapter 18. We note that the data in Fig. 1.4

Figure 1.3. Standing wave patterns obtained using scanning tunneling microscopy from a circular "corral" of radius ∼70 Å, constructed from 48 iron atoms on a copper surface. (Photo courtesy of IBM Almaden.)

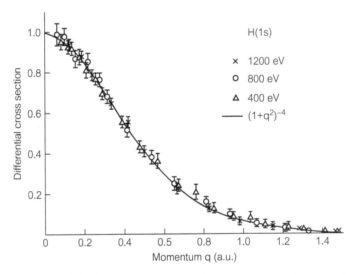

Figure 1.4. Electron probability density obtained by scattering with three different energy probes, compared with the theoretically calculated momentum-space probability density for the hydrogen-atom ground state, from Lohmann and Weigold (1981). The data are plotted again the scaled momentum in atomic units (a.u.), $q = a_0 p/\hbar$.

was obtained via scattering processes, and the importance of scattering methods in quantum mechanics is emphasized in both one-dimension (Chapter 11) and three-dimensions (Chapter 19). The fact that spin-1/2 particles must satisfy the Pauli principle has profound implications for the way that matter can arrange

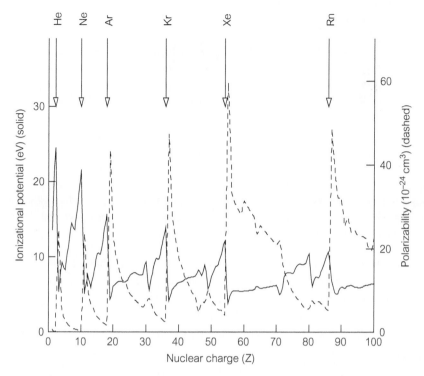

Figure 1.5. Plots of the ionization energy (solid) and atomic polarizability (dashed) versus nuclear charge, showing the shell structure characterized by the noble gas atoms, arising from the filling of atomic energy levels as mandated by the Pauli principle for spin-1/2 electrons.

itself, as shown in the highly correlated values of physical parameters shown in Fig. 1.5 for atoms of increasing size and complexity. While it is illustrated here in a numerical way, this should also be reminiscent of the familiar periodic table of the elements, and the Pauli principle has similar implications for nuclear structure. We discuss the role of spin in multiparticle systems described by quantum mechanics in Chapters 7, 14, and 17.

We remind the reader that similar dramatic manifestations of quantum phenomena (including all of the effects mentioned above) are still being discovered, as illustrated in Fig. 1.6. In a justly famous experiment,[1] two highly localized and well-separated samples of sodium atoms are cooled to sufficiently low temperatures so that they are in the ground states of their respective potential wells (produced by laser trapping.) The trapping potential is removed and the resulting coherent Bose–Einstein condensates are allowed to expand and overlap, exhibiting the quantum interference shown in Fig. 1.6 (the solid curve, showing

[1] From the paper entitled *Observation of interference between two Bose condensates* by Andrews *et al.* (1997).

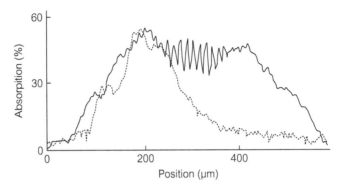

Figure 1.6. Data (from Andrews *et al.* (1997)) illustrating the interference of two Bose condensates as they expand and overlap (solid curve), compared to a single expanding Bose condensate (dotted curve).

regular absorption variations across the central overlap region), while no such interference is observed for a single expanding quantum sample (dotted data.) Many of the salient features of this experiment can be understood using relatively simple ideas outlined in Chapters 3, 4, and 9.

The ability to use the concepts and mathematical techniques of quantum mechanics to confront the wide array of experimental realizations that have come to characterize modern physical science will be one of the focuses of this text. Before proceeding, however, we reserve the remainder of this chapter for brief reviews of some of the essential aspects of both relativity and standard results from quantum theory.

1.2 **Essential Relativity**

While we will consider nonrelativistic quantum mechanics almost exclusively, it is useful to briefly review some of the rudiments of special relativity and the fundamental role played by the *speed of light, c*.

For a free particle of rest mass m moving at speed v, the total energy (E), momentum (p), and kinetic energy (T) can be written in the relativistically correct forms

$$E = \gamma mc^2, \quad p = \gamma mv, \quad \text{and} \quad T \equiv E - mc^2 = (\gamma - 1)mc^2 \tag{1.1}$$

where

$$\gamma \equiv \frac{1}{\sqrt{1 - v^2/c^2}} = \left(1 - \frac{v^2}{c^2}\right)^{-1/2} \tag{1.2}$$

The nonrelativistic limit corresponds to $v/c << 1$, in which case we can use the series expansion

$$(1+x)^n = 1 + nx + \frac{n(n-1)}{2!}x^3 + \frac{n(n-1)(n-2)}{3!}x^3 + \cdots \tag{1.3}$$

for $x = v^2/c^2$ small to show that

$$p \approx mv \quad \text{and} \quad T \approx \left(1 + \frac{1}{2}\frac{v^2}{c^2} + \cdots - 1\right)mc^2 \approx \frac{1}{2}mv^2 \tag{1.4}$$

which are the familiar nonrelativistic results for motion at speeds slow compared to the speed of light.

In quantum mechanics the momentum is a more natural variable than v, and a useful relation can be obtained from Eqn. (1.1), namely

$$E^2 = (pc)^2 + (mc^2)^2 \tag{1.5}$$

This form stresses the fact that E, pc, and mc^2 all have the same dimensions (namely energy), and we will often use these forms when convenient. As an example, the rest energies of various atomic particles will often be quoted in energy units; for the electron and proton we have

$$m_e c^2 = 0.511 \text{ MeV} \quad \text{and} \quad m_p c^2 = 938.3 \text{ MeV} \tag{1.6}$$

Recall that the *electron volt* or *eV* is defined by

$$1 \ eV = \text{the energy gained by a fundamental charge } e$$
$$\text{which has been accelerated through 1 V}$$
$$= (1.6 \times 10^{-19} \text{ C})(1 \ V) = 1.6 \times 10^{-19} \text{ J} \tag{1.7}$$

Atomic "masses" are often quoted in *unified atomic mass units* (formerly *amu*) which are given by $1 \ u = 931.5$ MeV.

The nonrelativistic limit of Eqn. (1.5), where $pc << mc^2$, is easily seen to be

$$E = mc^2 \left(1 + \left(\frac{pc}{mc^2}\right)^2\right)^{1/2} = mc^2 + \frac{p^2}{2m} - \frac{p^4}{8m^3 c^2} + \cdots \tag{1.8}$$

Since the rest energy is "just along for the ride" in most of the problems we consider, we will ignore its contribution to the total energy; thus a phrase such as "... a 2 *eV* electron ..." should be taken to mean that the electron has a kinetic energy $T = E - mc^2 \approx p^2/2m \approx 2$ eV. We will often write $pc = \sqrt{2(mc^2)T}$ in this limit.

At the other extreme, in the ultrarelativistic limit when $E \gg mc^2$ (or $v \lesssim c$), we can write

$$E = pc \left(1 + \left(\frac{mc^2}{pc} \right)^2 \right)^{1/2} \approx pc + \frac{1}{2} \frac{(mc^2)^2}{pc} + \cdots \tag{1.9}$$

which is also seen to be consistent with the energy–momentum relation for truly massless particles (such as photons), namely $E = pc$.

We list below several typical quantum mechanical systems and the order-of-magnitudes of the energies involved:

- **Electrons in atoms:** For the inner shell electrons of an atom with nuclear charge $+Ze$, the kinetic energy is of order $T \approx Z^2$ 13.6 eV. We can say, somewhat arbitrarily, that relativistic effects become nonnegligible when $T \gtrsim 0.05 \, mc^2$ (i.e. a 5% effect). This condition is satisfied when $Z \gtrsim 43$, implying that the effects of relativity must certainly be considered for heavy atoms.

- **Deuteron:** The simplest nuclear system is the bound state of a proton and neutron where the typical kinetic energies are $T \approx 2$ MeV; this is to be compared with $m_p c^2 \approx m_n c^2 \approx 939$ MeV so that the deuteron can be considered as a nonrelativistic system to first approximation.

- **Quarks in the proton and pion:** The constituent quark model of elementary particles postulates that three quarks of effective mass roughly $m_q c^2 \approx 350$ MeV form the proton; this implies binding energies and kinetic energies of the order of $1 - 10$ MeV which is consistent with "nonrelativity." The pion, on the other hand, is considered a bound state of two such quarks, but has rest energy $m_\pi c^2 \approx 140$ MeV, so that binding energies (and hence kinetic energies) of order several hundred MeV are required and relativistic effects dominate.[2]

- **Compact objects in astrophysics:** The electrons in white dwarf stars and neutrons in neutron stars have kinetic energies $T_e \approx 0.08$ MeV and $T_n \approx 140$ MeV respectively, so these objects are "barely" nonrelativistic.

1.3 **Quantum Physics: $\hbar$ as a Fundamental Constant**

Just as the speed of light, c, sets the scale for when relativistic effects are important, quantum physics also has an associated fundamental, dimensionful parameter,

---

[2] The pion is really a quark–antiquark system. Bound states of heavier quarks and antiquarks, which are more slowly moving, can be more successfully described using nonrelativistic quantum mechanics.

namely *Planck's constant*. Its first applications came in the understanding of some of the quantum aspects of the electromagnetic (EM) field and the particle nature of EM radiation.

- In his investigations of the black body spectrum emitted from heated objects (so-called cavity radiation), Planck found that he could only fit the observed intensity distribution if he made the (then radical) assumption that the EM energy of a given frequency f was quantized and given by

$$E_n(f) = nhf \quad \text{where} \quad n = 0, 1, 2, 3 \ldots \tag{1.10}$$

The constant of proportionality, h, was derived from a "fit" to the experimental data, and has been found to be

$$h = 6.626 \times 10^{-34} \text{ J} \cdot \text{s} \tag{1.11}$$

and is called Planck's constant; we will far more often use the related form

$$\hbar \equiv \frac{h}{2\pi} = 1.054 \times 10^{-34} \text{ J} \cdot \text{s} = 6.582 \times 10^{-16} \text{ eV} \cdot \text{s} = 6.582 \times 10^{-22} \text{ MeV} \cdot \text{s} \tag{1.12}$$

which is to be read as "h-bar".

- Einstein assumed the energy quantization of Eqn. (1.10) was a more general characteristic of light, and proposed that EM radiation was composed of *photons*[3] or "bundles" of discrete energy $E_\gamma = hf$. He used the photon concept to explain the *photoelectric effect*, and predicted that the kinetic energy of electrons emitted from the surface of metals after being irradiated should be given by

$$\tfrac{1}{2} m v_{\text{max}}^2 = E_\gamma - W = hf - W \tag{1.13}$$

where W is called the *work function* of the metal in question. Subsequent experiments were able to confirm this relation, as well as providing another, complementary measurement of h (P1.5) which agreed with the value obtained by Planck.

- The relativistic connection between energy and momentum for a massless particle such as the photon could be used to show that it has a momentum given by

$$p_\gamma c = E_\gamma = hf = \frac{hc}{\lambda} \quad \text{or} \quad p_\gamma = \frac{h}{\lambda} \tag{1.14}$$

---

[3] We use the notation γ (for gamma ray) to indicate a property corresponding to a photon of any energy or frequency.

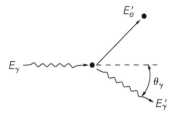

Figure 1.7. Geometry for Compton scattering. The incident photon scatters from an electron, initially at rest, at an angle θ_γ.

where λ is the wavelength. Arthur Compton noted that the scattering of X-rays by free electrons at rest could be considered as a collision process where the incident photon has an energy and momentum given by Eqn. (1.14), as shown in Fig. 1.7. Conservation of energy and momentum (P1.6) can then be applied to show that the wavelength of the scattered photon, λ', is given by the *Compton scattering formula*

$$\lambda' - \lambda = \frac{h}{m_e c}(1 - \cos(\theta_\gamma)) \tag{1.15}$$

where θ_γ is the angle between the incident and scattered photon directions; X-ray scattering experiments confirmed the validity of Eqn. (1.15).

The connection of Planck's constant to the properties of material particles, such as electrons, came later:

• Using yet another experimental "fit" to spectroscopic data, in this case the Balmer–Ritz formula for the frequencies in the spectrum for hydrogen, Bohr used semiclassical arguments to deduce that the angular momentum of the electron was quantized as

$$L = n\frac{h}{2\pi} = n\hbar \quad \text{with } n = 1, 2, 3 \ldots \tag{1.16}$$

• Motivated by the dual wave-particle nature exhibited by light, for example, in Compton scattering, de Broglie suggested that matter, specifically electrons, would exhibit wave properties. He postulated that the relation

$$\lambda_{\text{dB}} = \frac{h}{p} \tag{1.17}$$

apply to material particles as well as to photons, thereby defining the *de Broglie wavelength*. He could show that Eqn. (1.17) reproduced the Bohr condition of Eqn. (1.16), and thus explain the hydrogen atom spectrum.

Example 1.1. de Broglie Wavelength of a Truck?

Over the roughly 80 years since the Davisson–Germer experiment[4] directly demonstrated the wave nature of electrons by the observation of the diffraction of electron beams from nickel crystals, with a wavelength consistent[5] with Eqn. (1.17), the quantum mechanical *wave-particle duality* of objects of increasing size and complexity has been observed.

Only 3 years after the prediction by de Broglie, Davisson and Germer accelerated electrons through voltages of order $\mathcal{V} \sim 50$ V to speeds given by

$$\frac{1}{2}m_e v^2 = e\mathcal{V} \longrightarrow v = \sqrt{\frac{2e\mathcal{V}}{m_e c^2}}\, c = \sqrt{\frac{100 \text{ eV}}{0.51 \text{ MeV}}}\, c \approx 0.015\, c \qquad (1.18)$$

which is still nonrelativistic and gives a de Broglie wavelength of $\lambda = h/mv \approx 1.7$ Å, which nicely matched the atomic spacings in their sample (already determined by X-ray scattering experiments).

It is sometimes useful to compare the quantum mechanical wavelength of a particle to other physical dimensions, including its own size. While many particles which play a crucial role in determining the structure of matter have finite and measurable sizes, all ultrahigh energy scattering experiments involving electrons (which therefore probe ultra-small distance scales) are so far consistent with the electrons having no internal structure; various experiments can be interpreted as putting upper limits on an electron "size" of order 10^{-10} Å $= 10^{-5}$ F or roughly 50, 000 times smaller than a proton or neutron. This justifies the assumption of a "point-like" electron.

Sixty years after the Davisson and Germer experiments with electrons, single- and double-slit diffraction of slow neutrons was observed, giving the "*most precise realization hitherto for matter waves.*"[6] In this case, the neutrons have a physical size measured (in other experiments) to be of order 1 F $= 10^{-5}$ Å and ultracold neutrons with $\lambda = 15 - 30$ Å were utilized, so that the spatial extent of the particle is still orders-of-magnitude less than its quantum mechanical wave length. In the last decade or so, however, advances in atom interferometry have led to the observation of interference or diffraction phenomena for small atoms (helium, He), larger atoms (atomic sodium, Na), diatomic molecules (sodium dimer or Na_2), small clusters of molecules (of H, He_2, and D_2), and most recently C_{60} molecules (buckeyballs), all of atomic dimensions, and with increasingly small de Broglie wavelengths. Representative data (and references) are collected below.

---

[4] See Davisson and Germer (1927).
[5] Their exact words are "*The equivalent wave-lengths of the electron beams may be calculated from the diffraction data in the usual way. These turn out to be in acceptable agreement with the values of h/mv of the undulatory mechanics.*"
[6] Zeilinger *et al.* (1988).

(Continued)

Particle	Mass (atomic units)	approx. size (diameter)	λ_{dB} (in Å)	Reference
electron (e⁻)	5×10^{-4}	$<10^{-10}$Å	1–2	Davisson/Germer (1927)
neutron (n)	1	$\sim 10^{-5}$ Å	15–30	Zeilinger *et al.* (1988)
helium (He)	4	~ 1 Å	0.5–3	Carnal *et al.* (1991)
sodium (Na)	23	~ 4 Å	0.16	Keith *et al.* (1991)
helium (He$_2$)	8	~ 50 Å	1–2	Schollkopf *et al.* (1994)
sodium (Na$_2$)	$2 \cdot 23 = 46$	~ 8 Å	0.1	Chapman *et al.* (1995)
buckeyballs (C$_{60}$)	$60 \cdot 12 = 720$	7 Å	0.025	Arndt *et al.* (1999)

It is clear that objects of an increasing classical nature (like C$_{60}$, with a large number of internal degrees of freedom involved in the many bonds) are seen to exhibit quantum mechanical behavior. The possibility that the quantum mechanical wavelength of an object can be much smaller than its physical size (hence the question in the title of this Example) has been amply demonstrated.

The de Broglie relation contains the seeds of the *position–momentum uncertainty principle*, namely

$$\Delta x \, \Delta p \geq \frac{\hbar}{2} \tag{1.19}$$

where Δx and Δp are the uncertainties in a measurement of x and p respectively. Equation (1.19) puts fundamental limitations on one's ability to simultaneously measure the position and momentum of a particle; it also leads to the notion of *zero-point energy*, a minimum unavoidable energy of a particle confined to a localized region of space.

The example of a particle in a one-dimensional box illustrates this most simply. A particle of mass m confined to a one-dimensional box of length L will satisfy the "standing wave" condition for de Broglie waves if $n(\lambda_n/2) = L$ (compare this to Eqn. (1.35) below and explain any differences) with $n = 1, 2, 3 \ldots$ This corresponds to quantized momenta $p_n = n\hbar\pi/L$ and energies given by

$$E_n = \frac{p_n^2}{2m} = \frac{n^2 \hbar^2 \pi^2}{2mL^2} \quad \text{where } n = 1, 2, 3 \ldots \tag{1.20}$$

In contrast to the classical case, the particle cannot just "sit quietly in the box," but has a minimum energy. More generally, a particle localized to a region of spatial extent $\Delta x \sim L$ will have a corresponding uncertainty in momentum of order $p_{min} \sim \Delta p \sim \hbar/L$ or minimum kinetic energy

$$T_{min} \propto \frac{p_{min}^2}{2m} \propto \frac{\hbar^2}{mL^2} \tag{1.21}$$

While such "back-of-the-envelope" calculations should be used with care, they can often provide insight into the ground state of a quantum system.

- **Electrons in atoms?** The electrons in atoms and molecules are confined to a region of size $\Delta x \sim 1$ Å. The corresponding unavoidable spread in momentum is therefore roughly

$$\Delta pc \sim pc \sim \frac{\hbar c}{\Delta x} \approx \frac{2000 \text{ eV Å}}{1 \text{ Å}} \approx 2 \text{ keV} \tag{1.22}$$

Since this is much smaller than the electron rest energy, $mc^2 \approx 0.5$ MeV, the zero-point energy can be treated nonrelativistically and is roughly

$$E_0^{(e)} \approx \frac{p^2}{2m} = \frac{(pc)^2}{2mc^2} \approx \frac{(2000 \text{ eV})^2}{2(0.5 \times 10^6 \text{ eV})} \approx 4 \text{ eV} \tag{1.23}$$

which is, of course, exactly the order-of-magnitude for the hydrogen atom and other atomic systems.

- **Photons in atoms?** On the other hand, the photons emitted in the radiative decays of such atoms, cannot have been "stored" in the atom beforehand. To see this, we note that a photon "bouncing around" in an atomic-sized box will have the same $\Delta p \sim p$ as in Eqn. (1.22). Because massless photons are necessarily relativistic, the corresponding kinetic energy is then given by

$$E_0^{(\gamma)} \approx pc \approx 2000 \text{ eV} \tag{1.24}$$

which is much larger than the 1–10 eV observed in typical transitions.

- **Alpha particles in nuclei?** Radioactive nuclei emit α particles ($m_\alpha \approx 4m_p$) with kinetic energies of a few MeV. The minimum momentum in a heavy nucleus of radius $R \approx 5$ F is roughly

$$pc \approx \frac{\hbar c}{R} \approx \frac{200 \text{ MeV F}}{5 \text{ F}} \approx 40 \text{ MeV} \tag{1.25}$$

which corresponds to a (nonrelativistic) zero-point energy of

$$E_0^{(\alpha)} \approx \frac{(pc)^2}{2m_\alpha c^2} \approx \frac{(40 \text{ MeV})^2}{8 \times 940 \text{ MeV}} \approx 0.2 \text{ MeV} \tag{1.26}$$

which is consistent with observations.

- **Electrons from neutron β decay?** Neutrons decay via the process $n \rightarrow pe\bar{\nu}_e$ where the electron (anti-)neutrino is often not directly detected, but the emitted electron of a few MeV is easily observed. A "pre-existing" electron in the neutron, "waiting around" to decay, would have a value of pc roughly five times *larger* than in Eqn. (1.25) (since the neutron is roughly five times *smaller* than a heavy nucleus) implying relativistic electrons with kinetic energies of order $E_0^{(e)} \approx pc \approx 200$ MeV; the decay electrons must therefore be created *ex nihilo* at the time of decay. Arguments such as these were among the first pieces of evidence used to predict that a new force beyond those known classically, the so-called *weak interaction*, was required to explain such decays.

For single particles, it is often clear when wave mechanical effects are important. For example, in electron scattering from crystal planes, the Bragg condition for constructive interference can be written in the form $n\lambda = D\sin(\phi)$ where D is the interatomic spacing and ϕ is the scattering angle; clearly λ must be comparable to the other spatial dimensions in the problem for the wave properties of matter to be visible. Many problems have some other natural length scale against which to compare the de Broglie wavelength.

Example 1.2. Systems of Particles: Classical or Quantum Mechanics?

At high temperatures and/or low densities, the behavior of a gas can be described by classical statistical mechanics; the atoms, to a good approximation, move along classical trajectories. At low temperatures and/or high densities, quantum effects become important. The classical approximation will break down when the de Broglie wavelength of a typical particle becomes comparable to the average interparticle distance; if the number density is n, this distance is roughly $d \sim n^{-1/3}$ which can then be compared to $\lambda = h/p$. For a system in thermal equilibrium, the thermal energy is $E = p^2/2M = k_B T/2$ where k_B is Boltzmann's constant and T is the temperature; this gives

$$\lambda = \frac{h}{p} = \frac{2\pi\hbar}{\sqrt{Mk_B T}} \tag{1.27}$$

For air at typical room condition, one can estimate that $M \sim 28\ u$ (for diatomic nitrogen, N_2) and $T \sim 300\ K$ to find $\lambda \sim 0.45$ Å; at one atmosphere of pressure, one has $P_{atm} \sim 10^5$ N/m^2 giving[7] $n \sim 2.4 \times 10^{25}$ m^{-3} or $d = n^{-1/3} \sim 35$ Å. Since $d \gg \lambda$ the system can be considered classically.

On the other hand, the conduction electrons in a metal (which for many purposes can be considered as a gas) have a de Broglie wavelength which is $\sqrt{M/m_e} \approx 225$ times *larger* than for gas atoms, so that $\lambda_e \sim 100$ Å. The larger densities of solid matter, however, imply much *smaller* interparticle distances; with a few conduction electrons per atom, electron densities of $n_e \approx (1–10) \times 10^{28}$ m^{-3} are typical, so that $d = n_e^{-1/3} \approx 2–5$ Å giving $\lambda \gg d$ in this case.

---

[7] One uses the ideal gas law, $P = nk_B T$.

1.4 **Semiclassical Model of the Hydrogen Atom**

An essentially classical approach to the bound-state dynamics of the electron–proton (hydrogen atom) system can be extended by using the wave mechanics idea of de Broglie to derive many of the most important features of the hydrogen spectrum, as in the Bohr model. We note that other systems can be usefully described with the same techniques, including (i) multiply ionized atoms where all but one electron has been removed and (ii) states where an outer valence electron is in a highly excited state, and far from the ionic core so that it appears hydrogenic, so-called *Rydberg atom states.*

The *Coulomb force* between the two particles can be written in the form

$$\mathbf{F}(r) = -\frac{1}{4\pi\epsilon_0}\frac{e^2}{r^2}\hat{\mathbf{r}} = -\frac{Ke^2}{r^2}\hat{\mathbf{r}} \tag{1.28}$$

where we will conventionally write (in the units used in this book) the fundamental constant of electrostatics in the form $K = 1/4\pi\epsilon_0$. This force can be derived from the *Coulomb potential,* namely

$$V(r) = -\frac{Ke^2}{r} \tag{1.29}$$

- Before proceeding, let us pause and make a few comments about the dimensionful constants that appear in this and other atomic and nuclear physics systems involving electromagnetism. The combination of constants which determines the electrostatic force between two fundamental charges can be written in the form

$$Ke^2 = \left(\frac{Ke^2}{\hbar c}\right)\hbar c = \alpha\hbar c \quad \text{where} \quad \alpha \equiv \frac{Ke^2}{\hbar c} \approx \frac{1}{137} \tag{1.30}$$

and α is dimensionless and is called the *fine-structure constant.* The combination $\hbar c$ has dimensions and numerical values given by

$$\hbar c \approx 1973 \text{ eV Å} \approx 197.3 \text{ MeV } F \approx 0.1973 \text{ GeV } F \tag{1.31}$$

which are useful for atomic/molecular, nuclear, and elementary particle physics problems, respectively. Together, these give

$$Ke^2 \approx 14.4 \text{ eV Å} \approx 1.44 \text{ MeV } F \approx 2.31 \times 10^{-28} \text{ J} \cdot \text{m}. \tag{1.32}$$

Despite focusing on nonrelativistic systems, we will often manipulate factors of c to make use of these combinations. Now back to the hydrogen atom.

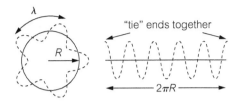

Figure 1.8. Standing wave pattern in circular orbits for deBroglie waves.

For circular orbits in which the electron (mass m) is assumed to orbit around the (stationary and infinitely heavy) proton, Newton's law implies that

$$m\frac{v^2}{r} = ma_C = F(r) = \frac{Ke^2}{r^2}$$ (1.33)

where we have used the appropriate centrifugal acceleration; this relation is classically consistent with any value of r.

If we wish to incorporate the wave properties of the electron via the de Broglie relation

$$\lambda = \frac{h}{p} = \frac{2\pi\hbar}{p}$$ (1.34)

then we must presumably insist that the appropriate number of de Broglie wavelengths "fit" into the circular orbit, as in Fig. 1.8, that is, that

$$n\lambda = 2\pi R \quad \text{where } n = 1, 2, 3 \ldots.$$ (1.35)

When combined with Eqn. (1.34), this implies that

$$n\hbar = pR = mvR = L$$ (1.36)

á la Bohr, and the orbital angular momentum must be quantized. This additional constraint, along with Eqn. (1.33), gives

$$r_n = \left(\frac{\hbar^2}{mKe^2}\right) n^2 = a_0 n^2$$ (1.37)

where we have defined the *Bohr radius* as

$$a_0 = \frac{\hbar^2}{mKe^2} = \frac{\hbar c}{mc^2\alpha} \approx \frac{(197.3 \text{ eVÅ}) (137)}{(0.511 \times 10^6 \text{ eV})} \approx 0.53 \text{ Å}$$ (1.38)

The corresponding speeds in the Bohr model are also quantized and given by

$$v_n = \frac{n\hbar}{mr_n} = \left(\frac{Ke^2}{\hbar}\right)\frac{1}{n} = \frac{\alpha c}{n} << c$$ (1.39)

which reminds us that the electron orbital motion (in hydrogen at least) is non-relativistic. The period (τ) of the classical orbit and the corresponding frequency

(f) are given by

$$\frac{1}{\tau_n} = f_n = \frac{v_n}{2\pi r_n} = \frac{1}{2\pi}\left(\frac{m(Ke^2)^2}{\hbar^3}\right)\frac{1}{n^3} = \frac{1}{2\pi}\left(\frac{mc^2\alpha^2}{\hbar}\right)\frac{1}{n^3} \qquad (1.40)$$

which will prove useful for comparisons to other timescales in atomic systems. For example, the periods can be written in the form

$$\tau_n = \tau_0 n^3 \quad \text{where } \tau_0 \equiv \frac{2\pi\hbar^3}{m(Ke^2)^2} = \frac{2\pi}{\alpha}\frac{a_0}{c} \approx 1.5 \times 10^{-16}\,\text{s} \qquad (1.41)$$

Even more importantly, the bound-state energies are also quantized since

$$\begin{aligned}
E_n &= \frac{1}{2}mv_n^2 - \frac{Ke^2}{r_n} = -\frac{1}{2}\frac{m(Ke^2)^2}{\hbar^2}\frac{1}{n^2} \\
&= -\left[\frac{1}{2}mc^2\alpha^2\right]\frac{1}{n^2} \\
&\approx -\frac{(0.51 \times 10^6\,\text{eV})}{2}\left(\frac{1}{137}\right)^2\frac{1}{n^2} \\
E_n &\approx \frac{-13.6\,\text{eV}}{n^2}
\end{aligned} \qquad (1.42)$$

While this has been derived assuming circular orbits, it can be shown that elliptical orbits, when properly quantized, are also described by this relation. The agreement of this simple result with experimental data on the hydrogen spectrum was one of the early successes of quantum theory.

Example 1.3. "Sieve" for Rydberg Atoms

While for small values of n, the typical sizes of atoms, as exemplified by Eqn. (1.37), are small, for large values of the quantum number (the Rydberg atom limit), the spatial extent of the state can easily fall into the micron range or larger. The large sizes of such Rydberg atoms make possible "slit" experiments in which the results do not depend on the quantum mechanical wave properties of the system but, rather, on their classical physical size. In one such experiment,[8] beams of Rydberg atoms (in this case highly excited sodium atoms) were produced with specified quantum numbers in the range $23 < n < 65$. These were allowed to drift toward an array of rectangular, micrometer size slits in gold foil; the average slit size was $2 \times 10\,\mu$m. If one assumes that the Rydberg atoms have an effective classical radius given by ka_0n^2, where k is a dimensionless constant, one can argue from Fig. 1.9 that if the center of the atom is more than $d = l/2 - ka_0n^2$ from the center of the slit, the atom will not pass through it (being ionized instead upon contact with the foil). The transmission probability, T,

---

[8] See Fabre *et al.* (1983) for details.

(Continued)

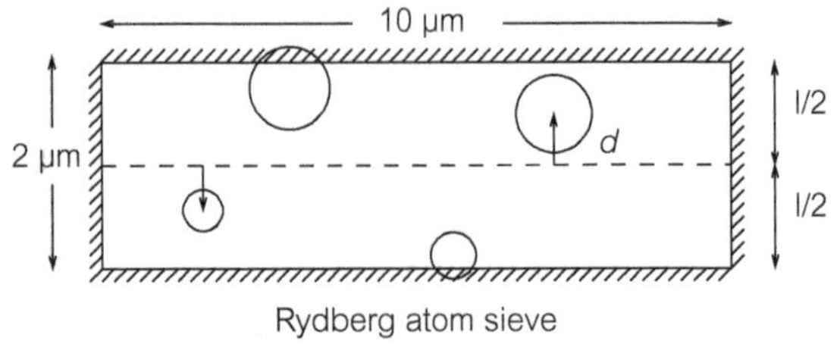

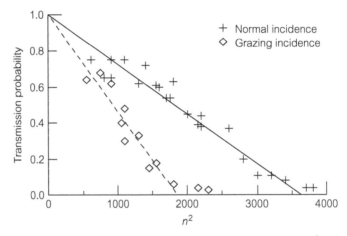

Rydberg atom sieve

Figure 1.9. Geometry of the Rydberg "sieve" in Example 1.3.

Figure 1.10. Transmission probability versus principal quantum number square, n^2, for the Rydberg "sieve"; the data are taken from Fabre *et al.* (1983).

is then determined solely by geometry, and is given by

$$T = \frac{d}{l/2} = 1 - k \frac{2l}{a_0} \frac{1}{n^2} \tag{1.43}$$

which predicts an $A - B/n^2$ dependence of the transmission and a "cut-off" size for the atoms. The data are plotted in Fig. 1.10 where the crosses correspond to normal incidence of the atomic beam (beam perpendicular to foil), while the diamonds are for incidence at an angle such that the effective width of the slit, l, is reduced by a factor of two. The predictions of the simple "hard-sphere" model described above are indicated by the straight lines for the two cases.

An important limit is suggested by Eqn. (1.36) where we note that $n \gg 1$ is required to obtain macroscopically large values of the angular momentum; the fact that quantum systems approach (in an average sense which we will discuss in later chapters) their classical counterparts in this limit is called the *correspondence principle*, and was used heavily by Bohr in his analyses. For example, he noted that the photons emitted in transitions between the quantized energy levels in Eqn. (1.42) satisfy the Balmer formula, written here in the form

$$2\pi\hbar f_\gamma = hf_\gamma = E_\gamma = E_n - E_{n'} = \frac{m(Ke^2)^2}{2\hbar^2}\left(\frac{1}{(n')^2} - \frac{1}{n^2}\right) \tag{1.44}$$

For transitions between neighboring states, that is of the form $n' = n - 1$, in the large n limit the emitted radiation is of frequency

$$f_\gamma = \frac{1}{2\pi\hbar}(E_n - E_{n-1}) \xrightarrow{n\gg 1} \frac{m(Ke^2)^2}{2\pi\hbar^3}\frac{1}{n^3} \tag{1.45}$$

A classical particle undergoing circular acceleration would emit radiation at its orbital frequency, f, which from Eqn. (1.40) is given by exactly the limit above. The connections and interpolations between the quantum mechanical and classical descriptions of the physical world are stressed in this book.

It is interesting to note in this context the role that wave mechanics and Coulomb's law (via $\hbar$ and e) play in determining the densities of "ordinary" solid matter.[9] The mass of atoms is due mostly to their nuclear constituents (the protons and neutrons), while their size is determined by the quantum properties of their electrons. For example, an order of magnitude estimate of the density of atomic hydrogen can be obtained by assuming that there is one proton mass in a cube of size $2a_0 \approx 1$ Å on a side; this gives a density of roughly

$$\rho \sim \frac{m_p}{(2a_0)^3} \approx 1.6 \times 10^3 \; kg/m^3 \sim 1.6 \; gr/cm^3 \tag{1.46}$$

which is in the right "ball-park."

A number of such useful results were derived using such early quantum mechanical ideas, but in order to utilize the full predictive power of modern quantum mechanics, we will incorporate important notions of classical wave physics (Chapter 2) into the Schrödinger equation formulation (Chapter 3 and beyond) of quantum physics.

1.5 **Dimensional Analysis**

Most problems in physics are ultimately to be related to measureable quantities in the "real world", and therefore have answers that carry dimensions. For purely

---

[9] The seemingly commonplace observation that "*matter held together by Coulomb forces is stable*" is a remarkable and rather subtle consequence of many aspects of quantum theory; for a nice discussion, see Lieb (1976). The corresponding classical "no-go" theorem of Earnshaw (see Jones 1980 for a brief history) goes something like "*No system of charged particles can be in stable static equilibrium.*"

mechanical problems, any constant or variable representing a physical property (call it $\mathcal{X}$) will have dimensions, which can be constructed from various powers of fundamental units of mass (M), length (L), and time (T). We can formalize this statement by using the notation

$$[\mathcal{X}] \equiv \text{dimensions of } \mathcal{X} = M^a L^b T^c \tag{1.47}$$

where a, b, c are real, possibly fractional, exponents. Familiar examples such as force (F), pressure (P), and density (ρ) correspond to

$$[F] = MLT^{-2}, \quad [P] = ML^{-1}T^{-2}, \quad \text{and} \quad [\rho] = ML^{-3} \tag{1.48}$$

Specific conventions giving the *units* of physical observables (such as the MKS or meter-kilogram-second system) rely on this observation, but it is more general. It can often be used to "solve" for the dependence of the physical quantity in question on the dimensionful parameters of the problem.

Example 1.4. The Dimensions of the Quantum Harmonic Oscillator

The only dimensional inputs to the classic problem of a mass and spring system are the mass, m, and spring constant, K, which have dimensions

$$[m] = M \quad \text{and} \quad [K] = MT^{-2} \tag{1.49}$$

The period of the oscillatory motion, τ, should presumably depend on these parameters, plus additional dimensionless constants; we therefore expect that $\tau \propto m^\alpha K^\beta$, so we write

$$T = [\tau] = [m^\alpha K^\beta] = M^\alpha (MT^{-2})^\beta \tag{1.50}$$

Comparing the powers of M, L, T on each side of the equation, we find

$$M : 0 = \alpha + \beta$$
$$L : 0 = 0 + 0 \tag{1.51}$$
$$T : 1 = 0 - 2\beta$$

which gives $\alpha = 1/2$ and $\beta = -1/2$ or

$$\tau \propto \sqrt{\frac{m}{K}} \tag{1.52}$$

The "exact" answer obtained from the solution of the equations of motion is, of course,

$$\frac{2\pi}{\omega} = \tau = 2\pi\sqrt{\frac{m}{K}} \tag{1.53}$$

This result is not atypical in that the dimensionless constant that is left unspecified by dimensional analysis is often within 1–2 orders-of-magnitude (either bigger or smaller) of unity.

(Continued)

For the classical oscillator, there is no combination of m and K, which naturally gives a length scale, or preferred amplitude. In the quantum version of this problem, however, we have another dimensionful constant at our disposal, namely Planck's constant or $\hbar$, which has dimensions

$$[\hbar] = ML^2T^{-1} \tag{1.54}$$

In contrast to the classical case, we can now also construct a fundamental length or amplitude by writing $A \propto m^\alpha K^\beta \hbar^\gamma$ and proceeding as above to find that $\gamma = 1/2$ and $\alpha = \beta = -1/4$ or

$$A \propto \left(\frac{\hbar^2}{mK}\right)^{1/4} \tag{1.55}$$

The energy of the oscillator should also have typical dimensions given by

$$[E] = [KA^2] \propto \hbar\sqrt{\frac{K}{m}} \propto \hbar\omega \tag{1.56}$$

For problems involving electricity and magnetism, an additional fundamental dimensionful quantity is introduced, often the dimension of charge[10] while thermodynamics problems require a temperature dimension, standardly taken as the degree Kelvin when used as a unit. We discuss in Appendix A the dimensions of many physical observables in an MKS-type system of units.

1.6 **Questions and Problems**

Q1.1. Everyone comes to any text with some idea of what they want to get out of it. What one question about, or important aspect of, quantum mechanics interests you the most? Look in the index, and see if that topic is covered in this book. If it is, find the reference, and see what you will have to learn in order to understand it. If it is not, do a library or web search (or ask someone) until you learn what you really want to know.

Q1.2. Try to imagine a world [11] in which the fundamental constants of relativity and quantum mechanics were on a more "human scale," namely $c = 10$ m/s and $\hbar = 0.1$ J · s. For example, how would the famous "twin paradox" of relativity

---

[10] Recall that the MKSA system of units actually uses the Ampere, that is, current, as the defining unit for EM.

[11] For an entertaining version of this "what-if exercise," see *Mr. Tompkins in Wonderland* by George Gamow (1946).

work for a trip to the neighborhood store? What would the zero-point energy of the keys in your pocket be? How about the probability that they could quantum mechanically tunnel out of your pocket? Could you walk through a doorway and be sure of which way you were going afterward (i.e. would you diffract?)

Q1.3. Discuss to what extent areas of physics other than special relativity and quantum mechanics have a "fundamental dimensionful constant"; consider, for example, gravitation (G?), statistical mechanics (k_B?), electricity (e and/or ϵ_0?), and magnetism (e and/or μ_0?).

Q1.4. Compton scattering experiments were first carried out by scattering X-rays from the valence electrons of carbon, which are bound. Why is the assumption of an initially free electron at rest still used?

Q1.5. How would the densities of "ordinary" solid matter change if the value of $\hbar$ were somehow suddenly doubled in magnitude? How about if the electron were 200 times heavier?

Q1.6. The probability density shown in Fig. 1.4 is plotted against "*Momentum q (a.u.)*," where (a.u.) stands for *atomic units*. Given what we know about dimensions for quantities in the hydrogen atom problem, what units do you think these would be?

P1.1. Relativistic decay kinematics:

(a) A particle of mass M (at rest) decays into a lighter particle of mass m and a massless photon, that is, $M \to m + \gamma$. Using conservation of energy and momentum, find the magnitude of the momentum and the energy for both final-state particles. Hint: Remember to use the relativistic formulae for E and pc.

(b) Show that m moves off with a speed given by

$$\frac{v}{c} = \frac{M^2 - m^2}{M^2 + m^2} \tag{1.57}$$

and comment on any obvious limiting cases.

(c) For the decay of an excited state of an atom, one has $Mc^2 = mc^2 + \Delta E$ where $\Delta E = E_\gamma$ is the energy of the photon emitted in the transition. Estimate the recoil speed of the hydrogen atom when it emits a photon in the $n = 2 \to n = 1$ transition. Compare this to the speed of the electron in the $n = 1$ state. Compare the recoil speed with the thermal speed of a hydrogen atom at room temperature, using $E = mv^2/2 = k_B T/2$.

P1.2. Ultrahigh energy cosmic ray interactions with the cosmic microwave background: Ultrahigh energy protons ($E_p >> m_p c^2$) can interact with the individual photons remaining from the "Big Bang", namely the cosmic microwave background (CMB) photons, which are characterized by a temperature of $T_{CMB} \approx 3$ K. For protons of sufficiently high energy, the reaction $p + \gamma_{CMB} \to p + \pi^0$ can produce pions (of mass $m_\pi \approx 140$ MeV), which then reduces the energy of the incident proton kinetic energy. This type of process limits the maximum energy of cosmic ray protons which should propagate through space.[12]

(a) One mechanism for this process is the resonant production of a Δ resonance by $p + \gamma_{CMB} \rightarrow \Delta \rightarrow p + \pi_0$. The mass of the proton and Δ are given by $m_p c^2 \approx 940$ MeV and $m_\Delta \approx 1230$ MeV. Using conservation of energy and momentum, show that the minimum incident proton energy for which such a reaction is kinematically possible and is given by

$$E_p^{(min)} = \frac{(M_\Delta c^2)^2 - (m_p c^2)^2}{4 E_{CMB}} \tag{1.58}$$

and evaluate this energy numerically. Hint: Use the following assumptions: (i) The most favorable kinematic situation is when the directions of the incident proton and photon are opposite, (ii) assume that the energy of the incident photon is $E_{CMB} = k_B T_{CMB}$ with $T = 3$ K.

(b) The average energy of a CMB photon is actually $\langle E_{CMB} \rangle \approx 2.7 k_B T_{CMB}$, while the minimum "effective mass" needed for pion production is actually $M_{min} = m_p + m_\pi$. What is the minimum proton energy needed for pion production under these assumptions?

(c) What is the proton speed for the minimum energies derived in parts (a) and (b)? What is the wavelength of the CMB photons used in parts (a) and (b)?

P1.3. **"Stopping atoms with laser light"**[13] or **Doppler cooling:** Laser light can be used to exert a force on atoms, as shown in the schematic diagram in Fig. 1.11. Photons which are absorbed to excite the atom will transfer a net momentum in the $+z$ direction, while the photons which are emitted in the subsequent decay are radiated isotropically (no preferred direction), so that there is a net momentum transfer.

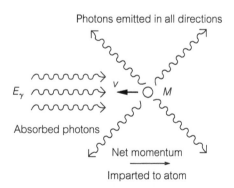

Figure 1.11. Schematic diagram of absorption of laser light by atom leading to net retarding force.

---

[12] Giving an "*End to the cosmic-ray spectrum*," as discussed by Griesen (1966).
[13] This is the title of a paper by Prodan *et al.* (1985) which discusses this effect.

(a) Recalling the Doppler effect, show that the photon E_γ energy required to excite the atomic transition of energy E_A is $E_\gamma = (1 - v/c)E_A$ where v is the speed of the atom.

(b) Show that each absorption results, on average, in a reduction in speed given by $\Delta v = \hbar k/M$ where M is the atom mass and $k = 2\pi/\lambda$.

(c) For a sodium atom ($M = 23$ amu) with initial speed $v_0 \sim 600$ m/s, how many such absorptions, N, are required to bring it to rest if $\lambda = 5890$ Å?

(d) What is the approximate spread in longitudinal (i.e. perpendicular to the laser beam) speed after this many absorptions? Hint: This is a statistical process, so estimate the error in N in part (c) by assuming that $\Delta N = \sqrt{N}$.

(e) If the time between photon absorptions is limited by the radiative lifetime of the atomic state to be $\Delta t \sim 10^{-8}$ s, estimate the average force on the atom using $F = \Delta p/\Delta t$ and compare it to the force of gravity.

P1.4. Laser trapping of atoms[14] **or "Optical Molasses."** Imagine an atom as shown in Fig. 1.12, which is now irradiated by two equal intensity lasers with frequencies just below a resonant frequency of the atom, namely $E_\gamma \leq E_A$. The probability of absorption of each beam has the so-called Lorentzian form, and is given by

$$\frac{1}{(E_\gamma - E_A)^2 + (\hbar/2\tau)^2} = \frac{1}{(\Delta E)^2 + (\hbar/2\tau)^2} \tag{1.59}$$

where τ is the lifetime of the stated excite at E_A.

(a) Show that if the atom moves to the right, say, with speed v that there is a net force on the atom proportional to

$$F_{\text{net}} \propto -\frac{|\Delta E|}{(\Delta E)^2 + (\hbar/2\tau)^2}\left(\frac{v}{c}\right) \propto -v \tag{1.60}$$

(b) For what value of ΔE is this force maximized?

(c) A linear restoring force proportional to $-x$ is like a spring, but one proportional to $-v$ is like a viscous damping force, hence the name "optical molasses." Solve the classical equation of motion for such an object with initial conditions $x(0) = 0$ and $v(0) = v_0$.

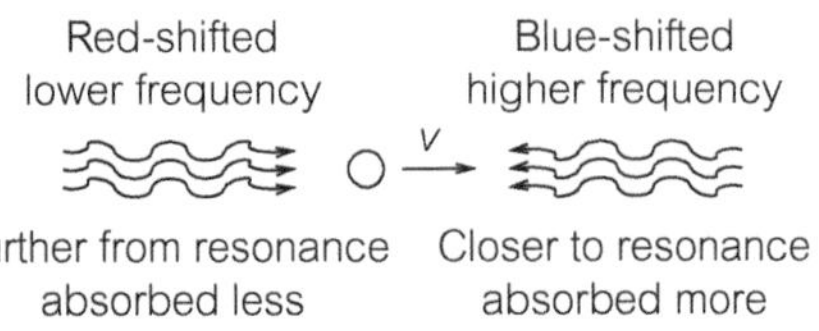

Red-shifted lower frequency   Blue-shifted higher frequency

Figure 1.12. Geometry of opposing laser beams leading to "optical molasses" or velocity-dependent damping force giving atom trapping.

Further from resonance absorbed less   Closer to resonance absorbed more

[14] For a discussion, see Cohen-Tannoudji (1990).

P1.5. Photoelectric effect data analysis. Experiments measuring the photoelectric effect often determine the *stopping potential, V_0,* required to bring the emitted photoelectrons to rest; this implies the relation

$$eV_0 = hf - W = \frac{hc}{\lambda} - W \qquad (1.61)$$

The following data (taken from Millikan (1916)) give the stopping potential for various wavelengths of incident light for a sodium–copper interface

λ (Å)	V_0 (Volts)
5461	-2.05
4339	-1.48
4047	-1.30
3650	-0.92
3216	-0.38

(Note that the values of V_0 are negative because of the effect of the additional metal interface.) Use these values to test the Einstein relation, Eqn. (1.13), and to extract a value for h. Try to estimate the errors for your value if you can.

P1.6. Compton scattering:

(a) Derive Eqn. (1.15) by using conservation of energy and momentum for the photon–electron collision in Fig. 1.5. Hint: Conservation of energy and momentum in this case look like

$$\text{momentum:} \quad \mathbf{p}_\gamma = \mathbf{p}'_\gamma + \mathbf{p}'_e \qquad (1.62)$$

$$\text{energy:} \quad p_\gamma c + m_e c^2 = p'_\gamma c + \sqrt{(p'_e c)^2 + (m_e c^2)^2} \qquad (1.63)$$

since the electron is initially at rest.

(b) Evaluate the fractional change in wavelength, $\Delta\lambda/\lambda$, for incident blue light ($\lambda = 4500$ Å) and X-rays ($\lambda = 0.7$ Å).

P1.7. Backscattered laser photons. Since the derivation in P1.6 is a staple of most modern physics texts, here is a different version of a photon–electron collision problem.[15] At the Stanford Linear Accelerator Center (SLAC), laser light (i.e. with visible wavelengths) has been backscattered ($\theta_\gamma = 180°$) from $E_e \approx 50$ GeV (and hence ultrarelativistic) electrons to obtain very high-energy photons, as in Fig. 1.13. Using any approximations suggested by this description of the problem,

---

[15] See Milburn (1963) for an early discussion.

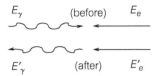

Figure 1.13. Geometry for light backscattered from electrons.

show that the energy of the backscattered photons is given by

$$E'_\gamma \approx E_e \left(1 + \frac{(m_e c^2)^2}{E_e E_\gamma} \right)^{-1} \tag{1.64}$$

Evaluate this energy for a "blue" laser and the electron energy above.

P1.8. Wave mechanical interference.

(a) The interference patterns in Fig. 1.2 were obtained using electrons that had been accelerated through a potential difference of 80 000 V and an "effective" slit width of $D \sim 6\,\mu$m. Calculate the de Broglie wavelength of the electrons and estimate the lateral size of the diffraction pattern on a screen 10 cm from the slit.

(b) Consider the diffraction pattern, due to wave mechanics effects, of a ball thrown through an open window. Using "everyday" values for all quantities, show *very roughly* that the first minimum of the diffraction pattern would be at a distance of one atomic radius from the central peak on a screen located in a neighboring galaxy.

P1.9. Discrete quantities in quantum physics.   Quantized energies in bound-state systems and quantized values of the angular momentum are common examples of physical quantities, which have discrete values in quantum mechanics, but there are a number of others.

(a) **Quantized circulation in low-temperature liquid helium.**   The circulation (line integral of the velocity field about any closed curve) of a sample of superfluid helium (mass M_4) at sufficiently low temperatures is known to be quantized in the form

$$\oint \mathbf{v} \cdot d\mathbf{l} = n \left(\frac{h}{M_4} \right) = n\kappa \tag{1.65}$$

Evaluate the *quantum of circulation*, κ, in MKS units. For a sample of superfluid helium obeying this relation, confined to a narrow annular region of average radius 3 cm, what would be the minimum angular velocity needed to observe a nontrivial rotation of the fluid?[16]

(b) **Quantum Hall effect.** In the classical Hall effect, a magnetic field is applied to sample, in a direction perpendicular to the direction of current flow. The

---

[16] For related experiments, see Hess and Fairbank (1967).

resulting voltage can be measured and the *Hall resistance* is measured as the ratio of the induced voltage to the observed current. For two-dimensional electron systems, at low temperature, von Klitzing (1980) observed that the Hall resistance exhibited a series of incredibly well-defined (flat to one part in 10^9!) steps, given by $R_i = (1/i)R_K$, corresponding to fractional values of $R_K \equiv h/e^2$. Calculate the value of this unit of resistance (in ohms or Ω), which is sometimes called the von-Klitzing constant.[17]

(c) **Flux quantum.** In superconductors, the allowed magnetic flux through a sample is quantized in units of $\Phi_B = h/(2e)$. Evaluate this flux in units of Tesla·m^2. If the earth's magnetic field of roughly 0.5 *gauss* (where 1 *gauss* = 10^{-4} Tesla) is perpendicular to a page of this book, estimate the magnetic flux through this one page, in terms of fundamental flux quanta.

(d) **Degeneracy of Landau levels.** The energy levels (so-called Landau levels) of a two-dimensional electron gas subject to an external magnetic field are quantized, with energy eigenvalues proportional to

$$E_L = \hbar \left(\frac{eB_0}{2m_e} \right) \qquad (1.66)$$

while the number of states with the same energy level (the degeneracy, N_d) is given by

$$N_d = \frac{eB_0 A}{\pi \hbar} \qquad (1.67)$$

where A is the area of the two-dimensional sample. First, show that these formulae are dimensionally correct. Then, evaluate E_L for an applied field of $B_0 = 5\,\text{T}$. The quantity N_d/A is the maximum number of energy levels per unit area which can be filled and this can be compared to the actual two-dimensional density, n_e. Evaluate N_d/A for the same external field, and compare it a typical two-dimensional electron density of $n_e \sim 10^{22}$ cm^{-2}.

P1.10. **Quantum version of Kepler's Third Law.** A standard exercise in introductory mechanics is to show that a light planet (of mass m) orbiting a heavy star (mass $M_S \gg m$) in a circular orbit satisfies Kepler's Third Law, namely that the period, τ, and orbital radius, r (or more generally the semimajor axis) are related by

$$\tau^2 = \frac{4\pi^2}{GM_S} r^3 \qquad (1.68)$$

For our solar system, if the periods are measured in years, $\tilde{\tau}$, and the distances in astronomical units (A.U.), $\tilde{r}$, this reduces to the simple relation $\tilde{\tau}^2 = \tilde{r}^3$. Derive a quantum version of Eqn. (1.68) for the Bohr model, and show that it also simplifies to $\tilde{\tau}^2 = \tilde{r}^3$ if appropriate atomic units, as in Eqns (1.37) and (1.41), are used.

---

[17] This quantity now forms the basis for international standards of resistance.

P1.11. Simple-minded scaling laws. Rederive all of the results for the Bohr model of the hydrogen atom (as in Section 1.4), where the potential is given by $V(r) \propto 1/r$, but here for the general power law radial potential and corresponding force

$$V(r) = Ar^s \quad \text{giving} \quad F(r) \propto Ar^{s-1} \tag{1.69}$$

For $s = -1$ and $A = -Ke^2$, this gives the Coulomb potential, while for $s = +2$ and $A = K/2$, it is represents a three-dimensional spring (or harmonic oscillator) potential; you should also be able to show that for $A = V_0/L^s$ and $s \to \infty$ one obtains the three-dimensional infinite well of radius L. Use these limits to check your general results against familiar special cases.

(a) Show that the period scales as

$$\tau_n \propto \left(\frac{m}{n\hbar}\right)\left(\frac{n^2\hbar^2}{Am}\right)^{2/(s+2)} \tag{1.70}$$

(b) Show that the quantized energies scale as

$$E_n \propto A\left(\frac{n^2\hbar^2}{Am}\right)^{s/(s+2)} \tag{1.71}$$

(c) Repeat the correspondence principle limit argument to show that the frequency of the photons emitted in the $n \to n-1$ transition scales in the same way as the classical rotation frequency when $n \gg 1$.

P1.12. Quantum numbers for macroscopic systems.

(a) Estimate the angular momentum quantum number n for the following macroscopic rotating systems, assuming that $L = n\hbar$: (i) a spinning compact disc (which rotates at approximately $200-500$ rev/s), (ii) a twirling ice skater (who might rotate at 3 rev/s), (iii) the earth rotating about its axis, and (iv) the earth rotating about the sun.

(b) The energies of a harmonic oscillator (i.e. mass and spring) are quantized via $E_n = (n+1/2)\hbar\omega$ where $\omega = \sqrt{K/m}$ is the natural frequency of oscillation. If you sit down on the back end of a car, and then jump off, it springs back and oscillates. Estimate the quantum number associated with this classical motion.

(c) The earth (M_e) and moon (M_m) system is similar to a hydrogen atom with the electrostatic force (i.e. Ke^2) replaced by the gravitational force (GM_eM_m) and with the moon playing the role of the electron. The earth–moon system is then quantized as in Section 1.4, and you should first rederive all of the relevant formulae for the energy, period, and orbital radius of the system. Using, for example, the length of the month, estimate the value of the quantum number n. (See also P1.10.)

P1.13. Quantum or classical systems.

(a) At one point, the "world's coldest gas" was a sample of ^{133}Cs atoms with a number density $N \sim 10^{10}$ cm^{-3} at a temperature of $T \sim 700$ nK. Is this a quantum or classical system? There are hopes to go to $T \sim 200$ nK and $n \sim 10^{14}$ cm^{-3}; how about that system? Hint: Use the ideas of Example 1.2.

(b) The electrons in a white dwarf star can be considered as a gas. There might be roughly 2×10^{57} of them in a volume about the size of the earth, with internal temperatures of order 10^{6-7} K. Is this a quantum or classical system?

P1.14. Zero-point energy using the uncertainty principle. Use the uncertainty principle to estimate the ground state energy (zero-point energy) of a mass (m) and spring (K) system. Hint: Write the energy in the form

$$E = \frac{p^2}{2m} + \frac{1}{2}Kx^2 \sim \frac{\hbar^2}{2mx^2} + \frac{1}{2}Kx^2 \sim E(x) \tag{1.72}$$

where one approximates $\Delta x \, \Delta p \sim x\,p \sim \hbar$; then find the minimum value of E as a function of x. What is the smallest amplitude of motion that a mass and spring system can have? Do your answers agree with the dimensional analysis result of Example 1.4?

P1.15. Repeat P1.14, but for a particle of mass m in a linear confining potential given by $V(x) = C|x|$. Does your result agree with the general scaling expression in P1.11, and with dimensional analysis arguments?

P1.16. Correspondence principle and the classical period.

(a) Show that the correspondence principle arguments used in Eqns (1.44) and (1.45) can be generalized to show that the classical periodicity, τ, of a quantum system in the large n limit would be given by

$$\tau_n = \frac{2\pi\hbar}{|dE_n/dn|} \tag{1.73}$$

(b) Using the expression for the quantized energies of a particle in a box of length L from Eqn. (1.20), find the classical period τ_n in state n and compare it to your expectations based on the classical motion.

(c) The quantized energies for a mass-spring (harmonic oscillator) system are given by $E_n = (n + 1/2)\hbar\omega$ where $\omega = \sqrt{k/m}$. Find the classical period τ_n using Eqn. (1.73) and compare it to the expected result.

P1.17. Dimensional analysis for mechanics: Gravitational bound states of neutrons. The physical problem of a particle bound to a horizontal surface by the (very weak) force of gravity has been explored experimentally in "*Quantum states of neutrons in the Earth's gravitational Field,*" V. V. Nesvizhevsky et al., Nature 415, 297–299 (2002). For this problem, there are only three relevant physical parameters, the mass of the neutron, m, the (very familiar) acceleration of gravity,

$g = 9.8 \text{ m/s}^2$, and Planck's constant, $\hbar$. The mass of the neutron is only slightly more than that of the proton, namely $m_n \approx 940 \text{ MeV}/c^2$ or 1.7×10^{-27} kg.

(a) Using only dimensional analysis, find the product of powers of $m, g, \hbar$ which give an energy, E. Namely, write $E \sim m^{\alpha} g^{\beta} \hbar^{\gamma}$ and solve for α, β, γ as in Section 1.5. This would then be the order-of-magnitude of the ground state energy of the system.

(b) Repeat part (a) to find those combinations that combine to describe the appropriate length, time, and speed (l, t, v) scales for this problem.

(c) Evaluate the values of E, l, t, v you obtained numerically, using the physical values for $m, g, \hbar$. About how far above the horizontal surface will the particle be found? (Use your estimate of the length scale l.)

(d) Compare your values (in tabular form) to the corresponding physical quantities for the ground state of the hydrogen atom, as outlined in Section 1.4. For example, the magnitude of the ground state energy of hydrogen is 13.6 eV, the length scale is of order of the Bohr radius, $a_0 \sim 0.5$ Å, and so forth.

P1.18. **Dimensions for EM and thermal quantities.** Table A2 in Appendix A lists dimensional equivalents for many standard quantities used in EM and thermal problems, in terms of a basic set of MKS-type units, namely the kilogram (kg), meter (m), second (s), and Coulomb (C). Reproduce as many of those as you can by using simple relationships familiar from introductory physics. For example, the connection between potential energy (U) and electric potential (V) for a particle of charge q is $U = qV$, which implies that the dimensions are related by $[V] = [U/q] = J/C$ or Joule/Coulomb.

P1.19. **More dimensional analysis for electromagnetism.** For charged particles, the combination Ke^2 appears frequently in problems involving EM interactions, and provides another natural dimensionful constant in addition to its mass, m.

(a) Using these, as well as $\hbar$ and c, show that the most general combination of these fundamental constants, which gives a length is

$$L = \alpha^n \left(\frac{\hbar}{mc} \right) \tag{1.74}$$

where $\alpha = Ke^2/\hbar c$ is the fine-structure constant and n is any power. Hint: Write

$$L = (Ke^2)^{n_1} m^{n_2} \hbar^{n_3} c^{n_4} \tag{1.75}$$

and solve for n_2, n_3, n_4 in terms of $n = n_1$

(b) Show that $n = 0$ corresponds to the Compton wavelength of the particle while $n = -1$ gives the Bohr radius. Do the cases with $n = +1$ and $n = -2$ look familiar as well?

(c) The lifetimes for the radiative decay of excited states of hydrogen atoms are known to have the dimensionful form

$$T_{\text{life}} = \frac{2\pi \hbar^6 c^3}{m(Ke^2)^5} \tag{1.76}$$

Show that this timescale can be written in the form $T_{\text{life}} = \tau_0 \alpha^{-3}$ where τ_0 is defined in Eqn. (1.41), evaluate it numerically and compare it to the classical period timescale.

P1.20. Quantum gravity?

(a) The natural length scale at which gravity, quantum mechanics, and relativity are all simultaneously important is called the *Planck length*, L_P. Using dimensional analysis, find the combination of powers of G (Newton's constant of gravitation), $\hbar$, and c, which make a length.

(b) Evaluate L_P numerically, and compare to a typical scale for nuclear or particle physics, namely $1\ F = 10^{-15}$ m.

(c) Repeat to find the *Planck mass*, M_P, evaluate it numerically, and compare to a typical mass of a nuclear constituent, like the proton mass.

(d) Repeat to find the *Planck time*, T_P, evaluate it numerically, and compare it to the light travel time across a nucleus, which is also a typical nuclear reaction time. Note that these values are *very* different from those encountered in P1.17!

P1.21. Dimensions in particle physics.

Practitioners of elementary particle physics often become so accustomed to the various factors of $\hbar$ and c in their calculations that they use a shorthand notation in which they simply do not bother to write them down; this is sometimes denoted by saying that "$\hbar = c = 1$." For example, the decay rate (λ) and lifetime (τ) of the muon, calculated theoretically, are sometimes written in the form

$$\lambda_\mu = \frac{1}{\tau_\mu} = \frac{G_F^2 m_\mu^5}{192\pi^3} \tag{1.77}$$

where $G_F = 1.166 \times 10^{-5}$ GeV^{-2} (Fermi's constant) and $m_\mu = 105.6$ MeV$/c^2$; this is obviously dimensionally wrong as it stands. Supply enough factors of $\hbar$ and c to make it dimensionally correct, evaluate τ_μ numerically, and compare it to the experimental value of $\tau_\mu = 2.2\ \mu$s.

TWO

Classical Waves

2.1 The Classical Wave Equation

As we initially approach quantum physics through the introduction of wave mechanics, it is useful to begin by recalling some of the standard results obtained from the study of the classical wave equation. For example, using Newton's laws, one can derive the equation of motion for the transverse displacement of a small piece of a stretched one-dimensional string; we obtain the equation

$$\rho \frac{\partial^2 A(x, t)}{\partial t^2} = T \frac{\partial^2 A(x, t)}{\partial x^2} \tag{2.1}$$

where T, ρ are the tension and linear mass density of the string, and $A(x, t)$ is the amplitude of the string at position x and time t. This can be written in the form

$$\frac{\partial^2 A(x, t)}{\partial t^2} = \frac{T}{\rho} \frac{\partial^2 A(x, t)}{\partial x^2} = v^2 \frac{\partial^2 A(x, t)}{\partial x^2} \tag{2.2}$$

where $v = \sqrt{T/\rho}$ is the wave velocity; this is one version of the *classical wave equation*.

In a similar way, Maxwell's equations (in vacuum) can be combined to yield the wave equation, this time in three dimensions, for the components of the electric (or magnetic) field, giving, for example,

$$\frac{\partial^2 \mathbf{E}(\mathbf{r}, t)}{\partial t^2} = \frac{1}{\sqrt{\epsilon_0 \mu_0}} \nabla^2 \mathbf{E}(\mathbf{r}, t) = c^2 \nabla^2 \mathbf{E}(\mathbf{r}, t) \tag{2.3}$$

where the fundamental constants of electricity (ϵ_0) and magnetism (μ_0) combine to yield the speed of light. (Because of the quantum connection between classical electromagnetic waves and their corresponding particle-like quanta, namely photons, we will be especially interested in this version.)

Both of these cases yield the *classical wave equation* which we will study in the form

$$\frac{\partial^2 \phi(x, t)}{\partial t^2} = v^2 \frac{\partial^2 \phi(x, t)}{\partial x^2} \tag{2.4}$$

for the time- and space-dependent wave amplitude $\phi(x, t)$. The most familiar solutions of Eqn. (2.4) are functions of the form

$$\sin(kx \pm \omega t) \quad \text{or} \quad \cos(kx \pm \omega t) \tag{2.5}$$

where the wave number, k, and angular frequency, ω, are related to the familiar wavelength and frequency/period via

$$k \equiv \frac{2\pi}{\lambda} \quad \text{and} \quad \omega \equiv 2\pi f = \frac{2\pi}{\tau} \tag{2.6}$$

In order to solve Eqn. (2.4), ω and k must be related via

$$\omega = vk \tag{2.7}$$

Any such relation between the oscillation rates in space (k) and time (ω) is called a *dispersion relation*. The two choices of sign correspond respectively to waves traveling at constant speed to the right $(-)$ or left $(+)$; this can be seen by examining a point of constant phase, $\theta = kx \pm \omega t = \theta_0$, implying that the same point on the wave satisfies $x(t) = \mp vt + \theta_0/k$. A linear relationship (dispersion relation) between ω and k is then a signal of constant speed wave motion.

While perhaps not as familiar, these traveling wave solutions can also be written compactly in complex notation; for right-moving waves one has

$$e^{+i(kx-\omega t)} \quad \text{and} \quad e^{-i(kx-\omega t)} \tag{2.8}$$

where we have used the fact that

$$e^{\pm iz} = \cos(z) \pm i\sin(z) \tag{2.9}$$

(See Appendix C for a review of complex numbers and functions.)

One of the most important features of Eqn. (2.4) is that it is a *linear differential equation*, defined by the property that if $\phi_1(x, t)$ and $\phi_2(x, t)$ are both solutions, then the linear combination

$$\Phi(x, t) = a_1\phi_1(x, t) + a_2\phi_2(x, t) \tag{2.10}$$

is also a solution. This notion can be generalized to show that infinite (discrete) sums

$$\Phi(x, t) = \sum_{n=0}^{\infty} a_n\phi_n(x, t) \tag{2.11}$$

and infinite (continuous) sums (i.e. integrals)

$$\Phi(x, t) = \int_{-\infty}^{+\infty} a(k)\phi_k(x, t)\,dk \tag{2.12}$$

of solutions are also solutions. This is the *principle of superposition.*

For wave problems with boundaries (and hence boundary conditions which must be satisfied), it is often more useful to consider linear combinations of plane waves traveling in opposite directions, for example,

$$A\sin(kx - \omega t) + A\sin(kx + \omega t) = 2A\sin(kx)\cos(\omega t) \tag{2.13}$$

which gives rise to standing waves. For problems with no boundaries, say infinitely long strings or regions of space with no conductors present (on which the electric field would have to vanish), all values of k (and hence ω) are allowed, and one has a continuous "spectrum."

If, however, one has to impose boundary conditions, there are additional constraints on the allowed solutions of the wave equations; these often require that the wave amplitude must vanish at the "edges." For a finite length string, fixed at both ends, say at $x = 0$ and $x = L$, with a standing wave solution of the form $A(x, t) = A\sin(kx)\cos(\omega t)$ we require that

$$A(0, t) = 0 = A(L, t) \quad \text{for all } t \tag{2.14}$$

This implies that $\sin(k_n L) = 0$ or equivalently $k_n = n\pi/L$ for $n = 1, 2, \ldots$. Thus, the imposition of boundary conditions on solutions of the wave equation can give rise to quantized values of the wave number k, and hence for the frequency ω. These quantization effects are a property of solutions to the wave equation and will necessarily appear in quantum mechanics as well, where the origin of quantized quantities can typically be traced to the imposition of boundary conditions.

2.2 Wave Packets and Periodic Solutions

2.2.1 General Wave Packet Solutions

Plane wave solutions, characterized by a well-defined wave number and frequency, are useful constructs for analyzing the wave properties of a system. They are, however, far from the most general solution of the classical wave equation so we will have to extend our analysis for several reasons:

- Because plane wave solutions imply a nonzero amplitude over all space, they can only be idealizations corresponding to wave pulses or wave trains of very long but finite duration. As we will see, this implies that any realizable waveform will have a finite spread in wavelength or wave number.

- The most general solution of the wave equation (e.g. see P2.1) can be shown to be given by any (suitably differentiable) function of the form $\phi(x, t) = f(x \pm vt)$ since it satisfies

$$(\pm v)^2 f''(x \pm vt) = \frac{\partial^2 \phi(x, t)}{\partial t^2} = v^2 \frac{\partial^2 \phi(x, t)}{\partial x^2} = v^2 f''(x \pm vt) \qquad (2.15)$$

This implies that any appropriate initial waveform, $f(x)$, can be turned into a solution of Eqn. (2.4), $f(x \pm vt)$, which propagates to the right ($-$) or left ($+$) with no change in shape, as in Fig. 2.1. Such a solution can be called a *wave packet*.

- While such wave packets are useful for describing the space-time evolution of localized wave phenomena (wave pulses traveling down a string, thunderclaps, laser pulses, etc.), they do not make the "wave content" (k or ω dependence) obvious.

To understand how to extract the wave number dependence of a given wave packet solution of the wave equation, it is instructive to ask the question in reverse and see how to construct localized wave packets using familiar plane wave solutions. Three effects will make this possible:

(1) The linearity of the wave equation ensures that one can add as many solutions as desired and still have a solution;

(2) The possibility of constructive and destructive interference allows us to imagine building up a localized solution;

(3) The fact that all plane wave components have the same common velocity guarantees that the entire wave packet will not disperse as it travels, consistent with the general solution of Fig. 2.1.

It is sufficient to study the problem of obtaining localized spatial waveforms, $f(x)$, as the complete time-dependent solution will then simply be given by $f(x - vt)$.

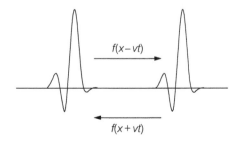

Figure 2.1. Left- and right-moving wave packet solutions to the classical wave equation.

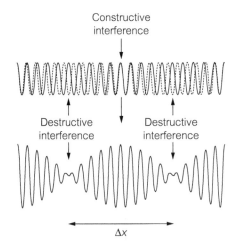

Figure 2.2. Interference of sinusoidal solutions giving "beats."

The simplest example that demonstrates the use of linearity and interference ideas is the phenomenon of *beats*, which is familiar from acoustics. We write the sum of two plane wave solutions as

$$f(x) = A\cos(k_1 x) + A\cos(k_2 x)$$

$$= 2A\cos\left[\frac{(k_1 + k_2)x}{2}\right]\cos\left[\frac{(k_1 - k_2)x}{2}\right] \tag{2.16}$$

which is illustrated in Fig. 2.2. For $k_1 \approx k_2 \approx k$ the resulting waveform is a plane wave, $\cos(kx)$, modulated by the "beat envelope", $2A\cos(\Delta kx)$, where $\Delta k \equiv (k_1 - k_2)/2$. This factor gives complete destructive interference when $x = (2n + 1)\pi/2\Delta k$; this implies that successive interference ("beat") minima will be separated by $\Delta x = \pi/\Delta k$ or $\Delta x \Delta k \approx \pi \geq \mathcal{O}(1)$. This is our first example of a quite general feature, namely that

- The degree of localization of a wave packet in space (Δx) making use of interference effects is inversely correlated with the spread in available k values (Δk).

2.2.2 Fourier Series

A more complex waveform, generated by selective constructive and destructive interference effects, can be obtained by the use of a *Fourier series expansion.*[1] Any (appropriately smooth) periodic function, satisfying $f(x) = f(x + 2L)$ or, equivalently, $f(x - L) = f(x + L)$, can be expanded in a linear combination of

---

[1] See any standard reference on mathematical physics for more details, for example, Butkov (1968) or Mathews and Walker (1970).

wave solutions via

$$f(x) = \frac{a_0}{2} + \sum_{n=1}^{\infty} \left[a_n \cos \left(\frac{n\pi x}{L} \right) + b_n \sin \left(\frac{n\pi x}{L} \right) \right] \qquad (2.17)$$

This solution will then contain varying contributions (given by the a_n, b_n) from plane wave solutions with wave numbers $k_n = n\pi/L$. The expansion coefficients a_n, b_n can be obtained by multiplying both sides of Eqn. (2.17) by $\cos(m\pi x/L)$, $\sin(m\pi x/L)$ respectively and integrating over one cycle. Making use of the relations

$$\int_{-L}^{+L} \cos \left(\frac{n\pi x}{L} \right) \cos \left(\frac{m\pi x}{L} \right) dx = L\delta_{n,m}$$

$$\int_{-L}^{+L} \sin \left(\frac{n\pi x}{L} \right) \sin \left(\frac{m\pi x}{L} \right) dx = L\delta_{n,m}$$

$$\int_{-L}^{+L} \cos \left(\frac{n\pi x}{L} \right) \sin \left(\frac{m\pi x}{L} \right) dx = 0 \qquad (2.18)$$

we find that

$$a_n = \frac{1}{L} \int_{0}^{2L} f(x) \cos \left(\frac{n\pi x}{L} \right) dx = \frac{1}{L} \int_{-L}^{+L} f(x) \cos \left(\frac{n\pi x}{L} \right) dx$$

$$b_n = \frac{1}{L} \int_{0}^{2L} f(x) \sin \left(\frac{n\pi x}{L} \right) dx = \frac{1}{L} \int_{-L}^{+L} f(x) \sin \left(\frac{n\pi x}{L} \right) dx \qquad (2.19)$$

The a_n (b_n) thus measure the "overlap" of the desired waveform with one of the basic cosine or sine solutions.

Example 2.1. Square waveform

As an example, consider the periodic square waveform, defined for one cycle $(-L, +L)$, by

$$\text{(square wave)} \quad f(x) = \begin{cases} +A & \text{for } |x| < L/2 \\ -A & \text{for } |x| > L/2 \end{cases} \qquad (2.20)$$

The Fourier coefficients a_n and b_n can be easily obtained, and we find

$$\text{(square wave)} \quad b_n = 0, \quad a_0 = 0, \quad \text{and} \quad a_n = 4A \frac{\sin(n\pi/2)}{n\pi} \qquad (2.21)$$

where $n = 1, 2, \ldots$ and so forth. In this case, the odd $\sin(N\pi x/L)$ terms cannot contribute to an even function, so that the b_n vanish for symmetry reasons. We plot the *partial* Fourier sums

$$F_N(x) = \sum_{n=1}^{N} a_n \cos \left(\frac{n\pi x}{L} \right) \qquad (2.22)$$

(Continued)

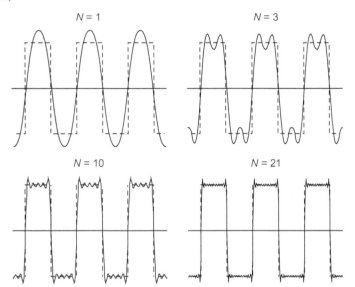

Figure 2.3. Fourier series approximations, $F_N(x)$, to a square wave pulse for $N = 1, 3, 5, 21$.

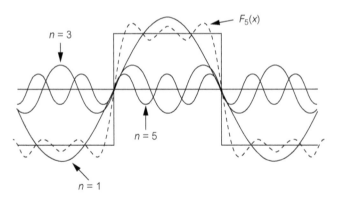

Figure 2.4. Partial Fourier sum, $F_5(x)$, for square wave and component terms.

for increasing values of N in Fig. 2.3; then, in Fig. 2.4 we show how the component waves contribute to a particular partial sum ($N = 5$). The Fourier series does seem to converge pointwise to the corresponding function where it should, that is, at points where the function is appropriately smooth.[2] One useful measure of the overall convergence can be defined via

$$\Delta_N = \frac{\int_{-L}^{+L} [f(x) - F_N(x)]^2 \, dx}{\int_{-L}^{+L} [f(x)]^2 \, dx} \tag{2.23}$$

---

[2] The "overshoots" at the discontinuities persist even in the complete sum. These "Gibbs peaks" are discussed in many textbooks on mathematical physics, for example, Mathews and Walker (1970).

(Continued)

where $F_N(x)$ is the Nth partial sum; this measures the overall deviation of the Nth partial sum from the function. For the square wave, one can show that it is given by

$$\text{(square wave)} \quad \Delta_N = 1 - \frac{8}{\pi^2} \sum_{n=1}^{N} \frac{\sin(n\pi/2)^2}{n^2} \qquad (2.24)$$

If we use the fact that

$$1 + \frac{1}{3^2} + \frac{1}{5^2} + \cdots = \sum_{k=1}^{\infty} \frac{1}{(2k-1)^2} = \frac{\pi^2}{8} \qquad (2.25)$$

(discussed in Appendix C under the topic "Zeta functions"), we see that $\Delta_N \to 0$ as $N \to \infty$; we can more and more closely approach the desired waveform except at isolated points.

Example 2.2. Triangle waveform

For comparison, we also consider a (slightly) better-behaved function, namely, a periodic triangle waveform, defined via

$$\text{(triangle wave)} \quad f(x) = A(L - |x|) \quad \text{for } -L < x < L \qquad (2.26)$$

over one cycle. This function is at least continuous at all points. We leave it as an exercise (P2.4) to show that the nonvanishing Fourier coefficients are given by

$$\text{(triangle wave)} \quad a_n = (1 - \cos(n\pi)) \frac{2AL}{n^2\pi^2} = 4AL \left(\frac{\sin(n\pi/2)}{n\pi} \right)^2 \qquad (2.27)$$

while the corresponding measure of the overall convergence is given by

$$\text{(triangle wave)} \quad \Delta_N = 1 - \frac{96}{\pi^4} \sum_{n=1}^{N} \frac{(1 - \cos(n\pi))^2}{n^4} \qquad (2.28)$$

This example implies that the convergence is faster (as a function of N) for smoother functions; this is clear from the form of the Fourier coefficients since $a_n \to 1/n^2$ $(1/n)$ for the triangle (square) wave which is continuous (discontinuous).

We see that with Fourier series, we can produce any desired *periodic* waveform and extract its wave number content (via the a_n and b_n). But even though we have used an infinite number of plane wave components, we still evidently do not have enough "degrees of freedom" to produce a truly localized wave packet.

The combination of large numbers of different wave numbers can, however, be accomplished in other ways.

For example, consider the linear combination of cosine solutions with wave numbers,

$$k_n = nK/N \quad \text{where} \quad n = -N, -(N-1), \dots, (N-1), N \qquad (2.29)$$

We can consider these to be wave numbers sampled uniformly (i.e. each with identical weight $1/N$) from the interval $(-K, K)$; we will eventually consider the limiting case with increasingly fine spacing, that is, letting $N \to \infty$.

We then have the solution

$$\psi_N(x) = \frac{1}{N} \sum_{n=-N}^{n=+N} \cos\left(\frac{nKx}{N}\right) = \frac{1}{N}\left(1 + 2\sum_{n=1}^{N} \cos\left(\frac{nKx}{N}\right)\right) \qquad (2.30)$$

This summation can actually be obtained in closed form (P2.9) giving

$$\psi_N(x) = \frac{1}{N}\left(1 + \frac{2\cos(Kx/2(1+1/N))\sin(Kx/2)}{\sin(Kx/2N)}\right) \qquad (2.31)$$

This function is also periodic in x, but now with period $2\pi N/K$ (see also P2.9); this implies that it can be localized by letting $N \to \infty$. We plot this summation for increasing N in Fig. 2.5, and we note that:

- The periodic "recurrences" are indeed pushed further away from the origin as N increases, giving a truly localized waveform in that limit.

- The wave packet still has an intrinsic width, even when $N \to \infty$, due to the finite range of k values used. This can also be seen analytically by noting that

$$\psi(x) = \lim_{N\to\infty} \psi_N(x) = \lim_{N\to\infty} \frac{2\cos(Kx/2(1+1/N))\sin(Kx/2)}{N\sin(Kx/2N)}$$
$$= \frac{2\sin(Kx)}{Kx} \qquad (2.32)$$

This limiting form is instructive as it clearly shows that $\psi(x) \to 0$ as $|x| \to \infty$.

- In order to accomplish the subtle destructive interference between plane wave components for arbitrarily large values of $|x|$, it seems that we must use an uncountably (continuous) large set of wave numbers.

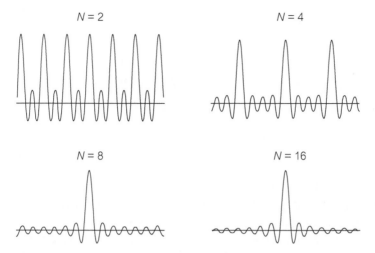

Figure 2.5. Linear superposition solution, $\psi_N(x)$, from Eqn. (2.31) for $N = 2, 4, 8, 16$ showing increasing localization as $N \to \infty$.

2.3 **Fourier Transforms**

This limit of a continuous summation over wave numbers, as well as the extension to include more general plane wave solutions, is formalized in the so-called *Fourier integral* or *Fourier transform*

$$f(x) = \frac{1}{\sqrt{2\pi}} \int_{-\infty}^{+\infty} dk\, A(k) e^{ikx} \tag{2.33}$$

The $A(k)$ gives the amplitude of each plane wave contribution to the resulting wave packet and is the continuous analog of the discrete Fourier coefficients, a_n, b_n. The seemingly arbitrary normalization factor $(1/\sqrt{2\pi})$ will be discussed in Section 2.4. As always, the final solution of the wave equation is obtained by letting $f(x) \to f(x \pm vt)$.

Example 2.3. Fourier transform with "flat" k values

The Fourier integral representation corresponding to the example at the end of Section 2.2.2 can be written by considering

$$A(k) = \begin{cases} 0 & \text{for } k > |K| \\ 1/\sqrt{2K} & \text{for } k < |K| \end{cases} \tag{2.34}$$

(Continued)

which has the resulting waveform

$$f(x) = \frac{1}{\sqrt{4\pi K}} \int_{-K}^{+K} e^{ikx}\, dk$$

$$= \frac{1}{\sqrt{4\pi K}} \left. \frac{e^{ikx}}{ix} \right|_{-K}^{+K}$$

$$= \sqrt{\frac{K}{\pi}} \frac{\sin(Kx)}{Kx} \tag{2.35}$$

We plot both $A(k)$ and $f(x)$ in Fig. 2.6 for two different values of K and note that the widths of the k and x distributions are inversely correlated. This arises because an increasingly large sample of wave numbers (larger Δk) can be more efficient in the destructive interference necessary to produce a smaller, more localized (smaller Δx) wave packet. In fact, if we make the identification of $\Delta k \approx 2K$ and estimate the spread in position by the location of the first set of nodes of $f(x)$, that is, $\Delta x \approx 2\pi/K$, we find that $\Delta k\, \Delta x \approx 4\pi$ or once again that

$$\Delta k\, \Delta x > \mathcal{O}(1) \tag{2.36}$$

independent of K.

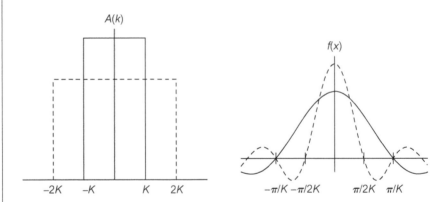

Figure 2.6. "Square" $A(k)$ and its Fourier transform $f(x)$ from Example 2.3 for two values of K.

Example 2.4. Fourier transform of exponential

Another example of a Fourier transform pair is obtained by considering

$$A(k) = \frac{1}{\sqrt{K}} e^{-|k|/K} \tag{2.37}$$

(Continued)
which gives

$$f(x) = \frac{1}{\sqrt{2\pi K}} \int_{-\infty}^{+\infty} e^{-|k|/K} e^{ikx}\, dk$$

$$= \frac{1}{\sqrt{2\pi K}} \left(\int_{-\infty}^{0} dk\, e^{k(1/K+ix)} + \int_{0}^{+\infty} dk\, e^{-k(1/K-ix)} \right)$$

$$= \sqrt{\frac{2K}{\pi}} \left(\frac{1}{1 + (Kx)^2} \right) \tag{2.38}$$

both of which we plot in Fig. 2.7; we see that the widths also satisfy Eqn. (2.36).

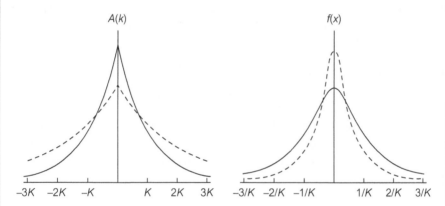

Figure 2.7. Exponential $A(k)$ and its Fourier transform $f(x)$ from Example 2.4 for two values of K.

This unavoidable constraint on the spatial extent and wave number content of a localized wave packet, $\Delta k \Delta x \geq 1$, is a fundamental limitation on physical systems. It restricts one's ability to make measurements or to produce physical phenomena in the same way as, for example, do the laws of thermodynamics or the limiting value of the speed of light.

Consider, for example, a laser pulse of finite duration (in both space and time) with the wave number distribution $(A(k))$ shown in Fig. 2.8(a); a long wave train with a correspondingly narrow k distribution has a relatively well-defined wavelength and so could resolve the two emission/absorption lines shown as dashed lines; if one wished to gain more "real time" information on the system by exciting it with laser pulses of very short duration (Fig.2.8(b)), the corresponding wave number distribution would accordingly *broaden*, making it no longer possible to resolve various spectral features.

Recognizing the importance of this result, we can make an immediate connection to quantum mechanics by using the de Broglie relation $p = h/\lambda = \hbar k$

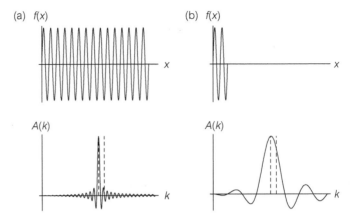

Figure 2.8. Schematic plot of wave amplitude for laser pulse, $f(x)$ versus x, for (a) long and (b) short pulses. The corresponding wave number amplitudes, $A(k)$ versus k, are shown and the dashed lines indicate two possible spectral lines.

to argue that any wave description of matter will necessarily satisfy

$$\Delta x \Delta p \approx \Delta x (\hbar \Delta k) \gtrsim \hbar \tag{2.39}$$

which is the content of the Heisenberg uncertainty principle. This limit on the ability to measure simultaneously the position and momentum of a quantum mechanical particle can thus be traced (in this language at least) to a wave description of mechanics.

2.4 Inverting the Fourier transform: the Dirac δ-function

We have seen that a truly localized wave packet can be constructed from plane wave solutions via the Fourier transform, that is,

$$f(x) = \frac{1}{\sqrt{2\pi}} \int_{-\infty}^{+\infty} A(k) \, e^{ikx} \, dk \tag{2.40}$$

If, however, we are given a spatial waveform, $f(x)$, we would also like to be able to extract the wave number or wavelength "components" by somehow inverting Eqn. (2.40) to obtain $A(k)$. Knowledge of $A(k)$ allows one to determine the behavior of a wave packet in, say, a diffraction or interference experiment, where one has simple rules for the behavior of each individual wavelength component. We devote this section to the question of how such an inversion is obtained, at the same time developing the mathematical properties of a new function which will be of continuing use, the so-called *Dirac δ-function*.

The final result we will obtain is quite simple, namely that

$$A(k) = \frac{1}{\sqrt{2\pi}} \int_{-\infty}^{+\infty} f(x) \, e^{-ikx} \, dx \tag{2.41}$$

This can be taken to mean that $A(k)$ and $f(x)$ are, in a sense, inverses of each other under the Fourier transform, and the similarity in form argues for the conventional use of the common $1/\sqrt{2\pi}$ factors.

In order to derive Eqn. (2.41) we multiply both sides of Eqn. (2.40) by $\exp(-ik'x)/\sqrt{2\pi}$ and integrate over x. Doing this, we obtain

$$\frac{1}{\sqrt{2\pi}} \int_{-\infty}^{+\infty} f(x) \, e^{-ik'x} \, dx = \frac{1}{2\pi} \int_{-\infty}^{+\infty} dx \int_{-\infty}^{+\infty} A(k) e^{ikx} e^{-ik'x} dk$$

$$= \int_{-\infty}^{+\infty} dk \, A(k) \left[\frac{1}{2\pi} \int_{-\infty}^{+\infty} e^{i(k-k')x} dx \right]$$

$$= \int_{-\infty}^{+\infty} dk \, A(k) \, \delta(k - k')$$

$$\stackrel{?}{=} A(k') \tag{2.42}$$

where we have implicitly defined a new function called the *Dirac δ-function* via

$$\delta(k - k') \equiv \frac{1}{2\pi} \int_{-\infty}^{+\infty} e^{i(k-k')x} \, dx. \tag{2.43}$$

Thus, if Eqn. (2.41) is to be true, we must have

$$\int_{-\infty}^{+\infty} A(k)\delta(k - k')dk = A(k') \tag{2.44}$$

that is, $\delta(k - k')$ must "pick" out the value of $A(k)$ only at $k = k'$ from the continuous integral. In this regard, it is similar in function (and name) to the *discrete* or *Kronecker δ-function* defined as

$$\delta_{n,m} = \begin{cases} 0 & \text{if } n \neq m \\ 1 & \text{if } n = m \end{cases} \tag{2.45}$$

which has the property that

$$\sum_{k=-\infty}^{+\infty} A_k \, \delta_{n,k} = A_n \tag{2.46}$$

namely, that it picks out a specific term in a discrete summation.

To study the properties of the Dirac δ-function, it suffices to consider the special case where $k' = 0$, that is,

$$\delta(k) = \frac{1}{2\pi} \int_{-\infty}^{+\infty} e^{ikx} \, dx \tag{2.47}$$

We can then argue (not necessarily prove) that

$$\delta(k) = \begin{cases} \frac{1}{2\pi} \int_{-\infty}^{+\infty} 1 \, dx = \infty & \text{for } k = 0 \\ \frac{1}{2\pi} \int_{-\infty}^{+\infty} (\cos(kx) + i\sin(kx)) \, dx = 0 & \text{for } k \neq 0 \end{cases} \tag{2.48}$$

where the vanishing of the integrals of the $\sin(kx)$ and $\cos(kx)$ functions occurs because of their oscillatory behavior, leading to cancellations. This heuristic definition of $\delta(k)$, namely, that it vanishes everywhere except at $k = 0$, where it is infinite, shows that it is is an extremely poorly behaved function[3] and has to be handled carefully.

We can study it a bit more rigorously by considering the family of auxiliary functions

$$\delta_\epsilon(k) \equiv \frac{1}{2\pi} \int_{-\infty}^{+\infty} e^{-\epsilon x^2} e^{ikx} \, dx = \frac{1}{2\pi} \sqrt{\frac{\pi}{\epsilon}} \, e^{-k^2/4\epsilon} \tag{2.49}$$

so that

$$\lim_{\epsilon \to 0} \delta_\epsilon(k) = \delta(k) \tag{2.50}$$

(We have simply included a convergence factor so that the integral can be performed in closed form using results in Appendix D.1.) Using this limiting representation, it is easier to argue that

$$\delta_\epsilon(k) \propto \begin{cases} 1/\sqrt{\epsilon} & \to \infty \quad \text{for } k = 0 \\ e^{-k^2/2\epsilon}/\sqrt{\epsilon} & \to 0 \quad \text{for } k \neq 0 \end{cases} \tag{2.51}$$

as $\epsilon \to 0$; we can also visualize the approach to the singular limit in Fig. 2.9.

This form also allows us to investigate the degree of "infiniteness" at $k = 0$ by considering

$$\int_{-\infty}^{+\infty} \delta_\epsilon(k) \, dk = \frac{1}{2\pi} \sqrt{\frac{\pi}{\epsilon}} \int_{-\infty}^{+\infty} e^{-k^2/4\epsilon} \, dk = \frac{1}{\sqrt{\pi}} \int_{-\infty}^{+\infty} e^{-q^2} \, dq = 1 \tag{2.52}$$

so that the total area under the δ-function family of curves is always normalized to unity. Thus we also take

$$\int_{-\infty}^{+\infty} \delta(k) \, dk = \lim_{\epsilon \to 0} \int_{-\infty}^{+\infty} \delta_\epsilon(k) \, dk = 1 \tag{2.53}$$

---

[3] It is in the class of mathematical objects called distributions or generalized functions for which the standardly cited reference is Lighthill (1958) in which he discusses, for example, *Good functions and fairly good functions*.

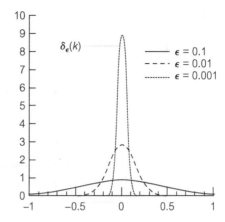

Figure 2.9. Limiting behavior of $\delta_\epsilon(k)$ in Eqn. (2.49); the dashed (solid, dotted) curves correspond to $\epsilon = 0.1$ (0.01, 0.001) respectively.

in much the same way that

$$\sum_{i=-\infty}^{+\infty} \delta_{i,j} = 1 \tag{2.54}$$

(One can (very loosely) say that $\delta(k = 0) = 1/dk$ where dk is the infinitesimal unit of measure.) We can also derive results such as

$$\int_{-\infty}^{+\infty} k\,\delta(k)\,dk = \lim_{\epsilon \to 0} \int_{-\infty}^{+\infty} k\,\delta_\epsilon(k)\,dk = 0 \tag{2.55}$$

and related ones involving higher powers of k. Using these results we can now argue that

$$\int_{-\infty}^{+\infty} A(k)\,\delta(k - k')\,dk$$

$$= \int_{-\infty}^{+\infty} A(q + k')\,\delta(q)\,dq$$

$$= \int_{-\infty}^{+\infty} \left(A(k') + qA'(k') + \frac{q^2}{2}A''(k') + \cdots \right) \delta(q)\,dq$$

$$= A(k') \left[\int_{-\infty}^{+\infty} \delta(q)\,dq \right] + A'(k) \left[\int_{-\infty}^{+\infty} q\,\delta(q)\,dq \right] + \cdots$$

$$= A(k') \tag{2.56}$$

where we have changed variables to $q = k - k'$ and expanded $A(k)$ in a Taylor expansion around $k = k'$. This is the desired property of the Dirac δ-function and shows that $A(k)$ and $f(x)$ are indeed related by Eqns (2.40) and (2.41).

The similarity of a spatial waveform and its Fourier transform in terms of their information content can be seen in other ways. Since we often consider complex waveforms, we will find useful the fact that

$$\int_{-\infty}^{+\infty} |f(x)|^2 \, dx = \int_{-\infty}^{+\infty} f^*(x) f(x) \, dx$$

$$= \int_{-\infty}^{+\infty} dx \, f^*(x) \left(\frac{1}{\sqrt{2\pi}} \int_{-\infty}^{+\infty} A(k) \, e^{ikx} \, dk \right)$$

$$= \int_{-\infty}^{+\infty} A(k) \, dk \left(\frac{1}{\sqrt{2\pi}} \int_{-\infty}^{+\infty} f^*(x) \, e^{ikx} \, dx \right)$$

$$= \int_{-\infty}^{+\infty} A(k) \, dk \left(\frac{1}{\sqrt{2\pi}} \int_{-\infty}^{+\infty} f(x) e^{-ikx} \, dx \right)^*$$

$$= \int_{-\infty}^{+\infty} A(k) \, A^*(k) \, dk$$

$$\boxed{\int_{-\infty}^{+\infty} |f(x)|^2 \, dx = \int_{-\infty}^{+\infty} |A(k)|^2 \, dk} \tag{2.57}$$

This result is sometimes called *Parseval's theorem*. We note that both examples considered in Section 2.3 were chosen so that they satisfied $\int_{-\infty}^{+\infty} |A(k)|^2 \, dk = 1$ (check this in P2.15) so that the corresponding spatial waveforms are guaranteed to yield $\int_{-\infty}^{+\infty} |f(x)|^2 \, dx = 1$ as well. A similar relation can also be proved for the *overlap* of two different waveforms, namely

$$\int_{-\infty}^{+\infty} f_1^*(x) f_2(x) \, dx = \int_{-\infty}^{+\infty} A_1^*(k) A_2(k) \, dk \tag{2.58}$$

While the Dirac δ-function is used here as a tool in proving the important physical connection between $A(k)$ and $f(x)$, its usefulness in mathematical physics will become more obvious, and we make some additional comments here; some of its other properties are discussed in Appendix E.8 to which we will refer the interested reader from time to time.

- The arguments of $\delta(k - k')$ are arbitrary, as we could equally well have considered the integral definition

$$\delta(x - x') = \frac{1}{2\pi} \int_{-\infty}^{+\infty} e^{ik(x-x')} \, dk \tag{2.59}$$

- The *dimensions* of the δ-function thus depend on its argument: namely, $[\delta(z)] = 1/[z]$, where $[z]$ gives the dimensions of z. This can be inferred from the dimensionlessness of the integral $\int_{-\infty}^{+\infty} \delta(z) \, dz = 1$.

- The definite integral of the δ-function, defined via

$$\theta(x) \equiv \int_{-\infty}^{x} \delta(x') \, dx' \tag{2.60}$$

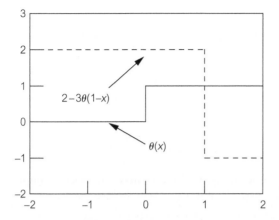

Figure 2.10. The Heaviside or step function $\theta(x)$ (solid curve) and $2 - 3\theta(1-x)$ (dashed curve) versus x.

is easily seen to satisfy

$$\theta(x) = \begin{cases} 0 & \text{for } x < 0 \\ 1 & \text{for } x > 0 \end{cases} \tag{2.61}$$

depending on whether the range of integration includes the singular point or not. This function (often called the *step function* or *Heaviside function*) can describe the instantaneous (and hence idealized) "turn-on" (or "turn-off" if one uses $1 - \theta(x)$) of some phenomenon or function, as shown in Fig. 2.10.

2.5 Dispersion and Tunneling

2.5.1 Velocities for Wave Packets

We have seen that wave packets constructed from plane waves satisfying the simple dispersion relation

$$\omega = \omega(k) = kv \tag{2.62}$$

propagate with no change in shape due to the constant speed, v, of each component. Many classical physical systems are characterized by dispersion relations for which this is not true and which exhibit a variety of new phenomena that have analogs in quantum mechanics, specifically *dispersion* and *tunneling*; we review these aspects of classical wave physics in this section using one model system as an example.

An important system in which both of these phenomena occur arises in the study of the propagation of electromagnetic (hereafter EM) radiation in a region of ionized gas (i.e. a plasma). The self-consistent application of Maxwell's equations (for the EM waves) and Newton's laws (for the motion of the charged particles, in this case the electrons) implies[4] that the dispersion relation is

$$\omega^2 = (kc)^2 + \omega_p^2 \tag{2.63}$$

Here ω_p is the *plasma frequency*,

$$\omega_p^2 \equiv \frac{n_e e^2}{\epsilon_0 m_e} \tag{2.64}$$

and n_e is the number density of electrons. The plasma frequency is the natural frequency of oscillation of the plasma system, arising, for example, when the positive ions and negative electrons are separated slightly and oscillate around their equilibrium neutral configuration.

This dispersion relation can be rewritten in the form

$$w(k) = \frac{c}{\sqrt{1 - \omega_p^2/\omega^2}} \, k = v_\phi k \tag{2.65}$$

where we have defined a *phase velocity*, $v_\phi \equiv \omega(k)/k$. If we restrict ourselves to the case where $\omega > \omega_p$, it is clear that $v_\phi > c$ and this velocity exceeds the speed of light in vacuum. This is not, however, in conflict with the tenets of special relativity, as the phase velocity measures a property of one single plane wave component, which, by definition, has a trivial, sinusoidal space- and time-dependence. We have already argued that such a waveform is an abstraction; it is, on average, uniform over all space and time and so can lead to no transfer of information. Any *changes* in the wavefield, which might carry a signal, will be described by *modulations* in the plane wave signals, and such disturbances will propagate at speeds less than c.

To see this, consider again a linear combination of two traveling waves (as done for beats) of differing wavenumbers and frequencies, governed by a general dispersion relation, $\omega = \omega(k)$. In this case,

$$\phi(x, t) = A \cos(k_1 x - \omega_1 t) + A \cos(k_2 x - \omega_2 t)$$

$$= 2A \cos\left(\frac{(k_1 - k_2)x}{2} - \frac{(\omega_1 - \omega_2)t}{2}\right) \cos\left(\frac{(k_1 + k_2)x}{2} - \frac{(\omega_1 + \omega_2)t}{2}\right)$$

$$= A_{\text{eff}}(x, t) \cos(\bar{k}x - \bar{\omega}t) \tag{2.66}$$

where

$$\bar{k} \equiv \frac{k_1 + k_2}{2} \quad \text{and} \quad \bar{\omega} \equiv \frac{\omega_1 + \omega_2}{2} \tag{2.67}$$

[4] See, for example, Kittel (1971) for a derivation.

are something like the "average" wave numbers and frequencies. We have defined an effective amplitude

$$A_{\text{eff}}(x, t) = 2A \cos(\Delta k\, x - \Delta\omega\, t) \tag{2.68}$$

with

$$\Delta k \equiv \frac{k_1 - k_2}{2} \quad \text{and} \quad \Delta\omega \equiv \frac{\omega_1 - \omega_2}{2} \tag{2.69}$$

If $k_1 \approx k_2$, we can consider Eqn. (2.66) to be a plane wave, $\cos(\bar{k}x - \bar{\omega}t)$, modulated by the amplitude $A_{\text{eff}}(x, t)$; information will be carried along the modulation wave crest, at a rate given by the condition

$$\theta_{\text{eff}} = \Delta k\, x - \Delta\omega\, t = \text{constant} \tag{2.70}$$

that is, examining a point of constant phase. This implies that the speed of information propagation is

$$\left(\frac{dx}{dt}\right)_{\text{information}} = \frac{\Delta\omega}{\Delta k} \longrightarrow \frac{d\omega}{dk} \tag{2.71}$$

We are thus led to define the *group velocity*,

$$v_g(\omega) = \frac{d\omega(k)}{dk} \tag{2.72}$$

and note that information contained in modulated waveforms or wave packets, as well as the rate of flow of energy density in a wave, are all governed by the group velocity. As an example, for EM waves in vacuum we have

$$w(k) = kc \quad \text{so that} \quad v_g = \frac{d\omega}{dk} = c \tag{2.73}$$

as expected. For the case of propagation in plasma, however, we find

$$v_g = \frac{d\omega(k)}{dk} = c\sqrt{1 - \omega_p^2/\omega^2} \le c \tag{2.74}$$

consistent with relativity. Note that if $\omega/\omega_p \gg 1$, the background plasma cannot "keep up" with the wave, and the radiation propagates at the same rate as in vacuum; as $\omega/\omega_p \to 1$ the effective speed can become arbitrarily small, and can even become imaginary when the ratio is less than unity.

2.5.2 **Dispersion**

For the dispersion relation in Eqn. (2.63), v_g clearly depends on frequency; this reflects the fact that different plane wave components of a wave packet will travel at different speeds due to different *phase velocities*. Any initially localized wave packet will necessarily contain a range of wavenumbers k (and hence frequencies ω); since the faster components will outpace the slower ones, the wave will necessarily spread or disperse as it propagates. Using Eqn. (2.74) as an example, a spread in frequencies, $\Delta\omega$, implies a range in group velocities

$$\Delta v_g = c \left(1 - \omega_p^2/\omega^2\right)^{-1/2} \frac{\omega_p^2}{\omega^3}\Delta\omega \approx c\frac{\omega_p^2}{\omega^3}\Delta\omega \qquad (2.75)$$

assuming that $\omega \gg \omega_p$; note that higher frequencies travel faster. If an initial pulse travels a fixed distance D, one should observe a difference in arrival times due to this effect, the two being related by $\Delta v_g = -c^2\Delta t/D$ (since higher-velocity components arrive earlier). This implies that the higher-frequency components of the pulse will arrive earlier, satisfying the relation

$$\Delta\omega = -\frac{c}{D}\frac{\omega^3}{\omega_p^2}\Delta t \qquad (2.76)$$

One of the most famous examples of this phenomenon is evident in the observed dispersion of pulses of microwave radiation emitted by the Crab Nebula which must travel through a region of ionized space. Figure 2.11 illustrates one experimental realization of this effect[5]; clearly the higher frequencies do arrive earliest. From this data, the plasma frequency (and average electron density) can be estimated (P2.22).

This phenomenon of dispersion can be examined in more mathematical detail by constructing wave packets consisting of a superposition of plane wave solutions, each satisfying Eqn. (2.63); these would be the analogs of the Fourier

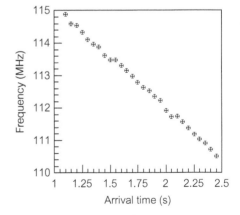

Figure 2.11. Plot of the frequency (MHz) versus arrival time (sec) for radio pulses from the Crab Nebula showing dispersion; the highest frequencies arrive earliest. The figure uses the original data of Staelin and Reifenstein (1968).

[5] Data take from Staelin and Reifenstein (1968).

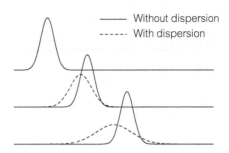

Figure 2.12. Electromagnetic wave packet in vacuum (solid) and in plasma (dashed) showing decrease in speed and spreading due to the dispersion relation in Eqn. (2.63).

integral solutions of Eqn. (2.40). For example, we can write

$$f(x,t) = \frac{1}{\sqrt{2\pi}} \int_{-\infty}^{+\infty} dk \, A(k) \, e^{i(kx-\omega(k)t)} \tag{2.77}$$

Even if we can calculate the Fourier transform explicitly at $t = 0$ to obtain $f(x,0)$, we can no longer assume that $f(x,t) = f(x - ct)$ unless $\omega = ck$. For a general dispersion relation, the integral must usually be done numerically at each desired time. We show in Fig. 2.12 the results of such a calculation; the dashed curve corresponds to a wave packet propagating with $\omega_p \neq 0$, and we compare it to a wave packet with the same initial shape but which propagates in vacuum (where $\omega_p = 0$); the increasing spread and the slower speed ($v_g < c$) are both evident.

If we choose $A(k)$ to be peaked around $k = k_0$, then the wave packet

$$f(x,t) = \frac{1}{\sqrt{2\pi}} \int_{-\infty}^{+\infty} dk \, A(k - k_0) \, e^{i(kx-\omega(k)t)} \tag{2.78}$$

will contain wavenumbers centered around k_0, with some spread Δk. If we change variables to $q = k - k_0$, it is natural to expand the exponent around k_0 to obtain

$$kx - \omega t = (k_0 x - \omega(k_0)t) + q\left(x - \frac{d\omega}{dk}(k_0)t\right) + \frac{1}{2}q^2 \frac{d^2\omega}{dk^2}(k_0)t + \cdots$$

$$= (k_0 x - \omega_0 t) + q(x - v_g t) + q^2 \beta t + \cdots \tag{2.79}$$

where $\beta \equiv d^2\omega/dk^2/2$. For EM waves in vacuum where $\omega = kc, \beta = 0$, the wave packet can be written as

$$f(x,t) = e^{i(k_0 x - \omega_0 t)} \frac{1}{\sqrt{2\pi}} \int_{-\infty}^{+\infty} dq \, A(q) \, e^{iq(x-v_g t)}$$

$$= e^{i(k_0 x - \omega(k_0)t)} f(x - v_g t) \tag{2.80}$$

so that $|f(x,t)|^2 = |f(x - v_g t)|^2$ as expected.

For a general dispersion relation, $\beta \neq 0$, and β measures the spread or range in propagation speeds; we can see this by writing

$$q(x - v_g t) + q^2 \beta t + \cdots = q(x - v_g t - q\beta t) + \cdots = q(x - (v_g - q\beta)t + \cdots \tag{2.81}$$

We know that the integral over q will be dominated by values in the range $(-\Delta k, \Delta k)$, so we associate

$$q\beta \approx \pm \Delta k \frac{d^2 \omega}{dk^2} \approx \pm \Delta v_g \tag{2.82}$$

as the spread in velocities, implying a dispersive wave.

The rate at which the wave packet spreads can be studied by noting that the peak of the packet is dominated by parts of the Fourier integral where $x - v_g t \approx 0$, so that as long as the next term in the expansion satisfies

$$q^2 \beta t \approx \Delta k^2 \beta t \ll 1 \tag{2.83}$$

the wave packet will not spread significantly. This condition can be rephrased in terms of a spreading time, defined by

$$t_0 \equiv \frac{1}{\beta \Delta k^2} \tag{2.84}$$

and noting that when $t \gtrsim t_0$ the spreading will become important.

2.5.3 Tunneling

So far we have assumed that $\omega > \omega_p$, but we can see that when $\omega < \omega_p$ we can write

$$k = \pm \frac{1}{c} \sqrt{\omega^2 - \omega_p^2} \longrightarrow \kappa = \pm \frac{i}{c} \sqrt{\omega_p^2 - \omega^2} \tag{2.85}$$

and the plane wave solutions now have the forms

$$\phi_k(x, t) = e^{\pm \kappa x} e^{-i\omega t} \tag{2.86}$$

Instead of oscillatory behavior, these solutions exhibit exponential decay (or growth).

If, for example, an EM wave with $\omega < \omega_p$ propagating to the right encountered a semi-infinite wall of plasma at $x = 0$ the possible solutions for $x > 0$ are shown in Fig. 2.13. Clearly the exponentially growing solution is unacceptable if the plasma extends forever to the right and must be discarded; this leaves only the solution $e^{-\kappa x} e^{-i\omega t}$. In this case we do not have propagation, but rather exponential damping or attenuation of the wave amplitude, $\phi(x, t)$. The corresponding

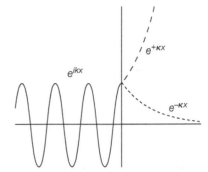

Figure 2.13. Electromagnetic wave in vacuum (on the left) impinging on a region of plasma (on the right); exponentially growing (dash) and decaying solutions (dotted) are allowed in the plasma.

energy density, which goes like $|\phi(x, t)|^2$, will penetrate into the plasma with a spatial dependence $e^{-2\kappa x} = e^{-x/d}$.

For static/time-independent ($\omega = 0$ or DC) or very slowly varying fields (for which we can assume that $\omega << \omega_p$), the penetration depth is set by the distance scale $d = c/2\omega_p$. An example is the plasma in fusion reactors where $n_e \approx 10^{12} - 10^{16}$ electrons/cm^3 giving $\omega_p \approx 6 \times 10^{10} - 6 \times 10^{12}$ s^{-1} or $d \approx 2$ mm $- 20\,\mu$m; electric and magnetic fields are effectively expelled from such plasmas. Similar exponential damping occurs for EM waves in conductors (P2.25) where the fields in the nonpropagating region are sometimes called "evanescent" waves.

In this context, it is interesting to note that such waves, although exponentially attenuated, can yield a finite wave amplitude if allowed to propagate through a *finite* thickness of plasma or conductor.[6] This is the classical wave analog of quantum mechanical tunneling, which is discussed in Chapters 8 and 11.

2.6 **Questions and Problems**

Q2.1. Suppose you had waves satisfying Eqn. (2.4) for $x \leq 0$, but which are reflected from an infinite wall located at $x = 0$. At the infinite wall, the wave amplitude must satisfy $\phi(x, t) = 0$ for all values of t. If an initial highly localized wave form, given by $f(x)$, is located far to the left of the wall, moving to the right, show that an appropriate solution $\phi(x, t)$ is given by

$$\phi(x, t) = \begin{cases} f(x - vt) - f(-x - vt) & \text{for } x \leq 0 \\ 0 & \text{for } 0 \leq x \end{cases} \tag{2.87}$$

Discuss the form of the reflected wave and compare to discussions in introductory textbooks on the behavior of wave pulses on "ropes with tied ends." What

---

[6] For a description of *Microwave experiments on electromagnetic evanescent waves and tunneling effect,* see Albiol *et al.* (1993).

would be the appropriate boundary condition on $\phi(x, t)$ if the infinite wall at $x = 0$ were replaced by a "free end," say a rope tied to a ring free to move up and down a pole. What would the solution corresponding to Eqn. (2.87) above be?

Q2.2. Suppose you have a function, $f(x)$, which has a Fourier *series* as in Eqn. (2.17). What would the Fourier *transform* of $f(x)$ look like?

Q2.3. In Examples 2.1 and 2.2, one can see that $a_n = 0$ for *even* values of n (except possibly for $n = 0$). Can you sketch (either literally or figuratively) a proof of why this should be so?

Q2.4. Is there any kind of "uncertainty principle" for *Fourier series*?

Q2.5. Show that the time and frequency variables t and $\omega = 2\pi f$ form a Fourier transform pair, that is, $f(t) \longleftrightarrow A(\omega)$. What is the corresponding classical uncertainty condition? How about its quantum analog? In this context, why would the Fourier transform pairs of Example 2.3 have anything to do with "ringing"? Why would the a_0 component of a Fourier series for $f(t)$ be called the DC component?

Q2.6. Why do functions with "sharp edges" have Fourier transforms which have "lots of wiggles"?

Q2.7. How would the experimental data shown in Fig. 2.11 change if the plasma density between us and the Crab Nebula were doubled? If the distance between us were doubled?

Q2.8. Assume an EM wave in vacuum is incident from the right on a semi-infinite region of plasma at $x = 0$. What is the appropriate amplitude for $x < 0$? If the wave is incident from the right on a finite thickness of plasma, what is the most general solution in that region?

P2.1. **Factoring the wave equation.** Show that the wave equation in one-dimension, Eqn. (2.4), can be written in the factored form

$$0 = \left(\frac{\partial^2}{\partial t^2} - v^2 \frac{\partial^2}{\partial x^2} \right) \phi(x, t) = \hat{D}_+ \hat{D}_- \phi(x, t) = \hat{D}_- \hat{D}_+ \phi(x, t) \qquad (2.88)$$

where $\hat{D}_\pm \equiv \partial/\partial t \pm v\partial/\partial x$. Use this to show that $\hat{D}_\pm \phi(x, t) = 0$ gives the most general solutions in Section 2.2.1. Factorization methods are discussed in more detail in Chapter 13.

P2.2. **Reflection and transmission on a string.** Consider an infinitely long string with constant tension T which has linear mass density ρ_1 for $x < 0$ and ρ_2 for $x > 0$.

(a) Assume a solution of the form

$$A(x, t) = \begin{cases} Ie^{i(k_1 x - \omega_1 t)} + Re^{i(-k_1 x - \omega_1 t)} & \text{for } x < 0 \\ Te^{i(k_2 x - \omega_2 t)} & \text{for } x > 0 \end{cases} \qquad (2.89)$$

which represents an incident and reflected wave for $x < 0$ and a transmitted wave for $x > 0$. Use the fact that the string amplitude should be continuous at $x = 0$ to show that $\omega_1 = \omega_2$ and that $I + R = T$.

(b) If the tension in both halves of the string is the same, there will be no "kink" at $x = 0$, that is, the slope $\partial A(x, t)/\partial x$, will also be continuous. Show that this implies that $k_1(I - R) = k_2 T$.

(c) Solve for R and T in terms of I and $\rho_{1,2}$. Does your solution make sense for the cases $\rho_1 = \rho_2$? for $\rho_2 \gg \rho_1$? for $\rho_2 \ll \rho_1$?

P2.3. Verify Eqns (2.21) and (2.24) for the Fourier series of the square waveform in Example 2.1.

P2.4. Verify Eqns (2.27) and (2.28) for the Fourier series of the triangular waveform in Example 2.2.

P2.5. (a) Find the Fourier series for the waveform defined in the interval $(-L, +L)$ by

$$f(x) = \begin{cases} -A & \text{for } -L < x < 0 \\ +A & \text{for } 0 < x < +L \end{cases} \tag{2.90}$$

(b) Compare the coefficients to those obtained for Example 2.1.

(c) Calculate the Δ_N for this series and show that $\Delta_N \to 1$ as $N \to \infty$.

P2.6. Find the Fourier series for the periodic waveform defined in the interval $(-L, +L)$ by

$$f(x) = A(L^2 - x^2)^2. \tag{2.91}$$

This function is continuous at $x = \pm L$ and has a continuous derivative there as well. Compare the convergence (perhaps by calculating the Δ_N) for this case, and compare to that found in Examples 2.1 and 2.2.

P2.7. Derive an expression for the Δ_N of a Fourier series in terms of its expansion coefficients, a_n and b_n.

P2.8. **Derivatives of Fourier series.** (a) If one periodic function is the derivative of another, how are their Fourier coefficients related?

(b) Using the definitions of the two functions in Examples 2.1 and 2.2, show that they satisfy

$$f'_{\text{triangle}}(x) = f_{\text{square}}\left(x + \frac{L}{2}\right). \tag{2.92}$$

Differentiate the Fourier series for $f_{\text{triangle}}(x)$ and show that it also satisfies this relation.

P2.9. (a) Consider the summation in Eqn. (2.30) and show that

$$S_N \equiv \sum_{n=1}^{N} \cos\left(\frac{nKx}{N}\right) = \text{Re}\left(\sum_{n=1}^{N} e^{inKx/N}\right) = \text{Re}\left(\sum_{n=1}^{N} (e^{iKx/N})^n\right) \equiv \text{Re}(T_N) \tag{2.93}$$

where $\text{Re}(z)$ denotes the real part of z.

(b) Show that

$$\sum_{n=1}^{N}(e^z)^n = \frac{(e^{(N+1)z} - e^z)}{(e^z - 1)} \tag{2.94}$$

and use this to show that

$$T_N = e^{iKx(1+1/N)/2}\frac{(e^{iKx/2} - e^{-iKx/2})}{(e^{iKx/2N} - e^{-iKx/2N})} \tag{2.95}$$

so that taking the real part gives

$$S_N = \frac{\cos(Kx(1+1/N)/2)\sin(Kx/2)}{\sin(Kx/2N)} \tag{2.96}$$

P2.10. Show that $\psi_N(x)$ in Eqn. (2.30) is periodic with period $2\pi K/N$ and verify the limit in Eqn. (2.32).

P2.11. Show that the waveform defined by

$$\tilde{\psi}_N(x) \equiv \frac{1}{N^2}\sum_{n=-N}^{n=+N} n\sin\left(\frac{nKx}{N}\right) \tag{2.97}$$

can also be evaluated in closed form. Hint: Show that

$$\tilde{\psi}_N(x) = -\frac{1}{K}\frac{\partial\psi_N(x)}{\partial x} \tag{2.98}$$

Find and then sketch the limiting waveform as $N \to \infty$ and verify that it is also localized. Is there an uncertainty relation of the form $\Delta k\,\Delta x > \mathcal{O}(1)$?

P2.12. Consider a Gaussian wavenumber distribution given by

$$A(k) = \frac{1}{\sqrt{K\sqrt{\pi}}}e^{-k^2/2K^2} \tag{2.99}$$

(a) Show that

$$\int_{-\infty}^{+\infty}|A(k)|^2 dk = 1 \tag{2.100}$$

(b) Find the Fourier transform $f(x)$ and confirm that it is normalized in the same way, that is

$$\int_{-\infty}^{+\infty}|f(x)|^2 dx = 1 \tag{2.101}$$

(c) *Estimate* the widths of the two distributions, Δk and Δx and show that they obey the $x - k$ uncertainty principle.

P2.13. Consider the localized triangle waveform defined by

$$f(x) = \begin{cases} N(a - |x|) & \text{for } |x| < a \\ 0 & \text{for } |x| > a \end{cases} \qquad (2.102)$$

(a) Find N such that $\int_{-\infty}^{+\infty} |f(x)|^2 dx = 1$.

(b) Find the Fourier transform $A(k)$ and confirm explicitly that it is normalized as well.

(c) Make estimates of Δx and Δk and show that the $x - k$ uncertainty principle holds.

P2.14. The integral in Example 2.3 was performed using complex exponential notation. Obtain the same result by writing e^{ikx} in terms cosines and sines and perform two "standard" integrals.

P2.15. Verify that the Fourier transform pairs in Examples 2.3 and 2.4 satisfy Parseval's theorem, namely

$$\int_{-\infty}^{+\infty} dx \, |f(x)|^2 = \int_{-\infty}^{+\infty} dk \, |A(k)|^2 \qquad (2.103)$$

P2.16. **Properties of Fourier transform pairs.** If $A(k)$ and $f(x)$ are related by Eqns. (2.40) and (2.41), verify the following relations. Note that we must generally assume that $f(x)$ and $A(k)$ are complex.

(a) If $A(k)$ is real, then $f(-x) = f^*(x)$. Show that this implies that $|f(x)|$ is an even function

(b) If $A(k)$ is imaginary, then $f(-x) = -f^*(x)$

(c) If $A(k)$ is even, then $f(-x) = f(x)$ i.e. $f(x)$ is even

(d) If $A(k)$ is odd, then $f(-x) = -f(x)$ i.e. $f(x)$ is odd

(e) If $A(k)$ and $f(x)$ are transform pairs, then so are $A(\alpha k)$ and $f(x/\alpha)/|\alpha|$; this "scaling" relation is sometimes useful

(f) If $f(x)$ and $A(k)$ are transform pairs, then so are $f(x + a)$ and $e^{ika} A(k)$.

(g) Show that the Fourier transform of $f'(x)$ is $ikA(k)$.

(h) What are the statements corresponding to cases (a) - (g) above, if any, about Fourier *series*?

P2.17. Consider the class of functions defined via

$$\delta_\epsilon(x) = \begin{cases} 0 & \text{for } |x| > \epsilon \\ 1/2\epsilon & \text{for } |x| < \epsilon \end{cases} \qquad (2.104)$$

(a) Show that this is an appropriate family of functions whose limit is a Dirac δ-function.

(b) Calculate $\theta_\epsilon(x) = \int_{-\infty}^{x} \delta_\epsilon(x) \, dx$ and show that it approaches a step-function.

P2.18. Consider the class of functions defined via

$$\delta_\epsilon(x) = \frac{\epsilon}{\pi} \frac{\sin^2(x/\epsilon)}{x^2} \tag{2.105}$$

Show that $\delta_\epsilon(x)$ has all of the desired properties of a Dirac δ-function as $\epsilon \to 0$.

P2.19. Overlap integrals. If we define

$$A(k; K) = \frac{1}{\sqrt{K}} e^{-|k|/K} \tag{2.106}$$

so that its Fourier partner is

$$f(x; K) = \sqrt{\frac{2K}{\pi}} \frac{1}{(1 + (Kx)^2)} \tag{2.107}$$

show that the overlap integrals satisfy

$$\int_{-\infty}^{+\infty} dk \, A(k; K_1) \, A(k; K_2) = \int_{-\infty}^{+\infty} dx \, f(x; K_1) \, f(x; K_2) \tag{2.108}$$

consistent with Eqn. (2.57). Confirm that the overlap is maximized at the value 1 when $K_1 = K_2$ as it should.

P2.20. It is sometimes useful to use a *finite well potential*, defined as

$$V(x) = \begin{cases} 0 & \text{for } |x| > a \\ -V_0 & \text{for } |x| < a \end{cases} \tag{2.109}$$

(a) Write $V(x)$ in terms of θ functions.

(b) Calculate the corresponding force via $F(x) = -dV(x)/dx$ and discuss its physical significance.

P2.21. The dispersion relation for gravitational water waves (ignoring surface tension effects) in a fluid of depth h is

$$\omega^2 = gk \frac{(1 - e^{-2kh})}{(1 + e^{-2kh})} \tag{2.110}$$

Calculate the phase and group velocities for both deep water (defined via $h \gg \lambda$) and shallow water ($h \ll \lambda$) waves. In which limit are the waves nondispersive?

P2.22. Dispersion data from the Crab Nebula. (a) Use the data in Fig. 2.11 to estimate $\Delta\omega$ and Δt; recall that $\omega = 2\pi f$. Assume that the distance to the Crab Nebula is roughly 2 kpc (where 1 kpc is one kiloparsec with a conversion given by $1 \, \text{pc} \approx 3.1 \times 10^{16}$ m.) Use these values to show that the plasma frequency is approximately $\omega_p = 9 \times 10^3$ rad/s. As a cross-check, compare this to the frequencies of the radio pulses detected in the experiment to confirm that $\omega \gg \omega_p$ so that the approximations leading to Eqn. (2.75) are appropriate.

(b) Show that the data for these experiments only give information on the product $n_e D$. Assuming that the electrons density is uniform, and using the value of D above (obtained from other measurements), estimate the *average* electron density in interstellar space; if most of the ionized gas is near the source, the interstellar electron density is even less than this.

P2.23. (a) Commercial radio stations broadcast signals in both the AM band ($f = 500 - 1500\,\text{kHz}$) and in the FM band ($88$–$108\,\text{MHz}$). Compare the corresponding frequencies, ω, to the plasma frequency of the ionosphere, using the fact that daytime plasma densities in the ionosphere are of the order $n_e \approx 10^{12} - 10^{13}$ electrons/m^3. Will either band of frequencies propagate through the ionosphere or reflect from it?

(b) At night, one can often hear radio stations which one does not hear during the day. Why? (There are both physics and non-physics reasons!)

(b) How far would DC ($f = 0$) electric and magnetic fields penetrate into the ionosphere?

P2.24. **Electromagnetic waveguides.** A standard problem in the study of EM waves is the propagation of EM waves in long metallic tubes or *waveguides*. For such a waveguide with a rectangular cross-section of $a \times b$, the dispersion relation between frequency and wavenumber is

$$\omega^2 = (kc)^2 + (\pi c)^2 \left[\left(\frac{m}{a} \right)^2 + \left(\frac{n}{b} \right)^2 \right] \tag{2.111}$$

where $m, n = 0, 1, 2, 3, \ldots$ and at least one of $m, n > 0$.

(a) What is the longest wavelength which can propagate freely in such a waveguide? (Assume that $a > b$ for definiteness.)

(b) If you drive your car into a tunnel (or parking structure), from which radio stations would you likely lose reception first, AM ($f = 540 - 1650\,\text{kHz}$) or FM ($f = 88 - 108\,\text{MHz}$)?

P2.25. **Electromagnetic waves in conductors.** The dispersion relation for the propagation of EM waves in a conducting medium is

$$\omega^2 = (kc)^2 - i\frac{\omega g}{\epsilon} \tag{2.112}$$

where g is the electrical conductivity.

(a) Find the wave number k as a function of ω and show that it corresponds to exponentially damped waves.

(b) In the limit of good conduction, that is, $g \gg \epsilon\omega$, show that the attenuation factor (for the field amplitude) reduces to the form

$$e^{-\beta x} = e^{-x/\delta} \quad \text{where } \beta = \sqrt{\frac{\omega g \mu}{2}} = \frac{1}{\delta} \tag{2.113}$$

The value of $\delta = 1/\beta$ is often called the "skin-depth."

(c) Calculate the frequency and wavelength of an EM wave in water for which the "skin-depth" is 1 meter. Use the values

$$\mu = \mu_0 = 4\pi \times 10^{-7} \frac{N \cdot s^2}{C^2} \quad \text{and} \quad g = 4.3 \frac{1}{\Omega \cdot m} \tag{2.114}$$

This problem has some relevance to the problem of communicating with submerged submarines[7].

P2.26. **"Culvert whistlers."**[8] Sound waves can be sent down a long tube (say of length L, assumed to be much greater than the effective width W, such as a culvert) where they are reflected. The sound heard upon reflection has been nicely described as follows:

Simply clap your hands at one end of the culvert and listen. You will hear a sharp echo of the hand clap at the expected delay time $t_0 = 2L/c$, where L is the length of the tube and $c = 340$ m/s is the velocity of sound in air. But, surprisingly, the sharp echo is followed immediately by a 'whistler,' a sound that starts at a very high pitch and then descends swiftly to a long lingering final note at frequency $f_{min} = 1/T_{max}$, where $T_{max} = 2W/c$ is the time it takes sound to travel directly back and forth across the tube of effective width W. (Crawford (1988))

(a) Assuming that the sound waves satisfy the dispersion relation

$$\omega^2 = (kc)^2 + \omega_0^2 \quad \text{where } \omega_0 = \frac{\pi c}{W} = 2\pi f_0 \tag{2.115}$$

evaluate the group velocity $v_g = d\omega(k)/dk$, and then the time it takes for one echo, $t = 2L/v_g$. Show that the frequency heard as a function of time is given by

$$f(t) = \frac{f_0}{\sqrt{1 - (t/t_0)^2}} \tag{2.116}$$

and discuss how your results describe the experimentally observed situation.

---

[7] See, for example, Reitz, Milford, and Christy (1993)
[8] See the similarly titled article by Crawford (1971).

THREE

The Schrödinger Wave Equation

3.1 The Schrödinger Equation

Just as it is impossible to *derive* Newton's equations of motion in classical mechanics or Maxwell's equations for electricity and magnetism (EM) from first principles, neither can we demonstrate the validity of the Schrödinger equation approach (or any other equivalent one) to quantum mechanics *a priori*. We can, however, make use of early quantum ideas and the connection between the classical wave equation and the photon concept to help in understanding its structure.

Let us consider one component (in one space dimension) of the classical EM wave equation, namely

$$\frac{\partial^2 \phi(x, t)}{\partial t^2} = c^2 \frac{\partial \phi(x, t)}{\partial x^2} \tag{3.1}$$

with a plane wave solution of the form

$$\phi(x, t) = A \, e^{i(kx - \omega t)} \tag{3.2}$$

We can use the photon concept by identifying the photon energy with its frequency (*à la* Einstein), namely

$$E = h\nu = \hbar\omega \tag{3.3}$$

and its momentum with its wavelength (as de Broglie did for material particles)

$$p = \frac{h}{\lambda} = \frac{\hbar 2\pi}{\lambda} = \hbar k \tag{3.4}$$

The plane wave solutions then take the form

$$\phi(x, t) = A \, e^{i(px - Et)/\hbar} \tag{3.5}$$

which satisfies the original wave equation, Eqn. (3.1), provided

$$E^2\phi(x, t) = (pc)^2\phi(x, t) \quad \text{or} \quad E^2 = (pc)^2 \tag{3.6}$$

We recognize this condition as the energy–momentum relation appropriate for a massless particle (namely, the photon). We can formalize this "wave–particle connection," not only for the solutions, but at the deeper level of the wave equation itself, provided we make the identifications

$$p \to \hat{p} = \frac{\hbar}{i}\frac{\partial}{\partial x} \quad \text{and} \quad E \to \hat{E} = i\hbar\frac{\partial}{\partial t} \tag{3.7}$$

In doing so, we have introduced two new *differential operators*, $\hat{p}$ and $\hat{E}$, representing the classical momentum and energy observables, p and E. (We will often distinguish quantum mechanical operators, $\hat{O}$, from their classical counterparts, O, by the use of this notation; $\hat{O}$ can then be read as *Oh-hat*.) When acting on plane wave solutions, these operators have a simple effect,

$$\hat{p}\phi(x, t) = \frac{\hbar}{i}\frac{\partial}{\partial x}\left[e^{i(px-Et)/\hbar}\right] = p\,\phi(x, t)$$

$$\hat{E}\phi(x, t) = i\hbar\frac{\partial}{\partial t}\left[e^{i(px-Et)/\hbar}\right] = E\,\phi(x, t) \tag{3.8}$$

namely, they return the classical numerical values. Using this formalism, we can write

$$\frac{\partial^2\phi(x, t)}{\partial t^2} = c^2\frac{\partial^2\phi(x, t)}{\partial x^2} \implies \hat{E}^2\phi(x, t) = \left(\hat{p}c\right)^2\phi(x, t) \tag{3.9}$$

which is a new *operator* version of the classical wave equation.

We might then be tempted to generalize this result to material particles (i.e. particles with mass) by associating

$$E^2 = (pc)^2 + (mc^2)^2 \implies \hat{E}^2\phi(x, t) = (\hat{p}c)^2\phi(x, t) + (mc^2)^2\phi(x, t) \tag{3.10}$$

or

$$\frac{\partial^2\phi(x, t)}{\partial t^2} = c^2\frac{\partial\phi(x, t)}{\partial x^2} - \left(\frac{mc^2}{\hbar}\right)^2\phi(x, t) \tag{3.11}$$

to obtain a (relativistically correct) wave equation for massive particles. This procedure yields the so-called *Klein–Gordon equation* which is, in fact, a useful dynamical equation for a certain class of particles. A problem arises, however, in the probabilistic interpretation of its solutions as representing a *single* particle. Very loosely speaking, the difficulty comes from the two possible signs of the square root operation when we write $E = \pm\sqrt{(pc)^2 + (mc^2)^2}$; this can be shown to give rise to antiparticles which must be included for self-consistency. (See Chapter 4 and especially P4.11 for more details.)

As we are more interested in generating a wave equation based on the *nonre-lativistic* connection between E and p for particles, we are led instead to write (for free particles)

$$E = \frac{p^2}{2m} \implies \hat{E}\psi(x,t) = \frac{\hat{p}^2}{2m}\psi(x,t) \tag{3.12}$$

or

$$i\hbar\frac{\partial\psi(x,t)}{\partial t} = -\frac{\hbar^2}{2m}\frac{\partial^2\psi(x,t)}{\partial x^2}. \tag{3.13}$$

This is the *time-dependent Schrödinger equation for a free particle*. We note that:

- As we will see, the free particle Schrödinger equation, along with its solutions, gives the quantum mechanical analog of Newton's first law of motion.

- It is a linear wave equation and so supports superposition and interference effects. This will allow us to construct localized wave packet solutions.

To allow for the possibility of interactions, we assume that any force, $F(x,t)$, felt by the particle is derivable from a potential energy function, $V(x,t)$, given by

$$F(x,t) = -\frac{\partial V(x,t)}{\partial x} \tag{3.14}$$

in which case we generalize $E = p^2/2m + V(x,t)$ to write

$$i\hbar\frac{\partial\psi(x,t)}{\partial t} = -\frac{\hbar^2}{2m}\frac{\partial^2\psi(x,t)}{\partial x^2} + V(x,t)\psi(x,t). \tag{3.15}$$

Equation (3.15) is the *time-dependent Schrödinger equation for an interacting particle*; it is the basic dynamical equation of quantum mechanics which generalizes Newton's second law, $F = ma$. Unlike Newton's law, which has two time derivatives in the acceleration term, Eqn. (3.15) is linear in the time-derivative; this implies that knowledge of a solution at $t = 0$, that is, $\psi(x,0)$, is sufficient to determine the wavefunction $\psi(x,t)$ at all later times.

3.2 Plane Waves and Wave Packet Solutions

3.2.1 Plane Waves and Wave Packets

It is easy to find plane wave solutions of the free particle Schrödinger equation, as one can show that

$$\psi_p(x,t) = e^{i(px - p^2 t/2m)/\hbar} \tag{3.16}$$

satisfies Eqn. (3.13). If we want solutions which can represent "particle-like" states, we can construct localized wave packets using linearity, superposition, and interference ideas as before.

Because the momentum label p is more natural than the wave-number k for a particle state, we choose to write the general linear combination solution in the form

$$\psi(x, t) = \frac{1}{\sqrt{2\pi\hbar}} \int_{-\infty}^{+\infty} dp\, \phi(p)\, \psi_p(x, t)$$

$$= \frac{1}{\sqrt{2\pi\hbar}} \int_{-\infty}^{+\infty} dp\, \phi(p)\, e^{i(px - p^2 t/2m)/\hbar} \tag{3.17}$$

where the normalization factor $(1/\sqrt{2\pi\hbar})$ is discussed below.

Before studying an explicit example of such a wave packet, we make the following observations:

- Each component wave, $\psi_p(x, t)$, is weighted by a different "momentum amplitude", $\phi(p)$; this function will eventually provide us with information on the momentum content of the solution, just as $A(k)$ encoded knowledge about the wave number dependence of a wave packet.

- Because each plane wave solution labeled by p now corresponds to a different classical velocity, $v = p/m$, the wave packet will be dispersive and will necessarily spread as it propagates. If these packets are meant to represent particle-like solutions in the macroscopic limit, we will have to make sure this spreading is consistent with our classical intuition and observations.

- A free, classical particle undergoes uniform, constant velocity motion, and a wave packet description should presumably yield this in the macroscopic limit as well.

For the case of a free particle wave packet, we can also write

$$\psi(x, t) = \frac{1}{\sqrt{2\pi\hbar}} \int_{-\infty}^{+\infty} dp\, \phi(p)\, e^{i(px - p^2 t/2m)/\hbar}$$

$$= \frac{1}{\sqrt{2\pi\hbar}} \int_{-\infty}^{+\infty} dp\, \left[\phi(p) e^{-ip^2 t/2m\hbar}\right] e^{ipx/\hbar}$$

$$\psi(x, t) = \frac{1}{\sqrt{2\pi\hbar}} \int_{-\infty}^{+\infty} dp\, \phi(p, t)\, e^{ipx/\hbar} \tag{3.18}$$

where

$$\phi(p, t) \equiv \phi(p) e^{-ip^2 t/2m\hbar} \tag{3.19}$$

Noting the analogy with Fourier transforms, we can invert this to obtain $\phi(p, t)$ as usual by writing

$$\frac{1}{\sqrt{2\pi\hbar}} \int_{-\infty}^{+\infty} dx\, \psi(x, t)\, e^{-ip'x/\hbar} = \int_{-\infty}^{+\infty} dp\, \phi(p, t) \left[\frac{1}{2\pi\hbar} \int_{-\infty}^{+\infty} e^{i(p-p')x/\hbar}\, dx \right]$$

$$= \int_{-\infty}^{+\infty} dp\, \phi(p, t)\, \delta(p - p')$$

$$= \phi(p', t) \tag{3.20}$$

where we have extended the definition of the Dirac δ-function. This important relation implies that

- $\psi(x, t)$ and $\phi(p, t)$ are Fourier transforms of each other and motivates the common $1/\sqrt{2\pi\hbar}$ normalization factor.

If an initial wave packet, $\psi(x, 0)$, is given, one can obtain the corresponding momentum distribution, $\phi(p) \equiv \phi(p, 0)$, required to produce it via Eqn. (3.20); the subsequent time-dependence of the spatial wavefunction, $\psi(x, t)$, is then given through Eqn. (3.18). This is one method of solving the *initial value problem* defined by the one-dimensional Schrödinger equation.

On the other hand, if $\psi(x, t)$ is somehow known, one can translate this information into knowledge of the momentum amplitude at any time t via Eqn. (3.20). For the case of a free particle only, the momentum amplitude has a trivial time-dependence which implies that

$$|\phi(p, t)|^2 = |\phi(p) e^{-ip^2 t/2m\hbar}|^2 = |\phi(p, 0)|^2 \tag{3.21}$$

and the momentum distribution does not change in time. This is consistent with Newtonian mechanics where a particle feeling no force would have a constant momentum (since $F = dp/dt = 0$).

This Fourier transform connection (and Eqns (2.57) and (2.58)) immediately implies that many integrals involving $\psi(x, t)$ and $\phi(p, t)$ are related, that is,

$$\int_{-\infty}^{+\infty} dx\, |\psi(x, t)|^2 = \int_{-\infty}^{+\infty} dp\, |\phi(p, t)|^2 \tag{3.22}$$

$$\int_{-\infty}^{+\infty} dx\, \psi^*(x, t)\psi(x, 0) = \int_{-\infty}^{+\infty} dp\, \phi^*(p, t)\phi(p, 0) \tag{3.23}$$

$$\int_{-\infty}^{+\infty} dx\, \psi_1^*(x, t)\psi_2(x, t) = \int_{-\infty}^{+\infty} dp\, \phi_1^*(p, t)\phi_2(p, t) \tag{3.24}$$

all of which will prove useful later.

3.2.2 **The Gaussian Wave Packet**

For most forms of $\phi(p)$ the integral in Eqn. (3.18) must be done numerically, and we will present some examples of such calculations in Section 3.4. In the case of a Gaussian p distribution, however, defined here by

$$\phi(p) = \sqrt{\frac{\alpha}{\sqrt{\pi}}}\, e^{-\alpha^2(p-p_0)^2/2} \tag{3.25}$$

the integrals can be done in closed form, giving an easy-to-analyze, analytic result. This distribution selects positive momentum components with values centered around p_0; this would correspond to a classical particle with speed $v_0 = p_0/m$, but there is a spread in momentum components of roughly $\Delta p \approx 1/\alpha$. We can then write

$$\psi(x,t) = \sqrt{\frac{\alpha}{2\pi\hbar\sqrt{\pi}}} \int_{-\infty}^{+\infty} dp\, e^{-\alpha^2(p-p_0)^2/2}\, e^{i(px - p^2 t/2m)/\hbar}$$

$$= \sqrt{\frac{\alpha}{2\pi\hbar\sqrt{\pi}}}\, e^{i(p_0 x - p_0^2 t/2m)/\hbar} \quad (\text{using } p - p_0 = q)$$

$$\times \int_{-\infty}^{+\infty} dq\, e^{-q^2(\alpha^2/2 + it/2m\hbar)}\, e^{iq(x - p_0 t/m)/\hbar} \tag{3.26}$$

The integral is of the form

$$\int_{-\infty}^{+\infty} e^{-ax^2 - bx} = \sqrt{\frac{\pi}{a}}\, e^{b^2/4a} \tag{3.27}$$

where

$$a \equiv \frac{\alpha^2}{2} + \frac{it}{2m\hbar} \quad \text{and} \quad b \equiv -i(x - p_0 t/m)/\hbar \tag{3.28}$$

which can be evaluated using the results in Appendix D.1. We find that

$$\psi(x,t) = \frac{1}{\sqrt{\alpha\hbar F\sqrt{\pi}}}\, e^{i(p_0 x - p_0^2 t/2m)/\hbar}\, e^{-(x - p_0 t/m)^2/2\alpha^2\hbar^2 F} \tag{3.29}$$

where we define

$$F \equiv 1 + \frac{it}{t_0} \quad \text{with } t_0 = m\hbar\alpha^2 \tag{3.30}$$

To better analyze this complex waveform, we evaluate its modulus squared and find

$$|\psi(x,t)|^2 = \psi^*(x,t)\psi(x,t)$$

$$= \frac{1}{\sqrt{\pi}\alpha\hbar}\frac{1}{\sqrt{(1+it/t_0)(1-it/t_0)}}\, e^{-(x-p_0t/m)^2/2\alpha^2\hbar^2(\frac{1}{1+it/t_0}+\frac{1}{1-it/t_0})}$$

$$= \frac{1}{\alpha\hbar\sqrt{\pi}\sqrt{1+t^2/t_0^2}}\, e^{-(x-p_0t/m)^2/\alpha^2\hbar^2(1+t^2/t_0^2)}$$

$$|\psi(x,t)|^2 = \frac{1}{\beta_t\sqrt{\pi}}\, e^{-(x-p_0t/m)^2/\beta_t^2} \tag{3.31}$$

where

$$\beta_t \equiv \alpha\hbar\sqrt{1+t^2/t_0^2} \tag{3.32}$$

This result illustrates several notable features:

- The central value of the wave packet (interpreted by looking at $|\psi(x,t)|^2$) is located at $x - p_0t/m = 0$, so that the peak moves at constant speed p_0/m, consistent with a particle of fixed momentum p_0 and mass m. One can also show quite generally (P3.2) that if we let

$$\phi(p) \quad \longrightarrow \quad \phi(p)\, e^{-ipx_0/\hbar} \tag{3.33}$$

the corresponding Schrödinger wavefunction satisfies

$$\psi(x,t) \quad \longrightarrow \quad \psi(x-x_0,t) \tag{3.34}$$

so that we can change the initial central position, and the peak location is given by $x = x_0 + p_0t/m$. This is the equivalent of fixing all of the initial conditions for a free particle. The most general, free-particle time-dependent Gaussian wave packet solution of this type is then given by

$$\psi_{(G)}(x,t) = \frac{1}{\sqrt{\alpha\hbar F}\sqrt{\pi}}\, e^{i(p_0(x-x_0)-p_0^2t/2m)/\hbar}\, e^{-(x-x_0-p_0t/m)^2/2\alpha^2\hbar^2 F} \tag{3.35}$$

with momentum–space counterpart

$$\phi_{(G)}(p,t) = \sqrt{\frac{\alpha}{\sqrt{\pi}}}\, e^{-\alpha^2(p-p_0)^2/2}\, e^{-ipx_0/\hbar}\, e^{-ip^2t/2m\hbar} \tag{3.36}$$

and we will frequently make use of both expressions as tractable analytic examples.

- The width of the wave packet, which is roughly given by β_t, does increase with time due to the variation in speed of the component plane waves. The timescale for dispersion is set by the "spreading time," defined as $t_0 = m\hbar\alpha^2$. For long times (defined as $t \gg t_0$), the spread in position goes as

$$\Delta x_t \sim \beta_t \longrightarrow \alpha\hbar\left(\frac{t}{t_0}\right) = \frac{t}{m\alpha} \tag{3.37}$$

Using $m\Delta v = \Delta p \approx 1/\alpha$ as the spread in velocity components in the wave packet, we see that

$$\Delta x_t \longrightarrow \Delta v\, t \sim \left(\frac{1}{m\alpha}\right) t \tag{3.38}$$

which provides a very intuitive (classical) explanation for the spreading behavior.

- The initial spread of the position wave packet is given by

$$\Delta x_0 \approx \beta_0 = \alpha\hbar \tag{3.39}$$

so the spreading time can be estimated as

$$t_0 = m\hbar\alpha^2 \approx \frac{m\hbar}{(\Delta p)^2} \approx \frac{m(\Delta x_0)^2}{\hbar} \tag{3.40}$$

These relations make clear the correlations (or anti-correlations) between

— the small (or large) initial spatial extent of the wave packet,

— the large (or small) spread in momentum required to achieve the necessary destructive interference to form the packet,

— the large (or small) spread in speed of the component plane waves, and

— the short (or long) time to exhibit the dispersion.

- We see that spatially small wave packets spread faster than those with larger initial widths and we visualize this effect in Fig. 3.1.

- We note the subtle time-dependence of both the real (Re) and imaginary (Im) parts of $\psi(x,t)$, which is required to maintain the Gaussian shape of the wave packet (if not its width) as it moves. This is shown in Fig. 3.2 where we plot $\text{Re}(\psi(x,t)), \text{Im}(\psi(x,t))$ as well as $|\psi(x,t)|$. We stress that only for the Gaussian wave packet does the wave packet remain in the same "family" of shapes for later times, and other examples are considered in Section 3.4.

- We will come to associate the "wiggliness" (or rate of spatial variation) of the wavefunction with the local kinetic energy (see Section 4.3.3); Fig. 3.2 then

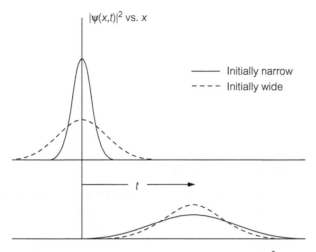

Figure 3.1. Spreading Gaussian wave packets illustrated by plots of $|\psi(x,t)|^2$ versus x at two different times. Note that the initially narrower packet (the solid curve) spreads faster.

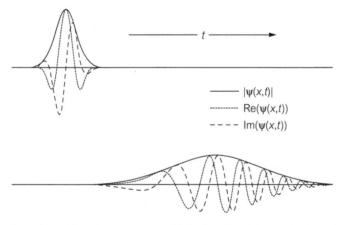

Figure 3.2. The real part (dotted), imaginary part (dashed), and modulus or absolute value (solid) of a spreading Gaussian wave packet: note how the "wiggliness" is concentrated in the leading edge.

shows that the "leading edge" of the wave packet has a larger "local kinetic energy" (it's wigglier) than the "trailing edge"; compare this behavior to the spreading wave pulse discussed in Section 2.5.2 where the faster components outpace the slower ones.

To verify that this wave packet spreading arises from a (quantum) wave mechanics description we also note that:

- In the spirit of Section 1.3, we find for a fixed initial spread or uncertainty in position, Δx_0, that $t_0 \to \infty$ as $\hbar \to 0$, so that spreading effects would be unobservable, consistent with its being a purely quantum effect.

• The spreads (or uncertainties) in x and p satisfy

$$\Delta x_t \, \Delta p_t \approx (\alpha\hbar\sqrt{1 + t^2/t_0^2}) \cdot \left(\frac{1}{\alpha}\right) \geq \hbar \qquad (3.41)$$

as expected from the uncertainty principle. The detailed definitions of Δx and Δp will be presented in the next chapter.

Example 3.1. Spreading wave packets

To see if the spreading of such wave packets has any observable consequences, we can make some numerical estimates of the spreading time in various cases. For a typical macroscopic particle with $m = 1$ kg and an initial uncertainty in position of, say, $\Delta x_0 \approx 0.1$ mm, using Eqn. (3.40) we find that $t_0 \approx 10^{26}$ s $\approx 3 \times 10^{16}$ years, which is roughly a million times the age of the universe. Once again, the smallness of $\hbar$ on a macroscopic scale can make quantum effects unobservably small.

On the other hand, we can consider an electron in a circular Bohr orbit in a hydrogen atom, with quantum number n, as described in Section 1.4. (This is not a perfectly valid comparison as we have dealt here only with straight line motion, but it makes the point.) The classical period, τ_n, is given by Eqn. (1.41) and can be written in the form

$$\tau_n = \frac{2\pi r_n}{v_n} = \left(\frac{2\pi a_0}{c\alpha}\right)n^3 \approx (1.5 \times 10^{-16}\,\text{s})n^3 \qquad (3.42)$$

If we were to imagine an electron wave packet localized to within one quadrant of its classical orbit (with radius $r_n = a_0 n^2$ as in Eqn. (1.37), that is, with a spatial uncertainty of $\Delta x_0^{(n)} = (2\pi r_n)/4$), the spreading time implied by Eqn. (3.40) would be

$$t_0^{(n)} \sim \frac{m(\Delta x_0^{(n)})^2}{\hbar} = \left(\frac{\pi^2 m a_0^2}{4\hbar}\right)n^4 \approx (6 \times 10^{-17}\,\text{s})n^4 \qquad (3.43)$$

The ratio of the classical period to the spreading time in a Bohr orbit is then roughly

$$\frac{\tau_n}{t_0^{(n)}} \approx \frac{8}{\pi n} \qquad (3.44)$$

In order for classical behavior to be observable over many periods, we require that $t_0^{(n)} >> \tau_n$ which, in turn, requires that $n >> 1$ in Eqn. (3.44); this is another example of the correspondence principle. Experiments using Rydberg electron wave packets with high n (40–70) states have observed 20 or more orbital periods.[1]

---

[1] For example, see Yeazell et al. (1990) and Wals et al. (1994).

3.3 **"Bouncing" Wave Packets**

Another case of a quasi-free particle whose quantum mechanical solution can be easily obtained is that corresponding to a free particle, incident from the left, which hits an impenetrable wall at $x = 0$. We can argue that the quantum mechanical particle is subject to a potential of the form

$$V(x) = \begin{cases} 0 & \text{for } x < 0 \\ \infty & \text{for } x > 0 \end{cases} \tag{3.45}$$

To solve this problem, we can look for plane wave solutions of the Schrödinger equation for $x < 0$ in the usual way, but then also enforce the boundary condition that

$$\psi(x, t) = 0 \quad \text{for } x \geq 0 \text{ for all } t \tag{3.46}$$

The complex exponential solutions $\exp(\pm ipx/\hbar)$ no longer satisfy the boundary condition at the origin, but the linear combination

$$\left(e^{ipx/\hbar} - e^{-ipx/\hbar} \right) \propto \sin(px/\hbar) \tag{3.47}$$

does. The relevant plane wave solutions, $\tilde{\psi}(x, t)$, are now proportional to

$$\tilde{\psi}_p(x, t) = \begin{cases} \left(e^{ipx/\hbar} - e^{-ipx/\hbar} \right) e^{-ip^2 t/2m\hbar} & \text{for } x \leq 0 \\ 0 & \text{for } x \geq 0 \end{cases} \tag{3.48}$$

We therefore have to evaluate the integral

$$\tilde{\psi}(x, t) = \frac{1}{\sqrt{2\pi\hbar}} \int_{-\infty}^{+\infty} dp\, \phi(p)\, \tilde{\psi}_p(x, t) \tag{3.49}$$

to obtain the resulting wave packet. The integrals over p can be done in exactly the same way as in the free-particle case, which allows us to write the solution of the "bouncing" wave packet as

$$\tilde{\psi}(x, t) = \begin{cases} \psi(x, t) - \psi(-x, t) & \text{for } x \leq 0 \\ 0 & \text{for } x \geq 0 \end{cases} \tag{3.50}$$

One can also derive these solutions by noting that if $\psi(x, t)$ is a solution of the time-dependent free-particle Schrödinger equation, then so is $\psi(-x, t)$, because the second derivative operator "brings down two minus signs"; then any linear combination of the two is as well, and this antisymmetric "difference" solution also satisfies the boundary condition at $x = 0$. Such a solution has been called

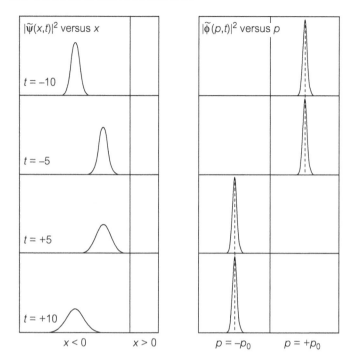

Figure 3.3. 'Bouncing' wave packet before and after a collision with an impenetrable wall at $x = 0$. The left sides shows $|\tilde{\psi}(x, t)|^2$ versus x, while the right side illustrates $|\tilde{\phi}(p, t)|^2$ versus p.

a "mirror" solution[2] and its use here is akin to the *method of images* techniques used in electrostatics to satisfy similar boundary conditions, which, in turn, are inspired by tricks from optics involving multiple mirrors. Because of the lack of symmetry imposed by the boundary conditions, the momentum–space partner of $\tilde{\psi}(x, t)$ cannot be evaluated using standard Gaussian integral tricks, and must be evaluated numerically.

As an example of the more complex dynamical behavior possible in this system, we show in Fig. 3.3(a) plots of $|\tilde{\psi}(x, t)|^2$ versus x and the corresponding $|\tilde{\phi}(p, t)|^2$ versus p for times before and after the classical "bounce" at the infinite wall where we use the general free-particle Gaussian solution $\psi_{(G)}(x, t)$ in Eqn. (3.35) with $x_0 < 0$ and $p_0 > 0$ corresponding to a classical particle impinging on the wall from the left. The classical reversal of direction, accompanied by the quantum mechanical spreading is apparent in the position-space plots, while the the change in momentum (from $+p_0$ to $-p_0$ values) before and after the impulsive collision with the wall is also apparent in the $|\tilde{\phi}(p, t)|^2$ plots. In Fig. 3.4(b), we show similar images for times closer to the classical collision.

---

[2] See, for example, Andrews (1998) or Doncheski and Robinett (1999) or Belloni *et al.* (2005).

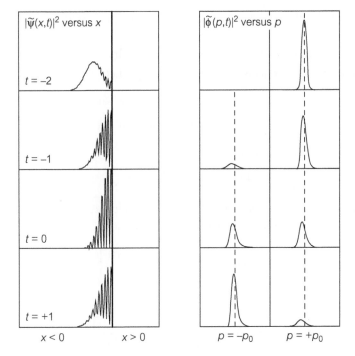

Figure 3.4. Same as Fig. 3.3, except for times closer to the 'impact'; note that the momentum distribution changes only over the duration of the impulsive collision. The vertical dashed lines correspond to the central value of the wave packet momentum, $\pm p_0$.

These illustrate the interference effects between the $\psi(x, t)$ and $\psi(-x, t)$ terms in Eqn. (3.50) near the wall, close to the classical collision time.

3.4 **Numerical Calculation of Wave Packets**

While it is possible to evaluate the wave packet integral of Eqn. (3.18) in closed form for only a few special cases, one can always perform the p integral numerically at each point of x, t. A conceptually simple method is to approximate

$$\psi(x, t) = \frac{1}{\sqrt{2\pi\hbar}} \int_{-\infty}^{+\infty} dp \, \phi(p) \, e^{i(px - p^2 t/2m)/\hbar} \tag{3.51}$$

$$\approx \frac{1}{\sqrt{2\pi\hbar}} \sum_{n=-N}^{n=N} \Delta p \, \phi(p_n) \, e^{i(p_n x - p_n^2 t/2m)/\hbar}$$

where $p_n = n\Delta p$ is evaluated on a set of discretized values ranging from $p_{max} = N\Delta p$ to $p_{min} = -N\Delta p$. (We make no claims that this method is the most computationally efficient way of calculating the time-dependence of a wave packet.[3])

In order to examine the generality of the results inferred from the analytic Gaussian example, we choose two initial wave packets given by

$$(a) \quad \psi_1(x,0) = \begin{cases} 0 & \text{for } |x| > a \\ 1/\sqrt{2a} & \text{for } |x| < a \end{cases} \tag{3.52}$$

that is, a square wave pulse, and

$$(b) \quad \psi_2(x,0) = e^{-|x|/a}/\sqrt{a} \tag{3.53}$$

These waveforms can be used in Eqn. (3.20) to calculate the initial momentum amplitude, $\phi(p)$, required to reproduce them, namely

$$\phi_1(p) = \sqrt{\frac{a}{\pi \hbar}} \left[\frac{\sin(ap/\hbar)}{ap/\hbar} \right] \tag{3.54}$$

and

$$\phi_2(p) = \sqrt{\frac{2a}{\pi \hbar}} \left[\frac{1}{1 + (ap/\hbar)^2} \right] \tag{3.55}$$

Once the initial momentum amplitudes are known, the resulting time-dependent wavefunctions can be obtained through Eqn. (3.18). The corresponding values of $|\psi(x,t)|^2$ are plotted in Fig. 3.5(a) and (b) for illustration. In each case, we use $\phi(p - p_0)$ so that the packets translate to the right. In addition, a Gaussian wave packet (dashed curve) is also plotted for comparison, and the values of a in each case are chosen to give the same initial "spread." We note that:

- Each wave packet spreads in a way similar to the analytically obtained Gaussian wave packet.

- The subtle interplay (constructive and destructive interference) between the component plane waves required to form the particular initial waveforms (e.g. the square bump) are rapidly altered due to the dispersive propagation; this results in rapid oscillations and changes in shape with time.

- The relative phases between the various components tend to "randomize" in some sense so that the long-term shape of the waveform seemingly approaches a Gaussian shape.

It seems that the Schrödinger wave equation prescription discussed in this chapter does incorporate wave mechanical effects into a dynamical equation

---

[3] See, for example, Press *et al.* (2002) for a comprehensive discussion of numerical Fourier transform methods.

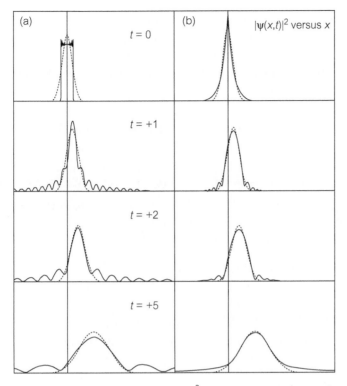

Figure 3.5. Spreading wave packets given by $|\psi(x, t)|^2$ versus x at increasing times for non-Gaussian momentum amplitudes: (a) is for an initial "square" wavepulse, $\psi_1(x, 0)$ and (b) is for an initial "exponential" wavepulse, $\psi_2(x, 0)$. In each case, the dashed curve is a Gaussian wave packet of the same initial width.

consistent with Newton's first law. Information on the desired "wave-like" and "particle-like" phenomena appear to be contained in both $\psi(x, t)$ and $\phi(p, t)$; we have not discussed, however, the precise physical interpretation of either wave amplitude. Because both $|\psi(x, t)|^2$ and $|\phi(p, t)|^2$ will be interpreted as probability densities, we begin the next chapter with a short review of probability concepts.

3.5 **Questions and Problems**

Q3.1. Discuss why the de Broglie relation (Eqn. (3.4)) is consistent with the statement in Section 3.2.2 that the spatial "wiggliness" of the Schrödinger wavefunction provides information on the kinetic energy of the associated particle.

Q3.2. Compare the expression for the spreading time, t_0, in Eqn. (3.40), for the special case of the Gaussian wave packet, to the expression given in Eqn. (2.84) for a general dispersive wave packet. What are the analogous quantities in each case?

Q3.3. How can the wavefunction $\tilde{\psi}_p(x, t)$ in Eqn. (3.48) contain information on the particle first moving to the right and then to the left after the bounce? Can you argue why this wave packet construction makes it clear that the wave packet will continue to spread after the collision in the same way it would have in the absence of the wall?

Q3.4. Could you construct wave packet solutions to the problem of a particle confined between two infinite walls (the familiar problem of the infinite well) by using "mirror solutions" as in Section 3.3? (See Kleber 1994 for a discussion.)

Q3.5. In Fig. 3.3 the dotted lines in the $|\tilde{\phi}(p, t)|^2$ plot correspond to the values of p_0 used in the calculation, and the momentum–space wavefunctions are indeed peaked near $\pm p_0$. In Fig. 3.4, however, during the collision, the peaks are clearly "off-center" from those central values. Why would this be so?

P3.1. Free-particle Schrödinger equation solutions.

    (a) Verify explicitly that the wave packet solution of Eqn. (3.29) satisfies the free-particle Schrödinger equation, and then repeat for the more general case in Eqn. (3.35).

    (b) The "phase" piece of the solution disappears when we evaluate $|\psi(x, t)|^2$. Show that $\psi(x, t)$ *without* this term is *not* a solution of the Schrödinger equation in each case from part (a).

P3.2. (a) Show generally that if one lets

$$\phi(p) \rightarrow \phi(p)\, e^{-ipx_0/\hbar} \tag{3.56}$$

    in Eqn. (3.17), then

$$\psi(x, t) \rightarrow \psi(x - x_0, t) \tag{3.57}$$

    illustrating how one derives Eqn. (3.35).

    (b) Analyze the effect on $\psi(x, t)$ of letting

$$\phi(p) \rightarrow \phi(p)\, e^{iAp^2} \tag{3.58}$$

    where A is a constant.

P3.3. Show that

$$\lim_{\beta_t \to 0} \frac{1}{\beta_t \sqrt{\pi}} e^{-(x - p_0 t/m)^2/\beta_t^2} = \delta(x - p_0 t/m) \tag{3.59}$$

$$\lim_{1/\alpha \to 0} \frac{\alpha}{\sqrt{\pi}} e^{-\alpha^2 (p - p_0)^2} = \delta(p - p_0) \tag{3.60}$$

Discuss how your results might bear on a classical limit of quantum mechanics.

P3.4. Other Gaussian wave packets I. One can construct other examples of closed-form wave packets using Eqn. (3.17) in certain special cases. Evaluate $\psi(x, t)$ for the momentum amplitude

$$\phi(p) = \sqrt{\frac{2\alpha^3}{\sqrt{\pi}}} (p - p_0) e^{-\alpha^2(p-p_0)^2/2} \qquad (3.61)$$

and find the resulting expression for $|\psi(x,t)|^2$. Does it spread in a manner consistent with Eqn. (3.32)?

P3.5. Other Gaussian wave packets II. Find the time-dependent solution to the Schrödinger equation, $\psi(x,t)$, corresponding to the initial momentum amplitude given by

$$\phi(p) = \sqrt{\frac{\alpha}{\sqrt{\pi}}} e^{-\alpha^2(p-p_0)^2(1+iC)/2} \qquad (3.62)$$

where C is a real parameter. Evaluate $|\psi(x,t)|^2$, find the analog of β_t in Eqn. (3.32), and discuss how the time-dependence of the position of the peak and the spread in position depend on the magnitude, and especially the sign, of C. Under what circumstances, if any, does Δx_t actually shrink with time, at least initially? What is it about the form in Eqn. (3.62) when $C \neq 0$ that allows one to "construct" an initial position-space wave packet so that this could happen?

P3.6. Show that the Gaussian wave packet solution from **P3.4**, in the special limit that $p_0, x_0 = 0$, also satisfies the Schrödinger equation (and boundary conditions) for the "bouncing" wave packet geometry in Section 3.3.

P3.7. Overlap integrals.

(a) Show that the general Gaussian wave packet

$$\psi_{(G)}(x,t) = \frac{1}{\sqrt{\alpha \hbar F \sqrt{\pi}}} e^{i(p_0(x-x_0)-p_0^2 t/2m)/\hbar} e^{-(x-x_0-p_0 t/m)^2/2\alpha^2 \hbar^2 F} \qquad (3.63)$$

satisfies

$$\int_{-\infty}^{+\infty} dx \, |\psi_{(G)}(x,t)|^2 = 1 \qquad (3.64)$$

for all values of t, p_0, and x_0. Show also that the corresponding Gaussian momentum amplitude

$$\phi_{(G)}(p,t) = \sqrt{\frac{\alpha}{\sqrt{\pi}}} e^{-\alpha^2(p-p_0)^2/2} e^{-ipx_0/\hbar} e^{-ip^2 t/2m\hbar} \qquad (3.65)$$

satisfies

$$\int_{-\infty}^{+\infty} dp \, |\phi_{(G)}(p,t)|^2 = 1 \qquad (3.66)$$

for all values of t, p_0, and x_0.

(b) Consider the overlap integral of two general Gaussian wave packet solutions of the form

$$\mathcal{O}^{(A,B)} = \int_{-\infty}^{+\infty} [\psi_{(G)}^{(B)}(x,t)]^* \, \psi_{(G)}^{(A)}(x,t) \, dx$$

$$= \int_{-\infty}^{+\infty} [\phi_{(G)}^{(B)}(p,t)]^* \, \phi_{(G)}^{(A)}(p,t) \, dp \qquad (3.67)$$

for two different sets of initial positions and momenta, $(x_0, p_0) = (x_A, p_A)$ and (x_B, p_B); the two forms are guaranteed to be equal from Eqn. (3.24). Show that it is *much* easier to evaluate this overlap integral using the momentum–space amplitudes with the result that

$$\mathcal{O}^{(A,B)} = e^{-(x_B-x_A)^2/4\beta^2} \, e^{-\alpha^2(p_B-p_A)^2/4} \, e^{i(p_A+p_B)(x_B-x_A)/2/\hbar} \qquad (3.68)$$

independent of time. Thus, these wave packet solutions can have "arbitrarily little in common" (exponentially so) if their initial positions and/or momenta differ significantly.

(c) Redo the evaluation of $\mathcal{O}^{(A,B)}$ using position-space wavefunctions, first specializing to $t = 0$ (where the calculation is easy) and then for the general $t \neq 0$ case (where it is much lengthier) and confirm the result of Eqn. (3.68).

P3.8. Autocorrelation functions. Another type of overlap integral which finds use in quantum mechanics is the *autocorrelation function*, defined by

$$A(t) \equiv \int_{-\infty}^{+\infty} \psi^*(x,t)\,\psi(x,0)\,dx = \int_{-\infty}^{+\infty} \phi^*(p,t)\,\phi(p,0)\,dp \qquad (3.69)$$

which measures the overlap between a wavefunction as it evolves in time with its initial value. Data on quantities which are related to $|A(t)|^2$ is sometimes available from experiments. The two forms give the same results from Eqn. (3.23).

(a) For the general Gaussian wave packet in Eqns (3.35) and (3.36), evaluate $A(t)$ using both the position–space and momentum–space wavefunctions and show that

$$A(t) = \frac{1}{\sqrt{1 - it/t_0}} \exp\left[\frac{i\alpha^2 p_0^2 t}{2t_0(1 - it/2t_0)}\right] \qquad (3.70)$$

so that

$$|A(t)|^2 = \frac{1}{\sqrt{1 + (t/2t_0)^2}} \exp\left[-2\alpha^2 p_0^2 \frac{(t/2t_0)^2}{(1 + (t/2t_0)^2)}\right] \qquad (3.71)$$

(b) Discuss how the pre-factor is due only to wave packet spreading (as it is present even when $p_0 = 0$.) Discuss why the exponential suppression "saturates" to a fixed values for long times as the wave packet "moves" away from its initial location.

P3.9. If two wave packets (representing the same mass particle) have different initial widths (determined by $\alpha_1 \neq \alpha_2$), and hence different spreading times, $t_0^{(1)}$ and $t_0^{(2)}$, show that the narrower one will spread faster, so that after a time $t = \sqrt{t_0^{(1)} t_0^{(2)}}$ the two packets will (temporarily) have the same width.

P3.10. (a) Assume that the electrons in a TV set are accelerated to 30 kV and travel 10 cm before they hit the back of the screen. If each electron is associated with a wave packet of initial size $\sim$0.1 mm, estimate the amount of spreading that occurs before the electron hits the screen. How far would the electron have to travel before the size of its wave packet would have grown to roughly 10 times its initial size?

(b) Electrons at the Stanford Linear Collider are accelerated to energies of $\sim$50 GeV. If their initial wave packets are localized to say 10 μm, what is their spatial extent at the time of their collisions when they have traveled 1 km? Hint: Are these particles nonrelativistic or not? If they are, what do you do?

P3.11. Show that the momentum amplitude of Eqn. (3.54) (or (3.55)) is indeed the Fourier transform of the initial spatial wavefunction in Eqn. (3.52) (or (3.53)).

FOUR

Interpreting the Schrödinger Equation

In stark contrast to classical Newtonian mechanics, the standard interpretation of quantum mechanics associates the Schrödinger wavefunction, not in a deterministic way with definite predictions about a specific, individual experimental result, but only in a probabilistic manner, with $|\psi(x,t)|^2$ acting as a probability density, in this case, for position measurements. Before analyzing this approach in more detail, we will find it profitable to briefly review some of the basic concepts of discrete and continuous probability theory.

4.1 Introduction to Probability

4.1.1 Discrete Probability Distributions

For a given experimental situation for which the outcomes are determined by random chance, the collection of the probabilities of all the possible outcomes defines a *probability distribution*. Such a collection of probabilities for which the outcomes form a discrete set, and hence can be classified by an integer label, is called a *discrete probability distribution*. We can therefore label the possible outcomes as $x_i, i = 1, 2, \ldots, N$ and the upper limit, N, can be either finite or infinite; the corresponding probabilities are labeled $P(x_i)$.

Trivial examples which are obviously labeled by integers are the number on the ith face in one roll of a die, or more prosaically, the number of raisins found in a single slice of raisin bread; in both cases the upper limit on N is finite. An example in the context of quantum mechanics is the set of quantized bound state energy levels, such as the allowed energy states of a hydrogen atom, $E_n = -E_0/n^2$. The x_i in this case are the energies and are not themselves integers (or even dimensionless numbers), but a measurement of the energy of an individual atom will yield only those discrete values; the upper limit for the integer label in this case is infinity.

A simple example which exhibits many of the most general features of a discrete probability distribution is the set of probabilities corresponding to the throw of a pair of dice. In this case, we can define $x_1 = 2, \ldots, x_{11} = 12$ as the possible outcomes, with probabilities given by

$$P(x_1 = 2) = \frac{1}{36} \cdots P(x_6 = 7) = \frac{6}{36} \cdots P(x_{11} = 12) = \frac{1}{36} \tag{4.1}$$

These *a priori* probabilities, $P(x_i)$ versus x_i, are shown in Fig. 4.1; they are clustered, as can often happen, about a central value, but with a characteristic spread.

Because in each measurement, *something* will be measured, the probabilities $P(x_i)$ must satisfy the obvious constrain

$$\sum_{i=1}^{N} P(x_i) = 1 \tag{4.2}$$

which is often called a *normalization* condition. While the complete set of the $P(x_i)$ encodes all of the available information concerning the system, specific combinations of the $P(x_i)$ are often useful. The *average value* or *expectation*

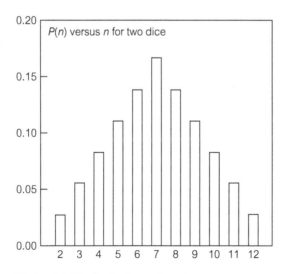

Figure 4.1. The predicted probabilities for the throw of two dice.

value of the discrete variable x is defined by

$$\langle x \rangle \equiv \sum_{i=1}^{N} x_i \, P(x_i) \tag{4.3}$$

as is the average value of any function of x,

$$\langle f(x) \rangle = \sum_{i=1}^{N} f(x_i) \, P(x_i) \tag{4.4}$$

Both give information on the expected results of many measurements of quantities depending on the variable. These *a priori* predictions can, of course, be compared to the experimental values given by repeated measurements of x or $f(x)$; if the values of a set of N_T *measurements* of the variable x are labeled $\overline{x}_s, s = 1, \ldots, N_T$, we define

$$\langle x \rangle_{\text{exp}} \equiv \frac{1}{N_T} \sum_{s=1}^{N_T} \overline{x}_s \quad \text{or} \quad \langle f(x) \rangle_{\text{exp}} \equiv \frac{1}{N_T} \sum_{s=1}^{N_T} f(\overline{x}_s) \tag{4.5}$$

While the values of the variable x often (but not always) cluster around the average value, a given measurement will often find x within a rather well-defined range about $\langle x \rangle$. The variable $y_i \equiv x_i - \langle x \rangle$ naturally measures the deviation of an individual value from the average, but it satisfies

$$\langle y \rangle = \langle x - \langle x \rangle \rangle = \langle x \rangle - \langle x \rangle = 0 \tag{4.6}$$

because the *average* deviation away from the mean vanishes. A better estimator of the *spread* or *uncertainty* in x that arises very naturally is the so-called *root mean square deviation* or *RMS deviation* or *standard deviation* defined by

$$\Delta x_{\text{RMS}} = \Delta x \equiv \left(\langle (x - \langle x \rangle)^2 \rangle \right)^{1/2} = \left(\sum_{i=1}^{N} (x_i - \langle x \rangle)^2 \, P(x_i) \right)^{1/2} \tag{4.7}$$

The name itself is a useful mnemonic for this formula as it requires one to

(1) first take the *deviation* (of each x_i from the average value $\langle x \rangle$);
(2) *square* it;
(3) then evaluate the *mean* (or average);
(4) and finally take the (square) *root*.

A more calculationally useful form of this relation is obtained by noting that

$$(\Delta x)^2 = \langle (x - \langle x \rangle)^2 \rangle$$
$$= \langle x^2 - 2x\langle x \rangle + \langle x \rangle^2 \rangle$$
$$= \langle x^2 \rangle - \langle x \rangle^2 \qquad (4.8)$$

so that an evaluation of $\langle x \rangle$ and $\langle x^2 \rangle$ suffice to determine Δx.

Example 4.1. Expectation values for two dice

For the case of the throw of two dice, we easily find that

$$\langle x \rangle = 7 \quad \text{and} \quad \langle x^2 \rangle = \frac{1974}{36} = 54.833 \quad \text{so that} \quad \Delta x = 2.415. \qquad (4.9)$$

If we sum the probabilities corresponding to results which fall within one standard deviation of the mean, that is, in the interval $(\langle x \rangle - \Delta x, \langle x \rangle + \Delta x)$, we find $24/36 = 2/3 \sim 0.67$; we similarly find that $34/36 \sim 0.94$ of the total (unit of) probability is found within two standard deviations.

This pattern is rather typical of probability distributions that are roughly centered and somewhat peaked around their average values (see also P4.1 – P4.4). Thus, a knowledge of $\langle x \rangle$ and Δx allows one to make reasonable predictions for the likely outcome of a given measurement, as well as for the reliability or uncertainty of the result.

4.1.2 Continuous Probability Distributions

If the random variable x can take on continuous values (e.g. heights of a population, location of a particle along a line segment, etc.) we can generalize the discrete probability distribution above by noting that the $P(x_i)$ can be trivially rewritten as $P(x_i)\Delta x_i$ if we define the "distance" between integrally labeled measurements to be $\Delta x_i = 1$, that is, a unit sized bin centered at x_i. The continuous case can then be obtained by the analogy

$$P(x_i) \sim P(x_i)\Delta x_i \implies P(x)\,dx \qquad (4.10)$$

where dx is an infinitesimally small unit of measure of the variable x. With this identification, we are led to the interpretation of $P(x)$:

- $P(x)\,dx$ is the probability that a measurement of the variable x will find it in the interval $(x, x + dx)$, implying that

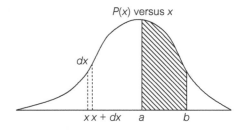

P(x) versus x

Figure 4.2. A generic continuous probability distribution. The predicted probability that a measurement will yield a value in the range (a, b) is given by the shaded area.

- the probability that a measurement of x will be in the finite interval (a, b) is given by

$$\text{Prob}(a < x < b) = \int_a^b P(x)\, dx \tag{4.11}$$

which we illustrate in Fig. 4.2.

Assuming for generality that the variable can take on values anywhere in the interval $(-\infty, +\infty)$, the normalization condition on the probability distribution $P(x)$ becomes

$$\sum_{i=1}^{N} P(x_i) = 1 \implies \int_{-\infty}^{+\infty} P(x)\, dx = 1 \tag{4.12}$$

The definitions of average values and RMS deviations are easily generalized to

$$\langle x \rangle = \int_{-\infty}^{+\infty} x\, P(x)\, dx \quad \text{or} \quad \langle f(x) \rangle = \int_{-\infty}^{+\infty} f(x)\, P(x)\, dx \tag{4.13}$$

and

$$(\Delta x)^2 = \int_{-\infty}^{+\infty} (x - \langle x \rangle)^2\, P(x)\, dx = \langle x^2 \rangle - \langle x \rangle^2 \tag{4.14}$$

Example 4.2. The Gaussian distribution

One of the most frequently occurring continuous probability distributions is the *Gaussian* (or *normal* or "bell-shaped") distribution, given by

$$P(x; \mu, \sigma) = \frac{1}{\sigma\sqrt{2\pi}} e^{-(x-\mu)^2/2\sigma^2} \tag{4.15}$$

which is shown in Fig. 4.3, and characterized by two constants μ, σ. This distribution arises in a fundamental manner in the theory of statistics (via the so-called Central Limit Theorem[1]),

---

[1] See, for example, Mathews and Walker (1970).

(Continued)

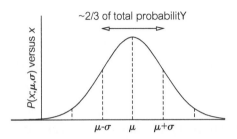

~2/3 of total probabilitY

$P(x;\mu,\sigma)$ versus x

$\mu{-}\sigma$ μ $\mu{+}\sigma$

Figure 4.3. Gaussian probability distribution: the mean value μ and the one-standard deviation $(\pm\sigma)$ values are shown.

but also appears in an important and natural way in quantum theory as well. Using the (appropriately named) Gaussian integrals discussed in Appendix D.1, it is easy to show that $P(x; \mu, \sigma)$ is appropriately normalized for any values of μ, σ, and that

$$\langle x \rangle = \mu \quad \text{and} \quad \langle x^2 \rangle = \mu^2 + \sigma^2, \quad \text{so that} \quad \Delta x = \sigma \quad (4.16)$$

Thus, the parameters μ and σ very directly characterize the average values and RMS deviations. The probability in any finite interval (i.e. the area under $P(x)$) can only be calculated numerically (see Appendix B.3) and one finds

$$\text{Prob}(|x - \mu| \leq \sigma) \approx 0.6826 \approx 68.3\%$$

$$\text{Prob}(|x - \mu| \leq 2\sigma) \approx 0.9544 \approx 95.5\%$$

$$\text{Prob}(|x - \mu| \leq 3\sigma) \approx 0.9974 \approx 99.7\% \quad (4.17)$$

Thus, the measurement of a quantity described by a Gaussian distribution found to have a value more than three standard deviations away from its average is rare (less than 0.3% of the time) but not impossible, while we expect roughly two-third of the measured events to be less than 1σ away from the mean.

Several points concerning $P(x)$ should be kept in mind:

- $P(x)$ is, by itself, not a probability but rather a *probability density* as it can be crudely defined as

$$P(x) \equiv \frac{d\text{Prob}(x)}{dx} \quad (4.18)$$

or the probability per unit x interval. The need for the infinitesimal unit of measure, dx, should always be kept in mind.

- Because of this, $P(x)$ has nontrivial dimensions, namely

$$[P(x)] = \left[\frac{1}{dx}\right] \tag{4.19}$$

whatever the units of x happen to be (height, etc.). (Recall that $[z]$ denotes the dimensions of the variable z.) This is clear from the specific example of the Gaussian distribution in Eqn. (4.15), where $[x] = [\sigma]$ and $[P(x)] = [1/\sigma]$.

The experimental measurement of a continuous probability distribution can be accomplished in a limiting procedure similar to that for discrete distributions. We can "discretize" the problem by defining "bins" of finite width in x, of some reasonable size, δx, centered at the desired value of x. The ratio of the number, of successes to total trials in that small bin, $\delta\text{Prob} = N_S/N_T$, divided by the "bin width" estimates the local probability density

$$P(x) = \frac{N_S/N_T}{\delta x} \longrightarrow \frac{\delta\text{Prob}}{\delta x} \tag{4.20}$$

One can decrease the bin width appropriately as the number of events collected increases. This is shown in Fig. 4.4 where a Gaussian probability is "built up"

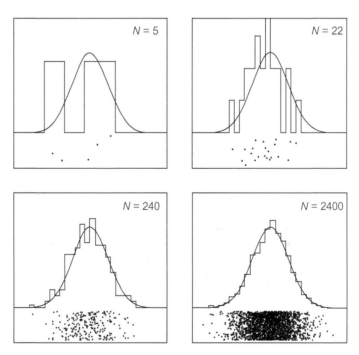

Figure 4.4. "Experimental" (i.e. computer) determination of a Gaussian probability distribution for increasingly large numbers of measurements. The individual measurements are shown as dots, while the smooth Gaussian curve is approximated by the binned data, using Eqn. (4.20).

by repeated measurements. The distribution of measured points is shown below each figure; the gradual accumulation of data points leading to the emergence of the probabilistic pattern should be reminiscent of Fig. 1.2 and is one of the hallmarks of many quantum mechanical experiments.

4.2 **Probability Interpretation of the Schrödinger Wavefunction**

One of our stated goals in understanding quantum physics has been to incorporate the experimentally observed wave properties of matter in a self-consistent manner with the basic (nonrelativistic) dynamical laws of motion for particles; this has led us to the time-dependent Schrödinger equation

$$i\hbar\frac{\partial\psi(x,t)}{\partial t} = -\frac{\hbar^2}{2m}\frac{\partial^2\psi(x,t)}{\partial x^2} + V(x,t)\psi(x,t) \tag{4.21}$$

The derivations leading to other wave equations make it clear that the solutions represent various physical observables (displacements of a string, an electric or magnetic field, or more generally the amplitude for some wave phenomenon), but the arguments leading to Eqn. (4.21) do not make obvious the appropriate interpretation of $\psi(x,t)$. Aside from the fact that we know that $\psi(x,t)$ (or rather $|\psi(x,t)|$ since ψ is complex) for wave packets is correlated with the position of the particle, the Schrödinger equation itself provides no obvious guidance. At the same time, we also wish to incorporate into our description of microscopic phenomena, the statistical nature of the measurement process mentioned in Section 1.1.

These two ideas come together in a natural way in the standard interpretation of the wavefunction solutions of the Schrödinger equation, namely that we are to interpret $|\psi(x,t)|^2$ as a *probability density* for position measurements. Specifically, if we define

$$P(x,t) = |\psi(x,t)|^2 = \psi^*(x,t)\psi(x,t) \tag{4.22}$$

then the so-called *Born interpretation*[2] states that

- $P(x,t)\,dx$ is the probability that a measurement of the position of the particle described by $\psi(x,t)$, at time t, will find it in the region $(x, x+dx)$.

In this view, it seems at this point that the wavefunction itself does not make a direct confrontation with experimental measurements of position, but does so

---

[2] Named after M. Born, 1928.

only through $|\psi(x, t)|^2$. This type of identification is not unique to quantum mechanics; the squared amplitude of a field is often used in the description of other wave phenomena where it appears, for example, in expressions for the energy density stored in the electric field given by $u_E(\mathbf{r}, t) = \epsilon_0|\mathbf{E}(\mathbf{r}, t)|^2/2$ or the magnetic field by $u_B(\mathbf{r}, t) = |\mathbf{B}(\mathbf{r}, t)|^2/2\mu_0$.

A similar and related *ensemble interpretation* of $\psi(x, t)$, which is perhaps more closely related to experimental test, is obtained by considering a (presumably large) number, N_0, of identically prepared particles, all described by the same wavefunction $\psi(x, t)$. Then

- $dN(x, t) = N_0 P(x, t) dx = N_0|\psi(x, t)|^2 dx$ is the number of particles found with position in the interval $(x, x + dx)$ at time t.

In this way, one can imagine measuring the probability density by binning measurements of the position in small increments δx, each bin having $\delta N(x, t)$ measured values, so that

$$\frac{\delta N(x, t)}{\delta x} = N_0|\psi(x, t)|^2 \tag{4.23}$$

gives $|\psi(x, t)|^2$.

Just as with any other continuous probability distribution, we then argue that

$$\text{Prob}[x \in (a, b)] = \int_a^b P(x, t)\, dx = \int_a^b |\psi(x, t)|^2 dx \tag{4.24}$$

is the probability of finding the particle in the *finite* interval (a, b); this in turn implies that

$$\int_{-\infty}^{+\infty} P(x, t)\, dx = \int_{-\infty}^{+\infty} |\psi(x, t)|^2\, dx = 1 \quad \text{for all } t \tag{4.25}$$

since the probability of finding the particle *somewhere* in its one-dimensional universe (i.e. the real line) must be unity.

This last constraint that the wavefunction be properly normalized is very important, as not all solutions of the Schrödinger equation will automatically satisfy this requirement; for example, the plane wave solutions for a free particle give

$$\int_{-\infty}^{+\infty} |\psi_p(x, t)|^2\, dx = \int_{-\infty}^{+\infty} \left|e^{i(px - p^2 t/2m)/\hbar}\right|^2 dx = \int_{-\infty}^{+\infty} 1\, dx = \infty \tag{4.26}$$

As long as the solutions are at least *square-integrable*, namely

$$\int_{-\infty}^{+\infty} |\psi(x, t)|^2 dx = C = \text{constant} < \infty \tag{4.27}$$

then we can use the linearity of the Schrödinger equation to write

$$\tilde{\psi}(x, t) = \frac{1}{\sqrt{C}} \, \psi(x, t) \tag{4.28}$$

Then $\tilde{\psi}(x, t)$ will still be a solution (because of linearity) with the same physical properties, but it will now satisfy Eqn. (4.25) and be properly normalized. This probability constraint is the motivation for the normalization factors in all of our examples throughout Chapters 3 and 4. We stress that the two steps of solving the Schrödinger equation (implementing the wave physics) and initially normalizing the solutions (ensuring a consistent probability interpretation), are independent of one another.

The requirement of square-integrability implies strong constraints on $\psi(x, t)$:

- $|\psi(x, t)|$ must tend to zero sufficiently rapidly as $x \to \pm\infty$ so that the integral of $|\psi(x, t)|^2$ will converge.

- We will usually assume that ψ is well enough behaved that various spatial derivatives of ψ also vanish at infinity.

- For a probability interpretation to be valid, we must also require that $\psi(x, t)$ be continuous in x as a discontinuous ψ would lead to ambiguous predictions for probabilities near the "jump", as in Fig. 4.5.

- Unless the potential energy function is extremely poorly behaved (see Section 8.1.1 for an example of just how bad it would have to be), we can also assume that $\psi'(x, t)$ and higher spatial derivatives are everywhere continuous.

We have argued that the requirement that $\psi(x, t)$ be properly normalized at, say, $t = 0$, is independent of the fact that it is a solution of the wave equation. It is not obvious, therefore, that once normalized it will continue to be so at future

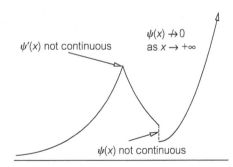

Figure 4.5. Unacceptable position-space wavefunction. $\psi(x)$ is assumed real, so it can be plotted easily.

times. We are thus naturally led to ask whether

$$\int_{-\infty}^{+\infty} |\psi(x, t_0)|^2 dx = 1 \qquad\qquad \int_{-\infty}^{+\infty} |\psi(x, t)|^2 dx = 1$$

$$\Longrightarrow \qquad\qquad\qquad (4.29)$$

at $t_0 = 0$ for all later times t.

We want to know if the time-evolution of the wavefunction, dictated by the Schrödinger equation, respects the normalization imposed by a probability interpretation.

To confirm this, we assume that $\psi(x, t)$ is a solution of Eqn. (4.21) (so that $\psi^*(x, t)$ satisfies the complex conjugated version of Eqn. (4.21)) and we note that

$$\frac{\partial P(x, t)}{\partial t} = \left(\frac{\partial \psi^*}{\partial t}\psi + \psi^*\frac{\partial \psi}{\partial t}\right)$$

$$= \left(\frac{i}{\hbar}\left(-\frac{\hbar^2}{2m}\frac{\partial^2 \psi^*}{\partial x^2} + V^*\psi^*\right)\right)\psi + \psi^*\left(\frac{-i}{\hbar}\left(-\frac{\hbar^2}{2m}\frac{\partial^2 \psi}{\partial x^2} + V\psi\right)\right)$$

$$\frac{\partial P(x, t)}{\partial t} = -\frac{i\hbar}{2m}\left(\frac{\partial \psi^*}{\partial x^2}\psi - \psi^*\frac{\partial \psi}{\partial x^2}\right) + \frac{i}{\hbar}(\psi^*\psi)(V^* - V) \qquad (4.30)$$

Classically, any potential energy function with which we are familiar is real, so that $V^*(x, t) - V(x, t) = 0$ and we can then write

$$\frac{\partial P(x, t)}{\partial t} = -\frac{\partial}{\partial x}\left[\frac{\hbar}{2mi}\left(\psi^*\frac{\partial \psi}{\partial x} - \frac{\partial \psi^*}{\partial x}\psi\right)\right] = -\frac{\partial}{\partial x}j(x, t) \qquad (4.31)$$

We have defined

$$j(x, t) \equiv \frac{\hbar}{2mi}\left(\psi^*(x, t)\frac{\partial \psi(x, t)}{\partial x} - \frac{\partial \psi^*(x, t)}{\partial x}\psi(x, t)\right) \qquad (4.32)$$

which can be interpreted as a *probability current* or *flux*. This relation, relating the *time rate of change* of a probability density to the *spatial change* in a flux,

$$\frac{\partial P(x, t)}{\partial t} = -\frac{\partial j(x, t)}{\partial x} \qquad (4.33)$$

is called the *equation of continuity*. It is based on conservation of probability in the same way that the similarly named equation for the flow of incompressible fluids, namely

$$\frac{\partial \rho(x, t)}{\partial t} = -\frac{\partial}{\partial x}[\rho(x, t)v_x(x, t)] \qquad (4.34)$$

or, in three dimensions,

$$\frac{\partial \rho(\mathbf{r}, t)}{\partial t} = -\nabla \cdot [\rho(\mathbf{r}, t)\mathbf{v}(\mathbf{r}, t)] \tag{4.35}$$

(where $\rho(x, t)$ is the fluid density), is based on the conservation of fluid mass or the similar statement in electromagnetism based on conservation of electric charge. Integrating Eqn. (4.33) over a finite region of space, we find

$$\frac{d}{dt}\left[\int_a^b P(x, t)\, dx\right] = -\int_a^b \frac{\partial j(x, t)}{\partial x}\, dx = j(a, t) - j(b, t) \tag{4.36}$$

which can be interpreted as saying

• The time rate of change of the probability in a finite interval (a, b) at any given time t (the left-hand side) is given by the difference in the rates of probability "flow" into $(j(a, t))$ and out of $(j(b, t)$ that interval (the right-hand side).

Most importantly, if we specialize to $(a, b) = (-\infty, +\infty)$, that is, the entire one-dimensional universe, we have

$$\frac{dP(t)}{dt} \equiv \frac{d}{dt}\left[\int_{-\infty}^{+\infty} P(x, t)\, dx\right] = j(-\infty, t) - j(+\infty, t) \longrightarrow 0 \tag{4.37}$$

because

$$\lim_{x \to \pm\infty} (\psi(x, t)) = 0 \tag{4.38}$$

if the wavefunction is to be normalizable. Here

$$P(t) \equiv \int_{-\infty}^{+\infty} P(x, t)\, dx \tag{4.39}$$

is the probability of finding the particle *somewhere* as a function of time. Thus, $P(t) = C$ is a constant for all times, one which we have already set to unity to implement probability conservation.

This shows that the total probability is indeed constant in time provided that

(1) the solutions satisfy the Schrödinger equation;

(2) the solutions are localized so that the wavefunction can be initially normalized;

(3) the potential energy function is real.

Under these conditions, the initial normalization factor is preserved by the subsequent time-evolution dictated by the Schrödinger equation; we can "*set it and forget it.*"

While the normalization of the wavefunction necessary for a probability interpretation is initially separate from the implementation of the wave equation, the Schrödinger equation guarantees that such an identification is preserved. This connection is a further assurance that the association of $|\psi(x, t)|^2$ with a probability density is a very natural one.

If we relax the constraint that the potential be real and allow it to have, for example, a constant, negative imaginary part $V(x) = V_R(x) - iV_I$, we can repeat the analysis above (P4.8). We find that

$$\frac{d\mathcal{P}(t)}{dt} = -\frac{2V_I}{\hbar}\mathcal{P}(t) = -\lambda\mathcal{P}(t) = -\frac{1}{\tau}\mathcal{P}(t) \tag{4.40}$$

where $\lambda \equiv 2V_I/\hbar \equiv 1/\tau$; this has the trivial solution

$$\mathcal{P}(t) = \mathcal{P}(0)\,e^{-\lambda t} = \mathcal{P}(0)\,e^{-t/\tau} \tag{4.41}$$

Viewed in the "ensemble" interpretation where

$$N(t) = N(0)e^{-\lambda t} = N_0 e^{-t/\tau} \tag{4.42}$$

it becomes clear that this represents the loss of probability or particles with an exponential decay law familiar from radioactivity. The decay rate, λ, and mean lifetime, τ, of an unstable particle (or ensemble thereof) can be described in quantum mechanical language at the cost of a nonintuitive (or at least nonclassical) complex potential energy. The relationship between the complex energy, V_I, and the lifetime, τ, namely, $V_I \cdot \tau = \hbar/2$, is reminiscent of an energy-time uncertainty principle.

4.3 Average Values

4.3.1 Average Values of Position

Given a probability density for position, we can immediately evaluate the expectation values of any x-related quantity; for example, the average value resulting from many position measurements is predicted to be

$$\langle x \rangle_t = \int_{-\infty}^{+\infty} x\,P(x, t)\,dx = \int_{-\infty}^{+\infty} x\,|\psi(x, t)|^2\,dx \tag{4.43}$$

where we stress that

- the time-dependence of $\langle x \rangle_t$ comes solely from the information contained in the wavefunction, $\psi(x, t)$.

The average value defined in this way is as close as one can come in quantum mechanics to the concept of a classical trajectory, $x(t)$. In a similar way, one can evaluate

$$\langle x^n \rangle_t = \int_{-\infty}^{+\infty} x^n \, P(x, t) \, dx \tag{4.44}$$

as well as

$$\langle f(x) \rangle_t = \int_{-\infty}^{+\infty} f(x) \, P(x, t) \, dx \tag{4.45}$$

for an arbitrary function of position.

Using these, we can return to the example of a spreading Gaussian wave packet considered in Section 3.2.2 where we studied

$$|\psi(x, t)|^2 = \frac{1}{\beta_t \sqrt{\pi}} e^{-(x - p_0 t/m)^2/\beta_t^2} \tag{4.46}$$

with $\beta_t = \alpha \hbar \sqrt{1 + t^2/t_0^2}$; we can now evaluate expectation values of any power of x. Using Appendix D.1 we find that

$$\langle x \rangle_t = \int_{-\infty}^{+\infty} dx \, x \, |\psi(x, t)|^2$$

$$= \int_{-\infty}^{+\infty} dx \, (x - p_0 t/m + p_0 t/m) \left[\frac{1}{\beta_t \sqrt{\pi}} e^{-(x - p_0 t/m)^2/\beta_t^2} \right]$$

$$= \int_{-\infty}^{+\infty} dx \, (x - p_0 t/m) \left[\frac{1}{\beta_t \sqrt{\pi}} e^{-(x - p_0 t/m)^2/\beta_t^2} \right]$$

$$\quad + p_0 t/m \int_{-\infty}^{+\infty} dx \left[\frac{1}{\beta_t \sqrt{\pi}} e^{-(x - p_0 t/m)^2/\beta_t^2} \right]$$

$$= \left[\frac{1}{\sqrt{\pi}} \int_{-\infty}^{+\infty} z \, e^{-z^2} \, dz \right] + (p_0 t/m) \left[\frac{1}{\sqrt{\pi}} \int_{-\infty}^{+\infty} e^{-z^2} \, dz \right]$$

$$\langle x \rangle_t = \frac{p_0 t}{m} \tag{4.47}$$

as expected, where we have used the fact that the odd Gaussian integrand ($z \, e^{-z^2}$) vanishes when integrated over all space. Using similar integrals, we find

$$\langle x^2 \rangle_t = \left(\frac{p_0 t}{m} \right)^2 + \frac{\beta_t^2}{2} \tag{4.48}$$

so that the spread in position does indeed increase with time via

$$\Delta x_t = \sqrt{\langle x^2 \rangle_t - \langle x \rangle_t^2} = \frac{\alpha \hbar}{\sqrt{2}} \sqrt{1 + t^2/t_0^2} = \frac{\beta_t}{\sqrt{2}} \tag{4.49}$$

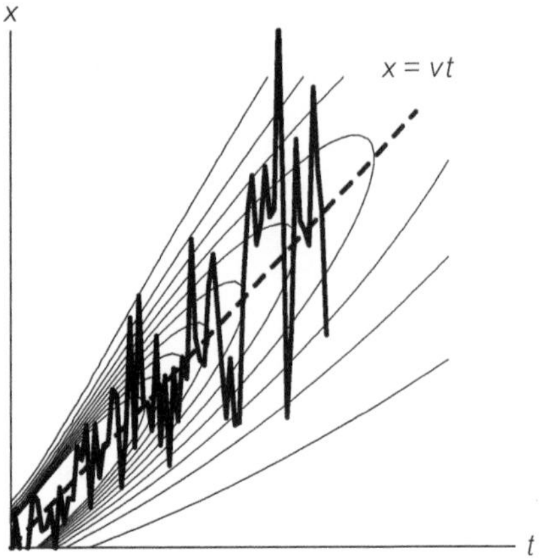

Figure 4.6. Contour plot of $|\psi(x,t)|^2$ versus (x,t) for a Gaussian free-particle wave packet. Superimposed is the classical straight-line trajectory, $x(t) = vt$ (dashed line), as well as random measurements of the particle position for increasing times (connected by the 'erratic' solid lines) to illustrate wave packet spreading.

We can "sample" this wavefunction by randomly measuring the position of a particle at various times and comparing these measurements to the classical, straight line trajectory in Fig. 4.6. We note that the measurements tend to cluster around the classical path, but with increasingly large excursions due to the spreading.

4.3.2 **Average Values of Momentum**

It is initially far from obvious how to extract further information from $\psi(x,t)$ on other physically observable quantities, such as momentum and energy, which are now represented by operators. For example, in Section 3.1 we identified the momentum with an operator via $\hat{p} = (\hbar/i)\partial/\partial x$. Does one then define

$$\langle \hat{p} \rangle_t \stackrel{?}{=} \int_{-\infty}^{+\infty} dx \, \hat{p} \, |\psi(x,t)|^2 \stackrel{?}{=} \int_{-\infty}^{+\infty} dx \, \psi^*(x,t) \, \hat{p} \, \psi(x,t) \qquad (4.50)$$

or in some other way? No such ambiguity arises, of course, for the position variable itself as

$$\langle x \rangle_t = \int_{-\infty}^{+\infty} dx \, x \, |\psi(x,t)|^2 = \int_{-\infty}^{+\infty} dx \, \psi^*(x,t) \, x \, \psi(x,t) \qquad (4.51)$$

To gain some guidance, we note that classically the trajectory $x(t)$ would satisfy

$$\frac{dx(t)}{dt} = \frac{p(t)}{m} \qquad (4.52)$$

so we examine the time-dependence of its quantum analog, $\langle x \rangle_t$, more fully. We can write

$$
\frac{d}{dt}\langle x \rangle_t = \frac{d}{dt}\left[\int_{-\infty}^{+\infty} x|\psi(x,t)|^2 dx\right]
$$

$$
= \int_{-\infty}^{+\infty} dx\, x\left(\frac{\partial \psi^*}{\partial t}\psi + \psi^*\frac{\partial \psi}{\partial t}\right)
$$

$$
= \frac{\hbar}{2mi}\int_{-\infty}^{+\infty} dx\left(\frac{\partial^2 \psi^*}{\partial x^2}x\psi - \psi^* x\frac{\partial^2 \psi}{\partial x^2}\right) \tag{4.53}
$$

where we have once again used the fact that $\psi(x,t)$ satisfies the Schrödinger equation (and ψ^* its conjugate) and have assumed that $V(x,t)$ is real. To simplify this, consider

$$
\int_{-\infty}^{+\infty} dx\, \frac{\partial^2 \psi^*}{\partial x^2}x\psi \stackrel{\text{IBP}}{=} \left(\frac{\partial \psi^*}{\partial x}(x\psi)\right)_{-\infty}^{+\infty} - \int_{-\infty}^{+\infty} dx\frac{\partial \psi^*}{\partial x}\frac{\partial}{\partial x}(x\psi)
$$

$$
= -\int_{-\infty}^{+\infty} dx\frac{\partial \psi^*}{\partial x}\left(\psi + x\frac{\partial \psi}{\partial x}\right) \tag{4.54}
$$

where we have used an integration by parts (IBP) and dropped the "surface" terms (those evaluated at $\pm\infty$), using the fact that the wavefunction vanishes sufficiently rapidly at infinity. We can repeat this trick once more to obtain

$$
\int_{-\infty}^{+\infty} dx\, \frac{\partial^2 \psi^*}{\partial x^2}x\psi = 2\int_{-\infty}^{+\infty} dx\, \psi^*\frac{\partial \psi}{\partial x} + \int_{-\infty}^{+\infty} dx\, \psi^* x\frac{\partial^2 \psi}{\partial x^2} \tag{4.55}
$$

so that substitution back in Eqn. (4.53) gives

$$
\frac{d\langle x \rangle_t}{dt} = \frac{\hbar}{2mi}\int_{-\infty}^{+\infty} dx\left(2\psi^*\frac{\partial \psi}{\partial x} + \psi^* x\frac{\partial^2 \psi}{\partial x^2} - \psi^* x\frac{\partial^2 \psi}{\partial x^2}\right)
$$

$$
= \frac{1}{m}\int_{-\infty}^{+\infty} dx\, \psi^*(x,t)\left(\frac{\hbar}{i}\frac{\partial}{\partial x}\right)\psi(x,t)
$$

$$
= \frac{1}{m}\int_{-\infty}^{+\infty} \psi^*(x,t)\,\hat{p}\,\psi(x,t)
$$

$$
\frac{d\langle x \rangle_t}{dt} \equiv \frac{\langle \hat{p}_t \rangle}{m} \tag{4.56}
$$

This identification of the average values, which is similar to the classical one for the trajectory variables, is valid provided we adopt the following general definition:

- The average value resulting from a large number of measurements of a physically observable quantity, O, corresponding to some quantum mechanical operator, $\hat{O}$, in a state described by a wavefunction $\psi(x, t)$, is

$$\langle \hat{O} \rangle_t \equiv \int_{-\infty}^{+\infty} dx \, [\psi^*(x, t)] \, \hat{O} \, [\psi(x, t)] \tag{4.57}$$

so that the operator is "sandwiched" between ψ^* and ψ, but acts "only to the right". We will speak of $\langle \hat{O} \rangle_t$ as the *average* or *expectation value* of that operator in the state $\psi(x, t)$. Clearly this definition reduces to the standard one for any function of the position "operator" x.

Given this general result, we can now extract some (but not all) of the information contained in $\psi(x, t)$ about any other physical observable O for which we have an associated quantum operator, $\hat{O}$. The expectation value of any power of an arbitrary operator, $\hat{O}^n$, is defined in a similar way, and we can generalize this further to any function of an operator, $f(\hat{O})$, provided we have a well-defined series representation for the function $f(y)$. Thus, if

$$f(y) = f(0) + f'(0)y + \frac{f''(0)y^2}{2} + \cdots = \sum_{n=0}^{\infty} \frac{f^{(n)}(0)}{n!} y^n \tag{4.58}$$

then

$$\langle f(\hat{O}) \rangle_t = \sum_{n=0}^{\infty} \frac{f^{(n)}(0)}{n!} \langle \hat{O}^n \rangle_t \tag{4.59}$$

which is sometimes useful. The need for well-defined values of moments of the momentum operator, implied by Eqn. (4.59), helps justify our continued assumptions that spatial derivatives of $\psi(x, t)$ exist and are well behaved at infinity.

4.3.3 **Average Values of Other Operators**

We can generalize the definition of $\langle \hat{p} \rangle_t$ to include any power of the momentum operator via

$$\langle \hat{p}^n \rangle_t = \int_{-\infty}^{+\infty} \psi^*(x, t) \, \hat{p}^n \, \psi(x, t) = \int_{-\infty}^{+\infty} \psi^*(x, t) \left(\frac{\hbar}{i} \frac{\partial}{\partial x} \right)^n \psi(x, t) \tag{4.60}$$

so, for example, the RMS spread in momentum measurements will be given by

$$\Delta p_t = \sqrt{\langle \hat{p}^2 \rangle_t - \langle \hat{p} \rangle_t^2} \tag{4.61}$$

Returning to the Gaussian wave packet of Eqn. (3.29), we can evaluate $\langle \hat{p} \rangle_t$ (P4.13) and find

$$\langle \hat{p} \rangle_t = \int_{-\infty}^{+\infty} \psi^*(x, t) \left(\frac{\hbar}{i} \frac{\partial}{\partial x} \right) \psi(x, t)$$

$$= \int_{-\infty}^{+\infty} dx \left(p_0 - \frac{2(x - p_0 t/m)}{2\hbar i \alpha F} \right) |\psi(x, t)|^2$$

$$= p_0 \int_{-\infty}^{+\infty} |\psi(x, t)|^2$$

$$\boxed{\langle \hat{p} \rangle_t = p_0} \tag{4.62}$$

where the two terms arise from differentiating the "phase" and the "Gaussian" term, respectively. In the case of position measurements, the "phase" term played no significant role, since only $|\psi(x, t)|^2$ appeared. For more general average value calculations where the operator must act on $\psi(x, t)$ *before* one squares, its effects can obviously be important. In this sense, $\psi(x, t)$, with all its phase information intact, is clearly more fundamental than $|\psi(x, t)|^2$.

A similar (and only slightly lengthier) calculation shows that

$$\boxed{\langle \hat{p}^2 \rangle_t = p_0^2 + \frac{1}{2\alpha^2}} \quad \text{so that} \quad \boxed{\Delta p_t = 1/\sqrt{2}\alpha} \tag{4.63}$$

is indeed constant in time, consistent with Eqn. (3.21). Combining this result and that of Eqn. (4.49) we find that

$$\boxed{\Delta x_t \, \Delta p_t = \frac{\hbar}{2} \sqrt{1 + t^2/t_0^2} \geq \frac{\hbar}{2}} \tag{4.64}$$

and at $t = 0$ this wave packet actually attains the minimum product of spreads in x and p allowed by the uncertainty principle.

The evaluation of $\langle \hat{p}^2 \rangle$ is especially interesting as it is related to the kinetic energy operator $\hat{T}$, and we can write rather generally

$$\langle \hat{T} \rangle_t = \frac{1}{2m} \langle \hat{p}^2 \rangle_t = -\frac{\hbar^2}{2m} \int_{-\infty}^{+\infty} dx \, \psi^* \frac{\partial^2 \psi(x, t)}{\partial x^2}$$

$$\overset{\text{IBP}}{=} -\frac{\hbar^2}{2m} \left(\psi^*(x, t) \frac{\partial \psi(x, t)}{\partial x} \right)_{-\infty}^{+\infty} + \frac{\hbar^2}{2m} \int_{-\infty}^{+\infty} dx \, \frac{\partial \psi^*}{\partial x} \frac{\partial \psi}{\partial x}$$

$$\boxed{\langle \hat{T} \rangle_t = \frac{\hbar^2}{2m} \int_{-\infty}^{+\infty} dx \left| \frac{\partial \psi(x, t)}{\partial x} \right|^2} \tag{4.65}$$

This not only simplifies the calculation of $\langle \hat{p}^2 \rangle$ somewhat (one derivative instead of two), but it also shows that the kinetic energy associated with a quantum mechanical wavefunction can be related to its spatial variation, that is, its "wiggliness."

This justifies the statement made in discussing Fig. 3.2. The quantity

$$T(x, t) \equiv \frac{\hbar^2}{2m} \left| \frac{\partial \psi(x, t)}{\partial x} \right|^2 \quad \text{where} \quad \int_{-\infty}^{+\infty} T(x, t)\, dx = \langle \hat{T} \rangle_t \quad (4.66)$$

can be associated with a *kinetic energy distribution*. While it is not as funda-mentally important as $|\psi(x, t)|^2$, $T(x, t)$ is sometimes of use in visualizing the distribution of kinetic energy in quantum wavefunctions. A similar quantity is the *potential energy distribution*, which we can define to be

$$V(x, t) \equiv |\psi(x, t)|^2 V(x, t) \quad (4.67)$$

whose integral gives the expectation value of $V(x, t)$.

Finally, the expectation value of the total energy, represented by the average value of the operator $\hat{E} = i\hbar \partial/\partial t$, can be evaluated for any state via

$$\langle \hat{E} \rangle_t = \int_{-\infty}^{+\infty} \psi^*(x, t) \left(i\hbar \frac{\partial}{\partial t} \right) \psi(x, t) \quad (4.68)$$

and for the Gaussian wave packet an explicit calculation gives

$$\langle \hat{E} \rangle_t = \frac{1}{2m} \left(p_0^2 + \frac{1}{2\alpha^2} \right) = \frac{\langle \hat{p}_t^2 \rangle}{2m} \quad (4.69)$$

consistent with Eqn. (4.63) and the fact that the wave packet is a solution of the free particle Schrödinger equation.

4.4 Real Average Values and Hermitian Operators

While we have introduced a well-defined operational procedure for the calcu-lation of expectation values of quantum mechanical operators, many questions about the connection between classical observable quantities and their quantum mechanical operator counterparts remain to be answered. For example, for $\langle x \rangle_t$ and related position averages, it is clear from Eqns (4.43) and (4.45) that we will always find real values, as we should if we are to confront the results of measure-ments of observable quantities. In contrast, the forms for the momentum and energy operators, for example,

$$\hat{p} = \frac{\hbar}{i} \frac{\partial}{\partial x} \quad \text{and} \quad \hat{E} = i\hbar \frac{\partial}{\partial t} \quad (4.70)$$

with their explicit factors of i, make it far from obvious that their expectation values will not be complex, and hence have no connection with real measure-ments. We presumably wish to restrict ourselves to operators, $\hat{O}$, for which we can guarantee that the expectation values will satisfy

$$\langle \hat{O} \rangle = \langle \hat{O} \rangle^* \tag{4.71}$$

or

$$\int_{-\infty}^{+\infty} dx \, \psi^*(x,t) \, \hat{O} \, \psi(x,t) = \left[\int_{-\infty}^{+\infty} dx \, \psi^*(x,t) \, \hat{O} \, \psi(x,t) \right]^* \tag{4.72}$$

for any physically admissible wavefunction, $\psi(x,t)$. Operators which satisfy Eqn. (4.72) are called *Hermitian*, which we can take to be an extension of the notion of "realness" to operators. The similar statement for general complex numbers would, of course, be that a complex number, z, is real provided $z = z^*$. Thus, a first test of any identification of an operator with a classical observable will be to check whether Eqn. (4.72) is satisfied.

We note that this definition can be easily extended (P4.20) to show that a Hermitian operator, $\hat{O}$, will actually satisfy the more general condition

$$\int_{-\infty}^{+\infty} dx \, \psi^*(x,t) \, \hat{O} \, \phi(x,t) = \left[\int_{-\infty}^{+\infty} dx \, \phi^*(x,t) \, \hat{O} \, \psi(x,t) \right]^* \tag{4.73}$$

for any two admissible wavefunctions $\psi(x,t)$, $\phi(x,t)$.

While Eqn. (4.56) shows implicitly that $\langle \hat{p} \rangle_t$ is real (because $\langle x \rangle_t$ is manifestly real), it is instructive to demonstrate this in a more explicit way. We can write

$$\langle \hat{p} \rangle^* = \left[\int_{-\infty}^{+\infty} dx \, \psi^* \left(\frac{\hbar}{i} \frac{\partial}{\partial x} \right) \psi \right]^*$$

$$= \int_{-\infty}^{+\infty} dx \, \psi \left(-\frac{\hbar}{i} \frac{\partial \psi^*}{\partial x} \right)$$

$$\stackrel{\text{IBP}}{=} -\frac{\hbar}{i} \left[(\psi^* \psi)_{-\infty}^{+\infty} - \int_{-\infty}^{+\infty} dx \, \psi^* \frac{\partial \psi}{\partial x} \right]$$

$$= \int_{-\infty}^{+\infty} dx \, \psi^* \left(\frac{\hbar}{i} \frac{\partial}{\partial x} \right) \psi$$

$$\langle p \rangle^* = \langle \hat{p} \rangle \tag{4.74}$$

One explicit factor of -1 arising from the complex conjugation has been canceled by a similar one from an IBP. This trick can be extended to show that $\langle \hat{p}^n \rangle$ is real for any power of the momentum operator if one assumes that all the relevant surface terms generated by the various integrations by parts vanish due to the behavior of ψ and its spatial derivatives at $x = \pm\infty$. Using Eqn. (4.59) then shows that an arbitrary function of $\hat{p}$ will also have real expectation values. Thus, we have found that x and $f(x)$ (trivially) and $\hat{p}$ and $f(\hat{p})$ are all Hermitian operators.

The proof that the energy operator is Hermitian is somewhat different, as we examine

$$\langle \hat{E} \rangle - \langle \hat{E} \rangle^* = \int_{-\infty}^{+\infty} dx\, \psi^* \left(i\hbar \frac{\partial}{\partial t} \right) \psi - \left[\int_{-\infty}^{+\infty} dx\, \psi^* \left(i\hbar \frac{\partial}{\partial t} \right) \psi \right]^*$$

$$= i\hbar \int_{-\infty}^{+\infty} dx \left(\psi^* \frac{\partial \psi}{\partial t} + \frac{\partial \psi^*}{\partial t} \psi \right)$$

$$= i\hbar \frac{d}{dt} \left[\int_{-\infty}^{+\infty} dx\, \psi^*(x,t) \psi(x,t) \right]$$

$$= i\hbar \frac{d}{dt} \left[\int_{-\infty}^{+\infty} dx\, P(x,t) \right]$$

$$= i\hbar \frac{d}{dt} [\mathcal{P}(t)]$$

$$\langle \hat{E} \rangle - \langle \hat{E} \rangle^* = 0 \tag{4.75}$$

so that $\hat{E}$ is a Hermitian operator *provided* the total probability, $\mathcal{P}(t)$, is constant in time. We have seen that the only situation in which this is not true is when the potential energy function, $V(x,t)$, does, in fact, have an imaginary part; in that case we might expect the classical energy to not be well-defined.

4.5 The Physical Interpretation of $\phi(p)$

We are now able to calculate average values (and higher moments) of the momentum operator, but even more detailed information on the "momentum content" of the wavefunction $\psi(x,t)$ is available. In order to extract it in a simple way, we will analyze the role played by $\phi(p)$ more carefully. So far, we have constructed localized wave packets from plane wave solutions by using

$$\psi(x) = \frac{1}{\sqrt{2\pi\hbar}} \int_{-\infty}^{+\infty} dp\, \phi(p)\, e^{ipx/\hbar} \tag{4.76}$$

where $\phi(p)$ simply played the role of a weighting function, the amplitude associated with each plane wave component of definite momentum, p. Given a solution of the Schrödinger equation, $\psi(x,t)$ we can invert this to obtain

$$\phi(p,t) = \frac{1}{\sqrt{2\pi\hbar}} \int_{-\infty}^{+\infty} dx\, \psi(x,t)\, e^{-ipx/\hbar} \tag{4.77}$$

so that this momentum–amplitude can be obtained from any solution of the Schrödinger equation. Recall that the normalizations of $\psi(x,t)$ and $\phi(p,t)$ are completely correlated as

$$\int_{-\infty}^{+\infty} dp\, |\phi(p,t)|^2 = \int_{-\infty}^{+\infty} dx\, |\psi(x,t)|^2 = 1 \qquad (4.78)$$

This is true provided that we have enforced the normalization for $\psi(x,t)$ as a consequence of its interpretation as giving a probability density for position measurements. This fact strongly suggests that we make the additional association that

$$P_{QM}(p,t) = |\phi(p,t)|^2 = \phi^*(p,t)\,\phi(p,t) \qquad (4.79)$$

is a *probability density* for momentum measurements, and that

- $|\phi(p,t)|^2\, dp$ is the probability that a measurement of the *momentum* of a particle described by $\phi(p,t)$ (obtained possibly via the Fourier transform of $\psi(x,t)$) will find a value in the interval $(p, p+dp)$ at time t.

We can call $\phi(p,t)$ the *momentum-space wavefunction* by analogy with $\psi(x,t)$ which is the *position-* or *configuration-space wavefunction*. This then allows one to make more detailed predictions about the distribution of momentum values using, for example,

$$\text{Prob}[p \in (p_a, p_b)] = \int_{p_a}^{p_b} dp\, |\phi(p,t)|^2 \qquad (4.80)$$

being the probability of measuring the momentum to be in the finite interval (p_a, p_b).

This identification is made more compelling by the observation that

$$\langle \hat{p} \rangle_t = \int_{-\infty}^{+\infty} dx\, \psi^*(x,t) \left(\frac{\hbar}{i}\frac{\partial}{\partial x}\right) \psi(x,t)$$

$$= \int_{-\infty}^{+\infty} \psi^*(x,t) \left(\frac{\hbar}{i}\frac{\partial}{\partial x}\right) \left[\frac{1}{\sqrt{2\pi\hbar}} \int_{-\infty}^{+\infty} dp\, \phi(p,t) e^{ipx/\hbar}\right]$$

$$= \left[\frac{1}{\sqrt{2\pi\hbar}} \int_{-\infty}^{+\infty} dx\, \psi^*(x,t) \int_{-\infty}^{+\infty} dp\, p\, \phi(p,t) e^{ipx/\hbar}\right]$$

$$= \int_{-\infty}^{+\infty} dp\, p\, \phi(p,t) \left[\frac{1}{\sqrt{2\pi\hbar}} \int_{-\infty}^{+\infty} dx\, \psi(x,t)\, e^{-ipx/\hbar}\right]^*$$

$$= \int_{-\infty}^{+\infty} dp\, p\, \phi(p,t)\phi^*(p,t)$$

$$= \int_{-\infty}^{+\infty} dp\, p\, |\phi(p,t)|^2$$

$$\langle \hat{p} \rangle_t \equiv \langle p \rangle_t \qquad (4.81)$$

where we have written the average value now as $\langle p \rangle_t$ (i.e. without the "hat" or operator symbol) when evaluated using the momentum–space wavefunction. In this representation of the quantum mechanical solution, the momentum observable is represented by a "trivial" operator, p. Position information is now less directly obtainable, as we can write

$$\langle x \rangle_t = \int_{-\infty}^{+\infty} dx \, \psi^*(x,t) \, x \, \psi(x,t)$$

$$= \int_{-\infty}^{+\infty} \left[\frac{1}{\sqrt{2\pi\hbar}} \int_{-\infty}^{+\infty} dp \, \phi^*(p,t) \, e^{-ipx/\hbar} \right] x \, \psi(x,t) \, dx$$

$$= \int_{-\infty}^{+\infty} dp \, \phi^*(p,t) \left[\frac{1}{\sqrt{2\pi\hbar}} \int_{-\infty}^{+\infty} dx \, x \, \psi(x) \, e^{-ipx/\hbar} \right]$$

$$= \int_{-\infty}^{+\infty} dp \, \phi^*(p,t) \left(i\hbar \frac{\partial}{\partial p} \right) \left[\frac{1}{\sqrt{2\pi\hbar}} \int_{-\infty}^{+\infty} dx \, \psi(x) \, e^{-ipx/\hbar} \right]$$

$$= \int_{-\infty}^{+\infty} \phi^*(p,t) \left(i\hbar \frac{\partial}{\partial p} \right) \phi(p,t)$$

$$= \int_{-\infty}^{+\infty} \phi^*(p,t) \, \hat{x} \, \phi(p,t)$$

$$\langle x \rangle_t = \langle \hat{x} \rangle_t \tag{4.82}$$

where we now identify the *position operator* in the momentum space representation as

$$\hat{x} = i\hbar \frac{\partial}{\partial p} \tag{4.83}$$

The probability interpretation of $\phi(p,t)$ also implies that:

- $\phi(p,t)$ should be square-integrable (as a function of p) and continuous in p and

- $\phi(p,t)$ (and its spatial derivatives) should be continuous and must vanish sufficiently rapidly as $p \to \pm\infty$ so that it is square-integrable.

The expectation values of higher powers or functions of x are also easily obtained, generalizing ideas in Sections 4.3.2 and 4.3.3. For example, a sometimes useful expression for the average value of x^2 is given by

$$\langle \hat{x}^2 \rangle_t = -\hbar^2 \int_{-\infty}^{+\infty} \phi^*(p,t) \frac{\partial^2 \phi(p,t)}{\partial p^2} \, dp = +\hbar^2 \int_{-\infty}^{+\infty} \left| \frac{\partial \phi(p,t)}{\partial p} \right|^2 dp \tag{4.84}$$

where the second form is obtained by an IBP, as in Eqn. (4.65).

4.6 **Energy Eigenstates, Stationary States, and the Hamiltonian Operator**

In much the same way that Newton's second law relates the time-dependence of a particle's trajectory to the external force, the time-dependent Schrödinger equation

$$i\hbar \frac{\partial}{\partial t} \psi(x, t) = -\frac{\hbar^2}{2m} \frac{\partial^2 \psi(x, t)}{\partial x^2} + V(x, t)\psi(x, t) \tag{4.85}$$

dictates the time-development of the wavefunction of a particle in the presence of an external potential. If $V(x, t)$ is truly time-dependent, the resulting partial differential equation can be difficult to solve, but we can often consider the special, but very important, case of a time-independent potential, that is, one for which

$$V(x, t) = V(x) \tag{4.86}$$

only. In this case, the Schrödinger equation can be separated in the form

$$i\hbar \frac{\partial}{\partial t} \psi(x, t) = \left[-\frac{\hbar^2}{2m} \frac{\partial^2}{\partial x^2} + V(x) \right] \psi(x, t) = \hat{H}\psi(x, t) \tag{4.87}$$

where we have introduced the *Hamiltonian operator*, which can be written as

$$\hat{H} \equiv -\frac{\hbar^2}{2m} \frac{\partial^2}{\partial x^2} + V(x) = \frac{\hat{p}^2}{2m} + V(x) \tag{4.88}$$

and is seen to be a function of position coordinates only. This operator is the quantum mechanical version of the corresponding classical Hamiltonian function.[3]

Equation (4.87) is now a *separable differential equation* and a standard method of solution is to assume a product wavefunction of the form

$$\psi(x, t) = \psi(x) T(t). \tag{4.89}$$

If we substitute this form into Eqn. (4.87) and divide by $\psi(x, t)$ we find

$$\left[i\hbar \frac{dT(t)}{dt} \right] \Big/ T(t) = \left[\hat{H}\psi(x) \right] \Big/ \psi(x) \tag{4.90}$$

which must be true, of course, for all possible values of x and t. The constraint that two different functions of independent variables be identical, that is,

$$F(t) = G(x) \qquad \text{for all } x \text{ and } t \tag{4.91}$$

---

[3] See many undergraduate texts, for example, Marion and Thornton (2003), for an introduction to the Hamiltonian formulation of classical mechanics; a brief review is also contained in Appendix G.

can be satisfied only if both functions are equal to a constant. This is easily shown by noting that

$$0 = \frac{\partial F(t)}{\partial x} = \frac{\partial G(x)}{\partial x} \Longleftrightarrow G(x) = \text{constant}$$

$$\frac{\partial F(t)}{\partial t} = \frac{\partial G(x)}{\partial t} = 0 \Longrightarrow F(t) = \text{same constant} \tag{4.92}$$

Noting the dimensions, we can then write the common constant as E giving

$$\frac{[i\hbar(dT(t)/dt)]}{T(t)} = \frac{\left[\hat{H}\psi(x)\right]}{\psi(x)} = E \tag{4.93}$$

so that the time-dependence is easily found to be

$$T(t) = e^{-iEt/\hbar} \tag{4.94}$$

The complete wavefunction is then

$$\psi(x, t) = \psi_E(x) e^{-iEt/\hbar} \tag{4.95}$$

where $\psi_E(x)$ now satisfies the *time-independent Schrödinger equation*

$$\hat{H}\psi_E(x) = \left(-\frac{\hbar^2}{2m}\frac{d^2}{dx^2} + V(x)\right)\psi_E(x) = E\psi_E(x) \tag{4.96}$$

We will devote much of the rest of the book to examining the mathematical properties and physical meaning of solutions of the time-independent Schrödinger equation. We note that:

- The *number E* can certainly be identified as the uniquely defined energy of the state since application of the energy *operator*, $\hat{E}$, gives

$$\hat{E}\psi_E(x, t) = i\hbar\frac{\partial}{\partial t}\left(\psi_E(x)e^{-iEt/\hbar}\right) = E\psi_E(x, t) \tag{4.97}$$

Moreover, calculations of expectation values of powers of the energy operator give

$$\langle\hat{E}^n\rangle = \int_{-\infty}^{+\infty} dx\, \psi^*(x, t)\, \hat{E}^n\, \psi(x, t)$$

$$= \int_{-\infty}^{+\infty} dx\, \left(\psi_E^*(x)e^{iEt/\hbar}\right)\left(i\hbar\frac{\partial}{\partial t}\right)^n\left(\psi_E(x)e^{-iEt/\hbar}\right)$$

$$= E^n \int_{-\infty}^{+\infty} dx\, |\psi_E(x)|^2$$

$$= E^n \tag{4.98}$$

This implies that the uncertainty in energy of this state is

$$\Delta E = \sqrt{\langle \hat{E}^2 \rangle - \langle \hat{E} \rangle^2} = 0 \tag{4.99}$$

Such a state, with a precisely defined value of energy, can be called an *energy eigenstate* with *energy eigenvalue* given by E. (The use of the German "eigen" meaning characteristic or "belonging to" or, colloquially, "own" is appropriate here.)

- The energy eigenvalue appears as a parameter in the time-independent Schrödinger equation, so that separate solutions must be found for each different value of E, hence the label $\psi_E(x)$. For the case of a free particle, where $V(x) = 0$, for example, we solve

$$\hat{H}\psi_E(x) = -\frac{\hbar^2}{2m}\frac{d^2\psi_E(x)}{dx^2} = E\psi_E(x) \tag{4.100}$$

to obtain

$$\psi_E(x) = e^{\pm i\sqrt{2mE}x/\hbar} \tag{4.101}$$

or

$$\psi_E(x,t) = e^{\pm i\sqrt{2mE}x/\hbar}e^{-iEt/\hbar} = e^{i(px-p^2t/2m)/\hbar} = \psi_p(x,t) \tag{4.102}$$

which are the standard plane wave solutions, with the identification $E = p^2/2m$ for the *parameters E, p*.

- An equation such as $\hat{H}\psi_E(x) = E\psi_E(x)$, of the form

$$\text{operator acting on function} = \text{number times a function} \tag{4.103}$$

is called an *eigenvalue problem*. Similar problems arise in matrix algebra and elsewhere; in that context, they are often of the form

$$\mathbf{M} \cdot \mathbf{v} = \lambda \cdot \mathbf{v} \tag{4.104}$$

where $\mathbf{M}$ is a matrix and $\mathbf{v}$ is a vector.

- The trivial time-dependence of such states implies that the corresponding probability densities are independent of time, since

$$P(x,t) = |\psi_E(x,t)|^2 = |\psi_E(x)|^2 e^{-iEt/\hbar}e^{+iEt/\hbar} = |\psi_E(x)|^2 \tag{4.105}$$

Therefore, the expectation values of most operators, $\hat{O}$, for such states will satisfy

$$
\langle \hat{O} \rangle_t = \int_{-\infty}^{+\infty} dx \left(\psi_E^*(x) e^{+iEt/\hbar} \right) \hat{O} \left(\psi_E(x) e^{-iEt/\hbar} \right)
$$

$$
= \int_{-\infty}^{+\infty} dx \, \psi_E(x)^* \, \hat{O} \, \psi_E(x)
$$

$$
\langle \hat{O} \rangle_t = \langle \hat{O} \rangle_{t=0} \tag{4.106}
$$

Such states for which the probability density and other observables are "frozen in time" are called *stationary states.*

- Because of the linearity of the Schrödinger equation, the most general solution will consist of linear combinations of such stationary state or energy eigenstate solutions, that is,

$$
\psi(x, t) = \left(\sum_E + \int dE \right) \psi_E(x) e^{-iEt/\hbar} \tag{4.107}
$$

where the sum is over all possible discrete and continuous values of E. Because this function contains solutions of different energies, it will no longer be an energy eigenstate and will, in general, have $\Delta E \neq 0$. In addition, because of the possibility of "cross-terms" in $|\psi(x, t)|^2$, the probability density (and physical observables) can have nontrivial time-dependence, and so it is not a stationary state either. Examples of this type include the Gaussian wave packet constructed from free-particle solutions in Section 3.2.2 or the accelerating wave packet considered in the next section.

An especially simple case of such time-dependence is a solution consisting of a linear combination of two normalized energy eigenstates, namely

$$
\psi(x, t) = \frac{1}{\sqrt{2}} \left(\psi_{E_1}(x) \, e^{-iE_1 t/\hbar} + \psi_{E_2}(x) \, e^{-iE_2 t/\hbar} \right) \tag{4.108}
$$

where we assume that $\psi_{E_1}(x)$ and $\psi_{E_2}(x)$ are real for simplicity. We will also assume that the two states have a vanishing overlap integral (to be proved quite generally in Section 6.3). In this case we have

$$
\langle \hat{E} \rangle = \frac{E_1 + E_2}{2} \quad \text{and} \quad \langle \hat{E}^2 \rangle = \frac{E_1^2 + E_2^2}{2} \tag{4.109}
$$

so that

$$
\Delta E = \sqrt{\langle \hat{E}^2 \rangle - \langle \hat{E} \rangle^2} = \frac{|E_1 - E_2|}{2} \tag{4.110}
$$

while

$$P(x, t) = |\psi(x, t)|^2 = \psi_{E_1}^2(x) + \psi_{E_2}^2(x)$$

$$+ 2\psi_{E_1}(x)\psi_{E_2}(x)\cos(|E_1 - E_2|t/\hbar) \qquad (4.111)$$

- This example serves to remind us that the actual value of E itself is not important, as for individual eigenstates the effect of the time-dependent phase, $e^{-iEt/\hbar}$, is irrelevant. In mixed states, only *energy differences* appear, and this fact is a reflection of the arbitrariness inherent in choosing the potential energy function, $V(x)$; in classical mechanics, letting $V(x) \rightarrow V(x) + V_0$ makes no change in the applied force (and hence the physics), and the choice of V_0 can change the overall energy scale, but not energy differences (P6.4).

- Energy eigenstates, characterized by $\Delta E = 0$, in order to be consistent with the energy–time uncertainty principle, $\Delta E \, \Delta t \geq \hbar/2$ require that $\Delta t = \infty$ in some sense; this is plausible given the static character of stationary states. For the two-state system above, the characteristic periodicity of the system from Eqn. (4.111) is

$$\tau = 2\pi \frac{\hbar}{|E_1 - E_2|} \qquad (4.112)$$

which indeed is perfectly consistent with

$$\Delta E \Delta t \approx \frac{|E_1 - E_2|}{2} \tau \approx \pi \hbar \qquad (4.113)$$

4.7 The Schrödinger Equation in Momentum Space

4.7.1 Transforming the Schrödinger Equation into Momentum Space

We will often concentrate on the solution of the time-dependent Schrödinger equation in position- or configuration-space, namely, solving

$$-\frac{\hbar^2}{2m} \frac{\partial^2 \psi(x, t)}{\partial x^2} + V(x)\psi(x, t) = i\hbar \frac{\partial \psi(x, t)}{\partial t} \qquad (4.114)$$

for $\psi(x, t)$ and then obtaining $\phi(p, t)$, if desired, by the appropriate Fourier transform,

$$\phi(p, t) = \frac{1}{\sqrt{2\pi\hbar}} \int_{-\infty}^{+\infty} \psi(x, t)e^{-ipx/\hbar}dx. \qquad (4.115)$$

It is possible, however, to transform the Schrödinger equation itself into momentum space and solve for $\phi(p, t)$ directly, and this strategy sometimes yields a simpler and more directly interpretable solution. (For another variation on this approach, see P4.22.) To this end, we multiply both sides of Eqn. (4.114) by $\exp(-ipx/\hbar)/\sqrt{2\pi\hbar}$ and integrate over dx to obtain

$$\frac{1}{\sqrt{2\pi\hbar}} \int_{-\infty}^{+\infty} i\hbar \frac{\partial \psi(x, t)}{\partial t} e^{-ipx/\hbar} dx$$

$$= \frac{1}{\sqrt{2\pi\hbar}} \int_{-\infty}^{+\infty} \left(-\frac{\hbar^2}{2m} \frac{\partial^2 \psi(x, t)}{\partial x^2} \right) \left(e^{-ipx/\hbar} \right) dx$$

$$+ \frac{1}{\sqrt{2\pi\hbar}} \int_{-\infty}^{+\infty} V(x)\psi(x, t) e^{-ipx/\hbar} dx \tag{4.116}$$

The order of integration and differentiation with respect to time on the right-hand side can be exchanged so that

$$\frac{1}{\sqrt{2\pi\hbar}} \int_{-\infty}^{+\infty} i\hbar \frac{\partial \psi(x, t)}{\partial t} e^{-ipx/\hbar} dx = i\hbar \frac{\partial}{\partial t} \left(\frac{1}{\sqrt{2\pi\hbar}} \int_{-\infty}^{+\infty} \psi(x, t) e^{-ipx/\hbar} dx \right)$$

$$= i\hbar \frac{\partial \phi(p, t)}{\partial t} \tag{4.117}$$

The spatial derivative term can be rewritten as

$$\frac{1}{\sqrt{2\pi\hbar}} \int_{-\infty}^{+\infty} \left(-\frac{\hbar^2}{2m} \frac{\partial^2 \psi(x, t)}{\partial x^2} \right) e^{-ipx/\hbar} dx$$

$$\stackrel{\text{IBP}}{=} -\frac{\hbar^2}{2m} \frac{1}{\sqrt{2\pi\hbar}} \int_{-\infty}^{+\infty} \psi(x, t) \left(\frac{d^2}{dx^2} \left[e^{-ipx/\hbar} \right] \right) dx$$

$$= \frac{p^2}{2m} \left(\frac{1}{\sqrt{2\pi\hbar}} \int_{-\infty}^{+\infty} \psi(x, t) e^{-ipx/\hbar} dx \right)$$

$$= \frac{p^2}{2m} \phi(p, t) \tag{4.118}$$

where we have used two integrations by parts to move the derivatives onto the exponential term. Finally, the potential energy function can formally be expanded in a Taylor series to yield

$$\frac{1}{\sqrt{2\pi\hbar}} \int_{-\infty}^{+\infty} V(x)\psi(x, t) e^{-ipx/\hbar} dx$$

$$= \sum_{n=0}^{\infty} \frac{V^{(n)}(0)}{n!} \frac{1}{\sqrt{2\pi\hbar}} \int_{-\infty}^{+\infty} x^n \psi(x, t) e^{-ipx/\hbar} dx \tag{4.119}$$

and we can use

$$\frac{1}{\sqrt{2\pi\hbar}} \int_{-\infty}^{+\infty} \psi(x,t) x^n e^{-ipx/\hbar} dx$$

$$= \frac{1}{\sqrt{2\pi\hbar}} \int_{-\infty}^{+\infty} \psi(x,t) \left(i\hbar\frac{\partial}{\partial p}\right)^n \left(e^{-ipx/\hbar}\right) dx$$

$$= \left(i\hbar\frac{\partial}{\partial p}\right)^n \phi(p,t) \tag{4.120}$$

in which case the potential term becomes

$$\left[\sum_{n=0}^{\infty} \frac{V^{(n)}(0)}{n!} \left(i\hbar\frac{\partial}{\partial p}\right)^n\right] \phi(p,t) = V\left(i\hbar\frac{\partial}{\partial p}\right) \phi(p,t) \tag{4.121}$$

where we once again have identified $\hat{x} = i\hbar\partial/\partial p$. Thus, the *time-dependent Schrödinger equation in momentum-space* can be written as

$$\frac{p^2}{2m}\phi(p,t) + V\left(i\hbar\frac{\partial}{\partial p}\right)\phi(p,t) = i\hbar\frac{\partial\phi(p,t)}{\partial t} \tag{4.122}$$

where we implicitly use the series expansion for $V(\hat{x}) = V(i\hbar\partial/\partial p)$. If the potential energy function does not depend explicitly on time, we can write

$$\phi(p,t) = \phi(p) e^{-iEt/\hbar} \tag{4.123}$$

and obtain the *time-independent Schrödinger equation in momentum-space*

$$\frac{p^2}{2m}\phi(p) + V\left(i\hbar\frac{\partial}{\partial p}\right)\phi(p) = E\phi(p) \tag{4.124}$$

in the usual way.

For the simple case of a free particle ($V(x) = 0$), we have

$$\frac{\partial\phi(p,t)}{\partial t} = -i\hbar\frac{p^2}{2m}\phi(p,t) \tag{4.125}$$

which is easily integrated to yield

$$\phi(p,t) = \phi_0(p) e^{-ip^2 t/2m\hbar} \tag{4.126}$$

where $\phi_0(p)$ is an arbitrary initial momentum distribution; this yields the standard position-space wave packet via the Fourier transform since

$$
\begin{aligned}
\psi(x,t) &= \int_{-\infty}^{+\infty} dp\, \phi(p,t)\, e^{ipx/\hbar} \\
&= \int_{-\infty}^{+\infty} \left(\phi_0(p) e^{-ip^2 t/2m\hbar} \right) e^{ipx/\hbar}\, dp \\
&= \frac{1}{\sqrt{2\pi\hbar}} \int_{-\infty}^{+\infty} \phi_0(p)\, e^{i(px - p^2 t/2m)\hbar}\, dp
\end{aligned}
\tag{4.127}
$$

as in Section 3.2.1.

4.7.2 Uniformly Accelerating Particle

A model system (with a very familiar classical analog) where the solutions of the Schrödinger equation are actually more easily obtained and analyzed using momentum-space methods is the case of a particle acted upon by a uniform or constant force. We take the force to be given by $F(x) = F$, so that the potential energy function is $V(x) = -Fx$. We can assume for definiteness that $F > 0$, corresponding in the classical case to uniform acceleration to the right. The time-dependent Schrödinger equation in p-space from Eqn. (4.122) has the form

$$
\frac{p^2}{2m}\phi(p,t) - F \cdot \left[i\hbar \frac{\partial}{\partial p} \right] \phi(p,t) = i\hbar \frac{\partial \phi(p,t)}{\partial t}
\tag{4.128}
$$

or

$$
i\hbar \left(F \frac{\partial \phi(p,t)}{\partial p} + \frac{\partial \phi(p,t)}{\partial t} \right) = \frac{p^2}{2m}\phi(p,t)
\tag{4.129}
$$

We note that the simple combination of derivatives guarantees that a function of the form $\Phi(p - Ft)$ will make the left-hand side vanish, so we assume a solution of the form $\phi(p,t) = \Phi(p - Ft)\tilde{\phi}(p)$, with $\Phi(p)$ arbitrary and $\tilde{\phi}(p)$ to be determined. Using this form, Eqn. (4.129) reduces to

$$
\frac{\partial \tilde{\phi}(p)}{\partial p} = -\frac{i\hbar p^2}{2m\hbar F}\tilde{\phi}(p)
\tag{4.130}
$$

with the solution

$$
\tilde{\phi}(p) = e^{-ip^3/6mF\hbar}
\tag{4.131}
$$

We can then write the general solution as

$$
\phi(p,t) = \Phi(p - Ft)\, e^{-ip^3/6mF\hbar}
\tag{4.132}
$$

or, using the arbitrariness of $\Phi(p)$, as

$$\phi(p,t) = \phi_0(p - Ft)\, e^{i((p-Ft)^3 - p^3)/6mF\hbar} \tag{4.133}$$

where now $\phi_0(p)$ is the initial momentum–space amplitude, since $\phi(p, t = 0) = \phi_0(p)$. If the initial momentum distribution is characterized by $\langle p \rangle_0 = p_0$ then (using an obvious change of variables) we find that

$$\langle p \rangle_t = \int_{-\infty}^{+\infty} p\, |\phi(p,t)|^2\, dp$$

$$= \int_{-\infty}^{+\infty} p\, |\phi_0(p - Ft)|^2\, dp \qquad \text{(and using } q = p - Ft)$$

$$= \int_{-\infty}^{+\infty} q|\phi_0(q)|^2\, dq + Ft \int_{-\infty}^{+\infty} |\phi_0(q)|^2\, dq$$

$$\langle p \rangle_t = \langle p \rangle_0 + Ft \tag{4.134}$$

Thus the average momentum value increases linearly with time, consistent with the classical result for a constant force, $F = dp/dt$; the momentum distribution simply 'translates' uniformly to the right with no change in shape since, from Eqn. (4.133)

$$|\phi(p,t)|^2 = |\phi_0(p - Ft)|^2 \tag{4.135}$$

The corresponding position-space wavefunction can be written as

$$\psi(x,t) = \frac{1}{\sqrt{2\pi\hbar}} \int_{-\infty}^{+\infty} \phi_0(p - Ft)\, e^{i((p-Ft)^3 - p^3)/6mF\hbar}\, e^{ipx/\hbar}\, dp \tag{4.136}$$

and, because the p^3 terms cancel in the exponent, this transform can be done analytically (P4.24) for a Gaussian momentum distribution. In that case, we have

$$\phi_0(p) = \sqrt{\frac{\alpha}{\sqrt{\pi}}}\, e^{-\alpha^2 p^2/2} \tag{4.137}$$

so that

$$\phi(p,t) = \sqrt{\frac{\alpha}{\sqrt{\pi}}}\, e^{-\alpha^2(p-Ft)^2/2}\, e^{i((p-Ft)^3 - p^3)/6mF\hbar} \tag{4.138}$$

and

$$\psi(x,t) = \frac{1}{\sqrt{\alpha\hbar\sqrt{\pi}(1 + it/t_0)}}\, e^{iFt(x - Ft^2/6m)/\hbar}\, e^{-(x - Ft^2/2m)^2/2\hbar^2\alpha^2(1 + it/t_0)} \tag{4.139}$$

where the spreading time is defined by $t_0 \equiv m\hbar\alpha^2$, just as in the free-particle wave packet case of Section 3.2.2. The corresponding probability density is then

$$|\psi(x,t)|^2 = \frac{1}{\beta_t \sqrt{\pi}} e^{-(x-Ft^2/2m)^2/\beta_t^2} \qquad (4.140)$$

where $\beta_t = \hbar\alpha\sqrt{1+t^2/t_0^2}$, also as before. It is easy to see (P4.25) that

$$\langle x \rangle_t = Ft^2/2m \quad \text{and} \quad \langle p^2 \rangle_t = (Ft)^2 + \frac{1}{2\alpha^2} \qquad (4.141)$$

so that the uncertainty principle product is given by

$$\Delta x \Delta p = \frac{\hbar}{2}\sqrt{1+t^2/t_0^2} \qquad (4.142)$$

as before. A position-space wave packet with arbitrary initial position (x_0) and momentum (p_0) can then be obtained by letting

$$\phi_0(p) \longrightarrow e^{-ipx_0/\hbar} \phi_0(p-p_0) \qquad (4.143)$$

which gives the most general such Gaussian solution with arbitrary initial conditions. Just as in Section 3.2.2, the integral in Eqn. (4.136) can be performed numerically for other initial momentum distributions. For $\phi(p,0)$ other than a Gaussian, the position-space wave packet will change shape, but Eqn. (4.135) guarantees that the momentum distribution will not disperse.

4.8 Commutators

In the analytic wave packet examples considered so far, the free-particle and accelerating Gaussian packets, we have explicitly demonstrated the validity of the position–momentum uncertainty principle,

$$\Delta x \, \Delta p \geq \frac{\hbar}{2} \qquad (4.144)$$

We have previously understood this as arising from a fundamental limitation imposed on wave packets formed by constructive/destructive interference by the relation $\Delta x \Delta k > \mathcal{O}(1)$ from Section 2.3 and the identification (via de Broglie) $p = \hbar k$, that is, as a basic constraint arising from a wave mechanics description of particle dynamics. We can approach this relation in a more formal way, using the notion of quantum mechanical operators, as a preview of the more rigorous proof of the uncertainty principle in Chapter 12.

If we choose to work in a position-space representation, we note that because $\hat{p}$ is associated with a nontrivial operator, the result of the application of p and x to a wavefunction will depend on their ordering, specifically

$$x\hat{p}\,\psi(x) \neq \hat{p}\,x\,\psi(x) \tag{4.145}$$

We can show quite generally that

$$
\begin{aligned}
\left(x\hat{p} - \hat{p}\,x\right)\psi(x) &= x\left(\frac{\hbar}{i}\frac{d\psi(x)}{dx}\right) - \left(\frac{\hbar}{i}\frac{d}{dx}\right)(x\,\psi(x)) \\
&= x\left(\frac{\hbar}{i}\frac{d\psi(x)}{dx}\right) - x\left(\frac{\hbar}{i}\frac{d\psi(x)}{dx}\right) - \frac{\hbar}{i}\psi(x) \\
&= i\hbar\psi(x) \neq 0.
\end{aligned} \tag{4.146}
$$

Since this is true for an arbitrary $\psi(x)$, we can write an *operator* identity, namely

$$[x,\hat{p}] \equiv x\hat{p} - \hat{p}\,x = i\hbar \tag{4.147}$$

where we have introduced the *commutator* of two operators, defined by

$$[\hat{A},\hat{B}] \equiv \hat{A}\hat{B} - \hat{B}\hat{A} \tag{4.148}$$

It is easy to show (P4.28) that this same result is also obtained in momentum-space where $\hat{x} = i\hbar\partial/\partial p$ is now the "nontrivial" operator, so that this relation is not dependent on a specific representation.

To preview the connection between this lack of commutativity (as measured by the nonvanishing commutator) and the uncertainty principle, we can look again at Eqn. (4.146). If measurements of x and p gave the same result done in either ordering, we could argue that the measurement procedures for these two physical quantities did not "interfere" with each other, and independent measurements of both are possible. One could then imagine making increasingly precise measurements of both quantities until the uncertainty principle product $\Delta x \Delta p$ was as small as desired. The content of Eqn. (4.145), however, is to say that a measurement of, say, p will necessarily alter some of the information regarding the position x of the particle, or vice versa. This formal statement is important, as it codifies the results of a large variety of "gedanken" experiments (thought experiments) designed to violate Eqn. (4.146) (e.g. see P4.29) but which do not after careful consideration.

4.9 **The Wigner Quasi-Probability Distribution**

We have examined the probability interpretation of both $\psi(x, t)$ and $\phi(p, t)$, their connection via the Fourier transform, and how to extract information on expectation values from either representation of the quantum wave function. In classical mechanics, one often asks about the behavior of the motion of a particle using *both* the x *and* p variables simultaneously, via a phase-space description,[4] as discussed in Appendix G; such an approach has proved especially useful in the analysis of systems exhibiting classical chaos. We also encounter this idea in statistical mechanics[5] where phase-space is a common topic, and related concepts play an important role in quantum optics.[6]

It is therefore natural to consider if a quantum mechanical analog of a phase-space probability distribution is a useful construct, despite the obvious problems raised by the $x - p$ uncertainty principle (discussed in the last section) and the resulting restriction on one's ability to make simultaneous measurements of both x and p.

The closest that one can seemingly come to a joint (x, p) probability distribution is a version proposed by Wigner (1932), who defined a *quasi-probability distribution* via

$$P_{\text{W}}(x, p; t) \equiv \frac{1}{\pi \hbar} \int_{-\infty}^{+\infty} \psi^*(x + y, t)\, \psi(x - y, t)\, e^{2ipy/\hbar}\, dy \qquad (4.149)$$

using position-space wavefunctions. Using the properties of the Fourier transform, this can also be written in an entirely equivalent form in terms of momentum-space wavefunctions as

$$P_{\text{W}}(x, p; t) = \frac{1}{\pi \hbar} \int_{-\infty}^{+\infty} \phi^*(p + q, t)\, \phi(p - q, t)\, e^{-2iqx/\hbar}\, dq \qquad (4.150)$$

We stress that in both equations x and p are simply *variables* and have no operator properties.

---

[4] See, for example, Marion and Thornton (2003) or Cassiday and Fowles (1999).
[5] See, for example, Reif (1965).
[6] See Scully and Zubairy (1997) and especially Schleich (2001).

The Wigner distribution is easily shown to be real, since

$$\left[P_W(x,p;t)\right]^* = \frac{1}{\pi\hbar} \int_{-\infty}^{+\infty} \psi(x+y,t)\,\psi^*(x-y,t)\,e^{-2ipy/\hbar}\,dy$$

$$= \frac{1}{\pi\hbar} \int_{-\infty}^{+\infty} \psi^*(x+\bar{y},t)\,\psi(x-\bar{y},t)\,e^{+2ip\bar{y}/\hbar}\,d\bar{y}$$

$$= P_W(x,p;t) \tag{4.151}$$

by using a simple change of variables ($\bar{y} = -y$). This is, of course, one of the desired properties of any probability distribution constructed from complex wavefunctions.

The integration of $P_W(x,p;t)$ over one variable or the other is seen to give the correct *marginal probability distributions* for x and p separately, since

$$\int_{-\infty}^{+\infty} P_W(x,p;t)\,dp = |\psi(x,t)|^2 = P_{QM}(x,t) \tag{4.152}$$

$$\int_{-\infty}^{+\infty} P_W(x,p;t)\,dx = |\phi(p,t)|^2 = P_{QM}(p,t) \tag{4.153}$$

where one uses the definition of the Dirac δ-function in Eqns (4.149) or (4.150). This property of $P_W(x,p)$ is clearly another necessary condition for a joint probability density. Thus, in the limit that we consider only one variable or the other, all results we have obtained so far are recovered.

However, one can also easily show that the Wigner distributions for two distinct quantum states, $\psi(x,t)$ and $\chi(x,t)$,

$$P_W^{(\psi)}(x,p;t) = \frac{1}{\pi\hbar} \int_{-\infty}^{+\infty} \psi^*(x+y,t)\,\psi(x-y,t)\,e^{2ipy/\hbar}\,dy \tag{4.154}$$

$$P_W^{(\chi)}(x,p;t) = \frac{1}{\pi\hbar} \int_{-\infty}^{+\infty} \chi^*(x+z,t)\,\chi(x-z,t)\,e^{2ipz/\hbar}\,dz \tag{4.155}$$

satisfy the relation

$$\int_{-\infty}^{+\infty} dx \int_{-\infty}^{+\infty} dp\, P_W^{(\psi)}(x,p;t)\, P_W^{(\chi)}(x,p;t)$$

$$= \frac{2}{\pi\hbar} \left| \int_{-\infty}^{+\infty} \psi^*(x,t)\,\chi(x,t)\,dx \right|^2 \tag{4.156}$$

So, for example, if ψ and χ are orthogonal states, so that their overlap integral vanishes, it cannot be true that the corresponding Wigner distributions, $P_W^{(\psi)}$ and $P_W^{(\chi)}$, can be positive-definite everywhere, as there must be cancellations in the integral on the left side of Eqn. (4.156). The fact that $P_W(x, p; t)$ can be negative in parts of (x, p) space is also easily confirmed by direct calculation (P4.37, P5.23, P9.16) of specific cases.

This feature is, after all, hardly surprising because of the noncommutativity of x and p encoded in the uncertainty principle. Despite this obvious drawback, the Wigner function is still useful for the visualization of the correlated position- and momentum-space behavior of both quantum eigenstates and time-dependent wave packets.

Example 4.3. Wigner distribution for the free-particle Gaussian wave packet

A useful benchmark example of the Wigner function is that of a Gaussian free-particle wave packet, where the time-dependent momentum- and position-space wavefunctions (for arbitrary initial x_0 and p_0) are given in Eqns (3.35) and (3.36). The corresponding Wigner function is easily obtained from either Eqn. (4.149) or (4.150) using standard Gaussian integrals, and is given by

$$P_W^{(G)}(x, p; t) = \frac{1}{\pi \hbar} e^{-\alpha^2 (p-p_0)^2} e^{-(x-x_0-pt/m)^2/\beta^2} \qquad (4.157)$$

where $\beta \equiv \hbar\alpha$. In this case, the ultrasmooth Gaussian solution does give a positive-definite $P_W(x, p; t)$ and it is known that such solutions are the only forms which give rise Wigner functions which are everywhere nonnegative. Note that in the second exponential term, it is the constant β that appears and *not* the time-dependent $\beta_t = \beta\sqrt{1 + (t/t_0)^2}$; similarly, the term pt/m that appears there is also correct, and *not* $p_0 t/m$ as one might have expected.

To see how this form allows us to visualize wave packet spreading, we plot $P_W(x, p; t)$ versus x, p in Fig. 4.7 for $t = 0, t_0$, and $2t_0$ (where t_0 is the spreading time) and note how it develops with time, especially the 'tilting' of the shape of P_W. A projection of $P_W(x, p)$ back onto the (vertical) p-axis, via Eqn. (4.153), is seen to give the same $P_{QM}(p, t)$ for each time

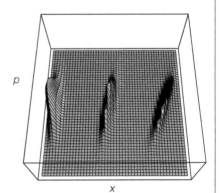

Figure 4.7. Plot of the Wigner distribution, $P_W(x, p; t)$ versus (x, p), for the free-particle Gaussian wave packet, from Eqn. (4.157), at $t = 0$, t_0, and $2t_0$.

(Continued)

shown, consistent with no change in the shape of $|\phi(p, t)|^2$. A similar projection downward onto the (horizontal) x-axis not only illustrates the constant speed motion (to the right), but also does give rise to an increasingly spread out $|\psi(x, t)|^2$. The large (small) momentum components of the packet are increasingly correlated with the front (back) of wave packet (as shown more quantitatively in P4.18) so that the fast (slow) parts are ahead (behind) the center of the packet, and therefore in the upper left (lower right) of the P_W peak, giving rise to the observed "tilt."

The Wigner distribution also provides a useful context in which to visualize the constraints placed on measurements of x and p by the uncertainty principle. In this picture, the smallest possible "area" over which a quantum wavefunction can be spread in (x, p) space is $\Delta x \cdot \Delta p \geq \hbar/2$; if ones squeezes the wavefunction down in one direction (say x), it must necessarily stretch out along the other (p) one. While we have focused on the use of $P_W(x, p; t)$ as a visualization tool, many of the formal results discussed in this chapter regarding expectation values and operators can be obtained in a natural way through its use, as in P4.37.

4.10 **Questions and Problems**

Q4.1. Would $\Delta x \equiv \langle |x - \langle x \rangle| \rangle$ be a good choice for the spread or uncertainty in x?

Q4.2. Can you show that $\langle x^2 \rangle \geq \langle x \rangle^2$ quite generally? Under what conditions is the equality possible? Can you find a probability distribution, $P(x)$, for which $\Delta x = 0$?

Q4.3. What is the statement of the equation of continuity in electromagnetism? What is the conserved quantity in that case?

Q4.4. What are the dimensions of probability flux in Eqn. (4.32)?

Q4.5. What does the conservation of flux equation look like in momentum-space? For simplicity consider only the case of a free-particle and one with a constant force, using the momentum-space equations in Eqns (4.125) and (4.128), respectively.

Q4.6. Consider a particle moving in an imaginary potential $V(x) = -iV_I$. What are the possible plane wave solutions of the Schrödinger equation? What would happen to a wave packet which impinged on a region with this potential? What does such a potential have to do with absorption?

Q4.7. Why do we more often solve the Schrödinger equation in position space than in momentum space?

Q4.8. For the free-particle and accelerating wave packets, the spread in position increases with time. Is it possible that the spread in position of a wave packet (not necessarily for a free particle) could *shrink* with time? What would that imply about the spread in momentum? How about the case of the 'bouncing' wave packet in Section 3.3? Or the alternate Gaussian wave packet of **P3.5**?

Q4.9. In Chapter 2, it was discussed how solutions of the classical wave equation of the form $\psi(x, t) = \psi(x - vt)$ propagated with no distortion or change in shape. Show how to calculate Δx_t for such a wave and discuss how the spread in position remains constant in time. Can such solutions satisfy the Schrödinger equation?

Q4.10. Sketch the contour plot and "trajectory" picture for $|\phi(p, t)|^2$ corresponding to a free particle; that is, what does Fig. 4.6 look like for the momentum space distributions? How about the corresponding plots for the accelerating wave packet in Section 4.7.2?

P4.1. Probabilities for dice. Consider the throw of a single die.

(a) Enumerate all of the possible outcomes, and evaluate the *a priori* probabilities, $P(x_i)$.

(b) Evaluate $\langle x \rangle$, $\langle x^2 \rangle$, and Δx.

(c) How much of the probability is found in the interval $(\langle x \rangle + \Delta x, \langle x \rangle - \Delta x)$? Within $2\Delta x$ of the average?

(d) Repeat for two dice thrown at the same time, and confirm the results from Example 4.1 and Eqn. (4.9),

(e) Are your results for the one- or two-dice problem similar to those expected for the Gaussian or normal distribution in Eqn. (4.17)?

P4.2. Poisson probability distribution. The Poisson probability distribution is defined by

$$P(n; \lambda) \equiv \frac{\lambda^n}{n!} e^{-\lambda} \tag{4.158}$$

where λ is a constant. This distribution can correspond, for example, to the probability of observing n independent events in a time interval t when the counting rate is r so that the *expected* number of events is $rt = \lambda$; this is especially relevant when $\lambda = rt$ is not too large.

(a) Show that the $P(n; \lambda)$ are properly normalized, namely that

$$\sum_{n=0}^{\infty} P(n; \lambda) = 1 \tag{4.159}$$

(b) Evaluate $\langle n \rangle$ and Δn for this distribution in terms of λ. Hint: One can write

$$\langle n \rangle = \sum_{n=0}^{\infty} n\, P(n; \lambda) = \sum_{n=0}^{\infty} n \frac{e^{-\lambda}\lambda^n}{n!} = e^{-\lambda} \left(\lambda \frac{\partial}{\partial \lambda} \right) \sum_{n=0}^{\infty} \frac{\lambda^n}{n!} \tag{4.160}$$

(c) Plot $P(n; \lambda)$ versus n for several values of λ, indicating the average value and the 1σ limits. Is the distribution symmetric around the average?

(d) Assume that a book is 600 pages long and has a total of 1200 typographical errors. What is the average number of errors per page? What is the probability that a single page has no errors? How many pages would be expected to have less than 3 errors? What is the probability that a single page has 4 or more errors?

P4.3. Binomial probability distribution. Consider a trial which only has two outcomes, the desired one with probability p, and anything else, with probability $q = 1 - p$. In N independent trials, the probability of n successful outcomes is

$$P(n; N) = \frac{N!}{n!(N-n)!} p^n q^{N-n} \tag{4.161}$$

This result depends on two factors; the last two terms correspond to the probability that the first n trials are successful while the last $N - n$ ones are not; the first term simply counts the number of different distinguishable ways in which one can obtain the n possible outcomes.

(a) Show that this distribution is normalized, namely that

$$\sum_{n=0}^{N} P(n; N) = (p+q)^N = 1 \tag{4.162}$$

Hint: Use the binomial theorem!

(b) Evaluate $\langle n \rangle$, $\langle n^2 \rangle$, and show that $\Delta n = \sqrt{Np(1-p)}$. Hint: Use a differentiation trick of the form

$$n^K P(n; N) = \left(p \frac{\partial}{\partial p} \right)^K P(n; N) \tag{4.163}$$

which "brings down" the requisite number of powers of n.

(c) Why does Δn behave as it does when $p \to 0$ or 1?

P4.4. Exponential probability distribution. A continuous probability distribution has the form

$$P(x) = Ne^{-|x|/a} \tag{4.164}$$

(a) Find N so that $P(x)$ is properly normalized.

(b) Evaluate $\langle x \rangle$, $\langle x^2 \rangle$, and Δx.

(c) What is the probability that a measurement would find x in the interval $(0, +\infty)$? $(-a, 2a)$? $(0, 0.0001a)$?

(d) What is the probability that a measurement would find $x = 1.5$ exactly?

(e) What is the probability that $|x - \langle x \rangle| \leq \Delta x$? $\leq 2\Delta x$? $\leq 3\Delta x$? Compare these values to the Gaussian distribution in Example 4.2.

(f) Plot the distribution above for a given value of a along with a Gaussian distribution with its parameters (μ and σ) chosen to give the same average value of x and spread Δx. Show how your plot illustrates the results of part (e).

P4.5. Classical probabilities for a particle in a box. Consider a particle which is located with equal probability anywhere along a one-dimensional segment of length L.

(a) Find $P(x; L)$, that is, the normalized probability of finding the particle at position x.

(b) Evaluate $\langle x \rangle$ and Δx.

(c) What are the probabilities that the particle would be found within one standard deviation of its mean value? Why is this value lower than for the Gaussian distribution?

(d) What are the dimensions of $P(x; L)$?

(e) This is an excellent problem for which you could try to generate "measurements" of the probability distribution by using a random number generator on a computer. Write a short program that generates a random number in the range $(0, 1)$ (i.e. let $L = 1$), and put it into bins, δx, of your own choosing, counting how many fall into each bin. Use this to evaluate $P(x) = (\Delta N/N_{\text{tot}})/\Delta x$ as a function of x, plotting your results.

P4.6. Consider a particle described by a wavefunction

$$\psi(x, t) = Ae^{-|x|/L - iEt/\hbar} \tag{4.165}$$

(a) Find A so that $\psi(x, t)$ is normalized properly.

(b) Evaluate $\langle x \rangle$, $\langle x^2 \rangle$, and Δx.

(c) What is the probability of finding a particle in the region $(-L, +L/2)$? How about in the interval $(+L, +\infty)$? and in the interval $(0.99999L, L)$? What is the probability of finding the particle exactly at $x = 3L$?

(d) Try to evaluate $\langle \hat{p} \rangle$ and see what you get. Now try $\langle \hat{p}^2 \rangle$. If you have problems doing this, to what do you trace your difficulties?

P4.7. Consider the same position-space wavefunction in P4.6.

(a) Using the Fourier transform, find $\phi(p, t)$.

(b) Evaluate $\langle p \rangle$, $\langle p^2 \rangle$, and Δp. Why do you think you got the results you did?

(c) What fraction of the particles would be moving with momentum in the range $p \in (\hbar/L, 2\hbar/L)$?

P4.8. Assuming that the potential has a negative imaginary part, that is $V(x) = V(x) - iV_I$, derive Eqns (4.40) and (4.41).

P4.9. Lorentzian line-shape. Consider the time-dependence of a wavefunction, which has an energy with both real and imaginary parts, namely

$$T(t) = e^{-i(E_A - iV_I)t/\hbar} \tag{4.166}$$

Use this wavefunction to evaluate $\langle t \rangle$ and $\langle t^2 \rangle$ using

$$\langle f(t) \rangle = \int_0^\infty dt \ |T(t)|^2 \ f(t) \ \bigg/ \int_0^\infty dt \ |T(t)|^2 \tag{4.167}$$

Find the analog of the Fourier transform of $T(t)$, namely

$$A(E) = \int_0^\infty dt \ T(t) \ e^{iEt/\hbar} \tag{4.168}$$

and show that

$$|A(E)|^2 = \frac{N}{(E - E_A)^2 + (\hbar/2\tau)^2} \tag{4.169}$$

and find the normalization factor N (by integrating over all possible energies, E.) Evaluate $\langle E \rangle$ and estimate the value of the spread in energy. This functional form for $A(E)$ is called a *Lorentzian* line shape and is relevant for describing the spectral line shapes of unstable quantum states.

P4.10. What is the probability flux for a plane wave of the form

$$\psi_p(x, t) = Ae^{i(\pm px - p^2 t/2m)/\hbar} \tag{4.170}$$

P4.11. The Klein–Gordon equation. Consider the relativistic wave-equation

$$\frac{\partial^2 \phi(x, t)}{\partial t^2} = c^2 \frac{\partial^2 \phi(x, t)}{\partial x^2} - \left(\frac{mc^2}{\hbar}\right)^2 \phi(x, t) \tag{4.171}$$

(a) Define the probability density for position to be

$$P(x, t) \equiv \phi^*(x, t) \frac{\partial \phi(x, t)}{\partial t} - \frac{\partial \phi^*(x, t)}{\partial t} \phi(x, t) \tag{4.172}$$

and show that the corresponding equation of continuity can be written as

$$\frac{\partial P(x, t)}{\partial t} = \frac{\partial j(x, t)}{\partial x} \tag{4.173}$$

where

$$j(x, t) \equiv \frac{\hbar c^2}{2mi} \left(\phi^*(x, t) \frac{\partial \phi(x, t)}{\partial x} - \frac{\partial \phi^*(x, t)}{\partial x} \phi(x, t)\right) \tag{4.174}$$

(b) Evaluate the probability density for a solution of the form

$$\phi(x, t) = u(x) \ e^{-iEt/\hbar} \tag{4.175}$$

Is the probability density positive definite no matter what the sign of E? Compare this to the probability density defined via Eqn. (4.22) for the Schrödinger equation.

P4.12. Show that the overlap of two solutions of the Schrödinger equation,

$$O_t^{(1,2)} = \int_{-\infty}^{+\infty} \psi_1^*(x, t)\, \psi_2(x, t)\, dx = O_0^{(1,2)} \tag{4.176}$$

is independent of time, provided $V(x)$ is real.

P4.13. Gaussian free-particle wave packets in position-space. Using the explicit form of the most general position-space free-particle Gaussian wave packet of Eqn. (3.35) (with arbitrary x_0, p_0), calculate

(a) $\langle x \rangle_t$ and $\langle x^2 \rangle_t$

(b) $\langle \hat{p} \rangle_t$ and $\langle \hat{p}^2 \rangle_t$

(c) Δx_t and Δp_t

(d) $\langle \hat{E} \rangle_t$, $\langle \hat{E}_t^2 \rangle$, and ΔE.

P4.14. Gaussian free-particle wave packets in momentum-space. Repeat P4.13, but use the general momentum space Gaussian wavefunction in Eqn. (3.36), namely

$$\phi(p, t) = \sqrt{\frac{\alpha}{\sqrt{\pi}}}\, e^{-\alpha^2(p-p_0)^2/2}\, e^{-ipx_0/\hbar}\, e^{-ip^2 t/2m\hbar} \tag{4.177}$$

Are all your answers the same?

P4.15. Kinetic energy calculations. Consider the wavefunction, $\psi(x)$, defined over the range $(0, a)$, given by

$$\psi(x) = \begin{cases} 0 & \text{for } x \leq 0 \\ Ax/c & \text{for } 0 \leq x \leq c \\ A(a-x)/(a-c) & \text{for } c \leq x \leq a \\ 0 & \text{for } a \leq x \end{cases} \tag{4.178}$$

(a) Find the value of A for which this is properly normalized.

(b) Evaluate the kinetic energy for this wavefunction using the two different forms discussed in Eqn. (4.65), namely

$$\langle \hat{T} \rangle = \frac{1}{2m}\langle \hat{p}^2 \rangle = -\frac{\hbar^2}{2m}\int_{-\infty}^{+\infty} dx\, \psi^* \frac{\partial^2 \psi(x, t)}{\partial x^2}$$

$$= \frac{\hbar^2}{2m}\int_{-\infty}^{+\infty} dx\, \left| \frac{\partial \psi(x, t)}{\partial x} \right|^2 \tag{4.179}$$

and show that you get

$$\langle \hat{T} \rangle = \frac{3\hbar^2}{2mc(a-c)} \tag{4.180}$$

For the first form, you should be careful when evaluating $d^2\psi(x)/dx^2$, by first writing $d\psi(x)/dx$ in terms of Θ-functions (as in Section 2.4), and *then* differentiating a second time.

(c) Why does $\langle \hat{T} \rangle$ diverge as $c \to 0$ or $c \to a$?

P4.16. Which of the following operators are Hermitian and which are not?

(a) $3 - 4i$

(b) $\partial/\partial x$

(c) $x \cdot \hat{p}$

(d) $x\hat{p}x$

P4.17. Show that the operator $(x\hat{p} + \hat{p}x)/2$ is Hermitian.

P4.18. Position-momentum correlation in quantum mechanics. The "mixed" expectation value involving x and $\hat{p}$ that measures a correlation between position and momentum in a quantum wave function, called the *covariance* (or *cov* for short) can be written as

$$\text{cov}(x, p) = \frac{1}{2} \langle (x - \langle x \rangle)(\hat{p} - \langle \hat{p} \rangle) + (\hat{p} - \langle \hat{p} \rangle)(x - \langle x \rangle) \rangle. \qquad (4.181)$$

(a) Show that this is equal to

$$\text{cov}(x, p) = \frac{1}{2} \left(\langle x\hat{p} + \hat{p}x \rangle \right) - \langle x \rangle \langle \hat{p} \rangle. \qquad (4.182)$$

(b) Evaluate $\text{cov}(x, p)$ for the general Gaussian wave packet in position space of the form in Eqn. (3.35).

(c) Repeat the calculation using the general Gaussian wave packet in momentum space of the form in Eqn. (3.36), now using $\hat{x}$ as the nontrivial operator

(d) The analog of a classical *correlation coefficient* for two quantum mechanical quantities, which may not commute, can be defined as

$$\rho(x, p; t) \equiv \frac{\text{cov}(x, p; t)}{\Delta x_t \cdot \Delta p_t} \qquad (4.183)$$

Evaluate $|\rho(x, p; t)|^2$ for the standard Gaussian wave packet, and discuss its time-dependence. Why would such a free-particle wave packet develop a nonzero correlation as it evolves in time?

P4.19. If the potential energy function is real, show that the Hamiltonian operator,

$$\hat{H} = \frac{\hat{p}^2}{2m} + V(x) \qquad (4.184)$$

is Hermitian.

P4.20. Show that any operator, $\hat{O}$, which satisfies Eqn. (4.72) (and which is therefore Hermitian) also satisfies Eqn. (4.73). Hint: You might wish to follow the following steps.

(a) Consider a wavefunction $\zeta(x) = \psi(x) + \lambda\phi(x)$ where λ is an arbitrary *complex* number.

(b) Since ζ is an admissible wavefunction (why?), we must have

$$I(\lambda) \equiv \int_{-\infty}^{+\infty} dx \, \zeta^*(x) \, \hat{O} \, \zeta(x) = [I(\lambda)]^* \qquad (4.185)$$

so that $I(\lambda) - I(\lambda)^* = 0$.

(c) Use this result and the fact that λ is arbitrary to complete the proof.

P4.21. Consider the *angular momentum operator*, defined by

$$\hat{L}_z = \frac{\hbar}{i} \frac{\partial}{\partial\phi} \qquad (4.186)$$

which acts on angular wavefunctions of the form $Z(\phi)$ only over the range $(0, 2\pi)$. To be consistently defined, the wavefunctions should satisfy $Z(\phi+2\pi) = Z(\phi)$. Explain why this is so and then use this fact to show that the angular momentum operator is Hermitian.

P4.22. (a) Show that the Schrödinger equation in momentum-space can be written in terms of an integral equation involving a "nonlocal" potential energy, namely

$$\left(E - \frac{p^2}{2m}\right) \phi(p) = \int_{-\infty}^{+\infty} V(p - \bar{p}) \, \phi(\bar{p}) \, d\bar{p} \qquad (4.187)$$

where

$$\overline{V}(q) = \frac{1}{2\pi\hbar} \int_{-\infty}^{+\infty} e^{-iqx/\hbar} \, V(x) \, dx \qquad (4.188)$$

is essentially the Fourier transform of $V(x)$.

(b) Find the form of $\overline{V}(p - \bar{p})$ in the case of the uniformly accelerated particle.

P4.23. **Position- and momentum-space wavefunctions for the harmonic oscillator.** Consider the Schrödinger equation for the harmonic oscillator, for which the potential energy is typically written as $V(x) = m\omega^2 x^2/2$.

(a) Write down the time-independent Schrödinger equation in position-space for $\psi_E(x)$.

(b) Do the same for the time-independent Schrödinger equation in momentum-space for $\phi_E(p)$.

(c) Show that *both* equations can be written in the form

$$-\frac{d^2 f_\epsilon(\zeta)}{d\zeta^2} + \zeta^2 f_\epsilon(\zeta) = \epsilon f_\epsilon(\zeta) \qquad (4.189)$$

where $\zeta = x/\sigma = p/\rho$ with σ, ρ carrying the appropriate dimensions. What is ϵ in terms of E and the other parameters of the problem? This implies that the stationary state or energy eigenstates of the harmonic oscillator have the same functional form in both position- and momentum-space. Why would you expect this to be the only case for which this might happen?

P4.24. Evaluate the integral in Eqn. (4.136) explicitly, using the Gaussian $\phi_0(p)$ in Eqn. (4.137), resulting in Eqn. (4.139).

P4.25. Gaussian accelerating wave packet in position space. Using the explicit form of the position-space accelerating Gaussian wave packet of Eqn. (4.139), evaluate

(a) $\langle x \rangle_t$ and $\langle x^2 \rangle_t$

(b) $\langle \hat{p} \rangle_t$ and $\langle \hat{p}^2 \rangle_t$

(c) Δx and Δp

(d) $\langle \hat{E} \rangle_t$, $\langle \hat{E}^2 \rangle_t$, and ΔE

(e) $\langle \hat{T} \rangle$ and $\langle V(x) \rangle$. Compare their sum to $\langle \hat{E} \rangle$

P4.26. Gaussian accelerating wave packet in momentum-space. Repeat P4.25, but use the momentum space wavefunction, namely

$$\phi(p, t) = \sqrt{\frac{\alpha}{\sqrt{\pi}}} \, e^{-\alpha^2 p^2/2} \, e^{i((p-Ft)^3 - p^3)/6mF\hbar} \qquad (4.190)$$

P4.27. General Gaussian accelerating wave packet. Calculate the position space wavefunction for the accelerating particle using the general Gaussian momentum distribution of Eqn. (4.133) and show that the probability density is localized at the point

$$x(t) = \frac{F}{2m} t^2 + \frac{p_0}{m} t + x_0 \qquad (4.191)$$

P4.28. Commutators.

(a) Show directly that $[x^n, \hat{p}] = i\hbar n x^{n-1}$ by acting with both sides on an arbitrary function $\psi(x)$ using the position-space form of $\hat{p} = (\hbar/i)(d/dx)$. (This takes one line.)

(b) Generalize this result to show that $[f(x), \hat{p}] = i\hbar(df(x)/dx)$

(c) Show that $[x, x\hat{p}] = [x, \hat{p}x] = i\hbar x$.

(d) Show that $[x, \hat{p}^2] = 2i\hbar\hat{p}$ and then generalize to show that $[x, \hat{p}^n] = i\hbar n \hat{p}^{n-1}$. (The last part takes n lines.)

(e) Show that $[x^2, \hat{p}^2] = 4x\hat{p}$.

(f) Repeat parts (a) - (e) using the momentum-space description where p is trivial and $\hat{x} = i\hbar(d/dp)$ is an operator and confirm that you get the same answers.

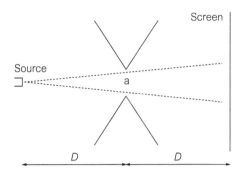

Figure 4.8. Schematic experimental apparatus for testing the uncertainty principle.

(g) Use this method to show that $[\hat{x}, g(p)] = i\hbar(dg(p)/dp)$.

(h) Evaluate $[\hat{E}, t]$ and $[\hat{E}, x]$.

P4.29. Outwitting the uncertainty principle? Consider the schematic experiment shown in Fig. 4.8 where electrons are emitted from a source, allowed to pass through an aperture of lateral width a, and are detected at a screen a distance D away. A naive analysis of this "gedanken" experiment might come to the conclusion that

$$\Delta p_y \sim \left(\frac{a}{D}\right)p \quad \text{and} \quad \Delta y \sim a \tag{4.192}$$

so that the uncertainty principle product $\Delta y \, \Delta p_y$ could be made arbitrarily small by letting $a \to 0$ and $D \to \infty$. Discuss what is wrong with such a analysis.

P4.30. Simple model of "interfering Bose condensates." In a well-known experiment,[7] sodium atoms are trapped and cooled to form two highly localized and well-separated samples, at which point the trapping potential is released and the two sets of atoms expand freely, overlap, and exhibit interference effects. A simple one-particle quantum wavefunction for a particle of mass m, which illustrates this type of behavior, is given by the initial state

$$\psi(x,0) = \frac{N}{\sqrt{\beta\sqrt{\pi}}}\left[e^{-(x-d/2)^2/2\beta^2} + e^{i\phi}e^{-(x+d/2)^2/2\beta^2}\right] \tag{4.193}$$

where the two samples are separated by a distance $d \gg \beta$ and ϕ describes a constant relative phase.

(a) Evaluate the normalization constant N exactly for the general case, and approximately in the limit when $d \gg \beta$.

(b) What is the resulting time-dependent wavefunction, $\psi(x,t)$, which solves the free-particle time-dependent Schrödinger equation? Evaluate the time-dependent probability density $P(x,t)$ and discuss why interference effects arise in the overlap region.

[7] See Andrews *et al.* (1997).

(c) In terms of β, d, m, and $\hbar$, how long does it take for the two localized samples to begin to overlap. Compare this time to the natural spreading time, t_0, for a Gaussian wave packet.

(d) For timescales for which the overlap is apparent, show that the spatial wavelength of the interference fringes goes like $\lambda = ht/md$.

(e) Show that the momentum-space probability density is time-independent and is given by

$$P(p, t) = \frac{4N^2\alpha}{\sqrt{\pi}} \cos^2\left(\frac{pd}{2\hbar}\right) e^{-\alpha^2 p^2} \tag{4.194}$$

where $\alpha \equiv \beta/\hbar$.

P4.31. **Expectation values and uncertainties for general free-particle wave packet.** We can evaluate the most general form for the expectation values of position and momentum, as well as the corresponding uncertainties, using a general time-dependent momentum-space solution as in Eqn. (4.126), namely

$$\phi(p, t) = \phi_0(p)\, e^{-ip^2 t/2m\hbar} \tag{4.195}$$

where $\phi_0(p)$ is any acceptable momentum-space wavefunction.[8]

(a) Show that the expectation values for p and p^2 are time-independent, namely that

$$\langle p \rangle_t = \langle p \rangle_0 \quad \text{and} \quad \langle p^2 \rangle_t = \langle p^2 \rangle_0 \qquad \text{so that} \quad \Delta p_t = \Delta p_0 \tag{4.196}$$

which is also consistent with Eqn. (3.21), since the momentum-space probability density $|\phi(p, t)|^2 = |\phi_0(p)|^2$ does not change with time.

(b) Use the momentum-space form of the position operator, $\hat{x} = i\hbar(d/dp)$, to show that

$$\langle \hat{x} \rangle_t = \langle \hat{x} \rangle_0 + \langle p \rangle_0 t/m \tag{4.197}$$

which is consistent with special cases we have considered, and classical expectations.

(c) Finally, show that

$$\langle \hat{x}^2 \rangle_t = \langle \hat{x}^2 \rangle_0 + \frac{t}{m}\langle p\hat{x} + \hat{x}p \rangle_0 + \langle p^2 \rangle_0 t^2/m^2 \tag{4.198}$$

so that the most general form for the time-dependent width of a free-particle wave packet is

$$(\Delta x_t)^2 = (\Delta x_0)^2 + (\Delta p_0)^2 t^2/m^2$$
$$+ \frac{t}{m}\langle (p - \langle p \rangle_0)(\hat{x} - \langle \hat{x} \rangle_0) + (\hat{x} - \langle \hat{x} \rangle_0)(p - \langle p \rangle_0) \rangle \tag{4.199}$$

[8] The general expressions here are nicely derived, using a different approach, by Styer (1990).

Hint: You might find it useful to use the commutation relation $[\hat{x}, p] = i\hbar$. The first two terms are familiar from the case of the Gaussian example in Section 4.3.1 and Eqn. (4.48). The term linear in t arises only in cases where there is a nontrivial correlation in the momentum-position construction of the wave packet. For example, if the higher momentum components $(p - \langle p \rangle_0 > 0)$ are, in general, farther back $(\hat{x} - \langle \hat{x} \rangle_0 < 0)$ in the packet, then the linear term will be nonvanishing.

Comments: (i) While these results have all been obtained using momentum-space methods, all of the $t = 0$ expectation values in Eqns (4.196–4.197) can be evaluated using either $\phi_0(p)$ or $\psi(x, 0)$. (ii) The covariance discussed in **P4.18** is clearly useful here.

P4.32. Expectation values and uncertainties for the "other" Gaussian wave packet. Evaluate $\langle x \rangle_t$ and Δx_t for the modified Gaussian wave packet in P3.5. If you have not already evaluated the appropriate $\psi(x, t)$ for that problem, you can do so now. Compare your results for that special case with the general expression in Eqn. (4.199) above.

P4.33. Properties of the Wigner distribution.

(a) Using the Fourier transform connection between $\psi(x, t)$ and $\phi(p, t)$, show that the two forms of the Wigner function in Eqns (4.149) and (4.150) are indeed equivalent.

(b) Using properties of the Dirac δ-function, confirm Eqn. (4.156).

P4.34. Wigner distribution for the free-particle Gaussian wave packet.

(a) Using both Eqns (4.149) and (4.150), and the free-particle Gaussian wave packet solutions in Eqns (3.35) and (3.36), evaluate $P_W(x, p; t)$ in two ways to confirm the result in Eqn. (4.157).

(b) Integrate $P_W(x, p; t)$, first over x, and then over p, to confirm that you obtain the appropriate single-particle probability densities, $|\psi_{(G)}(x, t)|^2$ and $|\phi_{(G)}(p, t)|^2$, respectively. Note especially how the time-dependent β_t comes to appear in the position-space probability distribution in the appropriate places.

P4.35. Wigner distribution for the uniformly accelerating wave packet. Using either the position-space or momentum-space Gaussian wavepackets in Eqns (4.138) or Eqns (4.139) representing a particle undergoing uniform acceleration, evaluate the Wigner distribution and show that you obtain

$$P_W(x, p; t) = \frac{1}{\pi\hbar} e^{-\alpha^2(p-p_0-Ft)^2} e^{-(x-x_0-pt/m+Ft^2/2m)^2/\beta^2} \qquad (4.200)$$

What do plots of $P_W(x, p; t)$ versus (x, p) look like?

P4.36. Wigner distribution for an exponential wavefunction. Recall the position-space and momentum-space wavefunctions of Section 3.4, given by

$$\psi(x) = \frac{1}{\sqrt{a}} e^{-|x|/a} \quad \text{and} \quad \phi(p) = \sqrt{\frac{2a}{\hbar\pi}} \frac{1}{(1 + (pa/\hbar)^2)} \quad (4.201)$$

(a) Use $\psi(x)$ to evaluate the corresponding Wigner distribution. Plot $P_W(x, p)$ in the easiest way you can think of to show that there is a small fraction of (x, p) space where it is negative. This provides a simple example of why $P_W(x, p)$ cannot be used as a true joint probability distribution.

(b) Integrate $P_W(x, p)$ over p and confirm that you obtain the correct position-space probability density, as in Eqn. (4.152); repeat for the integration over x, as in Eqn. (4.153).

P4.37. Expectation values, operators, and the Wigner distribution. A natural definition of the expectation value of momentum using the Wigner distribution would be

$$\langle p \rangle_t = \int_{-\infty}^{+\infty} \int_{-\infty}^{+\infty} p \, P_W(x, p; t) \, dx \, dp \quad (4.202)$$

Show that this gives

$$\langle p \rangle_t = \int_{-\infty}^{+\infty} \psi^*(x, t) \left(\frac{\hbar}{i} \frac{\partial}{\partial x} \right) \psi(x, t) \, dx \quad (4.203)$$

$$\langle p \rangle_t = \int_{-\infty}^{+\infty} p \, |\phi(p, t)|^2 \, dp \quad (4.204)$$

depending on which representation (Eqn. (4.149) or (4.150)) of the Wigner function one uses. (Hint: One case is straightforward, while the other requires several IBP tricks, but both rely on the definition of the Dirac δ-function.) Repeat for the expectation value of x in both representations.

FIVE

The Infinite Well: Physical Aspects

5.1 The Infinite Well in Classical Mechanics: Classical Probability Distributions

The problem of a particle moving in an infinite well or one-dimensional box, defined by the potential

$$V(x) = \begin{cases} 0 & \text{for } 0 < x < +a \\ +\infty & \text{for } x < 0 \text{ or } +a < x \end{cases} \tag{5.1}$$

is one of the most familiar and easiest to solve of all problems in introductory quantum mechanics. It is the simplest case in which to study the phenomenon of quantized energy levels in bound states, one of the "smoking guns" of wave mechanics. In this chapter and the next, we use this problem (and some related ones) as a tool in understanding the formalism used in solving the time-independent Schrödinger equation for bound states in a potential, understanding their physical interpretation, as well as some of the formal properties of its solutions. In Chapter 7, we also use the infinite well as a model system in which to examine the role that spin and the Pauli exclusion principle play in multiparticle quantum systems.

We also wish to develop some intuition about quantum mechanical wavefunctions, both in the 'quantum' limit of small quantum number and especially in the quasi-classical limit of large quantum number. To that end, we first discuss the problem of a particle moving in this potential well, treated classically, and introduce the notion of a classical probability distribution.

Classically, a particle in such a potential would move at constant speed inside the box, experiencing elastic collisions with the walls. The speed (v_0) and period (τ) of the motion are easily determined in terms of the energy to be

$$v_0 = \sqrt{2E/m} \quad \text{and} \quad \tau = \frac{2a}{v_0} = 2a\sqrt{\frac{m}{2E}} \tag{5.2}$$

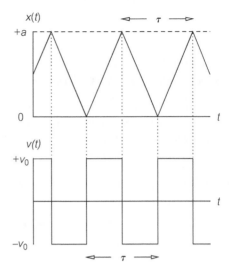

Figure 5.1. Classical trajectories in the infinite well, $x(t)$ and $v(t)$ versus t.

The position, $x(t)$, and velocity, $v(t)$, of a particle in such a well[1] are shown in Fig. 5.1. We note that measurements of the velocity of the particle at any time will only yield values of $\pm v_0$.

Although we know the exact trajectory, $x(t)$, for all times, we can still ask for the probability that a measurement of the position of the particle (using, for example, a large number of stroboscopic photographs of the system taken at random times) will find it in a given region inside the well. Since the particle moves at constant speed, and therefore spends equal amounts of time in all regions of the well, we must have

$$P_{\mathrm{CL}}(x)\,dx \equiv \text{Probability}\,[(x,x+dx)] = C\,dx \tag{5.3}$$

where C is a constant, and we have introduced the notion of a *classical probability distribution*, $P_{\mathrm{CL}}(x)$. Since the particle must be found somewhere in the box, $P_{\mathrm{CL}}(x)$ must be normalized in the usual way so that

$$1 = \int_0^a P_{\mathrm{CL}}(x)\,dx = aC \implies P_{\mathrm{CL}}(x) = C = \frac{1}{a} \tag{5.4}$$

Using this classical probability distribution, we can calculate average values as usual and find, for example,

$$\langle x \rangle_{\mathrm{CL}} = \int_0^a x\,P_{\mathrm{CL}}(x)\,dx = \frac{a}{2} \tag{5.5}$$

[1] See Styer (2001) for a discussion of the mathematical representation of these classical trajectories, and their quantum analogs.

Figure 5.2. Generic one-dimensional potential with bound states. For energy E_1, a and b are the classical turning points for bound motion. For energy E_2, c and d are turning points for unbound motion; particles incident from the left (right) will rebound at point c (d). For energy E_3, classical particles will slow down (speed up) as they travel over the "bump" ("well") in the potential, but will not rebound.

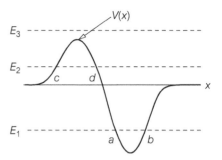

and

$$\langle x^2 \rangle_{\text{CL}} = \int_0^a x^2 P_{\text{CL}}(x)\, dx = \frac{1}{a} \int_0^a x^2\, dx = \frac{a^2}{3} \tag{5.6}$$

so that $\Delta x_{\text{CL}} = a/\sqrt{12}$.

One can also define a classical probability distribution for momentum, $P_{\text{CL}}(p)$, and in this simple case where only the values $p = \pm p_0 = \pm v_0 m$ are allowed, with equal probability, such a distribution might be written as

$$P_{\text{CL}}(p) = \frac{1}{2}[\delta(p - p_0) + \delta(p - p_0)] \tag{5.7}$$

and it is straightforward to show (P5.1) that $\langle p \rangle_{\text{CL}} = 0$, $\langle p^2 \rangle_{\text{CL}} = p_0^2$, and $\Delta p_{\text{CL}} = p_0$.

The notion of a classical probability distribution for position[2] can be generalized to any bound state problem with a potential $V(x)$. Consider a particle of mass m and energy E moving in a general confining potential such as in Fig. 5.2 (for the case E_1 at least). The motion in such a potential is periodic, the particle bouncing back and forth between the classical turning points a and b; the time for one traversal of the well (from, say, left to right) is half the period, $\tau/2$. The amount of time, dt, the particle spends in the small region of space, dx, near the point x is given by the speed there, $v(x)$, via

$$dt = \frac{dx}{dx/dt} = \frac{dx}{v(x)} \tag{5.8}$$

so that the the classical period is given by summing all of these infinitesimal times for one traversal, namely

$$\frac{\tau}{2} = \text{one back-and-forth time} = \int_{t_a}^{t_b} dt = \int_a^b \frac{dx}{v(x)} \tag{5.9}$$

[2] The comparison of classical and quantum probability distributions, in both position- and momentum-space, for many simple systems is discussed in Robinett (1995).

The probability of finding the particle in the small region $(x, x + dx)$ is simply the ratio of the time spent there, dt, to the total time for one traversal, that is

$$P_{CL}(x)\, dx \equiv \text{Probability}[(x,x+dx)] = \frac{dt}{(\tau/2)} = \frac{2}{\tau}\frac{dx}{v(x)} \qquad (5.10)$$

so that the classical probability density is

$$P_{CL}(x) = \frac{2}{\tau}\frac{1}{v(x)} \qquad (5.11)$$

which is normalized appropriately since

$$\int_a^b P_{CL}(x)\, dx = \frac{2}{\tau}\int_a^b \frac{dx}{v(x)} = \left(\frac{2}{\tau}\right)\left(\frac{\tau}{2}\right) = 1 \qquad (5.12)$$

from Eqn. (5.9). This definition shows that the particle will spend more time (and hence be found more often) in regions where the classical speed is low; this is especially true at classical turning points, where the particle is changing direction, with the velocity changing sign, implying that $v(x) \to 0$ at such points.

Since the classical speed is related to the kinetic energy by $T(x) = mv(x)^2/2$ and hence to the potential energy, we can write

$$P_{CL}(x) = \frac{2}{\tau}\sqrt{\frac{m}{2T(x)}} = \frac{2}{\tau}\sqrt{\frac{m}{2(E - V(x))}} \quad \text{(for a general bound state)}$$
$$(5.13)$$

Thus the classical probability density is large (small) where the kinetic energy is small (large) or the potential energy is large (small). For the case of the infinite well, where the potential vanishes inside the well, this reduces to

$$P_{CL}(x) = \frac{2}{\tau}\sqrt{\frac{m}{2E}} = \frac{1}{a} \quad \text{(for the infinite well)} \qquad (5.14)$$

as expected.

5.2 Stationary States for the Infinite Well

5.2.1 Position-Space Wavefunctions for the Standard Infinite Well

We now discuss the solutions of the quantum mechanical problem of a particle in an infinite well potential. We begin by focusing on the potential well of the form

$$V(x) = \begin{cases} 0 & \text{for } 0 < x < +a \\ +\infty & \text{for } x < 0 \text{ or } +a < x \end{cases} \tag{5.15}$$

which we will describe as the "standard" infinite well problem. The time-independent Schrödinger equation inside the potential well (where $V(x) = 0$) becomes

$$-\frac{\hbar^2}{2m}\frac{d^2\psi(x)}{dx^2} + V(x)\psi(x) = E\psi(x) \quad \longrightarrow \quad -\frac{\hbar^2}{2m}\frac{d^2\psi(x)}{dx^2} = E\psi(x) \tag{5.16}$$

This is of the form

$$\frac{d^2\psi(x)}{dx^2} = -k^2\psi(x) \quad \text{where} \quad k = \sqrt{\frac{2mE}{\hbar^2}} \tag{5.17}$$

The solutions of Eqn. (5.17), which are most like standing waves (and hence relevant for bound state problems) are

$$\psi(x) = A\sin(kx) + B\cos(kx) \tag{5.18}$$

where A, B are (at the moment) arbitrary constants. Since the particle is not allowed outside (i.e. $\psi(x) = 0$ for $x < 0$, $a < x$), and the wavefunction should be continuous, we must also implement the requirements that $\psi(0) = \psi(a) = 0$; the application of these boundary conditions there then requires that

$$\psi(0) = A\sin(0) + B\cos(0) = 0$$
$$\psi(a) = A\sin(ka) + B\cos(ka) = 0 \tag{5.19}$$

or

$$B = 0 \quad \text{and} \quad A\sin(ka) = 0 \tag{5.20}$$

If $A = 0$ as well, then $\psi(x)$ vanishes identically (the uninteresting case of zero total probability, corresponding to no particle in the well), so we must have

$$\sin(ka) = 0 \quad \text{or} \quad k_n a = n\pi \quad \text{where } n = 1, 2, 3 \dots \tag{5.21}$$

giving the quantized energies

$$E_n = \frac{\hbar^2 k_n^2}{2m} = \frac{\hbar^2 n^2 \pi^2}{2ma^2} \quad \text{(standard infinite well)} \tag{5.22}$$

This result can also be obtained from simple "fitting deBroglie wave" ideas, as in Sec. 1.3. The stationary state wavefunctions, here written in the form $\psi(x) = u_n(x)$, must satisfy the normalization condition that

$$1 = \int P_{QM}(x)\, dx = \int_0^a |u_n(x)|^2\, dx = 1 \tag{5.23}$$

and the appropriately normalized wavefunctions are conventionally written in
the form

$$u_n(x) = \sqrt{\frac{2}{a}} \sin\left(\frac{n\pi x}{a}\right) \quad \text{(standard infinite well)} \quad (5.24)$$

We note that this normalization has the appropriate dimensionality for a one-
dimensional wavefunction, but that the sign (or more generally the phase) of
the normalization constant is purely conventional, and $-\sqrt{2/a}$ or $\exp(i\theta)\sqrt{2/a}$
would serve just as well. The first few spatial wavefunctions, $u_n(x)$, are shown
in Fig. 5.3 (along with those for the related symmetric infinite well discussed in
Section 5.2.3.) We note the general feature that the number of nodes increases
with energy, starting with a nodeless ground state.

These results exemplify a standard analytic approach to the solutions of a
quantum mechanical problem of a particle in a potential well:

1. One solves the time-independent Schrödinger equation in position-space
   which gives the appropriate functional forms for $\psi(x)$.

2. One applies the boundary conditions, which usually involve the condition,
   either explicitly or implicitly, that the wavefunction vanish sufficiently rapidly
   at infinity, and this restriction gives rise to quantized energy levels; this is the
   analog of "fitting de Broglie waves" (as in Section 1.3) for a general potential.

3. Finally, one normalizes the wavefunctions to ensure that a probability interpretation is valid. Other information, such as average values of various
   operators or the corresponding momentum-space wavefunction, $\phi(p)$, are
   then easily calculated.

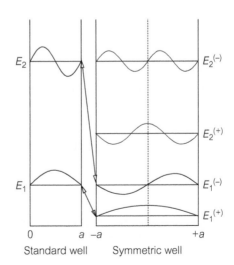

Figure 5.3. Energy eigenvalues and eigenfunctions
for the standard and symmetric infinite well.

Each step is independent of the others, and all are required to obtain the full information possible from the stationary state solutions.

The most general solutions will be linear combinations of these energy eigenstates with their associated time-dependence, namely

$$\psi(x, t) = \sum_n a_n \psi_n(x, t) = \sum_n a_n u_n(x) \, e^{-iE_n t/\hbar} \tag{5.25}$$

We stress again that, because the $u_n(x)$ are stationary state solutions, the squared moduli of the individual wavefunctions, $|\psi_n(x, t)|^2 = |u_n(x)|^2$, are time-independent.

5.2.2 Expectation Values and Momentum-Space Wavefunctions for the Standard Infinite Well

To make contact with the classical probability distributions in Section 5.1, we evaluate the quantum mechanical expectation values

$$\langle x \rangle = \int x \, |\psi(x)|^2 \, dx = \int_0^a x \, [u_n(x)]^2 \, dx = \frac{a}{2} \tag{5.26}$$

and

$$\langle x^2 \rangle = \int x^2 \, |\psi(x)|^2 \, dx = \int_0^a x^2 \, [u_n(x)]^2 \, dx = \frac{a^2}{3} \left(1 - \frac{3}{2(n\pi)^2} \right) \tag{5.27}$$

We note that the second result agrees with the classical expectation only in the limit of large quantum numbers ($n \to \infty$). This effect is illustrated in Fig. 5.4 where we note that the quantum probability densities, $P_{QM}^{(n)}(x) = |u_n(x)|^2$, corresponding to individual energy eigenstates, do not approach the classical limit $P_{CL}(a) = 1/a$ in the usual sense of smooth convergence, but rather oscillate increasingly rapidly about the classical result as n increases (as discussed in P5.5), locally averaging to the classical result.

The expectation value of momentum in an infinite well stationary state is given by

$$\langle \hat{p} \rangle = \int_0^a [u_n(x)]^* \, \hat{p} \, [u_n(x)] \, dx = \left(\frac{\hbar}{i} \right) \int_0^a u_n(x) \frac{du_n(x)}{dx} \, dx$$

$$= \frac{\hbar}{2i} \left[u_n(x)^2 \right]_0^a = 0 \tag{5.28}$$

since the eigenfunctions vanish at the boundaries. Thus, the average value of the momentum vanishes in each eigenstate. While this specific calculation provides

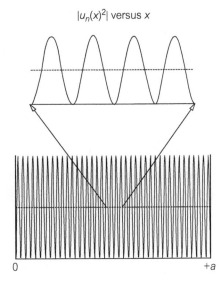

$|u_n(x)^2|$ versus x

Figure 5.4. Probability density, $P_{QM}^{(n)}(x) = |u_n(x)|^2$, for the standard infinite well for large quantum number ($n = 40$). $P_{QM}^{(n)}(x)$ averages locally to the flat classical probability distribution, $P_{CL}(x) = 1/a$.

0 $+a$

the correct answer, this important (and more general) result can be understood more intuitively and physically in several ways:

1. The energy eigenstates in this (or any) bound state system can be chosen to be purely real functions (or real with multiplicative complex phases that are independent of position). As the calculation in Eqn. (5.28) illustrates, this implies that the expectation value $\langle \hat{p} \rangle$, which we know must be real, can be written as a purely real integral times i. For consistency, this implies that the integral must vanish, which it manages to do by integrating to be $[u_n(x)]^2/2$, evaluated at the appropriate limits, where the wavefunction vanishes. For the simple case of the infinite well, these limits are $x = 0, a$, while for a more general (and realistic) case, the limits would be $x = -\infty, +\infty$ where $\psi(x)$ must vanish in order for the solution to be physically acceptable. We note that the inclusion of the exponential time-dependence $\exp(-iE_n t/\hbar)$ does not affect the argument. One can also show that the probability flux or current, $j(x, t)$, for such purely real energy eigenstates also vanishes for similar reasons.

2. The fact that the eigenstates can be chosen as real functions can also be understood in terms of classical results for standing waves, where (complex) traveling wave solutions of the form $\exp(\pm ikx)$ can be combined to form the real $\sin(kx), \cos(kx)$ solutions used in Eqn. (5.18).

3. The vanishing of the average momentum value can also be thought of heuristically in classical terms as the symmetry in velocity during subsequent "back

and forth" traversals of a potential during its periodic motion, as the same *speed* is achieved twice during each period. The classical velocity is related to the energy and potential at a point via

$$v(x) = \pm \sqrt{\frac{2}{m}(E - V(x))} \tag{5.29}$$

so that the classical velocity distribution is necessarily symmetric between $+v(x)$ and $-v(x)$, implying that $\langle p \rangle_{CL} = m\langle v(x) \rangle_{CL} = 0$.

4. While this result is valid for energy eigenstates, linear combinations of such eigenstates forming a general solution will contain nontrivial time-dependent phases between eigenstates and can yield nonvanishing values of $\langle \hat{p} \rangle_t$. Examples include the two-state systems discussed in Section 5.4.1 or wave packets as in Section 5.4.2.

In a similar way, one can evaluate the expectation value of powers of momentum, including the useful case of

$$\langle \hat{p}^2 \rangle = \int_0^a [u_n(x)]^* \, \hat{p}^2 \, [u_n(x)] \, dx = \left(\frac{n\pi\hbar}{a}\right)^2 \equiv p_n^2 \tag{5.30}$$

which is consistent with the fact that the kinetic energy for this system

$$\langle \hat{T} \rangle = \frac{1}{2m} \langle \hat{p}^2 \rangle = \frac{\hbar^2 \pi^2 n^2}{2ma^2} = \frac{p_n^2}{2m} = E_n \tag{5.31}$$

should be equal to the total energy.

To make connections with classical concepts of velocity, we can evaluate the momentum-space wavefunctions corresponding to the $u_n(x)$ via

$$\phi_n(p) = \frac{1}{\sqrt{2\pi\hbar}} \int_{-\infty}^{+\infty} dx \, \psi(x) \, e^{-ipx/\hbar}$$

$$\frac{1}{\sqrt{2\pi\hbar}} \int_0^a dx \, u_n(x) \, e^{-ipx/\hbar} = \frac{-i}{\sqrt{\pi\hbar a}} e^{-ipa/2\hbar} \left\{ e^{+in\pi/2} \left(\frac{\sin[(n\pi - pa/\hbar)/2]}{(n\pi/a - p/\hbar)} \right) \right.$$

$$\left. - e^{-in\pi/2} \left(\frac{\sin[(n\pi + pa/\hbar)/2]}{(n\pi/a + p/\hbar)} \right) \right\} \tag{5.32}$$

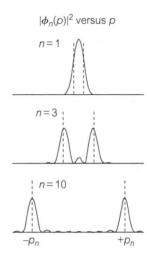

$|\phi_n(p)|^2$ versus p

$n = 1$

$n = 3$

$n = 10$

$-p_n$ $+p_n$

Figure 5.5. Momentum-space probability density, $|\phi_n(p)|^2$ versus p, for the standard infinite well for values of $n = 1, 3$, and 10. The dotted lines correspond to values of p given by $\pm p_n = \sqrt{2mE_n}$.

The momentum-space probability distribution, $P_{\mathrm{QM}}^{(n)}(p) = |\phi_n(p)|^2$, is easily evaluated and we plot in Fig. 5.5 the results for $n = 1, 3, 10$. We note that even for rather small values of the quantum number there are two well-defined peaks at $p = \pm p_n$, consistent with classical expectations where one would find only values $\pm v_n = \pm p_n/m = \pm\sqrt{2E_n/m}$ as the result of velocity measurements. To facilitate comparison with the classical probability distribution in Eqn. (5.7), we can rewrite the momentum-space wavefunction in the equivalent form

$$\phi_n(p) = \frac{-i}{\sqrt{2\pi\,\Delta p}} e^{-ip/\Delta p} \left\{ e^{+in\pi/2} \left(\frac{\sin[(p_n - p)/\Delta p]}{(p_n - p)/\Delta p} \right) \right.$$
$$\left. - e^{-in\pi/2} \left(\frac{\sin[(p_n + p)/\Delta p]}{(p_n + p)/\Delta p} \right) \right\} \qquad (5.33)$$

where $\Delta p \equiv 2\hbar/a$. If we take the corresponding momentum distribution, $|\phi_n(p)|^2$, in the limit that $\Delta p \to 0$ (either $\hbar \to 0$ or $a \to \infty$) and use the representation of the δ-function in Appendix D.8, we find (P5.10) that

$$\lim_{\Delta p \to 0} P_{\mathrm{QM}}^{(n)}(p) = \lim_{\Delta p \to 0} |\phi_n(p)|^2 = \frac{1}{2}\left[\delta(p - p_n) + \delta(p + p_n)\right] = P_{\mathrm{CL}}(p_n) \qquad (5.34)$$

We also note that the quantum mechanical momentum distributions reflect the same symmetry as their classical counterparts, as we have

$$|\phi_n(-p)|^2 = P_{\mathrm{QM}}^{(n)}(-p) = P_{\mathrm{QM}}^{(n)}(+p) = |\phi_n(+p)|^2 \qquad (5.35)$$

which is consistent with our observations about the speed distributions arising from the equivalence of the "back-and-forth" motions in a bound state system. This fact can also be used to show that

$$\langle p \rangle = \int_{-\infty}^{+\infty} p \, |\phi_n(p)|^2 \, dp = 0 \tag{5.36}$$

using the momentum-space representation, since the integrand is now manifestly odd in the variable p. Other properties of the momentum-space representation are explored in the problems (P5.6 and P5.8.)

5.2.3 The Symmetric Infinite Well

A variation on the standard infinite well problem, which we will find useful is the symmetric infinite well, defined by

$$V(x) = \begin{cases} 0 & \text{for } |x| < a \\ +\infty & \text{for } |x| > a \end{cases} \tag{5.37}$$

that is, a one-dimensional box of width $2a$, centered at the origin. We discuss this quite similar case for several reasons:

1. The two potentials, and their solutions (in both position- and momentum-space) can be obtained from each other by simple scaling arguments and symmetry relationships.

2. The symmetric infinite well introduces us to the notion of parity, which we will study further in Section 6.6.

3. It is a limiting case of the asymmetric infinite well considered in Section 5.3.

4. Finally, the symmetric well eigenfunctions are very similar to those used in standard Fourier series analyses where we have experience in their use in the expansion of arbitrary functions.

For this case we find the same Schrödinger equation and solutions as in Eqns (5.16) and (5.18), but the boundary conditions are implemented in a slightly different way. Now because $\psi(x) = 0$ for $|x| \geq a$, we must impose the continuity conditions at the walls

$$\psi(+a) = A \sin(ka) + B \cos(ka) = 0$$
$$\psi(-a) = -A \sin(ka) + B \cos(ka) = 0 \tag{5.38}$$

or

$$A \sin(ka) = 0 \quad \text{and} \quad B \cos(ka) = 0 \tag{5.39}$$

Once again, if $A = B = 0$ then $\psi(x)$ vanishes identically, so we can consider two cases separately:

A=0, even solutions: In this case, only the $\cos(kx)$ (i.e. the even, $\cos(-kx) = \cos(kx)$) solutions survive. We therefore set $\cos(ka) = 0$, which has solutions

$$k_n^{(+)} = \frac{(n-1/2)\pi}{a} \quad \text{with } n = 1, 2, 3, \ldots \tag{5.40}$$

so that the energy eigenvalues are

$$E_n^{(+)} = \frac{\hbar^2 (k_n^{(+)})^2}{2m} = \frac{\hbar^2 (2n-1)^2 \pi^2}{8ma^2} \quad \text{(symmetric infinite well)} \tag{5.41}$$

where the $(+)$ superscript denotes the even states. The corresponding normalized eigenfunctions are

$$u_n^{(+)}(x) = \frac{1}{\sqrt{a}} \cos\left(\frac{(n-1/2)\pi x}{a}\right) \quad \text{(symmetric infinite well)} \tag{5.42}$$

B=0, odd solutions: In this case, where only the $\sin(kx)$ (or the odd, $\sin(-kx) = -\sin(kx)$) solutions are used, we set $\sin(ka) = 0$, which has solutions

$$k_n^{(-)} = \frac{n\pi}{a} \quad \text{with } n = 1, 2, 3, \ldots \tag{5.43}$$

with energy eigenvalues

$$E_n^{(-)} = \frac{\hbar^2 n^2 \pi^2}{2ma^2} \quad \text{(symmetric infinite well)} \tag{5.44}$$

and normalized eigenfunctions

$$u_n^{(-)}(x) = \frac{1}{\sqrt{a}} \sin\left(\frac{n\pi x}{a}\right) \quad \text{(symmetric infinite well)} \tag{5.45}$$

We note that these solutions can be obtained from those for the standard well (in Eqn. (5.24) by first letting $a \to 2a$ (doubling the width of the well, so it covers the range $(0, 2a)$) and then letting $x \to x - a$ (shifting the center to the origin); see Fig. 5.3 for a comparison. Note that the solutions for the standard and symmetric wells are necessarily normalized differently, as they are defined over different length boxes.

These solutions possess the additional property of evenness and oddness, namely

$$u_n^{(+)}(-x) = +u_n^{(+)}(+x) \quad \text{and} \quad u_n^{(-)}(-x) = -u_n^{(-)}(+x) \tag{5.46}$$

which are obviously related to the symmetry of the well, $V(-x) = V(+x)$; we will introduce the important notion of parity in just this context in the next chapter.

5.3 **The Asymmetric Infinite Well**

An asymmetric infinite well potential can be defined by

$$V(x) = \begin{cases} +\infty & \text{for } |x| > a \\ 0 & \text{for } -a < x < 0 \\ V_0 & \text{for } 0 < x < +a \end{cases} \tag{5.47}$$

corresponding to a change in the potential in the right half of the symmetric infinite well; for definiteness, we will assume that $V > 0$ in what follows. This small variation is interesting as it provides an opportunity to study several new features, namely:

1. For the case where $E > V_0$, a variation in potential corresponds to a variation in speed so that the classical probability of finding the particle will not be uniformly distributed (P5.11) across the potential, providing another example of the correlation between the potential energy function and the probability distribution (both classical and quantum mechanical).[3]

2. For $E < V_0$, we will find nonclassical solutions, which are nonvanishing in the classically disallowed region $(0 < x < +a)$, corresponding to *quantum tunneling*, which is discussed further in Chapter 8.

An example of the potential (and representative energy levels for both cases) is shown in Fig. 5.6 and we consider the $E > V_0 > 0$ and $V_0 > E > 0$ cases separately.

The $E > V_0 > 0$ case. For the left-hand side of the well, where $V(x) = 0$, the problem can be analyzed as above, but for the right-hand side where there is a nonzero potential, the Schrödinger equation becomes

$$-\frac{\hbar^2}{2m}\frac{d^2\psi(x)}{dx^2} + V_0\psi(x) = E\psi(x) \tag{5.48}$$

or

$$\frac{d^2\psi(x)}{dx^2} = -q^2\psi(x) \quad \text{where} \quad \sqrt{\frac{2m(E - V_0)}{\hbar^2}} \equiv q < k = \sqrt{\frac{2mE}{\hbar^2}} \tag{5.49}$$

---

[3] The asymmetric well is discussed in Doncheski and Robinett (2000) and Gilbert *et al.* (2005).

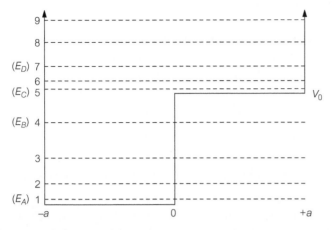

Figure 5.6. Potential energy function, $V(x)$, for the asymmetric infinite well, with representative energy eigenvalues for a particular choice of well parameters. The highlighted states, E_A, E_B, E_C, E_D, corresponding to $n = 1, 4, 5, 7$ are visualized in more detail in Fig. 5.7.

with obvious $\sin(qx), \cos(qx)$ solutions. The general solution for both sides of the well can be written in the form

$$\psi(x) = \begin{cases} 0 & \text{for } |x| > a \\ A' \sin(kx) + C' \cos(kx) & \text{for } -a < x < 0 \\ B' \sin(qx) + D' \cos(qx) & \text{for } 0 < x < +a \end{cases} \quad (5.50)$$

The boundary conditions on the wavefunction at $x = -a, +a$ (namely $\psi(-a) = 0 = \psi(+a)$) results in a considerable simplification, giving

$$\psi(x) = \begin{cases} A \sin(k(x + a)) & \text{for } -a < x < 0 \\ B \sin(q(x - a)) & \text{for } 0 < x < +a \end{cases} \quad (5.51)$$

Recalling the discussions in Section 4.2, we insist that $\psi(x)$ (and higher derivatives) should be continuous even at discontinuous boundaries[4]; matching both $\psi(x)$ and $\psi'(x)$ at $x = 0$ imposes the additional constraints

$$A \sin(ka) = -B \sin(qa) \quad (5.52)$$

$$Ak \cos(ka) = qB \cos(qa) \quad (5.53)$$

which can then be combined to yield the energy eigenvalue condition, namely

$$k \cos(ka) \sin(qa) + q \sin(ka) \cos(qa) = 0 \quad (5.54)$$

independent of the normalization constants A, B.

---

[4] The discontinuity in $V(x)$ at $x = 0$ is not "bad enough" to make $\psi'(x)$ discontinuous; see Section 8.1.1.

In the limit $V_0 \to 0$, where $q \to k$, this reduces to the condition $2k \sin(ka) \cos(ka) = 0$, consistent with our discussion of the symmetric well and Eqn. (5.39). However, for $V > 0$, the equation no longer has simple closed form or analytic solutions as encountered early, and the values of E (giving k, q), which satisfy Eqn. (5.54) must be obtained numerically. For example, for fixed values of $\hbar, m, a$, the $k(E)$ and $q(E)$ are functions of the energy and one can define

$$F(E) \equiv k(E) \cos[k(E)a] \sin[q(E)a] + q(E) \sin[k(E)a] \cos[q(E)a] \quad (5.55)$$

then plot $F(E)$ as a function of the energy, looking for zeroes, and finally use root-finding algorithms to locate values of E, which satisfy Eqn. (5.54).

Picking a representative set of parameters, we have followed this approach, and in Fig. 5.6 we have indicated a number of such energy eigenvalues for the $E > V_0$ case (states labeled 5 and higher); we also plot in Fig. 5.7 (left-hand side) the corresponding (normalized) probability distributions $|\psi_n(x)|^2$ versus x for states with $n = 5, 7$ (labeled E_C and E_D.) The dashed vertical line indicates the middle of the well, while the horizontal dotted line indicates the classical probability density for a symmetric well of width $2a$, namely $P_{\rm CL}(x) = 1/2a$. We note several features of these states, which are exemplary of the rather general

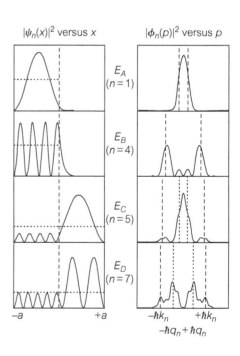

Figure 5.7. Plots of the position-space probability density $|\psi_n(x)|^2$ versus x (left) and momentum-space probability density $|\phi_n(p)|^2$ versus p (left) for four energy eigenvalues in the asymmetric infinite well corresponding to values of $E < V_0$ (E_A ($n = 1$) and E_B ($n = 4$) and $E > V_0$ (E_C ($n = 5$) and E_D ($n = 7$).) For the plots on the left, the center of the potential well is indicated by the vertical dashed lines, while the horizontal dotted lines correspond to the classical probability density in an infinite well of width a ($n = 1, 4$) and $2a$ ($n = 5, 7$). For the plot on the right, the vertical dashed lines indicate the values of $\pm \hbar k_n$ (motion in the left side of the well) while the vertical dotted lines indicate $\pm \hbar q_n$ (motion on the right.)

correlations between the magnitude and "wiggliness" of quantum wavefunctions, namely:

1. The wavefunction (and hence the quantum probability density) is larger (smaller) in magnitude in regions where the particle moves more slowly (more quickly), as in Eqn. (5.11), or where the kinetic energy is smaller (larger) or the potential energy is larger (smaller), as in Eqn. (5.13). The probability density in the right half of the well is larger than that for the symmetric potential with no potential step, since the classical particle would be going more slowly on that side.

2. The wavefunction is "wigglier" (less "wiggly"), with a smaller (larger) local deBroglie wavelength in regions where the classical speed or kinetic is larger (smaller), consistent with $\lambda = 2\pi\hbar/p$.

3. For bound state wavefunctions therefore, we typically expect to see solutions with

   - large magnitude and few "wiggles" ...or...
   - small magnitude and more "wiggles."

The dramatic difference between the wavefunctions on the left and right sides of the potential seen here will become less noticeable as the energy eigenvalue increases, since for $E_n \gg V_0$ the presence of the potential step will have increasingly less effect.

The momentum-space wavefunctions corresponding to these two cases, obtained by Fourier transform as usual, are also shown in Fig. 5.7 (right-hand side), where the symmetric character of $|\phi_n(p)|^2$ is again apparent. For the E_D ($n = 7$) case, there are obvious features corresponding to both the back-and-forth motion in the right side of the well, with momenta $p = \pm\hbar q$ (location indicated by the vertical dotted lines), as well as larger momentum values corresponding to $p = \pm\hbar k$ (dashed lines) and motion ("back" or "forth") in the left half of the well. The correlation between their heights, namely that the $\pm\hbar q$ peaks are higher than the $\pm\hbar k$ peaks, is again consistent with finding the particle more frequently with "motion" in the right side of the well. For the $n = 5$ case, the two $\pm\hbar q$ peaks have coalesced, similar to the $n = 1$ case in Fig. 5.5.

The $V_0 > E > 0$ case. For this case, the Schrödinger equation and its solutions in the left side of the well are still given by Eqn. (5.18), while for $0 < x < +a$, we have

$$\frac{d^2\psi(x)}{dx^2} = +\left(\frac{2m(V_0 - E)}{\hbar^2}\right)\psi(x) = \tilde{q}^2\,\psi(x) \qquad (5.56)$$

where

$$\tilde{q} \equiv \sqrt{\frac{2m(V_0 - E)}{\hbar^2}} \tag{5.57}$$

The solutions are now qualitatively different than the oscillatory $\sin(qx)$, $\cos(qx)$ functions, as they can be written in the form

$$\psi(x) = \tilde{B}\cosh(\tilde{q}x) + \tilde{C}\sinh(\tilde{q}x) \quad \text{for } 0 < x < +a \tag{5.58}$$

corresponding to linear combinations of exponential $(\exp(\pm\tilde{q}x))$ functions. In contrast to the classical case, where the particle would not be allowed in a region where $E < V_0$, there is a nonvanishing quantum wavefunction (and probability) in the classically disallowed region. (This same type of behavior is seen in the reflection of electromagnetic waves from boundaries where the index of refraction in the first medium is larger than in the second ($n_1 > n_2$) or optically dense to optically thin. In this case there are still nonzero fields[5] in medium 2, which are exponentially attenuated, so-called *evanescent waves*.

The matching at boundary/continuity conditions at $x = +a$ and $x = 0$ proceeds as before, with the new eigenvalue condition

$$k\cos(ka)\sinh(\tilde{q}a) + \tilde{q}\sin(ka)\cosh(\tilde{q}a) = 0 \quad \text{for } E < V_0 \tag{5.59}$$

which can also be solved numerically. The energy eigenvalues for this case are also shown in Fig. 5.6 (those labeled $n = 1 - 4$) and plots of $|\psi_n(x)|^2$ versus x corresponding to $n = 1$ (E_A) and 4 (E_C) are shown in Fig. 5.7. In this region, the behavior of the quantum wavefunctions is very similar to those for the standard infinite well (oscillating about the classical $P_{CL}(x) = 1/a$ prediction relevant for a particle restricted to the left-half of the well), but with a clear exponential "tail" in the classically disallowed region.

The behavior of the quantum wave function in this region can be approximated from the solution here as

$$\frac{\psi(x)}{\psi(0)} = \frac{\tilde{B}\sinh(\tilde{q}(a-x))}{\tilde{B}\sinh(\tilde{q}a)} = \frac{\exp(\tilde{q}(a-x)) - \exp(-\tilde{q}(a-x))}{\exp(\tilde{q}a) - \exp(-\tilde{q}a)} \approx e^{-\tilde{q}x} \tag{5.60}$$

or an exponential suppression as one penetrates into the nonclassical region. This behavior can be written in the form

$$\frac{\psi(x)}{\psi(0)} \approx e^{-\tilde{q}x} = e^{-x/L} \quad \text{where} \quad L \equiv \frac{1}{\tilde{q}} = \frac{\hbar}{\sqrt{2m(V_0 - E)}} \tag{5.61}$$

---

[5] For example, see Griffiths (1998).

and L is a penetration depth. This effect is seen to be quantum mechanical in origin in a variety of ways:

- This form immediately implies that if we formally let $\hbar \to 0$ as an effective limit of classical physics, then $L \to 0$, and the particle is constrained to stay in the left side of the well.

- The form of the penetration depth can be understood from a simple heuristic argument using the energy-time uncertainty principle. A particle with energy $E < V_0$ can "fluctuate" to one having energy $E \gtrsim V_0$, which is then "above" the potential step; this is only plausible provided it does so over a time interval consistent with the uncertainty principle, $\Delta E \cdot \Delta t \gtrsim \hbar$. Given this "excursion," the particle will necessarily have an uncertainty in its measured energy of $\Delta E = |V_0 - E|$, so the fluctuation can only last a time $\Delta t \sim \hbar/|V_0 - E|$. It can then, *very roughly*, go "into" the classically disallowed region and "back" again for half this time, that is, $\Delta t_{\text{out}} \sim \Delta t_{\text{back}} \sim \hbar/2|V_0 - E|$. If we associate a classical velocity of $|V_0 - E| = mv^2/2$ during this "motion," giving $v = \sqrt{2|V_0 - E|/m}$, the particle could travel a distance

$$L \sim \Delta x_{\text{out}} \sim \Delta t_{\text{out}}\, v \sim \frac{\hbar}{\sqrt{2m|V_0 - E|}} \sim \frac{1}{\tilde{q}} \qquad (5.62)$$

While not to be taken as a rigorous proof in any sense, this argument (or mnemonic device) does point out one useful way of thinking about the length scale in the quantum evanescent wavefunction.

We note that the penetration into the classically disallowed region in Fig. 5.7 is clearly larger for the $n = 4$ (smaller $\tilde{q}$ value state, hence less "cheating" required) than for the $n = 1$ ground state.

5.4 Time-Dependence of General Solutions

5.4.1 Two-State Systems

In all of the variations of the infinite well problem we have discussed, we have focused on the stationary state solutions and their physical interpretation. Because of the linearity of the Schrödinger equation, the most general time-dependent solution will consist of a linear combination of any, or all, such solutions. In the case of the standard well, for example, this implies that

$$\psi(x, t) = \sum_{n=1}^{\infty} a_n u_n(x)\, e^{-iE_n t/\hbar} \qquad (5.63)$$

while for the symmetric well,

$$\psi(x, t) = \sum_{n=1}^{\infty} \left(a_n^{(+)} u_n^{(+}(x) \, e^{-iE_n^{(+)}t/\hbar} + a_n^{(-)} u_n^{(-}(x) \, e^{-iE_n^{(-)}t/\hbar} \right) \tag{5.64}$$

where the a_n or $a_n^{(\pm)}$ are, at the moment, simply arbitrary (possibly complex) numbers.

As a simple example of the more complicated time-development possible with such general solutions, we consider a simple two-state (2S) system consisting an equally weighted combination of the ground state and first excited state wavefunctions of the symmetric well, $\psi_{2S}(x, t)$, that is,

$$\psi_{2S}(x, t) = \frac{1}{\sqrt{2}} \left(u_1^{(+)}(x) e^{-iE_1^{(+)}t/\hbar} + u_1^{(-)}(x) e^{-iE_1^{(-)}t/\hbar} \right) \tag{5.65}$$

that is, $a_1^{(+)} = a_1^{(-)} = 1\sqrt{2}$. The consistency of this choice of normalization will become clear shortly. We note that such states can be produced experimentally when two closely spaced energy levels of a system are simultaneously excited (say by a laser pulse.)

Recall first that

$$E_1^{(+)} = \frac{\hbar^2 \pi^2}{8ma^2}, \qquad E_1^{(-)} = \frac{4\hbar^2 \pi^2}{8ma^2} \tag{5.66}$$

and

$$u_1^{(+)}(x) = \frac{1}{\sqrt{a}} \cos\left(\frac{\pi x}{2a}\right) \qquad u_1^{(-)}(x) = \frac{1}{\sqrt{a}} \sin\left(\frac{\pi x}{a}\right) \tag{5.67}$$

The probability density for this solution has nontrivial time-dependence, since

$$P_{2S}(x, t) = |\psi_{2S}(x, t)|^2$$

$$= \frac{1}{2} \left\{ [u_1^{(+)}(x)]^2 + [u_1^{(-)}(x)]^2 + 2[u_1^{(+)}(x)][u_1^{(-)}(x)] \cos\left(\frac{\Delta E t}{\hbar}\right) \right\} \tag{5.68}$$

where $\Delta E \equiv E_1^{(-)} - E_1^{(+)} = 3\hbar^2\pi^2/8ma^2$ so this it not a stationary state solution, but one with periodicity given by $\tau = 2\pi\hbar/\Delta E$.

It is easy to show that this state is correctly normalized for all times, since

$$\int_{-a}^{+a} P_{2S}(x, t) \, dx = \frac{1}{2} \int_{-a}^{+a} [u_1^{(+)}(x)]^2 \, dx + \frac{1}{2} \int_{-a}^{+a} [u_1^{(-)}(x)]^2 \, dx$$

$$+ \cos\left(\frac{\Delta E t}{\hbar}\right) \int_{-a}^{+a} [u_1^{(+)}(x) \, u_1^{(-)}(x)] \, dx$$

$$= \frac{1}{2} + \frac{1}{2} + 0 = 1 \tag{5.69}$$

since the two eigenfunctions are individually normalized, and the integral containing the cross-term is found to vanish; this last fact could, in this case, be argued as being due to the integrand being an odd function integrated over a symmetric interval, but we will see in the next chapter that it has a deeper significance.

A straightforward calculation (P5.16) shows that the average position of the particle described by this wavefunction oscillates, alternating between being more or less localized on the right and left sides of the well, specifically

$$\langle x \rangle_t = \int_{-a}^{+a} x \, |\psi_{(2S)}(x,t)|^2 \, dx = a \left(\frac{32}{9\pi^2} \right) \cos \left(\frac{\Delta E t}{\hbar} \right) \tag{5.70}$$

which arises solely from the cross-term in Eqn. (5.68). The corresponding average value of the momentum operator can be calculated in the standard way, and one finds

$$\langle \hat{p} \rangle_t = \int_{-\infty}^{+\infty} dx \, \psi_{2S}^*(x,t) \, \hat{p} \, \psi_{2S}(x,t) = - \left(\frac{4}{3} \frac{\hbar}{a} \right) \sin \left(\frac{\Delta E t}{\hbar} \right) \tag{5.71}$$

which is also consistent $\langle \hat{p} \rangle_t = m d\langle x \rangle_t / dt$, as it must. The corresponding momentum space probability density for this state, obtained via taking the Fourier transform of each position-space $u_n^{(\pm)}(x)$, shows similar structure, since

$$\phi_{2S}(p,t) = \frac{1}{\sqrt{2}} \left(\phi_1^{(+)}(x) \, e^{-iE_1^{(+)}t/\hbar} + \phi_1^{(-)}(x) \, e^{-iE_1^{(-)}t/\hbar} \right) \tag{5.72}$$

which gives

$$P_{2S}(p,t) = |\phi_{2S}(p,t)|^2$$
$$= \frac{1}{2} \left[|\phi_1^{(+)}(p)|^2 + |\phi_1^{(-)}(p)|^2 + 2\mathrm{Re}[\phi_1^{(+)}(p)^* \phi_1^{(-)}(p)] \sin \left(\frac{\Delta E t}{\hbar} \right) \right] \tag{5.73}$$

Finally, the probability flux can be calculated using Eqn. (4.32), and one finds

$$j_{2S}(x,t) = F_{2S}(x) \sin \left(\frac{\Delta E t}{\hbar} \right) \tag{5.74}$$

where

$$F_{2S}(x) = - \frac{\hbar \pi}{2ma^2} \left\{ \cos \left(\frac{\pi x}{2a} \right) \cos \left(\frac{\pi x}{a} \right) + \frac{1}{2} \sin \left(\frac{\pi x}{2a} \right) \sin \left(\frac{\pi x}{a} \right) \right\} \tag{5.75}$$

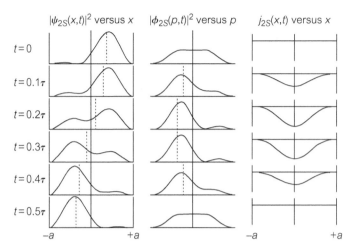

Figure 5.8. Time-development of two-state wavefunction in Eqn. (5.65), in the symmetric infinite well. Position-space probability density ($|\psi(x, t)|^2$ versus x); momentum-space probability density ($|\phi(p, t)|^2$ versus p); and probability flux ($j(x, t)$ versus x) are shown for various times during the first half-cycle. The time-dependent expectation values of $\langle x \rangle_t$ and $\langle \hat{p} \rangle_t$ are shown (as vertical dotted lines) in the respective figures.

The resulting time-dependent position-space and momentum-space probability densities, $P_{2S}(x, t)$ versus x and $P_{2S}(p, t)$ versus p, as well as the flux, are shown in Fig. 5.8 for various times during one half-cycle. The dotted lines indicate the location of the average position ($\langle x \rangle_t$) and momentum ($\langle \hat{p} \rangle_t$) values on the respective plots. As the probability density "sloshes" from right to left, the flux is everywhere negative (corresponding to probability flow to the left), while the average value of momentum (the vertical dotted line) is also indicating "motion" to the left, since $\langle \hat{p} \rangle_t \leq 0$; this behavior then reverses itself over the second half-cycle. Clearly, more dynamic behavior is allowed for general time-dependent states than for stationary states.

5.4.2 Wave Packets in the Infinite Well

In earlier discussions of the free particle, we made heavy use of the concept of a wave packet, a localized solution of the Schrödinger equation with a behavior similar to that of a classical free particle, but which inherently included the concept of spreading due to dispersion. One can construct a similar time-dependent solution in the standard infinite well[6] constructed from the solutions of Eqn. (5.24).

---

[6] For a early pedagogical example, see Segre and Sullivan (1976).

Motivated by the Gaussian wave packet of Section 3.2.2, we can write a solution involving stationary states of the standard infinite well given by

$$\psi_{WP}(x, t) = \sum_{n=1}^{\infty} a_n u_n(x)\, e^{-iE_n t/\hbar} \tag{5.76}$$

where we choose coefficients of the form

$$a_n = e^{-((p_n - p_0)^2 \alpha^2/2)}\, e^{-ip_n x_0/\hbar} \tag{5.77}$$

Here $p_n \equiv \hbar n\pi/a$ is the analog of the continuous momentum variable and $p_0 \equiv \hbar n_0 \pi/a$ defines some central value of n, while this form also allows for arbitrary values of the central peak position, x_0. (For a more complete discussion, see Example 6.4 below.) The solution in Eqn. (5.76), with these a_n, can be evaluated numerically (cutting off the infinite sum at some large but finite value of n) for various values of the parameters to illustrate the time-dependence. We note that the largest contribution to the summation will come from states with $E_n \approx p_0^2/2ma^2 = \hbar^2 n_0^2/2ma^2$, so that the classical speed of the wave packet is roughly $v_0 = p_0/m \approx \hbar n_0/ma$ with a corresponding classical period $\tau = 2a/v_0 = 2ma^2/\hbar n_0$. The spreading time (by analogy with Eqn. (3.40)) is $t_0 = m\hbar\alpha^2$; in the examples that follow, we will use parameters for which τ and t_0 are comparable.

We first show in Fig. 5.9 (left side) the time-dependence of $|\psi_{WP}(x, t)|^2$ over a single classical period, τ; the initial wave function is constructed so as to be slightly off-center. The "back-and-forth" motion is obvious (to be compared to the classical trajectory indicated by the dashed line), while the "bounces" at the wall (with the same quantum interference seen in Fig. 3.4) are also consistent with earlier discussions. As with the free-particle Gaussian wave packet, quantum spreading should also be present, and with $\tau \sim t_0$, we see significant spreading even after one classical period. The subsequent longer time behavior of $|\psi_{WP}(x, t)|^2$ is also illustrated in Fig. 5.9 (right side), now over 10 classical periods. Because of the confined nature of the potential, the spreading wavefunction increasingly fills the well, seemingly approaching the classical probability distribution, $P_{CL}(x) = 1/a$, shown as the horizontal solid line and arrows for the times $t = (5 - 10)t_0$.

However, because of the quantized nature of the energy eigenvalues, this spreading behavior can actually become "undone" and structure can reappear in the quantum probability density, as suggested for the latest times shown in Fig. 5.9. Most dramatically, after a time (called the *revival time*) defined by

$$T_{rev} \equiv \frac{4ma^2}{\hbar\pi} \tag{5.78}$$

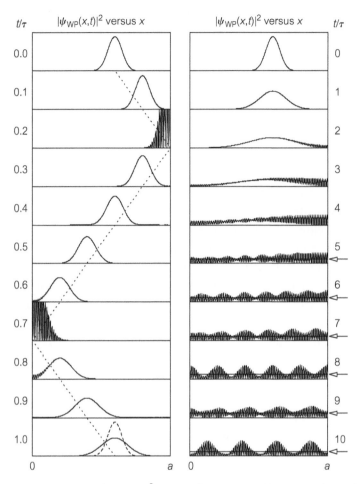

Figure 5.9. Time-development of $|\psi(x, t)|^2$ versus x for a Gaussian-like wave packet in the standard infinite well, over one classical period (left) and 10 classical periods (right); the classical periodicity and the spreading time (t_0) are chosen to be comparable. The classical trajectory is shown as the dotted line. The classical probability density, $P_{CL}(x) = 1/a$ is shown on the right as the horizontal solid line, indicated by the arrows for later times.

the initial wave packet is exactly reformed, since

$$\psi_{WP}(x, t + T_{rev}) = \sum_{n=1}^{\infty} a_n u_n(x) e^{-iE_n(t+T_{rev})/\hbar}$$

$$= \sum_{n=1}^{\infty} a_n u_n(x) e^{-iE_n/\hbar} e^{-2\pi i n^2}$$

$$\psi_{WP}(x, t + T_{rev}) = \psi_{WP}(x, t) \tag{5.79}$$

since the $\exp(-2\pi i n^2)$ phases are all unity. The same argument (using an expansion in momentum-space eigenstates) can be shown to imply that $\phi(p, t + T_{\text{rev}}) = \phi(p, t)$ as well. This phenomena, most often described as a *quantum revival*[7] or *quantum recurrence*, is exact for the infinite well, but approximate revival behavior of wave packets produced by ultrashort laser pulses has been experimentally observed in many atomic and molecular systems.

5.4.3 **Wave Packets Versus Stationary States**

We have now been able to analyze (either analytically or numerically) examples of wave packet solutions to the Schrödinger equation for several simple cases, namely, the free and accelerating particle and the particle in a box. We have argued that these solutions are the closest representation of something like the classical motion of a particle that we will find in quantum mechanics; they exhibit all of the expected dynamics in the classical limit,[8] but still have quantum features such as spreading.

They have been constructed from energy eigenstate solutions, which are, in some sense, more natural in the context of quantum physics. The $\psi_E(x)\, e^{-iEt/\hbar}$ solutions give information on the quantized energies of the bound state system, one of the most important features of quantum mechanics. One might think that these solutions could contain no interesting information on the dynamics of the system as $|\psi_E(x, t)|^2$ for such states is independent of time for stationary states.

We have seen, however, that the shape of $\psi_E(x)$ does make contact with the classical dynamics of the particle; the local magnitude and "wiggliness" of the wavefunction is correlated with the local speed of the particle in a meaningful way. Furthermore the quantum probability density can be used to approach the classical distribution of probability so that information on the particle trajectory is obtained, in a time-averaged sense. Assured of these connections, we will henceforward concentrate on the physical meaning and mathematical properties of energy eigenstates.

5.5 **Questions and Problems**

Q5.1. If one made many measurements of the position of a particle in the standard infinite well, and binned the results to estimate $P_{\text{CL}}(x)$ (as in Section 4.2 and

---

[7] For a review of quantum wave packet revivals, see Robinett (2004). Saxon (1968) discusses this effect in an older textbook, but (correctly) argues that for macroscopic systems the revival time is immeasurably long.
[8] See the discussion by Brown (1973) on the classical limit of quantum wave packets.

Fig. 4.4), what would the result look like? What would it look like for a particle that was undergoing simple harmonic motion? or a ball bouncing vertically under the influence of gravity?

Q5.2. In the quantum version of the infinite well, the energy eigenvalues are, in principle, precisely determined. The energy is all in the form of kinetic energy, $E = p^2/2m$, and so, classically, the magnitude of the momentum should be exactly known as well. But the peaks in $\phi(p)$ have a finite spread (as seen in Fig. 5.5.) How can this happen?

Q5.3. Comment on the statement made in the text that it is usually the boundary conditions on $\psi(x)$ at $x = \pm\infty$ that determine the quantized energy levels. In what sense is this true for the infinite well examples in this chapter?

Q5.4. For the standard infinite well, one could try a solution of the Schrödinger equation corresponding to $E = 0$. What mathematical form do the resulting solutions have? Can they satisfy the boundary conditions?

Q5.5. Referring to Fig. 5.3, why is the energy of the first excited state of the symmetric well identical to the ground state energy of the "standard well," that is, why is $E_2 = E_1^{(-)}$? Will there be any other such "degeneracies"? What do "degeneracies" such as these have to do with the relative size (i.e. width) of the well? Would there be such patterns of symmetry if the symmetric well was of width $4a$?

Q5.6. To explain the physics behind wave packet revivals, as in Eqn. (5.79), some authors have invoked the picturesque image of numerous runners on a circular track, beginning at the same starting line, but with different speeds. Can you use this analogy to explain both wave packet spreading and wave packet revivals?

P5.1. Classical probability density for momentum in the standard infinite well. Using the classical $P_{CL}(p)$ in Eqn. (5.7), evaluate $\langle p \rangle_{CL}$, $\langle p^2 \rangle_{CL}$, and Δp_{CL}.

P5.2. Classical probability density for the harmonic oscillator. The harmonic oscillator potential is conventionally written in quantum mechanics in the form $V(x) = m\omega^2 x^2/2$.

(a) For a fixed total energy E, find the classical turning points of motion.

(b) Use the expression in Eqn. (5.9) to find the classical period as a function of E. Do you get the answer you expect?

(c) Show that the classical probability density in Eqn. (5.13) can be written in the form

$$P_{CL}(x) = \frac{1}{\pi}\frac{1}{\sqrt{A^2 - x^2}} \tag{5.80}$$

and identify A. Show that this distribution is properly normalized and plot this for two values of E, say E_a, E_b), satisfying $E_b = 4E_a$. Contrast this behavior to that of $P_{CL}(x)$ for the infinite well.

(d) Use $P_{CL}(x)$ to evaluate $\langle x \rangle_{CL}$, $\langle x^2 \rangle_{CL}$, and Δx_{CL}. Use this result to evaluate the classical expectation value of the potential energy, $\langle V(x) \rangle_{CL}$.

(e) Use the energy relation $T(x) + V(x) = E$ to evaluate $\langle T(x)\rangle_{\text{CL}}$ and show that

$$\langle V(x)\rangle_{\text{CL}} = \langle T(x)\rangle_{\text{CL}} = \frac{E}{2} \tag{5.81}$$

so that, on average, the kinetic and potential energies are shared equally.

(f) What is $\langle p\rangle_{\text{CL}}$? Use the fact that $T = p^2/2m$ to evaluate $\langle p^2\rangle_{\text{CL}}$ and then Δp_{CL}.

(g) Finally, evaluate the uncertainty principle product $\Delta x_{\text{CL}} \cdot \Delta p_{\text{CL}}$ and show that it can be made arbitrarily small in classical mechanics.

P5.3. We solved the Schrödinger equation for the standard and symmetric infinite wells using $\sin(kx)$ and $\cos(kx)$ solutions. A general solution of the form

$$\psi(x) = Ae^{+ikx} + Be^{-ikx} \tag{5.82}$$

is, of course, just as acceptable.

(a) Use this solution for the standard infinite well and re-derive the quantized energies and normalized wavefunctions in Eqns (5.22) and (5.24).

(b) Do the same thing for the symmetric infinite well to reproduce the results in Section 5.2.3.

P5.4. Placement of the infinite well. Consider a particle of mass m moving in an infinite well potential of width $2a$, centered not at the origin, but rather at $x = d$.

(a) Find the solutions of the Schrödinger equation, apply the appropriate boundary conditions and show that the resulting quantized energies are identical to those found in Section 5.2.3.

(b) Show how the wavefunctions in this well are related to those discussed above.

(c) Discuss why the positioning of the well at different locations should have no effect on any important physical observables.

P5.5. Classical limit of infinite well eigenfunctions. Consider a particle in the standard infinite well in an eigenstate $u_n(x)$.

(a) Calculate the probability that the particle will be found in a finite interval of length $b < a$ located at some arbitrary point, $x = c$, in the well, that is, in the interval $(c, c + b)$ within the well.

(b) Show that as $n \to \infty$, this reproduces the classical result we discussed in Section 5.1, that is, that the particle is equally likely to be found anywhere in the well.

P5.6. We have shown (in Section 4.4) that the average value of the momentum operator, $\langle \hat{p}\rangle$, vanishes when evaluated in any position-space wavefunction, $u(x)$, which is purely real. In this problem, we will show the same thing, but will use momentum space methods.

(a) Use the momentum space representation

$$\phi(p) = \frac{1}{\sqrt{2\pi\hbar}} \int_{-\infty}^{+\infty} dx\, u(x)\, e^{-ipx/\hbar} \tag{5.83}$$

to show that when $u(x)$ is real, one can write $\phi(p) = f(p) + ig(p)$ where $f(p)$ and $g(p)$ are real functions, which are also even and odd functions of p respectively. Show that this means that $|\phi(p)|^2$ will be symmetric under $p \to -p$, which, in turn, implies that $\langle p \rangle_{QM} = 0$.

(b) Discuss what happens to your results if $u(x)$ is multiplied by an arbitrary complex number (even a function of time such as the standard time-dependent factor, $\exp(-iEt/\hbar)$) so long as it has no spatial dependence.

(c) Discuss how either of these proofs fail if the spatial wavefunction, $u(x)$, is allowed to be complex, for example, a plane wave solution of the form $\exp(ip_0 x/\hbar)$.

(d) Show that the flux, $j(x, t)$, also vanishes for any purely real state.

P5.7. Show that the $u_n(x)$ eigenfunctions of the standard infinite well in Eqn. (5.24) have a generalized parity property, namely that they satisfy

$$u_n(a - x) = (-1)^{n+1} u_n(x) \tag{5.84}$$

corresponding to simple symmetries under reflection about the center of the well. Verify this for the first few eigenfunctions in Fig. 5.3.

P5.8. **Properties of momentum-space wavefunctions for the infinite well.** Consider the momentum-space wavefunctions, $\phi_n(p)$ given in Eqn. (5.32) corresponding to the standard infinite well solutions, $u_n(x)$, in Eqn. (5.24).

(a) Show that the $\phi_n(p)$ are appropriately normalized, namely that

$$\frac{1}{\sqrt{2\pi\hbar}} \int_{-\infty}^{+\infty} |\phi_n(p)|^2 \, dp = 1 \tag{5.85}$$

(b) Show that the Fourier transform of the $\phi_n(p)$ returns the standard infinite well position-space wavefunctions in Eqn. (5.24), namely that

$$\int_{-\infty}^{+\infty} \phi_n(p)\, e^{-ipx/\hbar}\, dp = \begin{cases} 0 & \text{for } x < 0 \text{ or } x > +a \\ \sqrt{2/a}\, \sin(n\pi x/a) & \text{for } 0 < x < +a \end{cases} \tag{5.86}$$

You will likely find the following integrals useful:

$$\int_{-\infty}^{+\infty} \frac{\sin^2(q)}{q^2} \, dq = \pi \quad \text{and} \quad \int_{-\infty}^{+\infty} \frac{\sin(q)\cos(mq)}{q} \, dq$$

$$= \begin{cases} 0 & \text{for } |m| > 1 \\ \pi/2 & \text{for } |m| = 1 \\ \pi & \text{for } |m| < 1 \end{cases} \quad (5.87)$$

P5.9. Momentum-space wavefunctions for the symmetric infinite well. Using the $u_n^{(+)}(x)$ and $u_n^{(-)}(x)$ solutions of the symmetric infinite well, evaluate their Fourier transform momentum-space counterparts, $\phi_n^{(+)}(p)$ and $\phi_n^{(-)}(p)$. How are they related to those in Eqn. (5.32) for the standard infinite well. Compare your answers here to the very general results for the Fourier transforms of even and odd functions discussed in P2.16.

P5.10. Classical limit of momentum-space probability distributions. Use the expression for the standard infinite well $\phi_n(p)$ in Eqn. (5.33) to show that the corresponding momentum-space probability density approaches

$$P_{CL}(p) = \frac{1}{2}[\delta(p - p_0) + \delta(p + p_0)] \quad (5.88)$$

in the limit that $\Delta p \to 0$. Hint: Use the representation of the δ-function in Appendix E.8.

P5.11. Classical probability distribution for the asymmetric infinite well. Consider a particle of mass m and energy E moving in the asymmetric infinite potential well in Eqn. (5.47), namely

$$V(x) = \begin{cases} +\infty & \text{for } |x| > a \\ 0 & \text{for } -a < x < 0 \\ V_0 & \text{for } 0 < x < a \end{cases} \quad (5.89)$$

for the case where $E > V_0$.

(a) Sketch graphs of the classical trajectory variables $x(t)$ and $v(t)$ versus t.

(b) Use the expression in Eqn. (5.9) to find the classical period as a function of E. How does $\tau(E)$ behave for $E \gtrsim V_0$ and $E >> V_0$? Show that it reduces to the expected values when $V_0 \to 0$.

(c) Use the expression in Eqn. (5.13) for the classical probability distribution to write an expression for $P_{CL}(x)$ and confirm that it is properly normalized. Use it to find the probability that a measurement of the position of this particle would find it with $0 < x < a$? Give a numerical answer in the cases $E = 1.1 V_0$, $E = 2 V_0$, and $E = 10 V_0$

(d) Sketch the classical probability density for $E = 1.1 V_0, 2 V_0, 10 V_0$.

(e) Using the classical probability density, evaluate $\langle x \rangle$ as a function of E.

(f) What would the *classical* distribution of velocities (or momenta) look like for a particle in this potential?

P5.12. Zero-curvature solutions for the asymmetric infinite well. In finding the solutions for the asymmetric infinite well in Section 5.3, we considered solutions with $E > V_0$ (oscillatory for $0 < x < +a$) and with $0 < E < V_0$ (exponentially decaying for $0 < x < +a$).

(a) Show that it is possible, with special values of the parameters of the problem, to have solutions with $E = V_0$. For what values of $V_0, m, \hbar, a$ can this happen?

(b) For such solutions, find the (unnormalized) solutions for $0 < x < +a$ in two ways, namely, (i) by taking the $q \to 0$ ($E \to V_0$) limit of the solutions in Eqn. (5.51) and (ii) by solving the appropriate Schrödinger equation (which is what?) directly for this special case.

(c) Can you sketch such a solution?

P5.13. Normalization of asymmetric infinite well wavefunctions. Consider the solutions of the asymmetric infinite well, for the case of $E > V_0$, as in Eqn. (5.51).

(a) How does one normalize these wavefunctions?

(b) Find explicit expressions for A and B in Eqn. (5.51). Show that they reduce to the appropriate limits when $V_0 \to 0$.

(c) Repeat for the case of $E < V_0$.

P5.14. Momentum-space wavefunctions for the asymmetric infinite well. For the $E > V_0$ position-space wavefunctions in Eqn. (5.51), evaluate the corresponding momentum-space wavefunctions. Discuss the correlations between the peaks at $p = \pm \hbar k, \pm \hbar q$ and the coefficients A, B.

P5.15. Tunneling solutions for the asymmetric infinite well. Consider the asymmetric infinite well in the case when $E < V_0$. Show that the solutions for $0 < x < +a$ can be obtained from those for the $E > V_0$ case by the substitution $q \to i\tilde{q}$. Namely, continue the standard trigonometric functions into hyperbolic functions, using, for example, the results of Appendix C.

P5.16. Properties of two-state systems I. Consider the two-state solution of the symmetric well in Eqn. (5.65).

(a) Explicitly confirm the results in Eqns (5.70) and (5.71) by direct calculation. You will find the necessary integrals in Appendix D.1, but you might find the following integral useful

$$\int_{-1}^{+1} dy \, y \cos\left(\frac{\pi y}{2}\right) \sin\left(\pi y\right) = \frac{32}{9\pi^2} \qquad (5.90)$$

(b) Confirm the result in Eqn. (5.74) for the time-dependent probability flux, $j_{2S}(x, t)$.

(c) Evaluate the time-dependent expectation value of the energy operator in this state, $\langle \hat{E} \rangle_t$.

P5.17. Properties of two-state systems II. Consider two variations on the two-state solution in the symmetric well in Eqn. (5.65), obtained by slightly different choices of the coefficients $a_1^{(+)}, a_1^{(-)}$.

(a) Let $a_1^{(+)} = \cos(\theta)$ and $a_1^{(-)} = \sin(\theta)$. Confirm that the wavefunction is still correctly normalized for all times. Evaluate $\langle x \rangle_t$, $\langle \hat{p} \rangle_t$, and the probability flux, $j_{2S}(x, t)$. Are your results consistent with expectations when $\theta = 0$ or $\theta = \pi/2$? To what limits do those cases correspond?

(b) Repeat part (a), but for the case $a_1^{(+)} = 1/\sqrt{2}$ and $a_1^{(-)} = e^{i\phi}/\sqrt{2}$. What role does the phase ϕ play in the time-development of the two-state system?

P5.18. Properties of two-state systems III. Consider the two-state system in the symmetric infinite well described by the initial wavefunction

$$\psi_{2S}(x, 0) = \frac{1}{\sqrt{2}} \left[u_1^{(+)}(x) + u_2^{(+)}(x) \right] \tag{5.91}$$

namely, a linear combination of the ground state and the *second*-excited state.

(a) What is $\psi_{2S}(x, t)$?

(b) What are $\langle x \rangle_t$ and $\langle p \rangle_t$? (Before you calculate, look at the symmetry properties of $\psi_{2S}(x, t)$).

(c) What is $\langle \hat{E} \rangle_t$?

P5.19. Timescales for quantum revivals. We can estimate the hierarchy for the various timescales for quantum mechanical bound states, comparing the spreading time (t_0), the classical period (τ), and the revival time (T_{rev}), using Eqn. (5.78).

(a) Consider a 1 kg mass bouncing elastically between walls 1 m apart, with its position determined to within an experimental error of roughly 1 μm. Estimate both t_0 and T_{rev} and compare them to the classical period. Is quantum mechanical spreading or revival behavior observable in such macroscopic systems? Assume a kinetic energy $T = 1$ J.

(b) Recall the discussion in Example 3.1 comparing the classical period and spreading time of wave packets in large n atomic orbits. As an approximation, let us use Eqn. (5.78) to estimate the revival time for this system, where we can associate $L = 2r_n = 2(a_0 n^2)$ with the size of the quantum "box." Compare the relative sizes of $t_0, \tau, T_{\text{rev}}$ for such microscopic systems.

P5.20. Mirror revivals in the infinite well. Consider a general wave packet state in the standard infinite well, as in Eqn. (5.76), evaluated at half a revival time, namely, $\psi_{\text{WP}}(x, t + T_{\text{rev}}/2)$.

(a) Using the results of P5.7, show that

$$\psi(x, t + T_{rev}/2) = -\psi(a - x, t)$$

so that $\quad |\psi(x, t + T_{rev}/2)|^2 = |\psi(L - x, t)|^2 \quad$ (5.92)

and the shape of the initial wave packet is reformed exactly, but mirrored about the center of the well. Hint: You might want to confirm to yourself that $(-1)^n = (-1)^{n^2}$ for integral values of n.

(b) Show that two applications of the equation above reproduces Eqn. (5.79), namely that

$$\psi(x, t + T_{rev}/2 + T_{rev}/2) = \psi(x, t) \quad (5.93)$$

(c)) Show that the momentum-space wavefunction satisfies

$$\phi(p, t + T_{rev}/2) = -e^{-ipa/\hbar}\phi(-p, t)$$

so that $\quad |\phi(p, t + T_{rev}/2)|^2 = |\phi(-p, t)|^2 \quad$ (5.94)

and the momentum-space probability density is also flipped in sign, so that the particle is at the "other corner of phase space," centered at $L - x$ and going in the opposite direction.

P5.21. Classical forces. A classical particle rattling around in a one-dimensional box would exert a force on the walls and the same is true in quantum mechanics. To evaluate it, consider a particle in the ground state of the standard well (with walls at $(0, L)$) with energy E and imagine moving the right-hand wall *very slowly* (sometimes called *adiabatically*) to the right by an amount dx. It can be shown that any such slow change will leave the particle in the ground state of the evolving system.

(a) Calculate the new ground state energy, E', and the difference in energies $\Delta E = E' - E$. This difference in energies is associated with the work done *by the wall on the particle*, which we can write $dW = F_{wall} \cdot dx$; it can be used to evaluate F_{wall}.

(b) The force *of the particle on the wall* is then the same in magnitude, but with the opposite sign by required by Newton's second law. Use this force to calculate the work done on a particle by the walls if the right wall is slowly moved from L to $2L$ and show that it agrees with the change in energy.

P5.22. Properties of the Wigner distribution for energy eigenstates. Assume that you have an energy eigenstate representing a bound state system, so that the position-space wavefunction is given by $\psi(x, t) = u_n(x) e^{-iE_n t/\hbar}$, where $u_n(x)$ is real. Show that the Wigner distribution, $P_W(x, p; t)$, defined in Eqn. (4.149), is independent of time so that $P_W(x, p; t) = P_W(x, p)$ and that it also satisfies $P_W(x, -p; t) = P_W(x, +p; t)$; discuss the significance of the last property.

P5.23. **Wigner distribution for infinite well eigenstates.** Evaluate the Wigner quasi-probability distribution defined in Eqn. (4.149) for the case of the standard infinite square well solutions, $u_n(x)$, in Section 5.2.1. (You have to think a bit carefully about how to split up the integration region appropriately to handle the limits of integration.) Plot your result for the $n = 1$ state in the simplest way possible to show that while it is mostly nonnegative, it is not completely positive-definite. Plot $P_W(x, p)$ versus (x, p) for a value of $n \gg 1$ and discuss the rather intricate structure you find.

SIX

The Infinite Well: Formal Aspects

In the previous chapter, we concentrated on the most important physical aspects of the solutions of the Schrödinger equation in various versions of the infinite well, namely, the quantization of energy levels in a confining potential, the connection between the quantum wavefunctions (in both position- and momentum-space) and their classical counterparts, and wave packets.

We now turn to an examination of the more formal properties of energy eigenstate solutions, namely, the eigenfunctions of a Hamiltonian operator. Many of these properties can be immediately generalized to include the eigenfunctions of other Hermitian operators. We begin, however, by introducing a simple but useful notation.

6.1 Dirac Bracket Notation

Motivated initially by the notation for average values used before, we define the *Dirac bracket* of a position-space wavefunction via

$$\langle \psi | \psi \rangle \equiv \int_{-\infty}^{+\infty} dx \, \psi^*(x, t) \, \psi(x, t) \tag{6.1}$$

A properly normalized wavefunction will then have $\langle \psi | \psi \rangle = 1$. This is easily extended to include the overlap integrals of two different wavefunctions, namely

$$\langle \psi_1 | \psi_2 \rangle \equiv \int_{-\infty}^{+\infty} dx \, \psi_1^*(x, t) \, \psi_2(x, t) \tag{6.2}$$

We note the simple connection

$$\langle \psi_1 | \psi_2 \rangle^* = \langle \psi_2 | \psi_1 \rangle \tag{6.3}$$

since such overlap integrals are not necessarily real; on the other hand, $\langle \psi | \psi \rangle$ is manifestly real and positive-definite for any $|\psi\rangle$. An equivalent definition for the momentum-space representation is

$$\langle \phi_1 | \phi_2 \rangle \equiv \int_{-\infty}^{+\infty} dp \, \phi_1^*(p, t) \, \phi_2(p, t) \tag{6.4}$$

and Parseval's theorem (Eqn. (3.24)) guarantees that the Dirac bracket of two different wavefunctions, whether evaluated in position- or momentum-space, will be equal since

$$\begin{aligned} \langle \phi_1 | \phi_2 \rangle &= \int_{-\infty}^{+\infty} dp \, \phi_1^*(p, t) \, \phi_2(p, t) \\ &= \int_{-\infty}^{+\infty} dx \, \psi_1^*(x, t) \, \psi_2(x, t) \\ &= \langle \psi_1 | \psi_2 \rangle \end{aligned} \tag{6.5}$$

Expectation values of operators are written as

$$\langle \hat{O} \rangle = \langle \psi | \hat{O} | \psi \rangle \equiv \int_{-\infty}^{+\infty} dx \, \psi^*(x, t) \, \hat{O} \, \psi(x, t) \tag{6.6}$$

so that, in this simplified form, an operator is Hermitian if

$$\langle \psi | \hat{O} | \psi \rangle = \langle \psi | \hat{O} | \psi \rangle^* \tag{6.7}$$

From Eqn. (4.73), we also know that Hermitian operators will satisfy

$$\langle \psi_1 | \hat{O} | \psi_2 \rangle^* = \langle \psi_2 | \hat{O} | \psi_1 \rangle \tag{6.8}$$

Besides being of typographical convenience, the Dirac bracket notation has a geometrical significance, which will be discussed in Chapter 12.

6.2 Eigenvalues of Hermitian Operators

The basic equation of quantum mechanics, the time-independent Schrödinger equation,

$$\hat{H} \psi_E(x) = E \psi_E(x) \tag{6.9}$$

requires one to find the eigenvalues of a Hermitian operator, namely the Hamiltonian (recall P4.19). The energy eigenvalues so obtained are real, which is an example of a general result:

- The eigenvalues of a Hermitian operator are real numbers.

This is easily proved by considering a Hermitian operator $\hat{A}$ satisfying

$$\hat{A}\psi_a(x) = a\psi_a(x) \tag{6.10}$$

where we work in a position-space presentation for definiteness. If we multiply both sides of Eqn. (6.10) by $\psi_a^*(x)$ (on the left) and integrate, we find that

$$\langle \hat{A} \rangle = \langle \psi_a | \hat{A} | \psi_a \rangle = \int_{-\infty}^{+\infty} dx\, \psi_a^*(x)\, \hat{A}\, \psi_a(x)$$

$$= \int_{-\infty}^{+\infty} dx\, \psi_a^*(x)\, a\, \psi_a(x)$$

$$= a\langle \psi_a | \psi_a \rangle \tag{6.11}$$

so that

$$a = \frac{\langle \psi_a | \hat{A} | \psi_a \rangle}{\langle \psi_a | \psi_a \rangle} \tag{6.12}$$

which is obviously real since $\hat{A}$ is Hermitian. Not surprisingly, then, the eigenvalues associated with operators corresponding to physical observables are real; familiar examples considered so far include the momentum and energy operators.

6.3 Orthogonality of Energy Eigenfunctions

Starting with the standard infinite well, we found that we had to normalize the energy eigenfunctions by hand, because the relation

$$\langle u_n | u_n \rangle = \int_0^a |u_n(x)|^2 dx = \frac{2}{a} \int_0^a \sin^2\left(\frac{n\pi x}{a}\right) dx = 1 \tag{6.13}$$

was not an automatic property of solutions of the Schrödinger equation. We now note that the overlap of two *different* solutions of this problem, that is,

$$\langle u_n | u_m \rangle = \int_0^a [u_n(x)]^* u_m(x)\, dx$$

$$= \frac{2}{a} \int_0^a \sin\left(\frac{n\pi x}{a}\right) \sin\left(\frac{m\pi x}{a}\right) dx$$

$$= \frac{\sin[(m-n)\pi]}{(m-n)\pi} + \frac{\sin[(m+n)\pi]}{(m+n)\pi}$$

$$= \delta_{n,m} \tag{6.14}$$

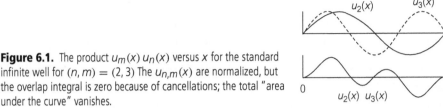

Figure 6.1. The product $u_m(x)\,u_n(x)$ versus x for the standard infinite well for $(n, m) = (2, 3)$ The $u_{n,m}(x)$ are normalized, but the overlap integral is zero because of cancellations; the total "area" under the curve" vanishes.

vanishes if the eigenfunctions correspond to different eigenvalues, that is, if $n \neq m$. In contrast to Eqn. (6.13), this fact *is* seemingly a consequence of the fact that the eigenfunctions satisfy the Schrödinger equation. We can visualize how this occurs by examining the integrands for a pair of states with $n = 2$, $m = 3$ in Fig. 6.1; this figure illustrates how the total "area" under the product $u_n(x)\,u_m(x)$ somehow conspires to vanish.

A similar result can be seen to hold for the symmetric infinite well,

$$\langle u_n^{(+)}|u_m^{(-)}\rangle = 0 \quad \text{for all } n, m, \quad \langle u_n^{(+)}|u_m^{(+)}\rangle = \delta_{n,m}, \quad \text{and}$$

$$\langle u_n^{(-)}|u_m^{(-)}\rangle = \delta_{n,m} \tag{6.15}$$

where the first condition also follows from the fact that the integrand is the product of an even times an odd function.

From Eqn. (3.24), we know that the equivalent overlaps of the momentum space wavefunctions will also satisfy this orthonormality condition; for example, for the standard infinite well we must have

$$\langle \phi_n|\phi_m\rangle = \int_{-\infty}^{\infty} [\phi_n(p)]^*\phi_m(p)\,dp$$

$$= \int_0^{+a} [u_n(x)]^* u_m(x)\,dx$$

$$= \langle u_n|u_m\rangle = \delta_{n,m} \tag{6.16}$$

The momentum space integrals can also be done analytically (P6.1) as a further check.

Finally, we also note that the infinite well is not special in this regard as we can show (P6.2), after some algebra, that the position-space wavefunctions of the "asymmetric" well corresponding to different energies also have vanishing overlap.

We can then say that

• Two functions are *orthogonal* if their overlap integral vanishes,

and we need not specify whether we work with position-space or momentum-space wavefunctions because of Eqn. (3.24). We are finding that

- The energy eigenfunctions form an *orthonormal* set, that is, they are mutually orthogonal and normalized (or can be made so) under a generalized form of dot- or inner-product (namely, the overlap integral).

This concept is very similar to the case of a set of unit vectors in an N-dimensional vector space, $\{\hat{e}_i, i = 1, \ldots, N\}$, where $\hat{e}_i \cdot \hat{e}_j = \delta_{ij}$ or the more familiar $\hat{x}, \hat{y}, \hat{z}$ mutually perpendicular unit vectors of three-dimensional geometry.

To see that this phenomenon is not restricted to solutions of the Schrödinger equation, that is, eigenfunctions of a Hamiltonian operator, consider the eigenfunctions of the *momentum operator*,

$$\hat{p}\psi_p(x) = \frac{\hbar}{i}\frac{d\psi_p(x)}{dx} = p\psi_p(x) \tag{6.17}$$

with solutions

$$\psi_p(x) = \frac{1}{\sqrt{2\pi\hbar}}e^{ipx/\hbar} \tag{6.18}$$

for any real value of the variable p. These solutions have overlap integrals given by

$$\langle\psi_p|\psi_{p'}\rangle = \int_{-\infty}^{+\infty}\left(\frac{1}{\sqrt{2\pi\hbar}}e^{ipx/\hbar}\right)^*\left(\frac{1}{\sqrt{2\pi\hbar}}e^{ip'x/\hbar}\right)dx$$

$$= \frac{1}{2\pi\hbar}\int_{-\infty}^{+\infty}e^{i(p'-p)x/\hbar}dx$$

$$= \delta(p - p') \tag{6.19}$$

which also vanishes if the eigenvalues are different. In this case, the normalization condition is appropriate for a continuous label.

Thus, this generalized orthogonality is seemingly not a specific property of either the position-space or momentum-space representation, or even of eigenfunctions of a Hamiltonian, but a more general result. The functions above all share the property that they are eigenfunctions of some Hermitian operator, and the most general result we can derive can be stated simply as

- The eigenfunctions of a Hermitian operator corresponding to *different* eigenvalues will be orthogonal,

which we can prove as follows: Consider two eigenfunctions of a Hermitian operator, $\hat{A}$, corresponding to two distinct eigenvalues,

$$\hat{A}\psi_a(x) = a\psi_a(x) \quad \text{and} \quad \hat{A}\psi_b(x) = b\psi_b(x) \tag{6.20}$$

where $a \neq b$ are both real (from Section 6.2) since $\hat{A}$ is Hermitian; we work with position-space functions for definiteness. We know from Eqn. (6.8) that any Hermitian operator will satisfy the relation

$$\langle \psi | \hat{A} | \phi \rangle = \langle \phi | \hat{A} | \psi \rangle^* \tag{6.21}$$

so we can write

$$\langle \psi_a | \hat{A} | \psi_b \rangle = \int_{-\infty}^{+\infty} dx \, \psi_a^*(x) \left[\hat{A} \psi_b(x) \right]$$

$$= \int_{-\infty}^{+\infty} dx \, \psi_a^*(x) \, b \, \psi_b(x) = b \langle \psi_a | \psi_b \rangle \tag{6.22}$$

while

$$\langle \psi_b | \hat{A} | \psi_a \rangle^* = \left(\int_{-\infty}^{+\infty} dx \, \psi_b^*(x) \left[\hat{A} \psi_a(x) \right] \right)^*$$

$$= \left(\int_{-\infty}^{+\infty} dx \, \psi_b^*(x) \, a \, \psi_a(x) \right)^*$$

$$= a^* \int_{-\infty}^{+\infty} dx \, \psi_a^*(x) \, \psi_b(x) = a \langle \psi_a | \psi_b \rangle \tag{6.23}$$

since a is real. Comparing these two quantities we see that

$$0 = \langle \psi_a | \hat{A} | \psi_b \rangle - \langle \psi_b | \hat{A} | \psi_a \rangle^* = (b - a) \langle \psi_a | \psi_b \rangle \tag{6.24}$$

so that $\langle \psi_a | \psi_b \rangle = 0$ if $b \neq a$ and the eigenfunctions are indeed orthogonal if their corresponding eigenvalues are different. (We note that if there is more than one eigenfunction corresponding to the same eigenvector, the set of such eigenfunctions can be made orthogonal "by hand.")

The use of Hermitian operators, which naturally correspond to physical observables, automatically induces a rich geometrical and algebraic structure on the solutions of the Schrödinger equation and other systems, and we will make extensive use of these properties.

6.4 Expansions in Eigenstates

Starting with our canonical example of the standard infinite well, we saw that the general, time-dependent solution of the Schrödinger equation could be written as a linear combination of energy eigenstates via

$$\psi(x, t) = \sum_{n=1}^{\infty} a_n u_n(x) \, e^{-iE_n t / \hbar} \tag{6.25}$$

where the a_n are (at the moment) arbitrary and possibly complex constants. We wish to explore the physical interpretation of the "expansion" coefficients, a_n. We first note that if the solution is properly normalized, we must have

$$1 = \langle \psi | \psi \rangle = \int_0^a \psi^*(x, t) \psi(x, t) \, dx$$

$$= \int_0^a \left(\sum_{n=1}^{\infty} a_n \, u_n(x) \, e^{-iE_n t/\hbar} \right)^* \left(\sum_{m=1}^{\infty} a_m \, u_m(x) \, e^{-iE_m t/\hbar} \right) dx$$

$$= \sum_{n=1}^{\infty} \sum_{m=1}^{\infty} a_n^* a_m e^{-i(E_m - E_n)t/\hbar} \int_0^a u_n(x) \, u_m(x) \, dx$$

$$= \sum_{n=1}^{\infty} \sum_{m=1}^{\infty} a_n^* a_m \, e^{-i(E_m - E_n)t/\hbar} \, \delta_{n,m}$$

$$1 = \sum_{n=1}^{\infty} |a_n|^2 \tag{6.26}$$

and the expansion coefficients must satisfy their own normalization condition. Thus, we see that

- For an expansion in eigenstates of the form in Eqn. (6.25) to be properly normalized, the $|a_n|^2$ appearing in the expansion must sum to unity.

While a particle in any particular energy eigenstate will have a uniquely defined energy, this general solution has an expectation value for the energy given by

$$\langle \psi | \hat{E} | \psi \rangle = \int_0^a \psi^*(x, t) \hat{E} \psi(x, t) \, dx$$

$$= \int_0^a \left(\sum_{n=1}^{\infty} a_n \, u_n(x) \, e^{-iE_n t/\hbar} \right)^* \left(i\hbar \frac{\partial}{\partial t} \right) \left(\sum_{m=1}^{\infty} a_m \, u_m(x) \, e^{-iE_m t/\hbar} \right) dx$$

$$= \sum_{n=1}^{\infty} \sum_{m=1}^{\infty} a_n^* a_m E_m \, e^{-i(E_m - E_n)t/\hbar} \int_0^a u_n(x) \, u_m(x) \, dx$$

$$= \sum_{n=1}^{\infty} \sum_{m=1}^{\infty} a_n^* a_m E_m \, e^{-i(E_m - E_n)t/\hbar} \, \delta_{n,m}$$

$$\langle \hat{E} \rangle = \sum_{n=1}^{\infty} E_n |a_n|^2 \tag{6.27}$$

independent of time. So, while many important properties of such a general state may vary in time, for example, average values of x and $\hat{p}$, the average value of the energy operator, $\hat{E}$, does not. Similar results can be obtained for any power of the energy operator,

$$\langle \hat{E}^k \rangle = \langle \psi | \hat{E}^k | \psi \rangle = \sum_{n=1}^{\infty} (E_n)^k |a_n|^2 \tag{6.28}$$

and this allows one to calculate the energy spread, given by

$$\Delta E = \sqrt{\langle \hat{E}^2 \rangle - \langle \hat{E} \rangle^2} \tag{6.29}$$

for any state. A general state such as Eqn. (6.25) will have $\Delta E \neq 0$, unless of course $a_n = \delta_{n,k}$, so that it is actually an energy eigenstate.

Taken together, these results imply that the squares of the expansion coefficients are related to the probability of "finding" the particle in one of the given energy eigenstates. Because the measurable quantity associated with the solutions of the Schrödinger equation is their energy eigenvalue, we have the more precise statement,

- In an expansion of a wavefunction $\psi(x, t)$ in terms of energy eigenstates, $|a_n|^2$ is the probability that a measurement of the energy of the particle described by $\psi(x, t)$ will yield the energy E_n as a result of such a measurement.

This definition is consistent with the expression for $\langle \hat{E} \rangle$ in Eqn. (6.27) as an average value of a discrete probability distribution. Since the $|a_n|^2$ are themselves actual probabilities (in contrast to, say, $|\psi(x, t)|^2$, which is a probability density), they must be dimensionless.

Example 6.1. Average energy in an eigenstate expansion

Consider the unnormalized state in the standard infinite well given by

$$\psi(x, t) = N \left(2u_1(x)\, e^{-iE_1 t/\hbar} + (1-2i)u_2\, e^{-iE_2 t/\hbar} - 3iu_3(x)\, e^{-iE_3 t/\hbar} \right) \tag{6.30}$$

where $E_n = \hbar^2 \pi^2 n^2 / 2ma^2$. The normalization factor N is determined by the fact that

$$1 = \sum_{n=1}^{\infty} |a_n|^2 = N^2(4 + 5 + 9) = 18N^2 \tag{6.31}$$

so $N = 1/\sqrt{18}$. Repeated measurements of the energy of an ensemble of such states would only find the values $E_1, E_2,$ and E_3 with probabilities $2/9, 5/18,$ and $1/2$, respectively; the average value of energy after many such measurements would be $\langle \hat{E} \rangle = (35/12)(\hbar^2 \pi^2)/ma^2$ or roughly $5.83E_1$.

This same interpretation for the expansion coefficients will hold for any expansion in terms of energy eigenstates (i.e. solutions of the Schrödinger equation) but, once again, is far more general. For example, we have seen that the eigenstates of the momentum operator can be written as

$$\psi_p(x) = \frac{1}{\sqrt{2\pi\hbar}} e^{ipx/\hbar} \tag{6.32}$$

and we can write a general function as a weighted (continuous) sum of such solutions via

$$\psi(x) = \int_{-\infty}^{+\infty} \phi(p)\,\psi_p(x)\,dp$$

$$= \frac{1}{\sqrt{2\pi\hbar}} \int_{-\infty}^{+\infty} \phi(p)\,e^{ipx/\hbar}\,dp \tag{6.33}$$

which is simply the Fourier transform. The expansion coefficients in this case are the $\phi(p)$ and have a continuous label, so that the probability of making a measurement of the physical observable (in this case momentum) is given by

$$\text{Prob}[(p, p + dp)] = |\phi(p)|^2 dp \tag{6.34}$$

as expected. We are thus led to argue:

• If we expand a function in terms of the eigenstates of a Hermitian operator, $\hat{A}$, i.e.

$$\psi(x) = \sum_a c_a \psi_a(x) \quad \text{where} \quad \hat{A}\psi_a(x) = a\psi_a(x) \tag{6.35}$$

the square of the expansion coefficients, $|c_a|^2$, give the probability of observing that state with the value of the observable, a; such expansion can be either discrete or continuous.

This discussion points up again the usefulness of having many different *representations* of the same quantum mechanical system. We have already argued that the position-space wavefunction, $\psi(x)$, and the momentum-space wavefunction, $\phi(p)$, are equivalent in their information content, and we can now add to that list the expansion coefficients of a quantum system in terms of some set of energy eigenstates, namely, the $\{a_n\}$. These three sets of numbers give complementary information on the probabilities of measuring the position, momentum, and energy, respectively, of the particle described by them.

The different descriptions have an equivalent geometrical structure, as their norms satisfy

$$\int_{-\infty}^{+\infty} |\psi_a(x)|^2 \, dx = \int_{-\infty}^{+\infty} |\phi_a(p)|^2 \, dp = \sum_{n=1}^{\infty} |a_n|^2 = 1 \qquad (6.36)$$

and generalized inner products are related by

$$\int_{-\infty}^{+\infty} \psi_a^*(x)\psi_b(x) \, dx = \int_{-\infty}^{+\infty} \phi_a^*(p)\phi_b(p) \, dp = \sum_{n=1}^{\infty} a_n^* b_n \qquad (6.37)$$

if

$$\psi_a(x) = \sum_{n=1}^{\infty} a_n u_n(x) \quad \text{and} \quad \psi_b(x) = \sum_{n=1}^{\infty} b_n u_n(x) \qquad (6.38)$$

We have explored in Chapter 4 how average values of the position and momentum operators can be obtained in the $\psi(x)$ and $\phi(p)$ representations; Section 10.4 explores how such information is obtained from the a_n.

6.5 Expansion Postulate and Time-Dependence

We have now seen that an arbitrary linear combination of energy eigenstate solutions (each with its trivial time dependence) will also be a solution, and we have found an interpretation of the corresponding coefficients. We wish to examine whether we can invert the process, namely, if we are given an arbitrary (but physically acceptable) initial state, $\psi(x, 0)$, whether we can expand it in terms of the energy eigenstates.

This procedure, if possible, would then allow us to solve the general initial value problem, as the resulting time-dependence would be simply that of Eqn. (6.25). For the infinite well, this is the quantum mechanical analog of plucking a stretched string in some initial configuration and asking about its future vibrations.

If we formally write

$$\psi(x, 0) = \sum_{n=1}^{\infty} a_n u_n(x) \qquad (6.39)$$

we can try to "invert" this to find the a_n by multiplying both sides by $u_m^*(x)\,dx$ and integrating. We find

$$\int_0^a u_m^*(x)\,\psi(x,0)\,dx = \sum_{n=1}^{\infty} a_n \int_0^a u_m(x)\,u_n(x)\,dx$$

$$= \sum_{n=1}^{\infty} a_n \delta_{n,m} = a_m \tag{6.40}$$

We have dropped the complex conjugation in this case because the $u_m(x)$ can be made real. Thus, the contribution of the mth eigenstate ($u_m(x)$) to the expansion of the initial wavefunction, $\psi(x,0)$, is given by their mutual overlap, defined by the integral in Eqn. (6.40).

Another immediate similarity with a *complete set* or *basis set* of unit vectors is thus apparent. If we write a general vector as $\mathbf{A} = \sum_i A_i\,\hat{e}_i$, we can extract the expansion coefficients via the inner product as

$$\mathbf{A} \cdot \hat{e}_j = \left(\sum_i A_i\,\hat{e}_i\right) \cdot \hat{e}_j = \sum_i A_i\,(\hat{e}_i \cdot \hat{e}_j) = \sum_i A_i\,\delta_{i,j} = A_j \tag{6.41}$$

With some trivial changes, this expansion also works for the symmetric well, where we write

$$\psi(x,0) = \sum_{n=1}^{\infty} \left(a_n^{(+)} u_n^{(+)}(x) + a_n^{(-)} u_n^{(-)}\right) \tag{6.42}$$

with

$$a_m^{(\pm)} = \int_{-a}^{+a} u_m^{(\pm)}(x)\,\psi(x,0)\,dx \tag{6.43}$$

This particular expansion can be seen to be formally identical to the *Fourier series* expansion discussed in Section 2.2.2. The only differences are that the constant term is not allowed due to the boundary conditions at the walls of the well, and that the series expansion is defined to vanish outside the well.

In general, the wavefunctions can be complex, so that for an arbitrary expansion

$$\psi(x,t) = \sum_n a_n \psi_n(x)\,e^{-iE_n t/\hbar} \tag{6.44}$$

the overlap integrals giving the expansion coefficients must be carefully written as

$$a_n = \int_{-\infty}^{+\infty} \psi_n^*(x)\,\psi(x,0)\,dx \tag{6.45}$$

Lastly, this inversion process is not limited to sums of energy eigenstates as the expansion in momentum eigenstates

$$\psi(x) = \frac{1}{\sqrt{2\pi\hbar}} \int_{-\infty}^{+\infty} \phi(p) \, e^{ipx/\hbar} dp \tag{6.46}$$

has the well-known inverse

$$\phi(p) = \frac{1}{\sqrt{2\pi\hbar}} \int_{-\infty}^{+\infty} \psi(x) \, e^{-ipx/\hbar} dx \tag{6.47}$$

Taken together, and generalized to include general Hermitian operators, these results are examples of the so-called *expansion postulate*, which states that

• The eigenfunctions of a Hermitian operator form a *complete set*

since any admissible wavefunction can be expanded in such eigenfunctions.

Example 6.2. Why bad wavefunctions are bad

Let us now turn to some examples. Consider first the initial waveform, defined inside the symmetric infinite well, $(-a, +a)$, by

$$\psi(x) = \begin{cases} 1/\sqrt{a} & \text{for } |x| < a/2 \\ 0 & \text{for } |x| > a/2 \end{cases} \tag{6.48}$$

While normalized appropriately, this is not an acceptable wavefunction due to its discontinuities, and we will use this opportunity to examine the physical consequences of such behavior. The expansion coefficients are easily calculated, giving

$$a_n^{(-)} = \int_{-a}^{+a} u_n^{(-)}(x) \, \psi(x, 0) \, dx = 0 \tag{6.49}$$

because of symmetry considerations (even function times odd function integrated over symmetric interval), while

$$a_n^{(+)} = \int_{-a}^{+a} u_n^{(+)}(x) \, \psi(x, 0) \, dx$$

$$= \frac{2}{a} \int_0^{+a/2} \cos\left(\frac{(n-1)/2\pi x}{a}\right) dx$$

$$= \frac{4 \sin[(2n-1)\pi/4]}{(2n-1)\pi} \tag{6.50}$$

which is dimensionless as it should be. The convergence is similar to that in Example 2.1 and is shown in Fig. 6.2, We can now use these expansion coefficients to address a more physical

(Continued)

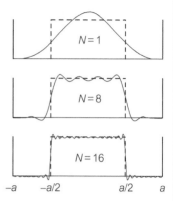

Figure 6.2. Generalized Fourier series expansion for a square wave in the symmetric infinite well for Example 6.2.

question, namely, to evaluate the energy of this state. We find

$$\langle \psi | \hat{E} | \psi \rangle = \sum_{n=1}^{\infty} \left\{ |a_n^+|^2 E_n^+ + |a_n^-|^2 E_n^- \right\}$$

$$= \sum_{n=1}^{\infty} \left(\frac{8}{(2n-1)^2 \pi^2} \right) \left(\frac{\hbar^2 \pi^2 (2n-1)^2}{8ma^2} \right)$$

$$= \frac{\hbar^2}{ma^2} \sum_{n=1}^{\infty} 1 = \infty \tag{6.51}$$

which is divergent. The kinetic energy corresponding to this state, measuring the "wiggliness" of ψ, is infinitely large due to the discontinuity. This can also be seen directly from the position- and momentum-space wavefunctions from a calculation of $\langle \hat{T} \rangle = \langle \hat{p}^2 \rangle / 2m$ (P6.8).

Example 6.3. The expanding box

As a further example, let us consider the case of a particle in the symmetric well, known somehow to be in its ground state. *Very suddenly*, at $t = 0$, the walls are pulled apart symmetrically to a new width $2b$ (where $b > a$). (This is in contrast to P5.21 where slow or adiabatic changes were discussed.)

The initial wavefunction of the particle in this new well (see Fig. 6.3 for the situation at $t = 0$), defined via

$$\psi(x,0) = \begin{cases} u_1^{(+)}(x;a) = \cos(\pi x/2a)/\sqrt{a} & \text{for } |x| < a \\ 0 & \text{for } a < |x| < b \end{cases} \tag{6.52}$$

(Continued)

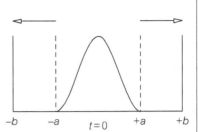

Figure 6.3. Initial state wavefunction of the "expanded well" state of Example 6.3 where the walls are suddenly moved from $\pm a$ to $\pm b$.

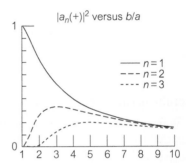

Figure 6.4. Expansion coefficients squared ($|a_n|^2$) for first three even levels versus b/a from Example 6.3.

can now be expanded in terms of the energy eigenstates of its new "universe" (the new well), that is, in terms of the $u_n^{(\pm)}(x; b)$. The result is that

$$a_n^{(-)} = 0 \quad \text{and} \quad a_n^{(+)} = \frac{4ab^2}{\pi\sqrt{ab}}\left(\frac{\cos((2n-1)\pi a/2b)}{(b^2-(2n-1)^2 a^2)}\right) \qquad (6.53)$$

Once again, the odd state expansion coefficients vanish because of symmetry considerations. We plot the probabilities of finding the particle in the ground state, and first and second excited *even* states of the new well versus b/a in Fig. 6.4; as $b/a \to 1$, only the original ground state is required.

Having once calculated the expansion coefficients for the given initial waveform, the future time-dependence of each term is dictated solely by the $\exp(-iE_n^{(\pm)}t/\hbar)$ factors. The wavefunction at later times is then given by

$$\psi(x, t) = \sum_{n=1}^{\infty} a_n^{(+)} \frac{1}{\sqrt{b}} \cos\left(\frac{(n-1/2)\pi x}{b}\right) e^{-iE_n^{(+)}(b)t/\hbar} \qquad (6.54)$$

and we plot in Fig. 6.5 the resulting probability density (given by $|\psi(x, t)|^2$) for various future times. Because of the simplicity of the infinite well (specifically its energy eigenvalues), the behavior is periodic, and we plot various times during the first half-cycle only; it need not be exactly periodic in a general potential.

(Continued)

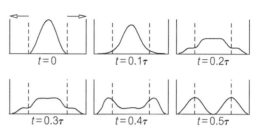

$t=0$ $t=0.1\tau$ $t=0.2\tau$

$t=0.3\tau$ $t=0.4\tau$ $t=0.5\tau$

Figure 6.5. Time-development of "expanded well" system illustrated by plots of $|\psi(x,t)|^2$ versus x for various times in the first half-cycle.

If this problem seems somewhat artificial (suddenly moving the walls of an infinite well in a microscopic system is admittedly somewhat difficult to imagine realizing experimentally), it is representative of a class of interesting real-life problems in which some parameter of the potential undergoes a sudden change; we will see a more realistic example in P17.8. We next show how a Gaussian wave packet can be constructed (approximately) in the standard infinite well using the expansion postulate.

Example 6.4. Gaussian wave packets in the standard infinite well

While the general Gaussian wave packet discussed in Chapter 3 (especially as defined in Eqn. (3.35)) is a very useful example, it is not strictly an allowable quantum state in the standard infinite well, since it does not satisfy the boundary conditions that $\psi(0,t) = \psi(a,t) = 0$. However, for a sufficiently narrow initial wave packet, far enough from either wall, the error made in neglecting the "tails" of the Gaussian wave packet outside the well can be made arbitrarily (exponentially) small. In order to extract the expansion coefficients of a Gaussian initial state (with $b \equiv \hbar\alpha$) given by

$$\psi_{(G)}(x,0) = \frac{1}{\sqrt{b\sqrt{\pi}}} e^{-(x-x_0)^2/2b^2} e^{ip_0(x-x_0)/\hbar} \tag{6.55}$$

placed inside the standard infinite well, we require the overlap integrals

$$a_n = \int_0^a u_n(x)\,\psi_G(x,0)\,dx \tag{6.56}$$

where we again use the fact that the $u_n(x)$ are real. Since the integral is assumed to be exponentially small outside the $(0,a)$ interval, we can extend the region of integration to $(-\infty, +\infty)$ with negligible error. This is important since the resulting Gaussian integrals can

(Continued)
be done in closed form, giving

$$a_n \approx \frac{1}{\sqrt{b\sqrt{\pi}}} \int_{-\infty}^{+\infty} \left[\sqrt{\frac{2}{a}} \sin\left(\frac{n\pi x}{a}\right) \right] e^{-(x-x_0)^2/2b^2} e^{ip_0(x-x_0)/\hbar} \, dx$$

$$= \left(\frac{1}{2i}\right) \sqrt{\frac{4b\pi}{a\sqrt{\pi}}} \left[e^{in\pi x_0/a} e^{-b^2(p_0+n\pi\hbar/a)^2/2\hbar^2} - e^{-in\pi x_0/a} e^{-b^2(p_0-n\pi\hbar/a)^2/2\hbar^2} \right]$$

$$(6.57)$$

(Note that in Section 5.4.2 we used an approximate, unnormalized version of this more precise result; the plots in Fig. 5.9 were obtained using the values in Eqn. (6.57).) The general time-dependent wavefunctions, in position- and momentum-space can then be written as

$$\psi_{WP}(x,t) = \sum_{n=1}^{\infty} a_n u_n(x) \, e^{-iE_n t/\hbar} \quad \text{and} \quad \phi_{WP}(p,t) = \sum_{n=1}^{\infty} a_n \phi_n(p) \, e^{-iE_n t/\hbar}$$

$$(6.58)$$

and while the sums are formally infinite, the contributions are peaked about values of n for which $p_n = n\hbar\pi/a \approx p_0$. We can also make use of this connection to estimate the number of states necessary to approximate the state by noting that

$$\frac{\hbar}{2\Delta x_0} = \frac{\hbar}{\sqrt{2b}} = \Delta p = \frac{\pi\hbar}{a} \Delta n \quad \text{or} \quad \Delta n = \frac{a}{2\pi \Delta x_0} \qquad (6.59)$$

so that the narrower the initial wave packet, the larger the number of states required to reproduce it faithfully.

6.6 Parity

We have concentrated our attention so far on Hermitian operators which have familiar classical analogs; the momentum, the energy, and the Hamiltonian operators all have recognizable classical counterparts. In this section, we study a somewhat more abstract operator which arises because of a symmetry.

We have seen that the energy eigenfunctions for the symmetric well can also be characterized in terms of their evenness or oddness, that is their symmetry properties under reflections; these properties are obviously connected to the symmetry of the potential itself. To formalize this notion, and to extend our experience with the properties of Hermitian operators, we are led to study the *parity operator*, defined via

$$\hat{P} f(x) \equiv f(-x) \qquad (6.60)$$

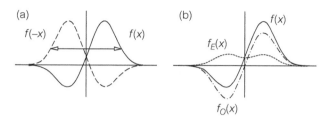

Figure 6.6. (a) Parity operation applied to a generic real function $f(x)$ and (b) its expansion in even ($f_E(x)$) and odd ($f_O(x)$) functions.

which has the effect of taking the mirror reflection of a function about the origin as illustrated in Fig. 6.6(a).

This operator is easily seen to be Hermitian since

$$
\left(\int_{-\infty}^{+\infty} dx\, \psi^*(x)\, \hat{P}\, \psi(x) \right) - \left(\int_{-\infty}^{+\infty} dx\, \psi^*(x)\, \hat{P}\, \psi(x) \right)^*
$$

$$
= \int_{-\infty}^{+\infty} dx\, \psi^*(x)\, \psi(-x) - \int_{-\infty}^{+\infty} dx\, \psi^*(-x)\, \psi(x)
$$

$$
= \int_{-\infty}^{+\infty} dx\, \psi^*(x)\, \psi(-x) - \int_{-\infty}^{+\infty} dy\, \psi^*(y)\, \psi(-y)
$$

$$
= 0 \tag{6.61}
$$

where we have simply changed variables $(x \to -y)$ in the second integral. This implies that the eigenvalues, λ_P, of the parity operator will be real (from Section 6.2). We can see that the definition

$$
\hat{P} f(x) = \lambda_P f(x) \tag{6.62}
$$

can be used twice to obtain

$$
f(x) = \hat{P} f(-x) = \hat{P}^2 f(x) = \hat{P} \lambda_P f(x) = \lambda_P \hat{P} f(x) = \lambda_P^2 f(x) \tag{6.63}
$$

implying that $\lambda_P = \pm 1$. Thus the eigenfunctions of the parity operator are simply even and odd functions with

$$
f_E(-x) = \hat{P} f_E(x) = +f_E(x) \quad \text{and} \quad f_O(-x) = \hat{P} f_O(x) = -f_O(x) \tag{6.64}
$$

The orthogonality of eigenfunctions belonging to different eigenvalues is trivially evident as

$$
\langle f_E | f_O \rangle = \int_{-\infty}^{+\infty} dx\, f_E^*(x)\, f_O(x) = 0 \tag{6.65}
$$

for any even and odd functions. The expansion postulate can also be confirmed in this case as we can always write a generic function in terms of even and odd solutions as

$$f(x) = \frac{f(x) + f(-x)}{2} + \frac{f(x) - f(-x)}{2}$$

$$= f_E(x) + f_O(x)$$

$$\equiv c_E \tilde{f}_E(x) + c_O \tilde{f}_O(x) \tag{6.66}$$

where $f_{E,O}(x)$ are obviously even and odd, respectively, but are not necessarily normalized properly. This expansion is illustrated in Fig. 6.6(b). If we write $f_{E(O)}(x) = c_{E(O)} \tilde{f}_{E(O)}(x)$ where $\tilde{f}_{E(O)}$ is normalized, then the expansion coefficients, namely, the $c_{E(O)}$, when squared, give the probability of finding the state with positive or negative parity (See P6.11).

6.7 Simultaneous Eigenfunctions

We have seen that the solutions for the symmetric infinite well problem are simultaneously eigenfunctions of both the Hamiltonian operator for that of system and of the parity operator; that is, they have precisely determined values of both the energy and the parity. On the other hand, the content of the standard uncertainty principle, $\Delta x \, \Delta p \geq \hbar/2$, is to say that it is not possible to know simultaneously the values of both the position and momentum to arbitrary precision.

We are naturally led to ask under what conditions two different Hermitian operators can share eigenfunctions. The general result can be stated as:

- Two Hermitian operators, $\hat{A}$ and $\hat{B}$, can have simultaneous eigenfunctions if and only if they commute with each other, that is,

$$[\hat{A}, \hat{B}] \equiv \hat{A}\hat{B} - \hat{B}\hat{A} = 0 \tag{6.67}$$

or their commutator vanishes.

Before proceeding with the proof, we note that the commutator satisfies

$$[\hat{A}, \hat{B}] = -[\hat{B}, \hat{A}] \tag{6.68}$$

as well as

$$[\alpha\hat{A} + \beta\hat{B}, \hat{C}] = \alpha[\hat{A}, \hat{C}] + \beta[\hat{B}, \hat{C}] \tag{6.69}$$

Let us first assume that $\hat{A}$ and $\hat{B}$ commute and that we have found the eigen-values and eigenfunctions of the operator $\hat{A}$, which we denote via $\hat{A}\psi_a = a\psi_a$. We assume for simplicity that the eigenvalue spectrum is not degenerate, that is that there is only one ψ_a for every eigenvalue a. (The case of degenerate eigenvalues is a straightforward extension.)

We then note that

$$\hat{A}\left(\hat{B}\psi_a\right) = \hat{B}\hat{A}\psi_a \quad \text{(since } \hat{A} \text{ and } \hat{B} \text{ commute)}$$

$$= \hat{B}\left(a\psi_a\right) \quad \text{(since } \psi_a \text{ is an eigenstate of} \hat{A})$$

$$= a\left(\hat{B}\psi_a\right). \tag{6.70}$$

Thus, $\hat{B}\psi_a$ is a function which, when acted upon by $\hat{A}$, returns the same function multiplied by the number a; this is just the definition of ψ_a so that $\hat{B}\psi_a$ must be, up to a multiplicative constant, the same as ψ_a, that is, $\hat{B}\psi_a \propto \psi_a$. But this, in turn, is just the definition of an eigenfunction since

$$\hat{B}\psi_a \propto \psi_a \implies \hat{B}\psi_a = b\psi_a \quad \text{or} \quad \hat{B}\psi_a^b = b\psi_a^b \tag{6.71}$$

so that ψ_a is also an eigenfunction of $\hat{B}$.

To complete the proof, we assume that we have found the simultaneous eigenfunctions of the operators $\hat{A}$ and $\hat{B}$ which satisfy

$$\hat{A}\psi_a^b = a\psi_a^b \quad \text{and} \quad \hat{B}\psi_a^b = b\psi_a^b \tag{6.72}$$

Clearly $\hat{A}$ and $\hat{B}$ commute on the common set of eigenfunctions since

$$[\hat{A}, \hat{B}]\psi_a^b = \left(\hat{A}\hat{B} - \hat{B}\hat{A}\right)\psi_a^b = \left(\hat{A}b - \hat{B}a\right)\psi_a^b = (ab - ba)\psi_a^b = 0 \tag{6.73}$$

Since we are dealing with Hermitian operators, we know that the ψ_a^b form a complete set so that an arbitrary wavefunction can be written as $\psi = \sum_a c_a \psi_a^b$; this implies that

$$\left(\hat{A}\hat{B} - \hat{B}\hat{A}\right)\psi = \sum_a c_a \left[\left(\hat{A}\hat{B} - \hat{B}\hat{A}\right)\psi_a^b\right] = \sum_a c_a [0] = 0 \tag{6.74}$$

and the commutator of $\hat{A}$ and $\hat{B}$ vanishes when acting on an arbitrary wavefunction.

Because $[x, \hat{p}] = i\hbar \neq 0$, we now understand why we can never have sim-ultaneous eigenfunctions of these two variables. For the case of the symmetric infinite well, the two operators are the Hamiltonian, $\hat{H} = \hat{p}^2/2m + V(x)$ and parity operator, $\hat{P}$. Their commutator, acting on a general state, can be written in two pieces,

$$[\hat{p}^2/2m, \hat{P}]\psi(x) = -\frac{\hbar^2}{2m}\left(\frac{d^2}{dx^2}\hat{P} + \hat{P}\frac{d^2}{dx^2}\right)\psi(x)$$

$$= -\frac{\hbar^2}{2m}\left(\frac{d^2}{dx^2}\psi(-x) + \frac{d^2\psi(-x)}{dx^2}\right)$$

$$= 0 \quad \text{(in general)} \tag{6.75}$$

and

$$[V(x), \hat{P}]\psi(x) = \left(V(x)\hat{P} - \hat{P}V(x)\right)\psi(x)$$

$$= (V(x) - V(-x))\psi(-x) \tag{6.76}$$

$$= 0 \quad \text{only if } V(x) = V(-x)$$

that is, if $V(x)$ is symmetric.

Because eigenfunctions provide such detailed information on the quantum state, one strategy is to search for

- A maximally large set of *mutually commuting observable operators*, that is, the largest possible set of operators, $\hat{A}_i$, which represent observables (i.e. Hermitian operators) and which all commute with each other, namely, $[\hat{A}_i, \hat{A}_j] = 0$. Such a set is guaranteed to have simultaneous eigenfunctions.

6.8 **Questions and Problems**

Q6.1. Give an example of an operator which is *not* Hermitian and show that its eigenvalues are not necessarily real.

P6.1. Consider the momentum-space wavefunctions, $\phi_n(p)$ for the standard infinite well, given by Eqn. (5.33). In P5.8 we showed that these were properly normalized.

(a) Show that $\phi_n(p)$ are actually orthonormal, namely that

$$\langle \phi_m | \phi_m \rangle = \int_{-\infty}^{+\infty} dp\, [\phi_m(p)]^* [\phi_m(p)] = \delta_{n,m} \tag{6.77}$$

Hints: (i) Use the integrals cited in P5.8 and (ii) use partial fractions to rewrite the products found in the denominators.

P6.2. **Orthogonality of asymmetric infinite well states.** Consider stationary state solutions of the asymmetric well discussed in Section 5.3 for the $E > V_0$ case. Show that two such solutions, $\psi_1(x), \psi_2(x)$, corresponding to *different* energy eigenvalues are mutually orthogonal, that is, their overlap vanishes

$$\langle \psi_1 | \psi_2 \rangle = \int_{-a}^{+a} dx \, \psi_1(x) \, \psi_2(x)$$

$$= \int_{-a}^{0} dx \, A_1 A_2 \sin(k_1(x+a)) \sin(k_2(x+a))$$

$$+ \int_{0}^{+a} dx \, B_1 B_2 \sin(q_1(x-a)) \sin(q_2(x-a))$$

$$\langle \psi_1 | \psi_2 \rangle = 0 \tag{6.78}$$

Hints: Make repeated use of the boundary conditions at $x = -a, a$ as well as the definitions of k, q. You may also find it useful to use the fact that $k_i^2 - k_2^2 = q_1^2 - q_2^2$ but you should prove it first. Do you need to have normalized the solutions first before you check whether they are orthogonal? Repeat for the $0 < E < V_0$ solutions.

P6.3. A particle in a symmetric infinite well is described by the wavefunction

$$\psi(x,0) = N[(3+2i)u_1^{(+)}(x) - 2u_1^{(-)}(x) + 3iu_2^{(+)}(x)] \tag{6.79}$$

where the $u_n^{(\pm)}(x)$ are the infinite well wavefunctions.

(a) Find N so that $\psi(x,0)$ is properly normalized.

(b) What is $\psi(x,t)$?

(c) Evaluate $\langle \hat{E} \rangle_t$ and ΔE.

(d) What is the probability that a measurement of the energy of the particle would find $E = \hbar^2\pi^2/8ma^2$? $E = 6\hbar^2\pi^2/8ma^2$? $E = 9\hbar^2\pi^2/8ma^2$?

P6.4. **Coordinate system choices in quantum mechanics.** Suppose that you have solved the Schrödinger equation for some arbitrary potential and found the generic eigenstates, for example, $u_n(x) \exp(-iE_n t/\hbar)$ so that the general solution is

$$\psi(x,t) = \sum_n a_n u_n(x) \exp(-iE_n t/\hbar) \tag{6.80}$$

(a) Assume that you suddenly decide to change what you call the *origin* of coordinates along the x-axis, i.e. letting $x \to x + d$, so that the new spatial solutions are $\tilde{\psi}(x,t) = \psi(x+d,t)$. Show that corresponding momentum space wavefunction is only changed by a phase factor, so that $|\phi(p,t)|^2$ is not changed at all. Discuss why this should be so.

(b) Suppose now that you decide to relabel the *zero* of your potential energy function via $V(x) \to V(x) + V_0$. Find $\tilde{\psi}(x,t)$ in terms of the original solution and show that observables depending on $P(x,t) = |\psi(x,t)|^2$ are unaffected by a different choice of zero of potential. How is $|\phi(p,t)|^2$ affected?

P6.5. Wavefunctions in the symmetric infinite well I. Consider an initial wavefunction in the symmetric infinite well defined by

$$\psi(x, 0) = \begin{cases} -N & \text{for } -a < x < 0 \\ +N & \text{for } 0 < x < +a \end{cases} \tag{6.81}$$

(a) Find N so that $\psi(x, 0)$ is properly normalized.

(b) Calculate the expansion coefficients, $a_n^{(\pm)}$, for this wavefunction.

(c) Use them to evaluate $\langle \hat{E} \rangle$. Is your answer consistent with what you know about this waveform?

P6.6. Wavefunctions in the symmetric infinite well II. An initial wavefunction in the symmetric infinite well is given by

$$\psi(x, 0) = N\left(1 - \frac{|x|}{a}\right) \tag{6.82}$$

(a) Answer the same questions as for P6.5. What is different about this waveform than that in P6.5 and Example 6.2.

(b) Try to evaluate ΔE in this state.

P6.7. Wavefunctions in the symmetric infinite well III. Consider two wavefunctions in the symmetric infinite well defined by

$$\psi_1(x) = \begin{cases} 1/\sqrt{a} & \text{for } -a < x < 0 \\ 0 & \text{for } 0 < x < +a \end{cases} \tag{6.83}$$

and

$$\psi_2(x) = \begin{cases} 0 & \text{for } -a < x < 0 \\ 1/\sqrt{a} & \text{for } 0 < x < +a \end{cases} \tag{6.84}$$

that is, waveforms which are completely localized within the left and right halves of the well.

(a) Show that the two states are normalized properly and that they are orthogonal, i.e.

$$\langle \psi_1 | \psi_2 \rangle = \int_{-a}^{+a} \psi_1^*(x)\, \psi_2(x)\, dx = 0 \tag{6.85}$$

(b) Calculate the corresponding momentum-space wavefunctions, $\phi_{1,2}(p)$, and show *explicitly* that

$$\int_{-\infty}^{+\infty} dp\, \phi_1^*(p)\, \phi_2(p) = 0 \tag{6.86}$$

and that they are normalized as well.

(c) Expand both wavefunctions in energy eigenfunctions and show that their overlap vanishes, that is,

$$\sum_{n=1}^{\infty} a_n^* b_n = 0 \tag{6.87}$$

Hint: Be sure to include both $a_n^{(\pm)}$ in the expansion and overlap summation.

P6.8. **Consider again the (unacceptable) wavefunction defined in Example 6.2.**

(a) Write the position-space wavefunction, $\psi(x)$, in terms of step-functions, $\theta(x)$ (as in Section 2.4). Evaluate the average kinetic energy (which is the total energy since the potential vanishes inside the well) using the form

$$\langle \hat{T} \rangle = \frac{\langle \hat{p}^2 \rangle}{2m} = \frac{\hbar^2}{2m} \int dx \left| \frac{d\psi(x)}{dx} \right|^2. \tag{6.88}$$

You may have to use some "seat-of-the-pants" mathematics to handle the Dirac δ-functions you encounter.

(b) Evaluate the momentum-space wavefunction, $\phi(p)$, and use

$$\langle T \rangle = \frac{\langle p^2 \rangle}{2m} = \frac{1}{2m} \int_{-\infty}^{+\infty} dp \, p^2 \, |\phi(p)|^2 \tag{6.89}$$

to find the kinetic energy.

P6.9. (a) In Example 6.3, show that the expansion coefficients are appropriately normalized, that is, that

$$\sum_{n=1}^{\infty} |a_n^{(+)}|^2 = 1 \tag{6.90}$$

in the case where $b = 2a$.

(b) The energy of the initial state is definitely known to be $E_1^{(+)} = \hbar^2 \pi^2 / 8ma^2$. Evaluate $\langle \hat{E} \rangle$ after the walls have expanded for a general value of b. Show explicitly for the case $b = 2a$ that your answer agrees with the initial value. Hint: The summations in Appendix D.2 may be useful.

P6.10. **The dissolving infinite well.** A particle is in an energy eigenstate, $u_n(x)$, of the standard infinite well. At $t = 0$, the walls of the well at $x = 0, a$ are suddenly removed, so that the wavefunction evolves in time as a free particle.

(a) What are the $t = 0$ position and momentum-space wavefunctions for this state, $\psi(x, 0)$ and $\phi(p, 0)$?

(b) What is the time-dependent momentum-space wavefunction, $\phi(p, t)$? What is $\psi(x, t)$?

(c) Using the results of P4.31, write down explicit formulae for the time-dependent $\langle x \rangle_t$ and $\langle x^2 \rangle_t$ for the state after the walls are dissolved. Hint: You can evaluate any $t = 0$ expectation values using either position- or momentum-space wavefunctions, whichever is easier.

(d) If you were to define a spreading time, t_0, for $\psi(x, t)$ using the free-particle result in Sections 3.2.2 or 4.3.1 as a model, how would it depend on the initial state, $u_n(x)$ used.?

P6.11. Expansion in parity eigenstates. Consider a real-valued function expanded in terms of even and odd functions as in Eqn. (6.66).

(a) Show that the probability that the wavefunction is in an even state is given by

$$c_E^2 = \frac{1}{2}\left(1 + \frac{\int_{-\infty}^{+\infty} dx\, f(x) f(-x)}{\int_{-\infty}^{+\infty} dx\, (f(x))^2}\right) \tag{6.91}$$

and an odd state by

$$c_O^2 = \frac{1}{2}\left(1 - \frac{\int_{-\infty}^{+\infty} dx\, f(x) f(-x)}{\int_{-\infty}^{+\infty} dx\, (f(x))^2}\right) \tag{6.92}$$

(b) Show that $c_E^2 = 1$ and $c_O^2 = 0$ for an even function.

(c) Consider the (unnormalized) function $f(x) = N \exp(-ax^2/2 - bx/2)$. First normalize $f(x)$ and then evaluate the c_E^2 and c_O^2 as a function of a, b? Are your results consistent with your expectations based on what this function looks like?

(d) Repeat part (c) for $f(x) = Nx \exp(-ax^2/2 - bx/2)$.

P6.12. Generalized parity operator. If one wants to "reflect" a function in a point other than the origin, show that the generalized parity operator, defined via

$$\hat{P}_a f(x) = f(2a - x) \tag{6.93}$$

reflects the function about the point $x = a$.

(a) Show that $\hat{P}_a$ is Hermitian, find its eigenvalues and eigenfunctions and interpret them.

(b) The standard infinite well potential is symmetric about the point $x = a/2$. Show that the energy eigenfunctions for this case are eigenfunctions of $\hat{P}_{(a/2)}$ and find their eigenvalues.

P6.13. Consider the initial wavefunction in P6.3.

(a) What is the average value of the parity operator, $\hat{P}$, for this state? Does it depend on time?

(b) What is the probability of measuring the particle described by this wavefunction to have positive parity? negative parity?

P6.14. The complex conjugation operator. Define the complex conjugation operator via

$$\hat{C}f(x) = f^*(x) \tag{6.94}$$

(a) Show that this operator is *not* Hermitian by finding a wavefunction for which $\langle \hat{C} \rangle = \langle \psi | \hat{C} | \psi \rangle$ is *not* real.

(b) Try to find the eigenvalues of this operator by modifying the derivation in Section 6.7.

(c) Show that *any* complex number $z = a + ib$ is an eigenfunction of $\hat{C}$ and find the corresponding eigenvalue.

P6.15. Which of the following pairs of operators can have simultaneous eigenfunctions?

(a) $\hat{p}$ and $\hat{T} = \hat{p}^2/2m$

(b) $\hat{p}$ and $V(x)$

(c) $\hat{E}$ and $\hat{p}$

P6.16. Consider the *translation operator*, $\hat{T}_a$ defined via

$$\hat{T}_a f(x) = f(x+a) \tag{6.95}$$

(a) Is $\hat{T}_a$ a Hermitian operator? Can you find an example where its expectation value is not real?

(b) What is $[\hat{T}_a, \hat{p}^2/2m]$?

(c) Under what conditions is $[\hat{T}_a, V(x)] = 0$?

P6.17. Gaussian wave packet at rest in the infinite well. Consider the expansion coefficients for the Gaussian wave packet in the standard infinite well in Eqn. (6.57) in the special limit of $p_0 = 0$. This is as close as one can come to trying to have a wave packet "at rest."

(a) For the case of $x_0 = a/2$ (center of the well) discuss which of the a_n are nonvanishing and explain why. For a general wave packet in the standard infinite well, we know from Section 5.4.2 that there will be wave packet revivals at a time $T_{rev} = 4ma^2/\hbar\pi$. Is the revival time different for the $p_0 = 0$ and $x_0 = a/2$ case?

(b) Repeat part (a) for the case of $x_0 = a/3, 2a/3$.

P6.18. Feynman–Hellman theorem. Suppose that the Hamiltonian of a system depends on a parameter λ in some well-defined manner, $\hat{H} = \hat{H}(\lambda)$, with energy eigenvalues satisfying $\hat{H}(\lambda)\psi = E(\lambda)\psi$. Remember that the eigenstates, $\psi(\lambda)$, will also depend on λ.

(a) Show that the dependence of the energy eigenvalue of a specific state on λ is then simply given by

$$\frac{\partial E(\lambda)}{\partial \lambda} = \left\langle \frac{\partial \hat{H}(\lambda)}{\partial \lambda} \right\rangle \qquad (6.96)$$

Hint: Use the fact that probability is conserved.

(b) Check this result explicitly for the infinite well energy eigenstates, using the mass m as the parameter.

SEVEN

Many Particles in the Infinite Well: The Role of Spin and Indistinguishability

7.1 The Exclusion Principle

The self-consistent inclusion of the wave properties of matter into the dynamical equations of motion, via the Schrödinger equation, is an important aspect of any attempt to understand the structure of matter. This aspect of microscopic physics, which we might dub "$\hbar$-physics," cannot be underestimated, as it accounts for much of the observable phenomena we attribute to quantum mechanics. The discrete sets of spectral lines in bound state systems (atomic, molecular, nuclear, quark/antiquark), for example, arise from quantization of energy levels. Even macroscopic quantities, such as the densities of ordinary matter, can be understood using simple quantum arguments as in Section 1.4.

Many physical systems, however, require us to understand the organization of collections of large numbers of seemingly indistinguishable particles. The "building blocks of matter," the electrons of atomic physics, the protons and neutrons (nucleons) which form nuclei, and the quarks which, in turn, are bound together to make the nucleons and other strongly interacting particles are all in this category. These particles all share a common feature, namely their *intrinsic angular momentum* or *spin*. The notion of "spin-up" and "spin-down" corresponding to such spin 1/2 particles will be an important additional degree of freedom which must be considered when constructing the multiparticle wavefunctions of such particles. The behavior of such particles is restricted by powerful constraints which do not arise from the machinery of wave mechanics but which, nonetheless, have a profound impact on the macroscopic properties of such systems. We will discuss this 'indistinguishability-physics' in much more detail in Chapters 14 and 17, but we wish to emphasize its impact as soon as possible.

The most familiar manifestation of such effects comes from the *Pauli exclusion principle*, which states roughly that

- No two electrons (or any two indistinguishable spin-1/2) particles can occupy the same quantum state.[1]

The fact that each atomic energy level can then accommodate only two electrons (one "spin-up" and one "spin-down") leads to the shell structure of atomic physics and is therefore arguably responsible for much of chemistry and biology; similar shell structure occurs in nuclei as well. In this chapter, using the infinite well potential as a model, we will examine the dramatic effects this restriction can have on macroscopic numbers of particles, applying it to condensed matter, to nuclear, and to astrophysical systems.

7.2 **One-Dimensional Systems**

We begin by considering N_e electrons in the standard infinite well (in one dimension, here with width L) with energy spectrum

$$E_n = \frac{\hbar^2\pi^2 n^2}{2mL^2}, \quad n = 1, 2, 3, \ldots \tag{7.1}$$

For particles not required to satisfy the exclusion principle, the total energy of such a system would simply be

$$E_{\text{tot}} = \left(\frac{\hbar^2\pi^2}{2mL^2}\right) N_e \quad \text{(without exclusion principle)} \tag{7.2}$$

since we would just sum up the ground state energy for each one. For electrons, however, to be consistent with the exclusion principle, one has to "fill up" the energy levels, two at a time, to a state characterized by $N_{\text{max}} = N_e/2$, as in Fig. 7.1. The total energy is then

$$E_{\text{tot}} = 2\sum_{n=1}^{N_{\text{max}}} E_n = \frac{\hbar^2\pi^2}{mL^2}\sum_{n=1}^{N_{\text{max}}} n^2 \tag{7.3}$$

---

[1] Like any other physical law, the Pauli principle should be amenable to experimental verification, and it is natural to ask "How well do we know that the exclusion principle is satisfied?." It turns out to be extremely difficult to construct logically self-consistent theories of quantum mechanics in which the Pauli principle is only violated by a small amount. Nonetheless, various experiments (e.g. see, Ramberg and Snow 1990) have been taken to imply that the probability that a multielectron system will be in a configuration which violates the exclusion principle is less than roughly 10^{-26}.

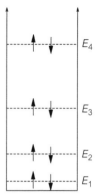

Figure 7.1. Filling of one-dimensional infinite well energy levels with spin-1/2 particles.

The summation can be done in closed form to yield

$$\sum_{n=1}^{N_{max}} n^2 = \frac{N_{max}(N_{max}+1)(2N_{max}+1)}{6} \approx \frac{N_{max}^3}{3} \qquad (7.4)$$

when $N_{max} \gg 1$. In this case, the labeling of states and energy summation is trivial, but in a more realistic three-dimensional example, the enumeration of states will be more complicated, so we also do the "counting" in a slightly more formal manner. We can write

$$N_e = 2\sum_{n=1}^{N_{max}} 1 = 2\sum_{n=1}^{N_{max}} \Delta n \approx 2\int_1^{N_{max}} dn \approx 2N_{max} \qquad (7.5)$$

and

$$\sum_{n=1}^{N_{max}} n^2 = \sum_{n=1}^{N_{max}} n^2 \Delta n \approx \int_1^{N_{max}} n^2 \, dn \approx \frac{N_{max}^3}{3} \qquad (7.6)$$

We have identified $\Delta n = 1$ with dn and used a simple version of the more general relation between a discrete sum and continuous integral, namely, the Euler–Maclaurin formula, which is discussed in Appendix D.2.

In either case we find

$$E_{tot} = \frac{\hbar^2 \pi^2}{mL^2}\left(\frac{N_{max}^3}{3}\right) = \frac{\hbar^2 \pi^2}{24mL^2}N_e^3 \quad \text{(with exclusion principle)} \qquad (7.7)$$

The average energy per particle, $\overline{E}$, is simply

$$\overline{E} = \frac{E_{tot}}{N_e} = \frac{\hbar^2 \pi^2}{24mL^2}N_e^2 \qquad (7.8)$$

while the energy of the last state to be filled, the so-called *Fermi energy*, is

$$E_{\text{Fermi}} = E_F = \frac{\hbar^2 \pi^2 N_{\text{max}}^2}{2mL^2} = \frac{\hbar^2 \pi^2}{8mL^2} N_e^2 \tag{7.9}$$

Taken together, the electrons in this configuration consistent with the exclusion principle are sometimes colloquially said to constitute the *Fermi sea*, so that E_F is the energy of an electron at the "top of the sea." From Eqns (7.8) and (7.9) we also note that the average energy is given by

$$\overline{E} = \frac{1}{3} E_F \tag{7.10}$$

The role of the exclusion principle can be seen by noting that the ratio of total energies with and without this constraint is roughly

$$\frac{E_{\text{tot}}(\text{with exclusion})}{E_{\text{tot}}(\text{without exclusion})} \approx \frac{N_e^2}{12} \tag{7.11}$$

which can be an enormous difference if $N_e \gg 1$.

We are accustomed to cases in which if we get all the dimensional factors right, then the estimate of the physical observable is usually wrong by less than an order-of-magnitude or so, in either direction. In this case, the "$\hbar$-physics" has predicted the dimensional factors correctly, but the exclusion principle can still play just as important a role in determining the actual state of the physical system.

7.3 Three-Dimensional Infinite Well

The one-dimensional example makes it clear that the exclusion principle can play a very important role. In order to make a plausible connection to a real system of spin-1/2 particles, however, we require a three-dimensional model, so we consider particles in a three-dimensional cubical box of volume $V = L^3$, that is, a potential given by

$$V(x) = \begin{cases} 0 & \text{for } |x|, |y|, |z| < L \\ \infty & \text{otherwise} \end{cases} \tag{7.12}$$

Either by "fitting de Broglie waves into the box" or via explicit solution of the three-dimensional Schrödinger equation, we can easily find the quantized energies, using

$$E = \frac{\mathbf{p}^2}{2m} = \frac{p_x^2 + p_y^2 + p_z^2}{2m} \tag{7.13}$$

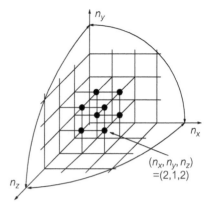

Figure 7.2. Filling of three-dimensional infinite well energy levels with spin-1/2 particles.

namely,

$$
E(\mathbf{n}) = E(n_x, n_y, n_z)
$$

$$
= \frac{\hbar^2 \pi^2}{2mL^2} \left(n_x^2 + n_y^2 + n_z^2 \right)
$$

$$
= \frac{\hbar^2 \pi^2}{2mL^2} \mathbf{n}^2 \quad \text{where } n_x, n_y, n_z = 1, 2, 3 \dots \tag{7.14}
$$

The quantity $\mathbf{n} \equiv (n_x, n_y, n_z)$ can be considered as a vector in the (abstract) three-dimensional number-space pictured in Fig. 7.2. Two electrons can be accommodated at each discrete point of the first quadrant (since $n_x, n_y, n_z \geq 1$) of this $\mathbf{n}$-space. The expression for the energy levels depends on $\mathbf{n}$ in a "spherically symmetric" way as it involves only $\mathbf{n}^2 = n_x^2 + n_y^2 + n_z^2$, so we must fill up the energy levels "radially" outward to a "radius" in $\mathbf{n}$-space which we call R_N. The enumeration of states is then best done using the integral approximations discussed above, and we find first that

$$
N_e = 2 \sum_{n_x, n_y, n_z = 1}^{|\mathbf{n}| = R_N} 1 = 2 \sum_{n_x, n_y, n_z = 1}^{|\mathbf{n}| = R_N} \Delta n_x \Delta n_y \Delta n_z
$$

$$
\approx 2 \int dn_x \, dn_y \, dn_z
$$

$$
= 2 \int_1^{R_N} n^2 \, dn \int d\Omega_n
$$

$$
= 2 \left[\frac{R_N^3}{3} \right] \left(\frac{4\pi}{8} \right)
$$

$$
\approx \frac{\pi}{3} R_N^3 \tag{7.15}
$$

where $d\Omega_n$ is the generalized solid angle in **n**-space, and we integrate only over the first octant (hence the $4\pi/8$ factor). This implies that

$$R_N = \left(\frac{3}{\pi}N_e\right)^{1/3} \tag{7.16}$$

The total energy of the system is then

$$E_{\text{tot}} = 2 \sum_{n_x,n_y,n_z=1}^{|n|=R_N} E(\mathbf{n})$$

$$= 2 \sum_{n_x,n_y,n_z=1}^{|n|=R_N} \frac{\hbar^2\pi^2}{2mL^2}(n_x^2 + n_y^2 + n_z^2)\Delta n_x \Delta n_y \Delta n_z$$

$$\approx \frac{\hbar^2\pi^2}{mL^2}\int |\mathbf{n}|^2 d^3n$$

$$= \frac{\hbar^2\pi^2}{mL^2}\int_1^{R_N} n^4 dn \left(\frac{4\pi}{8}\right)$$

$$\approx \frac{\hbar^2\pi^3}{10mL^2}R_N^5$$

$$E_{\text{tot}} = \frac{\hbar^2\pi^3}{10mL^2}\left(\frac{3}{\pi}\right)^{5/3}N_e^{5/3} \tag{7.17}$$

or

$$E_{\text{tot}} = \frac{\hbar^2\pi^3}{10mV^{2/3}}\left(\frac{3}{\pi}\right)^{5/3}N_e^{5/3} \tag{7.18}$$

where we have used $L^3 = V$. The average energy and Fermi energy are obtained as before and one finds

$$\overline{E} = \frac{\hbar^2\pi^3}{10mL^2}\left(\frac{3}{\pi}\right)^{5/3}N_e^{2/3} = \frac{3}{5}E_F \tag{7.19}$$

This form is useful as it implies that

$$\overline{E} = \frac{\pi^3}{10}\left(\frac{3}{\pi}\right)^{5/3}\frac{\hbar^2}{m_e}n_e^{2/3} \tag{7.20}$$

where $n_e \equiv N_e/L^3 = N_e/V$ is the number density of electrons in the system.

Once again, because of the exclusion principle, the total energy for a system of electrons is larger than for distinguishable particles, in this case by a factor of roughly $N_e^{2/3}$. For the electrons in a white dwarf star, for example, where

$N_e \approx 10^{57}$, forgetting about the constraints imposed by the exclusion principle would mean an error of roughly a factor of $(10^{57})^{2/3} \approx 10^{38}$ in an estimation of the total zero-point energy in the star due to electrons. There is a sense in which the universe covers roughly 60 orders-of-magnitude in length (from the smallest length scales to the largest). It is hard to imagine any other physical system in which leaving out a single physical effect, in this case the exclusion principle, could give rise to an error which is "2/3 the size of the universe."

7.4 Applications

In this section we will examine the role that the exclusion principle plays in some important physical systems.

7.4.1 Conduction Electrons in a Metal

Metals, with their observed large electrical and thermal conductivity, are characterized by the presence of quasi-free electrons which can move in response to electrical or thermal gradients. These *conduction electrons* can be modeled as a noninteracting gas in a three-dimensional box just as in Section 7.3.

Typically, one or two electrons per atom are available so that number densities of the order $n_e \sim 3 \times 10^{22}$ cm^{-3} are appropriate. This implies (P7.6) a Fermi energy of $E_F \sim 4$ eV; this in turn gives an average electron kinetic energy due to zero-point motion of roughly $\overline{E} \sim 2$ eV instead of the thermal energy of $k_B T \sim 1/40$ eV expected at room temperature on the basis of classical statistical mechanics.

This difference can be observed clearly in the behavior of the electronic contribution to the heat capacity of the metal. Classical kinetic theory would say that the thermal energy of N_e electrons would be $E \sim N_e k_B T$, so that the heat capacity due to electrons would be

$$C_{el} \sim \frac{\partial E}{\partial T} \sim N_e k_B \tag{7.21}$$

independent of temperature. This is inconsistent with experimental results which find an electronic heat capacity that varies linearly with T and which is smaller than this value by a factor of 0.01 or less.

In the quantum electron gas picture, only electrons in the Fermi sea in quantized states within roughly $k_B T$ of the Fermi energy can be thermally excited to new states (see Fig. 7.3) and so participate in the interactions required for

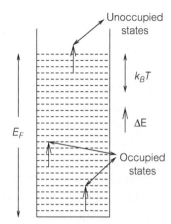

Figure 7.3. Filled energy levels in an electron gas picture of conduction electrons; electrons within a "distance" k_BT of the Fermi surface can be excited (with an energy increase ΔE) to unoccupied states and fully participate in thermal or electrical conduction.

conduction. Thus, only a fraction k_BT/E_F of the electrons are available, and the heat capacity is more properly

$$C_{el} \sim \frac{\partial}{\partial T}\left(\frac{k_BT}{E_f}N_e k_b T\right) \sim N_e k_B \frac{k_BT}{E_f} \tag{7.22}$$

The linear temperature dependence and resulting magnitude are then consistent with experiment.

Further experimental evidence for the electron gas picture comes from another macroscopic property of metals, namely, their compressibilities. The change in volume (ΔV) of a solid with applied pressure (ΔP) is often characterized by its *bulk modulus*, defined via

$$\Delta P = -B\frac{\Delta V}{V} \quad \text{or equivalently} \quad B = -V\frac{\partial P}{\partial V} \tag{7.23}$$

The appropriate pressure for an electron gas is

$$P(V) = -\frac{\partial E_{tot}}{\partial V} \tag{7.24}$$

so that from Eqn. (7.18), we find that (P7.6)

$$P_e = \frac{\pi^3}{15}\left(\frac{3}{\pi}\right)^{5/3}\frac{\hbar^2}{m_e}n_e^{5/3} \tag{7.25}$$

which implies that

$$B_e = \frac{5}{3}P_e \tag{7.26}$$

Using a value appropriate for sodium, $n_e = 2.6 \times 10^{22}$ cm^{-3}, we find that $B_e \approx 9 \times 10^9$ N/m^2 which is of the same order as the observed value of 6.4×10^{10} N/m^2; this suggests the important role that the exclusion principle plays in the compressibility of ordinary matter.

7.4.2 **Neutrons and Protons in Atomic Nuclei**

The nuclei of atoms are bound state systems of spin-1/2 protons and neutrons. The fact that protons are electrically charged, while neutrons are not, clearly makes them distinguishable from each other; the individual protons, however, are indistinguishable from each other, as are the neutrons, so the exclusion principle can play an important role in the structure of atomic nuclei.

One of the important problems of nuclear structure is to understand which combinations of Z protons and N neutrons will be stable for a given value of the atomic number, $A = Z + N$. Given the Coulomb repulsion between electrons, it would seem that a nucleus composed solely of neutrons, that is, $N = A$, $Z = 0$, would have the least energy; this is in marked contrast to the observed pattern where most light nuclei have $N \approx Z \approx A/2$.

If we model a nucleus as a three-dimensional infinite well, the total energy of a system of Z protons and N neutrons will be obtained by filling up energy levels, consistent with the exclusion principle with a Fermi sea for each species; this is shown in Fig. 7.4(a). If, for a given value of A, we were to have all neutrons instead of protons to minimize the Coulomb potential energy, the resulting quantum zero-point energy of the system would be much larger as in Fig. 7.4(b). The energy cost to do this is called the *symmetry energy* and is discussed in P7.7. The incorporation of this effect into the so-called *semiempirical mass formula* for nuclei is discussed in many textbooks.[2]

7.4.3 **White Dwarf and Neutron Stars**

We are used to thinking of the gravitational force, described via classical mechanics, as playing the dominant role in determining most of the structure of large astrophysical systems, that is, the solar system, galaxies, and beyond. The fact that stars use thermonuclear reactions as their energy source already implies

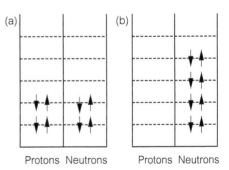

Figure 7.4. Filling of nuclear energy levels with protons and neutrons. Case (a) corresponds to $Z = N = A/$ and minimizes the zero-point energy, while (b) corresponds to $Z = 0$ and $N = A$ to attempt to reduce the electrostatic or Coulomb energy.

[2] See, for example, Krane (1988).

that the nuclear force, necessarily described by quantum mechanics, is also important in stellar evolution. The need to understand quantum tunneling effects (Chapter 11.4.4) and scattering and reaction cross-sections (Chapter 19) implies that "$\hbar$-physics" plays a key role in the structure of stars during the period when they are burning via fusion. It is perhaps not surprising then that "indistinguishability-physics" can have just as important an effect in the determination of the ultimate fate of stars.

In this section, we present an extended essay on the role played by the exclusion principle on the structure of compact astrophysical objects, namely, white dwarf and neutron stars.[3] To simplify matters, we will consider a highly simplified model of a star, namely a constant density, spherical object. We will also use the three-dimensional infinite well as a model for the quantum zero-point energy.

During its lifetime, a star is supported against gravitational collapse by the thermal pressure of the hot gas, with the energy being supplied by thermonuclear fusion reactions and radiated away in the form of electromagnetic radiation (and neutrinos). When the fusion reactions cease to be exothermic (no longer produce energy) the star will naturally begin to collapse. This can be seen by examining the total gravitational potential energy of a constant density, spherical object (P7.8) of mass M_*, namely

$$V_G(R) = -\frac{3}{5}\frac{GM_*^2}{R} \tag{7.27}$$

which obviously favors smaller radii. The corresponding gravitational pressure, P_G, can be obtained from the corresponding force via

$$P_G(R) = \frac{F_G(R)}{A} = \frac{-dV_G(R)/dR}{4\pi R^2} = -\frac{dV_G(\Omega)}{d\Omega} \tag{7.28}$$

where $\Omega = 4\pi R^3/3$ is the (spherical) volume. Using Eqn. (7.27), we find that

$$P_G(\Omega) = -\frac{1}{5}\left(\frac{4\pi}{3}\right)^{1/3}\frac{GM_*^2}{\Omega^{4/3}} \tag{7.29}$$

which is radially inward, favoring collapse. If thermal pressure is no longer sufficient to balance this, we need to look for a new source of energy or pressure, which increases with decreasing radius to compensate.

One such source is the zero-point energy of the (ionized and hence free) electrons in the star. We can estimate this using the infinite well model as

$$E_{\text{tot}} = \frac{\hbar^2\pi^3}{10m_e}\left(\frac{3}{\pi}\right)^{5/3}\frac{N_e^{5/3}}{\Omega^{2/3}} \tag{7.30}$$

---

[3] For more details, see the excellent survey by Shapiro and Teukolsky (1983).

where we have used Eqn. (7.18) and equated $V = \Omega$ with the volume of the star. (For sufficiently large numbers of particles, one can argue that the exact shape of the infinite well should be unimportant; see P7.4 for an example.)

As R (and hence Ω) decreases, this (positive) energy increases at a faster rate, proportional to $1/\Omega^{2/3}$, than the (negative) gravitational potential, which goes as $1/R \propto 1/\Omega^{1/3}$; for small enough radii, it can therefore be the dominant source of energy. Equivalently, we can calculate the corresponding pressure to find

$$P_{\text{EDP}} = -\frac{dE_{\text{tot}}(\Omega)}{d\Omega} = \frac{\hbar^2 \pi^3}{15 m_e} \left(\frac{3}{\pi}\right)^{5/3} \frac{N_e^{5/3}}{\Omega^{5/3}} \qquad (7.31)$$

We note that:

- We label this contribution as electron degeneracy pressure (EDP) as it arises from the zero-point energy of a degenerate electron gas.

- We ignore (for the moment) the similar contributions from protons, neutrons, and other heavier nuclear species as their masses are at least $m_p/m_e \approx 2000$ times heavier and give a much smaller contribution to the zero-point energy.

The total pressure, $P_{\text{TOT}}(\Omega) = P_G(\Omega) + P_{\text{EDP}}(\Omega)$, will vanish at some value of Ω as shown in Fig. 7.5. The value of R at which this happens will constitute a new stable configuration as a small decrease (increase) in R will cause a net positive (negative) pressure driving the system back to equilibrium. This can also be seen by examining the total energy, $V_G(\Omega) + E_{\text{tot}}(\Omega)$ versus Ω, and noting that it has a stable minimum (positive curvature or "concave up") when the total pressure (the derivative with respect to Ω) vanishes. The dependence of P_{EDP} on $\hbar$ implies that is a quantum effect balancing a macroscopic gravitational force. We might well then imagine that the resulting balance could only occur at microscopically small values of R.

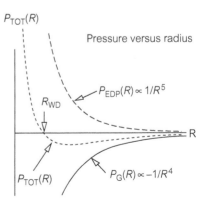

Figure 7.5. Gravitation pressure ($P_G(R)$), electron degeneracy pressure ($P_{EDP}(R)$), and total pressure ($P_{TOT}(R)$) versus stellar radius, R. The value of $P_{TOT}(R = R_{WD}) = 0$ gives a stable minimum of the total energy corresponding to a white dwarf star.

To find the radius of the new stable object, we simply set

$$P_G(\Omega) + P_{EDP}(\Omega) = 0 \tag{7.32}$$

or

$$\frac{1}{5}\left(\frac{4\pi}{3}\right)^{1/3}\frac{GM_*^2}{\Omega^{4/3}} = \frac{\hbar^2\pi^3}{15m_e}\left(\frac{3}{\pi}\right)^{5/3}\frac{N_e^{5/3}}{\Omega^{r/3}} \tag{7.33}$$

to find

$$R = \left(\frac{9\pi}{4}\right)^{2/3}\frac{\hbar^2 N_e^{5/3}}{m_e\,GM_*^2} \tag{7.34}$$

If we assume that the star was initially made of only hydrogen, that is, equal numbers of protons (N_p) and electrons (N_e), we can take $M_* = N_p m_p$ (since $m_e \ll M_p$) or $N_e = M_*/m_p$. Then Eqn. (7.34) can also be written as

$$R = \left(\frac{9\pi}{4}\right)^{2/3}\frac{\hbar^2}{m_e\,m_p^2\,GN_e^{1/3}} = \left(\frac{9\pi}{4}\right)^{2/3}\frac{\hbar^2}{m_e\,m_p^{5/3}\,GM_*^{1/3}} \tag{7.35}$$

showing that larger initial masses imply smaller final radii.

If we assume, for example, a two solar mass star, that is, $M_* = 2\,M_\odot \approx 4 \times 10^{30}$ kg, we find $N_p = N_e \approx 2.5 \times 10^{57}$ and (P7.9)

$$R_{WD} = 1.8 \times 10^7\,\text{km} \approx 1.8 \times 10^4\,\text{km} \tag{7.36}$$

to be compared with the radius of the earth, namely, $R_e = 0.6 \times 10^4$ km.

This new type of compact astrophysical object is called a *white dwarf star* (hence the label WD) and is characterized by the following properties:

- Its small size and the fact that it is still hot (typical surface (interior) temperatures of roughly 8000–10,000 K (10^{6-7} K)) justify its name.

- Typical densities are $\rho_{WD} \sim 2 \times 10^5$ gr/cm^3 compared to average stellar densities of 1–3 gr/cm^3.

- Such higher densities imply that the electrons are more closely packed than in "normal" matter; one can estimate the typical electron separation to be $d \approx 0.02$ Å $\approx 2000\,F$ which is roughly half way (on a logarithmic scale) between typical atomic and nuclear length scales.

- The average electron energy from Eqn. (7.19) is

$$\overline{E} = \frac{3}{5}E_F \approx 0.044\,\text{MeV} \tag{7.37}$$

which is significantly smaller than the electron rest energy of 0.51 MeV. This implies that

— With this energy, the electron speeds are roughly $v/c \approx 0.4$ or $(v/c)^2 \approx 0.15$ which helps justify, *a posteriori*, the use of a nonrelativistic approximation. For only somewhat heavier initial stellar masses, however, the white dwarf radii will be smaller, the zero-point energy larger, and the effects of relativity must be taken into account;

— The classical *thermal* energy of electrons in the core is $k_B T \approx 0.08$–0.8 keV $<< \overline{E}$ so that the electronic kinetic energy is indeed dominated by its zero-point motion.

• The important role of the exclusion principle in the determination of R_{WD} is now obvious, as the final numerical answer would have been a factor of $N_e^{2/3} \approx 2 \times 10^{38}$ *smaller* had we neglected its effects, giving

$$R_{WD} \approx 10^{-34} \text{ cm} \quad \text{(without the exclusion principle)} \tag{7.38}$$

instead.

These derivations seem to imply that every star will eventually collapse and stabilize as a white dwarf star. To see under what circumstances these arguments must be refined, we first note that we can combine Eqn. (7.19) and (7.34) to see that

$$\overline{E} = \frac{1}{5}\left[\frac{2}{3\pi^2}\right]^{1/3}\frac{m_e m_p^{8/3} G^2 M_*^{4/3}}{\hbar^2} \tag{7.39}$$

This implies that a roughly six-fold increase in M_* in our example will give $\overline{E} \approx 0.51$ MeV $\approx m_e c^2$ implying that the electrons are now relativistic and the original assumptions of "non-relativity" are invalid.

We should then use the most general energy–momentum relation

$$E^2 = (pc)^2 + (m_e c^2)^2 \tag{7.40}$$

but we can understand the qualitative changes by employing the ultrarelativistic limit where

$$E = |\mathbf{p}|c \tag{7.41}$$

In this case, the quantized energies are given by

$$E(\mathbf{n}) = \frac{\hbar c}{L}\sqrt{n_x^2 + n_y^2 + n_z^2} = \frac{\hbar c|\mathbf{n}|}{L} = \frac{\hbar c|\mathbf{n}|}{\Omega^{1/3}} \tag{7.42}$$

With this form, the total energy can be calculated in the same manner as Eqn. (7.18) (P7.10) and we find

$$E_{tot}^{rel}(\Omega) = \frac{\pi^2}{4} \left(\frac{3}{\pi}\right)^{4/3} \frac{\hbar c}{\Omega^{1/3}} N_e^{4/3} \tag{7.43}$$

which implies a degeneracy pressure of the form

$$P_{EDP}^{rel} = \frac{\pi^2}{12} \left(\frac{3}{\pi}\right)^{4/3} \frac{\hbar c}{\Omega^{4/3}} N_e^{4/3} \tag{7.44}$$

In this limit, the electron degeneracy pressure has the *same* Ω (volume) dependence as the gravitational pressure, so an equilibrium state can no longer be guaranteed by simply varying R. The fate of the star relies, instead, on the relative magnitudes of the *coefficients* of the $1/\Omega^{4/3}$ terms in Eqns (7.29) and (7.44), namely

$$C_{EDP} = \frac{\pi^2}{12} \left(\frac{3}{\pi}\right)^{4/3} \hbar c N_e^{4/3} \quad \text{versus} \quad C_G = \frac{1}{5} \left(\frac{4\pi}{3}\right)^{1/3} GM_*^2 \tag{7.45}$$

Since $C_{EDP} \propto N_e^{4/3} \propto M_*^{4/3}$ while $C_G \propto M_*^2$, the gravitational pressure will always win for sufficiently large initial stellar mass, causing continued collapse. The critical value beyond which a white dwarf star will not be stable can then be *roughly* estimated by letting $C_{EDP} = C_G$ to find

$$M_*^{crit} = \left(\frac{5}{4}\right)^{3/2} \left[\frac{9\pi}{4}\right]^{1/2} \left(\frac{\hbar c}{G}\right)^{3/2} \frac{1}{m_p^2} \quad \approx \quad 6.6 \times M_\odot \tag{7.46}$$

where $M_\odot$ is a solar mass.

This limit on the maximum mass of a white dwarf star is called the *Chandrasekhar limit* and a more sophisticated analysis finds that

$$M_*^{crit} \approx 1.4 \, M_\odot \tag{7.47}$$

The fact that we are within less than an order-of-magnitude from the "right answer" is now consistent with our belief that we have included much of the relevant physics, especially the exclusion principle, in however simplified a fashion. We note also that the Chandrasekhar mass depends only on the fundamental constants $\hbar$, c, G, and m_p in such a way that

$$M_*^{crit} \propto \frac{M_P^3}{m_p^2} \tag{7.48}$$

where M_P is the *Planck mass* discussed in P1.19,

$$M_P = \sqrt{\frac{\hbar c}{G}} \tag{7.49}$$

If $M_* > M_*^{\text{crit}}$, then the star will continue to collapse, but another stable configuration is possible. In this case, it can become energetically favorable for the electrons to interact with the protons via inverse β-decay, that is, via the reaction $e^- + p \rightarrow n + \nu_e$. The electrons and protons "disappear," leaving behind only the neutrons, as the weakly interacting neutrinos escape the star completely. This occurs when the average electron kinetic energy becomes greater than the energy cost of the reaction, namely, $Q = (m_n - m_p - m_e)c^2 \approx 0.78$ MeV which is (coincidentally) at roughly the same point at which the nonrelativistic calculation breaks down.

This leaves a gas of indistinguishable spin-1/2 neutrons, a *neutron star*, and the same analysis as for white dwarf stars can be performed. The radius of this new configuration is given by Eqn. (7.34), but with m_e replaced by m_n so that

$$R_{\text{NS}} = \frac{m_e}{m_n} R_{\text{WD}} \approx \frac{1}{1800} R_{\text{WD}} \approx 10 \text{ km} \tag{7.50}$$

Typical densities are now

$$\rho_{\text{NS}} \approx \left(\frac{m_n}{m_e}\right)^3 \rho_{\text{WD}} \approx 10^{15} \text{ gm/cm}^3 \tag{7.51}$$

which is similar to the density of atomic nuclei. We leave it to the interested reader to examine the phenomenology of such interesting astrophysical objects as well as the path leading to the most intriguing final configuration, namely, black holes.

7.5 **Questions and Problems**

Q7.1. We have applied a simplified picture of noninteracting spin-1/2 particles in a box to model various physical problems. In each case (conduction electrons in a metal, neutrons and protons in a nucleus, etc.) we have ignored the mutual interactions between the particles. How do the exclusion principle results and the Fermi sea picture help make this approximation more plausible? Hint: If two particles in the Fermi sea were to interact, thereby changing their momentum and energy, to which energy levels could they go?

Q7.2. Discuss why the exact shape of the infinite well potential should make no difference to the total zero-point energy if there are many particles.[4]

Q7.3. By how much would one side of a piece of metal one meter long increase in size if it were put into a vacuum so that atmospheric pressure were no longer acting on it? How does this depend on $\hbar$?

---

[4] See Balian and Bloch (1970).

Q7.4. Discuss the following statement: There are two regions of stable nuclei

- one with $A = 1$ to ≈ 200 with a balance between the strong attractive nuclear force and the Coulomb repulsion and

- one with $A \approx 10^{57}$ with a balance between the attractive gravitational force and quantum zero-point energy.

Q7.5. We estimated the cut-off mass beyond which a star no longer can become a white dwarf, the so-called Chandrasekhar limit. Real stars are initially rotating and often have nonnegligible magnetic fields. As the star collapses, the rotation rate must increase (to conserve angular momentum) and the energy density in the magnetic field also increases (same number of field lines in a smaller volume). What qualitative changes would the inclusion of such effects make on estimates of the upper limit on masses of white dwarf stars, that is, would the limit go up, down, or not change at all?

P7.1. Spin counting. A system of indistinguishable spin $S = 1/2$ particles can accommodate two states in each energy level, "spin-up" and "spin down," that is, with $S_z = \pm 1/2$. A particle with spin $S = 3/2$ can have four states with the same energy, $S_z = -3/2, -1/2, 1/2$, and $3/2$ in each level. Calculate the total, average, and Fermi energy of N such particles in a three-dimensional infinite well. Can you generalize your results to spin $5/2, 7/2, \ldots$ and so forth where the degeneracy per level is given by $2S + 1$?

P7.2. Fermi energy in two dimensions. Consider a two-dimensional system of electrons confined to a box of area $A = L^2$. Calculate the total energy of N_e such electrons as well as the average and Fermi energies. Can you show that in d dimensions that the total energy should scale as $E_{tot} \propto N_e^{(d+2)/d}$?

P7.3. Spin and energy levels for the harmonic oscillator.

(a) The quantized energy levels of a particle of mass m in a harmonic oscillator potential in one-dimension are given by

$$E_n = (n + 1/2)\hbar\omega \quad n = 0, 1, 2, \ldots \tag{7.52}$$

where $\omega = \sqrt{k/m}$ is the classical oscillation frequency. Find the total, average, and Fermi energy of N_e (noninteracting) electrons in this potential.

(b) Can you derive the same quantities for the three-dimensional harmonic oscillator for which the energy levels are

$$E(\mathbf{n}) = (n_x + n_y + n_z + 3/2)\hbar\omega \quad n_x, n_y, n_z = 0, 1, 2, \cdots \tag{7.53}$$

P7.4. Instead of assuming a cubical three-dimensional infinite well as in Sec. 7.3, assume that the electrons are in a box of dimensions L_x, L_y, L_z so that $V = L_x L_y L_z$.

(a) Find the allowed energy levels as in Eqn. (7.14).

(b) Calculate the total energy of the system and compare to the expression in Eqn. (7.18).

Hint: The volume of an ellipse governed by the formula

$$\frac{x^2}{a^2} + \frac{y^2}{b^2} + \frac{z^2}{c^2} = 1 \tag{7.54}$$

is given by $V = 4\pi \, abc/3$.

P7.5. Density of states.

(a) Using Eqn. (7.19), show that the number of states with energy between E and $E + dE$ is given by

$$\frac{dN}{dE} \propto \sqrt{E} = C\sqrt{E} \tag{7.55}$$

which is applicable in the interval $E \in (0, E_F)$.

(b) Normalize this to find C by using the fact that

$$N_e = \int_0^{E_F} \frac{dN}{dE} \, dE \tag{7.56}$$

(c) Use this expression to evaluate $\langle E \rangle$ and show that it agrees with Eqn. (7.19).

(d) What fraction of the particles have energy less than $0.1 E_F$? less than $0.5 E_F$?

P7.6. Electron gas model of metals. Consider the model of conduction electrons in Section 7.4.1.

(a) For an electron density $n_e = 3 \times 10^{22}$ cm^{-3}, evaluate the Fermi and average energies.

(b) Derive Eqn. (7.25).

(c) Confirm the numerical values for the bulk modulus for Na discussed in Section 7.4.1.

P7.7. Spin and energy levels in nuclear physics.

(a) Consider a heavy nucleus consisting of $Z = N_p$ protons and $N = N_n$ neutrons so the total atomic number is $A = Z + N$. If the nucleons have volume V_0 and are bound so that they are "just touching," show that the volume of such a nucleus will go as $V = V_0 A$ so that its radius will be given by $R = R_0 A^{1/3}$. With $R_0 \approx 1.2 - 1.4 \, F$. This form works well for many nuclei.

(b) Instead of considering all the many nucleon–nucleon interactions, model this system as A nucleons in a three-dimensional infinite well of radius R. Using Eqn. (7.18), find the total zero-point energy of a system of $A = Z + N$ nucleons, that is, $E(Z, N)$, recalling that both protons and neutrons are spin-1/2 particles and are distinguishable from each other. For simplicity, assume that the neutron and proton have the same mass, roughly $mc^2 \approx 940$ MeV.

(c) For fixed A, minimize this energy and show that equal numbers of neutrons and protons are favored.

(d) Assuming that $\Delta \equiv N - Z << A$, show that the zero-point energy can be approximated by an expression of the form

$$E(Z, N) \approx E_{min}(A) + E_{sym}\frac{\Delta^2}{A} + \cdots \qquad (7.57)$$

and find an expression for E_{sym}. Using the values above, show that the numerical value of E_{sym} is roughly 40 MeV. Note: This contribution to the total nuclear energy (or rest mass) is often called the *symmetry contribution* and is part of the well-known *semiempirical mass formula* of nuclear physics.

P7.8. Classical gravitational potential energy. Consider a uniform sphere of mass M and radius R. Find the total gravitational potential energy stored in this configuration, namely, Eqn. (7.27). Hint: Consider a uniform sphere of mass $M(r)$ of some intermediate radius r and calculate the work, dW, required to add a layer of thickness dr brought in from infinitely far away; then use $W = \int dW$.

P7.9. Verify the numerical values in Eqns (7.36) and (7.50).

P7.10. Relativistic electron degeneracy pressure. Repeat the calculation of the total electron energy in the extreme relativistic case and confirm Eqn. (7.43).

P7.11. We argued that the white dwarf configuration was stable against small perturbations in radius since the total energy was a minimum at $R = R_{WD}$. Expand $E_{tot}(R)$ as a function of R around this minimum and show that it is of the form

$$E_{tot}(R) = E_{tot}(R_{WD}) + \frac{1}{2}\frac{d^2 E_{tot}(R = R_{WD})}{dR^2}(R - R_{WD})^2 + \cdots \qquad (7.58)$$

which is like a harmonic oscillator potential. Use this connection to find the frequency of oscillations around the stable minimum configuration using $\tau = 2\pi\sqrt{m/K}$ where K is the effective spring constant in the problem.

EIGHT

Other One-Dimensional Potentials

Our study of the infinite well family of potentials has provided us with an array of insights into the physical meaning and mathematical structure of wave mechanics. There are other quantum mechanical properties which do not appear in such systems, and in this chapter we study several model potentials which illustrate many new aspects, both formal and intuitive, which have wide applications. We consider smoothness conditions on $\psi(x)$ and its derivatives, singular potentials, periodic potentials, and applications to models of band structure (Section 8.1), quantum mechanical tunneling and the large $|x|$ behavior of quantum wavefunctions (Section 8.2), and the application of one-dimensional problems to three-dimensions (Section 8.3).

8.1 Singular Potentials

8.1.1 Continuity of $\psi'(x)$

We have argued that the identification of $|\psi(x,t)|^2$ with an observable probability density requires that $\psi(x,t)$ be continuous in x. It is natural to ask about further smoothness conditions on the position-space wavefunction and their meaning. As an example we discuss below under what conditions we can demand that the spatial derivative of $\psi(x)$, namely, $\psi'(x)$ be continuous. We note that we can rewrite the Schrödinger equation in the form

$$\frac{d^2\psi(x)}{dx^2} = \frac{2m}{\hbar^2}\left(V(x) - E\right)\psi(x) \tag{8.1}$$

and in order to examine the derivative at a point $x = a$, we note that we can integrate Eqn. (8.1) over the narrow interval $(a - \epsilon, a + \epsilon)$ to obtain

$$\psi'(a^+) - \psi'(a^-) \equiv \frac{d\psi(a + \epsilon)}{dx} - \frac{d\psi(a - \epsilon)}{dx}$$

$$= \int_{a-\epsilon}^{a+\epsilon} dx \, \frac{d^2\psi(x)}{dx^2}$$

$$= \frac{2m}{\hbar^2} \int_{a-\epsilon}^{a+\epsilon} dx \, (V(x) - E) \, \psi(x) \qquad (8.2)$$

where $a^\pm \equiv a \pm \epsilon$ are x values which approach a infinitesimally closely from either side as $\epsilon \to 0$.

Since the $E\psi(x)$ term is everywhere finite, its contribution to the integral vanishes as $\epsilon \to 0$, and we need only worry about the behavior of $V(x)$ near a. Even if $V(x)$ is *discontinuous* at $x = a$ (say a step- or θ-function as in Fig. 8.1) the integral $\int_{a-\epsilon}^{a+\epsilon} dx \, V(x) \, \psi(x)$ will still vanish as ϵ is made smaller, as the "area" under $V(x) \, \psi(x)$ still decreases; this implies that $\psi'(a^+) = \psi'(a^-)$, that is, $\psi'(x)$ is continuous.

If, instead, the potential has a δ-function singularity at $x = a$, then the area under even an infinitesimally small region around a will be finite. Specifically, if $V(x) = \pm g\delta(x - a)$ (where g is a constant with the appropriate units), we find that the appropriate boundary condition on $\psi'(x)$ is

$$\psi'(a^+) - \psi'(a^-) = \pm\frac{2mg}{\hbar^2}\psi(a) \quad \text{when } V(x) = \pm g\delta(x - a) \qquad (8.3)$$

We can summarize these statements by saying that

- The spatial derivative, $\psi'(x)$, is everywhere continuous except at points where the potential is *singular*, where it then satisfies Eqn. (8.3).

The potential has to be this "badly behaved" in order for $\psi'(x)$ to be discontinuous. The smoothness of higher derivatives of $\psi(x)$ is explored in P8.1.

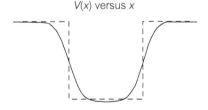

V(x) versus x

Figure 8.1. Simplified model of a smooth potential energy function (solid curve) using discontinuous step functions (dashed curve).

8.1.2 **Single δ-function Potential**

While a singular potential of the δ-function type is unphysical, it can provide a model system with a highly localized attractive ($g < 0$ or repulsive ($g > 0$) force; combinations of such potentials can then be used as simple models of both diatomic molecules and one-dimensional solids. Modern materials' fabrication techniques allow for the deposition of different atomic compounds providing "potential spikes" mimicking the effect of a δ-function[1] potential.

We first examine the bound state spectrum ($E = -|E| < 0$) for a single attractive δ-function potential of the form $V(x) = -g\delta(x)$ by looking for solutions of the Schrödinger equation

$$-\frac{\hbar^2}{2m}\frac{d^2\psi(x)}{dx^2} = E\psi(x) = -|E|\psi(x) \quad \text{for } x \neq 0 \tag{8.4}$$

since $V(x) = 0$ everywhere except at the origin. This differential equation has the general solution

$$\psi(x) = \begin{cases} Ae^{+Kx} + Be^{-Kx} & \text{for } x \leq 0 \\ Ce^{-Kx} + De^{+Kx} & \text{for } x \geq 0 \end{cases} \tag{8.5}$$

where $K = \sqrt{2m|E|/\hbar^2}$. Imposing the conditions that the wavefunction be square-integrable and continuous at $x = 0$ gives the physically acceptable solution

$$\psi(x) = \begin{cases} Ae^{+Kx} & \text{for } x \leq 0 \\ Ae^{-Kx} & \text{for } x \geq 0 \end{cases} \tag{8.6}$$

The final boundary condition available to determine the quantized energy eigenvalues is the "discontinuity" requirement in Eqn. (8.3), which implies that

$$\psi'(0^+) - \psi'(0^-) = -AK - AK = -\frac{2mg}{\hbar^2}A = -\frac{2mg}{\hbar^2}\psi(0) \tag{8.7}$$

which gives

$$K = \frac{mg}{\hbar^2} \quad \text{or} \quad E = -|E| = -\frac{mg^2}{2\hbar^2} \tag{8.8}$$

We note that:

• There is a single bound state energy, and the normalized wavefunction is

$$\psi(x) = \sqrt{K}e^{-K|x|} \tag{8.9}$$

---

[1] See, for example Salis *et al.* (1997)

which was discussed as an example in Section 3.4. The presence of the cusp at $x = 0$ in this wavefunction is now understood as arising from the singular potential. The fact that the expectation values of high powers of $\hat{p}$ are ill-defined (P4.6) is also plausible given the discontinuous behavior of higher derivatives of $\psi(x)$.

- The momentum-space wavefunction for this solution was also derived in Section (3.4) and in P4.7 by Fourier transform and has a *Lorentzian* form

$$\phi(p) = \sqrt{\frac{2p_0}{\pi}} \left(\frac{p_0}{p^2 + p_0^2} \right) \tag{8.10}$$

where $p_0 \equiv \hbar K$; this also shows that $\langle p^4 \rangle$ (and expectation values for higher powers of p) are not well defined.

- This problem can be also solved (P8.3) in momentum-space directly, giving the correct form of $\phi(p)$ first, and with the quantized energy level determined in a novel way.

- There is a discrete bound state spectrum for $E < 0$ (actually only one state), but there is also a continuum of unbound states with $E > 0$; this is similar to the spectrum of, say, the hydrogen atom. The discrete and continuous states, *taken together*, form a complete set from which any admissible wavefunction can be constructed; the bound state solution(s), by themselves, are not sufficient.

8.1.3 **Twin δ-function Potential**

The problem defined by a twin-δ-function potential, namely

$$V(x) = -g \left[\delta(x + a) + \delta(x - a) \right] \tag{8.11}$$

is a simple generalization of the problem above and can be used as a toy model for a one-dimensional diatomic ion[2]; it describes two attractive centers separated by a distance $2a$, which we can imagine being varied. As there is another natural length scale in the problem, namely, $L \equiv \hbar^2/mg$, we expect the physical results to depend qualitatively on the ratio of these two quantities.

Because of the symmetry of the potential, we can restrict ourselves to the study of even and odd solutions. For example, for even solutions we can write

$$\psi^{(+)}(x) = \begin{cases} Ae^{+Kx} & \text{for } x \leq -a \\ B\cosh(Kx) & \text{for } -a \leq x \leq +a \\ Ae^{-Kx} & \text{for } +a \leq x \end{cases} \tag{8.12}$$

[2] See Lapidus (1970).

where we have applied the boundary conditions at infinity, as well as the symmetry to take the even combination of e^{-Kx} and e^{+Kx}. Because of the symmetry of the wavefunction, the boundary conditions on ψ and ψ' at $x = -a$ give the same constraint as those at $x = a$, which we use to write

$$B\cosh(Ka) = Ae^{-Ka}$$

$$(-AKe^{-Ka}) - (BK\sinh(Ka)) = -\frac{2mg}{\hbar^2}Ae^{-Ka} \qquad (8.13)$$

arising from the continuity of $\psi(x)$ and "discontinuity" (Eqn. (8.3)) of $\psi'(x)$, respectively. These can be combined to yield the eigenvalue equation

$$\text{(even eigenvalue condition)} \quad f_E(y) \equiv y\left(1 + \tanh(y)\right) = \frac{2mga}{\hbar^2} \equiv \frac{a}{a_0} \quad (8.14)$$

where we have defined

$$y \equiv Ka \quad \text{and} \quad a_0 = \frac{\hbar^2}{2mg} \qquad (8.15)$$

The odd solutions, written as

$$\psi^{(-)}(x) = \begin{cases} -Ae^{+Kx} & \text{for } x \leq -a \\ B\sinh(Kx) & \text{for } -a \leq x \leq +a \\ +Ae^{-Kx} & \text{for } +a \leq x \end{cases} \qquad (8.16)$$

have the corresponding eigenvalue condition,

$$\text{(odd eigenvalue condition)} \quad f_O(y) \equiv y\left(1 + \coth(y)\right) = \frac{a}{a_0} \qquad (8.17)$$

The energy eigenvalues can be determined from Eqns. (8.14) and (8.17) by plotting the functions on the left-hand side versus y and looking for intersections with horizontal lines corresponding to values of a/a_0, as shown in Fig. 8.2. For comparison, in that diagram we also plot the eigenvalue condition for the *single* δ-function potential, namely

$$\text{(eigenvalue for } single \text{ } \delta\text{-function)} \quad K = \frac{mg}{\hbar^2} \qquad (8.18)$$

written in the form

$$f(y) \equiv 2y = 2Ka = 2\left(\frac{mg}{\hbar^2}\right)a = \frac{a}{a_0} \qquad (8.19)$$

for comparison. Plots of the position-space wavefunction for three values of a/a_0 are shown in Fig. 8.3. We note several important features:

- In the limit of large separation, $a \gg a_0$, there is one even and one odd solution, with energy identical to the single δ-function case; the two energy

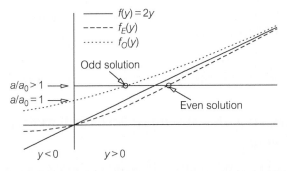

Figure 8.2. Energy eigenvalue condition for a twin δ-function potential. Intersections of the dotted (dashed) curves with horizontal lines of constant a/a_0 give the values of y for the odd (even) bound states. The solid line is the result for a single δ-function potential. For $a < a_0$ there are no odd solutions.

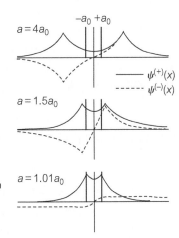

Figure 8.3. Position space wavefunctions for the twin δ-function potential. The solid (dotted) curves correspond to the even (odd) states, respectively. Results for a/a_0 corresponding to 4, 1.5, and 1.01 are shown from top to bottom. For $a/a_0 \gtrsim 1$, we can see the odd state begin to "unbind" as it ceases to be localized.

levels are said to be *degenerate* in energy. In this limit, the wavefunctions are simply related to those for the single δ-function case (P8.4).

- The odd solution (when it exists) always has a smaller value of y and hence is *less bound* than the corresponding even solution.

- For $a/a_0 < 1$, there is no odd solution, which we attribute to the increasing kinetic energy of the state as required by the node in the antisymmetric wavefunction at the origin.

- We can see in Fig. 8.3 the odd solution 'unbind' for $a/a_0 \to 1$ where the position-space wavefunction is becoming spatially uniform, indicating no localization near the origin; when $|E|$, and hence K, approach zero, the localization scale of the wavefunctions, $L = 1/K \to \infty$.

- The (unnormalized) momentum-space wavefunctions can be obtained by direct Fourier transform (or as in P8.6), and we find

$$\phi^{(+)}(p) \propto \frac{\cos(ap/\hbar)}{p^2 + (\hbar K)^2} \quad \text{and} \quad \phi^{(-)}(p) \propto \frac{\sin(ap/\hbar)}{p^2 + (\hbar K)^2} \tag{8.20}$$

8.1.4 Infinite Array of δ-functions: Periodic Potentials and the Dirac Comb

We can continue in this vein and discuss an infinitely long array of equally spaced attractive δ-function potentials, defined by

$$V_\infty(x) = -g \sum_{n=-\infty}^{+\infty} \delta(x - na) \tag{8.21}$$

which can be taken as a very simplified model of a one-dimensional solid.[3] The periodicity of this potential gives rise to many new features and we begin by first discussing the general form of solutions in the case when we have a periodic potential satisfying

$$V(x + a) = V(x) \tag{8.22}$$

We recall from P6.16 that the translation operator, $\hat{T}_a$, defined by

$$\hat{T}_a f(x) = f(x + a) \tag{8.23}$$

commutes with both the kinetic energy operator as well as any periodic potential of the form in Eqn. (8.22), and thus commutes with the Hamiltonian for this problem, $[\hat{H}, \hat{T}_a] = 0$. From the discussion in Section 6.7, we know that $\hat{H}$ and $\hat{T}_a$ can therefore have simultaneous eigenfunctions, and the allowed solutions of the Schrödinger equation for this problem can then also be assumed to satisfy

$$\psi(x + a) = \hat{T}_a \psi(x) = \lambda_a \psi(x) \tag{8.24}$$

Since $\hat{T}_a$ is not a Hermitian operator, its eigenvalues need not be real, but we can write quite generally that $\lambda_a = |\lambda_a| \exp(i\phi_a)$ for any complex number. If we repeatedly apply Eqn. (8.24) to $\psi(x)$, shifting either to the left or to the right, we obtain arbitrarily high powers of $|\lambda_a|^n$; these would diverge if $|\lambda_a| > 1$ and we go to the right, or if $|\lambda_a| < 1$ and we go to the left, so we clearly must have $|\lambda_a| = 1$ since, if not, $\psi(x)$ would not be square-integrable. Thus, we can assume that the λ_a eigenvalue is simply a complex phase, which we write as

$$\lambda_a = e^{iqa} \quad \text{so that} \quad \psi(x + a) = e^{iqa} \psi(x) \tag{8.25}$$

---

[3] This particular example of a periodic array of potential wells is often called a *Dirac comb*; this type of simplified model of a one-dimensional solid is most often associated with Kronig and Penney (1931).

This result is called *Bloch's theorem* and we see that it is certainly consistent with the expectation that

$$|\psi(x+a)|^2 = |e^{iqa}\psi(x)|^2 = |\psi(x)|^2 \tag{8.26}$$

We note that while no real solid actually extends to infinity, the finite boundaries of a macroscopic sample should have little effect on the interior many lattice spaces away. One standard method of taking this effect (or lack thereof) into account is to assume that the one-dimensional lattice actually satisfies periodic boundary conditions, namely that

$$\psi(x+Na) = \psi(x) \quad \text{for some } N >>> 1 \tag{8.27}$$

in which case we obtain the allowed values of q given by

$$\left(e^{iqa}\right)^N \psi(a) = \psi(x+Na) = \psi(x) \quad \text{or} \quad e^{iqNa} = 1 \tag{8.28}$$

so that

$$q = \left(\frac{2\pi}{a}\right)\left(\frac{n}{N}\right) \quad \text{where } n = 0, 1, 2, \ldots, N \tag{8.29}$$

Assuming a macroscopically large value for N, we see that the allowed q values range almost continuously from 0 to 2π.

Considering bound state $(E < 0)$ solutions as before, we first write the solutions in the interval $(0, a)$ in the form

$$\psi(x) = A\sinh(Kx) + B\cosh(Kx) \quad \text{for } 0 \le x \le +a \tag{8.30}$$

and using Eqn. (8.25), we can automatically write down the corresponding solution in the "unit cell" directly to the left, as

$$\psi(x) = e^{-iqa}\left[A\sinh(K(x+a)) + B\cosh(K(x+a))\right] \quad \text{for } -a \le x \le 0 \tag{8.31}$$

Application of the appropriate boundary conditions on $\psi(x)$ and $\psi'(x)$ need then only be done once, say at $x = 0$, giving the conditions

$$e^{-iqa}\left[A\sinh(Ka) + B\cosh(Ka)\right] = B \tag{8.32}$$

$$[AK] - [AK\cosh(Ka) + BK\sinh(Ka)] = -\frac{2mg}{\hbar^2}B \tag{8.33}$$

We can the eliminate A and B to obtain the energy eigenvalue condition

$$\cos(qa) = \cosh(Ka) - \left(\frac{a}{2a_0}\right)\frac{\sinh(Ka)}{Ka} \tag{8.34}$$

where $a_0 \equiv \hbar^2/2mg$ as before. We can write this in a form which is similar to the even and odd eigenvalue conditions for the twin δ-function case in Eqns (8.14) and (8.17), namely

$$f_\infty(y, z) \equiv \frac{2y(\cosh(y) - z)}{\sinh(y)} = \frac{a}{a_0} \tag{8.35}$$

where $y \equiv Ka$ and $z \equiv \cos(qa)$, with $-1 \leq z \leq +1$. We plot in Fig. 8.4 a number of representative example z values in the range $(-1, +1)$ of the eigenvalue function $f_\infty(y, z)$ versus y and can note the obvious similarities to Fig. 8.2. For $a/a_0 \gg 1$ (large separations) the solutions are all degenerate and equal to the single δ-function result. As a/a_0 is reduced, the $z = \cos(qa) = -1$ solutions unbind first, which is consistent with the fact that these are the most "odd" in the sense that $\psi(x + a) = \lambda_a \psi(x) = -\psi(x)$ and therefore has the most kinetic energy. Thus, as isolated bound states are combined to form a solid, we might expect the degenerate energy levels to split and form bands, with some higher-energy states in the band becoming unbound.

Motivated by this last fact, we should also discuss positive energy $(E > 0)$ solutions which would be relevant for any states which unbind. Instead of redoing

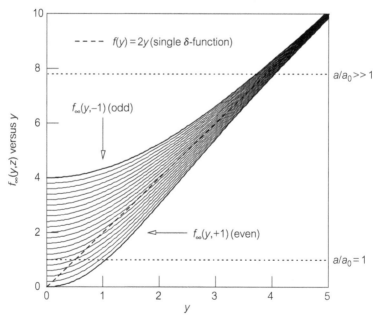

Figure 8.4. Plot of the energy eigenvalue function, $f_\infty(y, z)$ versus y in Eqn. (8.35), for values of $-1 \leq z \leq +1$ illustrating how the energy eigenvalues for the Dirac comb change as the attractive δ-function centers are brought together. For $a/a_0 \gg 1$, there is a huge degeneracy, and as a/a_0 is decreased the degeneracy is lifted and many states become unbound. The eigenvalue condition for a single, isolated δ-function potential from Eqn. (8.19), $f(y) = 2y$, is shown as the dashed line.

the problem from scratch, we note that we can simply make the changes

$$\text{for } E < 0 \quad K = \sqrt{\frac{-2mE}{\hbar^2}} = \sqrt{\frac{2m|E|}{\hbar^2}} \quad \longrightarrow \quad k = \sqrt{\frac{2mE}{\hbar^2}} = iK \quad \text{(for } E > 0)$$

(8.36)

and the relations

$$\cosh(iz) = \cos(z) \quad \text{and} \quad \sinh(iz) = i\sin(z)$$

(8.37)

so that the energy eigenvalue constraint for positive energies can be obtained from Eqn. (8.34) directly to be

$$\cos(qa) = \cos(y) - \left(\frac{a}{2a_0}\right)\frac{\sin(y)}{y} \equiv F(y)$$

(8.38)

where $y = ka$. For an otherwise free particle we would expect that any values of k would be allowed, but if we plot $F(y)$ on the right-hand side of Eqn. (8.38), we note that in intervals where $|F(y)| > 1$ that no solutions are possible since the fixed left-hand side of Eqn. (8.38) is bounded by the $-1 \leq \cos(qa) \leq +1$ term. The periodicity of the potential thus gives rise to allowed bands where free particle-like solutions are possible (shown as the hatched regions in Fig. 8.5) and disallowed gaps where such standing wave solutions are not supported.

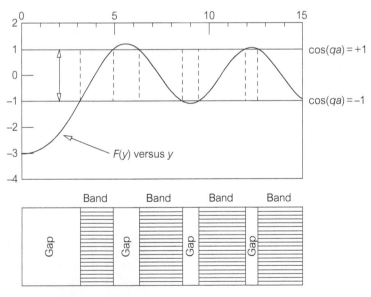

Figure 8.5. Plot of the energy eigenvalue function, $F(y)$, in Eqn. (8.38) showing that some values of y are not allowed since the function must fall in the range $-1 \leq F(y) \leq +1$. This gives rise to the allowed energy bands (hatched areas) with an almost continuous range of states available, and the disallowed energy gaps.

This simple soluble model usefully illustrates some of the physics responsible for the wide variety of properties which materials can exhibit when large numbers of atoms are brought together to make solids. A typical electronic energy level diagram for an isolated atom is shown in Fig. 8.6; when N such atoms are well-separated, we can think of each level shown as being N-fold degenerate where $N >>> 1$ for a macroscopic number of atoms. As the atoms are brought closer together, the degeneracy splits, with some levels becoming more or less strongly bound. When the atoms reach their equilibrium positions, the energy levels then can exhibit *bands* of energy levels, regions where the energy levels are very closely spaced, separated by well-defined *energy gaps* where no electronic states are allowed, as shown in Fig. 8.6. We note that overlapping energy bands are possible.

The allowed energy levels are then filled with electrons, consistent with the exclusion principle as in Chapter 7, and the position of the Fermi energy relative to the band gaps has a profound influence on the macroscopic properties of the material as shown in Fig. 8.7. If there are many unoccupied energy levels just

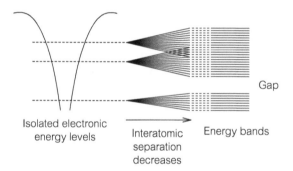

Isolated electronic energy levels

Interatomic separation decreases

Energy bands

Gap

Figure 8.6. Schematic representation of origin of energy band and gaps in solids. The large $N \gg 1$ degeneracy of the energy levels of isolated atoms is split as they are brought closer together.

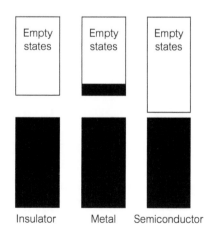

Empty states   Empty states   Empty states

Figure 8.7. Energy bands and gaps for various types of materials. The shaded regions indicated filled electron bands.

Insulator   Metal   Semiconductor

above the Fermi surface into which electrons can easily be scattered, they are free to respond to relatively small external electric fields and thermal gradients and thus exhibit metallic behavior. If, on the other hand, there is a large energy gap which would inhibit such transitions, one has an insulator. In this case a large electric field would be required to excite the electrons across the gap; this is the equivalent of a "spark" or "breakdown." The case of a very narrow energy gap corresponds to a semiconductor for which small but finite changes in the external parameters, such as the temperature or applied field, can cause excitations.

8.2 The Finite Well

8.2.1 Formal Solutions

While useful as a model system, the infinite well has at least one very unrealistic feature, namely, the fact that the particle cannot escape, that is, be "ionized" by the addition of a sufficiently large amount of energy to the system. A more realistic version, which also illustrates several new features of quantum physics, is the finite well, here defined as

$$V(x) = \begin{cases} -V_0 & \text{for } -a < x < +a \\ 0 & \text{otherwise} \end{cases} \tag{8.39}$$

and we wish to examine bound states with $E \equiv -|E| < 0$ as shown in Fig. 8.8. At first glance, it might seem more appropriate to use the form

$$\tilde{V}(x) = \begin{cases} 0 & \text{for } -a < x < +a \\ +V_0 & \text{otherwise} \end{cases} \tag{8.40}$$

which has the symmetric infinite well as its limit when $V_0 \to \infty$, but the latter choice is trivially related to the conventional one of Eqn. (8.39) via $\tilde{V}(x) = V(x) + V_0$. This simple relation ensures (P6.4) that the two systems will have the same observable physics. The standard choice, $V(x)$, is often used

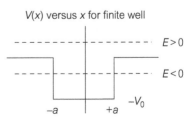

V(x) versus x for finite well

Figure 8.8. The finite potential well.

as it is conventional to pick the (arbitrary) zero of potential energy to vanish at infinite separations (e.g. the Coulomb potential, $V(r) = Kq_1q_2/r$). To facilitate comparisons with the infinite well limit, we note that the quantity $V_0 - |E|$ measures the difference between the energy levels and the bottom of the well; this then should equal the usual quantized energy levels $E_n^{(\pm)}$ of the symmetric infinite well in the limit that $V_0 \to \infty$.

With this choice of potential, the Schrödinger equation in the two regions is given by

$$|x| < a : -\frac{\hbar^2}{2m}\frac{d^2\psi(x)}{dx^2} - V_0\psi(x) = -|E|\psi(x) \tag{8.41}$$

$$|x| > a : -\frac{\hbar^2}{2m}\frac{d^2\psi(x)}{dx^2} = -|E|\psi(x) \tag{8.42}$$

or

$$|x| < a : \frac{d^2\psi(x)}{dx^2} = -q^2\psi(x) \tag{8.43}$$

$$|x| > a : \frac{d^2\psi(x)}{dx^2} = +k^2\psi(x) \tag{8.44}$$

where

$$k = \sqrt{\frac{2m|E|}{\hbar^2}} \quad \text{and} \quad q = \sqrt{\frac{2m(V_0 - |E|)}{\hbar^2}} \tag{8.45}$$

The most general solution can be written as

$$x \leq -a : De^{+kx} + Fe^{-kx}$$

$$-a \leq x \leq +a : A\cos(qx) + B\sin(qx)$$

$$+a \leq x : Ce^{-kx} + Ee^{+kx} \tag{8.46}$$

and one can immediately apply the boundary conditions at $x = \pm\infty$ (i.e. that the wavefunction be square-integrable) to insist that $E = F = 0$. The fact that the potential is symmetric can be used to infer that the energy eigenfunctions will also be eigenstates of parity, that is, even and odd functions, so we can specialize and note that for the even states we must have

$$\psi^{(+)}(x) = \begin{cases} Ce^{+kx} & \text{for } x \leq -a \\ A\cos(qx) & \text{for } -a \leq x \leq +a \\ Ce^{-kx} & \text{for } +a \leq x \end{cases} \tag{8.47}$$

while for odd states

$$\psi^{(-)}(x) = \begin{cases} -Ce^{+kx} & \text{for } x \le -a \\ B\sin(qx) & \text{for } -a \le x \le +a \\ +Ce^{-kx} & \text{for } +a \le x \end{cases} \qquad (8.48)$$

Despite the discontinuous nature of the potential at $x = \pm a$, the wavefunction and its derivative are still continuous there and these conditions provide the required boundary conditions to determine the quantized energies. For the even states, for example, these continuity conditions applied at $x = a$ give

$$\psi(a^-) = \psi(a^+) \Longrightarrow A\cos(qa) = Ce^{-ka}$$

$$\psi'(a^-) = \psi'(a^+) \Longrightarrow -qA\sin(qa) = -kCe^{-ka} \qquad (8.49)$$

and the symmetry ensures that the same conditions are obtained at $x = -a$. These can be combined to yield the condition

$$\text{even eigenvalue condition: } q\tan(qa) = k \qquad (8.50)$$

which depends on the energies (through the q, k), but not on the yet to be determined constants A, C; these coefficients can only be completely determined using the overall normalization condition and Eqn. (8.49) (See P8.9.) The appropriate energy eigenvalue condition for odd parity states is easily found to be

$$\text{odd eigenvalue condition: } -q\cot(qa) = k \qquad (8.51)$$

The equivalent eigenvalue conditions for the symmetric infinite well were

$$\text{even states: } \cos(ka) = 0$$

$$\text{odd states: } \sin(ka) = 0 \qquad (8.52)$$

respectively. These are also equations which involve transcendental functions, but which can be solved analytically. In the present case, however, Eqns (8.50) and (8.51) must be solved numerically, and a change to dimensionless variables is a useful first step. If we define

$$y \equiv qa \quad \text{and} \quad R \equiv \sqrt{\frac{2mV_0 a^2}{\hbar^2}} \qquad (8.53)$$

we can use the defining relations for q, k to write the eigenvalue conditions as

$$\text{even states} \quad \sqrt{R^2 - y^2} = +y\tan(y) \qquad (8.54)$$

$$\text{odd states} \quad \sqrt{R^2 - y^2} = -y\cot(y) \qquad (8.55)$$

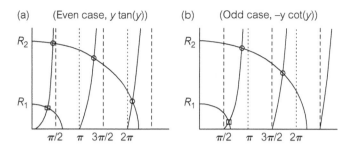

Figure 8.9. Energy eigenvalue conditions for the finite well for (a) even and (b) odd solutions. The vertical dashed (dotted) lines correspond to asymptotes of $\tan(y)$ $(\cot(y))$. Solutions corresponding to two different sets of model parameters, that is, values of $R = \sqrt{2mV_0a^2/\hbar^2}$ are shown; for the value R_2, there are three even and two odd solutions.

A standard method of visualizing the solution space of such equations is to plot both sides of say, Eqn. (8.54), versus y and look for points where the two curves intersect, as such points correspond to the discrete solutions y. We illustrate this in Fig. 8.9 for both the even and odd cases for two values of the dimensionless parameter R (which in turn depends on the dimensionful parameters of the problem). We note that:

- The number of intersections, and hence the number of bound states, is always finite but increases without bound as the values of V_0 and a increase. Thus, deeper (and hence more attractive) and wider potentials have larger numbers of bound states.

- For a fixed value of a, as $V_0 \to \infty$ the "radius" R increases without bound and the intersection points for *even* solutions approach the *asymptotes* of $\tan(y)$, that is, $y \to (n - 1/2)\pi$ where $n = 1, 2, 3 \ldots$. This implies that

$$V_0 - |E| \longrightarrow \frac{\hbar^2 \pi^2 (n - 1/2)^2}{2ma^2} \longrightarrow E_n^{(+)} \tag{8.56}$$

in agreement with the infinite well result. The normalized wavefunctions can also be shown to have the appropriate limit as well (P8.9); the odd solutions approach the $u_n^{(-)}(x)$ in this limit.

- The number of *even* bound states can be easily determined by comparing the value of R with the various *zeroes* of $\tan(y)$ and noting that there will be $n + 1$ *even* bound states if

$$n\pi < R < (n+1)\pi \tag{8.57}$$

There will be *n odd* solutions provided

$$(n - 1/2)\pi < R < (n + 1/2)\pi \tag{8.58}$$

- The values of

$$V_0 - |E| = \frac{\hbar^2 q^2}{2m} = \frac{\hbar^2 y^2}{2ma^2} \tag{8.59}$$

in the finite well are always smaller than the corresponding $E_n^{(\pm)}$ of the infinite well, and we discuss the origin of this effect below.

- No matter how shallow or narrow the well, there is always at least one intersection and hence at least one *even* bound state. It is a general feature that a purely attractive potential (carefully defined) *in one dimension* will always have at least one bound state (see P10.10), but this will not be true in higher dimensions. In contrast, there exists an odd parity bound state only if $R \geq \pi/2$.

The momentum space wavefunctions are also easily obtained (P8.12), and we will examine many of the physical interpretations of these solutions in the next section.

8.2.2 Physical Implications and the Large x Behavior of Wavefunctions

A particle in the finite well described by classical mechanics would still exhibit periodic oscillatory motion between the turning points, here the two walls, provided it was in the well to begin with, that is, if it has $|E| < 0$. Outside the classically allowed region, the kinetic energy of the particle,

$$mv^2/2 = T = E - V(x) = -|E| < 0 \tag{8.60}$$

would be *negative*, so that situation would not be kinematically allowed by energy conservation. In the quantum case, however, we have found explicitly that there is a nonvanishing probability of finding the particle *outside* the classically allowed region since the wavefunction is nonzero (albeit exponentially suppressed) in that region. This phenomenon will have important implications in the area of *quantum tunneling* in Chapter 11.

A remnant of this classical discrepancy can be seen by evaluating the average value of kinetic energy for, say, an even state using the standard definition, namely

$$
\begin{aligned}
\langle \hat{T} \rangle &= \int_{-\infty}^{+\infty} dx\, \psi^{(+)}(x)\, \frac{\hat{p}^2}{2m}\, \psi^{(+)}(x) \\
&= -\frac{\hbar^2}{2m} \int_{-\infty}^{+\infty} dx\, \psi^{(+)}(x)\, \frac{d^2\psi^{(+)}(x)}{dx^2} \\
&= \frac{\hbar^2 q^2 A^2}{m} \left[\int_0^a \cos^2(qx)\, dx \right] - \frac{\hbar^2 k^2 C^2}{m} \left[\int_a^{+\infty} dx\, e^{-2kx} \right]
\end{aligned} \tag{8.61}
$$

which shows that the contribution to $\langle \hat{T} \rangle$ from the region outside the well is indeed negative; the total value, which is what corresponds to a classical observable, is of course positive as it should be.

This behavior is similar to the tunneling wave solution found in Section 2.5.3 for plasma waves, and especially to the exponentially suppressed solution for the asymmetric square well in Section 5.3. where it was argued to be a general feature of quantum mechanical wavefunctions in the classically disallowed region. Similar methods can be used to determine the behavior of bound state wavefunctions for large $|x|$ in more realistic potentials. Imagine, for example, a particle bound in an arbitrary one-dimensional potential, $V(x)$. The Schrödinger equation can be written in the form

$$
\frac{d^2\psi(x)}{dx^2} = \frac{2m}{\hbar^2}(V(x) - E)\psi(x) \tag{8.62}
$$

which one can integrate *approximately* (twice) to give

$$
\psi(x) \sim \exp\left(\pm \sqrt{\frac{2m}{\hbar^2}} \int^x \sqrt{V(x) - E}\, dx \right) \tag{8.63}
$$

For the finite well case, where $V(x)$ is constant, this solution is, in fact, exact. Of most interest is the case when $|x|$ is very large and we have $V(x) \gg E$, in which case we find

$$
\psi \sim \exp\left(\pm \sqrt{\frac{2m}{\hbar^2}} \int^x \sqrt{V(x)}\, dx \right) \tag{8.64}
$$

which can be a useful approximate result for the large x behavior of a one-dimensional wavefunction.

Example 8.1. Large x behavior of the harmonic oscillator wavefunction

As an example, consider a particle moving in a harmonic oscillator potential defined as

$$V(x) = \frac{1}{2}kx^2 = \frac{m\omega^2}{2}x^2 \tag{8.65}$$

so that

$$\int^x \sqrt{V(x)}\,dx = \sqrt{\frac{m\omega^2}{2}}\int^x x\,dx = \sqrt{\frac{m\omega^2}{2}}\frac{x^2}{2} \tag{8.66}$$

so we expect

$$\psi_{SHO}(x) \stackrel{x\to\pm\infty}{\longrightarrow} \exp\left(-\frac{m\omega}{\hbar}\frac{x^2}{2}\right) \tag{8.67}$$

which is indeed what is obtained from an exact solution. These arguments are also useful in that they show the rapid convergence of most realistic wavefunctions, justifying *a posteriori* the assumptions made previously about the "good" behavior of quantum wavefunctions at infinity.

Example 8.2. Large x behavior of wavefunction for uniform accelerating particle

In Section 4.7.2 we considered time-dependent wave packet solutions corresponding to a particle undergoing uniform acceleration, subject to a constant force $F(x) = +F$ or potential $V(x) = -Fx$. We can also use the ideas above to obtain information on the energy eigenvalue or stationary state solutions of the Schrödinger equation for this problem, for large $x > 0$, now corresponding to *unbound* wavefunctions. The time-independent Schrödinger equation in position-space in this case is given by

$$-\frac{\hbar^2}{2m}\frac{d^2\psi(x)}{dx^2} - Fx\psi(x) = E\psi(x) \tag{8.68}$$

and, for simplicity, we will specialize to the case of $E = 0$ (but see Q8.3), namely

$$\frac{d^2\psi(x)}{dx^2} = -\left(\frac{2mF}{\hbar^2}\right)x\,\psi(x) \tag{8.69}$$

Integrating this twice we obtain the approximate solution

$$\psi^{(0)}(x) = \exp\left(\pm i\sqrt{\frac{2mF}{\hbar^2}}\frac{2x^{3/2}}{3}\right) \tag{8.70}$$

The increasingly oscillatory ("wiggly") behavior of this solution as $x \to +\infty$ is indicative of the increasing speed as the particle accelerates to the right, but no information on the relative

(Continued)

spatial probability has appeared in this form. To obtain information on the *magnitude* of the wavefunction, we try an improved solution of the form

$$\psi^{(1)}(x) = x^\alpha \, \psi^{(0)}(x) = x^\alpha \, \exp\left(\pm i \sqrt{\frac{2mF}{\hbar^2}} \frac{2x^{3/2}}{3}\right) \qquad (8.71)$$

and substituting this form into Eqn. (8.69) we find

$$\frac{d^2\psi^{(1)}(x)}{dx^2} = -\left(\frac{2mF}{\hbar^2}\right) x \left[\psi^{(1)}(x)\right] \pm i\sqrt{\frac{2mF}{\hbar^2}} (2\alpha + 1/2) x^{-1/2} \left[\psi^{(1)}(x)\right]$$
$$+ \alpha(\alpha - 1)x^{-2}\left[\psi^{(1)}(x)\right] \qquad (8.72)$$

The new solution will therefore satisfy Eqn. (8.69) to the next order, provided we choose $(2\alpha + 1/2) = 0$ or $\alpha = -1/4$. Thus, we have the approximate large $x > 0$ solution

$$\psi^{(1)}(x) \propto \frac{1}{\sqrt[4]{x}} \, e^{\pm i\sqrt{2mF/\hbar^2}(2x^{3/2}/3)} \qquad (8.73)$$

which corresponds to a spatial probability density

$$P_{QM}(x) = |\psi^{(1)}(x)|^2 \propto \frac{1}{\sqrt{x}} \qquad (8.74)$$

The corresponding classical probability distribution makes use of the classical trajectory information via

$$x(t) = \frac{at^2}{2} \quad \text{and} \quad v(t) = at \quad \implies \quad v(x) = \sqrt{2ax} \qquad (8.75)$$

so that from Section 5.1 we have

$$P_{CL} \propto \frac{1}{v(x)} \propto \frac{1}{\sqrt{x}} \qquad (8.76)$$

as well. While these probability densities are not normalizable (similar to those for plane wave solutions of the free particle Schrödinger equation) they do illustrate both the ability to extract information on the large x behavior of quantum wavefunctions (bound or unbound) and to be able to make comparisons with classical probability notions.

Returning now to the finite well, we illustrate in Fig. 8.10 the energy spectrum of the infinite well and a finite well which has the same width, but only three (two even and one odd) bound states. The finite well is scaled up in energy by adding V_0 for comparison and we note again that the corresponding energy levels are *lower* in energy in the finite well. A look at the corresponding (normalized) wavefunctions in Fig. 8.11 helps illustrate the origin of this effect. The finite well

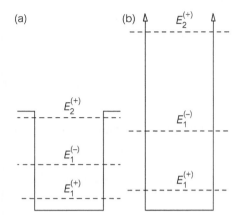

(a)   (b)

Figure 8.10. Comparison of energy levels for finite versus infinite wells of the same width.

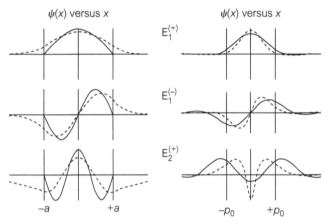

Figure 8.11. Position-space (left) and momentum-space (right) wavefunctions for the symmetric infinite well (solid) and finite well (dashed) energy levels in Fig. 8.10. On the left, the vertical solid lines show the edges of the well at $\pm a$. On the right, the vertical lines indicate the value of p corresponding to $\pm p_0 = \pm\sqrt{2mV_0}$; p values larger than this magnitude would classically allow the particle to escape the well.

solutions, while very similar in structure (i.e. numbers of nodes, parity, etc.), are allowed to tunnel into the classically disallowed region making for a "smoother" overall waveform and hence reducing its overall kinetic energy. Related effects are seen in the corresponding momentum-space wavefunctions (Fig. 8.11) where one sees that:

- The distributions are somewhat narrower (Δp smaller) consistent with the fact that finite well position-space wavefunctions are allowed to tunnel (giving Δx larger).

- The average values of $\langle p^2 \rangle$ for the finite well are correspondingly smaller (implied by the smaller spread in p values), consistent with less kinetic energy.

Using the fully normalized wavefunctions (P8.9), we can also calculate the probability that a position measurement would find the particle *outside* the potential, using

$$\text{Prob}(|x| > a) = \int_{|x|>a} dx \, |\psi_n^{(\pm)}(x)|^2 \tag{8.77}$$

For this particular example, we find that

$$\text{Prob}(|x| > a) = \begin{cases} 0.028 & \text{for } E_1^{(+)} \\ 0.127 & \text{for } E_1^{(-)} \\ 0.493 & \text{for } E_2^{(+)} \end{cases} \tag{8.78}$$

This is seen in Fig. 8.11 where the solid vertical lines indicate the position of the well boundaries; clearly the higher energy states "spend more of their time outside the well."

The corresponding calculation in momentum-space can be performed to give the probability that a measurement of $|p|$ would yield a value of $p^2/2m$ larger than necessary to "unbind" the particle, namely

$$\text{Prob}\left(|p| > \sqrt{2mV_0}\right) = \int_{|p|>\sqrt{2mV_0}} dp \, |\phi_n^{(\pm)}(p)|^2 \tag{8.79}$$

The numerical values corresponding to Fig. 8.11 are

$$\text{Prob}\left(|p| > \sqrt{2mV_0}\right) = \begin{cases} 0.00134 & \text{for } E_1^{(+)} \\ 0.031 & \text{for } E_1^{(-)} \\ 0.115 & \text{for } E_2^{(+)} \end{cases} \tag{8.80}$$

In this case, the solid vertical lines indicate the values of $\pm\sqrt{2mV_0}$.

8.3 Applications to Three-Dimensional Problems

8.3.1 The Schrödinger Equation in Three Dimensions

The extension of the one-particle Schrödinger equation to three dimensions is, in many ways, straightforward. A case of particular interest is when the interaction potential is a function only of the radial distance, that is, $V(\mathbf{r}) = V(r, \theta, \phi) = V(r)$ alone; that is for so-called *central potentials*. In this case, it is natural to use spherical coordinates and consider $\psi(\mathbf{r}) = \psi(r, \theta, \phi)$. Specializing even further to the instance where there is no angular momentum[4] we find (Chapter 16) that the wavefunction can be written in the form

$$\psi(r,\theta,\phi) = \psi(r) = \frac{u(r)}{r}\frac{1}{\sqrt{4\pi}} \tag{8.81}$$

where $u(r)$ satisfies a Schrödinger equation of the form

$$-\frac{\hbar^2}{2m}\frac{d^2u(r)}{dr^2} + V(r)u(r) = Eu(r) \tag{8.82}$$

which is identical to the one-dimensional Schrödinger equation. The difference arises in that the radial coordinate is only defined for $r > 0$, and, because the wavefunction $\psi(r)$ should be well-defined at the origin, one must also have $u(0) = 0$. Such solutions are then like the odd solutions of a typical one-dimensional problem with a symmetric potential.

Furthermore, the normalization condition, derived from an integration over the full three-dimensional position-space, is given by

$$
\begin{aligned}
1 &= \int d^3r\,|\psi(r,\theta,\phi)|^2 \\
&= \int_0^\infty r^2\,dr \int d\Omega\,\left|\frac{u(r)}{r\sqrt{4\pi}}\right|^2 \tag{8.83} \\
&= \int_0^\infty |u(r)|^2\,dr \tag{8.84}
\end{aligned}
$$

The obvious identification of $u(r)$ in three dimensions and $\psi(x)$ (especially odd solutions) in one-dimension, with their many similarities, is an example of how a relatively simple one-dimensional problem can have applications in a realistic three-dimensional system. We examine one specific example from nuclear physics in the next section.

8.3.2 Model of the Deuteron

The hydrogen atom, the two-body system consisting of an electron and proton interacting via the electromagnetic interaction, provides the simplest case in which to study realistically the effects of quantum mechanics in the atomic physics domain. The analogous system in nuclear physics, the deuteron, which consists of a neutron and proton bound by the strong nuclear force, plays somewhat the same role; it is experimentally known, however, to have only one bound state, so its spectrum is far less rich. Nonetheless, it provides some information

---

[4] The special case of no angular momentum ($l = 0$) is a highly nontrivial one, as the ground state solution, which falls into this category, is important for the determination of the ultimately stable configuration of most systems.

on the strength and range of the nucleon–nucleon potential and is amenable to study using the ideas of the last section.

This two-body system in three-dimensions can be described by an effective one-particle, one-dimensional Schrödinger equation provided that

- we use the *reduced mass*, μ, of the two-body system (to be discussed in Chapter 14) defined by

$$\mu = \frac{m_p m_n}{m_p + m_n} \approx \frac{m_p}{2} \approx 480 \text{ MeV}/c^2 \tag{8.85}$$

- and identify the coordinate $\mathbf{r} \equiv \mathbf{r}_1 - \mathbf{r}_2$ with the *relative coordinate* of the two bodies.

A model potential which captures some of the salient features of the nuclear force, especially its finite range, is the half-finite well, namely

$$V(r) = V(|\mathbf{r}_1 - \mathbf{r}_2|) = \begin{cases} -V_0 & \text{for } 0 < r < a \\ 0 & \text{for } r > a \end{cases} \tag{8.86}$$

where, *a priori*, the depth (V_0) and range (a) of the potential are unknown. Since we demand solutions which satisfy the boundary condition $u(r) = 0$, we focus on the relevant *odd* energy eigenfunctions of the symmetric finite well, which are simply given by Eqn. (8.48), namely,

$$u(r) \sim \psi^{(-)}(r) = \begin{cases} B \sin(qr) & \text{for } 0 < r < a \\ E e^{-kr} & \text{for } a < r \end{cases} \tag{8.87}$$

with appropriate matching conditions at $r = a$. We must also satisfy the eigenvalue constraint for odd states, namely

$$\sqrt{R^2 - y^2} = -y \cot(y) \tag{8.88}$$

where $R^2 = 2\mu V_0 a^2/\hbar^2$. We would then like to "fit" experimental data to determine the parameters of the potential; even in the context of a crude model, these give some indication of the properties of the true nuclear force.

Experiments find that it requires a gamma ray of energy roughly $E_\gamma = 2.23$ MeV to disassociate the deuteron, so that it has one bound state with $E = -|E| = -2.23$ MeV. With only this constraint, one cannot uniquely determine V_0 and a simultaneously, but some additional information is available. Scattering methods (Section 19.3) can be used to calculate the mean value of the radius-squared of the system, namely

$$\langle r^2 \rangle \equiv \int d\mathbf{r} \, |\mathbf{r}|^2 \, |\psi(r,\theta,\phi)|^2 = \int_0^\infty dr \, r^2 \, |u(r)|^2 \tag{8.89}$$

and the corresponding experimentally measured value is roughly $\langle r^2 \rangle \approx (4.2\ F)^2$. For the appropriate quasi-odd wavefunctions, this condition reduces to

$$\langle r^2 \rangle = B^2 \int_0^a dr\, r^2\, \sin^2(kr)r + E^2 \int_a^\infty dr\, r^2\, e^{-2kr} \tag{8.90}$$

where B, E are further constrained by the normalization condition

$$1 = B^2 \int_0^a dr\, \sin^2(kr)\, dr + E^2 \int_a^\infty dr\, e^{-2kr} \tag{8.91}$$

and the boundary conditions at $r = a$. These constraints can be fit numerically and one finds

$$a \approx 2.4\ F \quad \text{and} \quad V_0 \approx 27\ \text{MeV} \tag{8.92}$$

which do give a reasonable indication of the range and strength of the nuclear force as determined from other experiments. We note that:

- The depth of the potential also satisfies another plausible constraint, namely that the attractive nucleon–nucleon potential at these distances is much larger than the corresponding proton–proton repulsion due to their electrostatic interaction. Specifically, we find that

$$V_{\text{Coul}}(r = a) = \frac{Ke^2}{a} = \frac{1.44\ \text{MeV}\ F}{2.4\ F} \approx 0.6\ \text{MeV} \ll V_0 = 27\ \text{MeV} \tag{8.93}$$

- The fact that $|E|$ is so close to zero (compared to the value of V_0) implies that the state is extremely weakly bound. The two particles, in fact, spend roughly 60% of the time *outside* the range of the potential. This is also clear from the fact that $\sqrt{\langle r^2 \rangle} > a$. The approximate deuteron radial probability density, $|u(r)|^2$ versus r, is shown in Fig. 8.12.

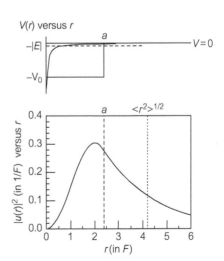

Figure 8.12. Radial probability density, $|u(r)|^2$ versus r, for the finite well model of the deuteron. The single bound state energy level is shown on the $V(r)$ versus r plot, as well as the Coulomb potential (solid curve) for comparison.

8.4 **Questions and Problems**

Q8.1. If you constructed a wave packet for the finite well representing a bound particle, that is, one with $\langle E \rangle < 0$, would you ever measure it with $E > 0$?

Q8.2. What is wrong with this picture? Figure 8.13 shows an energy level in a generic potential well. One of the position-space wavefunctions below it is the real solution. Identify the wrong solution and describe as many things wrong with the purported solution as you can; focus on the magnitude and "wiggliness" of $\psi(x)$ and its behavior in the classically disallowed region.

Q8.3. How would you generalize the results for the large x wavefunction for the uniformly accelerating particle in Example 8.2 for the case of $E \neq 0$.

Q8.4. How would you use the information in Fig. 8.12 to quantitatively confirm the statement made in Section 8.3.2 that the neutron and proton ... *spend roughly 60% of the time outside the range of the potential...*? Would having a mesh of grid lines added to the figure help?

P8.1. Show directly from Eqn. (8.1) that $\psi''(x)$ will be discontinuous if $V(x)$ is discontinuous. Can you calculate $\psi'''(x)$ and discuss the connection between its continuity and that of $V(x)$?

P8.2. For the single δ-function potential, show that the average values of kinetic and potential energy satisfy

$$ E = \frac{1}{2}\langle V(x) \rangle = -\langle \hat{T} \rangle \qquad (8.94) $$

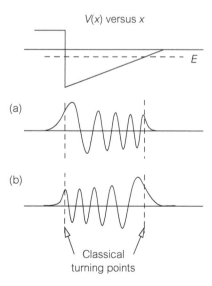

Figure 8.13. Energy levels in a generic potential well (top) and two purported solutions (a) and (b). Which one is right, and why?

using the position-space wavefunctions. You might wish to use the expression

$$\langle \hat{T} \rangle = \frac{\hbar^2}{2m} \int_{-\infty}^{+\infty} \left| \frac{d\psi(x)}{dx} \right|^2 dx \qquad (8.95)$$

for the kinetic energy averaging to avoid problems with higher derivatives.

P8.3. Single-delta function potential in momentum-space. One can solve the problem in Section 8.1.2 directly in momentum-space[5] in a way which illustrates, in a manner not seen so far, how boundary conditions give rise to quantized energy levels.

(a) As in Section 4.7.1, take the Schrödinger equation for this problem

$$\frac{\hat{p}^2}{2m}\psi(x) - g\delta(x)\psi(x) = -|E|\psi(x) \qquad (8.96)$$

multiply both sides by $(1/\sqrt{2\pi\hbar})\exp(-ipx/\hbar)$, and then integrate over x to directly obtain

$$\phi(p) = \frac{2mg}{\sqrt{2\pi\hbar}}\left(\frac{\psi(0)}{p^2 + (\hbar K)^2}\right) \qquad (8.97)$$

which immediately gives the Lorentzian form in Eqn. (8.10).

(b) To obtain the position-space wavefunction, Fourier transform back to obtain

$$\psi(x) = \frac{mg\psi(0)}{\hbar^2 K} e^{-|Kx|} \qquad (8.98)$$

which is also seen to be the correct form.

(c) Self-consistency demands that $\psi(0)$ be well-defined. Show that this condition gives the correct condition for the energy eigenvalue, namely

$$K = \frac{mg}{\hbar^2} \quad \text{or} \quad E = -\frac{mg^2}{2\hbar^2} \qquad (8.99)$$

in agreement with Eqn. (8.8). Does this approach imply that we must have normalized the wavefunction before determining the energy eigenvalue?

P8.4. (a) Find the normalized position-space wavefunctions for the even and odd state (when it exists) for the twin δ-function potential in Section 8.1.3. Show that for large separation (namely, $a \gg a_0$) they can be written as

$$\psi^{(\pm)}(x) \longrightarrow \frac{1}{\sqrt{2}}[\psi_1(x-a) \pm \psi_1(x+a)] \qquad (8.100)$$

where $\psi_1(x)$ is the solution for the single δ-function potential in Eqn. (8.9).

---

[5] For a similar derivation, see Lieber (1975).

(b) Say one tries to localize a particle around one or the other δ-function spikes by taking the initial wavefunction

$$\psi(x,0) = \frac{1}{\sqrt{2}} \left[\psi^{(+)}(x) + \psi^{(-)}(x) \right] \tag{8.101}$$

Does the particle stay localized for later times? What is the natural timescale for the problem? Hint: How does the energy difference, ΔE, between the two states scale when $a/a_0 \gg 1$?

P8.5. Three δ-function potential. Find the energy eigenvalue conditions for even and odd states for the three δ-function potential given by

$$V(x) = -g \left(\delta(x-a) + \delta(x) + \delta(x+a) \right) \tag{8.102}$$

Discuss the energy spectrum as a/a_0 is varied from large values to small. Is there a critical value of a/a_0 for which some states 'unbind'?

P8.6. Find the form of the momentum-space wavefunctions for the twin δ-function potential in Eqn. (8.11) by using the methods of P8.3. How do you obtain the energy eigenvalue condition? Would this work for a more complicated set of δ-function potentials as in P8.5?

P8.7. Consider the wavefunction in P4.15, defined over the range $(0, a)$, which has a cusp at $0 < x = c < a$. If this wavefunction is to be a solution of the Schrödinger equation in an infinite well, show that there must also be a δ-function potential present in the well, and find its strength, g. Do you need to normalize the wavefunction first in order to be able to extract g? With the appropriate strength δ-function added to the infinite well, what is the energy eigenvalue associated with the wavefunction in Eqn. (4.178)? Calculate the expectation values of both the kinetic and potential energies and confirm that they reproduce this energy eigenvalue.

P8.8. Infinite well plus δ-function potential. Consider the hybrid potential[6] consisting of the sum of symmetric infinite well potential

$$V_\infty(x) = \begin{cases} 0 & \text{for } |x| < a \\ \infty & \text{for } |x| > a \end{cases} \tag{8.103}$$

and a δ-function at the origin, namely

$$V(x) = V_\infty + g\delta(x) \tag{8.104}$$

where we make no assumption about the sign of g.

(a) Show that the potential is symmetric so that the solutions will be eigenstates of parity.

(b) Using that fact, solve the Schrödinger equation inside the well for both positive and negative parity states for *nonnegative* energies E. Apply the

---

[6] Some aspects of this problem are worked out in Lapidus (1982b).

appropriate boundary conditions (being especially careful at $x = 0$) and determine the energy eigenvalue conditions. Show that the negative energy states are unchanged by the presence of the potential.

(c) Repeat assuming negative energy states, that is, $E = -|E| < 0$.

(d) Plot the eigenvalue condition for even solutions, expressed as a function of the dimensionless parameter y and discuss how the even energies vary as the value of g is varied in the range $(-\infty, +\infty)$.

(e) Find the value of g for which there is a zero energy ground state. Find and sketch its wavefunction, and compare to P4.15.

(f) Show that for $g \to +\infty$, the even energies approach those of the odd states directly above them. How do their wavefunctions compare?

(g) Repeat for the case $g \to -\infty$ using the form appropriate for negative energy solutions.

P8.9. Normalization of the finite well eigenfunctions.

(a) Show that the appropriate normalization constants for the even states of the finite well are given by

$$A = \frac{1}{\sqrt{a}}\left(1 + \frac{\sin(2y)}{2y} + \frac{\cos^3(y)}{y\sin(y)}\right)^{-1/2}$$

$$= \frac{1}{\sqrt{a}}\left(1 + \frac{\sin(2y)}{2y} + \frac{\cos^2(y)}{ka}\right)^{-1/2} \tag{8.105}$$

$$C = A\cos(y)e^{ka} \tag{8.106}$$

(b) Show that the even finite well wavefunctions approach those of the infinite well in the limit $V_0 \to \infty$, that is, when $y \to (n - 1/2)\pi$.

(c) Calculate the probability that a measurement of the position of the particle will find it outside the classical disallowed region.

P8.10. (a) Evaluate the average value of kinetic energy for the *even* states in the finite well and show that your result can be expressed in the form

$$\langle \hat{T} \rangle = \frac{\hbar^2 y^2}{2m}\left[\frac{1}{1 + \sin(2y)/2y + \cos^2(y)/ak}\right] \tag{8.107}$$

How does this compare to the kinetic energy in the infinite well?

(b) Show that the ratio of the contribution of the kinetic energy from **outside** the well to the total is given by

$$\langle \hat{T} \rangle_{\text{out}}/\langle \hat{T} \rangle_{\text{total}} = -\frac{\sin(2y)}{2y} \tag{8.108}$$

Show that this increases in magnitude for less-bound states. Show that it has the appropriate limit for the infinite well.

P8.11. Suppose you tried to model a hydrogen atom in one-dimension by assuming that the electron of rest energy $m_e c^2 = 0.51$ MeV was in a finite well of width $2a$ with $a = 0.53$ Å (which is like the Bohr radius) and of depth $V_0 = 30$ eV (which is something like the value of the Coulomb potential evaluated at $r = a$). How many bound states would you find? Would this be a very realistic model?

P8.12. Calculate the momentum-space wavefunction, $\phi^{(+)}(p)$, for the finite well corresponding to the even solutions $\psi^{(+)}(x)$.

P8.13. Limit of a narrow and deep finite square well. The attractive δ-function potential, $V(x) = -g\delta(x)$, can be obtained from the finite well potential by taking the limits $V_0 \to \infty$ while $a \to 0$ in such a way that the "area" under the two potentials is kept the same, namely, $2V_0 a = g$.

(a) Show that in this limit, there is only one bound state for the finite well and that its energy reduces to that obtained in Eqn. (8.8).

(b) Show that the wavefunction approaches Eqn. (8.9) in this limit.

(c) In one experiment[7] a potential well of depth of roughly 10 meV and width of approximately 600 Å is constructed. A "potential spike" is added during the growth process to mimic a δ-function perturbation, with width ~ 9 Å and strength ~ 80 meV. Would it be appropriate to consider this addition as a perturbation to the original well?

P8.14. Large x behavior for the linear potential. Using Eqn. (8.64), find the large $|x|$ behavior of the wavefunction for a particle of mass m moving in a linear confining potential of the form $V(x) = C|x|$.

P8.15. How much lighter would the masses of the neutron and proton (assumed the same) have to be in order for there to be no deuteron at all, that is, no bound state? How much heavier would they have to be to have three bound states? Assume that the parameters of the potential stay the same.

---

[7] Salis et al. (1997).

NINE

The Harmonic Oscillator

9.1 The Importance of the Simple Harmonic Oscillator

The problem of a particle moving under the influence of a linear restoring force, $F(x) = -Kx$, or equivalently a quadratic potential, $V(x) = Kx^2/2$, is a problem which is studied at all levels of theoretical physics, from elementary classical mechanics through quantum field theory.[1] One of its most useful features is that, at every stage of development, it is exactly soluble and so can be easily used as a closed-form, analytic example. If it had no connection to real physical systems, however, that fact would be of only academic interest. In this section we mention two important applications of the harmonic oscillator problem to illustrate its potential wide-ranging usefulness. In Section 9.2, we will then derive its solution in nonrelativistic quantum mechanics, using a standard differential equation approach, and discuss its experimental realizations and classical limits in Sections 9.3 and 9.4. We also discuss wave packet solutions for the harmonic oscillator, but later in Section 12.6.2.

In classical mechanics, a conservative system can be described by a potential energy function, $V(x)$, and states of the system which will be in equilibrium will be found at the extrema of this potential, that is, places where

$$\left. \frac{dV(x)}{dx} \right|_{x=x_0} = -F(x_0) = 0 \tag{9.1}$$

or where the classical force vanishes as suggested by Newton's law. The *stability* of the equilibrium point is governed by the sign of the second derivative with

$$\left. \frac{d^2V(x)}{dx^2} \right|_{x=x_0} > 0 \quad \text{or} \quad \left. \frac{d^2V(x)}{dx^2} \right|_{x=x_0} < 0 \tag{9.2}$$

---

[1] For a comprehensive survey of the harmonic oscillator in classical and quantum mechanics, see Pippard (1978, 1983).

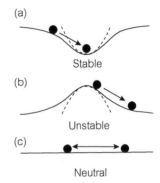

Figure 9.1. Generic potential with stable, unstable, and neutral equilibrium points.

implying stable, bounded, and periodic motion (for $V''(x_0) > 0$) or unstable (in fact exponentially increasing) and unbounded motion (for $V''(x_0) < 0$), respectively, as in Fig. 9.1.

We can expand the potential near the point of equilibrium as a power series,

$$V(x) = V(x_0) + \left.\frac{dV(x)}{dx}\right|_{x=x_0} (x - x_0) + \frac{1}{2}\left.\frac{d^2 V(x)}{dx^2}\right|_{x=x_0} (x - x_0)^2 + \cdots$$

$$= V_0 + \frac{1}{2}K(x - x_0)^2 + \cdots \tag{9.3}$$

where the effective "spring constant" is $K \equiv d^2 V(x)/dx^2|_{x=x_0}$. Since the constant part of the potential cannot affect the physics in any meaningful way, we see that:

- The simple harmonic oscillator (hereafter SHO) potential, $V(x) = Kx^2/2$, is often the "best first guess" for the potential near a point of stable equilibrium.

If the equilibrium point happens to be at $x_0 = 0$, we have,

$$m\ddot{x}(t) \approx -|K|x(t) \quad \text{when } K > 0 \tag{9.4}$$
$$m\ddot{x}(t) \approx +|K|x(t) \quad \text{when } K < 0 \tag{9.5}$$

with general solutions

$$x(t) = A\cos(\omega t) + B\sin(\omega t) \quad \text{for stable equilibrium} \tag{9.6}$$
$$x(t) = Ce^{\omega t} + De^{-\omega t} \quad \text{for unstable equilibrium} \tag{9.7}$$

where $\omega = \sqrt{K/m}$ in both cases; the constants are determined, of course, by the initial conditions. While the first case is by far the most important, we will also briefly discuss the quantum analog of unstable motion in Section 9.5.

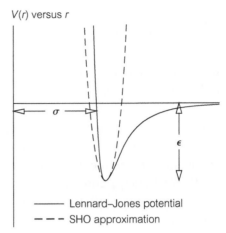

V(r) versus r

──── Lennard–Jones potential
── ── ── SHO approximation

Figure 9.2. $V(r)$ versus r for a Lennard–Jones potential (solid) and the corresponding SHO approximation (dashed).

A realization of this idea comes in the study of diatomic molecules[2] which interact via a potential of the generic type shown in Fig. 9.2. The particular form shown is often called a *Lennard–Jones* or 6–12 potential of the form

$$V(r) = 4\epsilon \left(\left(\frac{\sigma}{r} \right)^{12} - \left(\frac{\sigma}{r} \right)^{6} \right) \tag{9.8}$$

and is especially relevant for the interaction of spherical (i.e. noble gas) atoms. This functional form is not as good a representation of the potential for other types of bonding (i.e. ionic or covalent), but there will be a stable minimum in those systems as well. In this case, the potential is a function of the distance between the two molecules, $r \equiv |\mathbf{r}_1 - \mathbf{r}_2|$, and the reduction to relative coordinates (discussed already in Section 8.3.2) implies that the reduced mass of the system, defined as $\mu = m_1 m_2/(m_1 + m_2)$, should be used. A harmonic oscillator potential can be used to fit the potential near the stable equilibrium point with the classical result that such diatomic molecules will oscillate around their equilibrium separation with frequency

$$\omega = \sqrt{\frac{K}{\mu}} \quad \text{where } K = \left. \frac{d^2 V(r)}{dr^2} \right|_{r=r_0} \tag{9.9}$$

with any amplitude possible. The extent to which this approximation is a good one depends on the size of the anharmonic terms (the higher-order terms in the expansion of $V(r)$) as the amplitude of the motion increases. Classically, it is always possible to have oscillations with amplitude small enough that the SHO

---

[2] See any good text on modern physics for a further discussion of molecular bonding, for example, Eisberg and Resnick (1985).

approximation is a good one. Quantum mechanically, however, we know there will be quantized energy levels in such a confining potential; the extent to which an SHO approximation is useful now depends on the size of the energy level spacing and especially that of the lowest level given by the zero-point energy. We will discuss this in more detail in Section 9.3.

Another example, which at first seems very remote from classical considerations of vibrating masses and springs, comes from the desire to apply quantum ideas to the oscillations of the electromagnetic (EM) field. We give a preliminary discussion here[3] and return to the subject in Chapter 18.

The total energy contained in a configuration of EM fields is given by an integral over the respective electric $(u_E(\mathbf{r}, t))$ and magnetic $(u_B(\mathbf{r}, t))$ energy densities,

$$E_{\text{tot}} = \int d\mathbf{r} \, (u_E(\mathbf{r}, t) + u_B(\mathbf{r}, t)) = \int d\mathbf{r} \left(\frac{\epsilon_0}{2} |\mathbf{E}(\mathbf{r}, t)|^2 + \frac{1}{2\mu_0} |\mathbf{B}(\mathbf{r}, t)|^2 \right)$$

$$(9.10)$$

In a region of space where there is no charge density (so that the scalar potential, $\phi(\mathbf{r}, t)$, can be neglected), the electric and magnetic fields can be written in terms of the so-called vector potential, $\mathbf{A}(\mathbf{r}, t)$, via

$$\mathbf{E}(\mathbf{r}, t) = -\frac{\partial \mathbf{A}(\mathbf{r}, t)}{\partial t} \quad \text{and} \quad \mathbf{B}(\mathbf{r}, t) = -\nabla \times \mathbf{A}(\mathbf{r}, t) \tag{9.11}$$

We can write the vector potential in terms of its Fourier components via

$$\mathbf{A}(\mathbf{r}, t) = \frac{1}{(2\pi)^{3/2}} \int d\mathbf{k} \, \mathbf{A}(\mathbf{k}, t) \, e^{i\mathbf{k}\cdot\mathbf{r}} \tag{9.12}$$

and the total energy can be rewritten in the form

$$E_{\text{tot}} = \int d\mathbf{k} \left(\frac{\epsilon_0}{2} |\dot{\mathbf{A}}(\mathbf{k}, t)|^2 + \frac{k^2}{2\mu_0} |\mathbf{A}(\mathbf{k}, t)|^2 \right) \tag{9.13}$$

If we write the energy of a standard harmonic oscillator in terms of its amplitude, $x(t)$, as

$$E_{\text{tot}} = \frac{1}{2} m\dot{x}^2(t) + \frac{1}{2} Kx^2(t) \tag{9.14}$$

we can see that there is a very real sense in which the EM field can be considered as an infinite collection (specifically an integral over $d\mathbf{k}$) of harmonic oscillators, each with amplitude $\mathbf{A}(\mathbf{k}, t)$. Comparing the coefficients of the amplitude and

---

[3] For a more comprehensive discussion at this level, see Baym (1976).

derivative terms gives the appropriate frequency for both cases; for the mass and spring case we have

$$\omega = \sqrt{\frac{K/2}{m/2}} = \sqrt{\frac{K}{m}} \tag{9.15}$$

while for the EM field case we have

$$\omega_k = \sqrt{\frac{k^2/2\mu_0}{\epsilon_0/2}} = kc \quad \text{since } c = \frac{1}{\sqrt{\epsilon_0\mu_0}} \tag{9.16}$$

which is indeed the appropriate dispersion relation for photons. One intriguing implication of this result is that since there is a nonvanishing zero-point energy for *each* k mode, the total vacuum energy of the EM field will actually be divergent.[4]

9.2 Solutions for the SHO

9.2.1 Differential Equation Approach

We will now apply some standard techniques for the solution of ordinary differential equations to solve for the allowed energy eigenvalues and eigenfunctions, corresponding to the SHO potential in the standard form

$$V(x) = \frac{Kx^2}{2} = \frac{m\omega^2 x^2}{2} \tag{9.17}$$

We write the time-independent Schrödinger equation in position-space as

$$-\frac{\hbar^2}{2m}\frac{d^2\psi(x)}{dx^2} + \frac{m\omega^2 x^2}{2}\psi(x) = E\psi(x) \tag{9.18}$$

and we attempt a change of variables to make Eqn. (9.18) dimensionless. Specifically, we define $x = \rho y$ where ρ has the units of length and will be determined below. Substituting this above we find

$$\frac{d^2\psi(y)}{dy^2} - \frac{m^2\omega^2\rho^4}{\hbar^2}y^2\psi(y) = -\frac{2mE\rho^2}{\hbar^2}\psi(y) \tag{9.19}$$

---

[4] This property of the quantum version of the vacuum electric and magnetic fields has observable consequences in the so-called *Casimir effect*, a nice discussion of which is given by Elizalde and Romeo (1991).

which reduces to

$$\frac{d^2\psi(y)}{dy^2} - y^2\psi(y) = -\epsilon\psi(y) \tag{9.20}$$

provided we define

$$\rho \equiv \sqrt{\frac{\hbar}{m\omega}} \quad \text{and} \quad \epsilon = \frac{2E}{\hbar\omega} \tag{9.21}$$

where ϵ is now a dimensionless eigenvalue. We note that these are the same combinations of parameters for length and energy given simply by dimensional analysis in Example 1.4 and in P4.23.

Because the behavior of $\psi(y)$ at infinity is important for the existence of normalizable solutions, we first examine Eqn. (9.20) for large $|y|$,

$$\frac{d^2\psi(y)}{dy^2} \approx y^2\psi(y) \tag{9.22}$$

which has approximate solutions, $\psi(y) = e^{\pm y^2/2}$. Choosing only the square integrable solution leads us to try to "factor out" the behavior at infinity, once and for all, by writing

$$\psi(y) \equiv h(y)e^{-y^2/2} \tag{9.23}$$

Substitution into Eqn. (9.20) then gives

$$\frac{d^2h(y)}{dy^2} - 2y\frac{dh(y)}{dy} + (\epsilon - 1)h(y) = 0 \tag{9.24}$$

Since the potential $V(x)$ is symmetric, we know in advance that the solutions will be eigenfunctions of parity, that is, even and odd states; this implies that the $h(y)$ can be classified by their parity, and since this simplifies the analysis, we consider even solutions first.

A standard method of solution which can be applied to this problem is to assume a *power series expansion* for $h(y)$ so that in this even case we write

$$h^{(+)}(y) = \sum_{s=0}^{\infty} a_s y^{2s} \quad \text{(using only even powers of } y) \tag{9.25}$$

which must then satisfy

$$\sum_{s=0}^{\infty} 2s(2s-1)a_s y^{2s-2} + \sum_{s=0}^{\infty} (\epsilon - 1 - 4s) a_s y^{2s} = 0 \tag{9.26}$$

Using our freedom to relabel the dummy summation index in the first term by letting $s \to s + 1$, we can combine terms to obtain

$$\sum_{s=0}^{\infty} [2(s+1)(2s+1)a_{s+1} + (\epsilon - 1 - 4s)a_s] \, y^{2s} = \sum_{s=0}^{\infty} B_s y^{2s} = 0 \qquad (9.27)$$

This can only be true for every value of y if all of the coefficients B_s vanish separately, that is,

$$B_s = 0 \Longrightarrow a_{s+1} = a_s \left[\frac{4s+1-\epsilon}{2(s+1)(2s+1)} \right] \qquad (9.28)$$

Equation (9.28) is now a *recurrence* or *recursion relation*, which implicitly gives the solution because all of the coefficients are given in terms of the single parameter a_0 via

$$a_1 = a_0 \frac{(1-\epsilon)}{2}, \quad a_2 = a_1 \frac{(5-\epsilon)}{12} = a_0 \frac{(5-\epsilon)(1-\epsilon)}{24} \qquad (9.29)$$

and so forth. This gives

$$\psi^{(+)}(y) = a_0 \left[1 + \frac{a_1}{a_0} y^2 + \frac{a_2}{a_0} y^4 + \cdots \right] e^{-y^2/2} \qquad (9.30)$$

The fact that this solution of a second-order differential equation has only a single undetermined constant is explained by the fact that while a_0 can determine $\psi(0)$, $\psi'(0) = 0$ is already determined by the fact that $h^{(+)}(y)$ is an even function.

This solution for $h^{(+)}(y)$ is a nicely convergent series as a ratio test (Appendix D.2) gives

$$\frac{a_{s+1} y^{2s+2}}{a_s y^{2s}} \to \frac{y^2}{s} \to 0 \quad \text{as } s \to \infty \text{ for fixed } y \qquad (9.31)$$

We note, however, that the function e^{y^2}, which has the series expansion

$$e^{y^2} = 1 + y^2 + \frac{1}{2!}(y^2)^2 + \cdots = \sum_{s=0}^{\infty} \frac{1}{s!} (y^2)^s \qquad (9.32)$$

has the identical limiting behavior, so that

$$h^{(+)}(y) \to e^{y^2} \quad \text{and} \quad \psi^{(+)}(y) = h^{(+)}(y) \, e^{-y^2/2} \to e^{+y^2/2} \qquad (9.33)$$

So, while we have tried to eliminate the divergent behavior by hand, it has re-emerged in the series solution.

This argument relies, however, on the comparison of ratio tests, that is, on the assumption that the series solution for $h^{(+)}(y)$ has infinitely many terms. This problem is avoided if the series is, in fact, finite, that is, if it terminates for some finite value of s. If $a_n = 0$ for some value of n, then Eqn. (9.28) guarantees that all subsequent a_{n+k} will also vanish, so that $h^{(+)}(y)$ will be simply a polynomial in y^2. This can happen only if

$$a_n = 0 \Longrightarrow \epsilon = (4n + 1) \quad \text{for} \quad n = 0, 1, 2 \ldots \tag{9.34}$$

which implies the quantized energies

$$E_n^{(+)} = \hbar\omega(2n + 1/2) \quad \text{where } n = 0, 1, 2 \ldots \tag{9.35}$$

We plot in Fig. 9.3 the wavefunction corresponding to the first of these special even functions (i.e. $\epsilon = 1$), namely, the ground state, as well as solutions characterized by values of ϵ, which are only slightly different; each of these wavefunctions are solutions of the Schrödinger equation for the harmonic oscillator, but only one is square-integrable and therefore suitable as a physical solution. We note once again how the the requirement of normalizable wavefunctions leads to quantized energies for bound states. The acceptable even solutions are then

$$E_0^{(+)} = \frac{\hbar\omega}{2} \quad \psi_0^{(+)}(y) = a_0 \, e^{-y^2}$$

$$E_1^{(+)} = \frac{5\hbar\omega}{2} \quad \psi_1^{(+)}(y) = a_0(1 - 2y^2) \, e^{-y^2/2}$$

$$E_2^{(+)} = \frac{9\hbar\omega}{2} \quad \psi_2^{(+)}(y) = a_0(1 - 4y^2 + 4y^4/3) \, e^{-y^2/2} \tag{9.36}$$

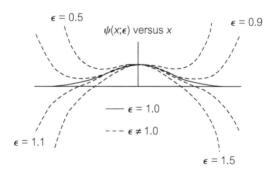

Figure 9.3. Square integrable ground state wavefunction (solid) for the SHO corresponding to $\epsilon = 1$ and divergent "near misses" (dashed curves) for various $\epsilon \neq 1$. Each of the wavefunctions are solutions of the Schrödinger equation for the harmonic oscillator problem, but only the $\epsilon = 1$ satisfies the necessary boundary conditions at $\pm\infty$ and is allowed.

The odd states can be obtained in an entirely similar way by assuming a series solution of the form

$$h^{(-)}(y) = \sum_{s=0}^{\infty} b_s y^{2s+1} \qquad \text{(using only odd powers of } y) \qquad (9.37)$$

which yields the recursion relation

$$b_{s+1} = b_s \left[\frac{4s + 3 - \epsilon}{2(s+1)(2s+3)} \right] \qquad (9.38)$$

and acceptable energies

$$E_n^{(-)} = (2n + 3/2)\hbar\omega \quad \text{with } n = 0, 1, 2 \ldots \qquad (9.39)$$

9.2.2 Properties of the Solutions

Collecting these results, we find that the two classes of solutions can be combined with a single label to write

$$E_n = (n + 1/2)\hbar\omega \quad n = 0, 1, 2, \ldots \qquad (9.40)$$

with corresponding normalized solutions, which are expressible in the form

$$\psi_n(x) = C_n h_n(y) \, e^{-y^2/2} \quad \text{where } y = \sqrt{\frac{m\omega}{\hbar}} x = \frac{x}{\rho} \qquad (9.41)$$

The $h_n(y)$ can be recognized as the *Hermite polynomials* (see Appendix E.3), and the normalization constants are found to be

$$C_n = \left(\frac{\sqrt{m\omega/\hbar\pi}}{2^n n!} \right)^{1/2} = \left(\frac{1}{\rho\sqrt{\pi}\,2^n n!} \right)^{1/2} \qquad (9.42)$$

Several comments can be made:

- The $h_n(y)$ are polynomials in y of degree n with parity $+1$ (-1) depending on whether n is even (odd). The first few can easily be obtained from the recursion relations (P9.2) and are given by

$$h_0(y) = 1 \quad h_1(y) = 2y$$
$$h_2(y) = 4y^2 - 2 \quad h_3(y) = 8y^3 - 12y$$
$$h_4(y) = 16y^4 - 48y^2 + 12 \quad h_5(y) = 32y^5 - 160y^3 + 120y \qquad (9.43)$$

- The first few energy eigenfunctions are plotted in Fig. 9.4 and show the usual pattern of increasing numbers of nodes and alternation of even and odd states.

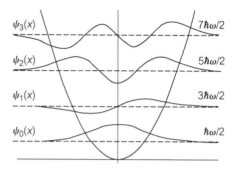

Figure 9.4. SHO wavefunctions for the lowest-lying energy levels, $n = 0, 1, 2, 3$.

- The nodeless ground state has a zero-point energy of $E_0 = \hbar\omega/2$ consistent with the uncertainty principle discussion of P1.14. In this case, $\psi_0(x)$ is a familiar Gaussian waveform implying that its Fourier transform, $\phi_0(p)$, is as well Gaussian. It can be shown that this wave packet has the minimum uncertainty principle product $\Delta x \Delta p = \hbar/2$ (Section 12.4).

- Because of parity, we have the expectation value

$$\langle \psi_n | x | \psi_n \rangle = \int_{-\infty}^{+\infty} dx\, x\, |\psi_n(x)|^2 = 0 \tag{9.44}$$

while

$$\langle \psi_n | \hat{p} | \psi_n \rangle = \int_{-\infty}^{+\infty} dx\, \psi^*(x)\, \hat{p}\, \psi(x) = 0 \tag{9.45}$$

as usual for energy eigenfunctions for bound states. Other average values will be more easily obtained using the operator formalism in Chapter 13; we list here for convenience the results

$$\langle \psi_n | x^2 | \psi_n \rangle = \frac{\hbar}{m\omega}(n+1/2) \tag{9.46}$$

$$\langle \psi_n | \hat{p}^2 | \psi_n \rangle = \hbar m\omega(n+1/2) \tag{9.47}$$

$$\langle \psi_n | x^4 | \psi_n \rangle = \frac{3}{4}\left(\frac{\hbar}{m\omega}\right)^2 (2n^2 + 2n + 1) \tag{9.48}$$

$$\langle \psi_n | \hat{p}^4 | \psi_n \rangle = \frac{3}{4}(\hbar m\omega)^2 (2n^2 + 2n + 1) \tag{9.49}$$

$$\langle \psi_n | x | \psi_k \rangle = \sqrt{\frac{\hbar}{2m\omega}}\left(\delta_{n,k-1}\sqrt{k} + \delta_{n,k+1}\sqrt{k+1}\right) \tag{9.50}$$

$$\langle \psi_n | \hat{p} | \psi_k \rangle = -i\sqrt{\frac{m\omega\hbar}{2}}\left(\delta_{n,k-1}\sqrt{k} - \delta_{n,k+1}\sqrt{k+1}\right) \tag{9.51}$$

- The solutions form a complete set of states so that any acceptable wavefunction can be expanded via $\psi(x) = \sum_{n=0}^{\infty} a_n \psi_n(x)$ in the usual way

Finally, the harmonic oscillator problem is the only one where the potential energy enters quadratically and this implies a complete symmetry between x and p representations; recall P4.23. For example, the Schrödinger equation in momentum-space reads

$$\frac{p^2}{2m}\phi(p) - \frac{\hbar^2 m\omega^2}{2}\frac{d^2\phi(p)}{dp^2} = E\phi(p) \qquad (9.52)$$

which is of the same form as Eqn. (9.18) in position-space. The solutions are easily obtained by the same methods, and one finds

$$\phi_n(p) = D_n h_n(q)e^{-q^2/2} \quad \text{where } q = \frac{p}{\sqrt{m\omega\hbar}} \qquad (9.53)$$

with normalization constants

$$D_n = \left(\frac{1}{n!2^n\sqrt{m\omega\hbar\pi}}\right)^{1/2} \qquad (9.54)$$

The momentum-space probability densities, plotted as a function of the scaled variable q, therefore have the same form as in Fig. 9.4.

9.3 **Experimental Realizations of the SHO**

Several aspects of these quantized vibrational states are clearly evident in the study of diatomic molecules; they include:

• Approximately evenly spaced energy levels
• Vibrational contribution to the heat capacity of diatomic gases
• The variation of the zero-point energy with the constituent masses

and we (briefly) discuss each in turn.

The validity of an SHO description of the quantized energy levels in a Lennard–Jones potential is illustrated in Fig. 9.5 for both relatively light (Ne–Ne) and heavy (Xe–Xe) molecules; clearly the approximation of the potential near the minimum is much better for the heavier molecules which sit 'further down' in the potential well. In the case of NaCl, for example, where the (ionic) binding potential is much larger, there are roughly 20 such levels which can be detected spectroscopically. As the quantum number is increased, however, the potential becomes more anharmonic, and the level spacings decrease somewhat (Can you argue why?). Since diatomic molecules which vibrate can also rotate,

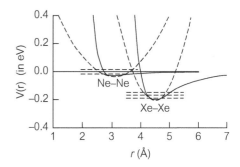

Figure 9.5. Lennard–Jones interaction potentials for pairs of noble gas atoms. The horizontal dashed lines indicate the first few energy levels using the SHO (dashed) approximation.

the energy spectrum also contains rotational levels, which we discuss in more detail in Chapter 16.

The vibrational motion can also contribute to the heat capacity of a system of diatomic molecules. Classically, each degree of freedom will contribute $k_B T/2$ to the thermal energy. For a monatomic gas, with only three translational degrees of freedom, this would yield a specific heat per mole given by

$$C_V = \frac{\partial E}{\partial T} = \frac{\partial}{\partial T}\left(\frac{3}{2}N_A k_B T\right) = \frac{3}{2}R \tag{9.55}$$

where $R = N_A k_B$ is the gas constant.

An additional vibrational degree of freedom would then contribute an extra R to this value. For this mode to contribute, however, the temperature must be large enough that such energy levels can actually be excited; thus this contribution becomes important only when $k_B T \gtrsim E_0 = \hbar\omega/2$. For H_2, for example, this temperature is so high that the molecule disassociates before the vibrational states can contribute fully.

Finally, the zero-point energy can be seen to scale with the reduced mass, μ, as

$$E_0 = \frac{1}{2}\hbar\omega = \frac{1}{2}\hbar\sqrt{\frac{K}{\mu}} \tag{9.56}$$

If we consider two atoms corresponding to different nuclear isotopes (i.e. the same nuclear charge but different mass), the effective interatomic potential between them should be identical (same K) since this is determined by the arrangements of the atomic electrons; the reduced mass, μ, which appears in Eqn. (9.56), however, will differ. This implies that the ground state energy (and hence the energy required to disassociate the molecule) will be different for different isotopes. This is most clearly seen using hydrogen and deuterium where the H–H, H–D, and D–D zero-point energies are in ratios $1 : \sqrt{4/3} : \sqrt{2}$; the energy required to disassociate H–H is then less than for D–D (as in Fig. 9.6) and this effect is observed experimentally.

Figure 9.6. Schematic description of interatomic potential for *H–H, H–D,* and *D–D* pairs (solid curve) and ground state energies (horizontal dotted lines) for molecules of hydrogen (*H*) and deuterium (*D*). The zero-point energy (dissociation energy) for the heavier *D–D* pair is smaller (larger) than for the lighter *H*-system, due to the dependence on the reduced mass of the system.

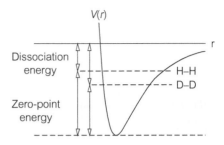

Advances in both materials science and atom trapping have allowed the construction of artificial harmonic traps, both in solids and for isolated atoms. Computer-controlled molecular beam epitaxy, for example, allowed the construction of *Remotely-doped graded potential well structures*[5] including symmetric parabolic potentials. With such techniques, even asymmetric wells, consisting of two half-parabolas with different curvatures have been studied; one such experiment[6] used a sample which was roughly 3000 Å wide and had effective left/right spring constants of $K_{L,R} = 5.1, 6.2 \times 10^{-5} \ meV/\text{Å}^2$, respectively.

In magnetic or optical traps, atoms or ions can be stored and even cooled to the ground state of the harmonic potential, at which point the *Generation of nonclassical motional states of a trapped atom*[7] is possible. This includes suddenly shifting the center of the trap to initiate wave packet motion, or even rapidly changing the effective spring constant (as discussed more abstractly in P9.6).

9.4 **Classical Limits and Probability Distributions**

The classical limit of the quantum oscillator can be approached in several ways; for example, wave packets constructed from the solutions can be shown to oscillate with the classical oscillation period, $\tau = 2\pi/\omega$, and we consider this in Section 12.6.2.

One can also examine the connection between the classical and quantum mechanical probability distributions discussed in Section 5.1. For the SHO, we found (P5.2) that

$$P_{CL}(x) = \frac{1}{\pi \sqrt{A^2 - x^2}} \qquad (9.57)$$

---

[5] The title of a paper by Sundaram *et al.* (1988); see also Shayegan *et al.* (1988).
[6] See Ying *et al.* (1992).
[7] The title of a paper by Meekhof *et al.* (1996).

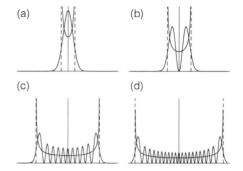

Figure 9.7. Classical (smooth, diverging curves) versus quantum probability distributions (oscillatory curves) for (a) $n = 0$, (b) $n = 1$, (c) $n = 10$, and (d) $n = 20$. The vertical dashed lines are turning points for the classical motion.

where A is the classical turning point defined by $E_{\text{tot}} = KA^2/2$. We plot in Fig. 9.7 the quantum and classical results for the position-space probability densities for the cases $n = 0, 1, 10$, and 20 and note that the quantum mechanical result, when locally averaged, does approach the classical result for large values of n. The form of the classical result is easy to understand, as the particle spends more time near the turning points where it must slow down and come to rest before reversing direction, while near the origin it has its largest kinetic energy (spring is unstretched) and so spends little time.

We can also inquire about the classical distribution of both kinetic and potential energies. We can consider the *potential energy density*, simply $V(x)$ weighted by the classical probability density, given by

$$
\mathcal{V}_{\text{CL}}(x) \equiv P_{\text{CL}}(x)\, V(x) = P_{\text{CL}}(x) \left(\frac{1}{2} Kx^2 \right) = \frac{K}{2\pi} \frac{x^2}{\sqrt{A^2 - x^2}} \tag{9.58}
$$

The similar weighted value of the kinetic energy, the *kinetic energy density*, is then

$$
\mathcal{T}_{\text{CL}}(x) \equiv P_{\text{CL}}(x)\, T(x) = P_{\text{CL}}(x) \left(\frac{1}{2} mv(x)^2 \right) = \frac{K}{2\pi} \sqrt{A^2 - x^2} \tag{9.59}
$$

where we have used the fact that $E = mv^2/2 + Kx^2/2$. The integrals of these quantities yield the average potential and kinetic energies, and we find

$$
\langle T(x) \rangle = \int_{-A}^{+A} P_{\text{CL}}(x)\, T(x)\, dx = \frac{KA^2}{4} \tag{9.60}
$$

$$
\langle V(x) \rangle = \int_{-A}^{+A} P_{\text{CL}}(x)\, V(x)\, dx = \frac{KA^2}{4} \tag{9.61}
$$

that is, $\langle V(x)\rangle = \langle T(x)\rangle$, so that the energy is, on average, shared equally. This result is familiar from classical mechanics where

$$V(x(t)) = \frac{Kx(t)^2}{2} = \frac{KA^2}{2}\cos^2(\omega t - \delta) \tag{9.62}$$

$$T(v(t)) = \frac{mv(t)}{2} = \frac{m\omega^2 A^2}{2}\sin^2(\omega t - \delta) \tag{9.63}$$

and when these are averaged over one period (P9.13), the same result is obtained.

In a position-space representation, the quantum mechanical probability density times the potential,

$$\mathcal{V}_{QM}(x) \equiv V(x)|\psi(x)|^2 \tag{9.64}$$

most closely resembles Eqn. (9.58). In addition, Eqn. (4.66) showed that

$$\mathcal{T}_{QM}(x) \equiv \frac{\hbar^2}{2m}\left|\frac{\partial\psi(x)}{\partial x}\right|^2 \tag{9.65}$$

when integrated over all space, gives the average kinetic energy, so that we can take this as a quantum mechanical equivalent of the classical kinetic energy density, the quantum version of Eqn. (9.59). In Fig. 9.8, we compare the classical and quantum versions of both the potential and kinetic energy densities. We see again that the quantum results, when locally averaged, approach the classical predictions.

The form of the quantum mechanical momentum-space probability density,

$$P_{QM}(p; n) = |\phi_n(p)|^2 \tag{9.66}$$

must necessarily have the same form as $P_{QM}(x; n)$, which at first might seem somewhat surprising. One might argue classically that since the particle spends most of its time near the turning points where the classical speeds are small, $P_{CL}(p)$ should be sharply peaked for *small* values of p. Such classical behavior would then seem to be in direct conflict with the $n \to \infty$ limit of the quantum probability densities, which have the same form as Fig. 9.7. The seeming contradiction arises from the mistaken assumption that we *first* "find" the particle near the classical turning points (i.e. make a position measurement) and *then* specify its momentum.

Figure 9.8. Classical (dashed) and quantum (solid) probability distributions for the potential energy (a) and kinetic energy (b) for the harmonic oscillator for $n = 10$.

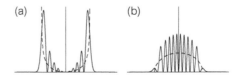

9.5 **Unstable Equilibrium: Classical and Quantum Distributions**

For the case of a particle near a point of *unstable* equilibrium, we have $V(x) = -Kx^2/2$, and the dimensionless Schrödinger equation becomes

$$\frac{d^2\psi(y)}{dy^2} + y^2\psi(y) = -\epsilon\psi(y) \tag{9.67}$$

The large y dependence can be obtained as done for Eqn. (9.23) giving

$$\psi(y) \to e^{\pm iy^2/2} \quad \text{or} \quad \psi(y) \to \sin(y^2/2) \,, \; \cos(y^2/2) \tag{9.68}$$

that is, oscillatory solutions consistent with an unbound particle. If we specialize to the special case of $\epsilon = 0$ for simplicity, we can get more information by assuming a solution of the form

$$\psi(y) = y^\alpha \sin(y^2/2) \tag{9.69}$$

and also noting that

$$\psi''(y) + y^2\psi(y) = (2\alpha + 1)y^\alpha \cos(y^2/2) + \alpha(\alpha - 1)y^{\alpha-2}\sin(y^2/2)$$
$$\approx 0 \quad \text{up to } O(y^{\alpha-2}) \tag{9.70}$$

provided $\alpha = -1/2$. Thus, the next better approximation gives

$$\psi(y) \propto \frac{\sin(y^2/2)}{\sqrt{y}} \quad \left(\text{or} \quad \frac{\cos(y^2/2)}{\sqrt{y}}\right) \tag{9.71}$$

which gives a probability density

$$P_{\text{QM}}(y) \propto \frac{\sin^2(y^2/2)}{y} \tag{9.72}$$

   To compare to the classical probability density, we note that for large time, any unstable solution will be dominated by the increasing exponential term in Eqn. (9.7), giving $x(t) \to De^{\omega t}$ so that

$$v(t) = \dot{x}(t) \to D\omega e^{\omega t} \propto x(t) \tag{9.73}$$

and

$$P_{\text{CL}}(x) \propto \frac{1}{v(x)} \propto \frac{1}{x} \tag{9.74}$$

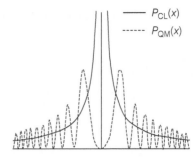

Figure 9.9. Classical (solid) and quantum (dotted) position space probability distribution $(|\psi(x)|^2)$ for the "unstable" oscillator, corresponding to the case of unstable equilibrium.

Changing this to dimensionless variables and noting that the $\sin(y^2/2)$ term averages to $1/2$ over many cycles, we can compare the quantum and classical probability distributions for an unstable particle, namely

$$P_{\text{CL}}(y) \propto \frac{1}{2y} \quad \text{and} \quad P_{\text{QM}}(y) \propto \frac{\sin^2(y^2/2)}{y} \tag{9.75}$$

which we plot in Fig. 9.9. Once again, the increasing wiggliness as $|y|$ increases is indicative of increasing kinetic energy, while the decreasing amplitude is consistent with less and less time spent in a given y interval, that is larger speeds. This, then, is the quantum mechanical "picture" of "falling off a log."

9.6 Questions and Problems

Q9.1. The vibrational degree of freedom in a diatomic gas does not contribute to the specific heat until $k_B T \gtrsim \hbar\omega$ because the energies are quantized. The translational degrees of freedom of a gas when confined to a box (as in a room) are also quantized. At what temperatures will the translational degrees of freedom be "frozen out" in the room you are sitting in right now?

Q9.2. In the context of the discussion in Section 9.5, compare $P_{\text{CL}}(x)$ for the cases of the completely free particle, a particle undergoing uniform acceleration (as discussed in Example 8.2, and the (exponentially) unstable particle considered above.

Q9.3. Most solids are characterized by the fact that they expand when heated. Show that if one models the interactions of the atoms in a solid by simple harmonic forces, any increase in energy of the individual atoms does not lead to an overall expansion. What is it about the real interactions of atoms which can lead to the observed expansion?

P9.1. Consider a particle of mass m, which moves in a potential of the form

$$V(x) = V_0 \left(\cosh\left(\frac{x}{a}\right) - 1\right) \tag{9.76}$$

(a) What is the "best fit" SHO approximation to this potential?

(b) What is the energy spectrum of the SHO fit?

(c) Is the approximation better for large or small a? (You should specify large or small compared to what.)

(c) For a given value of a, up to what values of n is the SHO spectrum a reasonable approximation?

P9.2. Using the recurrence relations in Eqn. (9.28), derive the first three even Hermite polynomial solutions. Do the same for the first three odd solutions using Eqn. (9.38).

P9.3. Linear combination of oscillator states. Consider a state which is a linear combination of the ground state and first excited state of the SHO, that is

$$\psi(x, 0) = \frac{1}{\sqrt{2}} \left(\psi_0(x) + \psi_1(x)\right) \tag{9.77}$$

(a) What is $\psi(x, t)$?

(b) Evaluate $\langle x \rangle_t$ and $\langle \hat{p} \rangle_t$ and show explicitly that $d\langle x \rangle_t/dt = \langle \hat{p} \rangle_t/m$.

(c) What is $\langle \hat{E} \rangle_t$?

P9.4. A particle is known to be in the ground state of the harmonic oscillator. What is the probability that a measurement of its position would find it outside of the classically allowed region? Hint: You can do the required integral numerically, or use the results for the Gaussian probability distribution in Appendix B.3 (with a suitable change of variables).

P9.5. Evaluate the position- and momentum-uncertainties, Δx_n and Δp_n in the nth SHO eigenstate, using the results in Eqns (9.46) and (9.47). What is the probability that a particle in the ground state of the harmonic oscillator would be measured to have a momentum larger than $+\Delta p_0$?

P9.6. Changing the spring constant. A particle of mass m is in the ground state of an SHO with spring constant K. Somehow, the spring constant is suddenly made four times smaller. What is the probability that the particle is in the ground state of the new system? in the first excited state? in the second excited state?

P9.7. A particle of mass m in a harmonic oscillator potential is described by the wavefunction

$$\psi(x, 0) = \frac{1}{\sqrt{a\sqrt{\pi}}} e^{-x^2/a^2} \tag{9.78}$$

What is the probability that a measurement of its energy would give the value $\hbar\omega/2$? $3\hbar\omega/2$? How do these values change with time?

P9.8. Dissolving oscillator. Assume that a particle is in the ground state of a harmonic oscillator potential. At a time $t = 0$, very suddenly, the spring is completely removed. What is the resulting time-dependent wavefunction, $\psi(x, t)$, of this suddenly free particle? What are the resulting time-dependent values of Δx_t and Δp_t?

P9.9. Harmonic oscillator in an external electric field. Consider a charged particle of mass m in an SHO potential, but which is also subject to an external electric field E. The potential for this problem is now given by

$$V(x) = \frac{1}{2}m\omega^2 x^2 - qEx \tag{9.79}$$

where q is the charge of the particle.

(a) Show that a simple change of variables makes this problem completely soluble in terms of the standard SHO solutions. Hint: Complete the square.

(b) Find the new eigenfunctions and energy eigenvalues.

(c) Show that for a particular value of E the ground state energy can be made to vanish. Does this mean that there is no zero-point energy in this case?

(d) Evaluate $\langle x \rangle$ and $\langle \hat{p} \rangle$ in each of the new eigenstates.

(e) What are the new momentum-space wavefunctions? Can you evaluate $\langle \hat{x} \rangle$ and $\langle p \rangle$ in this representation?

P9.10. Half-harmonic oscillator. (a) Evaluate the allowed energy eigenvalues and corresponding normalized wavefunctions for a particle of mass m moving in a potential given by

$$V(x) = \begin{cases} +\infty & \text{for } x < 0 \\ m\omega^2 x^2/2 & \text{for } 0 < x \end{cases} \tag{9.80}$$

Hint: Use the fact that the SHO solutions discussed above are eigenstates of parity.

(b) For the ground state of this system, evaluate $\langle x \rangle$, Δx, $\langle p \rangle$, and Δp.

P9.11. Lennard–Jones potentials. Consider the potential in Eqn. (9.8) for pairs of noble gas atoms.

(a) Show that the minimum of the potential is at $r_{min} = 2^{1/6}\sigma$ and that the depth of the potential there is $V(r_{min}) = -\epsilon$.

(b) "Fit" the potential near the minimum with a SHO potential and find the effective spring constant in terms of σ and ϵ. (From dimensional grounds alone, you know that $K \propto \epsilon/\sigma^2$.) What is the energy spectrum of the SHO potential in terms of these quantities.

(c) A measure of the "quantum-ness" of the system is given by the ratio E_0/ϵ. Calculate this in terms of the parameters of the problem.

(d) Using the numerical values below, evaluate E_0/ϵ in each case and comment.

Atom	μ (amu)	ϵ (eV)	σ (Å)
He	2	0.000875	2.56
Ne	10	0.00312	2.74
Ar	20	0.01	3.40
Kr	42	0.014	3.65
Xe	66	0.02	3.98

(e) An approximate criteria for the melting of a solid (assumed bound by harmonic potentials) is when the RMS amplitude ($\Delta r = \sqrt{(r - \langle r \rangle)^2}$) of vibration of atoms about their equilibrium position becomes larger than a certain fraction of the equilibrium spacing (d). This can be quantified to say that when the *Lindemann constant*, γ_L, is such that

$$\gamma_L \equiv \frac{\Delta r}{d} \gtrsim 0.15 \tag{9.81}$$

one finds classical melting. Calculate the Lindemann constant in terms of ϵ, μ, $\hbar$, and ϵ and evaluate it numerically for the ground states of the five systems above.

P9.12. Molecular vibrational data. The NaCl diatomic molecule has $\hbar\omega \approx 0.04\ eV$.

(a) Use the atomic weights of Na ($\sim$23) and Cl ($\sim$35) to estimate the effective spring constant K. Express this in eV/Å^2 and N/m.

(b) Estimate the amplitude of the vibrational motion in the ground state ($n = 0$) and in the last bound state ($n \sim 20$) by equating $E_n \sim KA^2/2$; express your answer in Å and compare to the equilibrium separation of the Na and Cl, namely, 2.4 Å.

(c) At roughly what temperature will the vibrational degree of freedom of NaCl contribute to the specific heat?

P9.13. Classical average values using trajectories.

(a) The average value of a function of position, $x(t)$, in a periodic system can be defined via

$$\langle f(x(t)) \rangle = \left[\int_0^\tau f(x(t))\, dt \right] \Big/ \left[\int_0^\tau dt \right] = \frac{1}{\tau} \int_0^\tau f(x(t))\, dt \tag{9.82}$$

Use the fact that $v(x) = dx/dt$ to show that this definition is equivalent the definition of average value used for functions of $x = x(t)$ in Eqn. (5.11), namely

$$\langle f(x) \rangle = \int_a^b P_{CL}(x) f(x)\, dx \quad \text{with } P_{CL}(x) = \frac{2}{\tau} \frac{1}{v(x)} \tag{9.83}$$

(b) Use the first method to calculate the average values of potential and kinetic energies using the forms in Eqn. (9.62) and (9.63).

P9.14. The classical anharmonic oscillator.

(a) Consider the *anharmonic oscillator* potential given by

$$V(x) = \frac{1}{2}Kx^2 - \lambda Kx^3 \tag{9.84}$$

and consider λ to be small. Show that the resulting classical equation of motion is

$$\ddot{x}(t) = -\omega^2 x(t) + 3\omega^2 \lambda x^2(t) \tag{9.85}$$

(b) Attempt a solution of the form

$$x(t) = A\sin(\omega t) + B_1 + [C_1\cos(2\omega t) + D_1\sin(2\omega t)] + \cdots \tag{9.86}$$

where A is fixed and B_1, C_1, D_1 are all of order λ and solve for the undetermined coefficients.

(c) In contrast to the (symmetric) harmonic oscillator, show that $\langle x \rangle = 3\lambda A^2/2 \neq 0$, where the average value is defined via

$$\langle f(t) \rangle = \frac{1}{\tau}\int_0^\tau f(t)\, dt \tag{9.87}$$

as in P9.13.

(d) Discuss Q9.3 in the context of your answer.

(e) Evaluate the "anharmonic" term in the SHO approximation to the Lennard–Jones potential by expanding Eqn. (9.8) to the appropriate order.

P9.15. Application of the Feynman–Hellmann theorem. Recall that the Feynman–Hellman theorem (from P6.18) required that

$$\frac{\partial E(\lambda)}{\partial \lambda} = \left\langle \frac{\partial \hat{H}(\lambda)}{\partial \lambda} \right\rangle \tag{9.88}$$

if the Hamiltonian depends on some parameter λ in some well-defined way, $\hat{H} = \hat{H}(\lambda)$ so that

$$\langle E \rangle(\lambda) = \langle \psi(\lambda)|\hat{H}(\lambda)|\psi(\lambda)\rangle \tag{9.89}$$

Confirm the Feynman–Hellman theorem for the harmonic oscillator eigenstates by using the mass m and the spring constant K as parameters.

P9.16. Wigner distribution for the harmonic oscillator eigenstates. Evaluate the Wigner quasi-probability distribution defined in Eqn. (4.149) for the $n = 0$ and $n = 1$ states of the harmonic oscillator. Show that for $n = 0$ one obtains a positive-definite result, while for $n = 1$ one does not.

TEN

Alternative Methods of Solution and Approximation Methods

It is a common practice to approach quantum mechanics through the study of a few, exactly soluble examples using the Schrödinger equation in position-space. The number of potential energy functions for which such closed-form solutions are available is, however, quite small. Luckily, many of them actually correspond reasonably well to actual physical systems; examples include the infinite well as a model of a free particle in a "box" (Chapters 5 and 7), the harmonic oscillator (Chapter 9), the rigid rotator (Chapter 16), and the Coulomb potential for the hydrogen atom (Chapter 17).

Nonetheless, it is important to recognize that other methods can be used to study the properties of a quantum system. Some of them are quite different from the Schrödinger equation approach, and many are amenable for use as numerical and approximation methods in problems for which analytic solutions are not available.

In this chapter, we focus on several methods which can be used to study the spectrum of energy eigenvalues and wavefunctions for time-independent systems and (more briefly) on the effects of time-dependent perturbations, not only as calculational tools for possible numerical analysis, but also as examples of very different ways of approaching quantum mechanical problems. We can make several general comments:

- Many (but by no means all) of the alternative approaches discussed here are most useful for the study of the ground state of the system. Because the structure of matter is ultimately determined by the lowest energy configuration, the determination of the properties of the ground state is arguably the most important; it is the very clearly "first amongst equals".

- Any method which is to be used as a numerical approximation technique should be capable of increased precision (usually at the cost of increased calculational difficulty) as well as providing an estimate of the errors made in the approximation. We will not focus extensively on these questions, but the reader should always keep in mind how each method can be extended in precision, as well as the possible effort involved in doing so.

- As our ultimate goal is to understand the physics behind the equations, we may well have to rethink what it means to "solve" a problem when we approach it numerically. For example, do we need an analytic functional form for $\psi(x)$ or is an array of numbers or an interpolating function enough? How precisely do we need to know the energy eigenvalues? When are we "done"?

- Finally, the use of numerical methods is often nicely complementary to the study of analytic examples. One often looks at a problem in a much different way when one approaches it expecting to write a computer program to "solve" it, and such new insights can be valuable. For example, the study of chaotic dynamics in classical mechanics owes much of its success to the application of numerical (as opposed to analytic) techniques to otherwise familiar problems.

In each section, we first discuss the formalism of each method and then give an example of its possible use as a computational tool.

10.1 **Numerical Integration**

Classical and quantum mechanics share the fact that their fundamental mathematical descriptions are given by second-order differential equations, Newton's law for a point particle

$$m\frac{d^2x(t)}{dt^2} = F(x) \tag{10.1}$$

and the time-independent Schrödinger equation

$$-\frac{\hbar^2}{2m}\frac{d^2\psi(x)}{dx^2} + V(x)\psi(x) = E\psi(x) \tag{10.2}$$

We are used to thinking of Eqn. (10.1) as being completely deterministic,[1] in that, if we are given the appropriate initial conditions, namely, $x_0 = x(0)$ and $v_0 = \dot{x}(0)$, the future time development of $x(t)$ is then predicted. To see how a particle "uses" Eqn. (10.1) to "know where it should be" at later times, we can

---

[1] We ignore any complications such as the extreme sensitivity to initial conditions present in chaotic systems.

use a conceptually simple method[2] to integrate Newton's law directly. We first approximate the acceleration (the second derivative) via

$$\frac{d^2 x(t)}{dt^2} = \ddot{x}(t) = \lim_{\delta \to 0} \left(\frac{\dot{x}(t+\delta) - \dot{x}(t)}{\delta} \right)$$

$$= \lim_{\delta \to 0} \left[\frac{\lim_{\delta \to 0}((x(t+2\delta) - x(t+\delta))/\delta) - \lim_{\delta \to 0}(((x(t+\delta) - x(t)))/\delta)}{\delta} \right]$$

$$= \lim_{\delta \to 0} \left(\frac{x(t+2\delta) - 2x(t+\delta) + x(t)}{\delta^2} \right)$$

$$\approx \frac{x(t+2\delta) - 2x(t+\delta) + x(t)}{\delta^2}. \tag{10.3}$$

With this approximation, Newton's law can be written as

$$x(t+2\delta) \approx 2x(t+\delta) - x(t) + \delta^2 \frac{F(x(t))}{m} \tag{10.4}$$

which is now a *difference equation* for $x(t)$, evaluated at the discretized times $t = n\delta$.

Since

$$v_0 = \dot{x}(0) \approx \frac{x(\delta) - x(0)}{\delta} \quad \text{we have} \quad x(\delta) \approx x(0) + \delta \dot{x}(0) \tag{10.5}$$

and the values of $x(t)$ at the first two of the discretized times, $n = 0, 1$, are fixed by the initial conditions; for later times, the $x(t = n\delta)$ with $n \geq 2$ are then determined by Eqn. (10.4).

Example 10.1. Numerical integration of the classical harmonic oscillator

The classical equation for a mass and a spring is of the form

$$\ddot{x}(t) = -\omega^2 x(t) \tag{10.6}$$

where $\omega = \sqrt{K/m}$. For any numerical problem, we must specialize to definite values, for both the physical parameters of the problem, and for the initial conditions; as an example, we choose

$$\omega = 2\pi, \quad x(0) = 1, \quad \text{and} \quad v(0) = \dot{x}(0) = 0 \tag{10.7}$$

---

[2] Much more powerful techniques, such as the Runge–Kutta method, are discussed in all textbooks dealing with numerical methods.

(Continued)

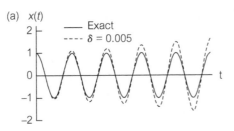

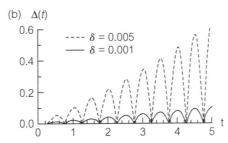

Figure 10.1. The exact (solid) and numerical (dashed) solutions of the harmonic oscillator differential equation are shown in (a). The differences between the numerical and exact solutions versus time for two different step sizes, δ, are plotted in (b).

which has the exact solution $x(t) = \cos(2\pi t)$. In Fig. 10.1(a), we show the result of a numerical solution of Eqn. (10.6) (dashed curve), using Eqn. (10.4), to be compared to the exact solution (solid curve). In Fig. 10.1(b) the difference between the numerical and exact solutions, $\Delta(t)$, is seen to increase with t, but it is also smaller for smaller step sizes, δ, as expected.

The same strategy can be used to solve the Schrödinger by approximating Eqn. (10.2) as

$$\psi(x + 2\delta) \approx 2\psi(x + \delta) - \psi(x) + \delta^2 \left[\frac{2m}{\hbar^2}(V(x) - E) \right] \psi(x) \qquad (10.8)$$

and using

$$\psi(0) \quad \text{and} \quad \psi(\delta) \approx \psi(0) + \psi'(0)\delta \qquad (10.9)$$

In this sense, Eqn. (10.2) is just as deterministic as Newton's laws; the chief differences are:

- The choice of $x = 0$ as the "initial" value is arbitrary.

- The differential equation can (and should) be integrated "to the left" as well to obtain $\psi(x)$ for $x < 0$.

- Most importantly, the Schrödinger equation can be integrated (solved) for any value of the energy eigenvalue, E; the solutions so obtained, however, will not necessarily be physically acceptable, that is, square-integrable.

To illustrate the usefulness of this approach to the isolation of energy eigenvalues and their corresponding eigenfunctions, we restrict ourselves to the special case of a symmetric potential for reasons that will become clear. In that case, we know that the solutions will also be eigenfunctions of parity and hence satisfy

$$\text{even solutions: } \psi(0) = \text{arbitrary} \quad \text{and} \quad \psi'(0) = 0 \qquad (10.10)$$

and

$$\text{odd solutions: } \psi(0) = 0 \quad \text{and} \quad \psi'(0) = \text{arbitrary} \qquad (10.11)$$

The arbitrariness in $\psi(0)$ or $\psi'(0)$ present at this point is eventually removed when the wavefunction is properly normalized, but that is separate from the solution of the Schrödinger equation itself. The overall normalization does not affect the *shape* of the solution.

We now focus on the behavior of the wavefunction at large $|x|$ for various values of E. For the even case, for example, we can start at $x = 0$ with an arbitrary value of $\psi(0)$, use the oddness of $\psi'(x)$ to determine $\psi(\delta) = \psi(0)$, and then use Eqn. (10.8) to numerically integrate to arbitrarily large values of $x = n\delta$; we find the generic behavior shown in Fig. 10.2(a). If we call the lowest even energy eigenvalue $E_1^{(+)}$, then for values of $E < E_1^{(+)}$, the solutions diverge as $\psi(x) \to +\infty$ as $x \to +\infty$. When $E \gtrsim E_1^{(+)}$, the solutions are still poorly behaved at infinity, but now diverge with the opposite sign. Clearly, the energy of the physically acceptable square-integrable ground state solution lies between E_a and E_b; this behavior is familiar from our study of the harmonic oscillator and Fig. 9.3.

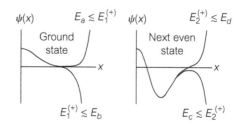

Figure 10.2. Numerical solutions of the Schrödinger equation for a symmetric potential. The energy parameters $E_a < E_1^{(+)} < E_b$ bracket the true ground state energy; $E_c < E_2^{(+)} < E_d$ bracket the first-excited even state.

Once such a pair of energy values which brackets the "acceptable" ground state solution is found, one can determine $E_1^{(+)}$ with increasing *precision* by a systematic exploration in the interval (E_a, E_b), finding values of E, which bracket the "true value" with decreasing error; the resulting *accuracy* of the estimated value of E_0, however, will still depend on the integration method used (Q10.2). As the energy parameter E is increased further, additional changes in sign of the wavefunction at infinity are encountered (Fig. 10.1(b)), and the energy spectrum can be systematically mapped out by finding pairs of energy values which bracket a "sign change."

Example 10.2. Energy eigenvalues for the harmonic oscillator

The numerical solution of the Schrödinger equation for the harmonic oscillator potential is easy to implement using Eqn. (10.8) provided the problem is put into dimensionless form as in Section 9.2.1, namely

$$\frac{d^2\psi(y)}{dy^2} = (y^2 - \epsilon)\psi(y) \tag{10.12}$$

where the dimensionless eigenvalues are $\epsilon_n = 2E_n/\hbar\omega = (2n + 1)$. The even states have $\epsilon_n = 1, 5, 9, \ldots$ and so forth. Values of E_a, E_b, which bracket the ground state $(E_1^{(+)})$ and first even excited state $(E_2^{(+)})$ energies for several values of δ are given by

δ	$E_a < E_1^{(+)} < E_b$	$E_a < E_2^{(+)} < E_b$
0.1	(1.191, 1.192)	(5.510, 5.511)
0.01	(1.0171, 1.0172)	(5.0431, 5.0432)
0.001	(1.00169, 1.00170)	(5.00423, 5.00424)

so that the effect of decreasing the step size on the reliability of the results is clear.

It is useful to keep in mind that before applying any numerical technique to a new problem, it is best to "test" it on a well-understood example if at all possible.

Once an approximate energy eigenvalue is found, the wavefunction for each energy eigenvalue is obtained from the numerical integration as the collection of points $\psi(x = n\delta)$, and can be fit to a smooth function using interpolation techniques if desired; in any case, it can be normalized and used to extract further information about the quantum system. The odd states are found in a similar way (Q10.4) by making use of Eqn. (10.11).

10.2 **The Variational or Rayleigh–Ritz Method**

Many branches of physics can be formulated in terms of a simple minimum principle using the methods of the calculus of variations. Examples include minimum surface problems (bubble problems and the like), Fermat's formulation of geometrical optics using a principle of least time and, perhaps most importantly, the principle of least action approach to classical mechanics.

In each case, the object of study is a *functional*, so-called because it takes as its argument a function and returns a number as its output. The *classical action* in mechanics, $S[x(t)]$, is just such an example; it takes any possible classical path, $x(t)$, and returns the numerical value

$$S[x(t)] = \int_{t_a}^{t_b} dt \, \left(\frac{1}{2} m \ddot{x}^2(t) - V(x(t)) \right) \tag{10.13}$$

and the trajectory realized in nature is the unique path which minimizes Eqn. (10.13).

It is perhaps then not surprising that quantum mechanics can also be formulated in such a manner. We will first discuss just such an approach and then discuss how it can be applied as a calculational tool to approximate energy eigenvalues and wavefunctions.

Consider a Hamiltonian, $\hat{H}$, defining the bound state spectra of some system. We assume that it will have a discrete spectrum of bound state energies, E_n, with corresponding, already normalized wavefunctions $\psi_n(x)$. We can define an *energy functional* for *any* trial wavefunction, $\psi(x)$, via

$$
\begin{aligned}
E[\psi] &\equiv \langle \psi | \hat{H} | \psi \rangle = \langle \psi | \hat{T} | \psi \rangle + \langle \psi | V(x) | \psi \rangle \\
&= \int_{-\infty}^{+\infty} dx \, \psi^*(x) \, \hat{H} \, \psi(x) \\
&= \frac{1}{2m} \int_{-\infty}^{+\infty} dx \, \psi^*(x) \, \hat{p}^2 \, \psi(x) + \int_{-\infty}^{+\infty} dx \, V(x) |\psi(x)|^2
\end{aligned}
\tag{10.14}
$$

This is defined whether $\psi(x)$ is an eigenfunction or not. It is often convenient to use the "alternative" form of the average value of kinetic energy (as in

Eqn. (4.65)), that is,

$$\langle\psi|\hat{T}|\psi\rangle = \langle\hat{T}\rangle = \frac{\hbar^2}{2m}\int_{-\infty}^{+\infty}dx\left|\frac{d\psi(x)}{dx}\right|^2 \tag{10.15}$$

If for some reason the trial wavefunction ψ is not already normalized, we can simply write

$$E[\psi] \equiv \frac{\langle\psi|\hat{H}|\psi\rangle}{\langle\psi|\psi\rangle} \tag{10.16}$$

It is easy to see that this functional simply returns an energy eigenvalue when its argument is a normalized eigenstate, since

$$E[\psi_n] = \int_{-\infty}^{+\infty}dx\,\psi_n^*(x)\,\hat{H}\,\psi_n(x) = \int_{-\infty}^{+\infty}dx\,\psi_n^*(x)\,E_n\,\psi_n(x) = E_n \tag{10.17}$$

For a general wavefunction, $\psi(x)$, we assume we can use the expansion theorem and write $\psi(x) = \sum_{n=0}^{\infty}a_n\,\psi_n(x)$ and we find that

$$E[\psi] = \langle\psi|\hat{H}|\psi\rangle = \sum_{n=0}^{\infty}\sum_{m=0}^{\infty}a_m^*a_n\langle\psi_m|\hat{H}|\psi_n\rangle$$

$$= \sum_{n=0}^{\infty}\sum_{m=0}^{\infty}a_m^*a_nE_n\delta_{n,m}$$

$$= \sum_{n=0}^{\infty}|a_n|^2E_n \tag{10.18}$$

This derivation is similar to that of Section 6.4 for the average value of the energy operator, $\hat{E}$, in a general state, but the quantity that appears in the energy functional here is the expectation value of the appropriate Hamiltonian for the problem, which in general acts only on spatial degrees of freedom.

We assume that the energy eigenvalues are ordered, that is, $\cdots \geq E_2 \geq E_1 \geq E_0$ so that

$$E[\psi] = \sum_{n=0}^{\infty}|a_n|^2E_n \geq \sum_{n=0}^{\infty}|a_n|^2E_0 = E_0 \quad\text{or}\quad E[\psi] \geq E_0 \tag{10.19}$$

because the expansion coefficients, when squared, sum to unity, since $\psi(x)$ is assumed normalized. The lower bound is only "saturated" when $\psi(x) = \psi_0(x)$ in which case $a_n = \delta_{n,0}$ and only the ground state energy term contributes.

This is then the desired minimum principle, namely that:

- The energy functional, $E[\psi]$, defined via Eqn. (10.14), always gives an energy at least as large as the true ground state energy, that is, $E[\psi] \geq E_0$ for all ψ.

To use this property as a calculational tool, we first note that if the wavefunction used in the functional has an arbitrary parameter, for example, $\psi(x) = \psi(x; a)$, then the energy functional yields a function of one variable, namely

$$E[\psi(x; a)] = E(a) \tag{10.20}$$

An example of this would be the family of Gaussian variational wavefunctions, $\psi(x; a) = \exp(-x^2/2a^2)/\sqrt{a\sqrt{\pi}}$, with a variable width.

Because the functional satisfies the minimum principle for each value of the parameter, one can minimize the variational function $E(a)$ and be assured that the resulting minimum is still greater than the true ground state energy. Thus, one can find the trial wavefunction, in the one parameter family considered, which has the lowest energy. The minimizing wavefunction accomplishes this by somehow "adjusting" to as similar as possible to the exact ground state solution. This approach is similar in spirit to the "zero-point energy" argument of P1.14, but is more powerful because:

- The guaranteed lower bound of Eqn. (10.19) provides a method of assessing the reliability of the approximations.

    — Of two variational estimates of the ground state energy, the lower one is always closer to the true value.

    In this context, "lower is always better" as we know that we can never "undershoot" E_0 on the negative side.

- It also provides an approximation to the wavefunction as well as to the energy; one can then use it to estimate expectation values and to find the approximate momentum-space wavefunction. As an aside, because the argument leading to Eqn. (10.18) is not specific to a position-space representation, one can also use the variational method with momentum-space wavefunctions (P10.5).

For illustrative purposes, we will sometimes calculate $|a_0|^2$ as a measure of the "overlap" of the trial solution with the exact ground state wavefunction (if known); it can be used as a quantitative measure of the similarity of any two functions. We reiterate, however, that the trial wavefunction of a given class which minimizes the energy is not necessarily the one which has the largest overlap with the true ground state wavefunction, that is, it does not necessarily maximize $|a_0|^2$ (see, for example, P10.9).

Example 10.3. Variational estimate for the harmonic oscillator I

As an example of the method, consider approximating the ground state energy and eigenfunction of the simple harmonic oscillator by using the family of trial wavefunctions $\psi(x; \alpha) = \exp(-x^2/2a^2)/\sqrt{a\sqrt{\pi}}$ mentioned above. Because the true ground state solution is also Gaussian, we expect to find the exact answer. We have to evaluate Eqn. (10.14) with $V(x) = m\omega^2 x^2/2$ and we find that

$$E[\psi(x; a)] = E(a) = \langle \hat{T} \rangle + \langle V(x) \rangle = \frac{\hbar^2}{4ma^2} + \frac{1}{4}m\omega^2 a^2 \qquad (10.21)$$

Minimizing this expression we find

$$\frac{dE(a)}{da} = -\frac{\hbar^2}{2ma^3} + \frac{1}{2}m\omega^2 a = 0 \qquad (10.22)$$

which yields $a_{min} = \sqrt{\hbar/m\omega}$ and $E(a_{min}) = \hbar\omega/2$ as expected.

Example 10.4. Variational estimate for the harmonic oscillator II

To illustrate the principle in the case where the form of the ground state wavefunction is not known, consider as a trial wavefunction for the simple harmonic oscillator (SHO) the wavefunction

$$\psi(x; a) = \begin{cases} 0 & \text{for } |x| > a \\ N(a^2 - x^2)^2 & \text{for } |x| < a \end{cases} \qquad (10.23)$$

where the variational parameter is again a and the normalization constant is given by $N = \sqrt{315/256a^9}$. A similar calculation to the one above shows (P10.3) that the energy function is

$$E[\psi] = E(a) = \frac{3\hbar^2}{2ma^2} + \frac{m\omega^2 a^2}{22}. \qquad (10.24)$$

This has a minimum value at $a_{min}^2 = \sqrt{33}\hbar/m\omega$ yielding

$$E(a_{min}) = \frac{\hbar\omega}{2}\sqrt{\frac{12}{11}} = (0.522)\hbar\omega \qquad (10.25)$$

which is only 4.4% greater than the exact value.

The trial wavefunctions, along with their corresponding energies for several choices of a are shown in Fig. 10.3 along with the value of $|a_0|^2$. We plot in Fig. 10.4 the fractional difference between the variational energy and the exact ground state value $(E(var) - E(exact))/E(exact)$ as well as the probability that the variational wavefunction is *not* in the ground state, that is, $1 - |a_0|^2$, versus the variational parameter a. We note

(Continued)

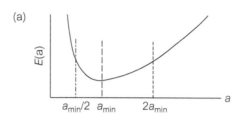

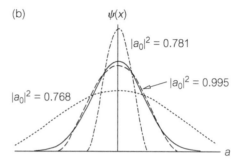

Figure 10.3. (a) The variational energy, $E(a)$ versus a, showing the value of a_{min} which minimizes the energy functional (vertical dashed line) and two other values (dotted and dot-dash). (b) The corresponding variational wavefunctions (same dashing) along with the exact ground state (solid curve). Values of the overlap, given by $|a_0|^2$, for each variational waveform are also shown.

Figure 10.4. The fractional energy error (solid curve) and the degree of "non-overlap" (dashed curve) versus variational parameter a for Example 10.4. This illustrates that first-order changes in the wavefunction give second-order changes in the energy functional.

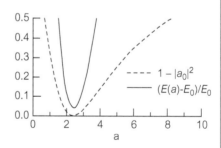

that variations in the parameter a seem to have a much larger effect on the energy functional than on the wavefunction itself.

To formalize this last observation further, let us imagine making small variations around the exact ground state wavefunction, $\psi_0(x)$, parameterized by $\psi_0(x) \rightarrow \psi_0(x) + \lambda\phi(x)$ so that $\phi(x)$ represents a first-order change in the wavefunction; we use λ to keep track of the expansion.

Consider then the energy functional (where we use Eqn. (10.16) since the new wavefunction is not properly normalized) and we find

$$
\begin{aligned}
E[\psi_0 + \lambda\phi] &= \frac{\langle \psi_0 + \lambda\phi | \hat{H} | \psi_0 + \lambda\phi \rangle}{\langle \psi_0 + \lambda\phi | \psi_0 + \lambda\phi \rangle} \\
&= \frac{E_0 \langle \psi_0 | \psi_0 \rangle + \lambda(\langle \psi_0 | \hat{H} | \phi \rangle + \langle \phi | \hat{H} | \psi_0 \rangle) + \lambda^2 \langle \phi | \hat{H} | \phi \rangle}{\langle \psi_0 | \psi_0 \rangle + \lambda \langle \psi_0 | \phi \rangle + \lambda \langle \phi | \psi_0 \rangle + \lambda^2 \langle \phi | \phi \rangle} \\
&= E_0 \left(\frac{1 + \lambda(\langle \psi_0 | \phi \rangle + \langle \phi | \psi_0 \rangle) + \lambda^2 \langle \phi | \hat{H} | \phi \rangle / E_0}{1 + \lambda(\langle \psi_0 | \phi \rangle + \langle \phi | \psi_0 \rangle) + \lambda^2 \langle \phi | \phi \rangle} \right) \\
&= E_0 \left(1 + \mathcal{O}(\lambda^2) \right)
\end{aligned}
\tag{10.26}
$$

since the zeroth-order and $\mathcal{O}(\lambda)$ terms are identical. This shows that, in general,

- *First-order changes* $(O(\lambda))$ in the trial wavefunction, away from the true ground state solution, give rise to *second-order changes* $(O(\lambda^2))$ in the corresponding energy functional.

This fact is reflected in Fig. 10.4 as the fractional change in energy does seems to vary quadratically with deviations away from the minimum value of the variational parameter, while the deviation in the wavefunction itself (as measured by $1 - |a_0|^2$) seems to vary much more weakly on a. This is a typical feature of problems involving the calculus of variations.

If the variational method is to be useful as an approximation method there should be some possibility of further refinement of the estimation of the ground state energy. This can be accomplished by simply taking as a trial wavefunction one with a larger number of variational parameters. For example, one might consider

$$
\psi(x; a, b) = e^{-x^2/2a^2} (1 + bx^2)
\tag{10.27}
$$

which has an additional parameter, b, but which reduces to the original choice in some limit (namely, $b = 0$). In this case, we are guaranteed to have

$$
E(a_{min}) \equiv E[\psi(x; a_{min})] > E(a_{min}, b_{min}) \equiv E[\psi(x; a_{min}, b_{min})] \geq E_0
\tag{10.28}
$$

because any variational energy must be larger than the true ground state and because the minimum with nonzero values of b will be at least as small as for

$b = 0$. The new minimum value will be determined by

$$\frac{\partial E(a, b)}{\partial a} = \frac{\partial E(a, b)}{\partial b} = 0 \qquad (10.29)$$

By adding more and more variational parameters, we can allow the trial wavefunction to conform as closely as possible to the exact ground state.

Example 10.5. Variational estimate for the harmonic oscillator III

We illustrate the improvement possible with multi-parameter trial wavefunctions by using the (unnormalized) function

$$\psi(x; a, b) = \begin{cases} 0 & \text{for } |x| > a \\ (a^2 - x^2)^2(1 + bx^2) & \text{for } |x| < a \end{cases} \qquad (10.30)$$

as a trial solution for the ground state of the SHO. We plot in Fig. 10.5 a contour plot of $E(a, b)$ versus a, b; the small star on the dotted line indicates the minimum for the $b = 0$ case, while the small $+$ indicates the new global minimum which does indeed have somewhat lower energy. The values of the exact, one-parameter, and two-parameter fits for various quantities are shown below:

quantity	exact	$\psi(x; a)$	$\psi(x; a, b)$
$E_0/(\hbar\omega/2)$	1	1.0445	1.0198
$|a_0|^2$	1	0.9951	0.9977
$\langle x^2 \rangle/\rho^2$	1/2	0.5222	0.5099
$\langle x^4 \rangle/\rho^4$	3/4	0.6923	0.6884

$$(10.31)$$

The energy is lower and the overall fit is better ($|a_0|^2$ is closer to 1) than in the one-parameter case; it is clear, however, that various higher moments (i.e. average values of x^{2n}) are never fit very well with this particular form which is not surprising given its lack of a realistic 'tail' for large $|x|$.

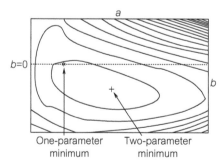

Figure 10.5. Contour plot of two-parameter variational energy $E(a, b)$ versus (a, b); the dotted line corresponds to the one-parameter family, $E(a)$ versus a. The minimum energy for the two-parameter family of trial wavefunctions is lower than for the one-parameter set.

The process can be continued with as many variational parameters as one can handle, presumably improving the agreement with experiment, if not providing much more useful insight into the basic physics.[3] One of the most famous *tour de force* calculations of this type are variational calculations of the ground state of the helium atom which use trial wavefunctions with hundreds of parameters.[4]

In some cases, it is also possible to extend the variational method to give rigorous lower bounds for excited states as well as the ground state. Suppose, for example, that one could choose a trial wavefunction which was somehow known to be orthogonal to the true ground state, that is, $\langle \psi | \psi_0 \rangle = 0$; this would guarantee that $a_0 = 0$ in the expansion theorem. The standard argument would then give

$$\psi = \sum_{n=1}^{\infty} a_n \psi_n \implies E[\psi] = \sum_{n=1}^{\infty} |a_n|^2 E_n \geq \sum_{n=1}^{\infty} |a_n|^2 E_1 = E_1 \qquad (10.32)$$

Thus, all the trial wavefunctions in this restricted class would have energies at least as great as the first excited state.

Various symmetries of the problem can often be used to restrict the form of the trial wavefunction so as to satisfy this constraint. For example, in a one-dimensional problem with a symmetric potential, $V(x) = V(-x)$, we know that the ground state will be an even function; therefore any odd trial wavefunction will have $a_0 = 0$ and hence satisfy Eqn. (10.32) and give a good estimate of the first excited state energy. Less prosaically, in three-dimensional problems with spherical symmetry, the ground state will have no angular momentum (i.e. $l = 0$) and excited states with higher values of l are automatically orthogonal to the ground state.

10.3 The WKB method

The variational method is best suited to evaluating the properties of the ground state solution, that is, for $n = 0$. It is useful to have a complementary approach, which is more appropriate for the quasi-classical regime where $n \gg 1$; we have argued that this limit is also attained, in some sense, when $\hbar \to 0$. Such an

[3] It is said that, when confronted with the result of an impressive numerical calculation, Eugene Wigner said "*It is nice to know that the computer understands the problem. But I would like to understand it too.*"; See Nussenzveig (1992).
[4] See Bethe and Jackiw (1968) or Park (1992) for discussions.

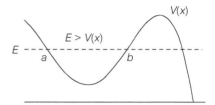

Figure 10.6. Generic potential with classical turning points for the WKB approximation.

approach was first discussed in the context of quantum mechanics by Wentzel, Kramers, and Brillouin[5] and is therefore often called the WKB method.

10.3.1 **WKB Wavefunctions**

Motivated by the simple form for a free-particle de Broglie wave, that is,

$$\psi(x) = A\, e^{i2\pi x/\lambda} = A\, e^{ikx} = A\, e^{ipx/\hbar} \tag{10.33}$$

we attempt a solution of the time-independent Schrödinger equation of the form

$$\psi(x) = A(x)\, e^{iF(x)/\hbar} \tag{10.34}$$

where $A(x)$ and $F(x)$ are an amplitude and phase term, respectively. We retain the explicit factor of $\hbar$ and will use it to parameterize the smallness of various terms. We also assume, for the moment, that we are in the classically allowed region so that $E > V(x)$, as in Fig. 10.6, so that $a < x < b$.

With the *ansatz*[6] in Eqn. (10.34), the Schrödinger equation becomes

$$0 = A(x)\left[\frac{1}{2m}\left(\frac{dF(x)}{dx}\right)^2 - (E - V(x))\right]$$
$$- \hbar\left(\frac{i}{2m}\right)\left[2\frac{dA(x)}{dx}\frac{dF(x)}{dx} + A(x)\frac{d^2F(x)}{dx^2}\right]$$
$$- \hbar^2\left[\frac{1}{2m}\frac{d^2A(x)}{dx^2}\right] \tag{10.35}$$

At this point, we can either consider $\hbar$ as an arbitrary small parameter and set the first two terms (of order $\mathcal{O}(\hbar^0)$ and $\mathcal{O}(\hbar^1)$, respectively) separately to zero or else we can require that both the real and imaginary parts of Eqn. (10.35) are satisfied; in either case, we neglect the last term (being of order $\mathcal{O}(\hbar^2)$) and discuss the validity of this approximation below.

[5] It was also studied independently by Jeffries; the name WKBJ approximation is therefore sometimes used.

[6] The German term *ansatz*, often defined as *formulation* or *setup*, is often taken to mean something like "assumed form of the solution" in the context of a physics or math problem.

The $\mathcal{O}(\hbar^0)$ equation (or real part) is easily written as

$$\frac{dF(x)}{dx} = \pm\sqrt{2m(E - V(x))} \equiv \pm p(x) \tag{10.36}$$

or

$$F(x) = \pm\int^x p(x)\,dx \tag{10.37}$$

where $p(x)$ is simply the classical momentum. The $\mathcal{O}(\hbar^1)$ (or imaginary part) then gives

$$2\frac{dA(x)}{dx}p(x) + A(x)\frac{dp(x)}{dx} = 0 \tag{10.38}$$

which can be multiplied on both sides by $A(x)$ to obtain

$$\left(2A(x)\frac{dA(x)}{dx}\right)p(x) + [A(x)]^2\frac{dp(x)}{dx} \equiv \frac{d}{dx}\left([A(x)]^2 p(x)\right) = 0 \tag{10.39}$$

or

$$[A(x)]^2 p(x) = C \tag{10.40}$$

where C is a constant. The two linearly independent solutions (corresponding to right $(+)$ and left-moving $(-)$ waves) are then given by

$$\psi_\pm(x) = \frac{C_\pm}{\sqrt{p(x)}}e^{\pm i\int^x p(x)dx/\hbar} \propto \frac{1}{\sqrt{v(x)}}e^{\pm i\int^x k(x)dx} \tag{10.41}$$

where $p(x) = \hbar k(x)$ defines the "local wavenumber" $k(x)$, and $v(x)$ is the local speed. This remarkably simple solution has several obvious features:

- The corresponding probability density, $|\psi(x)|^2$, satisfies

$$|\psi(x)|^2 \propto \frac{1}{p(x)} \propto \frac{1}{v(x)} \tag{10.42}$$

which is exactly of the form of the *classical probability distribution* first discussed in Section 5.1. This implies (recall Fig. 9.7) that the wavefunctions for the low-lying energy levels will be *poorly* described by the WKB solutions; the quantum wavefunctions for large quantum numbers will, however, approach these semiclassical solutions when suitably locally averaged.

- The phase of the wavefunction can be written as

$$\int^x k(x)\,dx = \int^x d\phi(x) \tag{10.43}$$

where

$$d\phi(x) = k(x)\, dx = \frac{2\pi}{\lambda(x)}\, dx \quad \text{or} \quad \frac{d\phi}{2\pi} = \frac{dx}{\lambda(x)} \tag{10.44}$$

Thus, as the particle moves a distance dx, or a fraction of a "local" wavelength, $df = dx/\lambda(x)$, through the potential, it acquires a phase $d\phi = 2\pi\, df$.

- The solutions can be easily extended to the case where $E < V(x)$, that is, in the classically disallowed regions by appropriate changes in sign giving

$$\psi_{\pm} = \frac{\tilde{C}_{\pm}}{\sqrt{p(x)}} \exp\left(\pm\sqrt{2m/\hbar^2} \int^x \sqrt{V(x) - E}\, dx\right) \tag{10.45}$$

which are the exponentially suppressed solutions discussed in Section 8.2.2; these give rise to quantum tunneling effects. The WKB wavefunction thus has features of both the classical probability distribution, arising from averaging over the trajectory, and the quantum wavefunction.

With this form of the solution, we can examine the effect of neglecting the $\mathcal{O}(\hbar^2)$ term in Eqn. (10.35). Taking the ratio of the last term to the first we find something of the order

$$\frac{\hbar^2}{F'(x)^2}\left(\frac{A''(x)}{A(x)}\right) \propto \frac{\hbar^2}{p(x)^2}\frac{1}{l^2} \propto \frac{1}{[lk(x)]^2} \propto \left(\frac{\lambda(x)}{2\pi l}\right)^2 \tag{10.46}$$

where l is a typical distance scale over which $E - V(x)$ changes. Thus, if the "local de Broglie wavelength," $\lambda(x)$, is much shorter than the distance scale over which the potential changes, the semiclassical approximation is a good one. This is obviously not the case near the classical turning points where the explicit $1/\sqrt{p(x)}$ factors in Eqn. (10.41) actually diverge, indicating that the solution is poorly behaved there.

To obtain a complete description of the wavefunction, the solutions inside and outside the well must be smoothly matched onto each other. The formalism for doing this is not beyond the level of this text, but we choose to only quote the results.[7] For example, we can take linear combinations of the complex exponential solutions near the left turning point to write

$$\psi_L(x) = \frac{A_L}{\sqrt{p(x)}} \cos\left(\int_a^x k(x)\, dx - C_L\pi\right) \tag{10.47}$$

---

[7] See, for example, Park (1992); I also like the discussion in Migdal and Krainov (1969).

For an infinite wall type boundary condition, it is easy to see that $C_L = 1/2$ since we require that

$$\psi_L(a) \propto \cos(-C_L \pi) = 0 \tag{10.48}$$

For a smoother potential, one for which one can approximate $V(x)$ near $x = a$ by a linear function,[8] the appropriate value of C_L turns out to be 1/4. This can be interpreted as saying that the quantum wavefunction penetrates $\pi/4 = 2\pi/8$ or $\sim 1/8$ of a local wavelength into the classically disallowed region.

10.3.2 **WKB Quantized Energy Levels**

One of the most useful results arising from the WKB method is a semiclassical estimate for the quantized energy levels in a potential. Matching the WKB wavefunctions at each of the two classical turning points yields two, presumably equivalent descriptions of $\psi(x)$ inside the well, namely

$$\psi_L(x) = \frac{A_L}{\sqrt{p(x)}} \cos\left(\int_a^x k(x)\,dx - C_L \pi\right) \tag{10.49}$$

and

$$\psi_R(x) = \frac{A_R}{\sqrt{p(x)}} \cos\left(\int_x^b k(x)\,dx - C_R \pi\right) \tag{10.50}$$

If these two solutions are to agree, we must clearly have $|A_L| = |A_R|$; then comparing the arguments of the cosines we find that

$$\int_a^b k(x)\,dx - (C_L + C_R)\pi = n\pi \quad \text{for } n = 0, 1, 2 \dots . \tag{10.51}$$

This implies that

$$\int_a^b k(x)\,dx = (n + C_L + C_R)\pi \quad \text{for } n = 0, 1, 2 \dots . \tag{10.52}$$

or

$$\int_a^b \sqrt{2m(E - V(x))}\,dx = (n + C_L + C_R)\pi \hbar \tag{10.53}$$

Recalling that $k(x) = 2\pi/\lambda(x)$, we see that Eqn. (10.52) is simply a more sophisticated version of "fitting an integral number of de Broglie half-wavelengths in a box" and generalizes the Bohr–Sommerfeld quantization condition. The value of n can be seen to count the number of nodes in the quantum wavefunction.

---

[8] In this case, the solution which interpolates between the inside and outside can be described by an Airy function (See Appendix E.2).

Example 10.6. Infinite well and harmonic oscillator

For the standard infinite well, we have $C_L = C_R = 1/2$ (infinite wall boundary conditions) and $k(x) = \sqrt{2mE/\hbar^2}$, so that the WKB quantization condition gives

$$\int_0^a \sqrt{2mE}\, dx = (n+1)\pi \tag{10.54}$$

or

$$E_n = \frac{\hbar^2 (n+1)^2 \pi^2}{2ma^2} \quad \text{for } n = 0, 1, 2 \ldots \tag{10.55}$$

which is the exact answer.

More interestingly, the WKB quantization also gives the correct answer for the harmonic oscillator. In that case we have

$$p(x) = \sqrt{2m(E - m\omega^2 x^2/2)} \quad \text{or} \quad k(x) = \frac{m\omega}{\hbar}\sqrt{A^2 - x^2} \tag{10.56}$$

where $E = m\omega^2 A^2/2$. Since $C_L = C_R = 1/4$ in this case (noninfinite walls) we find

$$\int_{-A}^{+A} k(x)\, dx = \frac{m\omega}{\hbar} \int_{-A}^{+A} \sqrt{A^2 - x^2}\, dx = (n + 1/2)\pi \tag{10.57}$$

or

$$E_n = (n + 1/2)\hbar\omega. \tag{10.58}$$

Once again, the harmonic oscillator problem can be solved exactly with seemingly every method brought to bear.

We have noted that we have dropped terms of order $\mathcal{O}(\hbar^2)$ or $1/(lk)^2$; since typically we find $k_n \propto n/l$, we expect the WKB estimates of the energies to have errors of order $\mathcal{O}(1/n^2)$. This is consistent with our keeping the C_L, C_R terms in Eqn. (10.52), which, in this language, are of order $\mathcal{O}(1/n)$.

10.4 Matrix Methods

The variational method relies on the expansion of a general quantum state in terms of energy eigenstates. In this section, we describe a matrix approach, which also uses the algebraic structure inherent in the Schrödinger equation, but in a rather different way.

Suppose that we have solved for the energy eigenstates of some Hamiltonian operator, $\hat{H}$. We call them $\psi_n(x)$ where we let the label n start with $n = 1$ for notational convenience; with this labeling, the ground state is $\psi_1(x)$, the first excited state $\psi_2(x)$, and so on.

We know that a general wavefunction can be expanded in such eigenstates via $\psi(x) = \sum_{n=1}^{\infty} a_n \psi_n(x)$ and from Section 6.4 we know that the information content in $\psi(x)$ and the expansion coefficients, $\{a_n\}$, is the same. We can write the collected $\{a_n\}$ as an (infinite-dimensional) vector **a**

$$\psi(x) \Longleftrightarrow \{a_n\} \Longleftrightarrow \begin{pmatrix} a_1 \\ a_2 \\ a_3 \\ \vdots \end{pmatrix} \Longleftrightarrow \mathbf{a} \tag{10.59}$$

where we demand that $\sum_{n}^{\infty} |a_n|^2 = 1$ for proper normalization. In this language, individual energy eigenstates are written as

$$\psi_1(x) \Longleftrightarrow \{a_1 = 1, a_{n>1} = 0\} \Longleftrightarrow \begin{pmatrix} 1 \\ 0 \\ 0 \\ \vdots \end{pmatrix} \Longleftrightarrow \mathbf{e}_1 \tag{10.60}$$

and so forth. The set of vectors, $\mathbf{e}_i$, corresponding to eigenfunctions are said to form a *basis* for the infinite-dimensional vector space; they are like the unit vectors of a more physical vector space. We then have $\mathbf{a} = \sum_i a_i \mathbf{e}_i$.

The Schrödinger equation $\hat{H}\psi = E\psi$ can be written in the form

$$\hat{H}\left(\sum_m a_m \psi_m(x)\right) = E\left(\sum_m a_m \psi_m(x)\right) \tag{10.61}$$

so that if we multiply both sides by $\psi_n^*(x)$ (on the left, as usual) and integrate we find that

$$\sum_m \langle \psi_n | \hat{H} | \psi_m \rangle a_m = E \sum_m \langle \psi_n | \psi_m \rangle a_m = E \sum_m \delta_{n,m} a_m = E a_n \tag{10.62}$$

We then choose to identify

$$\langle \psi_n | \hat{H} | \psi_m \rangle \equiv \mathsf{H}_{nm} \tag{10.63}$$

with the n, mth element of a *matrix* H in which case the Schrödinger equation takes the form of a matrix eigenvalue problem (see Appendix F.1), namely

$$\begin{pmatrix} \mathsf{H}_{11} & \mathsf{H}_{12} & \mathsf{H}_{13} & \cdots \\ \mathsf{H}_{21} & \mathsf{H}_{22} & \mathsf{H}_{23} & \cdots \\ \mathsf{H}_{31} & \mathsf{H}_{32} & \mathsf{H}_{33} & \cdots \\ \vdots & \vdots & \vdots & \ddots \end{pmatrix} \begin{pmatrix} a_1 \\ a_2 \\ a_3 \\ \vdots \end{pmatrix} = E \begin{pmatrix} a_1 \\ a_2 \\ a_3 \\ \vdots \end{pmatrix} \tag{10.64}$$

or

$$\mathbf{Ha} = E\mathbf{a} \tag{10.65}$$

for short. The H_{nm} are called the *matrix elements* of the Hamiltonian and are said to form a *matrix representation* of the operator $\hat{H}$.

We note that H can always be evaluated using a particular set of basis vectors, namely the eigenfunctions of $\hat{H}$ itself. In this particular case, the matrix takes an especially simple form namely, because

$$\mathsf{H}_{nm} = \langle \psi_n | \hat{H} | \psi_m \rangle = \langle \psi_n | E_m | \psi_m \rangle = E \delta_{nm} \tag{10.66}$$

so that the matrix H is *diagonal.* Thus, Eqn. (10.64) takes the form

$$\begin{pmatrix} E_1 & 0 & 0 & \cdots \\ 0 & E_2 & 0 & \cdots \\ 0 & 0 & E_3 & \cdots \\ \vdots & \vdots & \vdots & \ddots \end{pmatrix} \begin{pmatrix} a_1 \\ a_2 \\ a_3 \\ \vdots \end{pmatrix} = E \begin{pmatrix} a_1 \\ a_2 \\ a_3 \\ \vdots \end{pmatrix} \tag{10.67}$$

The only way Eqn. (10.67) can be satisfied is if

$$\det(\mathsf{H} - E\,\mathbf{1}) = \det \begin{pmatrix} E_1 - E & 0 & 0 & \cdots \\ 0 & E_2 - E & 0 & \cdots \\ 0 & 0 & E_3 - E & \cdots \\ \vdots & \vdots & \vdots & \ddots \end{pmatrix} = 0 \tag{10.68}$$

where $\mathbf{1}$ is the unit matrix. This is equivalent to

$$(E_1 - E)(E_2 - E)(E_3 - E)\cdots = \prod_{n=1}^{\infty}(E_n - E) = 0 \tag{10.69}$$

so that the energy eigenvalues are simply the E_n, as we knew; the corresponding eigenvectors of the matrix equation are then simply the $\mathbf{e}_i$ (Why?).

We then say that:

- The matrix representation of a Hamiltonian, when evaluated using its eigenfunctions as a basis, is diagonal and the diagonal entries are just its energy eigenvalues.

Matrix representations of other operators can also be generated. For example, the position and momentum operators, x and $\hat{p}$, have matrix counterparts denoted by x and p and are defined via

$$\mathsf{x}_{nm} = \langle \psi_n | x | \psi_m \rangle \quad \text{and} \quad \mathsf{p}_{nm} = \langle \psi_n | \hat{p} | \psi_m \rangle \tag{10.70}$$

Such matrix representations satisfy the usual rules of matrix algebra, namely

$$(x^2)_{nm} = \sum_k x_{nk}\, x_{km} \tag{10.71}$$

or more explicitly

$$\begin{pmatrix} x^2_{11} & x^2_{12} & \cdots \\ x^2_{21} & x^2_{22} & \cdots \\ \vdots & \vdots & \ddots \end{pmatrix} = \begin{pmatrix} x_{11} & x_{12} & \cdots \\ x_{21} & x_{22} & \cdots \\ \vdots & \vdots & \ddots \end{pmatrix} \cdot \begin{pmatrix} x_{11} & x_{12} & \cdots \\ x_{21} & x_{22} & \cdots \\ \vdots & \vdots & \ddots \end{pmatrix} \tag{10.72}$$

The matrix representation for the kinetic energy operator is, for example, $T = p^2/2m$ or

$$T_{nm} = \frac{1}{2m} \sum_k p_{nk}\, p_{km} \tag{10.73}$$

Example 10.7. Matrix representation of the harmonic oscillator

We can make use of the results of Chapter 9 to evaluate many of these matrix representations for the specific case of the harmonic oscillator. Using the standard energy eigenvalues we find that

$$H = \frac{\hbar\omega}{2} \begin{pmatrix} 1 & 0 & 0 & \cdots \\ 0 & 3 & 0 & \cdots \\ 0 & 0 & 5 & \cdots \\ \vdots & \vdots & \vdots & \ddots \end{pmatrix} \tag{10.74}$$

Using the results in Section 9.2.1, we then find that

$$x_{nm} = \sqrt{\frac{\hbar}{2m\omega}} \left(\delta_{n,m-1}\sqrt{m} + \delta_{n,m+1}\sqrt{m+1} \right) \tag{10.75}$$

and we also quote the result

$$x^2_{nm} = \frac{\hbar}{2m\omega} \left(\delta_{n,m+2}\sqrt{(m+1)(m+2)} + (2n+1)\delta_{n,m} \right.$$
$$\left. + \delta_{n,m-2}\sqrt{m(m-1)} \right) \tag{10.76}$$

We can check that the matrix equation

$$x^2_{nm} = \sum_k x_{nk}\, x_{km} \tag{10.77}$$

(Continued)

holds explicitly by comparing

$$
\frac{\hbar}{2m\omega}
\begin{pmatrix}
1 & 0 & \sqrt{1\cdot 2} & 0 & \cdots \\
0 & 3 & 0 & \sqrt{2\cdot 3} & \cdots \\
\sqrt{1\cdot 2} & 0 & 5 & 0 & \cdots \\
0 & \sqrt{2\cdot 3} & 0 & 7\cdots \\
\vdots & \vdots & \vdots & \vdots & \ddots
\end{pmatrix}
\tag{10.78}
$$

$$
\sqrt{\frac{\hbar}{2m\omega}}
\begin{pmatrix}
0 & \sqrt{1} & 0 & 0 & \cdots \\
\sqrt{1} & 0 & \sqrt{2} & 0 & \cdots \\
0 & \sqrt{2} & 0 & \sqrt{3} & \cdots \\
0 & 0 & \sqrt{3} & 0 & \cdots \\
\vdots & \vdots & \vdots & \vdots & \ddots
\end{pmatrix}
\cdot
\sqrt{\frac{\hbar}{2m\omega}}
\begin{pmatrix}
0 & \sqrt{1} & 0 & 0 & \cdots \\
\sqrt{1} & 0 & \sqrt{2} & 0 & \cdots \\
0 & \sqrt{2} & 0 & \sqrt{3} & \cdots \\
0 & 0 & \sqrt{3} & 0 & \cdots \\
\vdots & \vdots & \vdots & \vdots & \ddots
\end{pmatrix}.
$$

Similar results hold for p and p^2 and one can show (P10.16) that

$$
H_{nm} = \frac{1}{2m}p_{nm}^2 + \frac{m\omega^2}{2}x_{nm}^2
\tag{10.79}
$$

holds as a matrix equation.

The *average* or *expectation value* of an operator in any state can also be written in this language. For example, we have

$$
\langle x \rangle = \langle \psi | x | \psi \rangle = \left\langle \sum_n a_n \psi_n \left| x \right| \sum_m a_m \psi_m \right\rangle
$$

$$
= \sum_{n,m} a_n^* \langle \psi_n | x | \psi_m \rangle a_m
$$

$$
= \sum_{n,m} a_n^* x_{nm} a_m
\tag{10.80}
$$

with similar expressions for other operators. We can also easily include the time-dependence for any state via

$$
\mathbf{a}(t) \Longleftrightarrow \sum_i a_i \mathbf{e}_i\, e^{-iE_i t/\hbar} \Longleftrightarrow
\begin{pmatrix}
a_1\, e^{-iE_1 t/\hbar} \\
a_2\, e^{-iE_2 t/\hbar} \\
\vdots
\end{pmatrix}
\tag{10.81}
$$

The expectation value of the energy operator can be checked to satisfy

$$\langle \hat{E} \rangle_t = \sum_n |a_n|^2 E_n \tag{10.82}$$

independent of time because the energy matrix is diagonal; other average values have less trivial time-dependence (P10.17) in agreement with earlier examples.

Thus far we have only considered the case in which we already know the energy eigenfunctions and eigenvalues of the Hamiltonian operator $\hat{H}$. In this instance, the discussion above is interesting, but provides little new information; we have just provided yet another representation of the solution space. If, on the other hand, we did not know the stationary states we could still proceed as follows:

1. Pick a convenient set of energy eigenfunctions to *some* problem, called $\zeta_n(x)$; we immediately know that they form a complete set so that the expansion theorem will work.

2. Evaluate the Hamiltonian matrix using this set of basis functions, that is calculate

$$\mathsf{H}_{nm} \equiv \langle \zeta_n | \hat{H} | \zeta_m \rangle = \frac{1}{2m} \langle \zeta_n | \hat{p}^2 | \zeta_m \rangle + \langle \zeta_n | V(x) | \zeta_m \rangle \tag{10.83}$$

   In this case, H will no longer be diagonal.

3. The Schrödinger equation in matrix form is still an eigenvalue problem of the form in Eqn. (10.64); its eigenvalues are determined by the condition that $\det(\mathsf{H} - E\mathbf{1}) = 0$.

4. If the eigenvalues are labeled via E_i and the corresponding eigenvectors by $\mathbf{a}^{(i)}$, the position-space wavefunctions are given by $\psi_i(x) = \sum_n^\infty a_n^{(i)} \zeta_n(x)$.

Since finding the *exact* eigenvalues and eigenvectors of an infinite-dimensional matrix is only possible in very special cases, to use this method as a real calculational tool we most often restrict ourselves to a truncated version of the problem. More specifically, we try to diagonalize the $N \times N$ submatrix in the upper left-hand corner for some finite value of N. As N is made larger, we expect to obtain an increasingly good representation of the exact result. Because there exist powerful techniques for diagonalizing large matrices, especially if they happen to have large numbers of vanishing components (so-called sparse matrices), this technique is well suited for numerical computations.

Example 10.8. Infinite well plus δ-function I: Matrix methods

As an example of this method, consider the potential discussed in P8.8, namely, a symmetric infinite well defined via

$$V(x) = \begin{cases} 0 & \text{for } |x| < a \\ +\infty & \text{for } |x| > a \end{cases} \tag{10.84}$$

plus a δ-function potential spike at the origin,

$$V_g(x) = g\delta(x) \tag{10.85}$$

This problem can be solved exactly and hence is useful as a testing ground for various approximation techniques. (For a thorough discussion, see Lapidus (1987).)

We know that the odd states are unaffected by $V_g(x)$ since they all possess nodes at $x = 0$. We thus only consider the even states only for which the energy eigenvalue condition can be written as

$$\lambda = -2y \cot(y) \tag{10.86}$$

where

$$\lambda \equiv \frac{2mag}{\hbar^2} \quad \text{and} \quad E = \frac{\hbar^2 y^2}{2ma^2} \tag{10.87}$$

We naturally choose as a set of basis functions the even solutions of the symmetric well *without* the δ-function potential, that is,

$$\psi_n(x) = \frac{1}{\sqrt{a}} \cos\left(\frac{(n-1/2)\pi x}{a}\right) \tag{10.88}$$

Evaluating the Hamiltonian matrix with this basis set, we find

$$H = \begin{pmatrix} \hbar^2\pi^2/8ma^2 + g/a & g/a & g/a & \cdots \\ g/a & 9\hbar^2\pi^2/8ma^2 + g/a & g/a & \cdots \\ g/a & g/a & 25\hbar^2\pi^2/8ma^2 + g/a & \cdots \\ \vdots & \vdots & \vdots & \ddots \end{pmatrix} \tag{10.89}$$

or

$$H = \left(\frac{\hbar^2\pi^2}{8ma^2}\right) \begin{pmatrix} 1+\epsilon & \epsilon & \epsilon & \cdots \\ \epsilon & 9+\epsilon & \epsilon & \cdots \\ \epsilon & \epsilon & 25+\epsilon & \cdots \\ \vdots & \vdots & \vdots & \ddots \end{pmatrix} \tag{10.90}$$

where $\epsilon \equiv 4\lambda/\pi^2$. Since we rely on matrix diagonalization methods ("canned" packages exist in many programming languages which will find the eigenvalues and eigenvectors of

(Continued)

matrices), we must choose some specific numerical values. For $\lambda = 5$, the exact even energy eigenvalues (in terms of $\hbar^2\pi^2/8ma^2$) obtained from Eqn. (10.86) are

$$2.2969\,(1) \qquad 10.8048\,(9) \qquad 26.9303\,(25) \qquad 50.9743\,(49) \qquad (10.91)$$

where the terms in parentheses are the values without the δ-function term.

Using an available package (in this case *Mathematica*®), we find the eigenvalues for increasingly large $N \times N$ truncated basis sets:

$$
\begin{array}{lllll}
1 \times 1 & 3.02642 & -- & -- & -- \\
2 \times 2 & 2.54241 & 11.5104 & -- & -- \\
3 \times 3 & 2.44832 & 11.1412 & 27.4898 & -- \\
4 \times 4 & 2.40672 & 11.0335 & 27.2396 & 51.4298 \\
\vdots & \vdots & \vdots & \vdots & \vdots \\
10 \times 10 & 2.33857 & 10.8861 & 27.0277 & 51.07991 \\
\vdots & \vdots & \vdots & \vdots & \vdots \\
50 \times 50 & 2.30506 & 10.8204 & 26.9487 & 50.9937
\end{array}
\qquad (10.92)
$$

It does seem that the eigenvalues of the truncated set approach the exact values at N is increased. Such programs also give the eigenvectors as well; we display the components corresponding to the ground state solution below:

$$
\begin{array}{ll}
1 \times 1 & (1) \\
2 \times 2 & (0.97264, -0.23232) \\
3 \times 3 & (0.97451, -0.21543, -0.06258) \\
4 \times 4 & (0.97575, -0.20818, -0.06075, -0.02946)
\end{array}
\qquad (10.93)
$$

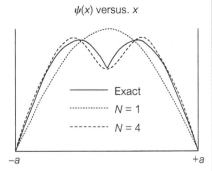

$\psi(x)$ versus. x

——— Exact
·············· $N = 1$
-------- $N = 4$

$-a$ $+a$

Figure 10.7. The exact (solid) and two approximate solutions for Example 10.8.

Using these values, we illustrate in Fig. 10.7 the approximations to the ground state wavefunction for the first four approximations, comparing them to the exact solution, with the cusp expected from the singular δ-function; the convergence to the exact solution is not particularly rapid.

10.5 **Perturbation Theory**

We now turn to what is undoubtedly the most widely used approximation method we will discuss, perturbation theory. We are certainly used to the notion of the systematic expansion of some quantity in terms of a small parameter; a familiar example is the series expansion of a function,

$$f(x) = f(0) + f'(0)x + \frac{1}{2!}f''(0)x^2 + \cdots \tag{10.94}$$

Such an expansion may well formally converge for all values of x (such as for the series for $\exp(x)$), but is often most useful as a calculational tool when $|x| \ll 1$.

Perturbation theory extends this notion to quantum mechanics in cases where the system under study can be described by an "unperturbed" Hamiltonian, $\hat{H}_0$, for which the energy eigenstates can be obtained exactly, that is,

$$\hat{H}_0 \psi_n^{(0)} = E_n^{(0)} \psi_n^{(0)} \tag{10.95}$$

We began this chapter with the observation that many important systems such as the hydrogen atom or the harmonic oscillator can actually be solved exactly. One can then imagine "turning on" an additional perturbing interaction, $\hat{H}'$, which will change the spectrum and wavefunctions; examples include the addition of an electric field acting on a charged particle (via a term $\hat{H}' = V(x) = -qEx$ in one dimension) or a magnetic field acting on a magnetic moment ($\hat{H}' = -\mu \cdot B$). (While we will most often consider the case where the perturbation is a (small) additional potential energy function, other cases are possible (P10.20).) We can then write

$$\hat{H} = \hat{H}_0 + \lambda \hat{H}' \tag{10.96}$$

where we introduce a dimensionless parameter λ (which can be set equal to unity at the end of the calculation) to act as an expansion parameter. Our goal is then to solve the Schrödinger equation for the complete system,

$$\hat{H}\psi_n = E_n \psi_n \tag{10.97}$$

as a series in λ.

We focus in the next two sections on time-independent problems, but briefly discuss problems involving perturbations which evolve in time in Section 10.5.3.

10.5.1 **Nondegenerate States**

We will begin by making the assumption that the energy levels of the unperturbed system are all distinct, that is, that there are no degeneracies where $E_n^{(0)} \approx E_l^{(0)}$

for some pair l, n. Then, as we imagine $\lambda \to 0$, we can unambiguously write

$$\psi_n \xrightarrow{\lambda \to 0} \psi_n^{(0)} \quad \text{and} \quad E_n \xrightarrow{\lambda \to 0} E_n^{(0)} \tag{10.98}$$

and make a unique identification of each perturbed state with its unperturbed counterpart. Motivated by these assumptions, we first write

$$E_n = E_n^{(0)} + \lambda E_n^{(1)} + \lambda^2 E_n^{(2)} + \cdots \tag{10.99}$$

as a series in λ. Then, since the unperturbed eigenstates form a complete set, we always have

$$\psi_n = \sum_{j=0}^{\infty} a_{nj} \psi_j^{(0)} = a_{nn} \psi_n^{(0)} + \sum_j{}' a_{nj} \psi_j^{(0)} \tag{10.100}$$

where $\sum_j{}'$ denotes the infinite sum with the $j = n$ term removed. The new (perturbed) eigenstates can always be written as a linear combination of the old (unperturbed) eigenstates.

The coefficients have slightly different expansions in λ,

$$a_{nn} = a_{nn}^{(0)} + \lambda a_{nn}^{(1)} + \lambda^2 a_{nn}^{(2)} + \cdots \tag{10.101}$$

$$a_{nj} = \lambda a_{nj}^{(1)} + \lambda^2 a_{nj}^{(2)} + \cdots \quad \text{for } j \neq n \tag{10.102}$$

because Eqn. (10.98) implies that

$$\lim_{\lambda \to 0} a_{nj} = \delta_{nj} \tag{10.103}$$

We can constrain the expansion coefficients of Eqn. (10.100) further by noting that the normalization condition

$$\sum_{j=1}^{\infty} |a_{nj}|^2 = 1 \quad \text{for all } n \tag{10.104}$$

implies that

$$1 = |a_{nn}|^2 + \sum_j{}' |a_{nj}|^2 = |a_{nn}|^2 + \sum_j{}' \left(\lambda a_{nj}^{(1)} + \cdots \right)^2 = |a_{nn}|^2 + \mathcal{O}(\lambda^2) \tag{10.105}$$

so that

$$a_{nn} \approx 1 \quad \text{to} \quad \mathcal{O}(\lambda^2) \quad \text{which implies that } a_{nn}^{(1)} = 0 \tag{10.106}$$

The Schrödinger equation can now be written (to $\mathcal{O}(\lambda^2)$) in the form

$$\left(\hat{H}_0 + \lambda\hat{H}'\right)\left(\psi_n + \lambda\sum_j{}'a_{nj}\psi_j^{(0)} + \cdots\right) =$$

$$(E_n^{(0)} + \lambda E_n^{(1)} + \lambda^2 E_n^{(2)} + \cdots)\left(\psi_n + \lambda\sum_j{}'a_{nj}\psi_j^{(0)} + \cdots\right) \qquad (10.107)$$

We first multiply Eqn. (10.107) by $(\psi_n^{(0)})^*$ on the left and integrate to obtain

$$\left\langle\psi_n^{(0}\left|\hat{H}_0 + \lambda\hat{H}'\right|\left(\psi_n + \lambda\sum_j{}'a_{nj}\psi_j^{(0)} + \cdots\right)\right\rangle$$

$$= (E_n^{(0)} + \lambda E_n^{(1)} + \lambda^2 E_n^{(2)} + \cdots)\left\langle\psi_n^{(0)}\left|\left(\psi_n + \lambda\sum_j{}'a_{nj}\psi_j^{(0)} + \cdots\right)\right.\right\rangle$$

$$(10.108)$$

Equating powers of λ and making extensive use of the orthogonality of the unperturbed wavefunctions, namely that $\langle\psi_n^{(0)}|\psi_j^{(0)}\rangle = \delta_{nj}$, we find

$$\mathcal{O}(\lambda^0) : E_n^{(0)} = \langle\psi_n^{(0)}|\hat{H}_0|\psi_n^{(0)}\rangle \qquad (10.109)$$

$$\mathcal{O}(\lambda^1) : E_n^{(1)} = \langle\psi_n^{(0)}|\hat{H}'|\psi_n^{(0)}\rangle \equiv \mathsf{H}'_{nn} \qquad (10.110)$$

$$\mathcal{O}(\lambda^2) : E_n^{(2)} = \sum_j{}'a_{nj}^{(1)}\langle\psi_n^{(0)}|\hat{H}'|\psi_j^{(0)}\rangle \equiv \sum_j{}'a_{nj}^{(1)}\mathsf{H}'_{nj} \qquad (10.111)$$

These expressions all require the *matrix elements* of the *perturbing Hamiltonian*, evaluated using the *unperturbed eigenfunctions*, H'_{nk}.

The $\mathcal{O}(\lambda^0)$ term $(E_n^{(0)})$ simply reproduces the unperturbed energy spectrum. The equation for $E_n^{(1)}$ in Eqn. (10.110) is a very important result as it states that:

- The first-order shift in the energy of level n due to a (small) perturbation is given by the diagonal matrix element of the perturbing Hamiltonian, H'_{nn}, evaluated with the unperturbed wavefunctions, that is,

$$E_n^{(1)} = \langle\psi_n^{(0)}|\hat{H}'|\psi_n^{(0)}\rangle = \mathsf{H}'_{nn} \qquad (10.112)$$

which we repeat because of its extreme importance.

Using Eqn. (10.112) we see that it can sometimes happen that the first-order energy shift vanishes identically because of symmetry. For example, a charged particle in the infinite symmetric well subject to a weak electric field, given by a potential of the form $V(x) = -qEx$, would have a first-order energy shift given by

$$E_n^{(1)} = \begin{cases} -qE\langle u_n^{(+)}|x|u_n^{(+)}\rangle & \text{for even states} \\ -qE\langle u_n^{(-)}|x|u_n^{(-)}\rangle & \text{for odd states} \end{cases} \qquad (10.113)$$

which vanishes for all states. In such cases, the second-order term $E_n^{(2)}$ is the leading correction.

The form of Eqn. (10.111) also suggests the more general result:

- The kth order correction to the energy levels requires knowledge of the $(k-1)$th order wavefunctions

so that to determine $E^{(2)}$ we require the leading-order expansion coefficients, $a_{nj}^{(1)}$.

To obtain information on the expansion coefficients, we multiply Eqn. (10.107) by $(\psi_j^{(0)})^*$ with $j \neq n$ and integrate. The $\mathcal{O}(\lambda^0)$ terms are absent, while the $\mathcal{O}(\lambda^1)$ terms require that

$$a_{nk}^{(1)} = \frac{\langle \psi_k^{(0)}|\hat{H}'|\psi_n^{(0)}\rangle}{(E_n^{(0)} - E_k^{(0)})} = \frac{H'_{nk}}{(E_n^{(0)} - E_k^{(0)})} \qquad (10.114)$$

The first-order wavefunction thus receives contributions from every state for which the *off-diagonal* matrix elements are nonvanishing, that is, $H'_{nk} \neq 0$. Combining Eqns. (10.114) and (10.111), we find that the second-order corrections to the energies are given by

$$E_n^{(2)} = \sum_j{}' a_{nk}^{(1)} H'_{nk}$$

$$= \sum_k{}' \left[\frac{\langle \psi_k^{(0)}|\hat{H}'|\psi_n^{(0)}\rangle}{(E_n^{(0)} - E_k^{(0)})} \right] \langle \psi_n^{(0)}|\hat{H}'|\psi_k^{(0)}\rangle$$

$$= \sum_k{}' \frac{|\langle \psi_n^{(0)}|\hat{H}'|\psi_k^{(0)}\rangle|^2}{(E_n^{(0)} - E_k^{(0)})}$$

$$E_n^{(2)} = \sum_k{}' \frac{|H'_{nk}|^2}{(E_n^{(0)} - E_k^{(0)})} \qquad (10.115)$$

In the last step, we have made use of the fact that $\hat{H}'$ is Hermitian to write $H'_{nk} = (H'_{kn})^*$; this form makes it clear that the second-order shift in energy is manifestly real. This result has many interesting consequences:

- The second-order shift depends on the off-diagonal matrix elements, but inversely weighted by the "distance in energy" to the state in question; thus, in general, states nearby in energy have a larger effect.

- This form also implies that the spacing in energy levels must be larger than the matrix elements of the perturbation for the expansion to be valid, that is, we demand that

$$H'_{nk} << |E_n^{(0)} - E_k^{(0)}| \tag{10.116}$$

This shows that degenerate energy levels must clearly be handled in a different way.

- States with energy below (above) a given level induce a second-order energy shift which is positive (negative); this effect is often referred to as "level repulsion".

- In particular, the second-order shift in the ground state energy is clearly always negative as all the other states lie above it. For many problems for which there are large numbers of levels, one can argue heuristically that the second-order shift for any fixed energy level will be negative due to the large number of states above it;[9] this can be motivated on more physical grounds and is often observed.

The second-order expansion coefficients (the $a_{nj}^{(2)}$) are too complicated to reproduce here but, for reference, we state without proof that the result for the third-order shift in energies is

$$E_n^{(3)} = {\sum_k}' {\sum_j}' \frac{H'_{nk} H'_{kj} H'_{jn}}{(E_n^{(0)} - E_k^{(0)})(E_n^{(0)} - E_j^{(0)})} - H'_{nn} {\sum_k}' \frac{|H'_{nk}|^2}{(E_n^{(0)} - E_k^{(0)})^2} \tag{10.117}$$

We see that the work required to continue the perturbation theory expansion increases rapidly, so that often only the first- and second-order corrections are calculated. We now turn to some examples.

---

[9] See the nice discussion by Saxon (1968).

Example 10.9. Harmonic oscillator with applied electric field

The problem of a charged oscillator in a constant electric field, described by the Hamiltonian

$$\hat{H} = \frac{1}{2}m\omega^2 x^2 - Fx \tag{10.118}$$

(where $F = qE$) was investigated in P9.9 where it was shown that it could be solved exactly; the resulting energy spectrum is

$$E_n = (n + 1/2)\hbar\omega - \frac{F^2}{2m\omega^2} \tag{10.119}$$

Let us approach this problem by considering the electric field interaction to be a small perturbation about the unperturbed oscillator, that is, $\hat{H}' = -Fx$; we can then use F as an expansion coefficient to count powers in perturbation theory. We have $E_n^{(0)} = (n + 1/2)\hbar\omega$, of course, while the first-order correction vanishes (because of symmetry) since

$$E_n^{(1)} = \langle\psi_n|\hat{H}'|\psi_n\rangle = -qE\langle\psi_n|x|\psi_n\rangle = 0 \tag{10.120}$$

The second-order correction is given by

$$E_n^{(2)} = F^2 {\sum_k}' \frac{|\langle\psi_n|x|\psi_k\rangle|^2}{(E_n^{(0)} - E_k^{(0)})} = F^2 {\sum_k}' \frac{|\langle n|x|k\rangle|^2}{(n-k)\hbar\omega} \tag{10.121}$$

Using the results in Section 9.2.2, we know that

$$\langle n|x|k\rangle = \sqrt{\frac{\hbar}{2m\omega}}\left(\sqrt{n}\,\delta_{k,n-1} + \sqrt{n-1}\,\delta_{k,n+1}\right) \tag{10.122}$$

Inserting this result into Eqn. (10.121), we find that

$$\begin{aligned}
E_n^{(2)} &= \frac{F^2}{\hbar\omega}\left(\frac{\hbar}{2m\omega}\right){\sum_k}' \frac{(n\delta_{k,n-1} + (n+1)\delta_{k,n+1})}{(n-k)} \\
&= \frac{F^2}{2m\omega^2}\left(\frac{n}{1} + \frac{(n+1)}{-1}\right) \\
&= -\frac{F^2}{2m\omega^2}
\end{aligned} \tag{10.123}$$

which reproduces the exact answer. One would then expect that all of the higher-order corrections to the energy would then vanish identically and one can confirm explicitly (P10.21) that the third-order correction in Eqn. (10.117) is indeed zero in this case. It is also an example where the second-order corrections are, in fact, negative for all energy levels.

This does not imply, however, that the expansion coefficients have a similarly simple series behavior. To see this, we can make use of the exact ground state solution to the complete problem (see P9.9 again)

$$\psi(x; F) = \frac{1}{\sqrt{\rho\sqrt{\pi}}} e^{-(x-x_0)^2/2\rho^2} \tag{10.124}$$

(Continued)

where $x_0 = F/m\omega^2$ and $\rho = \sqrt{\hbar/m\omega}$. The expansion coefficient a_{00} is then given by

$$a_{00} \equiv \int [\psi(x; F = 0)]^* \, \psi(x; F) \, dx = e^{-x_0^2/4\rho^2} = e^{-F^2/F_0^2} \tag{10.125}$$

where $F_0 \equiv 2\sqrt{\hbar m\omega}$. Expanding a_{00} in powers of F we find that

$$a_{00} = 1 - \frac{F^2}{F_0^2} + \frac{1}{2}\frac{F^4}{F_0^4} + \cdots \tag{10.126}$$

Thus, while the perturbation series for the energies terminates at second-order, the expansion coefficients require the full series to converge to the exact answer.

Example 10.10. Infinite well plus δ-function II: Perturbation theory

Consider the problem, discussed in P8.8, of the symmetric infinite square well potential plus a δ-function potential at the origin; in this case, let the δ-function constitute the perturbation so that $\hat{H}' = g\delta(x)$.

For the odd case, the explicit application of the boundary conditions for the full problem require that the $u_n^{(-)}(x)$ vanish at the origin and gives the same energy eigenvalue condition as for the infinite well alone. This can be confirmed to any order in perturbation theory since all of the relevant matrix elements in Eqns (10.112), (10.115), and (10.117) vanish explicitly. (See P10.22.)

For the even case, the exact eigenvalue condition was given by $\lambda = -2y \cot(y)$ where $\lambda \equiv 2mag/\hbar^2$ and $E = \hbar^2 y^2/2ma^2$. Focusing only on the ground state, this eigenvalue condition can be expanded to second order (see P10.23) to yield

$$y = \frac{\pi}{2} + \frac{\lambda}{\pi} - 2\frac{\lambda^2}{\pi^3} + \mathcal{O}(\lambda^3) = \frac{\pi}{2}\left(1 + \frac{2\lambda}{\pi^2} - \frac{4\lambda^2}{\pi^4}\right) + \cdots \tag{10.127}$$

or

$$E_1^{(+)} \approx \frac{\hbar^2}{2ma^2}\frac{\pi^2}{4}\left(1 + \frac{2\lambda}{\pi^2} - \frac{4\lambda^2}{\pi^4} + \cdots\right)^2$$

$$\approx \frac{\hbar^2\pi^2}{8ma^2} + \lambda\left(\frac{\hbar^2}{2ma^2}\right) - \lambda^2\left(\frac{\hbar^2}{2\pi^2 ma^2}\right) + \cdots \tag{10.128}$$

The first-order perturbation result for even states is simply

$$(E_n^{(+)})^{(1)} = \langle u_n^{(+)}|g\delta(x)|u_n^{(+)}\rangle = \frac{g}{a} = \lambda\left(\frac{\hbar^2}{2ma^2}\right) \tag{10.129}$$

(Continued)

independent of n for all even states; this obviously agrees with the explicit expansion of the eigenvalue condition for the ground state in Eqn. (10.128). The second-order correction is then

$$E_n^{(+)}{}^{(2)} = \sum_{k=2}^{\infty} \frac{|\langle u_n^{(+)}|g\delta(x)|u_k^{(+)}\rangle|^2}{(\hbar^2\pi^2/8ma^2)((1-(2k-1)^2))}$$

$$= -\lambda^2 \frac{2\hbar^2}{m\pi^2a^2} S \quad \text{where } S \equiv \sum_{k=2}^{\infty} \frac{1}{(2k-1)^2-1} = \frac{1}{4}$$

$$= -\lambda^2 \left(\frac{\hbar^2}{2\pi^2ma^2}\right) \tag{10.130}$$

which also agrees with the expansion of the exact result. (See Epstein (1960) for a nice discussion of the subtleties of this problem.)

While the technical details of the calculation are beyond our level, it is appropriate to note here that one of the most spectacularly successful predictions in all of physics makes use of perturbation theory. The magnetic moment of both the electron and the muon can be calculated in the theory of quantum electrodynamics,[10] using more advanced perturbation theory methods. A recent theoretical result for the electron magnetic moment (expressed as a dimensionless number) is

$$g_e(\text{theory}) = 2.0023193048\,(8) \tag{10.131}$$

where the uncertainty is indicated in the last significant digit. Amazingly, it can also be measured to a similar precision with the result

$$g_e(\text{experiment}) = 2.0023193048\,(4) \tag{10.132}$$

10.5.2 **Degenerate Perturbation Theory**

When two (or more) energy levels of the unperturbed system are degenerate, any linear combinations of the corresponding wavefunctions, $\psi_n^{(0)}(x)$ and $\psi_l^{(0)}(x)$, still give same energy eigenvalue. (Such combinations can still be made orthogonal to each other, of course.) This implies, however, that the unique identification of each perturbed state with an unperturbed counterpart as in Eqn. (10.98) is not possible. The breakdown of the perturbation method in this case is clearly signaled by the appearance of small energy denominators

---

[10] See Perkins (2000) and references therein for a discussion at an undergraduate level.

in Eqn. (10.114); states nearby in energy can play an important role and have to be considered on a more equal footing.

In this limit, it is convenient to return to the matrix formulation of the eigenvalue problem and consider the 2×2 submatrix involving the states in question, namely

$$
\begin{pmatrix} E_n^{(0)} + \lambda H'_{nn} & \lambda H'_{nl} \\ \lambda H'_{ln} & E_l^{(0)} + \lambda H'_{ll} \end{pmatrix} \begin{pmatrix} a_n \\ a_l \end{pmatrix} = E \begin{pmatrix} a_n \\ a_l \end{pmatrix}
\tag{10.133}
$$

where the a_n, a_l are the expansion coefficients. In general, for a case with N degenerate levels, the corresponding $N \times N$ submatrix must be considered.

This system of linear equations (for the $a_{n,l}$) will only have a nontrivial solution if the appropriate determinant vanishes, that is,

$$
\det \begin{pmatrix} E_n^{(0)} + \lambda H'_{nn} - E & \lambda H'_{nl} \\ \lambda H'_{ln} & E_l^{(0)} + \lambda H'_{ll} - E \end{pmatrix} = 0
\tag{10.134}
$$

The special case of an exact degeneracy where $E_n^{(0)} = E_l^{(0)} \equiv \mathcal{E}$ is easiest to treat; in this case, the energy eigenvalues are determined by the condition

$$
\left(E - [\mathcal{E} + \lambda H'_{nn}] \right) \left(E - [\mathcal{E} + \lambda H'_{ll}] \right) - \lambda^2 H'_{ln} H'_{nl} = 0
\tag{10.135}
$$

or

$$
E_{\pm} = \mathcal{E} + \frac{\lambda}{2} \left(H'_{nn} + H'_{ll} \right) \pm \frac{\lambda}{2} \sqrt{(H'_{nn} - H'_{ll})^2 + 4 H'_{ln} H'_{ln}}
\tag{10.136}
$$

The first term is obviously the (common) value of the unperturbed energy while the second is the average of the first-order energy shifts in each level, consistent with the non-degenerate case; the third term, however, can split the two levels and generally removes the degeneracy.

Substituting the result of Eqn. (10.136) into the matrix equation Eqn. (10.133), we find that the expansion coefficients are given by

$$
\frac{a_n^{(\pm)}}{a_l^{(\pm)}} = \frac{2 H'_{nl}}{(H'_{ll} - H'_{nn}) + \sqrt{(H'_{nn} - H'_{ll})^2 + 4 H'_{ln} H'_{ln}}}
\tag{10.137}
$$

so that the appropriate (unnormalized) eigenfunctions are given by

$$
\psi^{(\pm)}(x) \propto a_n^{(\pm)} \psi_n(x) + a_l^{(\pm)} \psi_l(x)
\tag{10.138}
$$

The actual energy splitting in Eqn. (10.136) clearly depends on λ, that is, on the *magnitude* of the perturbation; the appropriate linear combinations, however, do not, but are determined by the *form* of the perturbation, that is the *relative* sizes of the matrix elements H'_{nn}, H'_{ll}, and H'_{ln} (P10.26).

Because degeneracy of energy levels is far more common in multi-particle or multidimensional systems, we postpone presenting examples until later chapters.

10.5.3 **Time-Dependent Perturbation Theory**

The problem of determining the future time-development of a given quantum mechanical state, if one knows the exact energy eigenvalues and eigenfunctions of the system, is usually straightforward, since if we have $\hat{H}^{(0)}\psi_n^{(0)} = E_n\psi_n^{(0)}$, then

$$\psi^{(0)}(x,0) = \sum_{n=0}^{\infty} a_n\psi_n^{(0)} \longrightarrow \psi^{(0)}(x,t) = \sum_{n=0}^{\infty} a_n e^{-iE_nt/\hbar}\psi_n^{(0)} \qquad (10.139)$$

A similar problem arises in *time-dependent perturbation theory* when the system is subject to a (small) time-dependent change, so that the resulting Hamiltonian is given by

$$\hat{H} = \hat{H}^{(0)} + \lambda\hat{H}'(t), \qquad (10.140)$$

just as in Eqn. (10.96), but now with $\hat{H}'$ depending explicitly on time. We wish to see how the introduction of the perturbing potential changes the time-development of a quantum state, since we now must satisfy

$$\left[\hat{H}^{(0)} + \lambda\hat{H}'(t)\right]\psi(x,t) = \hat{H}\psi(x,t) = i\hbar\frac{\partial\psi(x,t)}{\partial t} \qquad (10.141)$$

Because the eigenfunctions of $\hat{H}^{(0)}$ still form a complete set, we can always write

$$\psi(x,t) = \sum_{n=0}^{\infty} a_n(t) e^{-iE_nt/\hbar}\psi_n^{(0)} \qquad (10.142)$$

where we now assume that the expansion coefficients, the $a_n(t)$, themselves depend on time. (We can choose this form without loss of generality, thereby defining the $a_n(t)$, but this form is convenient since for the case of no perturbation, it reduces to the standard result in Eqn. (10.139).) We can substitute a solution of the form in Eqn. (10.142) into the time-dependent Schrödinger equation (Eqn. (10.141)), giving the respective left- and right-hand sides

$$\hat{H}\psi(x,t) = \sum_n E_n a_n(t) e^{-iE_nt/\hbar}\psi_n^{(0)} + \sum_n a_n(t) e^{-iE_nt/\hbar}\left[\lambda\hat{H}'\psi_n\right]$$

$$(10.143)$$

$$\Updownarrow \qquad\qquad\qquad \Updownarrow$$

$$i\hbar\frac{\partial\psi(x,t)}{\partial t} = i\hbar\sum_n\left[\frac{da_n(t)}{dt} - \frac{iE_n}{\hbar}a_n(t)\right]e^{-iE_nt/\hbar}\psi_n \qquad (10.144)$$

Equating these two, cancelling the terms proportional to E_n, and multiplying (on the left, as usual) by $(\psi_m^{(0)})^*$ and integrating, we can make use of the orthonormality properties of the eigenstates to write

$$i\hbar \frac{da_m(t)}{dt} e^{-iE_m t/\hbar} = \sum_n a_n(t)\, e^{-iE_n t/\hbar}\, \langle \psi_m^{(0)} | \hat{H}'(t) | \psi_n^{(0)} \rangle \tag{10.145}$$

or

$$\frac{da_m(t)}{dt} = -\frac{i}{\hbar} \sum_n a_n(t)\, e^{i(E_m - E_n)t/\hbar} \langle \psi_m^{(0)} | \hat{H}'(t) | \psi_n^{(0)} \rangle$$

$$\equiv -\frac{i}{\hbar} \sum_n a_n(t)\, e^{i\omega_{m,n} t}\, H'_{mn} \tag{10.146}$$

where

$$\omega_{m,n} \equiv \frac{(E_m - E_n)}{\hbar} \quad \text{and} \quad H'_{mn} \equiv \langle \psi_m^{(0)} | \hat{H}'(t) | \psi_n^{(0)} \rangle \tag{10.147}$$

and H'_{nm} can be called in this context a *transition matrix element*. We recall that the frequency $\omega_{m,n}$ sets the characteristic timescale for any two-state system (Section 4.6). At this stage, the infinite set of coupled equations implied by Eqn. (10.146) is still completely equivalent to the time-dependent Schrödinger equation for the perturbed system, as no approximations have been made.

We now specialize to the case where the initial state of the system is that of an energy eigenstate of the unperturbed Hamiltonian, namely, we assume that $\psi(x, 0) = \psi_k^{(0)}$. We then expect that for a small perturbation the expansion coefficients, $a_m(t)$, for states with $m \neq k$ will be small (since $a_{m \neq k}(t = 0) = \delta_{m,k} = 0$ to begin with), while the single $a_k(t)$ corresponding to the original eigenstate will be of order unity, namely

$$a_m(t) = \begin{cases} \mathcal{O}(\lambda) \ll 1 & \text{for } m \neq k \\ \mathcal{O}(1) & \text{for } m = k \end{cases} \tag{10.148}$$

Using this approximation in Eqn. (10.146), we find that

$$\frac{da_m(t)}{dt} \approx -\frac{i}{\hbar} e^{i\omega_{m,k} t}\, H'_{mk} \tag{10.149}$$

or

$$a_m(t) = -\frac{i}{\hbar} \int_{t_0}^{t} e^{i\omega_{m,k} t}\, H'_{mk}(t)\, dt \tag{10.150}$$

for $m \neq k$, if we assume that the perturbation is "turned on" at time $t = t_0$. This form is important since it implies that the probability that the particle will be found in the mth state of the unperturbed system at a later time is given by

$$P_{k \to m}(t) = |a_m(t)|^2 = \frac{1}{\hbar^2} \left| \int_{t_0}^{t} e^{i\omega_{m,k}t} H'_{mk}(t) \, dt \right|^2 \qquad (10.151)$$

We expect that the chance of being excited (or decaying) into a different final state (the *transition probability*, $P_{k \to m}(t)$) will depend both on the transition matrix element (which includes information on the perturbation *and* its ability to "connect" the initial and final states) as well as on the detailed history of how the perturbation is applied in time, via the integral, which is weighted by the oscillatory exponential factor present in all two-state systems.

Example 10.11. Harmonic oscillator in a time-dependent electric field

Consider a particle of mass m and charge q in the ground state, $|0\rangle$, of the harmonic oscillator potential $V(x) = m\omega^2 x^2/2$. It is subject to an external time-dependent electric field of the form

$$\hat{H}'(t) = (-q\mathcal{E}_0 x) e^{-t^2/2\tau^2} \qquad (10.152)$$

The perturbation is allowed to act over the time interval $(-\infty, +\infty)$ and we wish to evaluate the probability that the particle is eventually found in any excited state, $|n\rangle$.

The expansion coefficient from Eqn. (10.150) in this case is given by

$$a_m(t) = \frac{i}{\hbar} \langle n|q\mathcal{E}_0 x|0 \rangle \int_{-\infty}^{+\infty} e^{i\omega_{n,0}t} e^{-t^2/2\tau^2} \, dt \qquad (10.153)$$

The off-diagonal oscillator matrix element is given by Eqn. (9.50) as

$$\langle n|x|0\rangle = \sqrt{\frac{\hbar}{2m\omega}} \, \delta_{n,1} \qquad (10.154)$$

while the integral over time is a standard Gaussian form. Combining these results, we find that the transition probability to any excited state is given by

$$P_{0 \to n}(t) = |a_n(t)|^2 = \frac{1}{\hbar^2} \left[\frac{q^2 \mathcal{E}_0^2 \hbar}{2m\omega} \delta_{n,1} \right] \left[2\pi\tau^2 e^{-(\omega_{n,0}\tau)^2} \right] \qquad (10.155)$$

This result does show that transition probabilities in time-dependent perturbation theory can depend on *what* the perturbation is, to *which* states it is trying to couple, and *how* it is applied.

- The explicit form of the electric field potential in this case, for example, has implied that only adjacent levels can be excited (to this order of perturbation theory, at least) which is reminiscent of the selection rules discussed in Section 16.3.3. For an initial state $|k\rangle$, only the final states $|k \pm 1\rangle$ would be populated by this perturbation. The transition probability does, of course, also depend on the strength of the external field.

(Continued)

- The transition probability in this case is peaked at a value of $\omega_{n,0}\tau = 1$ so that application of the external perturbing potential at a rate (given by the timescale τ) which matches the natural frequency or period of the system gives the biggest effect. For the special case of the oscillator, $\omega_{n,0}\,\delta_{n,1} = \omega = 2\pi/T_{cl}$ is precisely the classical periodicity of the problem, so perturbations which occur on timescales much longer than this allow the system to accommodate itself to the change. This connection to the classical periodicity is much more general as we recall (P1.16 and Section 12.7)) that the classical period of a quantum system can be written as $T_{cl} = 2\pi\hbar/|dE_n/dn| \sim 2\pi/|\Delta E_{n,m}|$.

We can examine the result of applying the perturbation slowly over a long timescale in some detail, by writing Eqn. (10.150) in a way which includes information on the rate at which $\hat{H}'$ is applied more directly. For example, using an identity and an integration-by-parts (IBP) trick, we have

$$
a_m(t) = -\frac{i}{\hbar}\int_{t_0}^{t} e^{i\omega_{m,k}t}\,\mathsf{H}'_{mk}(t)\,dt
$$

$$
= -\frac{1}{\hbar\omega_{m,k}}\int_{t_0}^{t}\frac{d}{dt}\left[e^{i\omega_{m,k}t}\right]\mathsf{H}'_{mk}(t)\,dt
$$

$$
\stackrel{\text{IBP}}{=} -\frac{1}{\hbar\omega_{m,k}}\left[-\int_{t_0}^{t}e^{i\omega_{m,k}t}\left(\frac{d\mathsf{H}'_{mk}}{dt}\right)dt + e^{i\omega_{m,k}t}\,\mathsf{H}'_{mk}(t)\right] \quad (10.156)
$$

If the rate at which the perturbation is applied, as encoded in the $d\mathsf{H}'_{mk}/dt$ term, is slow enough that that term can be neglected, we have the simplified result

$$
a_m(t) = \frac{\mathsf{H}'_{mk}(t)}{(E_k - E_m)}\,e^{i(E_m - E_k)t/\hbar} \quad (10.157)
$$

Since $a_k(t)\sim\mathcal{O}(1)\cdot e^{-iE_k t/\hbar}$, the time-dependence of the perturbed state is simply

$$
\psi(x,t) = e^{-iE_k t/\hbar}\,\psi_k^{(0)} + \sum_{m\neq k}\left[\frac{\mathsf{H}'_{mk}(t)}{(E_m - E_k)}\,e^{i(E_m - E_k)t/\hbar}\right]e^{-iE_m t/\hbar}\,\psi_m^{(0)}
$$

$$
= \left(\psi_k^{(0)} + \sum_{m\neq k}\frac{\mathsf{H}'_{mk}(t)}{(E_m - E_k)}\psi_m^{(0)}\right)e^{-iE_k t/\hbar} \quad (10.158)
$$

We stress that the time-dependence of this state is simply that of a single eigen-state, while the form of the (spatial) wavefunction is that due to a first-order perturbation theory treatment of $\hat{H}'(t)$, as in Eqn. (10.114), evaluated at time t.

This implies that in the limit of slow application of the perturbation, the individual eigenstates of the old system "morph" into the corresponding eigenstates of the new system, in a state-by-state or quantum-number by quantum-number manner; thus the ground state of the new system corresponds to the ground state of the old system, the first excited state to the first excited state, and so forth. This limit of a slowly acting perturbation is called the *adiabatic approximation* and was used (implicitly) in P5.21.

The opposite extreme, where the system undergoes a very rapid change, is called the *sudden approximation* and in this case we have a discontinuous change in the Hamiltonian of the system. For example, if we have $\hat{H}_1 \rightarrow \hat{H}_2$ at $t = 0$, the eigenfunctions of the system can be written as

$$\hat{H}_1 \psi_n = E_n \psi_n \quad t < 0 \text{ (original system)} \tag{10.159}$$

$$\hat{H}_2 \phi_n = \mathcal{E}_n \phi_n \quad t > 0 \text{ (new system)} \tag{10.160}$$

In this case, if the initial state was that of an eigenstate of the original system (ψ_n) for $t < 0$, then in the new universe of solutions we have the standard expansion theorem result that

$$\psi(x, 0) = \psi_n = \sum_k c_k \phi_k \quad \text{with expansion coefficient } c_k = \langle \phi_k | \psi_n \rangle \tag{10.161}$$

and the future time-dependence of the original eigenstate in the new system will be given by

$$\psi(x, t) = \sum_k c_k e^{-i\mathcal{E}_k t/\hbar} \phi_k \tag{10.162}$$

We have also made implicit use of this method in earlier problems (Example 6.3 and P9.6) with a more physical use discussed in P17.8.

10.6 **Questions and Problems**

Q10.1. If you are given a numerical solution of the Schrödinger equation in the form of a list of values at discrete points, that is, $\psi(x = n\epsilon)$, how would you normalize the solution? How would you find $\langle x \rangle$? How about $\langle p \rangle$? How would calculate the momentum-space wavefunction, $\phi(p)$?

Q10.2. Distinguish carefully between the *precision* and the *accuracy* of a measurement. For example, for a given approximation method, you can imagine determining the range over which the solutions change their sign at infinity more and more precisely; is this increased precision or increased accuracy? Do you think that decreasing the step size ϵ or using a better integration method results in increased accuracy or increased precision?

Q10.3. Assume that you have a program which numerically integrates the Schrödinger equation and that you have found two energy values, E_a and E_b, which bracket an acceptable (i.e. square-integrable) solution of the SE. Describe an efficient strategy to get arbitrarily close to the "real" energy. Hint: If someone tells you they have a number between 1 and 1000, what is the optimal strategy to find their number using the minimum number of "yes–no" questions.

Q10.4. If you have a program which solves the Schrödinger equation for the even solutions in a symmetric potential, what lines of code would you have to change to let it solve for the odd solutions?

Q10.5. Assume that you have a quantum mechanical system with quantized energies E_i and probability densities $|\psi_i(x)|^2$. Suppose that you can add a δ-function perturbation at an arbitrary location. How could you then "map out" the wavefunction using the observed shifts in energy. Assume that the δ-function strengths are small enough that first-order perturbation theory can be used. This approach was followed by Salis *et al.* (1997).

P10.1. Numerical integration—Newton's law. Pick some simple technique designed to numerically integrate second order differential equations, perhaps even the simple one used in Section 10.1. Write a short program (using a computer language, programmable calculator, or even a spread sheet program) to solve Newton's laws for a general potential or force law.

(a) Apply it to the differential equation

$$\frac{d^2x(t)}{dt^2} = -x(t) \quad \text{where } x(0) = 1 \quad \text{and} \quad \dot{x}(0) = 0 \qquad (10.163)$$

Compare your results for decreasing step size with the exact solution (which is, of course, $x(t) = \cos(t)$.) Try to reproduce Fig. 10.1.

(b) Try the same thing for the equation

$$\frac{d^2x(t)}{dt^2} = +x(t) \quad \text{where } x(0) = 1 \quad \text{and} \quad \dot{x}(0) = -1 \qquad (10.164)$$

and also for the initial conditions $x(0) = +1$ and $\dot{x}(0) = +1$. What are the exact solutions, how well does your program work in these two cases, and why?

P10.2. Numerical integration—The Schrödinger equation. Using your experience from P10.1, modify your program to solve the Schrödinger equation for a symmetric potential.

(a) Apply it to the case of the harmonic oscillator written in dimensionless coordinates as

$$\frac{d^2 \psi(y)}{dy^2} - y^2 \psi(y) = -\epsilon \psi(y) \tag{10.165}$$

where ϵ are dimensionless eigenvalues. Try to reproduce the values in Example 10.2. Repeat for the odd case where the eigenvalues are $\epsilon = 3, 7, 11, \ldots$.

(b) Apply your program to the case of a quartic potential, that is, $V(x) = Cx^4$. Write the Schrödinger equation in dimensionless variables and find the first two even and odd energy eigenvalues. Use your previous experience with the oscillator case to estimate the errors in your calculation.

P10.3. Show that the trial wavefunction in Example 10.4 yields the energy function in Eqn. (10.24). Try the problem with the nonzero piece of the wavefunction given by $N(a^2 - x^2)^n$ with $n = 1, 3, 4$ as well and compare your results.

P10.4. Estimate the ground state energy of the SHO by using the family of trial wavefunctions

$$\psi(x; a) = \sqrt{\frac{1}{a}} e^{-|x|/a} \tag{10.166}$$

Why is your answer so much worse than that using the cut-off polynomial expression of Example 10.4?

P10.5. The momentum-space wavefunction corresponding to P10.4 is

$$\phi(p) = \sqrt{\frac{2p_0}{\pi}} \left(\frac{p_0}{p^2 + p_0^2} \right) \tag{10.167}$$

where $p_0 \equiv \hbar/a$. Evaluate the energy functional in momentum space using this trial wavefunction for the SHO and show that you get the same result (for the energy and trial parameter) as in position space.

P10.6. Estimate the energy of the first excited state of the SHO potential by using a trial wavefunction of the form

$$\psi(x; a) = \begin{cases} 0 & \text{for } |x| > a \\ Nx(a^2 - x^2)^2 & \text{for } |x| > a \end{cases} \tag{10.168}$$

Is your answer guaranteed to be larger than the real answer?

P10.7. Use a Gaussian trial wavefunction to estimate the ground state energy for the quartic potential, $V(x) = gx^4$. Show that your answer is

$$E_{\min} = \left(\frac{3}{4} \right)^{4/3} \left(\frac{\hbar^4 g}{m} \right)^{1/3} \tag{10.169}$$

Compare this to the "exact" answer (determined by numerical integration) which has the prefactor 0.668.

P10.8. (a) Estimate the ground state energy of the symmetric infinite well using the family of trial wavefunctions

$$\psi(x) = \begin{cases} 0 & \text{for } |x| > a \\ N(a^\lambda - |x|^\lambda) & \text{for } |x| > a \end{cases} \tag{10.170}$$

where λ is the variational parameter and you must determine the normalization constant N

(b) Estimate the energy of the first excited state by using the wavefunction in (a) multiplied by x to make an appropriate odd trial wavefunction. You will, of course, have to renormalize the wavefunction,

P10.9. (a) Estimate the ground state energy of the symmetric infinite well by using the wavefunction

$$\psi(x) = \begin{cases} 0 & \text{for } |x| > a \\ N(a^2 - x^2) & \text{for } |x| < a \end{cases} \tag{10.171}$$

Evaluate $E(var)/E_0 - 1$ and $1 - |a_0|^2$ for this state. Note that this has no variational parameter.

(b) Now consider the family of trial functions

$$\psi(x; b) = \begin{cases} 0 & \text{for } |x| > a \\ N'(a^2 - x^2)(1 + bx^2/L^2) & \text{for } |x| < a \end{cases} \tag{10.172}$$

which does have an additional parameter. Calculate both the variational energy $E(b) = E[\psi(x; b)]$ and $1 - |a_0|^2$. Find the values of b, which minimize each of these two quantities and show that they are slightly different. Specifically, show that

$$b_{min} = \left(\frac{504 - 51\pi^2}{7\pi^2 - 72} \right) \approx -0.223216 \tag{10.173}$$

for the overlap maximum, while

$$b_{min} = \frac{(-98 + 8\sqrt{133})}{26} \approx -0.22075 \tag{10.174}$$

for the energy minimum. This demonstrates that while there is a strong correlation between the wavefunction which minimizes the variational energy and the one which maximizes the overlap with the ground state wavefunction, the two criteria are ultimately independent.

P10.10. Using the variational method, show that any purely attractive potential in one dimension has at least one bound state. By purely attractive, we mean that $V(x) \leq 0$ for all x. We also assume that $V(x) \to 0$ as $x \to \pm\infty$. Hint: Show that we can find a (perhaps very shallow and narrow) finite square well potential, $V_0(x)$, which satisfies $0 > V_0(x) > V(x)$ for all x and use the fact that a finite square well always has at least one bound state.

P10.11. Show that the matching of solutions leading to the WKB quantization condition implies that $C_L = C_R(-1)^n$.

P10.12. Apply the WKB quantization condition to the symmetric linear potential, $V(x) = F|x|$. The "exact" answers for the lowest lying even (+) and odd (−) states are given by

$$E_i^{(\pm)} = y_i^{(\pm)} \left(\frac{\hbar^2 F^2}{2m} \right)^{1/3} \tag{10.175}$$

where

$$y_1^{(+)} = 1.0188 \qquad y_1^{(-)} = 2.3381$$

$$y_2^{(+)} = 3.2482 \qquad y_2^{(-)} = 4.0879$$

$$y_3^{(+)} = 4.8201 \qquad y_3^{(-)} = 5.5206$$

$$y_4^{(+)} = 6.1633 \qquad y_4^{(-)} = 6.7867 \tag{10.176}$$

Does the agreement get better with increasing n as expected? Can you plot the WKB estimates and "exact" answers in such a way as to demonstrate that?

P10.13. Apply the WKB quantization condition to the "half-harmonic oscillator" potential, namely,

$$V(x) = \begin{cases} +\infty & \text{for } x < 0 \\ m\omega^2 x^2/2 & \text{for } x > 0 \end{cases} \tag{10.177}$$

What are the appropriate values of C_L, C_R and what are the WKB energies? What are the exact results for this problem? Hint: Recall P9.10.

P10.14. Apply the WKB quantization condition to estimate the bound state energies of the potential

$$V(x) = -\frac{V_0}{\cosh(x/a)^2} \tag{10.178}$$

(a) Show that your results can be written in the form

$$E_n = -\left(\sqrt{V_0} - (n + 1/2)\sqrt{\frac{\hbar^2}{2ma^2}} \right)^2 \tag{10.179}$$

Hint: You might use the integral

$$\int_0^A \frac{\sqrt{A^2 - u^2}}{1 + u^2} \, du = \frac{\pi}{2} \left(\sqrt{1 + A^2} - 1 \right) \tag{10.180}$$

(b) If $V_0 \gg \hbar^2/2ma^2$, show that your result approximates the harmonic oscillator approximation for this potential.

(c) One might think that one could take the limit $V_0 \to \infty$ and $a \to 0$ in such a way as to reproduce an attractive δ-function potential. Discuss the WKB approximation to the energy levels in this limit; if it works, does it reproduce the result of Section 8.1.2? If it does not, why?

P10.15. Harmonic oscillator matrix elements.

(a) Evaluate p_{nm} for the harmonic oscillator using the methods in Example 10.7. Using your result, show that the commutator $[x, p] = i\hbar$ holds as a matrix equation.

(b) Evaluate $p^2{}_{nm}$ and use your result to show that Eqn. (10.79) holds by evaluating both sides as matrices.

P10.16. Infinite well matrix elements.

(a) Evaluate the matrix elements p_{nm} using the 'standard' infinite well energy eigenstates as a basis. How would you show that

$$H_{nm} = \frac{1}{2m} \sum_k p_{nk}\, p_{km} \tag{10.181}$$

is the (diagonal) Hamiltonian matrix.

(b) Evaluate the matrix elements x_{nm}. Can you show that $[x, p] = i\hbar$ holds as a matrix equation?

P10.17. What is the expectation value, $\langle \hat{x} \rangle_t$, for a state vector in a matrix representation for general t, that is, how does Eqn. (10.80) generalize to $t \neq 0$?. What does your expression look like for a state with only two components?

P10.18. Show that the wavefunction to second order in perturbation theory (assuming no degeneracies) is given by

$$\psi_n^{(2)} = \sum_m{}' \sum_k{}' \frac{H'_{mk} H'_{kn}}{(E_n^{(0)} - E_k^{(0)})(E_n^{(0)} - E_m^{(0)})} \psi_m^{(0)} - \sum_m{}' \frac{H'_{nn} H'_{mn}}{(E_n^{(0)} - E_m^{(0)})^2} \psi_m^{(0)}$$
$$- \frac{1}{2} \psi_n^{(0)} \sum_m{}' \frac{|H'_{mn}|^2}{(E_n^{(0)} - E_m^{(0)})^2} \tag{10.182}$$

P10.19. We have seen in P6.4 that a constant shift in the potential energy function, that is, $V(x) \to V(x) + V_0$ can have no effect on the observable physics. Consider such a shift as a perturbation and evaluate (i) the first-, second-, and third-order changes in the energy of any state using Eqns. (10.112), (10.115), and (10.117) and (ii) the first-order shift in the wavefunction using Eqn. (10.114) and discuss your results. You can also use the results of P10.18 to check the second-order shift in the wavefunction.

P10.20. Relativistic effects in perturbation theory. The nonrelativistic series for the kinetic energy in Eqn. (1.8) is given by

$$T = \frac{p^2}{2m} - \frac{p^4}{8m^3c^2} + \cdots \qquad (10.183)$$

Using first-order perturbation theory, valuate the effect of the second term in this expansion (with p replaced by the operator $\hat{p}$) on the nth level of a harmonic oscillator.

P10.21. Referring to Example 10.9, use Eqn. (10.117) to show that the third-order energy shift to the energy levels vanishes, as expected since the exact result is of second order.

P10.22. Referring to Example 10.10, evaluate the first-, second-, and third-order shifts in energy for the odd states due to the $\delta(x)$ perturbation and show that they vanish.

P10.23. Derive the expansion in Eqn. (10.127) for the ground state solution in Example 10.10 by writing

$$y = \frac{\pi}{2} + a\lambda + b\lambda^2 + \cdots \qquad (10.184)$$

substituting this into the exact eigenvalue, $\lambda = -2y \cot(y)$ and equating powers of λ.

P10.24. A particle of mass m in a harmonic oscillator potential $V(x) = m\omega^2 x^2/2$ is subject to a small perturbing potential of the same type, namely, $V'(x) = \lambda x^2$.

(a) Show that the energy spectrum can be derived exactly with the result

$$E'_n = \left(n + \frac{1}{2}\right)\hbar\bar{\omega} \qquad (10.185)$$

where $\bar{\omega} = \omega\sqrt{1 + 2\lambda/m\omega^2}$. Expand this for small λ to $\mathcal{O}(\lambda^2)$ for comparison with part (b).

(b) Evaluate the first- and second-order shifts in energy using Eqns (10.112) and (10.115) and compare your results to the exact answer in part (a). You will find the matrix elements of $\langle n|x^2|k\rangle$ in Example 10.7 useful.

P10.25. Anharmonic oscillator in perturbation theory. Evaluate the effect of a small anharmonic term of the form

$$V'(x) = -\lambda k x^3 \qquad (10.186)$$

on the spectrum of the harmonic oscillator in first- and second-order perturbation theory. You may find the following matrix element useful:

$$\langle \psi_n|x^3|\psi_k\rangle = \left(\frac{\hbar}{2m\omega}\right)^{3/2}\left(\sqrt{(n+1)(n+2)(n+3)}\,\delta_{k,n+3} + 3(n+1)^{3/2}\,\delta_{k,n+1}\right.$$

$$\left. + 3n^{3/2}\,\delta_{k,n-1} + \sqrt{n(n-1)(n-2)};\delta_{k,n-3}\right) \qquad (10.187)$$

P10.26. Degenerate states in perturbation theory

    (a) Show that the first-order shifts in energy in Eqn. (10.136) are real as they should be. Do the $a_{n,l}$ have to be real?

    (b) Using Eqn. (10.137), show that the linear combinations $\psi^{(+)}(x)$ and $\psi^{(-)}(x)$ are always orthogonal.

    (c) Discuss the energy levels and mixing of eigenstates in the case where $H'_{nn} = H'_{ll} = 0$ but $H'_{ln} = (H'_{nl})^* \neq 0$; show that the eigenfunctions are "completely mixed."

    (d) Discuss the case where the degenerate states are not connected in the Hamiltonian to lowest order, that is for which $H'_{ln} = (H'_{nl})^* = 0$.

P10.27. Consider a particle of mass m and charge q in the ground state of the symmetric infinite well of Section 5.2.3. It is subject to a time-dependent electric field of the form

$$\hat{H}'(t) = (-q\mathcal{E}_0 x)e^{-|t|/\tau} \qquad (10.188)$$

Find the probability that the particle will be excited to the first excited state if the perturbation is allowed to act over the time range $(-\infty, +\infty)$. Repeat for the probability that it is excited to the second-excited state.

P10.28. Consider a system in a general eigenstate ψ_k, which is subject to a time-dependent harmonic perturbation of the form

$$\hat{H}'(t) = 2V(x)\cos(\omega t) \qquad (10.189)$$

If this perturbation is turned on at $t = 0$ and then removed at $t > 0$, find the probability that the system is in a new state, ψ_m. In the limit of long times, what states are most likely to be connected by this perturbation? Discuss what this might have to do with the emission or absorption of radiation.

ELEVEN

Scattering

11.1 Scattering in One-Dimensional Systems

11.1.1 Bound and Unbound States

Besides the bounded, periodic classical motion of particles in potentials as shown in Fig. 5.2 (for energy E_1), there is also the possibility of unbound states which are not localized in space and not repetitive in time; these correspond to particles incident on the potential and which subsequently "bounce" off (energy E_2 in Fig. 5.2) or temporarily change their speeds as they go over the potential (energy E_3). Classical mechanics, which solves for the exact trajectories in either case, makes little distinction between the two types of motion aside from some technical details, such as the use of Fourier series in the study of periodic motion. In quantum mechanics, on the other hand, the experimental realizations of these two classes of classical motion are quite distinct, so that the theoretical formalisms and methods used to analyze them are necessarily somewhat different.

One main source of experimental information on microscopic systems which allows us to test the ideas of quantum mechanics is spectroscopy, the study of bound states and their radiative decays. This corresponds most closely to bound states as studied in Chapters 5–10. The quantization of energy levels for particles bound in potentials gives rise to a discrete spectrum of photons (or other particles) when excited states decay into lower energy levels. A precise map of the photon energies can allow one to reconstruct the energy level diagram, which, of course, then conveys information on the nature of the bound particles and their interactions.

Quantum scattering experiments, however, make use of rather different experimental techniques. Typically, a beam of incident particles is directed toward a target and the scattered particles are collected (detected) and counted at various angular locations. A collection of classical trajectories corresponding to unbound motions would, as a function of say the particle energy and impact parameter, also allow for a mapping of the scattering force. Quantum mechanically, the particle trajectories are replaced, at best, by probabilistic wave packet motion, but variations of the probability of scattering at different angles and energies still gives information on the nature of the scattering potential.

We will discuss the formalism of scattering theory in much more detail in Chapter 19, but we find it useful here to introduce some of the ideas and notation of three-dimensional scattering before specializing to one-dimensional problems.

In a three-dimensional scattering experiment, a given intensity of incident particles

$$j_{inc}^{(3)}(\mathbf{r}, t) \equiv \frac{dN_{inc}}{dt \, dA} \tag{11.1}$$

that is, the number of particles (dN_{inc}) incident on a target per unit time (dt) per unit area (dA), can be directly associated with the probability flux, defined in three dimensions via

$$\mathbf{j}(\mathbf{r}, t) = \frac{\hbar}{2mi} \left[\psi^*(\mathbf{r}, t) \nabla \psi(\mathbf{r}, t) - \nabla \psi^*(\mathbf{r}, t) \psi(\mathbf{r}, t) \right] \tag{11.2}$$

which is an obvious generalization of the one-dimensional result in Eqn. (4.32). The number of particles scattered into a given small solid angle ($d\Omega$) at a specific angular location specified by (θ, ϕ), per unit time, as in Fig. 11.1, is described by

$$j_{scatter}^{(3)}(\theta, \phi) = \frac{dN_{scatter}}{dt \, d\Omega} \tag{11.3}$$

and will certainly depend on the incident intensity. An appropriate ratio which measures the probability of a scattering event is

$$\frac{j_{scatter}^{(3)}}{j_{inc}^{(3)}} = \frac{dN_{scatter}}{dt \, d\Omega} \bigg/ \frac{dN_{inc}}{dt \, dA} \equiv \frac{d\sigma(\theta, \phi)}{d\Omega} \tag{11.4}$$

which defines a *differential cross-section* for scattering, $d\sigma(\theta, \phi)/d\Omega$, which has the dimensions of an effective area; this quantity can be calculated from a knowledge of the scattering potential, $V(\mathbf{r})$, and can also be directly compared to experiment. For classical "specular" (equal angle reflective) scattering, idealized by scattering small masses (e.g. BBs or marbles) from larger, heavy shapes (e.g. billiard balls), the differential cross-section gives direct information on the size and shape of the scatterers, hence the notion of "cross-sectional area" or cross-section. In quantum mechanics, the wave properties of the scatterers will also be important and the analogs of such effects as interference and diffraction will be apparent.

For scattering in two dimensions, the similar quantity involves ratios of particles incident per unit time (dt) per unit length (dx) and numbers of particles

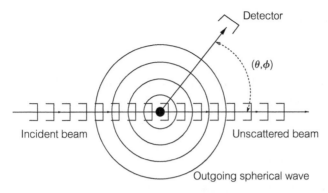

Figure 11.1. Geometry of a scattering experiment in three dimensions.

scattered per unit time per unit angle $(d\theta)$, that is,

$$\frac{j_{scatter}^{(2)}}{j_{inc}^{(2)}} = \frac{dN_{scatter}}{dt\, d\theta} \Bigg/ \frac{dN_{inc}}{dt\, dx} \equiv \frac{d\rho(\theta)}{d\theta} \tag{11.5}$$

This is an effective "width" or lateral size of the target for scattering through various angles θ in the plane.

In one dimension, which we consider in this chapter, the geometric situation is far simpler as there are only two possible directions. In this case, incident particles which continue forward are "transmitted," while those undergoing a back scatter are called "reflected." The incident particles will be described by the incoming number per unit time

$$j_{inc}^{(1)} = \frac{dN_{inc}}{dt} \tag{11.6}$$

with similar expressions for the scattered (i.e. reflected) and transmitted fluxes,

$$j_{ref}^{(1)} = \frac{dN_{ref}}{dt} \quad \text{and} \quad j_{trans}^{(1)} = \frac{dN_{trans}}{dt} \tag{11.7}$$

The ratio of reflected to incident flux, $j_{ref}^{(1)}/j_{inc}^{(1)}$ is then the analog of the scattering cross-section or size, but is dimensionless.

A description of scattering in one, two, or three dimensions involving wave packets (with obvious connections to classical particle trajectories) is possible, but following our discussion for free particles in Chapter 3, we will begin by considering the fluxes (and their ratios) corresponding to the scattering of plane wave solutions as it contains much of the physics involved.

11.1.2 **Plane Wave Solutions**

If we deal with plane wave solutions, we immediately come up against the problem of how to use nonnormalizable wavefunctions. We can avoid any such questions in a highly physical way by considering only the concept of probability flux or currents in Eqns (11.6) and (11.7). For a right- or left-moving plane wave of the form

$$\psi(x, t) = Ae^{i(\pm px - p^2 t/2m)/\hbar} = Ae^{i(\pm kx - \hbar k^2 t/2m)} \tag{11.8}$$

the probability flux is given by (P4.10)

$$j(x, t) = \pm \frac{\hbar k}{m}|A|^2 \tag{11.9}$$

independent of time. The factor $\pm \hbar k/m = \pm p/m = \pm v$ simply corresponds to the classical velocity of the particle beam. Since we expect (normalizable) one particle wavefunctions in one dimension to have dimensions given by $[A] = 1/\sqrt{\text{length}}$, the dimensions of flux can indeed be thought of as the number of particles per unit time; thus we take the probability flux as equivalent to Eqn. (11.6). The flux or incident intensity of a particle beam can be made larger by increasing A corresponding, say, increasing the intensity of the source, or by changing the speed of the particles, that is, increasing k (having them "come at you faster").

We stress that *ratios of fluxes* are then free from any ambiguities from the lack of normalizability and correspond most naturally to the real experimental situation where just such ratios are measured. The conservation of particle flux (no particles are assumed lost) should also follow immediately from the fact that we have introduced no possible absorption processes, and this can be used as a check in any calculation. We will implement these ideas in a series of examples involving plane waves in the next few sections; we will also exhibit some wave packet solutions for comparison.

11.2 **Scattering from a Step Potential**

The simplest one-dimensional potential which can be investigated analytically corresponds to scattering from a step potential. Any physical "step" will have smooth edges as in Fig. 11.2(a), but it is easiest to treat the discontinuous

(a)

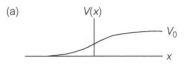

(b)

Figure 11.2. Models of a step potential in one dimension; (a) a physically acceptable smooth step, (b) an idealization of (a) as a discontinuous potential, and (c) a "linear step potential" where the potential is allowed to "turn on" over a distance a.

(c)

potential in Fig. 12.2(b). This is defined by

$$V(x) = \begin{cases} 0 & \text{for } x < 0 \\ V_0 & \text{for } x > 0 \end{cases} \tag{11.10}$$

where we consider both $V_0 > 0$, $V_0 < 0$, but allow only $E > 0$.

If a quantum mechanical free particle in one dimension corresponds to classical traveling waves on a string, this potential is the analog of two strings of differing mass density, and hence with different propagation velocities, "tied together" at the origin, as in P2.2. The corresponding classical particle picture would be described by a force of the form $F(x) = -V_0\delta(x)$ implying an impulsive "kick" each time the particle crosses the origin, changing the magnitude (and/or direction) of its momentum (Q11.1). For example, a particle incident on the step from the left with energy $E > V_0 > 0(V_0 > E > 0)$, would slow down, but continue over (bounce back from) such a potential step.

Considering first the case where $E > V_0$, the allowed plane wave solutions in the two regions are simply

$$\psi(x,0) = \psi(x) = \begin{cases} Ie^{ikx} + Re^{-ikx} & \text{for } x < 0 \\ Te^{iqx} + Se^{-iqx} & \text{for } 0 < x \end{cases} \tag{11.11}$$

where

$$k = \sqrt{\frac{2mE}{\hbar^2}} \quad \text{and} \quad q = \sqrt{\frac{2m(E - V_0)}{\hbar^2}} \tag{11.12}$$

The solutions for $x < 0$ are taken to correspond to an initial plane wave (Ie^{ikx}) incident from $-\infty$ (i.e. from the left) and a reflected component (Re^{-ikx}) going back to $-\infty$. For $x > 0$, the $T\exp(iqx)$ term certainly represents the right-moving transmitted wave; as we expect there to be no left-moving solution for

$x > 0$ if particles are incident on the step from the left, we require $S = 0$. The wavefunction must still satisfy the appropriate continuity conditions on $\psi(x)$ and $\psi'(x)$, so we insist that

$$\psi(0^-) = \psi(0^+) \Longrightarrow I + R = T \tag{11.13}$$

$$\psi'(0^-) = \psi'(0^+) \Longrightarrow ikI - ikR = iqT$$

which gives

$$R = I\left(\frac{k-q}{k+q}\right) \quad \text{and} \quad T = I\left(\frac{2k}{k+q}\right) \tag{11.14}$$

As expected, we cannot solve for R, T completely as they depend on the arbitrary incident amplitude I and we cannot normalize the plane wave solutions; ratios of fluxes, however, will be well-defined and independent of I.

A check on this (otherwise trivial) calculation is to confirm that the probability fluxes are also consistent with expectation. If we calculate the fluxes corresponding to $x < 0$ (j_L) and $x > 0$ (j_R) (see P11.1) we find that

$$j_L(x,t) = \frac{\hbar k}{m}|I|^2 - \frac{\hbar k}{m}|R|^2 = \frac{\hbar k}{m}|I|^2\frac{4kq}{(k+q)^2} \tag{11.15}$$

$$j_R(x,t) = \frac{\hbar q}{m}|T|^2 = \frac{\hbar k}{m}|I|^2\frac{4kq}{(k+q)^2} \tag{11.16}$$

are equal, as expected. The fact that, classically, the particle is moving at a different speed for $x > 0$ (and so q and not k appears in the transmitted flux) is obviously important to remember in this context.

The quantities which would be most similar to experimental observables are the ratios of reflected and transmitted to incident fluxes, that is,

$$\frac{j_{\text{ref}}}{j_{\text{inc}}} = \left|\frac{R}{I}\right|^2 = \left(\frac{k-q}{k+q}\right)^2 = \left(\frac{\sqrt{E} - \sqrt{V_0 + E}}{\sqrt{E} + \sqrt{V_0 + E}}\right)^2 \tag{11.17}$$

$$\frac{j_{\text{trans}}}{j_{\text{inc}}} = \frac{q}{k}\left|\frac{T}{I}\right|^2 = \frac{2kq}{(k+q)^2} = \frac{2\sqrt{E(V_0 + E)}}{(\sqrt{E} + \sqrt{V_0 + E})^2} \tag{11.18}$$

Using these expressions, it is easy to check that the probability of a backscatter goes to zero when $E \gg V_0$, and the effect of a small potential "bump" is negligible.

We plot in Fig. 11.3(a) and (b) plots of the real part of the incident, reflected, and transmitted wavefunctions corresponding to two signs of the potential step. These figures illustrate several aspects of both the wave and particle aspects of

Figure 11.3. Plane wave scattering from a discontinuous step potential; two cases where (a) $E > V_0 > 0$ and (b) $E > 0 > V_0$ are shown. The real part of the incident and reflected amplitude (solid and dotted curves for $x < 0$) and transmitted (dashed curve for $x > 0$) are illustrated.

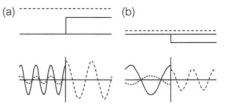

scattering:

- The wavefunction is wigglier (less wiggly) and smaller (larger) in magnitude when the kinetic energy is larger (smaller) as expected from our earlier discussions of the intuitive connections between classical and quantum probability distributions; this shows that such ideas are not restricted to bound state problems.

- A classical particle with energy larger than the step height would, however, never scatter back, so the reflected flux is purely a wave phenomenon. This is not at all apparent from Eqn. (11.17) for the ratio of scattered to incident flux as it depends only on E, V_0 and has no explicit factor of $\hbar$ as would be expected for a purely wave (and hence quantum) effect.

- The phase of the reflected wave relative to that of the incident one depends on the sign of V_0 (or equivalently the classical propagation speed) in a manner which is familiar from other wave phenomena; in going from a region of large to small speed, for example, there is a phase change on reflection.

For the case of $V_0 > E > 0$, we can either find the solutions for $x > 0$ directly or analytically continue the ones of Eqn. (11.11) by letting

$$q = \sqrt{\frac{2m(E - V_0)}{\hbar^2}} \quad \longrightarrow \quad \kappa = iq = \sqrt{\frac{2m(V_0 - E)}{\hbar^2}} \qquad (11.19)$$

so that, for example, the new solutions satisfy

$$\psi(x) = Te^{iqx} \quad \longrightarrow \quad \psi(x) = Te^{-\kappa x} \quad \text{for } x > 0 \qquad (11.20)$$

Such a real function has vanishing probability flux, so the particles must be completely reflected from this step potential. The wavefunction itself, however, is nonzero in the classically disallowed region as we expect from tunneling ideas. The reflection coefficient is then given by

$$R = I\left(\frac{k - i\kappa}{k + i\kappa}\right) \qquad (11.21)$$

which can be written in terms of a simple phase

$$R = Ie^{-2i\phi} \quad \text{where} \quad \tan(\phi) = \kappa/k \qquad (11.22)$$

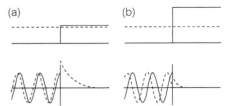

Figure 11.4. Same as for Fig. 11.3, but for two cases where $V_0 > E > 0$.

Using these, we indeed find that

$$\frac{j_{\text{ref}}}{j_{\text{inc}}} = \left|\frac{R}{I}\right|^2 = 1 \quad \text{and} \quad \frac{j_{\text{trans}}}{j_{\text{inc}}} = 0 \tag{11.23}$$

as expected. Figure 11.4 illustrates two such cases where we can observe the tunneling behavior as the amount of energy "cheating" increases. One obvious limit is when $V_0 \gg E$ in which case we have the "infinite wall" scattering case considered in Section 3.3. We note that in this limit we have $\kappa \gg k$, so that $R \to -I$ and the wave solution in the allowed region ($x < 0$) is then

$$\psi_L(x) = Ie^{ikx} + Re^{-ikx} \quad \longrightarrow \quad \left(Ie^{ikx} - Ie^{-ikx}\right) \sim \sin(px/\hbar) \tag{11.24}$$

as discussed previously.

To see the connections to wave packet scattering, we write the general time-dependent plane wave solution in the form

$$\psi_p(x, t) = \begin{cases} Ie^{i(px-Et)/\hbar} + I\left(\frac{p-p_q}{p+p_q}\right) e^{i(-px-Et)/\hbar} & \text{for } x < 0 \\ I\left(\frac{2p_q}{p+p_q}\right) e^{i(p_q x - Et)/\hbar} & \text{for } 0 < x \end{cases} \tag{11.25}$$

where $p = \hbar k$, $p_q = \hbar q$, and $E = p^2/2m = p_q^2/2m + V_0$. This can then be used with a Gaussian weighting distribution to numerically generate the time-dependent wave packet, as in Section 3.4. We illustrate the results for two cases in Fig. 11.5 and note many similarities with the plane wave results:

• The dotted curves correspond to unscattered (but spreading) wave packets and help illustrate the "slow-down" and "speed-up" of the classical particle when $V_0 > 0$ (a) and $V_0 < 0$ (b).

• The probability of reflection and transmission of such packets can be determined by the relative areas under the separate "bumps" of $|\psi(x, t)|^2$ calculated (numerically) for times long after the scatter; the complete wave packet, of course, remains properly normalized at all times. One should not think, therefore, of the particle somehow "splitting" into two separate "blobs"; as with all quantum measurements, a measurement of the particle will find it in a definite position, but with probability related to the relative sizes of the "bumps."

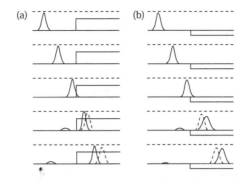

Figure 11.5. Gaussian wavepacket scattering (solid curves) from a discontinuous step potential for (a) $E > V_0 > 0$ and (b) $E > 0 > V_0$. The dashed curves show an unscattered Gaussian (no step potential) for comparison. For example, in (a), the transmitted "lump" of probability lags behind the "free Gaussian" as it slows down over the potential barrier, while for (b) it speeds up over the potential well and is ahead.

11.3 **Scattering from the Finite Square Well**

An important, analytically calculable, example of scattering in one dimension, which again illustrates classical connections, but which also introduces several new features, especially quantum tunneling, is the finite well or barrier defined by the potential

$$V(x) = \begin{cases} -V_0 & \text{for } |x| < a \\ 0 & \text{for } |x| > a \end{cases} \tag{11.26}$$

This potential satisfies the more realistic condition that $V(x) \to 0$ when $|x|$ becomes large. Please note that with this notation, the case where $V_0 > 0$ corresponds to an attractive well, while $V_0 < 0$ is a repulsive barrier.

11.3.1 **Attractive Well**

For the case $V_0 > 0$, we can easily solve the Schrödinger equation in the three appropriate regions. Assuming a plane wave solution incident from the left, we find

$$\psi(x) = \begin{cases} Ie^{ikx} + Re^{-ikx} & \text{for } x < -a \\ Ee^{iqx} + Fe^{-iqx} & \text{for } -a < x < +a \\ Te^{ikx} & \text{for } +a < x \end{cases} \tag{11.27}$$

where

$$k = \sqrt{\frac{2mE}{\hbar^2}} \quad \text{and} \quad q = \sqrt{\frac{2m(E + V_0)}{\hbar^2}} \tag{11.28}$$

as usual; inside the well, we expect both left- and right-moving waves due to reflections at the edges. In contrast to the bound state problem, the boundary conditions at $x = \pm a$ are not equivalent because of the asymmetric nature of the incident scattering. The four independent boundary conditions are given by

$$\text{match } \psi \text{ at } x = -a \quad Ie^{-ika} + Re^{ika} = Ee^{-iqa} + Fe^{iqa}$$

$$\text{match } \psi' \text{ at } x = -a \quad ik(Ie^{-ika} - Re^{ika}) = iq(Ee^{-iqa} - Fe^{iqa})$$

$$\text{match } \psi \text{ at } x = +a \quad Ee^{iqa} + Fe^{-iqa} = Te^{ika}$$

$$\text{match } \psi' \text{ at } x = +a \quad iq(Ee^{iqa} - Fe^{-iqa}) = ikTe^{ika} \tag{11.29}$$

We can then solve for R, T, E, F in terms of I; for the determination of the scattering and transmission probabilities, we only require R, T and one can obtain (P11.5)

$$R = Iie^{-2ika} \frac{(q^2 - k^2)\sin(2qa)}{2kq\cos(2qa) - i(q^2 + k^2)\sin(2qa)} \tag{11.30}$$

$$T = Ie^{-2ika} \frac{2kq}{2kq\cos(2qa) - i(q^2 + k^2)\sin(2qa)} \tag{11.31}$$

while E, F can also be determined if one wishes to see the wavefunction over the entire region.

We illustrate an example of the real parts of the various wavefunctions for a typical case in Fig. 11.6 and we note some general features:

- A classical particle speeds up as it goes over the (attractive) well, and this is consistent with the wavefunction shown in the figure, where the wave function is "wigglier," and has smaller amplitude.

- Another quantum remnant of this increase in velocity over the well is present in the *phase* of the transmitted wave. If we write the wavefunction for $x > a$ as

$$\psi(x) = |T| e^{i\phi} e^{ikx} \tag{11.32}$$

where ϕ is the phase of the (complex) transmission coefficient, we find that

$$\phi = \tilde{\phi} - 2ka \quad \text{where} \quad \tan(\tilde{\phi}) = \frac{k^2 + q^2}{2kq} \tan(2qa) \tag{11.33}$$

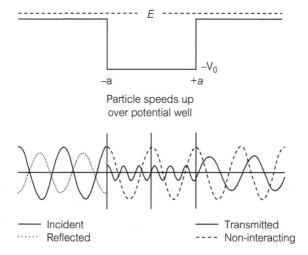

Figure 11.6. The real part of the incident, reflected, and transmitted plane wave solutions for a finite well. For $x < -a$ the dotted curve represents the reflected wave; for $x > -a$ the dashed curve simply continues the incident wave showing an unscattered wave for comparison.

In the limit that $E \gg V_0$ we have

$$\frac{k^2 + q^2}{2kq} = \frac{2E + V_0}{2\sqrt{E(E + V_0)}} \longrightarrow 1 + O(E^2/V_0^2) \approx 1 \qquad (11.34)$$

so that $\phi \to 2a(q-k)$; in this case, the transmitted wave is of the simple form

$$Ie^{i(kx+2a(q-k))} \qquad (11.35)$$

If we compare this form to an incident plane wave which does not interact (i.e. $V_0 = 0$), we find that *at a given time*, the position of a point *of the same phase* on the two waves satisfies

$$\text{scattered} \Longleftrightarrow \text{unscattered}$$

$$kx' + 2a(q - k) = kx$$

so that the difference in positions between the two cases is given by

$$\Delta x \equiv x' - x = \frac{2a}{k}(k - q) \qquad (11.36)$$

If we write these quantities in terms of the classical speeds in the two regions,

$$\hbar k = mv_0 \quad \text{and} \quad \hbar q = mv \qquad (11.37)$$

we find that

$$\Delta x = \frac{2a}{v_0}(v_0 - v) = t_{\text{across}}\Delta v \qquad (11.38)$$

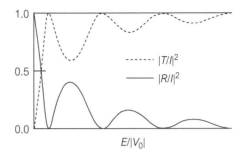

Figure 11.7. The ratio of reflected (dashed) and transmitted (solid) to incident flux versus energy for the attractive square well. The zeroes in $|R/I|^2$ correspond to transmission resonances.

where $\Delta v < 0$ since the particle speeds up ($v > v_0$) over the well; here t_{across} is just the time it would take the (unscattered) particle to cross the well. This is shown in Fig. 11.6 where the transmitted waves does indeed "lead" a wave (the dashed curve) which does not interact. This appearance of the classical "speed-up" in the phase of the transmitted wave is related to the so-called *phase-shift* of three-dimensional scattering.

We next plot the probabilities of both reflection and transmission versus incident energy in Fig. 11.7 (for a particular choice of V_0 and a) where we see that:

- In the limit $E \gg V_0$, one has $q^2 - k^2 \ll 2kq$, $q^2 + k^2$, so that there is little reflection as expected.

- A completely nonclassical phenomenon is readily apparent as there are special values of E for which there is no reflection, corresponding to so-called *transmission resonances*. These points can be traced to the vanishing of $\sin(2qa)$ for $2qa = n\pi$ i.e. for energies satisfying

$$E + V_0 = \frac{n^2 \hbar^2 \pi^2}{8ma^2} \quad \text{for } n = 1, 2, 3 \ldots \tag{11.39}$$

This effect is easily understood in wave terms as due to the complete destructive interference between waves scattered at the first "step" (for which there is a phase change on reflection) and the second (for which there is no phase change). In that case one requires that

$$\text{back and forth distance across the well} = 4a = n\lambda \tag{11.40}$$

so that the difference in path lengths between the two waves is an integral number of wavelengths, while one additional phase change from a reflection from an edge guarantees a minimum instead of a maximum. This effect is familiar from geometrical optics and is often used to minimize the amount of reflected light for optical instruments. A similar effect is seen in atomic systems (hence in three dimensions) where low energy electrons are scattered from atoms of certain noble gases (such as neon or argon); in this case, called the *Ramsauer–Townsend effect,* the scattering cross-section exhibits dramatic dips as a function of energy, indicating near perfect transmission.

11.3.2 **Repulsive Barrier**

All of the formulae for the attractive well can be taken over to the case of the repulsive barrier by simply letting $V_0 \to -V_0$, provided that the incident energy is larger than the step size, that is, $E > V_0$; for example,

$$q = \sqrt{\frac{2m}{\hbar^2}(E - V_0)} \tag{11.41}$$

and the particle now slows down classically over the barrier.

A more interesting case arises when the energy is less than the height of the barrier, $E < V_0$. The form of the solutions for $|x| > a$ are unchanged, but in the barrier region the Schrödinger equation now takes the form

$$\frac{d^2\psi(x)}{dx^2} = \kappa^2\psi(x) \tag{11.42}$$

where $\kappa = \sqrt{2m(V_0 - E)/\hbar^2}$. We can write the most general solution in the region $|x| < a$ as

$$\psi(x) = Ae^{-\kappa x} + Be^{+\kappa x} \tag{11.43}$$

but we can save considerable work by noting that the case under consideration can be easily obtained from an analytic continuation of earlier results since we have

$$q = \sqrt{\frac{2m}{\hbar^2}(E - V_0)} \quad \longrightarrow \quad i\kappa = i\sqrt{\frac{2m}{\hbar^2}(V_0 - E)} \tag{11.44}$$

The results of Eqns (11.30) and (11.31) can then be taken over by using the relations

$$\sin(iz) = i\sinh(z) \quad \text{and} \quad \cos(iz) = \cosh(z) \tag{11.45}$$

so that, for example, the transmission coefficient is given by

$$T = Ie^{-2ika} \frac{2k\kappa}{2k\kappa\cosh(2\kappa a) - i(k^2 - \kappa^2)\sinh(2\kappa a)} \tag{11.46}$$

We illustrate the corresponding wavefunctions in Fig. 11.8 for a generic case.

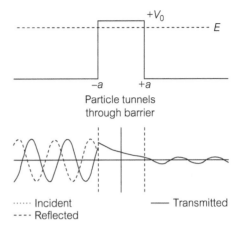

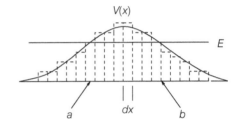

Figure 11.8. Representation of a plane wave incident on square barrier showing tunneling wavefunction.

Figure 11.9. Generic potential barrier, $V(x)$, approximated as a series of square barriers making connection to the WKB approximation for the tunneling wavefunction.

A particle incident on such a rectangular barrier would have a probability of penetration given by

$$\frac{|T|^2}{|I|^2} = \frac{(2\kappa k)^2}{(k^2 + \kappa^2)\sinh^2(2\kappa a) + (2k\kappa)^2} \approx \left(\frac{4k\kappa}{k^2 + \kappa^2}\right)^2 e^{-4\kappa a} \qquad (11.47)$$

We note that the dominant exponential piece of the tunneling probability in Eqn. (11.47) is just that given by the WKB formula of Section 10.3.1. We can make a useful connection between this square well result and the WKB formulae.

We derive this by noting that in order for a particle to tunnel through the general potential of Fig. 11.9, it must successfully tunnel through each of the thin rectangular barriers in turn, with probability $P_i(x_i)$ for each barrier; we make the following identifications.

$$\text{width, } 2a \longrightarrow \Delta x$$

$$\kappa \longrightarrow \kappa(x_i)$$

$$P = e^{-2(\kappa)2a} \longrightarrow P_i(x_i) = e^{-2(\kappa(x_i))\Delta x} \qquad (11.48)$$

Because each of the "successes" is an independent event, the total probability is the product of the $P_i(x_i)$; we thus have

$$P_T = \prod_i P_i(x_i)$$

$$\approx \prod_i e^{-2\kappa(x_i)\Delta x} \quad \text{(where} \quad \kappa(x) \equiv \sqrt{2m(V(x) - E)/\hbar^2}\text{)}$$

$$= e^{-2\sum_i \kappa(x_i)\Delta x}$$

$$\approx e^{-2\int_a^b \kappa(x)\,dx}$$

$$P_T = \exp\left(-2\sqrt{\frac{2m}{\hbar^2}} \int_a^b \sqrt{V(x) - E}\,dx\right) \tag{11.49}$$

where a, b are the classical turning points. We will use this approximate expression for the transmission probability for barrier penetration in the examples discussed in the next section.

11.4 **Applications of Quantum Tunneling**

11.4.1 **Field Emission**

A subject which is often discussed in modern physics courses is the *photoelectric effect* in which electrons are emitted from a metal by the absorption of sufficiently energetic photons. The effect is illustrated in Fig. 11.10(a) where the Fermi energy of the filled electron sea is still an energy W below the threshold for a free particle; W is often called the *work function* of the metal. A photon of energy E_γ can extract an electron from the sea provided that $E_\gamma \geq W$, with any remaining energy transferred to the electron as kinetic energy. Such experiments provide

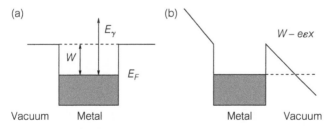

Figure 11.10. (a) Allowed electron states for a metal showing the filled Fermi sea and the photoelectric effect; (b) An external electric field is applied illustrating the triangular barrier giving rise to field emission.

evidence for the photon concept and the quantization of the electromagnetic field energy.

A completely different form of electron emission which relies instead on a purely classical electric field, but which makes use of quantum tunneling, is *field emission*. In this case, shown in Fig. 11.10(b), an external electric field $\mathcal{E}$ is applied to the sample; electrons at the top of the sea can now tunnel through the triangular-shaped potential barrier. In this simple approximation, the probability of tunneling corresponding to Eqn. (11.49) is (P11.8)

$$ P_T = \exp\left(-\frac{4}{3}\sqrt{\frac{2mW^3}{\hbar^2}}\frac{1}{e\mathcal{E}}\right) = \exp\left(-\frac{\mathcal{E}_0}{\mathcal{E}}\right) \tag{11.50} $$

This expression shows the strong dependence on the local value of the work function W at the surface. The resulting electron current due to quantum tunneling should be directly proportional to this probability, namely

$$ I = I_0 e^{-\mathcal{E}_0/\mathcal{E}} \quad \text{or} \quad \log(I) = \log(I_0) - \frac{\mathcal{E}_0}{\mathcal{E}} \tag{11.51} $$

We compare this prediction with some of the data from one of the original experiments[1] in Fig. 11.11.

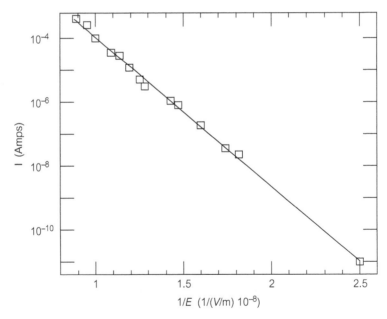

Figure 11.11. Semilog plot of tunneling current, I, versus $1/\mathcal{E}$ where $\mathcal{E}$ is the applied electric field, illustrating field emission. The data are taken from Millikan and Eyring (1926).

---

[1] We replot the data off Millikan and Eyring (1926) in the way suggested by Eqn. (11.51).

This effect is also used as the basis of an imaging device called the field ion microscope[2] (FIM), which was the first microscope to achieve atomic resolutions enabling one to "see" individual atoms. The device works roughly as follows:

- A sharp, metallic tip with radius of curvature in the range 100 to 200 Å is placed in a vacuum and charged to a large voltage, typically 1–20 kV; this process itself helps to smooth the surface by selective field ionization of the metal atoms to what can be called "atomic smoothness".

- A very dilute gas of noble gas atoms (often helium) is introduced; this is used as the imaging gas. These atoms are adsorbed onto the surface of the probe due to dipole–dipole attractions to the tip atoms (remember that both atomic species are initially neutral).

- The image gas atoms, once attached to a tip atom, can be ionized via field emission, losing an electron to the tip; the resulting positively charged *ions* are then accelerated by the electric field toward a phosphorescent screen some tens of centimeters, away forming the image. A schematic representation of the process is shown in Fig. 11.12 and an FIM image is shown in Fig. 11.13.

- In this device, the electron tunneling serves only to initiate ion formation and the electrons themselves do not participate in the image formation. In the original *field emission microscope*, the electrons emitted via field ionization were used to image the surface; this process relied on the local variations of

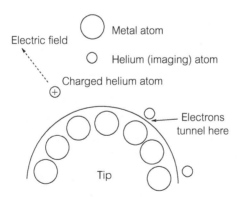

Figure 11.12. Schematic representation of field ion microscope. The metal atoms (large circles) forming the surface of the smooth probe tip attract (via dipole–dipole forces) the atoms of the imaging gas (small circles are helium). Once bound, electrons from the He can tunnel into the tip via field emission. The resulting charged ions travel along the electric field lines (the tip is held at a large electric potential) and form an image on the screen.

---

[2] For many technical details, see the excellent books by Tsong (1990) and Müller and Tsong (1969).

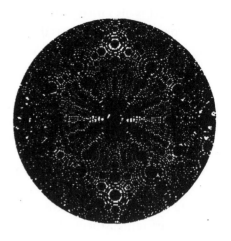

Figure 11.13. Field ion microscope image of a tungsten tip of radius ~ 400 Å. The original image had a magnification of 3 million. (Photo courtesy of T. Tsong.

the work function W on the surface which gives rise to large variations in the tunneling probability, and in turn the electron current. The large lateral velocity spread of the emitted electrons, as well as de Broglie wave diffraction effects, limited this technique to resolutions of the order 20–25 Å.

11.4.2 Scanning Tunneling Microscopy

A newer technique which has had great success in obtaining images of atomic structures on (typically graphite or silicon) surfaces is *scanning tunneling micro-scopy*[3] (STM). A schematic representation of the physics involved is shown in Fig. 11.14:

1. Two metal electrodes are placed close together (often only Å's apart), one being the sample while the other is the tip.

2. Their Fermi surfaces differ and electrical equilibrium is reached only when enough electrons have tunneled through the junction (from left to right in this case). The resulting charge separation results in an electric field in the vacuum region between the electrodes.

3. An external voltage difference is applied to the tip shifting the Fermi energies again and allowing electron tunneling to occur.

As the tip is scanned over a plane surface, feedback circuits monitor the tunneling current, adjusting the tip height to maintain it at a constant value. The resulting height profile provides a map of the surface.

---

[3] See the recent books by Stroscio and Kaiser (1993) and Chen (1993) for many details. The image in Fig. 1.3 were obtained using this technique.

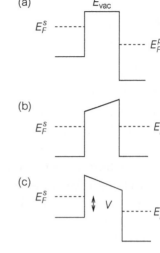

Figure 11.14. Schematic representation of the energy levels relevant for a scanning tunneling microscope (STM).

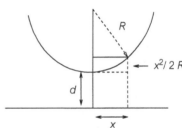

Figure 11.15. Typical geometry of the tip of an STM probe.

An estimate of the lateral resolution possible with this instrument can be made by assuming a simple shape for the STM tip probe as in Fig. 11.15. Assuming a parabolic probe with radius of curvature R of say 1000 Å, one finds a current profile as a function of distance away from the closest point d given by

$$I(x) \propto e^{-2(d+x^2/2R)\kappa} \quad \propto \quad e^{-x^2\kappa/R} \quad \propto \quad e^{-x^2/\rho^2} \tag{11.52}$$

where $\rho \equiv \sqrt{R/\kappa}$; typically $\kappa \approx \sqrt{2mW/\hbar^2} \approx 1\,\text{Å}^{-1}$. This familiar Gaussian distribution has a spread in lateral position given by $\Delta x = \rho/\sqrt{2} \approx 20\,\text{Å}$. As mentioned above, even smaller tip sizes are possible, but because of the exponential sensitivity of the tunneling current to the tip-to-surface distance d, it can well be the case that the best images arise from the tunneling from a very few surface atoms forming an atomic-scale 'dimple' closest to the surface.

11.4.3 α-Particle Decay of Nuclei

One of the most famous early (semiquantitative) successes of quantum tunneling theory in nuclear physics was the understanding of the process of α-particle tunneling. In this process, a heavy nucleus decays to a lighter one by the emission of an α-particle, that is, the nucleus He^4. Using a compact notation, the process can be written as

$$_N^Z X^A \longrightarrow {}_{(N-2)}^{(Z-2)} Y^{(A-4)} + {}_2^2 He^4 \tag{11.53}$$

where Z, N, and A are, respectively, the numbers of protons, neutrons, and total nucleons in the nuclear species denoted by X (the "parent") or Y (historically the "daughter" nucleus).

Because this is a two-body decay, the energy of the emitted α is determined uniquely from conservation of energy and momentum, and can be calculated from a knowledge of the masses of the parent and daughter nuclear species. For the nuclei for which α-decay is an important decay mechanism, the range in numerical values for the appropriate dimensionful parameters in the problem is not very large,

$$R \sim 2\text{--}4 \; F, \quad E_\alpha \sim 2\text{--}8 \; MeV, \quad \text{and} \quad Z \sim 50\text{--}100 \tag{11.54}$$

while the observed lifetimes have been measured over an incredibly large range

$$\tau \sim 10^{17} \, s - 10^{-12} \, s \tag{11.55}$$

A simple model for this process assumes that the α-particle moves in the potential of the *daughter* nucleus, modeled by a combination of an attractive square well (as in Section 8.2), along with the mutual Coulomb repulsion. This can be written as

$$V(r) = \begin{cases} -V_0 & \text{for } r < R \\ Z_1 Z_2 Ke^2/r & \text{for } R < r \end{cases} \tag{11.56}$$

We would then take $Z_1 = Z_\alpha = 2$ and $Z_2 = Z - 2$ where Z is the charge of the *parent* nucleus. This potential is illustrated in Fig. 11.16 and the α-particle is assumed to have positive energy E_α equal, to its observed final kinetic energy; the model pictures the α-particle as "rattling around" inside the nucleus with a small (exponentially so) quantum tunneling probability of escaping each time it "hits" the Coulomb barrier. The tunneling probability for this process is then given from Eqn. (11.49) by

$$P_T = \exp\left[-2\sqrt{\frac{2\mu}{\hbar^2}} \int_a^b dr \sqrt{\frac{Z_1 Z_2 Ke^2}{r} - E} \right] = e^{-2G} \tag{11.57}$$

where the factor in the exponential (G) is known as the *Gamow factor*. The classical turning points are taken to be

$$a = R \quad \text{and} \quad b = \frac{Z_1 Z_2 Ke^2}{E_\alpha} \tag{11.58}$$

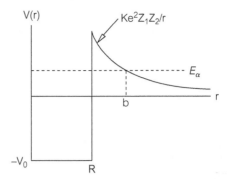

Figure 11.16. Simple model for the effective potential seen by an α-particle undergoing tunneling.

and we have used the reduced mass μ as is appropriate for a two-body problem; since the daughter nucleus is much heavier than the α-particle, however, one has $\mu \approx m_\alpha$. The Gamow factor can be written in the form

$$G = Z_1 Z_2 \alpha \sqrt{\frac{2\mu c^2}{E_\alpha}} \int_{\omega^2}^{1} d\eta \sqrt{\frac{1}{\eta} - 1} \tag{11.59}$$

where $\alpha = Ke^2/\hbar c$ as always and $\omega^2 \equiv R/b$. The integral can be done in closed form giving

$$\int_{\omega^2}^{1} d\eta \sqrt{\frac{1}{\eta} - 1} = \frac{\pi}{2} - \sin^{-1}(\omega) - \sqrt{\omega^2(1 - \omega^2)}$$

$$\downarrow$$

$$\frac{\pi}{2} \quad \text{for } \omega = \sqrt{b/R} << 1 \tag{11.60}$$

One then has very roughly that

$$2G \approx 4\frac{(Z - 2)}{\sqrt{E_\alpha(\text{MeV})}} \tag{11.61}$$

The decay *lifetime* itself can be estimated by noting that there is roughly an e^{-2G} probability of a tunneling "escape" every time the α "hits" the electrostatic barrier. The time between such 'escape attempts' can be approximated as

$$T_0 \approx \frac{2R}{v_\alpha} \quad \text{where} \quad v_\alpha \approx \sqrt{\frac{2E_\alpha}{m_\alpha}} \approx \frac{1}{20} c \tag{11.62}$$

and $R \approx 5$–$8\,F$ is a typical (heavy) nuclear radius; this gives $T_0 \approx 10^{-21}$ s, which is indeed a typical nuclear reaction time. The lifetimes in this simple

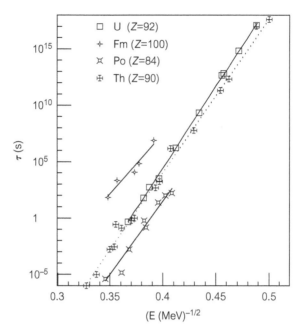

Figure 11.17. Semilog plot of α-decay lifetime (τ in seconds) versus $1/\sqrt{E_\alpha}$ (in MeV) for four different radioactive decay series, the so-called Geiger–Nutall plot. The data are taken from a recent edition of the Chart of the Nuclides (Walker (1983).)

picture then scale as

$$\tau = T_0 e^{+2G} \quad \text{or} \quad \log(\tau) \approx \log(T_0) + 4\frac{(Z-2)}{\sqrt{E_\alpha(\text{MeV})}} \tag{11.63}$$

This behavior is most easily studied by examining the α-decay lifetimes of different isotopes of the same element (so that the value of Z is fixed and only E_α varies). We plot the lifetimes for several such series (on a log scale) versus $1/\sqrt{E_\alpha}$ in Fig. 11.17 (a so-called *Geiger–Nuttall* plot) and note the reasonable straight line fits. The simple approximations made here can be refined,[4] but they provide convincing evidence for the importance of quantum tunneling effects in nuclear decay processes.

11.4.4 **Nuclear Fusion Reactions**

For α-decay of heavy nuclei, the repulsive Coulomb potential forms a barrier through which α-particles must tunnel to get out of the nucleus. For *fusion reactions*, we must consider the inverse process in which light nuclei must penetrate

---

[4] See, for example, Park (1992) for a more extensive discussion.

their mutual Coulomb barrier in order to get close enough to participate in strong nuclear or weak interactions, which are short-ranged forces.

For example, in the production of thermonuclear energy in stars, protons (^1H) are eventually converted to helium nuclei via the so-called *pp cycle.* The overall reaction can be written as

$$\text{pp cycle: } 4\,^1H \rightarrow \,^4He + 2e^+ + 2\nu_e \quad (Q = 26.7\,\text{MeV}) \tag{11.64}$$

where Q is the energy released per reaction. The net reaction is the result of a series of two-body interactions given by

$$^1H + \,^1H \rightarrow \,^2H + e^+ + \nu \quad (Q = 1.44\,\text{MeV})$$
$$^2H + \,^1H \rightarrow \,^3He + \gamma \quad (Q = 5.49\,\text{MeV})$$
$$^3He + \,^3He \rightarrow \,^4He + 2\,^1H + \gamma \quad (Q = 12.86\,\text{MeV}) \tag{11.65}$$

The reaction rate for such two-body processes is determined not only by the probability for the quantum mechanical tunneling event (given by the interaction cross-section), but also by the available number of initial particle pairs, and each of these factors has a very different dependence on the center-of-mass energy in the collision. The cross-section must include the Gamow factor, e^{-2G}, while the number density of particles in the hot gas is proportional to the Boltzmann factor, e^{-E/k_BT}. This means that the interaction rate is proportional to

$$\text{reaction rate} \propto e^{-2G - E/k_BT} = e^{-f(E)} \tag{11.66}$$

where

$$f(E) = Z_1 Z_2 K e^2 \pi \sqrt{\frac{2\mu c^2}{E}} + \frac{E}{k_BT} \tag{11.67}$$

and the *maximum* event rate corresponds to the *minimum* value of $f(E)$. The value of energy at this point is given (P11.10) by

$$\left(\frac{E_{\max}}{k_BT}\right)^{3/2} = Z_1 Z_2 \pi \alpha \sqrt{\frac{\mu c^2}{2k_BT}} \tag{11.68}$$

This implies that if one wishes to study the dynamics of the nuclear reactions responsible for stellar fusion at a given temperature, one should do nuclear scattering experiments in terrestrial accelerators at energies very near $E_{\max}$ as given by Eqn. (11.68); the interaction rate for values not too different from $E_{\max}$ will be hugely suppressed due to the exponential sensitivity in Eqn. (11.66).

11.5 **Questions and Problems**

Q11.1. Consider a classical particle of energy E incident from the left on the step potential of Eqn. (11.10). On the same graph, sketch the classical trajectory, $x(t)$ versus t, for the cases (i) $E = 4V_0 > 0$, (ii) $E = 1.1V_0 > 0$, (iii) $E = 0.5V_0 > 0$, and (iv) $E = -V_0 > 0$.

Q11.2. Consider a wave packet of mean energy $\overline{E}$ incident on a step potential of height $V_0 > 0$. What would $\langle x \rangle_t$, Δx_t, $\langle \hat{p} \rangle_t$, and Δp_t all look like as functions of time for the cases (i) $\overline{E} >> V_0$, (ii) $\overline{E} \gtrsim V_0$, and (iii) $V_0 >> \overline{E} > 0$?

Q11.3. If you wanted to construct a wave packet for the attractive finite well, what would be the complete set of states you would have to use?

Q11.4. To what temperature would you have to heat a typical metal to remove electrons by thermal excitations? Why is field emission sometimes referred to as "cold emission"?

Q11.5. Why are neutrons most often used in initiating fission processes in heavy nuclei? Would protons work as well? How about antiprotons?

Q11.6. Look up the *Arrhenius law* in any book on physical chemistry which is used to describe the rate of chemical reactions. Discuss the similarities to our simplified discussion of α-particle decay.

P11.1. Show that the probability flux corresponding to the wavefunction

$$\psi(x, t) = Ie^{i(kx - \hbar k^2 t/2m)} + Re^{i(-kx - \hbar k^2 t/2m)} \tag{11.69}$$

is given by

$$j(x, t) = \frac{\hbar k}{m}|I|^2 - \frac{\hbar k}{m}|R|^2 \tag{11.70}$$

P11.2. Scattering from δ-function potentials I. Consider plane waves of amplitude I incident on a single attractive δ-function potential, $V(x) = -g\delta(x)$.

(a) Match boundary conditions to calculate R and T in terms of I and the other parameters of the problem.

(b) Calculate the ratios of the reflected and transmitted flux to the incident flux and check conservation of flux.

(c) Express the reflected and transmitted fluxes in terms of E and sketch them versus E.

(d) For what value of E will 99% of the incident flux be reflected? How about 1%?

(e) What changes if we choose a repulsive potential instead?

P11.3. Absorptive scattering. Consider a potential similar to that of P11.2 except that the scattering center now has a negative imaginary part, that is,

$$V(x) = (g - ig_0)\delta(x). \tag{11.71}$$

(a) Repeat the analysis above and show that flux is *not* conserved in this case. This result is expected from the discussion of Section 4.2 where we noted that this type of potential corresponds to absorption.

(b) Calculate the rate of loss of particles 'into' the absorption potential and show that it approaches the incident particle rate when $E \ll E_0$, that is, effectively all of the particles are absorbed below some energy E_0.

P11.4. Scattering from δ-function potentials II. Consider scattering from a twin repulsive δ-function potential defined by

$$V(x) = g[\delta(x-a) + \delta(x+a)] \tag{11.72}$$

(a) Calculate R and T.

(b) Show that flux is conserved.

(c) Show that this potential exhibits transmission resonances in the same way as the square barrier, with the same physical condition being required. (For a discussion of this problem, see Lapidus 1982c.)

P11.5. Derive the expressions for R and T in Eqns. (11.30) and (11.31). Also solve for the E and F coefficients. Show that flux is conserved in this reaction, namely that the flux is the same in all three regions considered.

P11.6. The attractive δ-function potential, $-g\delta(x)$, can be thought of as the limiting case of a deep and narrow attractive well where the width and depth satisfy $2aV_0 = g$ as $a \to 0$ and $V_0 \to \infty$. Show that your results for $|T/I|^2$ in the last problem in this limit gives the same result as in P11.2.

P11.7. (a) Show that the smooth potential function defined via

$$V(x) = V_0 \left(\frac{1}{1 + e^{-x/L}} \right) \tag{11.73}$$

approaches the discontinuous step potential for $L \to 0$.

(b) One can show[5] that the ratio of reflected to incident flux is given by

$$\frac{|R|^2}{|I|^2} = \left(\frac{\sinh(\pi(k-q)L)}{\sinh(\pi(k+q)L)} \right)^2 \tag{11.74}$$

Show that the $L \to 0$ limit reduces to the result for the step potential.

(c) Discuss the $L \to \infty, \hbar \to 0$, and $m \to \infty$ limits.

---

[5] See Landau and Lifschitz (1965).

P11.8. Derive Eqn. (11.50).

P11.9. Geiger–Nuttall plot for thorium The lifetimes (in various units) and observed α-particle energies (in MeV) for decays of various isotopes of thorium ($Z = 90$) are given below:

Mass number	τ	E_α (MeV)
232	1.4×10^{10} yr	4.01
230	7.7×10^4 yr	4.69
229	7.34×10^3 yr	4.85
228	1.91 yr	5.42
227	18.7 day	6.04
226	31 min	6.34
225	8 min	6.48
224	1.04 s	7.17
223	0.66 s	7.29
222	2.9 ms	7.98
221	1.68 ms	8.15
220	10 μs	8.79
219	1.05 μs	9.34
218	0.11 μs	9.67

Plot $\log(\tau)$ versus $1/\sqrt{E_\alpha}$ and try to "fit" a straight line through the data points (either by eye or via some fitting routine if you know how). Also plot the theoretical expression in Eqn. (11.63) on your graph and compare the "slopes" and "intercepts" of your lines; this exercise gives some feel for the reliability of this simplest estimate of tunneling effects.

P11.10. Evaluate the optimal energy, E_{max}, for stellar fusion for the two situations

$$\text{H–H reactions: } Z_1 = Z_2 = 1 \qquad \mu \approx m_p/2 \qquad T = 10^7 \text{ K}$$
$$\text{C–C reactions: } Z_1 = Z_2 = 6 \qquad \mu \approx 6m_p \qquad T = 10^9 \text{ K}$$

What is the ratio of interaction rates as one varies $E = rE_{max}$ over the range $r = (0.5, 2)$. Over what range of energies around E_{max} does the interaction rate drop to half its peak value?

TWELVE
More Formal Topics

In this chapter, we investigate in more depth many of the formal properties of quantum mechanics introduced in Chapters 4 and 6, extending the discussions of Hermitian operators (Section 12.1), the vector structure of quantum wave-functions (12.2), and commutators (12.3). We use these techniques to prove important new results involving uncertainty principles (12.4) and the time-development of quantum states and its connection to conservation laws (12.5). We conclude with a discussion of the propagator approach to the time-evolution of quantum systems (12.6) and a discussion of timescales in bound state systems (12.7).

12.1 Hermitian Operators

We have stressed (in Section 4.4) that in order for a quantum operator, $\hat{O}$, to correspond to a classical observable quantity, O, it must be Hermitian; we have defined this by insisting that its expectation value in any quantum state be real, that is,

$$\langle\hat{O}\rangle^* = \left[\int \psi^*\hat{O}\,\psi\right]^* = \int \left(\hat{O}\psi\right)^*\psi \stackrel{?}{=} \int \psi^*\hat{O}\,\psi = \langle\hat{O}\rangle \qquad (12.1)$$

where any proof consists in verifying the equality in question. We also have, even more compactly, in bracket notation

$$\langle\psi|\hat{O}|\psi\rangle^* = \langle\psi|\hat{O}\psi\rangle^* = \langle\hat{O}\psi|\psi\rangle = \langle\psi|\hat{O}|\psi\rangle \qquad (12.2)$$

and we will often use whichever notation is more convenient. From Section 6.1, we know that Hermitian operators will also satisfy the more general requirement

$$\langle\chi|\hat{O}|\psi\rangle^* = \langle\hat{O}\chi|\psi\rangle = \langle\psi|\hat{O}|\chi\rangle \qquad (12.3)$$

It will turn out to be extremely useful to generalize the notion of complex conjugation used here for average values to the operators themselves. To this end, and motivated by Eqn. (12.1), we define the *Hermitian conjugate* of the operator $\hat{O}$, denoted by $\hat{O}^\dagger$ and read as "O-dagger," by the relation

$$\int \left(\hat{O}\psi\right)^* \psi \equiv \int \psi^* \hat{O}^\dagger \psi \tag{12.4}$$

or equivalently

$$\langle \hat{O}\psi | \psi \rangle \equiv \langle \psi | \hat{O}^\dagger | \psi \rangle \tag{12.5}$$

Using our established definition of Hermitian-ness, we see that:

- An operator is Hermitian provided

$$\langle \psi | \hat{O} | \psi \rangle^* = \langle \psi | \hat{O}\psi \rangle^* = \langle \hat{O}\psi | \psi \rangle \equiv \langle \psi | \hat{O}^\dagger | \psi \rangle = \langle \psi | \hat{O} | \psi \rangle \tag{12.6}$$

  for all ψ or, more simply, if

$$\hat{O}^\dagger = \hat{O} \tag{12.7}$$

This is then a test which is performed directly on the operator. The analogous condition for a complex number, c, to be real is simply

$$c = c^* \tag{12.8}$$

and it is easy to show for any complex constant

$$\int \psi^* c^\dagger \psi \equiv \int (c\psi)^* \psi = \int \psi^* c^* \psi \quad \text{so that } c^\dagger = c^* \tag{12.9}$$

This is then our first example of Hermitian conjugation and shows that this operation reduces to ordinary complex conjugation when applied to complex numbers. It is similarly easy to show that $x^\dagger = x$ in this language so that x is again confirmed to be Hermitian, as is any real function of x.

A less trivial example is the calculation of $(d/dx)^\dagger$ which gives

$$\int_{-\infty}^{+\infty} \psi^* \left(\frac{d}{dx}\right)^\dagger \psi \equiv \int_{-\infty}^{+\infty} \left(\frac{d}{dx}\psi\right)^* \psi$$

$$\overset{\text{IBP}}{=} -\int_{-\infty}^{+\infty} dx\, \psi^* \frac{d\psi}{dx} + (\psi^*\psi)_{-\infty}^{+\infty}$$

$$= \int_{-\infty}^{+\infty} \psi^* \left(-\frac{d}{dx}\right) \psi \tag{12.10}$$

whence

$$\left(\frac{d}{dx}\right)^\dagger = -\frac{d}{dx} \tag{12.11}$$

Combining these results we presumably then have (see below) for the momentum operator

$$\hat{p}^{\dagger} = \left(\frac{\hbar}{i}\frac{d}{dx}\right)^{\dagger} = \left(\frac{\hbar}{i}\right)^{*}\left(\frac{d}{dx}\right)^{\dagger} = \left(\frac{\hbar}{-i}\right)\left(-\frac{d}{dx}\right) = \frac{\hbar}{i}\frac{d}{dx} = \hat{p} \quad (12.12)$$

and $\hat{p}$ is Hermitian. This example demonstrates some of the utility of a definition of Hermitian-ness involving operators themselves; if one establishes, via direct calculation, a small class of known Hermitian operators, the Hermitian conjugate operation and its properties can be used to test new combinations of operators in a straightforward way.

Some of the most useful properties of Hermitian conjugation (whose proofs will be left to the problems) are:

- The Hermitian conjugate satisfies the more general relation

$$\langle \hat{A}\chi|\psi\rangle = \langle \chi|\hat{A}^{\dagger}|\psi\rangle \quad (12.13)$$

  for all states ψ, χ.

- This can be used to show that

$$(\hat{A}^{\dagger})^{\dagger} = \hat{A} \quad (12.14)$$

  in much the same way that $(c^{*})^{*} = c$ for complex conjugation.

- The Hermitian conjugate reverses the order of a product of operators, that is,

$$(\hat{A}\hat{B})^{\dagger} = \hat{B}^{\dagger}\hat{A}^{\dagger} \quad (12.15)$$

  This can be shown by considering

$$\begin{aligned}
\langle\psi|(\hat{A}\hat{B})^{\dagger}|\psi\rangle &\equiv \langle(\hat{A}\hat{B})\psi|\psi\rangle \\
&= \langle\hat{A}(\hat{B}\psi)|\psi\rangle \\
&= \langle\hat{B}\psi|\hat{A}^{\dagger}|\psi\rangle \\
&= \langle\psi|\hat{B}^{\dagger}|\hat{A}^{\dagger}\psi\rangle \\
&= \langle\psi|\hat{B}^{\dagger}\hat{A}^{\dagger}|\psi\rangle \quad (12.16)
\end{aligned}$$

- This implies that

  - If both $\hat{A}$ and $\hat{B}$ are separately Hermitian, their product, $\hat{A}\hat{B}$, is Hermitian provided that $\hat{A}$ and $\hat{B}$ also commute, that is, if $[\hat{A}, \hat{B}] = 0$.

  This observation helps justify the derivation in Eqn. (12.12).

- The combinations

$$\hat{A}\hat{A}^{\dagger}, \quad (\hat{A} + \hat{A}^{\dagger})/2, \quad \text{and} \quad -i(\hat{A} - \hat{A}^{\dagger})/2 \tag{12.17}$$

are all Hermitian, even if $\hat{A}$ itself is not; these correspond respectively to

$$cc^{*} = |c|^{2}, \quad \text{Re}(c), \quad \text{and} \quad \text{Im}(c) \tag{12.18}$$

which are the modulus squared, and the real and imaginary parts of an ordinary complex number, c.

- The combination

$$i[\hat{A}, \hat{B}] \tag{12.19}$$

is Hermitian provided both $\hat{A}$ and $\hat{B}$ are also; this will prove useful in our derivations of generalized uncertainty principles.

- For real functions of operators, we have

$$f(\hat{O})^{\dagger} = f(\hat{O}^{\dagger}) \tag{12.20}$$

which can be seen by using a power series expansion for $f(x)$.

Finally, we can make a connection with the matrix representation of operators in Section 10.4. If we define the matrix corresponding to some operator using a complete set of states, $u_n(x)$, as

$$\mathbf{O}_{nm} \equiv \langle u_n | \hat{O} | u_m \rangle \tag{12.21}$$

then the definition of Hermitian conjugate gives

$$\langle u_n | \hat{O}^{\dagger} | u_m \rangle \equiv \langle \hat{O} u_n | u_m \rangle^{*} = \langle u_m | \hat{O} | u_n \rangle^{*}. \tag{12.22}$$

This can be written in the form

$$\left(\mathbf{O}^{\dagger} \right)_{nm} = \left(\mathbf{O}^{*} \right)_{mn} = \left(\mathbf{O}^{T*} \right)_{nm} \tag{12.23}$$

where $\mathbf{O}^{T}$ defines the *transpose* of the matrix, that is, the matrix resulting from interchanging rows and columns (see Appendix F.1); thus

- The *Hermitian conjugate of a matrix* is obtained by taking its transpose and the complex conjugate of all its elements, that is, $\mathbf{O}^{\dagger} = \mathbf{O}^{T*}$. A *matrix* is then described as Hermitian if $\mathbf{O}^{\dagger} = \mathbf{O}$.

12.2 Quantum Mechanics, Linear Algebra, and Vector Spaces

We have made occasional reference to the algebraic structure of solutions of the Schrödinger equation, especially in position-space, noting similarities to the vectors of a linear vector space. In this section, we wish to extend and amplify this identification by collecting more examples of the apparent similarities. We know that the different representations of the Schrödinger wavefunction

$$\psi(x) \longleftrightarrow \phi(p) = \frac{1}{\sqrt{2\pi\hbar}} \int_{-\infty}^{+\infty} dx \, e^{-ipx/\hbar} \, \psi(x) \tag{12.24}$$

$$\psi(x) \longleftrightarrow \mathbf{a} = \left\{ a_n = \int_{-\infty}^{+\infty} ([u_n(x)]^* \psi(x) \, dx; \ n = 1, 2 \ldots \right\} \tag{12.25}$$

all have the same information content and satisfy the same normalization condition

$$1 = \int_{-\infty}^{+\infty} |\psi(x)|^2 \, dx = \int_{-\infty}^{+\infty} |\phi(p)|^2 \, dp = \sum_{n=1}^{\infty} |a_n|^2 \equiv \mathbf{a}^* \cdot \mathbf{a} \tag{12.26}$$

This connection of $\psi(x)$ and $\phi(p)$ with the discrete, but infinite-dimensional, complex vector $\mathbf{a} = \{a_n; n = 1, 2, \ldots\}$ motivates us to consider all three as simply different representations of the same *quantum state vector*. Extending the Dirac bracket notation, we often denote

$$\text{"ket" vector} \quad \longleftrightarrow \quad |\psi_a\rangle \quad \longleftrightarrow \quad \psi_a(x), \phi_a(p), \mathbf{a}$$
$$\text{"bra" vector} \quad \longleftrightarrow \quad \langle\psi_b| \quad \longleftrightarrow \quad \psi_b^*(x), \phi_b^*(p), \mathbf{b}^* \tag{12.27}$$

The various overlap integrals (and sums) can now be described as an *inner-product* or *generalized dot-product* of two such vectors; their common value is then given by

$$\int_{-\infty}^{+\infty} \psi_b^*(x)\psi_a(x) \, dx = \int_{-\infty}^{+\infty} \phi_b^*(p) \, \phi_a(p) \, dp = \mathbf{b}^* \cdot \mathbf{a} \equiv \langle\psi_b|\psi_a\rangle \tag{12.28}$$

where a "bra" vector "dotted into" a "ket" vector forms the familiar Dirac "bra"·"ket" or bracket representing the overlap of these two quantum states.

We can generalize the notion of a *linear vector space* to include the following features:

- In position-space, for example, the vectors are associated with square-integrable functions, $\psi(x)$, where the continuous label x generalizes the discrete label $\mathbf{a} = \{a_n; n = 1, 2, 3 \ldots\}$. Such an infinite-dimensional vector space, conventionally denoted as $\mathcal{H}$, is often called a *Hilbert space*.

- We assume that $\alpha \psi_a(x) + \beta \psi_b(x) \in \mathcal{H}$ for complex constants α, β provided ψ_a, ψ_b are as also. The space $\mathcal{H}$ is therefore closed, since linear combinations of vectors are still vectors, that is, linear combinations of square-integrable functions are still square-integrable.

- The (suitably normalized) eigenfunctions of a Hermitian operator, which we know form a complete set, can be thought as a set of orthonormal *unit vectors* since $\langle u_n | u_m \rangle = \delta_{nm}$.

- We will concentrate on *linear operators*, $\hat{O}$, on this space, namely, ones which satisfy

$$\hat{O}(\alpha \phi_a + \beta \psi_b) = \alpha \left(\hat{O} \phi_a \right) + \beta \left(\hat{O} \psi_b \right) \tag{12.29}$$

A very important set of operators are those which preserve the inner-product among vectors; these can be thought of as generalized rotations which keep the "length" of state vectors or functions fixed. We will denote such operators by $\hat{U}$, with the notation $|\psi'\rangle = \hat{U}|\psi\rangle$. If the dot-product is to remain unchanged under such transformations, we must have

$$\langle \psi_a | \psi_b \rangle = \langle \psi_a' | \psi_{b'} \rangle = \langle \hat{U} \psi_a | \hat{U} \psi_b \rangle \equiv \langle \psi_a | \hat{U}^\dagger \hat{U} | \psi_b \rangle \tag{12.30}$$

or

$$\hat{U}^\dagger \hat{U} = \hat{1} \tag{12.31}$$

where $\hat{1}$ is the unit operator and we have used the definition of Hermitian conjugate. Operators which satisfy Eqn. (12.31) are call *unitary operators*. Just as ordinary rotations preserve the length of standard vectors, unitary transformations preserve the norm of a vector in Hilbert space ($\sqrt{\langle \psi | \psi \rangle}$) and hence conserve probability.

In a matrix notation (P12.5), where the operator is defined via

$$\mathbf{a}' = \mathsf{U}\mathbf{a} \quad \text{or} \quad a_i' = \sum_j \mathsf{U}_{ij} a_j \tag{12.32}$$

we have the equivalent statement

$$\mathsf{U}^{T*}\mathsf{U} = \mathsf{U}^\dagger \mathsf{U} = 1 \quad \text{or} \quad \left(\mathsf{U}^{T*}\mathsf{U} \right)_{ij} = \sum_k \left(\mathsf{U}^{T*} \right)_{ik} \mathsf{U}_{kj} = \delta_{ij} \tag{12.33}$$

Example 12.1. Rotation matrices as unitary transformations

A simple example involving the rotation matrix for two-dimensional real vectors is a useful reminder. In this case we have

$$\mathbf{x}' = \mathsf{R}\mathbf{x} \quad \text{or} \quad \begin{pmatrix} x' \\ y' \end{pmatrix} = \begin{pmatrix} \cos(\theta) & \sin(\theta) \\ -\sin(\theta) & \cos(\theta) \end{pmatrix} \begin{pmatrix} x \\ y \end{pmatrix} \tag{12.34}$$

which rotates the vector $\mathbf{x}$ through θ radians. It is then easy to check that

$$\mathsf{R}^\dagger \mathsf{R} = \begin{pmatrix} \cos(\theta) & -\sin(\theta) \\ \sin(\theta) & \cos(\theta) \end{pmatrix} \begin{pmatrix} \cos(\theta) & \sin(\theta) \\ -\sin(\theta) & \cos(\theta) \end{pmatrix} = \begin{pmatrix} 1 & 0 \\ 0 & 1 \end{pmatrix} = 1 \tag{12.35}$$

In this case $\mathsf{R}^\dagger = \mathsf{R}^T$ since R is real.

Example 12.2. Fourier transform as a unitary transformation

An example of such a norm-preserving or unitary transformation of more relevance to quantum mechanics is the Fourier transform

$$\phi(p) = \frac{1}{\sqrt{2\pi\hbar}} \int_{-\infty}^{+\infty} dx\, \psi(x) e^{-ipx/\hbar} \tag{12.36}$$

which we already know satisfies $\langle\phi|\phi\rangle = \langle\psi|\psi\rangle$. If we write the Fourier transform in the form

$$\phi(p) = \int_{-\infty}^{+\infty} dx \left(\frac{e^{-ipx/\hbar}}{\sqrt{2\pi\hbar}}\right) \psi(x) = \int_{-\infty}^{+\infty} dx\, R_{px}\, \psi(x) \tag{12.37}$$

we can note the immediate similarity to Eqn. (12.32) with the discrete labels and summations replaced by continuous variables and integrations. The 'rotation matrix elements' are

$$R_{px} \equiv \frac{1}{\sqrt{2\pi\hbar}} e^{-ipx/\hbar} \tag{12.38}$$

and satisfy their own (continuous) version of Eqn. (12.33), namely

$$\int_{-\infty}^{+\infty} dx\, (R_{p'x})^* R_{xp} = \frac{1}{2\pi\hbar} \int_{-\infty}^{+\infty} dx\, e^{i(p-p')x/\hbar} = \delta(p - p') \tag{12.39}$$

This identification with norm-preserving transformations makes it even clearer that $\psi(x)$ and $\phi(p)$ share the same information content. A very similar set of results for the expansion coefficients for a complete set of states can also be shown; they follow from the definition

$$\psi(x) = \sum_n a_n u_n(x) \equiv \sum_n U_{xn} a_n \tag{12.40}$$

(Continued)
where

$$a_n = \int dx \, [u_n(x)]^* \, \psi(x) = \int dx \, \left(U^*_{nx}\right) \psi(x) \tag{12.41}$$

In this case, the norm-preserving condition of Eqn. (12.33) reads

$$\int_{-\infty}^{+\infty} dx \, (U_{nx})^* U_{xm} = \int_{-\infty}^{+\infty} dx \, u_n^*(x) u_m(x) = \delta_{nm} \tag{12.42}$$

which is familiar from the orthogonality of eigenfunctions belonging to different energy eigenvalues. We find similarly for the other (discrete) label that

$$\sum_n U^*_{x'n} U_{nx} = \sum_n u_n^*(x') u_n(x) = \delta(x - x') \tag{12.43}$$

which is useful in the study of propagators in Section 12.6.

The powerful analogy between the Hilbert space of quantum mechanics and ordinary vector spaces can be used to great benefit to prove new results which have an obvious geometrical analogy in the more familiar setting of ordinary dot products. An example is the so-called *Schwartz* or *triangle inequality* for vectors which reads

$$\mathbf{A}^2 \mathbf{B}^2 \geq (\mathbf{A} \cdot \mathbf{B})^2 \tag{12.44}$$

and which in two and three dimensions can be easily translated into the statement that

$$A^2 B^2 \geq A^2 B^2 \cos^2(\theta) \quad \text{or} \quad 1 \geq \cos^2(\theta) \tag{12.45}$$

where θ is the angle between $\mathbf{A}$ and $\mathbf{B}$; the equality is only achieved when $\mathbf{A}, \mathbf{B}$ are parallel or antiparallel, that is, $\cos(\theta) = \pm 1$.

A completely analogous result can be proved for complex wavefunctions, namely,

$$\left[\int_{-\infty}^{+\infty} dx \, |\psi(x)|^2 \right] \left[\int_{-\infty}^{+\infty} dx \, |\chi(x)|^2 \right] \geq \left| \int_{-\infty}^{+\infty} dx \, \psi^*(x) \chi(x) \right|^2 \tag{12.46}$$

or more generally in terms of quantum states,

$$\langle \psi | \psi \rangle \langle \chi | \chi \rangle \geq |\langle \psi | \chi \rangle|^2 \tag{12.47}$$

We will prove Eqn. (12.47) in a formal, abstract way by generalizing the standard argument for ordinary vectors. We begin by considering a general quantum state given by a linear combination of $|\psi\rangle$ and $|\chi\rangle$, whose "ket" vector is written as

$$|\zeta\rangle = \alpha|\psi\rangle + \beta|\chi\rangle \equiv |\psi\rangle\langle\chi|\chi\rangle - |\chi\rangle\langle\chi|\psi\rangle \tag{12.48}$$

where we recall that the overlaps $\alpha = \langle \chi | \chi \rangle$ and $\beta = -\langle \chi | \psi \rangle$ are inner-products of quantum states and hence simply (complex) numbers. The corresponding "bra" vector is then

$$\langle \zeta | = \langle \chi | \chi \rangle \langle \psi | - \langle \chi | \psi \rangle^* \langle \chi | = \langle \chi | \chi \rangle \langle \psi | - \langle \psi | \chi \rangle \langle \chi | \qquad (12.49)$$

The overlap of any generalized vector with itself must necessarily be nonnegative so that $\langle \zeta | \zeta \rangle \geq 0$, with the lower bound of 0 only being saturated if $| \zeta \rangle$ itself vanishes. This then implies that

$$
\begin{aligned}
0 \leq \langle \zeta | \zeta \rangle &= \{ \langle \chi | \chi \rangle \langle \psi | - \langle \psi | \chi \rangle \langle \chi | \} \{ | \psi \rangle \langle \chi | \chi \rangle - | \chi \rangle \langle \chi | \psi \rangle \} \\
&= \langle \chi | \chi \rangle \langle \chi | \chi \rangle \langle \psi | \psi \rangle - \langle \psi | \chi \rangle \langle \chi | \psi \rangle \langle \chi | \chi \rangle \\
&\quad - \langle \chi | \chi \rangle \langle \psi | \chi \rangle \langle \chi | \psi \rangle + \langle \chi | \chi \rangle \langle \psi | \chi \rangle \langle \chi | \psi \rangle \\
&= \langle \chi | \chi \rangle \langle \chi | \chi \rangle \langle \psi | \psi \rangle - \langle \psi | \chi \rangle \langle \chi | \psi \rangle \langle \chi | \chi \rangle \qquad (12.50)
\end{aligned}
$$

or

$$\langle \psi | \psi \rangle \langle \chi | \chi \rangle \geq \langle \psi | \chi \rangle \langle \chi | \psi \rangle = \langle \psi | \chi \rangle \langle \psi | \chi \rangle^* = | \langle \psi | \chi \rangle |^2 \qquad (12.51)$$

Once again, the equality is only achieved when the state $| \zeta \rangle$ vanishes identically which implies that

$$| \psi \rangle = \left(\frac{\langle \chi | \psi \rangle}{\langle \chi | \chi \rangle} \right) | \chi \rangle \quad \text{or} \quad \psi(x) \propto \chi(x) \qquad (12.52)$$

that is, $| \psi \rangle$ and $| \chi \rangle$ are proportional to each other, or "parallel" in a generalized sense.

12.3 **Commutators**

Since it is obvious that the commutator of two operators plays such a fundamental role in the formalism of quantum mechanics, we find it useful to collect below some basic results on commutators which can make the evaluation of complicated combinations easier. We can easily show by direct manipulation (P12.8) that

- The commutator of a nontrivial operator with any (complex) number vanishes trivially,

$$[c, \hat{A}] = 0 \quad \text{for any complex number } c. \qquad (12.53)$$

This is related to our assumption that we will only be considering linear operators, satisfying Eqn. (12.29).

• The commutator of an operator with itself vanishes, that is,

$$[\hat{A}, \hat{A}] = \hat{A}\hat{A} - \hat{A}\hat{A} = 0 \tag{12.54}$$

This statement is not as trivial as it may at first seem.

• The commutator "distributes" in that

$$[\hat{A} + \hat{B}, \hat{C}] = [\hat{A}, \hat{C}] + [\hat{B}, \hat{C}] \tag{12.55}$$

• The commutator of $\hat{A}$ and $\hat{B}$ is an antisymmetric function of its arguments, namely,

$$[\hat{A}, \hat{B}] = -[\hat{B}, \hat{A}] \tag{12.56}$$

so that the ordering of the operators is critically important.

• More involved commutators can often be simplified by repeated use of the relation

$$[\hat{A}\hat{B}, \hat{C}] = \hat{A}[\hat{B}, \hat{C}] + [\hat{A}, \hat{C}]\hat{B} \tag{12.57}$$

or its equivalent

$$[\hat{A}, \hat{B}\hat{C}] = \hat{B}[\hat{A}, \hat{C}] + [\hat{A}, \hat{B}]\hat{C} \tag{12.58}$$

We note that the last five relations (Eqns (12.54)–(12.58)) have a structure similar to the vector- or cross-product of two vectors (Q12.2).

• Commutators can be applied repeatedly so that, for example,

$$[\hat{A}, [\hat{B}, \hat{C}]] = \hat{A}[\hat{B}, \hat{C}] - [\hat{B}, \hat{C}]\hat{A} \tag{12.59}$$

which is to be distinguished from Eqn. (12.58).

• A sometimes useful relation amongst double commutators is the so-called *Jacobi identity* which states that

$$[\hat{A}, [\hat{B}, \hat{C}]] + [\hat{C}, [\hat{A}, \hat{B}]] + [\hat{B}, [\hat{C}, \hat{A}]] = 0 \tag{12.60}$$

Note the cyclic permutations.

As an example of the use of these relations, we can make use of the relation $[x, \hat{p}] = i\hbar$ to show that

$$[x, \hat{p}^2] = \hat{p}[x, \hat{p}] + [x, \hat{p}]\hat{p} = 2i\hbar\hat{p} \tag{12.61}$$

instead of evaluating such quantities directly, as in Section 4.8 or P4.16.

12.4 **Uncertainty Principles**

We have already seen several cases where it is possible to have simultaneous eigenfunctions of two different Hermitian operators, $\hat{A}$ and $\hat{B}$; for example, the energy and parity eigenfunctions for Hamiltonians with a symmetric potential.

In this section, we can combine several of our formal results to derive a general form for the minimum uncertainty principle product for any two observable quantities, A, B represented by quantum operators, $\hat{A}, \hat{B}$. To this end, we note that:

- The spread or RMS deviation in the measurement of any observable is given by

$$(\Delta A)^2 = \langle \psi | (\hat{A} - \langle \hat{A} \rangle)^2 | \psi \rangle \tag{12.62}$$

where $\langle \hat{A} \rangle = \langle \psi | \hat{A} | \psi \rangle$, with a similar expression for ΔB.

- Since we are dealing with observable quantities, we will assume that both $\hat{A}$ and $\hat{B}$ are Hermitian.

Using the same trick as in the proof of the Schwartz inequality, we define a "ket" state given by

$$|\zeta\rangle \equiv (\hat{A} - \langle \hat{A} \rangle) | \psi \rangle + i\lambda(\hat{B} - \langle \hat{B} \rangle) | \psi \rangle \tag{12.63}$$

where λ is an arbitrary real number. The corresponding "bra" vector is then

$$\langle \zeta | = \left\langle \left(\hat{A} - \langle \hat{A} \rangle\right) \psi \right| - i\lambda \left\langle \left(\hat{B} - \langle \hat{B} \rangle\right) \psi \right| \tag{12.64}$$

The inner-product $\langle \zeta | \zeta \rangle$ is then a function of λ and, of course, is nonnegative so that $\langle \zeta | \zeta \rangle \geq 0$. It can then be written as

$$
\begin{aligned}
0 \leq I(\lambda) \equiv \langle \zeta | \zeta \rangle = & \left\langle (\hat{A} - \langle \hat{A} \rangle)\psi \,\middle|\, (\hat{A} - \langle \hat{A} \rangle)\psi \right\rangle \\
& + \lambda^2 \left\langle (\hat{B} - \langle \hat{B} \rangle)\psi \,\middle|\, (\hat{B} - \langle \hat{B} \rangle)\psi \right\rangle \\
& + i\lambda \left[\langle (\hat{A} - \langle \hat{A} \rangle)\psi \,\middle|\, (\hat{B} - \langle \hat{B} \rangle)\psi \rangle \right. \\
& \left. - \langle (\hat{B} - \langle \hat{B} \rangle)\psi \,\middle|\, (\hat{A} - \langle \hat{A} \rangle)\psi \rangle \right]
\end{aligned}
\tag{12.65}
$$

We can make repeated use of the fact that $\hat{A}, \hat{B}$ are Hermitian, and that $\langle \hat{A} \rangle, \langle \hat{B} \rangle$ are therefore real, to move all the terms "out from under" the "bra" vector; we then write this as

$$I(\lambda) = \langle \psi \left| (\hat{A} - \langle \hat{A} \rangle)^2 \right| \psi \rangle + \lambda^2 \langle \psi \left| (\hat{B} - \langle \hat{B} \rangle)^2 \right| \psi \rangle \tag{12.66}$$

$$+ i\lambda \langle \psi \left| (\hat{A} - \langle \hat{A} \rangle)(\hat{B} - \langle \hat{B} \rangle) - (\hat{B} - \langle \hat{B} \rangle)(\hat{A} - \langle \hat{A} \rangle) \right| \psi \rangle$$

We note that the first two terms can be written as

$$(\Delta A)^2 + \lambda^2 (\Delta B)^2 \tag{12.67}$$

while the third term can be written in terms of a commutator, namely

$$(\hat{A} - \langle \hat{A} \rangle)(\hat{B} - \langle \hat{B} \rangle) - (\hat{B} - \langle \hat{B} \rangle)(\hat{A} - \langle \hat{A} \rangle) = \hat{A}\hat{B} - \hat{B}\hat{A} = [\hat{A}, \hat{B}] \tag{12.68}$$

with the simplification due to the fact that the commutator of an operator with any (complex) number vanishes, as in Eqn. (12.53). The resulting term has the form

$$i\lambda \left\langle \psi \left| [\hat{A}, \hat{B}] \right| \psi \right\rangle = \lambda \langle \psi | \hat{F} | \psi \rangle \tag{12.69}$$

where we know that the combination $\hat{F} \equiv i[\hat{A}, \hat{B}]$ is a Hermitian operator (P12.3) which must necessarily have real expectation values. We can thus write $I(\lambda)$ in terms of manifestly real quantities, namely

$$I(\lambda) = (\Delta A)^2 + \lambda^2 (\Delta B)^2 + \lambda \langle \psi | \hat{F} | \psi \rangle \geq 0 \tag{12.70}$$

Since this is, by construction, nonnegative for all values of λ, it will be so at the minimum, namely, for λ determined by

$$0 = \frac{dI(\lambda_{\min})}{d\lambda} = 2\lambda_{\min}(\Delta B)^2 + \langle \psi | \hat{F} | \psi \rangle \quad \text{or} \quad \lambda_{\min} = -\frac{\langle \psi | \hat{F} | \psi \rangle}{2(\Delta B)^2} \tag{12.71}$$

The inequality for this value of λ can then be written in the form of an uncertainty product bound, since we have

$$I(\lambda_{\min}) = (\Delta A)^2 + \left(-\frac{\langle \psi | \hat{F} | \psi \rangle}{2(\Delta B)^2} \right)^2 (\Delta B)^2 - \left(\frac{\langle \psi | \hat{F} | \psi \rangle}{2(\Delta B)^2} \right) \langle \psi | \hat{F} | \psi \rangle \geq 0 \tag{12.72}$$

which gives

$$(\Delta A)^2 (\Delta B)^2 \geq \frac{\langle \psi | \hat{F} | \psi \rangle^2}{4} = \frac{\langle \psi | i[\hat{A}, \hat{B}] | \psi \rangle^2}{4} \tag{12.73}$$

where the right-hand side is obviously real and positive (despite the explicit i) since $\langle \psi | \hat{F} | \psi \rangle$ is real. This general result shows that:

- It is, in principle, possible to have simultaneous eigenvalues, that is, states for which *both* ΔA and ΔB are vanishing, provided $[\hat{A}, \hat{B}] = 0$. The uncertainty principle product can be written as

$$\Delta A \, \Delta B \geq \frac{\left| \langle \psi \left| i[\hat{A}, \hat{B}] \right| \psi \rangle \right|}{2} \tag{12.74}$$

The famous Heisenberg uncertainty principle (for position-momentum) is now simply a special case with

$$\hat{A} = \hat{p} \quad \text{and} \quad \hat{B} = x \text{ which gives } \hat{F} = i[\hat{A}, \hat{B}] = \hbar \tag{12.75}$$

so that

$$\Delta x \, \Delta p \geq \frac{\hbar}{2} \langle \psi | \psi \rangle = \frac{\hbar}{2} \tag{12.76}$$

This derivation also allows us to calculate the minimum uncertainty product waveform for the pair x, p as we know that the lower bound $\Delta x \, \Delta p = \hbar/2$ will only be saturated when $|\zeta\rangle$ in Eqn. (12.63) vanishes identically. Using explicit position-space forms, this occurs when

$$0 = \zeta(x) = \left(\frac{\hbar}{i} \frac{d}{dx} - \langle \hat{p} \rangle \right) \psi(x) + i\lambda(x - \langle x \rangle)\psi(x) \tag{12.77}$$

or using

$$\lambda_{\min} = -\frac{\hbar}{2(\Delta x)^2} \tag{12.78}$$

we have

$$\frac{d\psi(x)}{dx} = \frac{ip_0}{\hbar} \psi(x) - \frac{(x - x_0)}{2(\Delta x)^2} \psi(x) \tag{12.79}$$

where $\langle x \rangle \equiv x_0$ and $\langle \hat{p} \rangle = p_0$. This has the simple solution

$$\psi(x) \propto e^{-(x-x_0)^2/4(\Delta x)^2} e^{ip_0 x/\hbar} \tag{12.80}$$

which is the familiar Gaussian wavefunction with arbitrary initial position and "speed"; this state is thus singled out not only because of the mathematical simplicity associated with its manipulation, but for important physical reasons.

12.5 **Time-Dependence and Conservation Laws in Quantum Mechanics**

In classical mechanics, Newton's laws yield not only particle trajectories, $x(t)$, but through them, the time behavior of all other observables. For example, the kinetic energy varies in time via

$$T(t) = \frac{1}{2}mv^2(t) = \frac{1}{2}m\left(\frac{dx(t)}{dt}\right)^2 \tag{12.81}$$

In a similar way, we have seen that the time-dependence of the expectation values of the quantum operator analogs of classical observables for most quantities arises solely through the time-dependence of the wavefunctions satisfying the Schrödinger equation, that is,

$$\langle \hat{O} \rangle_t = \int_{-\infty}^{+\infty} dx\, \psi^*(x,t)\, \hat{O}\, \psi(x,t) \tag{12.82}$$

Thus, the time-dependent kinetic energy in Eqn. (12.81), has the quantum analog

$$\langle \hat{T} \rangle_t = \int_{-\infty}^{+\infty} dx\, \psi^*(x,t)\, \hat{T}\, \psi(x,t) \overset{\text{IBP}}{=} \frac{\hbar^2}{2m} \int_{-\infty}^{+\infty} dx\, \left|\frac{\partial \psi(x,t)}{\partial x}\right|^2 \tag{12.83}$$

An important special case occurs in classical mechanics when the time-development of an observable is trivial, that is, it is constant in time. This results in so-called *conservation laws*; examples include the conservation of energy, momentum, and angular momentum under the appropriate circumstances. The use of such conservation laws and the study of the conditions under which they are valid is an important topic in classical mechanics, and in this section we extend these notions to the quantum arena by examining in detail the time-dependence of expectation values of quantum operators.

A deceptively simple result arises when one considers energy eigenstates or stationary states. In that case the time-dependence is trivial,

$$\psi(x,t) = \psi_n(x)\, e^{-iE_n t/\hbar} \tag{12.84}$$

so that the average values of operators satisfy

$$\langle \hat{O} \rangle_t = \int_{-\infty}^{+\infty} dx\, \left[\psi_n^*(x)\, e^{+iE_n t/\hbar}\right] \hat{O}\, \left[\psi_n(x) e^{-iE_n t/\hbar}\right]$$

$$= \int_{-\infty}^{+\infty} dx\, \psi_n^*(x)\, \hat{O}\, \psi_n(x) = \langle \hat{O} \rangle_0 \tag{12.85}$$

which is trivially time-independent and further justifies the name stationary state. For a more general state, $\psi(x, t) = \sum_n a_n \psi_n(x) e^{-iE_n t/\hbar}$, no such simple cancellation of time factors occurs, and in general one has $d\langle \hat{O} \rangle_t / dt \neq 0$.

Specializing, for familiarity, to a position-space representation, we can calculate the time-dependence of the expectation value of a generic operator $\hat{O}$ in some quantum state, $\psi(x, t)$, by writing

$$
\frac{d\langle \hat{O} \rangle_t}{dt} = \frac{d}{dt} \langle \hat{O} \rangle_t = \frac{d}{dt} \left[\int_{-\infty}^{+\infty} dx \, \psi^*(x, t) \hat{O} \, \psi(x, t) \right]
$$

$$
= \int_{-\infty}^{+\infty} dx \, \psi^* \left(\frac{\partial \hat{O}}{\partial t} \right) \psi
$$

$$
+ \int_{-\infty}^{+\infty} dx \left[\frac{\partial \psi^*}{\partial t} \hat{O} \psi + \psi^* \hat{O} \frac{\partial \psi}{\partial t} \right] \tag{12.86}
$$

where we have assumed, for generality, that $\hat{O}$ itself may have *explicit* time-dependence. Since $\psi(x, t)$ satisfies the Schrödinger equation, we can write

$$
i\hbar \frac{\partial \psi}{\partial t} = \hat{H} \psi \quad \text{and hence} \quad -i\hbar \frac{\partial \psi^*}{\partial t} = (\hat{H} \psi)^* \tag{12.87}
$$

and using these relations in Eqn. (12.86), and switching to the more compact bracket notation, we have

$$
\frac{d\langle \hat{O} \rangle_t}{dt} = \left\langle \psi \left| \frac{\partial \hat{O}}{\partial t} \right| \psi \right\rangle + \frac{i}{\hbar} \langle \hat{H} \psi | \hat{O} | \psi \rangle - \frac{i}{\hbar} \langle \psi | \hat{O} | \hat{H} \psi \rangle
$$

$$
= \left\langle \psi \left| \frac{\partial \hat{O}}{\partial t} \right| \psi \right\rangle + \frac{i}{\hbar} \left(\langle \psi | \hat{H} \hat{O} | \psi \rangle - \langle \psi | \hat{O} \hat{H} | \psi \rangle \right)
$$

$$
\frac{d\langle \hat{O} \rangle_t}{dt} = \left\langle \frac{\partial \hat{O}}{\partial t} \right\rangle + \frac{i}{\hbar} \left\langle [\hat{H}, \hat{O}] \right\rangle_t \tag{12.88}
$$

where we have used the fact that $\hat{H}$ is Hermitian to move it from the "bra" to the "ket" vector.

This is an extremely important result as it states that:

- If the quantum operator analog $\hat{O}$ of the classical observable O

  - is **not** itself an explicit function of time and

  - if it commutes with the Hamiltonian, that is $[\hat{H}, \hat{O}] = 0$,

then the expectation value of $\hat{O}$ in any quantum state is *independent of time*. This implies that the classical observable O is *conserved*.

Example 12.3. Time-dependent expectation values

Before discussing the meaning of this result in more detail, let us consider several examples. The quantum momentum operator $\hat{p}$ obviously has no explicit time-dependence and its commutator with a generic Hamiltonian is given by

$$[\hat{H}, \hat{p}] = [\hat{p}^2/2m + V(x), \hat{p}]$$

$$= \frac{1}{2m}[\hat{p}^2, \hat{p}] + [V(x), \hat{p}]$$

$$= 0 - \frac{\hbar}{i}\left(\frac{dV(x)}{dx}\right) = \frac{\hbar}{i}F(x) \tag{12.89}$$

so that

$$\frac{d}{dt}\langle\hat{p}\rangle_t = \frac{i}{\hbar}\langle[\hat{H}, \hat{p}]\rangle = \langle F(x)\rangle \tag{12.90}$$

where $F(x)$ is the classical force. We thus have the quantum version of the familiar classical condition on conservation of momentum, namely that $\langle\hat{p}\rangle_t$ will be constant in time provided the classical force, $F(x)$, vanishes. Similar conditions relating to the conservation of energy can be discussed (Q12.5).

Equation (12.88) can also be used to derive already familiar relations in a more compact form, such as

$$\frac{d\langle x\rangle_t}{dt} = 0 + \frac{i}{\hbar}\langle[\hat{H}, x]\rangle$$

$$= \frac{i}{\hbar}\left\langle \frac{1}{2m}[\hat{p}^2, x] + [V(x), x]\right\rangle$$

$$= \frac{i}{2m\hbar}\langle\hat{p}[\hat{p}, x] + [\hat{p}, x], \hat{p}\rangle$$

$$\frac{d\langle x\rangle_t}{dt} = \frac{\langle\hat{p}\rangle_t}{m} \tag{12.91}$$

This can be combined with the result of Eqn. (12.90) to give

$$m\frac{d^2\langle x\rangle_t}{dt^2} = m\frac{d}{dt}\left(\frac{\langle\hat{p}\rangle_t}{m}\right) = \langle F(x)\rangle = -\left\langle\frac{dV(x)}{dx}\right\rangle \tag{12.92}$$

This quantum version of Newton's second law is called Ehrenfests theorem.

If we choose $\hat{O} = \hat{p}^2$, we find that

$$\frac{d\langle\hat{p}^2\rangle}{dt} = \frac{i}{\hbar}\langle[\hat{p}^2 2m + V(x), \hat{p}^2]\rangle = \frac{i}{\hbar}\langle[V(x), \hat{p}^2]\rangle = \langle\hat{p}F(x) + F(x)\hat{p}\rangle \tag{12.93}$$

We can use this general result to note that the expectation value of the kinetic energy, $\langle\hat{p}^2\rangle/2m$ satisfies

$$\frac{d}{dt}\langle\hat{T}\rangle = \frac{1}{2m}\left[\frac{d}{dt}\langle\hat{p}^2\rangle\right] = \frac{1}{2}\left\langle\frac{\hat{p}}{m}F(x) + F(x)\frac{\hat{p}}{m}\right\rangle \tag{12.94}$$

(Continued)

which is the analog of the classical work-energy theorem relating the rate of change of the kinetic energy to the power done as work; note the symmetric combination of $\hat{p}/m$ and $F(x)$ which is the analog of $\mathbf{F} \cdot \mathbf{v}$. This result in Eqn. (12.93) is also useful in the evaluation of the time-dependence of the momentum-spread or uncertainty, Δp_t.

In a similar way, one can obtain the following general relations

$$\frac{d\langle x^2 \rangle}{dt} = \frac{1}{m} \langle x\hat{p} + \hat{p}x \rangle \tag{12.95}$$

$$\frac{d\langle x\hat{p} \rangle}{dt} = \frac{\langle \hat{p}^2 \rangle}{m} + \langle xF(x) \rangle = \frac{d\langle \hat{p}x \rangle}{dt} \tag{12.96}$$

which can be combined to give

$$\frac{d^2\langle x^2 \rangle}{dt^2} = \frac{2}{m} \left[\frac{\langle \hat{p}^2 \rangle}{m} + \langle xF(x) \rangle \right] \tag{12.97}$$

all of which are relevant for the calculation of $\Delta x_t{}^1$ (as in P12.16.)

Returning to Eqn. (12.88), it is easy to understand that a quantum operator must have no explicit time-dependence in order to be conserved. To appreciate why the commutator with the $\hat{H}$ operator arises, let us examine more fully the role that the Hamiltonian plays in quantum mechanics. The time-dependent Schrödinger equation can be written as

$$i\hbar \frac{\partial \psi(x, t)}{\partial t} = \hat{H}\psi(x, t) \quad \text{or} \quad \frac{\partial \psi(x, t)}{\partial t} = -\frac{i\hat{H}}{\hbar}\psi(x, t) \tag{12.98}$$

which we can *formally* integrate to yield

$$\psi(x, t) = e^{-i\hat{H}t/\hbar}\psi(x, 0) \tag{12.99}$$

thus solving the initial value problem; recall that the exponential of an operator is only defined via the series expansion of e^x, that is,

$$e^{\hat{O}} = 1 + \hat{O} + \frac{1}{2!}\hat{O}^2 + \cdots = \sum_{n=0}^{\infty} \frac{\hat{O}^n}{n!} \tag{12.100}$$

While this relation is of somewhat limited use as a calculational tool[2] it does show that:

- The Hamiltonian acts as the *time-development operator* in quantum mechanics.

---

[1] See Styer (1990) for a nice discussion of *The motion of wave packets through their expectation values and uncertainties*.

[2] See, however, Press *et al.* (2002).

For example, for small enough times, the approximation

$$\psi(x, t + dt) \quad \approx \quad \psi(x, t) - \frac{i}{\hbar}(\hat{H}\psi(x, t))dt \tag{12.101}$$

will be a good one, so that the operation of $-i\hat{H}dt/\hbar$ "translates" the wavefunction ahead in time by an infinitesimal amount dt.

Example 12.4. Time-development of free-particle Gaussian wave packet

One case in which the time-development operation in Eqn. (12.99) can be performed in closed form is for the free-particle propagation of the Gaussian wave packet.[3] To be consistent with the notation in Section 3.2.2, we write the initial wavefunction in the form

$$\psi(x, 0) = \frac{1}{\sqrt{\alpha\hbar\sqrt{\pi}}} e^{-x^2/2\alpha^2\hbar^2} = \sqrt{\frac{\alpha\hbar}{2\sqrt{\pi}}} \left[\frac{1}{\sqrt{w}} e^{-x^2/4w} \right] \tag{12.102}$$

where we have written $\alpha^2\hbar^2 = 2w$ for use in the "trick" we will employ below; this is the same form as in Section 3.2.2, but we choose $p_0 = 0$ for simplicity.

The free-particle Hamiltonian is simply

$$\hat{H} = -\frac{\hbar^2}{2m} \frac{\partial^2}{\partial x^2} \tag{12.103}$$

so the time-development operator is

$$e^{-i\hat{H}t/\hbar} = \sum_{n=0}^{\infty} \left(\frac{-i\hat{H}t}{\hbar} \right)^n = \sum_{n=0}^{\infty} \left(\frac{it\hbar}{2m} \frac{\partial^2}{\partial x^2} \right)^n \tag{12.104}$$

and we can use the identity

$$\frac{\partial^2}{\partial x^2} \left[\frac{1}{\sqrt{w}} e^{-x^2/4w} \right] = \frac{\partial}{\partial w} \left[\frac{1}{\sqrt{w}} e^{-x^2/4w} \right] \tag{12.105}$$

to write Eqn. (12.104) as

$$\sum_{n}^{\infty} \left(\frac{it\hbar}{2m} \frac{\partial}{\partial w} \right)^n = e^{c[\partial/\partial w]} \tag{12.106}$$

where $c \equiv it\hbar/2m$; we note that this is true only when acting on Gaussian functions. We can then use the results of P12.6 to show that

$$e^{c\partial/\partial w} f(w) = f(w + c) \tag{12.107}$$

---

[3] See Blinder (1968).

(Continued)

so that the net result of the time-development operator is to give

$$\psi(x, t) = e^{c[\partial/\partial w]} \psi(x, 0)$$

$$= \psi(x, 0; w \to w + (i\hbar t/2m))$$

$$= \sqrt{\frac{\alpha \hbar}{2\sqrt{\pi}}} \left[\frac{1}{\sqrt{w + it\hbar/2m}} e^{-x^2/4(w + it\hbar/2m)} \right]$$

$$= \frac{1}{\sqrt{\alpha \hbar F \sqrt{\pi}}} e^{-x^2/2\alpha^2 \hbar^2 F} \qquad (12.108)$$

where $F = 1 + it/m\hbar\alpha = 1 + it/t_0$, just as in Section 3.2.2.

Using the identification in Eqn. (12.101), we can now discuss the conditions under which a quantity might be conserved, that is, have a constant value in time. If such a quantity O is to be conserved, we should obtain the same value from a measurement at different times, and we can check this by comparing

$$\hat{H}\hat{O}\psi(x, t) \Longleftrightarrow \qquad (\textit{then evolve in time})(\text{measure } O \textit{ first})$$

$$? \qquad\qquad\qquad\qquad\qquad ?$$

$$= \qquad\qquad\qquad\qquad\qquad = \qquad (12.109)$$

$$? \qquad\qquad\qquad\qquad\qquad ?$$

$$\hat{O}\hat{H}\psi(x, t) \Longleftrightarrow (\textit{ then measure } O \textit{ later})(\text{evolve in time } \textit{first})$$

Thus, whether the operation of measuring O (via $\hat{O}$ acting on ψ) and the evolution of the wavefunction in time (via $\hat{H}$ acting on ψ) 'interfere' with each other, as measured by whether $[\hat{H}, \hat{O}] = 0$ or not, helps determine whether O will change in time.

Defining the time-development operator as $\hat{U}_t = e^{-i\hat{H}t/\hbar}$, we note that $\hat{U}_t$ is unitary since

$$\hat{U}_t^\dagger \hat{U}_t = e^{+i\hat{H}t/\hbar} e^{-i\hbar Ht/\hbar} = \hat{I} \qquad (12.110)$$

where we have used the fact (P12.20) that $e^{i\hat{O}}$ is unitary if $\hat{O}$ is Hermitian. This is the 'fancy' version of the statement that Schrödinger wavefunctions, if initially normalized, stay normalized at later times, the *set it and forget it* rule.

12.6 **Propagators**

The expansion theorem approach to solving for the time-dependence of quantum wavefunctions can be formally written in a way which is suggestive of yet another solution method familiar in classical mechanics, namely, the technique of Greens functions.[4]

12.6.1 **General Case and Free Particles**

The general solution

$$\psi(x,t) = \sum_n a_n \psi_n(x) e^{-iE_n t/\hbar} \tag{12.111}$$

with the initial conditions included by using

$$a_n = \int_{-\infty}^{+\infty} dx' \, \psi_n^*(x') \, \psi(x',0) \tag{12.112}$$

can be combined to give

$$\psi(x,t) = \sum_n \left[\int_{-\infty}^{+\infty} dx' \, \psi_n^*(x') \, \psi(x',0) \right] \psi_n(x) e^{-iE_n t/\hbar} \tag{12.113}$$

$$= \int_{-\infty}^{+\infty} dx' \left\{ \sum_n \psi_n^*(x') \, \psi_n(x) \, e^{-iE_n t/\hbar} \right\} \psi(x',0)$$

$$= \int_{-\infty}^{+\infty} dx' \, K(x,x';t,0) \, \psi(x',0) \tag{12.114}$$

where we have defined the *propagator*

$$K(x,x';t,0) = \sum_n \psi_n^*(x') \, \psi_n(x) \, e^{-iE_n t/\hbar} \tag{12.115}$$

As its name implies, the propagator dictates the time-dependence, or propagation in time, of the initial solution. An obvious property of $K(x',x;t,0)$ is that it must reproduce the initial state when $t \to 0$, so that one necessarily has

$$\delta(x'-x) = K(x',x;0,0) = \sum_m \psi_n^*(x') \, \psi_n(x) \tag{12.116}$$

as already shown in Eqn. (12.43), which is sometimes a useful relation.

---

[4] See, for example, Marion and Thornton (2004).

A generalization to states with a continuous energy eigenvalue label, such as for free particles, is straightforward. For that particular case, for example, we can write

$$K(x, x'; t, 0) = \int_{-\infty}^{+\infty} dp \, \psi_p^*(x') \, \psi_p(x) \, e^{-iE_p t/\hbar} \tag{12.117}$$

$$= \int_{-\infty}^{+\infty} dp \left(\frac{1}{\sqrt{2\pi\hbar}} e^{-ipx'/\hbar} \right)^* \left(\frac{1}{\sqrt{2\pi\hbar}} e^{ipx/\hbar} \right) e^{-ip^2 t/2m\hbar}$$

and the integrals can be evaluated using standard Gaussian techniques (P12.23) to give

$$K(x, x'; t, 0) = \sqrt{\frac{m}{2\pi i\hbar t}} \, e^{-im(x-x')^2/2\hbar t} \tag{12.118}$$

and one can check explicitly that

$$\lim_{x \to x'} K(x, x'; t, 0) = \delta(x - x') \tag{12.119}$$

as it should. The time-dependence of an initial Gaussian wave packet,

$$\psi(x, 0) = \frac{1}{\sqrt{b\sqrt{\pi}}} e^{-(x-x_0)^2/2b^2 - ip_0(x-x_0)/\hbar} \tag{12.120}$$

is then easily evaluated using Eqn. (12.114) (P12.23), and the result agrees with our earlier derivation in Section 3.2.2.

A similar expansion in momentum-space,

$$\phi(p, t) = \sum_n b_n \phi_n(p) e^{-iE_n t/\hbar} \tag{12.121}$$

implies a parallel formalism for propagators with

$$K_p(p', p; t, 0) = \sum_n \phi_n^*(p') \phi_n(p) e^{-iE_n t/\hbar} \tag{12.122}$$

giving

$$\phi(p, t) = \int_{-\infty}^{+\infty} dp' \, K_p(p, p'; t, 0) \phi(p', 0) \tag{12.123}$$

or suitable generalizations to continuous labels.

12.6.2 **Propagator and Wave Packets for the Harmonic Oscillator**

Any time-dependent state composed of a linear combination of harmonic oscillator eigenfunctions can easily be shown to exhibit a purely periodic behavior.

The general solution

$$\psi(x, t) = \sum_{n=0}^{\infty} a_n u_n(x) e^{-iE_n t/\hbar} \tag{12.124}$$

when evaluated after one classical period, $\tau = 2\pi/\omega$, is given by

$$\psi(x, t + \tau) = \sum_{n} a_n u_n(x) \, e^{-iE_n(t+\tau)/\hbar}$$

$$= \sum_{n} a_n u_n(x) \, e^{-iE_n t/\hbar} e^{-i(n+1/2)2\pi}$$

$$= e^{-i\pi} \psi(x, t) = -i\psi(x, t) \tag{12.125}$$

This implies that for integral multiples of the period we have

$$|\psi(x, t + m\tau)|^2 = |(-i)^m \psi(x, t)|^2 = |\psi(x, t)|^2 \tag{12.126}$$

so that the position probability density is indeed periodic, with similar results for the momentum-space distribution.

Localized wave packet-like linear combinations of energy eigenstates can be formed for the harmonic oscillator, which also indicate even more clearly the expected quasi-classical behavior. A very convenient formulation for their construction is the propagator method introduced in the last section. In this case we have

$$K(x, x'; t, 0) = \sum_{n=0}^{\infty} \psi_n^*(x') \, \psi_n(x) \, e^{-i(n+1/2)\omega t} \tag{12.127}$$

which gives the time-development of any initial state via

$$\psi(x, t) = \int_{-\infty}^{+\infty} dx' \, \psi(x', 0) \, K(x, x'; t, 0) \tag{12.128}$$

The periodicity of the classical harmonic oscillator can be obtained from Eqn. (12.127) by recalling that $\tau = 2\pi/\omega$, so

$$K(x, x'; t + m\tau) = (-i)^m K(x, x'; t, 0) \tag{12.129}$$

which gives the same result in Eqn. (12.125).

Using the formalism of raising and lowering operators of Chapter 13, one can actually calculate the summation in Eqn. (12.127) in closed form[5] with the result

$$K(x, x'; t, 0) = \sqrt{\frac{m\omega}{\pi\hbar 2i \sin(\omega t)}} \exp\left(\frac{im\omega((x^2 + x'^2)\cos(\omega t) - 2xx')}{2\hbar \sin(\omega t)}\right)$$

$$\tag{12.130}$$

---

[5] See, for example, the excellent discussion in Saxon (1968).

We note that:

- This expression explicitly exhibits the periodicity in time dictated by Eqn. (12.129).

- In the limit in which the value of the spring constant, and hence ω, vanishes, the system describes a free particle; it is easy to check that $K(x, x'; t, 0)$ approaches the free-particle propagator found in Eqn. (12.118) (P12.28).

Using this form and a Gaussian wave packet representing a particle with initial position and momentum equal to $x_0 = 0$ and p_0, respectively, namely

$$\psi(x, 0) = \frac{1}{\sqrt{\beta\sqrt{\pi}}} e^{-x^2/2\beta^2} e^{ip_0 x/\hbar} \tag{12.131}$$

we can perform the time-development integral in closed form and obtain the time-dependent position-space wavefunction (P12.30)

$$\psi(x, t) = \exp\left[\frac{im\omega x^2 \cos(\omega t)}{2\hbar \sin(\omega t)}\right] \frac{1}{\sqrt{A(t)\sqrt{\pi}}} \exp\left[-\frac{im\omega\beta}{2\hbar \sin(\omega t)} \frac{(x - x_s(t))^2}{A(t)}\right] \tag{12.132}$$

where we have defined

$$L(t) \equiv \beta \cos(\omega t) + i\left(\frac{\hbar}{m\omega\beta}\right) \sin(\omega t) \quad \text{and} \quad x_s(t) \equiv \frac{p_0 \sin(\omega t)}{m\omega} \tag{12.133}$$

This gives the time-dependent probability density

$$P(x, t) = |\psi(x, t)|^2 = \frac{1}{\sqrt{\pi}|A(t)|} e^{-(x-x_s(t))^2/|A(t)|^2} \tag{12.134}$$

where

$$|A(t)| = \sqrt{\beta^2 \cos^2(\omega t) + (\hbar/m\omega\beta)^2 \sin^2(\omega t)} \tag{12.135}$$

is the time-dependent width. The expectation value of position is easily evaluated, and one finds

$$\langle x \rangle_t = x_s(t) = \frac{p_0}{m\omega} \sin(\omega t) \tag{12.136}$$

which is consistent with the classical solution of Newton's equations for the same initial conditions. The time-dependent spread is more interestingly given by

$$\Delta x_t = \frac{|A(t)|}{\sqrt{2}} \tag{12.137}$$

which does not increase monotonically with time, but shrinks and expands periodically. The time-evolution of this wave packet is illustrated in Fig. 12.1.

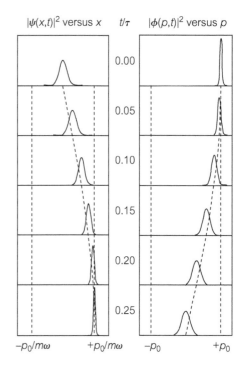

Figure 12.1. Plots of $|\psi(x, t)|^2$ versus x (left) and $|\phi(p, t)|^2$ versus p (right) for the harmonic oscillator wave packet in Eqn. (12.132) during the first quarter cycle. The dashed curves are the classical trajectories, $x(t) = p_0 \sin(\omega t)/m\omega$ and $p(t) = p_0 \cos(\omega t)$, corresponding to the initial conditions. Note the anticorrelation between the widths of the x and p distributions.

The oscillatory behavior of Δx_t is perfectly consistent with uncertainty principle arguments, as the corresponding momentum-space distributions show the same behavior with the appropriate phase relation; specifically, one can show that

$$\Delta p_t = p_L(t)/\sqrt{2} \tag{12.138}$$

where

$$p_L(t) = \sqrt{(\hbar/\beta)^2 \cos^2(\omega t) + (m\omega\beta)^2 \sin^2(\omega t)} \tag{12.139}$$

The propagator for the momentum-space wavefunctions is obtained as easily, with the result that

$$K_p(p', p; t) = \frac{1}{\sqrt{2\pi \hbar m\omega \sin(\omega t)}} \exp\left(\frac{i((p^2 + p'^2)\cos(\omega t) - 2pp')}{2\hbar m\omega \sin(\omega t)}\right) \tag{12.140}$$

and the momentum-space wave packets also be constructed.

12.7 **Timescales in Bound State Systems: Classical Period and Quantum Revival Times**

In general, the time-dependence of an arbitrary time-dependent bound state wavefunction, $\psi(x, t)$, with the expansion in eigenstates, $u_n(x)$, of the form

$$\psi(x, t) = \sum_n u_n(x)\, e^{-iE_n t/\hbar} \tag{12.141}$$

can be quite complicated. However, in many experimental realizations, a localized wave packet is excited with an energy spectrum which is tightly spread around a large central value of the quantum number, n_0, such that $n_0 \gg \Delta n \gg 1$. In that case, it is appropriate to expand the individual energy eigenvalues, $E(n) \equiv E_n$, about this value, giving the approximation

$$E(n) \approx E(n_0) + E'(n_0)(n - n_0) + \frac{E''(n_0)}{2}(n - n_0)^2 + \cdots \tag{12.142}$$

where $E'(n_0) = (dE_n/dn)_{n=n_0}$ and so forth. This gives the time-dependence of each individual quantum eigenstate through the factors

$$
e^{-iE_n t/\hbar} = \exp\left(-i/\hbar\left[E(n_0)t + (n - n_0)E'(n_0)t + \frac{1}{2}(n - n_0)^2 E''(n_0)t + \cdots\right]\right)
$$
$$
\equiv \exp\left(-i\omega_0 t - 2\pi i(n - n_0)t/T_{cl} - 2\pi i(n - n_0)^2 t/T_{rev} + \cdots\right) \tag{12.143}
$$

where each term in the expansion (after the first) defines an important characteristic timescale, and we have kept the first two terms, namely

$$T_{cl} = \frac{2\pi\hbar}{|E'(n_0)|} \quad \text{and} \quad T_{rev} = \frac{2\pi\hbar}{|E''(n_0)|/2} \tag{12.144}$$

which will be associated with the *classical period* and the *quantum revival time*, respectively.

The first term ($\omega_0 = E(n_0)/\hbar$) is an unimportant, n-independent overall phase, common to all terms in the expansion, and which therefore induces no interference between them; it is similar to the time-dependent phase for a single stationary state solution and has no observable effect in $|\psi(x, t)|^2$.

The second term in the expansion is familiar from correspondence principle arguments (as in Section 1.4 and P1.16) as being associated with the classical period of motion in the bound state. This connection is perhaps most easily seen using a semiclassical argument and the WKB quantization condition from Section 10.3.2.

For a particle of fixed energy E in a one-dimensional bound state potential, $V(x)$, we have $E = mv(x)^2/2 + V(x)$ and the short time, dt, required to traverse a distance dx can be obtained from this and integrated over the range defined by the classical turning points; this then gives *half* the classical period since

$$dt = \frac{dt}{v(x)} = \sqrt{\frac{m}{2}} \frac{dx}{\sqrt{E - V(x)}} \tag{12.145}$$

implies that

$$\frac{\tau}{2} = \int_a^b dt = \sqrt{\frac{m}{2}} \int_a^b \frac{dx}{\sqrt{E - V(x)}} \tag{12.146}$$

as discussed in Section 5.1. The WKB quantization condition (from Eqn. (10.53)) in this same potential (with the same classical turning points, a, b) can be written in the form

$$\sqrt{2m} \int_a^b \sqrt{E_n - V(x)}\, dx = (n + C_L + C_R)\pi\hbar \tag{12.147}$$

in terms of the appropriate matching coefficients C_L, C_R. Both sides of this expression can then be differentiated implicitly with respect to the quantum number, n, to obtain

$$\sqrt{2m} \int_a^b \frac{|dE_n/dn|\, dx}{2\sqrt{E_n - V(x)}} = \pi\hbar \tag{12.148}$$

This, in turn, can be related to the classical period in Eqn. (12.146) to give

$$\tau = \sqrt{2m} \int_a^b \frac{1}{\sqrt{E_n - V(x)}}\, dx = \frac{2\pi\hbar}{|dE_n/dn|} = T_{cl} \tag{12.149}$$

(via WKB methods) (via general expansion)

Example 12.5. Classical periods

The most obvious example of such a connection is for the harmonic oscillator, where the WKB condition gives the exact eigenvalues, $E_n = (n + 1/2)\hbar\omega$, and the classical period from Eqn. (12.144) is

$$T_{cl} = \frac{2\pi\hbar}{|dE/dn|} = \frac{2\pi\hbar}{\hbar\omega} = \frac{2\pi}{\omega} \tag{12.150}$$

as expected. For the special case of the oscillator, all wave packets are exactly periodic and all higher-order derivatives ($E'', E''', \ldots$) vanish, so no other longer timescales are present.

(Continued)

For the other standard example in one-dimensional quantum mechanics, namely, the infinite square well, we have

$$E_n = \frac{p_n^2}{2m} = \frac{n^2 \hbar^2 \pi^2}{2ma^2} \quad \text{so that } |E'_n| = \frac{n\hbar^2 \pi^2}{ma^2} \qquad (12.151)$$

giving

$$T_{cl} = \frac{2\pi \hbar}{|E'_n|} = \frac{2ma^2}{\hbar \pi n} \qquad (12.152)$$

which is consistent with entirely classical expectations for the period as given by

$$\tau = \frac{2a}{v_n} = \frac{2a}{p_n/m} = \frac{2ma^2}{\hbar \pi n} \qquad (12.153)$$

where we use $p_n = n\pi \hbar/a$.

For future reference, even in the presence of higher-order timescales, we can define the *classical* component of the wave packet to be

$$\psi_{cl}(x, t) \equiv \sum_{n=0}^{\infty} a_n u_n(x) e^{-2\pi i(n-n_0)E'_n t/\hbar} \equiv \sum_{k} a_k u_k(x) e^{-2\pi ikt/T_{cl}} \qquad (12.154)$$

where we define $k \equiv n - n_0$. This component can be used to describe the short-term ($t \approx T_{cl} << T_{rev}$) time-development and is especially helpful in discussing quantum revivals.

The next term in the expansion in Eqn. (12.143) is given by

$$T_{rev} = \frac{2\pi \hbar}{|E''(n_0)|/2} \qquad (12.155)$$

which will be associated with the quantum revival timescale. This timescale determines the relative importance of the $(n - n_0)^2$ term in the exponent for $t > 0$. It is responsible for the long-term ($t >> T_{cl}$) spreading of the wave packet in the same way that the difference in the p-dependence of the px and $p^2 t/2m$ terms in $\exp(i(px - p^2 t/2m)/\hbar)$ in the plane wave expansion of free-particle wave packets gives rise to dispersion.

More interestingly, for times of the order of T_{rev}, the additional $\exp(2\pi i(n - n_0)^2 t/T_{rev})$ phase terms all return to unity, giving the $t \approx 0$ time-dependence in Eqn. (12.154) and a return to approximate semiclassical behavior. This type of behavior was first discussed in Section 5.4.2 in the context of the infinite square well. Using the more general approach described here, we can write for the standard infinite well

$$|E_n''| = \frac{\hbar^2 \pi^2}{ma^2} \quad \longrightarrow \quad T_{\text{rev}} = \frac{4\pi \hbar}{|E_n''|} = \frac{4ma^2}{\hbar \pi} \qquad (12.156)$$

which is consistent with the result in Eqn. (5.78).

12.8 Questions and Problems

Q12.1. Using the infinite well eigenfunctions as an example, can you write a program which would test the result in Eqn. (12.43)? For example, can you generate a two-dimensional plot of the left-hand side (versus x, x') for increasingly large numbers of states in the summation, and see if it looks like the right-hand side?

Q12.2. What are the analogous expressions to Eqns (12.54) to (12.58) for the cross-products of ordinary vectors?

Q12.3. Is the complex conjugation operator, defined in P6.13, a linear operator as defined by Eqn. (12.29)

Q12.4. Would the operator $\hat{O} = x\hat{p}$ have an observable classical counterpart? How would you decide?

Q12.5. Using the ideas in Section 12.5, discuss under what conditions the average value of the energy operator will be constant in time.

Q12.6. Are there circumstances under which the Gaussian harmonic oscillator wave packets in Section 12.6.2 can propagate in time with no change in shape, either in position-space, or momentum-space, or both? Such solutions are called *coherent states*.

P12.1. Which of the following are Hermitian operators?

(a) $x\hat{p}$.

(b) $x\hat{p} + \hat{p}x$.

(c) $\hat{E}x$.

(d) $x\hat{p}x$.

P12.2. Show that $\left(\hat{A}^\dagger\right)^\dagger = \hat{A}$.

P12.3. Show that the combinations in Eqn. (12.17) and (12.19) are Hermitian.

P12.4. If $\hat{A}$ is Hermitian, show that $\langle \psi | \hat{A}^2 | \psi \rangle \geq 0$ for any $|\psi\rangle$. Why do we expect this kind of result?

P12.5. (a) Consider the matrix operator relation defined by Eqn. (12.32). Show that if $\mathbf{a}' \cdot \mathbf{a}' = \mathbf{a} \cdot \mathbf{a}$, then the matrix $\mathbf{U}$ satisfies Eqn. (12.33).

(b) Show that the rows and columns of a Hermitian matrix can be thought of as orthonormal vectors.

(c) Show that the matrix

$$U = \frac{1}{5\sqrt{2}} \begin{pmatrix} 4 - 4i & 3 - 3i \\ -3\sqrt{2}i & 4\sqrt{2}i \end{pmatrix} \tag{12.157}$$

is unitary.

(d) Find values of b and c for which the matrix

$$U = \frac{1}{5} \begin{pmatrix} 3 + i & \sqrt{15}i \\ b & ci \end{pmatrix} \tag{12.158}$$

is unitary. Are your choices unique?

P12.6. Consider a *translation operator* defined by

$$\hat{T}_a \equiv e^{ia\hat{p}/\hbar} = e^{a\partial/\partial x} \tag{12.159}$$

where a is a constant.

(a) Show that $\hat{T}_a f(x) = f(x + a)$. Hint: Expand both the operator and the function in a series expansion and compare.

(b) Show that $\hat{T}_a$ is a unitary operator and interpret this statement.

P12.7. Show that the set of translation operators, $\hat{T}_a$, where a is any real number, form a group. Hint: Show that this set satisfies all of the group requirements in Appendix F.2.

P12.8. Verify the commutator relations in Eqns. (12.53) to (12.60).

P12.9. Exponentials of operators

(a) Since we know that an operator commutes with itself, we would expect that

$$e^{a\hat{O}} e^{b\hat{O}} = e^{(a+b)\hat{O}} \tag{12.160}$$

since there would be no problems with operator ordering. Show this explicitly by comparing series expansions for both sides, that is, show that

$$\left(\sum_{n=0}^{\infty} \frac{(a\hat{O})^n}{n!} \right) \left(\sum_{m=0}^{\infty} \frac{(b\hat{O})^m}{m!} \right) = \sum_{J=0}^{\infty} \frac{[(a+b)\hat{O}]^J}{J!} \tag{12.161}$$

Hint: Relabeling the sums and using the binomial expansion helps.

(b) A more general identity, valid for any two operators $\hat{A}$ and $\hat{B}$, is given by the so-called *Campbell–Baker–Hausdorff formula*, namely

$$e^{\hat{A}} e^{\hat{B}} = e^{\hat{O}} \tag{12.162}$$

where

$$\hat{O} = \hat{A} + \hat{B} + \frac{1}{2}[\hat{A}, \hat{B}] + \frac{1}{12}\left([[\hat{A}, \hat{B}], \hat{B}] + [\hat{A}, [\hat{A}, \hat{B}]]\right)$$

$$+ \frac{1}{24}[[[\hat{B}, \hat{A}], \hat{A}], \hat{B}] + \cdots \tag{12.163}$$

is an infinite series of nested commutators. Verify this relation to second order.

P12.10. Using explicit representations of operators in momentum-space, find the minimum-uncertainty wavefunction corresponding to Eqn. (12.80) for the x, p pair.

P12.11. Consider a general quantum state which has been expanded in energy eigenfunctions, $\psi(x, 0) = \sum_n a_n u_n(x)$. What is the effect of operating on $\psi(x, 0)$ with the time-development operator, $\hat{U}_t = e^{-i\hat{H}t/\hbar}$?

P12.12. The quantum version of the virial theorem.

(a) Use the fact that average values of operators in energy eigenstates are time-independent to show that

$$\langle n|\hat{T}|n\rangle = \frac{1}{2}\left\langle n\left|x\frac{dV(x)}{dx}\right|n\right\rangle \tag{12.164}$$

when evaluated using energy eigenstates. Do this by considering the time-dependence of the expectation value of the operator $x\hat{p}$, that is, calculate

$$\frac{d}{dt}\langle n|x\hat{p}|n\rangle \tag{12.165}$$

as in Eqn. (12.96), but specialize to energy eigenstates.

(b) This relation often can give information on how the potential and kinetic energy in a quantum system are "shared." Show this by considering power law potentials of the form $V(x) = Kx^n$ and finding the fraction of the total energy which, on average, is in the form of potential and kinetic energy; specifically, show that

$$\frac{\langle V(x)\rangle}{E} = \frac{2}{n+2} \quad \text{and} \quad \frac{\langle \hat{T}\rangle}{E} = \frac{n}{n+2} \tag{12.166}$$

(c) Try to confirm your results in (b) by considering both the harmonic oscillator potential ($n = 2$) and infinite well potential ($n = \infty$).

(d) Compare this problem to the classical results of P5.2.

P12.13. Bound on quantum correlations. Recall the definition of the quantum versions of the *covariance* and *correlation coefficient* in P4.18, generalized here to any two quantum operators, $\hat{A}$ and $\hat{B}$, namely

$$\text{cov}(\hat{A}, \hat{B}) \equiv \frac{1}{2}\left[\langle(\hat{A} - \langle\hat{A}\rangle)(\hat{B} - \langle\hat{B}\rangle) + (\hat{B} - \langle\hat{B}\rangle)(\hat{A} - \langle\hat{A}\rangle)\rangle\right] \tag{12.167}$$

and

$$\rho(\hat{A}, \hat{B}) \equiv \frac{\text{cov}(\hat{A}, \hat{B})}{\Delta A \cdot \Delta B} \tag{12.168}$$

We will assume that both $\hat{A}$ and $\hat{B}$ are Hermitian, and that the expectation values can be evaluated in any acceptable quantum state.

(a) Use a variation on the proof of the general uncertainty principle in Section 12.4 to derive the bound on the quantum correlation coefficient,

$$[\rho(\hat{A}, \hat{B})]^2 \le 1 - \left[\frac{|\langle [\hat{A}, \hat{B}] \rangle|}{2\Delta A \cdot \Delta B} \right]^2 \tag{12.169}$$

Do this by using a general complex number $\kappa = c + id$ in Eqn. (12.63) instead of a pure imaginary number, $i\lambda$, used in that analysis.

(b) Recalling P4.18 (d), show that this bound is saturated (i.e. there is an equality) for the standard free-particle Gaussian wave packet in Eqn. (3.35).

P12.14. Assuming that the operator $\hat{O}$ it not itself time-dependent, show that the second derivative of the expectation value is given by

$$\frac{d^2}{dt^2} \langle \hat{O} \rangle_t = -\frac{1}{\hbar^2} \langle [\hat{H}, [\hat{H}, \hat{O}]] \rangle \tag{12.170}$$

What new term is introduced if $\hat{O}$ *does* depend on time?

P12.15. If $\hat{A}$ and $\hat{B}$ are arbitrary, perhaps time-dependent operators, show that

$$\frac{d}{dt} \langle \hat{A}\hat{B} \rangle = \left\langle \frac{\partial \hat{A}}{\partial t} \hat{B} \right\rangle + \left\langle \hat{A} \frac{\partial \hat{B}}{\partial t} \right\rangle + \frac{i}{\hbar} \langle [\hat{H}, \hat{A}] \hat{B} \rangle + \frac{i}{\hbar} \langle \hat{A}[\hat{H}, \hat{B}] \rangle \tag{12.171}$$

P12.16. **Time-dependent expectation values and uncertainties for free-particle wave packets.**

(a) For free particles, where $V(x) = 0$, use Eqns (12.90) and (12.91) to show that the expectation values for position and momentum are given very generally as

$$\langle \hat{p} \rangle_t = \langle \hat{p} \rangle_0 \tag{12.172}$$

$$\langle x \rangle = \langle p \rangle_0 t / m + \langle x \rangle_0 \tag{12.173}$$

(b) Using Eqn (12.93), show that $\langle \hat{p}^2 \rangle_t = \langle \hat{p}^2 \rangle_0$, so that the time-dependent spread in momentum for any free-particle wave packet is actually constant, namely, $\Delta p_t = \Delta p_0$.

(c) Using Eqns (12.95–12.97), show that

$$\langle x^2 \rangle_t = \frac{\langle \hat{p}^2 \rangle_0 t^2}{m^2} + \frac{t}{m} \langle x\hat{p} + \hat{p}x \rangle_0 + \langle x^2 \rangle_0 \tag{12.174}$$

which then gives the time-dependent spread in position as

$$(\Delta x_t)^2 = \langle x^2 \rangle_t - \langle x \rangle_t^2 = \left(\frac{\Delta p_0 t}{m}\right)^2 + (\Delta x_0)^2$$

$$+ \frac{t}{m}\langle (x - \langle x \rangle_0)(\hat{p} - \langle \hat{p} \rangle_0) + (\hat{p} - \langle \hat{p} \rangle_0)(x - \langle x \rangle_0) \rangle \quad (12.175)$$

Thus, in the most general case, the long-term uncertainty in position is dominated by the $\Delta x_0 \approx \Delta p_0 t/m$ term, just as argued in Section 3.2.2 for the Gaussian example.

(d) Compare this general result to that for the general Gaussian wave packet in Eqns (3.35) and 3.36). Are all of the terms in Eqn. (12.175) present? What is the origin of the term linear in t in Eqn. (12.175), and what does it mean?

(e) Consider the modified Gaussian wave packet in P3.5. Show that Δx_t for that packet includes the term linear in t.

P12.17. Repeat all parts of P12.16, but for the case of a constant force, namely, when $F(x) = +F$ and $V(x) = -Fx$. Specifically, show that the results for Δp_t and Δx_t are the same as for the free-particle results.

P12.18. (a) The use of Eqn. (12.99) is not restricted to position-space wavefunctions. What is the time-development operator for free-particle momentum-space wavefunctions?

(b) Generalize the result of Example 12.4 to include the case where the Gaussian wave packet has a nontrivial initial position (x_0) and momentum (p_0).

(c) Can you generalize Example 12.4 to find the explicit time-dependence of the initial wavefunction

$$\psi(x; 0) = \sqrt{\frac{2}{\sqrt{\pi}\beta^3}}\, x\, e^{-x^2/2\beta^2} \quad (12.176)$$

Hint: You will need to generate a new "tricky identity" to replace Eqn. (12.105).

P12.19. Use the Hamiltonian for a particle subject to a uniform force, corresponding to $V(x) = -Fx$, to calculate the effect of the time-development operator on both position- and momentum-space wavefunctions. Hint: Use the Hausdorff formula in P12.9.[6]

P12.20. Prove that $e^{i\hat{O}}$ is unitary is $\hat{O}$ is Hermitian.

P12.21. (a) If we let

$$|\psi\rangle_t = e^{-i\hat{H}t/\hbar}|\psi\rangle_0 \quad (12.177)$$

---

[6] And if you get stuck, look at Robinett (1996b).

show that the time-dependent expectation value of any operator can be written as

$$\langle \hat{O} \rangle_t = \langle \psi_0 | e^{+i\hat{H}t/\hbar} \hat{O} e^{-i\hat{H}t/\hbar} | \psi_0 \rangle \Rightarrow \langle \psi_0 | \hat{O}(t) | \psi_0 \rangle \qquad (12.178)$$

where $\hat{O}(t) \equiv e^{+i\hat{H}t/\hbar} \hat{O} e^{-i\hat{H}t/\hbar}$. In this view, the quantum state vectors, $|\psi_0\rangle$, are fixed once and for all, but the operators evolve in time; this is often referred to as the *Heisenberg picture* of quantum mechanics as opposed to the *Schrödinger picture* which we have adopted.

(b) Use these results to show that the time-derivative of a Heisenberg operator can be written as

$$\frac{d}{dt} \hat{O} = \frac{i}{\hbar} [\hat{H}, \hat{O}] \qquad (12.179)$$

in the case where $\hat{O}$ itself has no explicit time-dependence. Hint: Simply recall the definition of derivative as

$$\frac{d\hat{O}(t)}{dt} = \lim_{dt \to 0} \left(\frac{\hat{O}(t + dt) - \hat{O}(t)}{dt} \right) \qquad (12.180)$$

This result generalizes Eqn. (12.88) for the time-development of matrix elements of operators.

P12.22. (a) Using the results of the last problem, find the time-dependent operators, $\hat{x}(t)$ and $\hat{p}(t)$, in the Heisenberg picture in the free-particle case where the Hamiltonian is $\hat{H} = \hat{p}^2/2m$; comment on any classical analogs to your results. Hint: Use the operator identity

$$e^{\hat{A}} \hat{B} e^{-\hat{A}} = \hat{B} + [\hat{A}, \hat{B}] + \frac{1}{2}[\hat{A}, [\hat{A}, \hat{B}]] + \frac{1}{3!}[\hat{A}, [\hat{A}, [\hat{A}, \hat{B}]]] + \cdots \qquad (12.181)$$

(b) Repeat part (a), but use the Hamiltonian corresponding to the case of a constant force, F, in the $+x$ direction, namely

$$\hat{H}_F = \frac{\hat{p}^2}{2m} - Fx \qquad (12.182)$$

P12.23. **Free-particle propagators.**

(a) Do the necessary Gaussian integral (assuming that it converges, or adding a convergence factor) to obtain Eqn. (12.118).

(b) Apply it to the initial Gaussian wave packet in Eqn. (12.120) and show that you obtain the general, time-dependent Gaussian in Eqn. (3.35).

P12.24. (a) Show that the propagator for two arbitrary times $t_2 > t_1$ is given by

$$K(x, x'; t_2, t_1) = \sum_n \psi_n^*(x') \psi_n(x) e^{-iE_n(t_2 - t_1)/\hbar} \qquad (12.183)$$

(b) Show that the propagator satisfies the identity

$$K(x, x'; t_3, t_1) = \int_{-\infty}^{+\infty} dx'' \, K(x, x''; t_3, t_2) \, K(x'', x', t_2, t_1) \quad (12.184)$$

and interpret the result in terms of propagating from $t_1 \to t_2 \to t_3$

(c) Check the result of part (b) explicitly for the case of the free-particle propagator in Eqn. (12.118).

P12.25. Position-space versus momentum-space propagators.

(a) Show that the free-particle position-space propagator can be derived from the one in momentum-space via

$$K(x', x; t, 0) = \frac{1}{2\pi \hbar} \int_{-\infty}^{+\infty} dp' \, e^{ip'x'/\hbar} \int_{-\infty}^{+\infty} dp \, e^{-ipx/\hbar} \, K_p(p', p; t, 0) \quad (12.185)$$

Give the corresponding expression for $K_p(p', p; t, 0)$ in terms of $K(x', x; t, 0)$.

(b) Use the continuous generalization of Eqn. (12.122) to evaluate the momentum-space propagator for a free particle. Apply it as in Eqn. (12.123) to an initial Gaussian wavefunction $\phi(p', 0)$ and show that it reproduces the result of Eqn. (3.36).

P12.26. Propagators for uniform acceleration. Consider the accelerating particle from Section 4.7.2, described by the time-independent Schrödinger equation in momentum space

$$\frac{p^2}{2m} \phi_E(p) - i\hbar F \frac{d\phi_E(p)}{dp} = E\phi_E(p) \quad (12.186)$$

where $\phi(p, t) = \phi(p)e^{-iEt/\hbar}$.

(a) Show that an appropriate solution is

$$\phi_E(p) = Ce^{-ip^3/6m\hbar F} e^{-iEp/\hbar} \quad (12.187)$$

where C is an arbitrary constant.

(b) Use the fact that eigenstates corresponding to different eigenvalues should be properly normalized to evaluate C and show that $C = 1/\sqrt{2\pi \hbar F}$.
Hint: Use the fact that

$$\int_{-\infty}^{+\infty} dp \, \phi_{E'}^*(p) \, \phi_E(p) = \delta(E - E') \quad (12.188)$$

(c) Construct the momentum-space propagator by evaluating

$$K_p(p', p; t, 0) = \int dE \, \phi_E^*(p') \, \phi_E(p) \, e^{-iEt/\hbar} \quad (12.189)$$

where the summation is over the continuous E label. Show that the result, including the appropriate constant C, is

$$K_p(p', p; t, 0) = e^{i((p-Ft)^3 - p^3)/6m\hbar F} \delta(p - p' - Ft) \qquad (12.190)$$

and show that when applied to an initial momentum-space wavefunction it reproduces Eqn. (4.133). Show also that this propagator satisfies the appropriate initial condition and that it reduces to the free-particle result when $F \to 0$.

P12.27. Using the results of Problem P12.24 and P12.25, find the position-space propagator for the accelerating particle. Show that it reduces to the free-particle case when $F \to 0$.

P12.28. (a) Show that the $\omega \to 0$ limit of Eqn. (12.130) gives the free-particle propagator.

(b) Use Eqn. (12.130) to find the $\psi(x, t)$ for the initial Gaussian in Eqn. (12.131).

(c) Evaluate $\langle x \rangle_t$ to confirm Eqn. (12.136).

P12.29. Discuss the position-space and momentum-space probability densities for a general harmonic oscillator wave packet at $t \to t + \tau/2$ and show that

$$|\psi(x, t + \tau/2)|^2 = |\psi(-x, t)|^2 \quad \text{and} \quad |\phi(p, t + \tau/2)|^2 = |\phi(-p, t)|^2 \tag{12.191}$$

and interpret your results.

P12.30. Harmonic oscillator wave packets.

(a) Using the propagator in Eqn. (12.130), and the initial state in Eqn. (12.131), do the the necessary integral in Eqn. (12.128) to confirm Eqn. (12.132).

(b) Evaluate $\langle x \rangle_t$ and Δx_t for this state and confirm Eqns (12.136) and (12.137).

(c) Are there circumstances where the time-dependent spreads in position (Δx_t) and momentum Δp_t for the SHO wave packet do not oscillate in time?

(d) Use the propagator approach to evaluate $\psi(x, t)$ for the initial state

$$\psi(x, 0) = \frac{1}{\sqrt{\beta\sqrt{\pi}}} e^{-(x-x_0)^2/2\beta^2} \qquad (12.192)$$

corresponding to initial values $\langle x \rangle_0 = x_0$ and $\langle p \rangle_0 = 0$. Use your result to derive $|\psi(x, t)|^2$ and to evaluate $\langle x \rangle_t$ and Δx_t. Hint: To minimize the algebra, in the integral make the change of variables $y = x' - x_0$.

(e) Finally, show that the most general initial Gaussian given by

$$\psi(x, 0) = \frac{1}{\sqrt{\beta\sqrt{\pi}}} e^{-(x-x_0)^2/2\beta^2} e^{+ip_0 x/\hbar} \qquad (12.193)$$

gives a time-dependent wave packet with probability density

$$P(x, t) = \frac{1}{|L(t)|\sqrt{\pi}} \exp\left(-\frac{(x - \tilde{x}(t))^2}{|L(t)|^2}\right) \tag{12.194}$$

where

$$\tilde{x}(t) = x_0 \cos(\omega t) + \frac{p_0}{m\omega} \sin(\omega t). \tag{12.195}$$

P12.31. Momentum-space wave packets using propagators.

(a) Using the p-space propagator in Eqn. (12.140) and the initial state

$$\phi(p, 0) = \sqrt{\frac{\alpha}{\sqrt{\pi}}} e^{-\alpha^2 (p - p_0)^2 / 2} \tag{12.196}$$

evaluate $\phi(p, t)$ and show that

$$P(p, t) = |\phi(p, t)|^2 = \frac{1}{|p_L(t)|\sqrt{\pi}} \exp\left(-\frac{(p - p_0 \cos(\omega t)^2}{|p_L(t)|^2}\right) \tag{12.197}$$

and evaluate $\langle p \rangle_t$ and Δp_t.

P12.32. Wave packet for unstable equilibrium. The unstable quadratic potential, $V(x) = -Kx^2/2$, can be trivially obtained from the stable harmonic oscillator by the identification $K \to -K$.

(a) Show that this corresponds to the substitution $\omega \to i\omega$.

(b) Use this and the relations of Appendix C (for complex numbers and functions) to show that a wave packet satisfying the time-dependent Schrödinger equation

$$-\frac{\hbar^2}{2m} \frac{\partial^2 \psi(x, t)}{\partial x^2} + V(x) \psi(x, t) = i\hbar \frac{\partial \psi(x, t)}{\partial t} \tag{12.198}$$

with initial mean position and momentum fixed as $x_0 = 0$ and p_0 is given by

$$|\psi(x, t)|^2 = \frac{1}{\sqrt{\pi} \tilde{L}(t)} e^{-(x - p_0/m\omega \sinh(\omega t))^2 / |\tilde{L}(t)|^2} \tag{12.199}$$

where

$$|\tilde{L}(t)|^2 = \beta^2 \cosh^2(\omega t) + (\hbar/m\omega\beta)^2 \sinh^2(\omega t). \tag{12.200}$$

(c) Evaluate $\langle x \rangle_t$ and show that it is consistent with the classical solutions of Newton's equations with the same initial conditions.

P12.33. Classical period of a "bouncer."

(a) Show that the WKB approximation for the energies of a "quantum bouncer," defined by the potential

$$V(y) = \begin{cases} mgy & \text{for } y > 0 \\ \infty & \text{for } y < 0 \end{cases} \qquad (12.201)$$

gives

$$E_n = \left(\frac{9\pi^2}{8} mg^2 \hbar^2 \right)^{1/3} (n + 3/4)^{2/3} \qquad (12.202)$$

(b) Use the result in Eqn. (12.144) to evaluate the classical period in this system as a function of n.

(c) If we equate the quantized energies with the classical potential energy via $E_n = mgH$, we can relate the maximum height reached by the particle with the quantum number n. Use this association to rewrite the expression for T_{cl} in part (a) to reproduce the standard classical result that $T_{cl} = \sqrt{8H/g}$ for a ball bouncing without loss of energy.

(d) Evaluate the quantum revival time, T_{rev}, for this system and compare to T_{cl}. Compare T_{rev} for this case with the result for the infinite well in Eqn. (5.78).

P12.34. Classical period and revival time for hydrogen atom states.

(a) Use the Bohr result for the energy spectrum of the hydrogen atom in Eqn. (1.42), namely

$$E_n = -\frac{1}{2} \left(\frac{m(Ke^2)^2}{\hbar^2} \right) \frac{1}{n^2} \equiv -\frac{\epsilon_0}{n^2} \qquad (12.203)$$

to derive the classical period for this system, comparing it to the result in Eqn. (1.41).

(b) Using Eqn. (12.155), evaluate the revival time, T_{rev}, for hydrogen-atom wave packets and show that $T_{rev}/T_{cl} = 2n/3$. Compare this to the discussion in P5.19 where the revival time was estimated using the infinite square well result as a simple model. Recall that revival behavior on these timesscales has been observed with large n Rydberg atoms; see Yeazell et al.(1990) or Wals et al.(1994).

P12.35. Wigner distribution for a harmonic oscillator wave packet. Evaluate the Wigner quasi-probability distribution, $P_W(x, p; t)$ in Eqn. (4.149), for the time-dependent oscillator wave packet in Eqn. (12.132). For simplicity, assume that $\beta = \sqrt{\hbar/m\omega}$, which simplifies the algebra considerably. Is $P_W(x, p; t)$ everywhere positive-definite? Does its behavior in time mimic the corresponding classical phase-space description of a particle undergoing simple harmonic motion, as in Appendix G, especially Fig. G1?

THIRTEEN

Operator and Factorization Methods for the Schrödinger Equation

13.1 Factorization Methods

The special and recurring role that the simple harmonic oscillator (SHO) plays in quantum mechanics can be attributed both to its physical relevance and its simple solutions. The fact that the ground state solution is the minimum uncertainty wave packet (Section 12.4) and the highly constrained connection between the position-space and momentum-space wavefunctions (P4.23 and Section 9.2.2) also indicates that this problem occupies a special niche. The symmetry between x and p present in the solutions is obviously a reflection of the fact that only for the SHO is the potential energy function quadratic in x. It is often the case in physics where systems with a high degree of symmetry are amenable to solution in a variety of ways, sometimes quite unexpected. In this chapter, we discuss a powerful method of solving the harmonic oscillator problem involving the factorization of the differential equation using differential operators; we then (briefly) discuss extensions and applications of operator methods[1] to other physical problems.

Factorization methods are often used in the solution of linear differential equations with constant coefficients, that is, equations of the form

$$a_n \frac{d^n}{dx^n} y(x) + a_{n-1} \frac{d^{n-1}}{dx^{n-1}} y(x) + \cdots + a_1 \frac{d}{dx} y(x) + a_0 y(x) = 0 \quad (13.1)$$

One standard approach is to define a differential operator, $\hat{D} \equiv d/dx$, and write Eqn. (13.1) in operator form as

$$\left(a_n \hat{D}^n + a_{n-1} \hat{D}^{n-1} + \cdots + a_1 \hat{D} + a_0 \right) y(x) = 0 \quad (13.2)$$

---

[1] One of the most comprehensive treatment of operator methods in quantum mechanics is DeLange and Raab (1991).

which is similar to a polynomial equation in $\hat{D}$. If we can factor the associated polynomial, finding its real and imaginary roots, r_n, we can write this equation as

$$a_n \left[(\hat{D} - r_n)(\hat{D} - r_{n-1}) \cdots (\hat{D} - r_1) \right] y(x) = 0 \tag{13.3}$$

where the r_i are the n roots of the polynomial equation $a_n x^n + \cdots a_1 x + a_0 = 0$. The n independent solutions of Eqn. (13.1) are then obtained from

$$(\hat{D}_i - r_i) y_i(x) = 0 \quad \text{for } i = 1, \ldots, n \tag{13.4}$$

Each such simple differential equations has an exponential solution of the form

$$y_i(x) = C_i e^{-r_i x} \tag{13.5}$$

so that the general solution is

$$y(x) = \sum_{i=1}^{n} C_i e^{-r_i x} \tag{13.6}$$

Special care is necessary when there are multiple roots, but the extension to that case is mathematically straightforward (P13.1). This method then allows us to solve an nth order differential equation in terms of n first-order ones. A trivial example is the differential equation in Section 5.2 for the infinite well

$$\frac{d^2 \psi(x)}{dx^2} + k^2 \psi(x) = 0 \tag{13.7}$$

which we can write as

$$\left(\hat{D}^2 + k^2 \right) \psi(x) = (\hat{D} + ik)(\hat{D} - ik) \psi(x) = 0 \tag{13.8}$$

which has the solutions $\psi(x) = \exp(\pm ikx)$, usually written as $\sin(kx)$ and $\cos(kx)$, as expected.

13.2 **Factorization of the Harmonic Oscillator**

The form of the Hamiltonian for the harmonic oscillator problem

$$\hat{H} \psi(x) = \left(\frac{\hat{p}^2}{2m} + \frac{1}{2} m \omega^2 x^2 \right) \psi(x) = E \psi(x) \tag{13.9}$$

gives rise to a differential equation which does not have constant coefficients. Nonetheless, the form of $\hat{H}$ suggests that a similar factorization might be possible as classically one would have

$$\frac{p^2}{2m} + \frac{1}{2}m\omega^2 x^2 = \left(\frac{p}{\sqrt{2m}} + i\sqrt{\frac{m\omega^2}{2}}x \right)\left(\frac{p}{\sqrt{2m}} - i\sqrt{\frac{m\omega^2}{2}}x \right) \qquad (13.10)$$

and the lack of commutivity between the operators x and p in quantum mechanics will make things only somewhat less simple. We are therefore motivated to consider such a possible factorization, and to that end we first write

$$\hat{H} = \hbar\omega \left(\frac{\hat{p}^2}{2m\hbar\omega} + \frac{m\omega}{2\hbar}x^2 \right) \qquad (13.11)$$

as we know, if only on grounds of dimensional analysis, that the energies will be given in terms of $\hbar\omega$. We then define an operator

$$\hat{A} \equiv \sqrt{\frac{m\omega}{2\hbar}}x + i\frac{\hat{p}}{\sqrt{2m\hbar\omega}} \qquad (13.12)$$

from which one can immediately obtain

$$\hat{A}^\dagger \equiv \sqrt{\frac{m\omega}{2\hbar}}x - i\frac{\hat{p}}{\sqrt{2m\hbar\omega}} \qquad (13.13)$$

because both x and $\hat{p}$ are Hermitian. We first note that $\hat{A}, \hat{A}^\dagger$ are not Hermitian operators themselves (as $\hat{A}^\dagger \neq \hat{A}$) and so cannot represent physical observables. They do not commute either as

$$[\hat{A}, \hat{A}^\dagger] = \left[\sqrt{\frac{m\omega}{2\hbar}}x + i\frac{\hat{p}}{\sqrt{2m\hbar\omega}}, \sqrt{\frac{m\omega}{2\hbar}}x + i\frac{\hat{p}}{\sqrt{2m\hbar\omega}} \right]$$

$$= \frac{i}{2\hbar}\{-[x,\hat{p}] + [\hat{p},x]\} = 1 \qquad (13.14)$$

since $[x,\hat{p}] = -[\hat{p},x] = i\hbar$. Thus, we have

$$\hat{A}\hat{A}^\dagger = \hat{A}^\dagger\hat{A} + 1 \quad \text{and} \quad \hat{A}^\dagger\hat{A} = \hat{A}\hat{A}^\dagger - 1 \qquad (13.15)$$

which will prove useful.

To see if these operators factorize Eqn. (13.9), we can invert Eqns (13.12) and (13.13) and write

$$x = \sqrt{\frac{\hbar}{2m\omega}}(\hat{A} + \hat{A}^\dagger) \quad \text{and} \quad \hat{p} = i\sqrt{\frac{m\omega\hbar}{2}}(\hat{A}^\dagger - \hat{A}) \qquad (13.16)$$

If we substitute these into the Hamiltonian, we obtain

$$\hat{H} = \frac{1}{2m}\left(-\frac{m\hbar\omega}{2}\right)\left(\hat{A}^\dagger - \hat{A}\right)^2 + \frac{m\omega^2}{2}\left(\frac{\hbar}{2m\omega}\right)\left(\hat{A} + \hat{A}^\dagger\right)^2$$

$$= \frac{\hbar\omega}{2}\left(\hat{A}^\dagger\hat{A} + \hat{A}\hat{A}^\dagger\right)$$

$$= \hbar\omega\left(\hat{A}^\dagger\hat{A} + \frac{1}{2}\right) \quad \text{(using Eqn. (13.15))}$$

$$\equiv \hbar\omega\left(\hat{N} + \frac{1}{2}\right) \tag{13.17}$$

where we have defined a new *number operator* via

$$\hat{N} \equiv \hat{A}^\dagger\hat{A} \tag{13.18}$$

Because of the lack of commutativity of x, p we have not achieved a complete factorization, but we have shown that the Hamiltonian can be put into a form which is highly suggestive of the known result for the energy eigenvalues themselves. Thus, instead of considering the energy eigenvalue problem using the Hamiltonian, written as

$$\hat{H}\psi_n(x) = E_n\psi_n(x) \quad \text{in position-space} \tag{13.19}$$

$$\hat{H}\phi_n(p) = E_n\phi_n(p) \quad \text{in momentum-space} \tag{13.20}$$

we abstractly examine the eigenvalues of the number operator in Eqn. (13.18), that is,

$$\hat{N}|n\rangle = n|n\rangle \tag{13.21}$$

While we label the number eigenstates by n, we cannot assume (at least initially) that they are nonnegative integers; however, since $\hat{N}$ is Hermitian, because

$$\hat{N}^\dagger = \left(\hat{A}^\dagger\hat{A}\right)^\dagger = \hat{A}^\dagger\hat{A} = \hat{N} \tag{13.22}$$

we do know that the eigenvalues of $\hat{N}$ are real.

To examine the effects of $\hat{A}$ and $\hat{A}^\dagger$, and to provide constraints on the "number spectrum," we first note that

$$\hat{N}\left(\hat{A}|n\rangle\right) = \left(\hat{A}^\dagger\hat{A}\right)\left(\hat{A}|n\rangle\right)$$

$$= \left(\hat{A}\hat{A}^\dagger - 1\right)\left(\hat{A}|n\rangle\right)$$

$$= \hat{A}\left(\hat{N}|n\rangle\right) - \hat{A}|n\rangle$$

$$= (n-1)\left(\hat{A}|n\rangle\right) \tag{13.23}$$

where we have used the commutation relations between $\hat{A}$ and $\hat{A}^\dagger$. This can be interpreted as saying that the number operator, $\hat{N}$, acting on the state $\hat{A}|n\rangle$ yields the same eigenvalue, namely, $n - 1$, as when it acts on the state $|n - 1\rangle$. A similar derivation shows that

$$\hat{N}\left(\hat{A}^\dagger|n\rangle\right) = (n + 1)\left(\hat{A}^\dagger|n\rangle\right) \tag{13.24}$$

Taken together, these imply that:

- The operator $\hat{A}$ ($\hat{A}^\dagger$) acting on $|n\rangle$ must give, up to an arbitrary constant, the state $|n - 1\rangle, (|n + 1\rangle)$, that is,

$$\hat{A}|n\rangle \propto |n - 1\rangle \quad \text{and} \quad \hat{A}^\dagger|n\rangle \propto |n + 1\rangle \tag{13.25}$$

- The eigenvalues of the number operator are thus separated by integer differences and the $\hat{A}$ and $\hat{A}^\dagger$ operators act to move on up and down the ladder of possible eigenvalues, as shown in Fig. 13.1. For this reason, $\hat{A}$ and $\hat{A}^\dagger$ can be called *lowering* and *raising operators*, respectively; collectively they are called *ladder operators*.

It is still not clear (from this algebraic derivation) that the number eigenvalues are given by integers (and not say, $\pi, \pi + 1, \pi + 2 \ldots$) or even if the spectrum is bounded from below. To address the second question, using the explicit form of $\hat{N}$ and the definition of Hermitian conjugate, we can easily show that the expectation value of the number operator in the state $|n\rangle$ satisfies

$$n = n\langle n|n\rangle = \langle n|\hat{N}|n\rangle = \langle n|\hat{A}^\dagger\hat{A}|n\rangle = \langle \hat{A}n|\hat{A}n\rangle \geq 0 \tag{13.26}$$

Thus, the n are nonnegative numbers and therefore there must be a smallest one which we label n_{min}. By assumption, there can be no state with a lower value of n, so we must have

$$\hat{A}|n_{min}\rangle \propto |n_{min} - 1\rangle = 0 \tag{13.27}$$

$|n+2\rangle$ ——————— $\Big\uparrow$ $\hat{A}^\dagger$

$|n+1\rangle$ ———————

$|n\rangle$ ———————

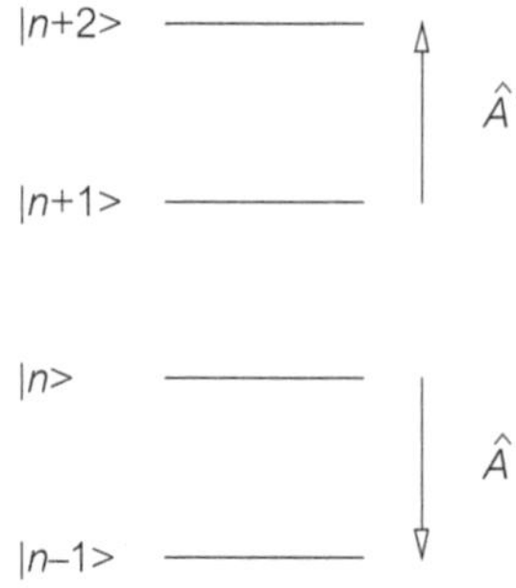

Figure 13.1. The operation of the raising (or creation) operator $\hat{A}^\dagger$ and lowering (or annihilation) operator $\hat{A}$ on number states for the harmonic oscillator.

$\Big\downarrow$ $\hat{A}$

$|n-1\rangle$ ———————

since $|n_{min} - 1\rangle$ does not exist. The number operator when acting on this lowest state then gives

$$n_{min}|n_{min}\rangle = \hat{N}|n_{min}\rangle = \left(\hat{A}^\dagger \hat{A}\right)|n\rangle = \hat{A}^\dagger \left(\hat{A}|n\rangle\right) = 0 \qquad (13.28)$$

so that $n_{min} = 0$; we sometimes say that $\hat{A}$ "annihilates the vacuum." Thus, the number and energy spectrum are given by

$$\hat{N}|n\rangle = n|n\rangle \quad n = 0, 1, 2 \ldots$$

$$\hat{H}|n\rangle = \hbar\omega \left(n + \frac{1}{2}\right)|n\rangle \quad n = 0, 1, 2 \ldots \qquad (13.29)$$

and these results have been obtained in a purely algebraic way.

While these techniques have yielded the energy spectrum in an elegant fashion, we might also wish to extract information from the position-space or momentum-space wavefunctions. The first step toward obtaining the wavefunctions is to note that any desired state can be obtained from the ground state by successive applications of the raising operator, namely

$$|n\rangle \propto \left(\hat{A}^\dagger\right)^n |0\rangle \qquad (13.30)$$

provided we have an explicit representation of the ground state. This representation can be obtained by using the fact that the lowering operator annihilates the lowest state, that is, $\hat{A}|0\rangle = 0$. Using an explicit position-space representation, we write

$$0 = \hat{A}\psi_0(x) = \left(\sqrt{\frac{m\omega}{2\hbar}}x + i\frac{\hat{p}}{\sqrt{2m\hbar\omega}}\right)\psi_0(x)$$

$$= \sqrt{\frac{\hbar}{2m\hbar\omega}}\left(\frac{m\omega}{\hbar}x + \frac{d}{dx}\right)\psi_0(x) \qquad (13.31)$$

giving

$$\frac{d\psi_0(x)}{dx} = -x\left(\frac{m\omega}{\hbar}\psi_0(x)\right) = -\frac{x}{\rho^2}\psi_0(x) \qquad (13.32)$$

or

$$\psi_0(x) = \frac{1}{\sqrt{\rho\sqrt{\pi}}}e^{-x^2/2\rho^2} \qquad (13.33)$$

when finally normalized. The corresponding momentum-space wavefunctions can be obtained from $(\hat{A}^\dagger)^n\phi_0(p)$ by solving $\hat{A}\phi_0(p) = 0$.

The nth eigenstate is obtained by repeated applications of the raising operator on the ground state as in Eqn. (13.30), and we can easily determine the precise normalization by considering

$$|n\rangle = c_n(\hat{A}^\dagger)^n|0\rangle \quad \text{or equivalently} \quad \langle n| = \langle 0|\hat{A}^n c_n \tag{13.34}$$

Assuming that all states are to be properly normalized, we note that

$$
\begin{aligned}
n = n\langle n|n\rangle &= \langle n|\hat{N}|n\rangle \\
&= \langle n|\hat{A}^\dagger\hat{A}|n\rangle \\
&= \langle n|\left(\hat{A}\hat{A}^\dagger - 1\right)|n\rangle \\
&= \langle n|\hat{A}\hat{A}^\dagger|n\rangle - 1
\end{aligned}
\tag{13.35}
$$

so that

$$
\begin{aligned}
n + 1 &= \langle n|\hat{A}\hat{A}^\dagger|n\rangle \\
&= c_n^2\langle 0|(\hat{A})^n\left(\hat{A}\hat{A}^\dagger\right)(\hat{A}^\dagger)^n|0\rangle \\
&= c_n^2\langle 0|\hat{A}^{n+1}(\hat{A}^\dagger)^{n+1}|0\rangle \\
&= c_n^2\frac{\langle n+1|n+1\rangle}{c_{n+1}^2} \\
&= \frac{c_n^2}{c_{n+1}^2}
\end{aligned}
\tag{13.36}
$$

Thus, the normalization constants are related via

$$c_{n+1} = \sqrt{n+1} \quad \text{which implies that} \quad c_n = \sqrt{n!} \tag{13.37}$$

since $c_0 \equiv 1$. Thus

$$|n\rangle = \sqrt{n!}(\hat{A}^\dagger)^n|0\rangle \tag{13.38}$$

and this can be used (P13.3) to show that

$$\hat{A}|n\rangle = \sqrt{n}|n-1\rangle \quad \text{and} \quad \hat{A}^\dagger|n\rangle = \sqrt{n+1}|n+1\rangle \tag{13.39}$$

These same techniques can be used to show explicitly that eigenfunctions corresponding to different eigenvalues are orthogonal (P13.4).

Example 13.1. Expectation values for the harmonic oscillator

The power of these general algebraic methods which is evident in generating the properly normalized wavefunctions can also be used in the efficient evaluation of expectation values of many operators. For example, the vanishing of $\langle n|x|n \rangle$ is physically obvious from the parity properties of the wavefunctions. In operator language, this relation arises because

$$\langle n|x|n \rangle = \left\langle n \left| \sqrt{\frac{\hbar}{2m\omega}}(\hat{A} + \hat{A}^\dagger) \right| n \right\rangle$$

$$= \sqrt{\frac{\hbar}{2m\omega}} \left\{ \langle n|\hat{A}|n \rangle + \langle n|\hat{A}^\dagger|n \rangle \right\}$$

$$\propto \sqrt{n}\langle n|n-1 \rangle + \sqrt{n+1}\langle n|n+1 \rangle = 0 \qquad (13.40)$$

because these states are orthogonal; this proof gives no new information, but does illustrate the method. The expectation value of the potential energy, $V(x) = m\omega^2 x^2/2$, requires

$$\langle n|x^2|n \rangle = \frac{\hbar}{2m\omega} \left\langle n \left| \hat{A}^2 + \hat{A}\hat{A}^\dagger + \hat{A}^\dagger\hat{A} + \hat{A}^{\dagger 2} \right| n \right\rangle$$

$$= \frac{\hbar}{2m\omega} \langle n|2\hat{N} + 1)|n \rangle$$

$$= \frac{\hbar}{m\omega}(n + \frac{1}{2}) \qquad (13.41)$$

so that

$$\langle n|V(x)|n \rangle = \frac{\hbar\omega}{2}(n + \frac{1}{2}) = \frac{1}{2}E_n \qquad (13.42)$$

with a similar result for $\langle n|\hat{T}|n \rangle = \langle n|\hat{p}^2|n \rangle/2m$.

13.3 **Creation and Annihilation Operators**

The importance of raising and lowering operators is not limited to the study of the quantum version of the classical oscillating particle. We have seen in Section 9.1 that the vibrations of the electromagnetic field can be represented by an ensemble of effective harmonic oscillators where the "amplitude" is essentially the vector potential $\mathbf{A}(\mathbf{k}, t)$. This similarity can be pushed even further if we define generalized ladder operators, $\hat{a}_{\mathbf{k}}$ and $\hat{a}_{\mathbf{k}}^\dagger$, for each wave number[2] mode $\mathbf{k}$.

---

[2] For simplicity, we neglect the additional labels which describe the photons state of polarization; see Baym (1976) for more complete details.

To proceed formally, we need only specify the appropriate commutation relations, that is,

$$[\hat{a}_k, \hat{a}_{k'}^\dagger] = \delta_{kk'} \tag{13.43}$$

while we also note that the $\hat{a}_k$, $\hat{a}_k^\dagger$ always commute with themselves, that is,

$$[\hat{a}_k, \hat{a}_k] = 0 \quad \text{and} \quad [\hat{a}_k^\dagger, \hat{a}_k^\dagger] = 0 \tag{13.44}$$

The number operator for each mode is simply $\hat{N}_k = \hat{a}_k^\dagger \hat{a}_k$ with corresponding integer eigenvalues n_k; the Hamiltonian is simply $\hat{H}_k = \left(\hat{N}_k + 1/2\right) \hbar \omega_k$. We then formally have relations such as

$$\hat{a}_k^\dagger |n_k\rangle = \sqrt{n_k + 1}|n_k\rangle \quad \text{and} \quad \hat{a}|n_k\rangle = \sqrt{n_k}|n_k - 1\rangle \tag{13.45}$$

corresponding to increasing or decreasing the energy of the system by one unit of $\hbar \omega_k$ where $\omega_k = |k|c$. These results are consistent with the following interpretation:

- The radiation field, quantized using such ladder operators, consists of an ensemble of photons, each of quantized energy $\hbar \omega_k$

- The number of photons of wave number k is given by n_k

- The shift operator $\hat{a}_k^\dagger (\hat{a}_k)$ increases (decreases) the number of photons by one; this justifies the use of the term *creation* (*annihilation*) operator for $\hat{a}_k^\dagger$ ($\hat{a}_k$).

Such operators, describing the particle-like quanta of the electromagnetic field are useful in many aspects of quantum optics, laser physics and beyond. It turns out that they are also the prototypes of the general class of creation and annihilation operators for an entire class of particles. Particles can be classified by the value of their intrinsic angular momentum or spin, J, with particles of integral spin, $J = 0, 1, 2 \ldots$, being called *bosons*, while those with half-integral spin $J = 1/2, 3/2, 5/2, \ldots$ called *fermions*. This last class includes the most familiar components of matter, the spin 1/2 electrons, protons, and neutrons which are known to satisfy the exclusion principle.

The creation and annihilation operators for bosons are obviously inappropriate for fermions as they allow arbitrarily many particles to be in the same quantum state. Remarkably, the corresponding operators for fermionic degrees of freedom satisfy a set of commutation relations, which are very similar in appearance to Eqns (13.43) and (13.44), namely

$$\{\hat{b}_k, \hat{b}_k^\dagger\} = \delta_{kk'} \tag{13.46}$$

and

$$\{\hat{b}_k, \hat{b}_k\} = 0 \quad \text{and} \quad \{\hat{b}_k^\dagger, \hat{b}_k^\dagger\} = 0 \tag{13.47}$$

The new symbol, $\{,\}$, denotes the so-called *anticommutator* of two operators and is defined by

$$\{\hat{A}, \hat{B}\} \equiv \hat{A}\hat{B} + \hat{B}\hat{A} \tag{13.48}$$

It is also sometimes written $\{\hat{A}, \hat{B}\} = [\hat{A}, \hat{B}]_+$.

While the formalism of anticommutation relations is superficially similar to that for commutators encountered earlier, the physical implications of these anticommutation relations are entirely different:

- The anticommutation relation of $\hat{b}^\dagger$ with itself implies that

$$\hat{b}_k^\dagger \hat{b}_k^\dagger = 0 \tag{13.49}$$

which may at first appear to be somewhat puzzling. This is simply the restatement of the exclusion principle that no two identical fermions can be put into the same quantum state; the state $|0\rangle$ with no quanta (the ground state or "vacuum") and $|k\rangle = \hat{b}_k^\dagger |0\rangle$ with one quantum are both allowed, but operating *twice* with $\hat{b}_k^\dagger$ always gives a vanishing result.

- The *number operator*, $\hat{N}_k = \hat{b}_k^\dagger \hat{b}_k$, satisfies (with the k label suppressed)

$$\hat{N}^2 = \left(\hat{b}^\dagger \hat{b}\right)\left(\hat{b}^\dagger \hat{b}\right)$$
$$= \hat{b}^\dagger \left(\hat{b}\hat{b}^\dagger\right) \hat{b}$$
$$= \hat{b}^\dagger \left(1 - \hat{b}^\dagger \hat{b}\right) \hat{b}$$
$$\hat{N}^2 = \hat{b}^\dagger \hat{b} = \hat{N} \tag{13.50}$$

This implies that the corresponding *number eigenvalues*, given by $\hat{N}|n\rangle = n|n\rangle$, satisfy $n^2 = n$, which has only the trivial (but appropriate) solutions $n = 0, 1$; again, *at most one* fermion is allowed per quantum state.

- Ordinary numbers or operators cannot satisfy the anticommutation relations of Eqns (13.46) and (13.47). One can check that an appropriate *matrix representation* of the $\hat{b}$ and $\hat{b}^\dagger$ is given by (P13.8)

$$\hat{b} = \begin{pmatrix} 0 & 1 \\ 0 & 0 \end{pmatrix} \quad \text{and} \quad \hat{b}^\dagger = \begin{pmatrix} 0 & 0 \\ 1 & 0 \end{pmatrix} \tag{13.51}$$

One of the most important implications of this formalism is that the wavefunctions for multiparticle quantum states are highly correlated. For example, if we label a state of two particles of similar type, say both photons or both electrons, as $|1; 2\rangle$ we can write

$$|1; 2\rangle = \hat{a}_1^\dagger \hat{a}_2^\dagger |0\rangle = +\hat{a}_2^\dagger \hat{a}_1^\dagger |0\rangle = +|2; 1\rangle \quad \text{for bosons} \tag{13.52}$$

while

$$|1; 2\rangle = \hat{b}_1^\dagger \hat{b}_2^\dagger |0\rangle = -\hat{b}_2^\dagger \hat{b}_1^\dagger |0\rangle = -|2; 1\rangle \quad \text{for fermions.} \tag{13.53}$$

Such wavefunctions are classified respectively as being *symmetric* or *antisymmetric* under the interchange of the two particles. This constraint on systems of similar particles imposes important new constraints on the Schrödinger wavefunction. This will be explored further in the next chapter.

We see that far from being only a clever way to solve a special problem in quantum mechanics, the method of factorization, with its resulting ladder operators, plays a central role in the description of systems of particles of all possible spins.

13.4 **Questions and Problems**

P13.1. Factoring differential equations I. Multiple roots. Consider the factorized differential equation

$$y''(x) - 2ry'(x) + r^2 y(x) = (\hat{D} - r)^2 y(x) = 0 \tag{13.54}$$

which has r as a double root.

(a) Show that $y(x) = e^{rx}$ is still a solution.

(b) To extract the second, linearly independent solution, imagine that there are two *different* roots, r, s. The linear combination given by

$$y(x) = \frac{e^{rx} - e^{sx}}{r - s} \tag{13.55}$$

will also be a solution in this case. Let $s \to r$ to obtain a second solution and explicitly check that it works.

(c) Extend these results to the case where there is a root with an N-fold degeneracy.

P13.2. Factoring differential equations II. Complex roots. In classical mechanics, the equation of motion for a damped oscillator is given by

$$m \frac{d^2 x(t)}{dt^2} + b \frac{dx(t)}{dt} + kx(t) = 0 \tag{13.56}$$

where $F_{damping} = -bv(t)$ is a velocity-dependent frictional force. In the study of circuits, there is an identical equation describing a damped LC oscillator, given by

$$L\frac{d^2q(t)}{dt^2} + R\frac{dq(t)}{dt} + \frac{1}{C}q(t) = 0 \tag{13.57}$$

with obvious terms representing the effects of the inductance (L), the capacitance (C), and the electrical resistance (R).

(a) Factor either equation to find two different roots. For small values of the damping/resistance, show that you obtain two complex conjugate roots.

(b) Use these to solve for the motion of a damped oscillator corresponding to the initial conditions $x(0) = x_0$ and $v(0) = 0$.

P13.3. Confirm that Eqn. (13.39) gives the proper normalization for the action of $\hat{A}$ and $\hat{A}^\dagger$ on a general state $|n\rangle$.

P13.4. Orthogonality of harmonic oscillator states. We can use purely algebraic techniques to show that the harmonic oscillator states are mutually orthogonal, namely that

$$\langle n|m\rangle = \delta_{n,m} \tag{13.58}$$

It is easiest to use a proof by induction as follows:

(a) Since $|n\rangle \propto (\hat{A}^\dagger)^n|0\rangle$, we also have $\langle n| \propto \langle 0|(\hat{A})^n$, so that

$$\langle n|0\rangle = \langle 0|(\hat{A})^n|0\rangle = \langle 0|(\hat{A})^{n-1}\hat{A}|0\rangle = 0 \tag{13.59}$$

since $\hat{A}$ "annihilates the vacuum state." This gives

$$\langle n|0\rangle = 0 \quad \text{for } n > 0 \tag{13.60}$$

(b) *Assume* that it has been shown that

$$\langle n|m\rangle = 0 \tag{13.61}$$

for *all* n values for *some* value of m. Show that

$$\langle n|m+1\rangle \propto (m+1)\langle n-1|m\rangle \tag{13.62}$$

Show that these two pieces, taken together, constitute a proof by induction.

P13.5. Consider a Hamiltonian written in terms of raising and lowering operators of the form

$$\hat{H} = \epsilon_1\left(\hat{A}^\dagger\hat{A}\right) + \epsilon_2\left(\hat{A} + \hat{A}^\dagger\right) \tag{13.63}$$

where $\epsilon_{1,2}$ are real constants with units of energy and $[\hat{A}, \hat{A}^\dagger] = 1$.

(a) Show that $\hat{H}$ is Hermitian.

(b) Find the energy eigenvalues of the Hamiltonian *in a purely algebraic way*, that is, do not transform back to $x, \hat{p}$ operators. Hint: Introduce *new* operators $\hat{D} = \alpha\hat{A} + \beta$ (which defines $\hat{D}^{\dagger}$) and choose the α, β so as to make the resulting problem look like the harmonic oscillator.

(c) Repeat for the Hamiltonian

$$\hat{H} = \epsilon_3 \left(\hat{A}^{\dagger}\hat{A} \right) + \epsilon_4 i \left(\hat{A} - \hat{A}^{\dagger} \right) \tag{13.64}$$

(d) Transform these Hamiltonians into position- or momentum-space differential operators, solve the differential appropriate equations, and show that the spectra are the same as given using purely operator methods.

P13.6. Use raising and lowering operators to show that

$$\langle n|x|k \rangle = \sqrt{\frac{\hbar}{2m\omega}} \left(\sqrt{n+1}\, \delta_{n+1,k} + \sqrt{n}\, \delta_{n-1,k} \right) \tag{13.65}$$

and

$$\langle n|x^2|k \rangle = \left(\frac{\hbar}{2m\omega} \right) \left[\sqrt{(n+1)(n+2)}\, \delta_{n+2,k} \right.$$
$$\left. + (2n+1)\, \delta_{n,k} + \sqrt{(n(n-1)}\, \delta_{n-2,k} \right] \tag{13.66}$$

Find the corresponding relations for the momentum operator, $\hat{p}$. These relations are useful both for the matrix formulation of the harmonic oscillator in Example 10.7 and for perturbation theory as in Example 10.9.

P13.7. Raising and lowering operators. (a) Show that the raising and lowering operators satisfy

$$[\hat{H}, \hat{A}] = -\hbar\omega\hat{A} \quad \text{and} \quad [\hat{H}, \hat{A}^{\dagger}] = \hbar\omega\hat{A}^{\dagger} \tag{13.67}$$

(b) In the Heisenberg picture (see P12.21), where operators have a nontrivial time-dependence, we can think of the harmonic oscillator states, $|n\rangle$ as being time-independent while the raising and lowering operators are given by

$$\hat{A}(t) = e^{+i\hat{H}t/\hbar}\, \hat{A}\, e^{-i\hat{H}t/\hbar} \tag{13.68}$$

and similarly for $\hat{A}^{\dagger}(t)$. Show that $[\hat{A}(t), \hat{A}^{\dagger}(t)] = 1$, independent of t. Show that the commutation relations of part (a) are also valid for the $\hat{A}(t)$.

(c) Use these facts, and the time-development equation for operators,

$$\frac{d\hat{O}(t)}{dt} = \frac{i}{\hbar}[\hat{H}, \hat{O}] \tag{13.69}$$

to show that $\hat{A}(t) = \hat{A}(0)e^{-i\omega t}$ and $\hat{A}^{\dagger}(t) = \hat{A}^{\dagger}(0)e^{+i\omega t}$. Use the defining relations for the $\hat{A}(0)$, $\hat{A}^{\dagger}(0)$ to find expressions for the time-dependent *operators*

$\hat{x}(t)$ and $\hat{p}(t)$, namely

$$\hat{x}(t) = \hat{x}(0) \cos(\omega t) + \hat{p}(0) \sin(\omega t)/m\omega \qquad (13.70)$$

$$\hat{p}(t) = -m\omega\hat{x}(0) \sin(\omega t) + \hat{p}(0) \cos(\omega t) \qquad (13.71)$$

P13.8. Show that the matrices in Eqn. (13.51) satisfy the anticommutation relations in Eqns (13.43) and (13.46).

FOURTEEN

Multiparticle Systems

Just as in classical mechanics, the study of single particle systems in one dimension provides invaluable experience in the use of quantum mechanical concepts and techniques. To extend these ideas to more realistic applications (to be able to enter the Quantum World of Part II) we also need to develop the formalisms necessary to handle multiparticle systems. In this chapter we will discuss separable systems (in Section 14.2), the important special case of two-body systems (Section 14.3), notation for spin-1/2 wavefunctions (Section 14.4), and lastly, in Section 14.5, the important constraints placed on quantum wavefunctions of multiparticle systems due to indistinguishability.

14.1 Generalities

In classical mechanics, Newton's laws for a multiparticle system have the form

$$m_i \frac{d^2 x_i(t)}{dt^2} = \mathcal{F}_i(x_i) + \sum_{j}^{j \neq i} F_{ij}(x_i - x_j) \quad \text{for } i = 1, 2, \ldots, N \quad (14.1)$$

where we have specialized to the case of external forces, $\mathcal{F}_i(x_i)$, which act on each particle separately and mutual two-body interactions, $F_{ij} = F_{ij}(x_i - x_j)$; the functional form of the F_{ij} is consistent with Newton's third law which requires that $F_{ij} = -F_{ji}$. When solved self-consistently, these equations predict the time-dependence of the coordinates of all of the particles, $x_i(t)$, once the initial conditions are specified.

The time-development of the corresponding quantum system is dictated by the *multiparticle Hamiltonian* operator

$$\hat{H} = \sum_{i} \frac{\hat{p}_i^2}{2m_i} + \sum_{i} \mathcal{V}_i(x_i) + \sum_{i,j}^{i>j} V_{ij}(x_i - x_j) \quad (14.2)$$

Here the external and two-body potentials give the corresponding forces via

$$\mathcal{F}_i(x_i) = -\frac{\partial V_i(x_i)}{\partial x_i} \quad \text{and} \quad F_{ij}(x_i - x_j) = -\frac{\partial V_{ij}(x_i - x_j)}{\partial x_i} \tag{14.3}$$

and the restriction $i > j$ on the double sum is to avoid double counting. The momentum operator corresponding to each coordinate is $\hat{p}_i = (\hbar/i)\partial/\partial x_i$ and one has $[x_j, \hat{p}_k] = i\hbar\delta_{jk}$.

This acts on a multiparticle wavefunction $\psi(x_1, x_2, \ldots, x_n; t)$ as the time-development operator and generalizes the Schrödinger equation to

$$\hat{H}\psi(x_1, x_2, \ldots, x_N; t) = i\hbar\frac{\partial}{\partial t}\psi(x_1, x_2, \ldots, x_n; t) \tag{14.4}$$

If none of the potentials actually depend on time, the usual exponential time-dependence is found so that

$$\psi(x_1, x_2, \ldots, x_N; t) = \psi_E(x_1, x_2, \ldots, x_n)\, e^{-iEt/\hbar} \tag{14.5}$$

where ψ_E satisfies the *time-independent multiparticle Schrödinger equation*

$$\hat{H}\psi_E(x_1, x_2, \ldots, x_N) = E\psi_E(x_1, x_2, \ldots, x_N) \tag{14.6}$$

The multiparticle wavefunction is then associated with a probability amplitude, so that

$$P(x_1, x_2, \ldots, x_N; t) = |\psi(x_1, x_2, \ldots, x_N; t)|^2 \tag{14.7}$$

is a *multivariable probability density*. A more concrete definition is:

- The quantity $|\psi(x_1, x_2, \ldots, x_n; t)|^2\, dx_1\, dx_2, \ldots, dx_n$ is the probability that a measurement of the positions of the N particles, at time t, would find

particle 1 in the interval $(x_1, x_1 + dx_1)$

and

particle 2 in the interval $(x_2, x_2 + dx_2)$

and

$$\vdots \tag{14.8}$$

and

particle N in the interval $(x_N, x_N + dx_N)$

Just as with any multivariable probability distribution, this implies that the wavefunction must be normalized so that

$$\int_{-\infty}^{+\infty} dx_1 \int_{-\infty}^{+\infty} dx_2 \cdots \int_{-\infty}^{+\infty} dx_n\, |\psi(x_1, x_2, \ldots, x_n; t)|^2 = 1 \tag{14.9}$$

and the total probability of measuring "something" is unity and not, for example, N since we have more than one particle; we emphasize that $|\psi|^2$ does not "count the number of particles," but rather specifies the probability of the entire system of particles being in a particular configuration.

All of the usual conditions on the smoothness and convergence of one-dimensional wavefunctions can be easily generalized, as can the expressions for expectation values; for example, one has

$$\langle \hat{O} \rangle_t = \int_{-\infty}^{+\infty} dx_1 \int_{-\infty}^{+\infty} dx_2 \cdots \int_{-\infty}^{+\infty} dx_n$$

$$\times \psi^*(x_1, x_2, \dots, x_N; t)\, \hat{O}\, \psi(x_1, x_2, \dots, x_N; t) \tag{14.10}$$

for the average value of any operator $\hat{O}$.

Example 14.1. Correlations in two-particle wavefunctions

Consider a two-particle wavefunction

$$\psi(x_1, x_2) = Ne^{-\left(ax_1^2 + 2bx_1x_2 + cx_2^2\right)} \tag{14.11}$$

where we must have $a, c > 0$ in order for the wavefunction to be normalizable. The normalization constant is determined by the condition that

$$1 = \int_{-\infty}^{+\infty} dx_1 \int_{-\infty}^{+\infty} dx_2\, |\psi(x_1, x_2)|^2$$

$$= N^2 \int_{-\infty}^{+\infty} dx_2\, e^{-(c-b^2/a)x_2^2} \int_{-\infty}^{+\infty} dx_1\, e^{-a(x_1 + bx_2/a)^2} \tag{14.12}$$

$$= N^2 \frac{\pi}{\sqrt{ac - b^2}}$$

This form (obtained by completing the square in the exponent) is useful in that it shows that one must also have $ac - b^2 > 0$ in order for the state to be acceptable. It is then easy to show that $\langle x_1 \rangle = \langle x_2 \rangle = 0$, while the *covariance* (measuring a correlation between the two particles) is given by

$$\langle (x_1 - \langle x_1 \rangle)(x_2 - \langle x_2 \rangle) \rangle = \langle x_1 x_2 \rangle = -\frac{b}{(ac - b^2)} \tag{14.13}$$

This shows that the positions of the coordinates are correlated with each other, so that a measurement of one provides nontrivial information on the other. When $b \to 0$, we note that the wavefunction becomes the (uncorrelated) product of two Gaussians, and is simply a product wavefunction.

14.2 **Separable Systems**

A great simplification occurs if the mutual interactions of the particles can be ignored ($V_{ij} = 0$), because the Hamiltonian then takes the form

$$\hat{H} = \sum_i \frac{\hat{p}_i^2}{2m_i} + \sum_i V(x_i) = \sum_i \left(\frac{\hat{p}_i^2}{2m_i} + V(x_i) \right) \equiv \sum_i \hat{H}_i \qquad (14.14)$$

With no mutual interactions present to give rise to dynamical correlations, it is natural to assume a product solution of the form

$$\psi_E(x_1, x_2, \dots, x_N) = \psi_1(x_1)\, \psi_2(x_2) \cdots \psi_N(x_N). \qquad (14.15)$$

The time-independent Schrödinger equation, Eqn. (14.6), can then be written as

$$E\left[\psi_1(x_1)\psi_2(x_2)\cdots\psi_N(x_N)\right] = \left[\hat{H}_1\psi_1(x_1)\right]\psi_2(x_2)\cdots\psi_N(x_N)$$

$$+ \psi_1(x_1)\left[\hat{H}_2\psi_2(x_2)\right]\cdots\psi_N(x_N) \quad (14.16)$$

$$+ \cdots + \psi_1(x_1)\psi_2(x_2)\cdots\left[\hat{H}_N\psi_N(x_N)\right]$$

Using the usual separation of variables trick, we divide both sides by Eqn. (14.15) and find that

$$\frac{[\hat{H}_1\psi_1(x_1)]}{\psi_1(x_1)} + \frac{[\hat{H}_2\psi_2(x_2)]}{\psi_2(x_2)} + \cdots + \frac{[\hat{H}_N\psi_N(x_N)]}{\psi_N(x_N)} = E. \qquad (14.17)$$

This is only consistent if

$$\hat{H}_i\psi_i(x_i) = E_i\psi_i(x_i) \quad \text{for } i = 1, 2, \dots, N \qquad (14.18)$$

where $E_1 + E_2 + \cdots + E_N = E$; we then have to solve N "versions" of the one-dimensional problem. Several comments can be made:

- If each component wavefunction is properly normalized, then the product solution is also, since

$$\int_{-\infty}^{+\infty} dx_1 \cdots \int_{-\infty}^{+\infty} dx_N |\psi(x_1, x_2, \dots, x_N)|^2 = \prod_i \left[\int_{-\infty}^{+\infty} dx_i\, |\psi_i(x_i)|^2 \right] = 1$$

$$(14.19)$$

- The overall time-dependence can also be factorized since

$$e^{-iEt/\hbar} = e^{-iE_1t/\hbar} \cdots e^{-iE_nt/\hbar} \qquad (14.20)$$

which implies that

$$\psi(x_1, x_2, \ldots, x_N; t) = \psi_1(x_1, t)\, \psi_2(x_2, t) \cdots \psi_N(x_N, t) \qquad (14.21)$$

This is potentially useful as wave packets for each particle can be constructed using superposition techniques, so that products of such wave packets will also be valid solutions for the noninteracting case.

Example 14.2. Degeneracy in two-particle systems

Consider two particles of the same mass m confined to the standard infinite well; for the moment, we neglect any mutual interactions.[1] The general solution to this two-particle system is

$$\psi_{(n_1, n_2)}(x_1, x_2) = u_{(n_1)}(x_1)\, u_{(n_2)}(x_2) \quad \text{where } u_n(x) = \sqrt{\frac{2}{a}} \sin\left(\frac{n\pi x}{a}\right) \qquad (14.22)$$

with the corresponding energy spectrum

$$E_{(n_1, n_2)} = \frac{\hbar^2 \pi^2}{2ma^2}\left(n_1^2 + n_2^2\right) \qquad (14.23)$$

The ground state energy is $\hbar^2\pi^2/ma^2$, corresponding to $(n_1, n_2) = (1, 1)$ and is unique. The first excited state, given by the two choices $(1, 2)$ and $(2, 1)$, is doubly degenerate, and the corresponding wavefunctions can be written as $\psi_\alpha = \psi_{(1,2)}(x_1, x_2)$ and $\psi_\beta = \psi_{(2,1)}(x_1, x_2)$; these two choices are not unique because we can invoke the linearity of the Schrödinger equation to show that any (appropriately orthogonal) linear combination of these two is also a solution with energy $E_{(1,2)} = 5\hbar^2\pi^2/2ma^2 = E_{(2,1)}$.

We can now use this example to illustrate the methods of degenerate perturbation theory, as outlined in Section 10.5.2. We add a small mutual interaction term given by $V'(x_1, x_2) = g\delta(x_1 - x_2)$ where positive (negative) g corresponds to a repulsive (attractive) interaction between the two particles. Referring to Eqn. (10.134), we require the various matrix elements of the perturbing interaction; for example,

$$H'_{\alpha\alpha} = \langle \psi_\alpha | V(x_1 - x_2) | \psi_\alpha \rangle$$

$$= \int_0^a dx_1 \int_0^a dx_2\, [u_1(x_1)u_2(x_2)]\, (g\delta(x_1 - x_2))\, [u_1(x_1)u_2(x_2)]$$

$$= g\left(\frac{2}{a}\right)^2 \int_0^a dx_1\, \sin^2\left(\frac{\pi x_1}{a}\right) \sin^2\left(\frac{2\pi x_1}{a}\right)$$

$$H'_{\alpha\alpha} = \frac{g}{a} \qquad (14.24)$$

---

[1] If they are both in the same one-dimensional well, this implies that they are somewhat "ghostlike" as they must be able to "pass through" each other.

(Continued)

with identical answers for $H'_{\beta\beta}$ and $H'_{\alpha\beta}$. The condition determining the (split) energy eigenvalues (Eqn. (10.134)) then reads

$$\det \begin{pmatrix} \mathcal{E} + g/a - E & g/a \\ g/a & \mathcal{E} + g/a - E \end{pmatrix} = 0 \tag{14.25}$$

where $\mathcal{E} = E_{(2,1)} = E_{(1,2)}$ is the initially degenerate energy level. The resulting polynomial equation is easily solved and yields

$$E^{(\pm)} = \mathcal{E} + \frac{g}{a} \pm \frac{g}{a} \tag{14.26}$$

The two energy eigenvalues and corresponding (normalized) eigenstates are given by

$$E^{(-)} = \mathcal{E} \qquad \psi^{(-)}(x_1, x_2) = \frac{1}{\sqrt{2}}\left(\psi_{(1,2)}(x_1, x_2) - \psi_{(2,1)}(x_1, x_2)\right) \tag{14.27}$$

$$E^{(+)} = \mathcal{E} + \frac{2g}{a} \qquad \psi^{(+)}(x_1, x_2) = \frac{1}{\sqrt{2}}\left(\psi_{(1,2)}(x_1, x_2) + \psi_{(2,1)}(x_1, x_2)\right) \tag{14.28}$$

The antisymmetric combination state, corresponding to $E^{(-)}$, is unshifted in energy because the $\psi^{(-)}$ wavefunction vanishes where the perturbation has any effect, namely, for $x_1 = x_2$. The symmetric solution has a larger probability of having $x_1 = x_2$ than do either ψ_α or ψ_β individually and it can "feel" the effect of the perturbation; the energy of the symmetric state is therefore increased or decreased depending on the sign of g.

14.3 **Two-Body Systems**

While much of classical mechanics is concerned with the motion of single particles under the influence of external forces, many standard problems, especially in gravitation, are concerned with the motion of two bodies subject only to their mutual interaction. While general methods of solution for the N-body problem[2] (with $N \geq 3$) do not exist, a simple change of variables is often enough to transform Newton's equations for two particles into an effective one-particle problem which can then be approached using a variety of familiar techniques.

Such techniques are perhaps even more important in quantum mechanics where many of the "textbook" examples are two-body systems; examples include diatomic molecules, the hydrogen atom, the deuteron (proton–neutron bound state), and quarkonia (quark–antiquark bound states). In these cases, we are often more interested in probing the (sometimes unknown) force between the particles, so that not having to deal with the complications of many particles

[2] See, for example, Symon (1971).

is extremely important. Even though it is implemented in a very different way, the same coordinate transformation "trick" works in both classical and quantum mechanics and we begin our study by reviewing the classical case.

14.3.1 Classical Systems

The classical equations of motion for a two-particle system with no external forces and only mutual two-body interactions are

$$m_1 \ddot{x}_1(t) = F_{21}(x_1 - x_2) \quad \text{and} \quad m_2 \ddot{x}_2(t) = F_{12}(x_1 - x_2) \tag{14.29}$$

and we recall that $F \equiv F_{21} = -F_{12}$ from Newton's third law. Two combinations of these equations then immediately suggest themselves and naturally select out a new set of variables. If, for example, we add the two equations in Eqn. (14.29), we obtain

$$0 = F_{21} + F_{12} = m_1 \ddot{x}_1(x) + m_2 \ddot{x}_2(t)$$

$$= (m_1 + m_2) \left(\frac{m_1 \ddot{x}_1(x) + m_2 \ddot{x}_2(t)}{m_1 + m_2} \right) \tag{14.30}$$

$$0 = M \ddot{X}(t)$$

where we define the *total mass*, $M = m_1 + m_2$, and the *center-of-mass coordinate*

$$X(t) = \frac{m_1 x_1(t) + m_2 x_2(t)}{m_1 + m_2} \tag{14.31}$$

We note that Eqn. (14.30) gives the standard result that if there are no net external forces, the center-of-mass of a system moves at constant speed. A related variable is the *total momentum*, given by

$$P(t) = M \dot{X}(t) = m_1 v_1(t) + m_2 v_2(t) = p_1(t) + p_2(t) \tag{14.32}$$

so that Eqn. (14.30) also shows that the total momentum is conserved.

If we now divide both sides of Eqns (14.29) by the respective masses, and then subtract, we find

$$F \left(\frac{1}{m_1} + \frac{1}{m_2} \right) = \frac{F_{21}}{m_1} - \frac{F_{12}}{m_2} = \ddot{x}_1(t) - \ddot{x}_2(t) \tag{14.33}$$

or

$$F(x) = \mu \, \ddot{x}(t) \tag{14.34}$$

where we have defined the *reduced mass* via

$$\frac{1}{\mu} \equiv \frac{1}{m_1} + \frac{1}{m_2} \quad \text{or} \quad \mu = \frac{m_1 m_2}{m_1 + m_2} \tag{14.35}$$

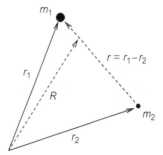

Figure 14.1. Center-of-mass (R) and relative ($r = r_1 - r_2$) coordinates for a two-body system in two-dimensions, which generalizes Eqn. (14.31). You should be able to estimate the ratio m_1/m_2 from the figure.

and the *relative coordinate* via

$$x(t) = x_1(t) - x_2(t) \tag{14.36}$$

while $F(x) \equiv F_{21}(x_1 - x_2)$. The nontrivial dynamics of the system is then described by Eqn. (14.34); the "interesting" physics is all contained in the relative coordinate which describes the motion of a fictitious particle of effective mass μ. The change of variables can also be inverted to give

$$x_1(t) = X(t) + \frac{m_2}{M}x(t) \quad \text{and} \quad x_2(t) = X(t) - \frac{m_1}{M}x(t) \tag{14.37}$$

so that the motion of each particle can be extracted if so desired. The same variable change works in more realistic two- and three-dimensional systems and we visualize the new coordinates in two-dimensions in Fig. 14.1.

14.3.2 Quantum Case

The quantum version of the two-body problem requires us to solve the two-particle Schrödinger equation given by

$$\hat{H}\psi(x_1, x_2) = E\psi(x_1, x_2) \tag{14.38}$$

where the Hamiltonian is given by

$$\hat{H} = \frac{\hat{p}_1^2}{2m_1} + \frac{\hat{p}_2^2}{2m_2} + V(x_1 - x_2)$$

$$= -\frac{\hbar^2}{2m_1}\frac{\partial^2}{\partial x_1^2} - \frac{\hbar^2}{2m_2}\frac{\partial^2}{\partial x_2^2} + V(x_1 - x_2) \tag{14.39}$$

if there are no external forces. The change to center-of-mass and relative coordinates,

$$X = \frac{m_1 x_1 + m_2 x_2}{m_1 + m_2} \quad \text{and} \quad x = x_1 - x_2 \tag{14.40}$$

(where we drop the t dependence, as these are no longer classical coordinates, but quantum mechanical labels) is trivially implemented for the potential energy term where $V(x_1 - x_2) = V(x)$. For the kinetic energy operators, we need to rewrite

$$\hat{p}_i = \frac{\hbar}{i} \frac{\partial}{\partial x_i} \quad \text{for } i = 1, 2 \tag{14.41}$$

in terms of the spatial derivatives of the X, x coordinates which give the momentum operators corresponding to those variables, namely

$$\hat{P} \equiv \frac{\hbar}{i} \frac{\partial}{\partial X} \quad \text{and} \quad \hat{p} \equiv \frac{\hbar}{i} \frac{\partial}{\partial x}. \tag{14.42}$$

This requires the chain rule relation

$$\frac{\partial}{\partial x_{1,2}} = \frac{\partial X}{\partial x_{1,2}} \frac{\partial}{\partial X} + \frac{\partial x}{\partial x_{1,2}} \frac{\partial}{\partial x} \tag{14.43}$$

which, using Eqn. (14.40), gives

$$\hat{p}_1 = \frac{m_1}{M} \hat{P} + \hat{p} \quad \text{and} \quad \hat{p}_2 = \frac{m_2}{M} \hat{P} - \hat{p} \tag{14.44}$$

We note that adding these two equations gives $\hat{P} = \hat{p}_1 + \hat{p}_2$, now as an operator relation. The kinetic energy operators in the Hamiltonian can now be written as

$$\begin{aligned}
\frac{\hat{p}_1^2}{2m_1} + \frac{\hat{p}_2^2}{2m_2} &= \frac{1}{2m_1} \left(\frac{m_1}{M} \hat{P} + \hat{p} \right)^2 + \frac{1}{2m_2} \left(\frac{m_2}{M} \hat{P} - \hat{p} \right)^2 \\
&= \frac{1}{2m_1} \left(\frac{m_1^2}{M^2} \hat{P}^2 + \frac{m_1}{M} \left(\hat{p}\hat{P} + \hat{P}\hat{p} \right) + \hat{p}^2 \right) \\
&\quad + \frac{1}{2m_2} \left(\frac{m_2^2}{M^2} \hat{P}^2 - \frac{m_2}{M} \left(\hat{p}\hat{P} + \hat{P}\hat{p} \right) + \hat{p}^2 \right)
\end{aligned} \tag{14.45}$$

$$\frac{\hat{p}_1^2}{2m_1} + \frac{\hat{p}_2^2}{2m_2} = \frac{\hat{P}^2}{2M} + \frac{\hat{p}^2}{2\mu}$$

While we have been careful with the ordering of $\hat{P}$ and $\hat{p}$, it is easy to see that $[\hat{P}, \hat{p}] = 0$. This can also be used to confirm that the total momentum of the system is a constant since $\hat{P}$ commutes with the Hamiltonian operator, that is,

$$[\hat{H}, \hat{P}] = \frac{1}{2M} [\hat{P}^2, \hat{P}] + \frac{1}{2\mu} [\hat{p}^2, \hat{P}] + [V(x), \hat{P}] = 0 \tag{14.46}$$

We emphasize that this is true only under our assumption that the potential is only a function of the relative coordinate; if there are external forces, an equivalent classical result is obtained (P14.3).

Most importantly, in the new coordinates, the Hamiltonian is now separable since

$$\hat{H} = \frac{\hat{P}^2}{2M} + \left(\frac{\hat{p}^2}{2\mu} + V(x) \right) = \hat{H}_X + \hat{H}_x \tag{14.47}$$

so a product wavefunction of the form

$$\left[\Psi(X)e^{-iE_X t/\hbar}\right]\left[\psi(x)e^{-iE_x t/\hbar}\right] \qquad (14.48)$$

satisfying

$$\hat{H}_X \Psi(X) = E_X \Psi(X) \quad \text{and} \quad \hat{H}_x \psi(x) = E_x \psi(x) \qquad (14.49)$$

is a solution. The center-of-mass equation has trivial plane wave solutions of the form

$$\Psi_P(X) = e^{i(PX - P^2 t/2M)/\hbar} \qquad (14.50)$$

(with P a number) from which wave packets, representing the constant velocity motion of the center-of-mass, can be constructed.

Example 14.3. Reduced mass effects in two-particle systems

A simple model of a one-dimensional diatomic molecule consists of two masses m_1, m_2 interacting via the potential $V(x_1 - x_2) = K(x_1 - x_2 - l)^2/2$ where l is the equilibrium separation of the two masses. The equation for the relative coordinate,

$$\left(\frac{\hat{p}^2}{2\mu} + V(x - l)\right)\psi(x) = E_x \psi(x) \qquad (14.51)$$

has the standard harmonic oscillator solutions $\psi(x) = \phi_n(x - l)$ with quantized energy levels given by $E_x^{(n)} = (n+1/2)\hbar\omega$ where $\omega = \sqrt{K/\mu}$. The dependence on μ of the zero-point energy of vibrational states in diatomic molecules was mentioned in Section 9.3.

In systems where $m_1 \approx m_2$, the reduced mass is roughly $\mu \approx m_1/2 \approx m_2/2$ and its effect is obviously important to include. In cases such as the hydrogen atom, where one has $m_1 = m_e << m_p = m_2$ and $\mu = m_e/(1 + m_e/m_p) \approx m_e$, the effect is much smaller (since $m_e/m_p \approx 1/2000$) and is sometimes not stressed sufficiently. The discrete energy spectrum of a hydrogen-like atom (a single electron interacting via a Coulomb force with a nucleus of charge Z) is given by Eqn. (1.42) as

$$E_n = -\frac{1}{2}\mu c^2 Z^2 \alpha^2 \frac{1}{n^2} \qquad (14.52)$$

where the reduced mass μ now properly appears. The frequencies of the photons emitted in a transition are

$$\hbar\omega_{nl} = \hbar 2\pi f_{nl} = E_n - E_l = \frac{1}{2}\mu c^2 Z^2 \alpha^2 \left(\frac{1}{n^2} - \frac{1}{l^2}\right) \qquad (14.53)$$

The dependence on μ implies that the corresponding lines in atoms with the same value of Z, but with different nuclear masses (i.e. isotopes) will have slightly differing wavelengths. This effect was utilized in the discovery of the "*hydrogen isotope of mass 2*," now known as deuterium. (See P14.6 for details.)

14.4 Spin Wavefunctions

In describing the quantum state of a particle, we have concentrated on the wavefunctions corresponding to observable quantities such as position ($\psi(x)$), momentum ($\phi(p)$), or energy eigenvalues ($\{a_n; n = 0, 1 \ldots\}$). The "spin-up" and "spin-down" label necessary to describe spin-1/2 particles must also be included in the multiparticle wavefunctions for such particles. A more comprehensive discussion of spin in quantum mechanics is given in the next chapter, but we introduce here, for convenience, some of the basic formalism for spin-1/2 particles.

A convenient (matrix) representation of the spin operator (quantized along some convenient direction, often the z-axis) is given by

$$S_z = \frac{\hbar}{2} \begin{pmatrix} 1 & 0 \\ 0 & -1 \end{pmatrix} \tag{14.54}$$

which acts on a (complex) *spinor wavefunction*

$$\chi = \begin{pmatrix} \alpha \\ \beta \end{pmatrix} \tag{14.55}$$

In order to be normalized, such spinors must satisfy

$$(\chi)^{\dagger}\chi = (\alpha^*, \beta^*) \begin{pmatrix} \alpha \\ \beta \end{pmatrix} = |\alpha|^2 + |\beta|^2 = 1 \tag{14.56}$$

The eigenvectors and eigenvalues of S_z are seen to be

$$\chi^+ = \begin{pmatrix} 1 \\ 0 \end{pmatrix} \quad \text{with} \quad S_z\chi^+ = +\frac{\hbar}{2}\chi^+ \tag{14.57}$$

$$\chi^- = \begin{pmatrix} 0 \\ 1 \end{pmatrix} \quad \text{with} \quad S_z\chi^- = -\frac{\hbar}{2}\chi^- \tag{14.58}$$

Since S_z is a Hermitian (matrix) operator (note that $S_z^{\dagger} = S_z$), it is not surprising that its eigenvalues, $\pm\hbar/2$, are real. For the same reason, the eigenfunctions (in this case eigenvectors) also satisfy the usual orthonormality conditions since

$$(\chi^{(+)})^\dagger \chi^{(+)} = 1 = (\chi^{(-)})^\dagger \chi^{(-)} \quad \text{and} \quad (\chi^{(+)})^\dagger \chi^{(-)} = 0 = (\chi^{(-)})^\dagger \chi^{(+)}$$
(14.59)

and there is a corresponding expansion theorem, written as

$$\chi = \begin{pmatrix} a^{(+)} \\ a^{(-)} \end{pmatrix} = a^{(+)} \begin{pmatrix} 1 \\ 0 \end{pmatrix} + a^{(-)} \begin{pmatrix} 0 \\ 1 \end{pmatrix} = a^{(+)} \chi^{(+)} + a^{(-)} \chi^{(-)}$$
(14.60)

This form makes it clear that $|a^{(+)}|^2 (|a^{(+)}|^2)$ is the probability that a measurement of the spin (projected onto the z-axis) will yield a value of $+\hbar/2$ ($-\hbar/2$); it is also consistent with the expectation value

$$\langle \chi | S_z | \chi \rangle = (a^{(+)*}, a^{(-)*}) \frac{\hbar}{2} \begin{pmatrix} 1 & 0 \\ 0 & -1 \end{pmatrix} \begin{pmatrix} a^{(+)} \\ a^{(-)} \end{pmatrix} = \frac{\hbar}{2} \left(|a^{(+)}|^2 - |a^{(+)}|^2 \right)$$
(14.61)

A spin-1/2 particle can then carry information on its spin state in its quantum wavefunction, $\psi(x, \chi)$, and inner products between different quantum states must be generalized. For example, the overlap "integral" of $\psi_a(x) \chi_a$ and $\psi_b(x) \chi_b$ will be

$$\langle \psi_a | \psi_b \rangle = \left[\int_{-\infty}^{+\infty} dx \, \psi_a^*(x) \, \psi_b(x) \right] \left[(\chi_a)^\dagger \chi_b \right]$$
(14.62)

and wavefunctions can be orthogonal because of different spin dependences. An expansion in energy and spin eigenstates might then have the form

$$\psi(x, \chi) = \sum_n \left(a_n^{(+)} u_n(x) \chi^{(+)} + a^{(-)} u_n(x) \chi^{(-)} \right)$$
(14.63)

Example 14.4. Expansion in energy and spin eigenstates

Consider a spin-1/2 particle in a harmonic oscillator potential described by the wavefunction

$$\psi(x, \chi) = N \left(3\psi_0(x) \chi^{(+)} - (2 + i)\psi_1(x) \chi^{(-)} + \sqrt{6}\psi_1(x) \chi^{(+)} \right)$$
(14.64)

The normalization constant can be determined by the requirement that

$$\sum_n (|a_n^{(+)}|^2 + |a_n(-)|^2) = 1 \quad \text{so that } N = 1/\sqrt{20}$$
(14.65)

The average value of the energy is

$$\langle \hat{E} \rangle = \left(\frac{9}{20} \right) \left(\frac{1}{2}\hbar\omega \right) + \left(\frac{5 + 6}{20} \right) \left(\frac{3}{2}\hbar\omega \right) = \frac{21}{20}\hbar\omega$$
(14.66)

(Continued)
while the expectation value of S_z is

$$\langle S_z \rangle = \frac{\hbar}{2}\left(\frac{9+6}{20}\right) - \frac{\hbar}{2}\left(\frac{5}{20}\right) = \frac{\hbar}{4} \tag{14.67}$$

The combined probability that a measurement will find $S_z = +\hbar/2$ and $E = 3\hbar\omega/2$ is $P = 3/10$.

For multiparticle wavefunctions, we have to specify the position and spin label for each particle,[3] for example, $\psi(x_1, \chi_1; x_2 \chi_2; \ldots; x_n, \chi_n)$. We will often denote all of the relevant labels for a given particle by simply specifying the common numerical index, that is $\psi(1; 2; \ldots; N)$.

For example, for two (noninteracting) electrons in an infinite well, the following wavefunctions will all turn out be physically acceptable:

$$\psi_A(1; 2) = \frac{1}{\sqrt{2}}\left(\chi_1^{(-)}\chi_2^{(+)} - \chi_1^{(-)}\chi_2^{(+)}\right) u_1(x_1) u_1(x_2) \tag{14.68}$$

$$\psi_B(1; 2) = \chi_1^{(+)}\chi_2^{(+)} \frac{1}{\sqrt{2}}\left(u_1(x_1) u_2(x_2) - u_2(x_1) u_1(x_2)\right) \tag{14.69}$$

$$\psi_C(1; 2) = \frac{1}{\sqrt{2}}\left(u_1(x_1) u_2(x_2)\chi_1^{(+)}\chi_2^{(-)} - u_2(x_1) u_1(x_2)\chi_1^{(-)}\chi_2^{(+)}\right) \tag{14.70}$$

The inner product for spin-states for different particles is generalized to be

$$\langle \chi_1, \chi_2 | \chi_1, \chi_2 \rangle = \langle \chi_1 | \chi_1 \rangle \langle \chi_2 | \chi_2 \rangle = (|\alpha_1|^2 + |\beta_1|^2)(|\alpha_2|^2 + |\beta_2|^2) = 1 \tag{14.71}$$

You should now be able to show that the wavefunctions in Eqns (14.68)–(14.70) are properly normalized and mutually orthogonal and be able to calculate the energy of each state.

14.5 **Indistinguishable Particles**

Thus far, we have focused for the most part on exploring the consequences of a wave description of particles, its implications for observable phenomena, and the connection between the classical and quantum limit; we have thus concentrated on what we have termed "$\hbar$ physics." But there is another dichotomy between the extreme quantum and classical limits of particles which was introduced

---

[3] We do not consider other possible degrees of freedom which might be labeled, such as "isospin" for nucleons and "color" for quarks.

in Chapter 7 which we can describe as "indistinguishability physics"; hereafter, indistinguishable will be abbreviated IND for convenience.

Unlike a set of billiard balls which have different colors and even distinct numeric labels, each of the electrons in an atom is seemingly equivalent to every other electron. In the same way, all protons are effectively identical, all neutrons are equivalent, and so forth; no experiment has yet been able to discern any measureable differences between individual electrons, individual protons, and so on. The same statement holds for other particles other than just the "building blocks of nature," such as photons and all other "elementary" particles.

We can roughly define:

- A set of indistinguishable (IND) particles is one in which the interchange of any two particles has no observable effect on any property of the system.

The notion of indistinguishability raises an interesting question when one considers the total number of wavefunctions which can have the same total energy, that is, the degeneracy. For simplicity, say we have a product wavefunction of N IND particles of the form

$$\psi(1; 2; \ldots; N) = \phi_a(1)\phi_a(2) \cdots \phi_a(N) \qquad (14.72)$$

where each particle is in the same quantum state; the total energy is simply $E_{\text{tot}} = NE_a$. The assumption of indistinguishability means any permutation of the indices will give a state with the same energy. Because of the special form of Eqn. (14.72), however, the resulting exchanges do not change the wavefunction and there is only one distinct wavefunction with this energy.

Contrast to this is the situation where all N particles are in totally different one-particle configurations,

$$\psi(1; 2; \ldots; N) = \phi_{a_1}(1)\phi_{a_2}(2) \cdots \phi_{a_N}(N) \quad \text{with } a_1 \neq a_2 \neq \cdots \neq a_N \qquad (14.73)$$

with energy $E_{\text{tot}} = E_{a_1} + E_{a_2} + \cdots + E_{a_N}$. One can use N different labels for the first state ϕ_{a_1}, leaving $N - 1$ for the second state ϕ_{a_2}, and so forth; there are thus $N!$ different permutations of the labels, each of which gives a state of the same total energy and, in this case, all $N!$ wavefunctions are distinct. For example, with three particles in the $n = 0, 1, 2$ levels of the harmonic oscillator, we might have (ignoring spin labels) the $3! = 6$ different states

$$\phi_0(x_1)\phi_1(x_2)\phi_2(x_3), \quad \phi_0(x_1)\phi_1(x_3)\phi_2(x_2), \quad \phi_0(x_2)\phi_1(x_1)\phi_2(x_3)$$
$$\phi_0(x_2)\phi_1(x_3)\phi_2(x_1), \quad \phi_0(x_3)\phi_1(x_1)\phi_2(x_2), \quad \phi_0(x_3)\phi_1(x_2)\phi_2(x_1) \qquad (14.74)$$

For *distinguishable* particles, each of these choices corresponds to a different physical system since one can, by definition, tell the particles apart; the probability density for a hydrogen atom with an electron "here" and a proton "there" is obviously different from the exchanged system. For IND particles, however, we have the possibility of $N!$ wavefunctions which supposedly all describe the same (presumably unique) physical system and the obvious question is:

• Which one (if any) of these $N!$ choices is the appropriate wavefunction?

To help answer this question, we first formalize the notion of interchange by defining the *exchange operator*, $\hat{\mathcal{E}}_{ij}$, which has the effect of exchanging particles i and j with *all* their appropriate labels. Recalling the notation

$$\psi(x_1, \chi_1; x_2, \chi_2; \ldots; x_N, \chi_N) \equiv \psi(1; 2; \ldots; N) \tag{14.75}$$

we define the exchange operator such that

$$\hat{\mathcal{E}}_{ij}\psi(1; 2; \ldots; i; \ldots; j; \ldots; N) = \psi(1; 2 \ldots; j; \ldots; i; \ldots; N) \tag{14.76}$$

so that $x_i \leftrightarrow x_j$ *and* $\chi_i \leftrightarrow \chi_j$.

If, for example, one has two IND spin-1/2 particles in the infinite well with wavefunction

$$\psi(1; 2) = u_4(x_1)u_7(x_2)\chi_1^{(+)}\chi_2^{(-)} \tag{14.77}$$

we will have

$$\psi(2; 1) = \hat{\mathcal{E}}_{12}\psi(1; 2) = u_4(x_2)u_7(x_1)\chi_1^{(-)}\chi_2^{(+)} \tag{14.78}$$

The two-particle wavefunctions in Eqns (14.68)–(14.70) are easily seen to be antisymmetric under the action of $\hat{\mathcal{E}}_{12}$.

There are $N(N-1)$ distinct $\hat{\mathcal{E}}_{ij}$ which, by themselves, exchange labels pairwise; the complete set of all the $N!$ permutations of the particle indices is generated by taking products of the individual $\hat{\mathcal{E}}_{ij}$. For example, when $N = 3$, we have

$$\psi(3; 2; 1) = \hat{\mathcal{E}}_{13}\psi(1; 2; 3) \tag{14.79}$$

while

$$\psi(2; 3; 1) = \hat{\mathcal{E}}_{12}\psi(1; 3; 2) = \hat{\mathcal{E}}_{12}\hat{\mathcal{E}}_{23}\psi(1; 2; 3) \tag{14.80}$$

The set of the $\hat{\mathcal{E}}_{ij}$ and their products forms a *permutation group* (P14.7).

We can establish several important properties of the exchange operators using two-particle systems as an example, for ease of notation:

- The exchange operator is Hermitian. We show this explicitly for the position degree of freedom by noting that

$$
\langle \hat{\mathcal{E}}_{12} \rangle^* = \left[\int_{-\infty}^{+\infty} dx_1 \int_{-\infty}^{+\infty} dx_2 \, \psi^*(x_1, x_2) \, \hat{\mathcal{E}}_{ij} \, \psi(x_1, x_2) \right]^*
$$

$$
= \int_{-\infty}^{+\infty} dx_1 \int_{-\infty}^{+\infty} dx_2 \, \psi^*(x_2, x_1) \, \psi(x_1, x_2)
$$

$$
= \int_{-\infty}^{+\infty} dy_2 \int_{-\infty}^{+\infty} dy_1 \, \psi^*(y_1, y_2) \, \psi(y_2, y_1)
$$

$$
= \int_{-\infty}^{+\infty} dy_2 \int_{-\infty}^{+\infty} dy_1 \, \psi^*(y_1, y_2) \, \hat{\mathcal{E}}_{12} \, \psi(y_1, y_2)
$$

$$
\langle \hat{\mathcal{E}}_{12} \rangle^* = \langle \hat{\mathcal{E}}_{12} \rangle \tag{14.81}
$$

where a simple relabeling of the dummy integration variables is used; the similar proof for spin wavefunctions is discussed in P14.8.

- The operator $\hat{\mathcal{E}}_{ij}$ certainly commutes with the many-body Hamiltonian because

$$
[\hat{H}, \hat{\mathcal{E}}_{12}] \psi(1; 2) = \hat{H} \hat{\mathcal{E}}_{12} \psi(1; 2) - \hat{\mathcal{E}}_{12} \hat{H} \psi(1; 2)
$$

$$
= \hat{H} \psi(2; 1) - \hat{\mathcal{E}}_{ij} E_{(12)} \psi(1; 2)
$$

$$
= (E_{(21)} - E_{(12)}) \psi(1; 2) = 0 \tag{14.82}
$$

since the energy is an observable which should be unchanged by interchange. We then know that the states of the system are simultaneous eigenfunctions of both the energy and *all* the exchange operators.

- The square of the exchange operator is just the identity since

$$
\left(\hat{\mathcal{E}}_{12} \right)^2 \psi(1; 2) = \hat{\mathcal{E}}_{12} \hat{\mathcal{E}}_{12} \psi(1; 2) = \hat{\mathcal{E}}_{12} \psi(2; 1) = \psi(1; 2) \tag{14.83}
$$

Just as with the parity operator (Section 6.6), this fact implies that the eigenvalues of the exchange operator are ± 1, corresponding to states which are symmetric $(+1)$ and antisymmetric (-1) under interchange of any two particles, that is

$$
\text{symmetric: } \psi(2; 1) = \hat{\mathcal{E}}_{12} \psi(1; 2) = +\psi(1; 2) \tag{14.84}
$$

or

$$
\text{antisymmetric: } \psi(2; 1) = \hat{\mathcal{E}}_{12} \psi(1; 2) = -\psi(1; 2). \tag{14.85}
$$

This is certainly consistent with the fact that the probability distribution of two exchanged wavefunctions should be the same under exchange, namely that

$$P(2;1) = |\psi(2;1)|^2 = |\pm\psi(1;2)|^2 = P(1;2) \tag{14.86}$$

- These last two points, taken together, imply a very powerful constraint on the total wavefunction (by which we mean both position and spin degrees of freedom), namely:

  The wavefunction of N IND particles must be either totally symmetric (S) or totally antisymmetric (A) under the exchange of any two of the IND particles, that is either

$$\text{totally symmetric: } \hat{\mathcal{E}}_{ij}\psi_S(1;2;\dots;N) = +\psi_S(1;2..;N) \tag{14.87}$$

or

$$\text{totally antisymmetric: } \hat{\mathcal{E}}_{ij}\psi_A(1;2;\dots;N) = -\psi_A(1;2..;N) \tag{14.88}$$

for all possible pairs (i,j).

This important result is the key ingredient in determining the correct form of the quantum wavefunction for a system of IND particles. Using the states in Eqn. (14.74) as an example, we see that none of them satisfy either (14.87) or (14.88) by themselves. One can see, however, that the linear combinations

$$\begin{aligned}
\psi_S(1;2;3) = C_S\,[&\phi_0(x_1)\phi_1(x_2)\phi_2(x_3) + \phi_0(x_1)\phi_1(x_3)\phi_2(x_2) \\
+ &\phi_0(x_2)\phi_1(x_1)\phi_2(x_3) + \phi_0(x_2)\phi_1(x_3)\phi_2(x_1) \qquad (14.89) \\
+ &\phi_0(x_3)\phi_1(x_1)\phi_2(x_2) + \phi_0(x_3)\phi_1(x_2)\phi_2(x_1)]
\end{aligned}$$

and

$$\begin{aligned}
\psi_A(1;2;3) = C_A\,[&\phi_0(x_1)\phi_1(x_2)\phi_2(x_3) - \phi_0(x_1)\phi_1(x_3)\phi_2(x_2) \\
- &\phi_0(x_2)\phi_1(x_1)\phi_2(x_3) + \phi_0(x_2)\phi_1(x_3)\phi_2(x_1) \qquad (14.90) \\
+ &\phi_0(x_3)\phi_1(x_1)\phi_2(x_2) - \phi_0(x_3)\phi_1(x_2)\phi_2(x_1)]
\end{aligned}$$

are respectively symmetric and antisymmetric under the interchange of any two labels; the constants C_S, C_A are determined, of course, by the overall normalization.

A general prescription for the construction of such properly symmetrized or antisymmetrized linear combinations is easy to generate. For the symmetric combination, one can take

$$\text{completely symmetric: } \psi_S(1;2;\dots;N) = C_S\sum_P \psi(1;2;\dots;N) \tag{14.91}$$

where $\sum_P$ denotes the sum over all possible permutations of the N indices; this form certainly reproduces Eqn. (14.89). While there can be as many as $N!$ terms in this sum, if the IND particles are not all in different quantum states, the number of terms can be far less. For example, if the N particles are all in the same state given by Eqn. (14.72), the permutations in (14.91) are all identical and we find

$$\psi_S(1; 2 \ldots; N) = C_S \, N! \, \psi(1; 2; \ldots; N). \tag{14.92}$$

In this language, the antisymmetric combination can be written schematically as

completely antisymmetric: $\psi_A(1; 2; \ldots; N) = C_A \sum_P (-1)^P \psi(1; 2; \ldots; N).$

$$\tag{14.93}$$

- Here n_P is the *number of two-particle permutations or exchanges* which are required to achieve the overall permutation denoted by P starting from the canonical ordering $(1; 2 \ldots; N)$; this factor gives the alternating signs required by the antisymmetry.

As an example of this last case, we note that

$$\psi(2; 1; 3) = \hat{\mathcal{E}}_{12} \psi(1; 2; 3) \implies n_P = 1, \ (-1)^{n_P} = -1$$
$$\psi(3; 1; 2) = \hat{\mathcal{E}}_{13} \hat{\mathcal{E}}_{23} \psi(1; 2; 3) \implies n_P = 2, \ (-1)^{n_P} = +1$$

$$\tag{14.94}$$

as in Eqn. (14.90). This form also implies that:

- No two particles described by a totally antisymmetric wavefunction can be in the same quantum state,

which we can see as follows. Suppose particles i and j were in the same quantum state, namely, $i = j$; then since the overall wavefunction must be antisymmetric under $\hat{\mathcal{E}}_{ij}$, we have

$$\psi(1; 2; \ldots; j; \ldots; i; \ldots; N) = -\psi(1; 2; ..; i; \ldots; j; \ldots; N) \ \text{by antisymmetry}$$

$$\Downarrow$$

$$\psi(1; 2; \ldots; i; \ldots; i; \ldots; N) = -\psi(1; 2; ..; i; \ldots; i; \ldots; N) \ \text{since } i = j, \text{implying}$$
$$\psi(1; 2; \ldots; i; \ldots; i; \ldots; N) = 0 \tag{14.95}$$

The more careful way of stating this result is that:

- The wavefunction (and hence the probability density) for two particles in a completely antisymmetric state to occupy the same quantum "niche" vanishes.

For the case of noninteracting particles where the wavefunction can be written in product form, the antisymmetric combination can be written in an especially simple form using a determinant, namely

$$\psi_A(1; 2; \ldots; N) = \frac{1}{\sqrt{N!}} \det \begin{pmatrix} \phi_{a_1}(1) & \phi_{a_2}(1) & \cdots & \phi_{a_N}(1) \\ \phi_{a_1}(2) & \phi_{a_2}(2) & \cdots & \phi_{a_N}(2) \\ \vdots & \vdots & \ddots & \vdots \\ \phi_{a_1}(N) & \phi_{a_2}(N) & \cdots & \phi_{a_N}(N) \end{pmatrix} \quad (14.96)$$

The "recipe" for constructing this matrix is to put the N (necessarily different) single-particle wavefunctions in succeeding *columns* while the particle state labels $1, 2, \ldots, N$ are then inserted in different *rows*. The overall antisymmetry of the wavefunction is guaranteed by the linear algebra result that the exchange of any two rows (or columns) of a matrix introduces a factor of (-1) in the determinant. The overall normalization constant is correct provided each ϕ_{a_i} is properly normalized (P14.9). This form is called a *Slater determinant* and is useful even when the particles interact with each other as it can be used as a trial wavefunction for a variational calculation.

A similar shorthand notation for the *symmetric* state is

$$\psi_S(1; 2; \ldots; N) = C_S \det \begin{pmatrix} \phi_{a_1}(1) & \phi_{a_2}(1) & \cdots & \phi_{a_N}(1) \\ \phi_{a_1}(2) & \phi_{a_2}(2) & \cdots & \phi_{a_N}(2) \\ \vdots & \vdots & \ddots & \vdots \\ \phi_{a_1}(N) & \phi_{a_2}(N) & \cdots & \phi_{a_N}(N) \end{pmatrix}_+ \quad (14.97)$$

where the $+$ subscript indicates that the determinant should be taken with all positive signs. For symmetric states, more than one particle can be in a given state, not all of the resulting terms will necessarily be different, and the normalization must be determined case by case; an example is Eqn. (14.92) and P14.9.

The requirement that IND particles have totally symmetric or antisymmetric wavefunctions has therefore reduced the possible ambiguity in the number of quantum states describing the same physics from being as large as $N!$ possible choices to only two. The physical property of the particles in question that determines which choice is actually realized in nature is their intrinsic angular momentum or spin, specifically whether the particles have integral spin ($J = 0, 1, 2 \ldots$) or half-integral spin ($1/2, 3/2, 5/2, \ldots$); the former are called *bosons* while the latter are known as *fermions*. This distinction is the content of the *spin-statistics theorem* which states that:

- The total wavefunction (including both spin and position information) of a system of indistinguishable bosons (fermions) must be symmetric (antisymmetric) under the interchange of any two particles.

The exclusion principle, as stated in Section 7.1 that "*no two electrons may be in the same quantum state*" is seen to be an immediate consequence of this result from Eqn. (14.95). The same result must then hold for neutrons and protons in nuclear systems, quarks inside nucleons, and all other particles with half-integral spin. Besides yielding a "no-go" theorem for what is not allowed, the spin-statistics theorem provides a prescription for the construction of the appropriate wavefunction for a system of IND particles via Eqns (14.91) and (14.93) and, as such, is a much more powerful statement about how nature organizes itself.

Example 14.5. Two electrons with spin in a box

Consider two noninteracting electrons in an infinite well potential. The ground state of the system is achieved when both particles are in the lowest allowed energy state with $E_{tot} = 2E_1$. This is allowed provided their spins are different (as in Fig. 14.2(a)), in which case the Slater determinant wavefunction is

$$\psi(1;2) = \frac{1}{\sqrt{2!}} \det \begin{pmatrix} u_1(x_1)\chi_1^{(+)} & u_1(x_1)\chi_1^{(-)} \\ u_1(x_2)\chi_2^{(+)} & u_1(x_2)\chi_2^{(-)} \end{pmatrix}$$

$$= u_1(x_1)u_2(x_2)\frac{1}{\sqrt{2}}\left(\chi_1^{(+)}\chi_2^{(-)} - \chi_2^{(+)}\chi_1^{(-)}\right) \qquad (14.98)$$

The antisymmetric wavefunction with both particles in the ground state *and* with spins aligned vanishes, as in Eqn. (14.95).

For the first excited state of the system, one electron can be "elevated" to the next energy level so that $E_{tot} = E_1 + E_2$ and both spin configurations in Fig. 14.2(b) and (c) are possible; the corresponding wavefunctions are then

$$\psi(1;2) = \frac{1}{\sqrt{2!}} \det \begin{pmatrix} u_1(x_1)\chi_1^{(+)} & u_2(x_1)\chi_1^{(-)} \\ u_1(x_2)\chi_2^{(+)} & u_2(x_2)\chi_2^{(-)} \end{pmatrix} \qquad (14.99)$$

$$= \frac{1}{\sqrt{2}}\left(u_1(x_1)u_2(x_2)\chi_1^{(+)}\chi_2^{(-)} - u_2(x_1)u_1(x_2)\chi_1^{(-)}\chi_2^{(+)}\right)$$

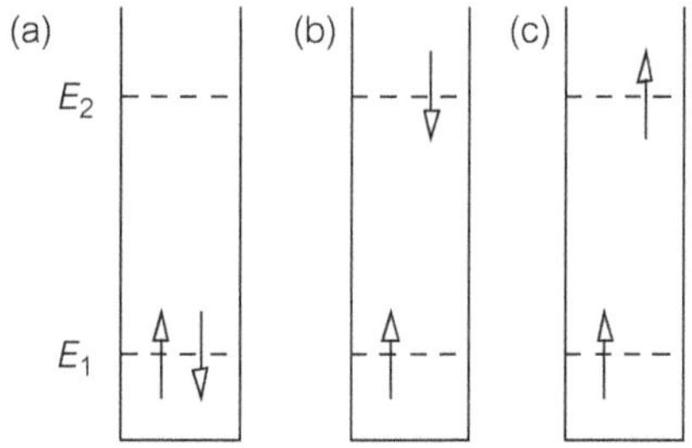

Figure 14.2. Allowed states of two indistinguishable spin-1/2 particles in an infinite well; Case (a) is the ground state, while (b) and (c) show two possible first excited states.

(Continued)

and

$$\psi(1;2) = \frac{1}{\sqrt{2!}} \det \begin{pmatrix} u_1(x_1)\chi_1^{(+)} & u_2(x_1)\chi_1^{(+)} \\ u_1(x_2)\chi_2^{(+)} & u_2(x_2)\chi_2^{(+)} \end{pmatrix} \tag{14.100}$$

$$= \chi_1^{(+)}\chi_1^{(+)} \frac{1}{\sqrt{2}} (u_1(x_1)u_2(x_2) - u_2(x_1)u_1(x_2))$$

with a similar state with both spins down also possible. These are just the wavefunctions of Eqns (14.68)–(14.70).

We know that the presence of mutual interactions between particles will induce dynamical correlations between them which are reflected in their quantum wavefunctions. What is more surprising is that IND particles exhibit such correlations even when they do not interact, simply due to the requirement of indistinguishability; these can be called an effective "Fermi repulsion" and "Bose attraction" which we illustrate in the next example.

Example 14.6. Correlations due to indistinguishability

Consider two particles of mass m in a harmonic oscillator potential; assume that we somehow know that there is one particle in the ground state (ψ_0) and one in the first excited state (ψ_1). We define the three wavefunctions

$$\psi_D(x_1, x_2) = \psi_0(x_1)\psi_1(x_2) \tag{14.101}$$

$$\psi_B(x_1, x_2) = \frac{1}{\sqrt{2}} (\psi_0(x_1)\psi_1(x_2) + \psi_1(x_1)\psi_0(x_2)) \tag{14.102}$$

$$\psi_F(x_1, x_2) = \frac{1}{\sqrt{2}} (\psi_0(x_1)\psi_1(x_2) - \psi_1(x_1)\psi_0(x_2)) \tag{14.103}$$

where

$$\psi_0(x) = \sqrt{\frac{1}{\rho\sqrt{\pi}}} e^{-x^2/2\rho^2} \quad \text{and} \quad \psi_1(x) = \sqrt{\frac{2}{\rho\sqrt{\pi}}} \left(\frac{x}{\rho}\right) e^{-x^2/2\rho^2} \tag{14.104}$$

are the appropriate SHO position-space wavefunctions, and the labels stand for distinguishable (D), boson (B), and fermion (F), respectively. We imagine for example that the bosons have no spin so that their wavefunction must be symmetric in the position coordinate, while the fermions have a symmetric spin wavefunction (which we do not exhibit) implying that their position-space wavefunction must be odd under exchange. There is, of course, another distinguishable wavefunction with $1 \leftrightarrow 2$.

(Continued)

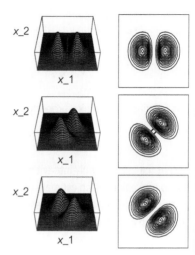

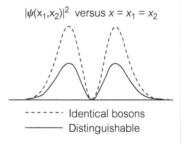

Figure 14.3. Three-dimensional plots (left) and contour plots (right) of $|\psi(x_1, x_2)|^2$ versus x_1, x_2 for the case of (top) distinguishable particles, (middle) indistinguishable bosons, and (bottom) indistinguishable fermions.

$|\psi(x_1,x_2)|^2$ versus $x = x_1 = x_2$

Figure 14.4. A "slice" through the contour plots of Fig. 14.3 along the $x_1 = x_2$ direction. The fermion probability distribution vanishes (Pauli principle), and the identical boson configuration is twice as likely as the indistinguishable particle state.

‑ ‑ ‑ ‑ ‑ ‑ Identical bosons
——— Distinguishable

To see the correlations contained in these wavefunctions, we calculate expectation values and note that $\langle x_1 \rangle = \langle x_2 \rangle = 0$ for all cases but:

$$D: \langle x_1^2 \rangle = \rho^2/2 \quad \langle x_2^2 \rangle = 3\rho^2/2 \quad \langle x_1 x_2 \rangle = 0 \qquad \langle (x_1 - x_2)^2 \rangle = 2\rho^2$$

$$B: \langle x_1^2 \rangle = \rho^2 \quad\;\; \langle x_2^2 \rangle = \rho^2 \quad\;\;\; \langle x_1 x_2 \rangle = +\rho^2/2 \quad \langle (x_1 - x_2)^2 \rangle = \rho^2 \quad (14.105)$$

$$F: \langle x_1^2 \rangle = \rho^2 \quad\;\; \langle x_2^2 \rangle = \rho^2 \quad\;\;\; \langle x_1 x_2 \rangle = -\rho^2/2 \quad \langle (x_1 - x_2)^2 \rangle = 3\rho^2$$

The two particles described by the fermion (boson) wavefunction are, on average, farther apart (closer together) than if they were distinguishable particles. This is illustrated in Fig. 14.3(a–c) for the three cases where we plot $|\psi(x_1, x_2)|^2$ versus x_1, x_2. A "slice" through these plots along the line $x_1 = x_2$ is shown in Fig. 14.4 where we see that the totally symmetric wavefunction is twice as probable to be found with the particles in the same state; the totally antisymmetric wavefunction, of course, vanishes identically in this case.

(Continued)

We can further explore the implications of these correlations by "turning on" a mutual interaction between the two particles of the form

$$V'(x_2 - x_2; d) \equiv \frac{\lambda}{d\sqrt{\pi}} \exp\left(-\frac{(x_2 - x_2)^2}{2d^2}\right)$$ (14.106)

This function has the nice property that

$$\lim_{d \to 0} [V(x_2 - x_2; d)] = \lambda \delta(x_2 - x_2)$$ (14.107)

so that the particles only interact when they are on "top of each other" in this limit. This form is also convenient as an estimate of the effect of this interaction can be made using first order perturbation theory and the necessary overlap integrals can all be done analytically (P14.10). The shift in energy due to this perturbation at this order can be written as

$$E_D^{(1)} = \Delta E_D = \frac{\lambda}{\sqrt{\pi}} \frac{\rho^2 + d^2}{(2\rho^2 + d^2)^{3/2}} = V_b \left[\frac{1 + z^2}{(2 + z^2)^{3/2}}\right]$$

$$E_B^{(1)} = \Delta E_B = \frac{\lambda}{\sqrt{\pi}} \frac{2\rho^2 + d^2}{(2\rho^2 + d^2)^{3/2}} = V_b \left[\frac{2 + z^2}{(2 + z^2)^{3/2}}\right]$$ (14.108)

$$E_F^{(2)} = \Delta E_F = \frac{\lambda}{\sqrt{\pi}} \frac{d^2}{(2\rho^2 + d^2)^{3/2}} = V_b \left[\frac{z^2}{(2 + z^2)^{3/2}}\right]$$

where $V_b \equiv \lambda/b\sqrt{\pi}$ and $z \equiv d/\rho$; we plot these results in Fig. 14.5. One sees that when $d << \rho$ ($z << 1$ or "range of mutual interaction" << "particle separation"), the perturbation "samples" the various wavefunctions in a region where the correlations are dramatic, and the resulting energy shifts are very different; when $d >> \rho$ ($z >> 1$) the effects of indistinguishability are less important.

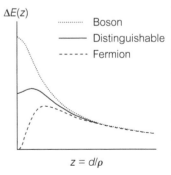

$\Delta E(z)$

.... Boson
——— Distinguishable
- - - - - Fermion

$z = d/\rho$

Figure 14.5. Energy shift, $\Delta E(z)$, versus the range of the mutual interaction, $z = d/\rho$. Results for the distinguishable (solid), indistinguishable boson (dotted), and indistinguishable fermion (dashed) states are shown.

14.6 **Questions and Problems**

Q14.1. What are the *dimensions* of an N-particle wavefunction in one dimension? in three dimensions?

Q14.2. Does the notion of probability flux generalize to multiparticle wavefunctions?

Q14.3. Referring to Fig. 14.1, show how to *estimate* the ratio m_1/m_2.

P14.1. **Multiparticle momentum-space wavefunctions.**
    (a) Show that the definition

$$\phi(p_1, p_2) = \frac{1}{2\pi\hbar} \int_{-\infty}^{+\infty} dx_1 \int_{-\infty}^{+\infty} dx_2 \, e^{-i(p_1 x_2 + p_2 x_2)/\hbar} \, \psi(x_1, x_2)$$

$$(14.109)$$

is appropriate for the momentum-space wavefunction corresponding to a two-particle position-space $\psi(x_1, x_2)$; specifically, show that $\phi(p_1, p_2)$ is normalized if $\psi(x_1, x_2)$ is, that the appropriate inverse relation holds, and anything else you think is important.

    (b) Evaluate the momentum-space wavefunction corresponding to $\psi(x_1, x_2)$ in Example 14.1 and show that it is proportional to

$$\phi(p_1, p_2) = \exp\left(-\frac{(cp_1^2 - 2bp_1 p_2 + ap_2^2)}{2\hbar^2(ac - b^2)}\right)$$

$$(14.110)$$

Normalize this wavefunction, show that this form has the right limit as $b \to 0$ and $ac - b^2 \to 0$, and interpret your results.

    (c) Find the covariance for the variables p_1, p_2 and show that it is opposite in sign to that for x_1, x_2 and interpret your result.

P14.2. Consider two *distinguishable* and noninteracting particles of mass m moving in the same harmonic oscillator potential.
    (a) What is the ground state energy E_0 and wavefunction $\psi_0(x_1, x_2)$ of the two-particle system? Is the ground state energy degenerate?

    (b) Show that the first excited state E_1 is doubly degenerate and write down the two wavefunctions, $\psi_1^{a,b}(x_1, x_2)$. Consider a δ-function interaction between the particles as in Example 14.2, $V'(x_1 - x_2) = g\delta(x_1 - x_2)$-as a small perturbation. Find the perturbed energies and eigenfunctions.

    (c) Show that the second excited state is triply degenerate and write down the possible wavefunctions. Show if a $g\delta(x_1 - x_2)$ mutual interaction is added that the resulting eigenstates are given by

$$\psi_2^a(x_1, x_2) = \frac{1}{\sqrt{2}} \left[\psi_0(x_1)\psi_2(x_2) - \psi_2(x_1)\psi_0(x_2) \right]$$

$$\psi^b(x_1, x_2) = \frac{1}{2} \left[\psi_0(x_1)\psi_2(x_2) - \sqrt{2}\psi_1(x_1)\psi_1(x_2) + \psi_2(x_1)\psi_0(x_1) \right]$$

$$\psi^c(x_1, x_2) = \frac{1}{2} \left[\psi_0(x_1)\psi_2(x_2) + \sqrt{2}\psi_1(x_1)\psi_1(x_2) + \psi_2(x_1)\psi_0(x_2) \right]$$

where the $\psi_n(x)$ are the one-particle SHO eigenfunctions. Show that these states are mutually orthogonal. If the mutual interaction is repulsive ($g > 0$), which state has the highest energy? the lowest energy? What are the energy eigenvalues? Hint: For this three-state system you have to diagonalize a 3×3 matrix. You presumably know one linear combination which would be unaffected by the perturbation and hence one eigenvector and eigenfunction. Extracting this one helps you solve for the other two.

P14.3. Consider the operator representing the total momentum of a multiparticle system, namely

$$\hat{P} = \hat{p}_1 + \hat{p}_2 + \cdots + \hat{p}_N \tag{14.111}$$

The time-dependence of the expectation value of this operator will be given (recall Section 12.5) by

$$\frac{d}{dt}\langle \hat{P} \rangle = \frac{i}{\hbar}\langle [\hat{H}, \hat{P}] \rangle \tag{14.112}$$

where $\hat{H}$ is now the multiparticle Hamiltonian in Eqn. (14.2). Show that this can be written as

$$\frac{d}{dt}\langle \hat{P} \rangle = \langle \mathcal{F}_1 + \mathcal{F}_2 + \cdots + \mathcal{F}_N \rangle = \langle \mathcal{F}_{\text{tot}} \rangle \tag{14.113}$$

where $\mathcal{F}_{\text{tot}}$ corresponds to the total external force; if this vanishes, the total momentum is conserved as in the classical case. Note: This result implies that the effects of the two-body mutual forces cancel and you should show that

$$\frac{\partial V_{ij}}{\partial x_i} + \frac{\partial V_{ij}}{\partial x_j} = 0 \tag{14.114}$$

P14.4. In changing to center-of-mass and relative coordinates for a two-body system, the wavefunction in terms of the original x_1, x_2 coordinates can be recovered by using

$$\Psi(X)\phi(x) = \Psi(X(x_1, x_2))\phi(x(x_1, x_2)) \longrightarrow \psi(x_1, x_2) \tag{14.115}$$

We have stressed that a probability interpretation requires not only the value of $|\psi|^2$, but also the measure, so it is more relevant to compare

$$|\Psi(X(x_1, x_2))\phi(x(x_1, x_2))|^2 \, dX \, dx \overset{?}{\longleftrightarrow} |\psi(x_1, x_2)|^2 \, dx_1 \, dx_2 \qquad (14.116)$$

Use the Jacobian of the transformation from x_1, x_2 to x, X to show that $dX \, dx = dx_1 \, dx_2$. Hint: The Jacobian of a coordinate transformation from (w, v) to (x, y) is given by

$$dw \, dv = \left| \det \begin{pmatrix} \partial w/\partial x & \partial w/\partial y \\ \partial v/\partial x & \partial v/\partial y \end{pmatrix} \right| dx \, dy \qquad (14.117)$$

(As a test, recall the change from Cartesian to polar coordinates and show that $dx \, dy = r \, dr \, d\theta$.)

P14.5. Consider two *distinguishable* particles of mass m moving in the potential

$$V(x_1, x_2) = \frac{1}{2} m\omega^2 x_1^2 + \frac{1}{2} m\omega^2 x_2^2 \qquad (14.118)$$

(a) Using the fact that the potential is separable in the x_1, x_2 coordinates, find the energy spectrum and ground state wavefunction. Since this is a product wavefunction, there can be no correlations between the the two particles; show that $\text{cov}(x_1, x_2)$ vanishes in any state. Recall that

$$\text{cov}(x_1, x_2) \equiv \langle (x_1 - \langle x_1 \rangle)(x_2 - \langle x_2 \rangle) \rangle \qquad (14.119)$$

(b) Show that the potential also separates when expressed in center-of-mass (X) and relative (x) coordinates and find the energy spectrum and ground state wavefunction in these coordinates. Show that the degeneracy is the same in each representation and that the ground state wavefunctions agree.

(c) Add an additional mutual interaction of the form

$$V'(x_1, x_1) = V'(x_1 - x_2) = \lambda(x_1 - x_2)^2 \qquad (14.120)$$

and find the energy spectrum exactly using center-of-mass and relative coordinates. Are the energy levels changed in the way you expect from the form of V'?

(d) When such a mutual interaction is present, we expect the positions of the particles to be correlated. Evaluate $\text{cov}(x_1, x_2)$ for the ground state wavefunction and show that it is proportional to λ when λ is small. Convince yourself that the correlation should be positive (negative) when $\lambda > 0$ (< 0). Hint: Use the coordinate transformations to show things like

$$\langle x_1 \rangle = \left\langle X + \frac{1}{2} x \right\rangle = 0 + 0 \qquad (14.121)$$

and

$$\langle x_1^2 \rangle = \langle X^2 \rangle + \langle Xx \rangle + \frac{1}{4} \langle x^2 \rangle \qquad (14.122)$$

P14.6. Reduced mass effects in "heavy hydrogen."

(a) The energy levels of hydrogen-like atoms with a proton nucleus (ordinary hydrogen or H^1) will be slightly different from those with a deuteron nucleus (so-called "heavy hydrogen" or deuterium, H^2) which has roughly twice the mass of a proton, due to reduced mass effects. Show that the shift in wavelength of a given line for H^2 relative to H^1 is roughly

$$\frac{\Delta\lambda}{\lambda} \approx m_e \left(\frac{1}{M_D} - \frac{1}{m_p} \right) \tag{14.123}$$

where $M_D \approx 2m_p$. Evaluate this fractional change numerically.

(b) The original discovery of deuterium was made by looking for such shifts in the *visible*, atomic Balmer spectra of hydrogen. The original paper[4] says that

When with ordinary hydrogen, the times of exposure required to just record the strong H^1 lines were increased 4000 times, very faint lines appeared at the calculated positions for the H^2 lines" ... "on the short wave-length side and separated from them by between 1 and 2 Å."

Using the result of part (a), quantitatively explain the wavelength shifts observed. Estimate the relative abundance of H^2 and H^1 in normal hydrogen.

P14.7. Permutation groups. We have seen that permutations play an important role in the physics of IND particles and in this problem you are asked to study some of the properties of the permutation group, using the case of three particles as an example. Consider three $(N = 3)$ objects, labeled a, b, and c which can be in the three positions 1,2, and 3; there are then $N! = 3! = 6$ different ways in which the labels can be placed. These permutations can all be obtained from one standard labeling, say (a, b, c) by the action of 6 permutation operators;

$$1(a, b, c) \longrightarrow (a, b, c)$$
$$(12)(a, b, c) \longrightarrow (b, a, c)$$
$$(13)(a, b, c) \longrightarrow (c, b, a)$$
$$(23)(a, b, c) \longrightarrow (a, c, b)$$
$$(231)(a, b, c) \longrightarrow (b, c, a)$$
$$(312)(a, b, c) \longrightarrow (c, a, b)$$

The element 1 is the identity operator. The natural multiplication on these group elements is obtained by letting the permutations act in order, for example,

$$(g_1 \cdot g_2)(a, b, c) \Longrightarrow g_1(g_2(a, b, c)) \tag{14.124}$$

---

[4] Urey, Brickwedde, and Murphy (1932).

so that, for example,

$$(12) \cdot (231) = (13) \quad \text{and} \quad (23) \cdot (13) = (312) \tag{14.125}$$

Complete the multiplication table below and show that these elements form a group satisfying all of the requirements in the definition of Appendix F.2.

	1	(12)	(13)	(23)	(231)	(312)
1	1	(12)	(13)	(23)	(231)	(312)
(12)	(12)				(13)	
(13)	(13)					
(23)	(23)		(312)			
(231)	(231)					
(312)	(312)					

P14.8. (a) Generalize the proof in Eqn. (14.81) to show that the exchange operator $\hat{\mathcal{E}}_{ij}$ is Hermitian by showing that $\langle \hat{\mathcal{E}}_{ij} \rangle$ is real when evaluated with any multiparticle position space wavefunction.

(b) Show that the expectation value of $\hat{\mathcal{E}}_{12}$ is real when evaluated between spin-states, that is, show that

$$\langle \chi_1; \chi_2 | \hat{\mathcal{E}}_{12} | \chi_1; \chi_2 \rangle = (\alpha_1^*, \beta_1^*)(\alpha_2^*, \beta_2^*) \hat{\mathcal{E}}_{12} \begin{pmatrix} \alpha_1 \\ \beta_1 \end{pmatrix} \begin{pmatrix} \alpha_2 \\ \beta_2 \end{pmatrix} = |\alpha_1 \alpha_2^* + \beta_1 \beta_2^*|^2 \tag{14.126}$$

is real.

P14.9. (a) Show that the Slater determinant in Eqn. (14.96) is properly normalized.

(b) Four spinless particles move in the same harmonic oscillator potential; two are in the ground state and two in the first excited state. Write down the normalized wavefunction for this system.

P14.10. Confirm the results in Eqn. (14.108).

P14.11. Consider two IND spin-1/2 particles which interact via the potential $V(x_1 - x_2) = k(x_1 - x_2)^2/2$.

(a) Ignoring spin for the moment, show that the position-space wavefunctions can be written as

$$\psi(x_1, x_2) \propto e^{iP(x_1+x_2)/2\hbar} \psi_n(x_1 - x_2) \tag{14.127}$$

where the ψ_n are the harmonic oscillator eigenstates.

(b) Show that under the exchange $1 \leftrightarrow 2$ that these solutions satisfy

$$\hat{\mathcal{E}}_{12} \psi(x_1, x_2) = (-1)^n \psi(x_1, x_2) \tag{14.128}$$

    (c) Add the appropriate symmetric or antisymmetric spin wavefunctions and find the allowed states of the system.

P14.12. Consider the following very simplified model of the lithium 7 nucleus (^7Li), consisting of 3 protons and 4 neutrons in a one-dimensional infinite well of width a. Assume that the protons and neutrons do not interact with each other.

    (a) What is the ground state energy for this system?

    (b) Write down the normalized wavefunction for the ground state.

    (c) What is the energy of the first excited state?

PART II

The Quantum World

FIFTEEN
Two-Dimensional Quantum Mechanics

One-dimensional (1D) systems provide examples of many of the most important features of quantum mechanics, but it is also instructive to consider two-dimensional (hereafter 2D or planar) systems for several reasons:

- Systems with two spatial degrees of freedom provide more opportunities to study multivariable probability concepts, separation of coordinates techniques, and new mathematical methods and special functions. They also allow for the visualization of many quantum phenomena which arise in more realistic three-dimensional (3D) systems, but which are obviously difficult to plot in 3D.

- Two-dimensional systems naturally exhibit symmetries not present in 1D systems, and provide a glimpse of the intimate connection between symmetries and the degeneracy of energy levels.

- Two-dimensional systems allow one to study rotational motion and its symmetries as well as the properties of angular momentum, both quantum mechanically and in its approach to the classical limit. For example, charged particles in a uniform magnetic field classically can undergo circular planar orbits (see Section 18.5); a similar quantum system of electrons in two-dimensions in a uniform **B** field has important implications for the understanding of the so-called quantum Hall effect in condensed matter physics.

- Finally, and perhaps most importantly, while often considered of pedagogical use only, 2D systems of particles are rapidly becoming of more practical importance as their realization in surface physics becomes increasingly easy. It is now possible, using modern crystal growth techniques such as molecular-beam epitaxy and other methods, to fabricate semiconductor nanostructures, artificially created patterns of atoms whose atomic composition and sizes are

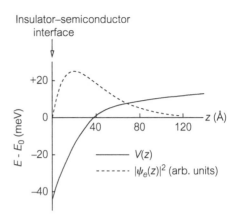

Figure 15.1. Semischematic representation of the potential energy, $V(z)$ (solid curve), and electron wavefunction, $|\psi_e(z)|^2$ (dashed curve), versus the distance from the surface, z, for a 2D electron gas near an insulator–semiconductor interface, indicating the approximate localization distances and energy scales. Adapted from von Klitzing (1987).

controllable at the nanometer scale, which is comparable to interatomic distances. At such length scales, quantum effects obviously become increasingly important. Even more dramatically, scanning tunnel microscopy (STM) techniques can now be used to manipulate individual atoms and molecules[1] with atomic scale precision. The quantum corral shown in Fig. 1.3 was constructed in this manner.

As an example of such a system, consider a 2D electron gas, bound to a surface or interface between surfaces by the potential shown in Fig. 15.1. The typical localization scale (determined, say, by the thickness of the interface layer) might be $L \sim 40$ Å; this implies quantized energies in the direction perpendicular to the surface (here the z direction) of the order of

$$E_n \sim \frac{\hbar^2 \pi^2 n^2}{2 m_e L^2} \sim n^2 25 \,\text{meV} \tag{15.1}$$

so that the energy required to excite the electrons in this direction would be roughly $\Delta E \sim 75$ meV. This can be compared to other typical energy scales for the two-dimensional electron gas which might be either:

- The thermal energy, $k_B T \lesssim 25$ meV, for temperatures at, or below, room temperature (300 K) or

- The Fermi energy of the system. Recall from P7.2 that

$$E_F^{(2D)} = \frac{\hbar^2 \pi}{m_e} n_e^{(2D)} \tag{15.2}$$

---

[1] See, for example, Stroscio and Eigler (1991).

where $n_e^{(2D)} = N_{tot}/L^2$ is the electron surface density in two dimensions. Using a typical value for $n_e^{(2D)}$, this can be written in the form

$$E_F^{(2D)} = 2.5 \text{ meV} \left(\frac{n_e^{(2D)}}{10^{12} \text{ e}^- /\text{cm}^2} \right)$$ (15.3)

and we see that typical 2D collisions will not be able to "excite" electrons in the z direction, provided the density is not too high; thus the electron wavefunctions will stay effectively localized within the interface or on the surface (as in Fig. 15.1), and one can study an effectively 2D problem.

15.1 **2D Cartesian Systems**

The simplest example of a quantum system in two-dimensions is one described by Cartesian coordinates with a Hamiltonian of the form

$$\hat{H} = \frac{1}{2m} \left(\hat{p}_x^2 + \hat{p}_y^2 \right) + V(x, y)$$ (15.4)

where

$$\hat{p}_x = \frac{\hbar}{i} \frac{\partial}{\partial x} \quad \text{and} \quad \hat{p}_y = \frac{\hbar}{i} \frac{\partial}{\partial y}$$ (15.5)

with a wavefunction

$$\psi(x, y; t) = \psi(x, y) \, e^{-iEt/\hbar}$$ (15.6)

satisfying the time-independent Schrödinger equation

$$\hat{H}\psi(x, y) = -\frac{\hbar^2}{2m} \left(\frac{\partial^2}{\partial x^2} + \frac{\partial^2}{\partial y^2} \right) \psi(x, y) + V(x, y)\psi(x, y) = E\psi(x, y)$$ (15.7)

The probability density (now in two-dimensions) is given by $|\psi(x, y, t)|^2$, which must satisfy

$$\int_{-\infty}^{+\infty} dx \int_{-\infty}^{+\infty} dy \, |\psi(x, y; t)|^2 = 1$$ (15.8)

and average or expectation values are calculated in the usual way via

$$\langle \hat{O} \rangle_t \equiv \int_{-\infty}^{+\infty} dx \int_{-\infty}^{+\infty} dy \, \psi^*(x, y; t) \, \hat{O} \, \psi(x, y; t)$$ (15.9)

If we consider *separable potentials* of the form

$$V(x, y) = V_x(x) + V_y(y) \tag{15.10}$$

we can assume a factorized form for the wavefunction

$$\psi(x, y; t) = X(x)\, Y(y)\, e^{-i(E_x + E_y)t/\hbar} \tag{15.11}$$

where each coordinate satisfies its own, 1D Schrödinger equation

$$\frac{\hat{p}_x^2}{2m} X(x) + V_x(x)X(x) = E_x X(x) \tag{15.12}$$

$$\frac{\hat{p}_y^2}{2m} Y(y) + V_y(y)Y(y) = E_y Y(y) \tag{15.13}$$

and the total energy of the system is given by $E = E_x + E_y$.

15.1.1 2D Infinite Well

A simple and instructive case is that of the 2D infinite well (or square box) with walls at $x = 0, L$ and $y = 0, L$; this is of the form above as we can define

$$V_{1D}(z; L) = \begin{cases} 0 & \text{for } 0 < z < L \\ \infty & \text{otherwise} \end{cases} \tag{15.14}$$

in which case the 2D potential is of the form

$$V_{2D}(x, y) = V_{1D}(x; L) + V_{1D}(y; L) \tag{15.15}$$

The fully normalized solutions can be written in the form

$$u_{(n,m)}(x, y) = u_n(x) u_m(y) = \frac{2}{L} \sin\left(\frac{n\pi x}{L}\right) \sin\left(\frac{m\pi y}{L}\right) \tag{15.16}$$

with

$$E_{(n,m)} = E_n + E_m = \frac{\hbar^2 \pi^2}{2mL^2}(n^2 + m^2) \tag{15.17}$$

and the spectrum is illustrated in Fig. 15.2. The wavefunctions for several sets of n, m are shown in Fig. 15.3 illustrating the wave properties of the system. We note that if the 1D infinite well corresponds to waves on a string with fixed ends, then this case corresponds to the vibrations of a square drumhead. Such wavefunctions are not only of pedagogical interest, as very similar patterns of "standing electron waves" can be directly observed on surfaces using STM techniques.[2]

[2] For one of the first experimental realizations, see Crommie, Lutz, and Eigler (1993).

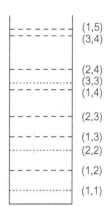

Figure 15.2. The spectrum of the 2D infinite square well. The values of (n, m) for the lowest-lying states are shown; states for which $n \neq m$ are doubly degenerate and shown dashed.

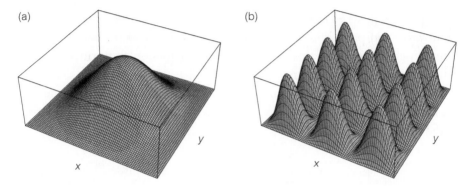

Figure 15.3. Plots of $|u_{(n,m)}(x, y)|^2$ versus (x, y) for the 2D infinite square well for (a) $(n, m) = (1, 1)$ and (b) $(n, m) = (3, 4)$.

Particle-like, wave packet solutions for this separable potential can be formed by using the linearity of the Schrödinger equation to write

$$\psi_{WP}(x, y; t) = \psi_{WP}(x; t)\psi_{WP}(y, t) \tag{15.18}$$

where one has

$$\psi_{WP}(x, t) = \sum_{n=1}^{\infty} a_n^{(x)} u_n(x) e^{-iE_n^{(x)} t/\hbar} \tag{15.19}$$

and similarly for y. For simplicity, we can use Gaussian weighting factors

$$a_n^{(x)} = e^{-(p_n - p_0^{(x)})^2 \alpha^2/2} e^{-ip_n x_0} \tag{15.20}$$

with $p_n = \hbar n/L$, as in Section 5.4.2, or the more rigorous version in Example 6.4. We can localize the packet initially such that it is centered at (x_0, y_0) with a central value of momentum $\mathbf{p}_0 = (p_0^{(x)}, p_0^{(y)})$. The ballistic propagation of such a wave packet with elastic collisions from the walls is illustrated in Fig. 15.4 where we

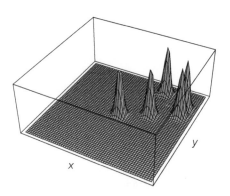

Figure 15.4. Series of snapshots (taken at equal time intervals) showing the propagation of a quasi-Gaussian wavepacket in a two-dimensional infinite square well; the packet is initially localized in the center of the well, with initial momenta $p_0^{(x)} = 2p_0^{(y)}$.

have chosen $(x_0, y_0) = (L/2, L/2)$ and $p_{(0)}^x = 2p_{(0)}^y$. Thus, the system can exhibit both wave- and particle-like behavior in a fashion similar to the 1D case. (2D infinite well potentials of this type, of arbitrary shapes or "footprints," are often called *quantum billiard* systems.)

The energy eigenfunction solutions also form a complete set in that the time-dependence of any allowable wavefunction in the 2D infinite well can be written in the form

$$\psi(x, y; t) = \sum_{n=1}^{\infty} \sum_{m=1}^{\infty} a_{(n,m)} u_{(n,m)}(x, y) e^{-iE_{(n,m)}t/\hbar} \tag{15.21}$$

where $|a_{(n,m)}|^2$ is the probability that a measurement of the energy associated with $\psi(x, y)$ will yield the value $E_{n,m}$. The form in Eqn. (15.18) is then a special case of the time-development of such a solution.

The most interesting new feature of this system is the fact that more than one independent energy eigenstate corresponds to the same energy level, at least for $n \neq m$, where the exchange $n \leftrightarrow m$ gives the same energy. Such a system is said to exhibit *degeneracy* and we say that:

- The quantum value of an observable quantity is *degenerate* when two (or more) independent eigenfunctions of an operator yield the *same* eigenvalue.

In this case, the degeneracy is easily traced to the symmetry of the potential (and kinetic energies) since the exchange of labels $x \leftrightarrow y$ has no observable effect on the system, leading naturally to a doubly degenerate set of levels (when $n \neq m$.) We can formalize this notion by introducing an *exchange operator*, $\hat{E}_{(x,y)}$, defined via

$$\hat{E}_{(x,y)} f(x, y) \equiv f(y, x) \tag{15.22}$$

which can be easily shown to be Hermitian even though its classical connection is far from obvious; this operator is very similar to the multiparticle exchange operator, $\mathcal{E}_{ij}$, used in Chapter 14.

Following the discussion of Section 12.5 on conserved quantities, we note that the statement above that the exchange $x \leftrightarrow y$ "has no effect on the system" can be associated with the fact that this operator commutes with the Hamiltonian, that is,

$$\left(\hat{H}\hat{E}_{(x,y)} - \hat{E}_{(x,y)}\hat{H} \right) \psi(x, y) = 0 \quad \text{or} \quad [\hat{H}, \hat{E}_{(x,y)}] = 0 \tag{15.23}$$

This implies that there will be simultaneous eigenfunctions of *both* $\hat{H}$ and $\hat{E}_{x,y}$. By invoking the same arguments used previously for both the parity and exchange operators (mostly the fact that $\left(\hat{E}_{(x,y)} \right)^2 = 1$), the eigenvalues (eigenfunctions) of the exchange operator can be seen to be ± 1 (even-odd functions under exchange), that is,

$$\hat{E}_{(x,y)}f^{(+)}(x, y) = +f^{(+)}(x, y) \quad \text{and} \quad \hat{E}_{(x,y)}f^{(-)}(x, y) = -f^{(-)}(x, y) \tag{15.24}$$

The $u_{(n,m)}(x, y)$ solutions individually do not, however, immediately satisfy this requirement. We note that any linear combination of degenerate energy eigenstates will also be an energy eigenstate with the same energy eigenvalue since

$$\hat{H}\left(\sum_E a_E \psi_E(x) \right) = \sum_E a_E \hat{H}\psi_E(x)$$

$$= \sum_E E a_E \psi_E(x)$$

$$= E\left(\sum_E a_E \psi_E(x) \right) \tag{15.25}$$

We are thus free to take appropriate linear combinations of degenerate solutions provided they remain orthogonal, and it is easy to see that the required combinations are

$$u_{(n,m)}^{(\pm)}(x, y) \equiv \frac{1}{\sqrt{2}} \left(u_{(n,m)}(x, y) \pm u_{(m,n)}(x, y) \right) \tag{15.26}$$

which do satisfy Eqn. (15.24). (See P15.8 for an application of these linear combinations.)

It is something of a "folk-theorem" (meaning roughly a statement which is universally accepted as being true, but difficult to state precisely and to prove in each case) that:

- Most degeneracies are necessarily a result of *some* symmetry (sometimes not obvious) of the system under consideration.

The study of symmetries in quantum mechanics (under the guise of group theory) has had profound applications in atomic, nuclear, and elementary particle physics.

15.1.2 **2D Harmonic Oscillator**

Another separable Cartesian system is described by an isotropic harmonic oscillator, that is, a 2D mass and spring, defined by the potential energy

$$V(x, y) = \frac{1}{2}K(x^2 + y^2) = \frac{1}{2}m\omega^2(x^2 + y^2). \tag{15.27}$$

The product wavefunctions are given by

$$\psi_{(n,m)}(x, y) = \psi_n(x)\psi_m(y) \tag{15.28}$$

where the $\psi_n(x)$ are the solutions of Section 9.2.2. The energy spectrum is given by

$$E_{n,m} = E_n + E_m = (n + m + 1)\hbar\omega = (N + 1)\hbar\omega \tag{15.29}$$

which is illustrated in Fig. 15.5. As before, the energy levels with $n \leftrightarrow m$ are degenerate. The total degeneracy, that is, the number of distinct states, N_s, with energy value labeled by N is $N_s = N$ this is much larger than expected solely on the basis of the $x \leftrightarrow y$ symmetry. The enlarged degeneracy is partly due to the fact that the system also exhibits a symmetry under *rotations*, since the potential can also be written in the circularly symmetric way

$$V(x, y) = \frac{1}{2}K(x^2 + y^2) = \frac{1}{2}Kr^2 = V(r) \tag{15.30}$$

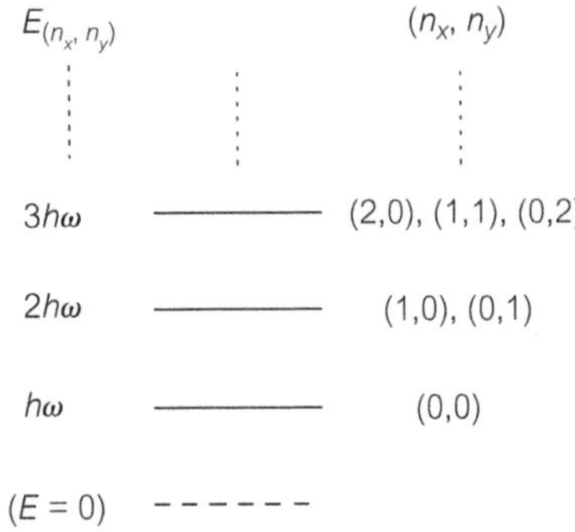

$4\hbar\omega \longrightarrow (3,0),(0,3)\ (2,1)\ (1,2)$

$E_{(n_x, n_y)}$		(n_x, n_y)
$\vdots$	$\vdots$	$\vdots$
$3\hbar\omega$	——	(2,0), (1,1), (0,2)
$2\hbar\omega$	——	(1,0), (0,1)
$\hbar\omega$	——	(0,0)
$(E = 0)$	— — — —	

Figure 15.5. Energy spectrum and degeneracies for the 2D simple harmonic oscillator potential using Cartesian coordinates. Values of (n_x, n_y) for each level are shown.

and the system can then be separated in cylindrical coordinates as well, as we will see in Section 15.3.3.

Two-dimensional wave packets, *a la* Eqn. (15.18), can also be constructed in this case using, for example, the special Gaussian packet of Section 12.6.2, and can be shown to undergo semiclassical motion (P15.12). This is especially interesting in the case of an *non-isotropic spring*, that is, a potential of the form

$$V(x, y) = \frac{1}{2}K_x x^2 + \frac{1}{2}K_y y^2 \tag{15.31}$$

which is still separable. In this case, the natural vibration frequencies are different, $\omega_{x,y} = \sqrt{K_{x,y}/m}$, and the "trajectories" of the wave packets will, in general, not be periodic. In the special case where the the frequencies are *commensurate*, namely, rational multiples of each other, that is

$$\frac{\omega_x}{\omega_y} = \frac{p}{q} \tag{15.32}$$

(with p, q integers) the classical motion *is* periodic and the quantum wave packets can reproduce the classical "Lissajous figures" discussed in many classical mechanics texts.[3]

15.2 **Central Forces and Angular Momentum**

15.2.1 **Classical Case**

Cartesian coordinates may not be the most natural set of variables for the study of many systems, and this is especially true for 2D systems described by a cylindrically symmetric potential of the form

$$V(\mathbf{r}) = V(r, \theta) = V(r) \tag{15.33}$$

In this case, the classical force is given by

$$\mathbf{F}(\mathbf{r}) = -\nabla V(r, \theta)$$

$$= -\frac{\partial V(r, \theta)}{\partial r}\hat{\mathbf{r}} - \frac{1}{r}\frac{\partial V(r, \theta)}{\partial \theta}\hat{\boldsymbol{\theta}}$$

$$= -\frac{dV(r)}{dr}\hat{\mathbf{r}}$$

$$\mathbf{F}(\mathbf{r}) = F(r)\hat{\mathbf{r}} \tag{15.34}$$

---

[3] See, for example, Marion and Thornton (2004).

where $\hat{\mathbf{r}}$ and $\hat{\boldsymbol{\theta}}$ are unit vectors in the radial and tangential directions, respectively. Thus, for a central potential, the force is directed radially toward (or away from) the origin.

The corresponding classical torque, $\boldsymbol{\tau} = \mathbf{r} \times \mathbf{F}$, then vanishes and the relation

$$\frac{d\mathbf{L}}{dt} = \boldsymbol{\tau} = 0 \tag{15.35}$$

guarantees that the classical angular momentum, L, is conserved, that is, is constant in time. Because of its importance as an additional conserved quantity (along with the total energy), we will discuss the quantum version of angular momentum extensively, in 2D in Section 15.2.2 and in 3D in Chapter 16.

The classical equations of motion for the particle in polar coordinates can be derived from those in Cartesian coordinates, namely,

$$\mathbf{F}(r) = F(r)\,\hat{\mathbf{r}} = m\mathbf{a}(t) \tag{15.36}$$

giving

$$x: F(r) \left(\frac{x(t)}{r(t)} \right) = m\ddot{x}(t) \tag{15.37}$$

$$y: F(r) \left(\frac{y(t)}{r(t)} \right) = m\ddot{y}(t) \tag{15.38}$$

and by using the relations

$$x(t) = r(t)\cos(\theta(t)) \quad \text{and} \quad y(t) = r(t)\sin(\theta(t)) \tag{15.39}$$

to obtain

$$\frac{F(r)}{m}\cos(\theta) = \ddot{r}\cos(\theta) - 2\dot{r}\sin(\theta)\dot{\theta} - r\cos(\theta)\dot{\theta}^2 - r\sin(\theta)\ddot{\theta} \tag{15.40}$$

$$\frac{F(r)}{m}\sin(\theta) = \ddot{r}\sin(\theta) + 2\dot{r}\cos(\theta)\dot{\theta} - r\sin(\theta)\dot{\theta}^2 + r\cos(\theta)\ddot{\theta}. \tag{15.41}$$

The linear combination of equations given by (15.41) $\times$ ($r\cos(\theta)$) $-$ (15.40) $\times$ ($r\sin(\theta)$) implies that

$$0 = 2\dot{r}r\dot{\theta} + r^2\ddot{\theta} = \frac{d}{dt}\left(r^2\dot{\theta}\right) = 0 \tag{15.42}$$

which is another statement of conservation of angular momentum as $L_z = rp = mrv = mr^2\dot{\theta}$ in polar coordinates. Using this identification, the other obvious combination, (15.41) $\sin(\theta)$ + (15.40) $\cos(\theta)$, then gives the dynamical equation of motion for the radial coordinate,

$$F(r) = m\ddot{r} - mr\dot{\theta}^2 = m\ddot{r} - \frac{L_z^2}{mr^3} \tag{15.43}$$

which is the standard Newton's law result, including the familiar "centrifugal force" term. The derivation makes clear that this term is solely due to the proper accounting of the rotational motion and not to any "fictitious force."

Most importantly for connections to quantum mechanics, the total energy will be constant for a conservative potential, so we can write

$$E = \frac{1}{2}mv^2(t) + V(x, y) \tag{15.44}$$

and the chain rule, and Eqn. (15.39), gives

$$v^2 = v_x^2 + v_y^2 = \dot{x}^2 + \dot{y}^2 = \dot{r}^2 + r^2\dot{\theta}^2 \tag{15.45}$$

so that the total energy can be written as

$$E = \frac{m}{2}(\dot{r} + r^2\dot{\theta}^2) + V(r, \theta) = \frac{m\dot{r}^2}{2} + \frac{L_z^2}{2mr^2} + V(r, \theta) \tag{15.46}$$

The second term can be put in the form

$$T_{rot} = \frac{L_z^2}{2mr^2} = \frac{1}{2}mr^2\dot{\theta}^2 = \frac{1}{2}I\omega^2 \tag{15.47}$$

where $I = mr^2$ is the rotational moment of inertia for a point mass; this makes it clear that it represents the rotational kinetic energy. It is this form for the energy which can be most easily generalized to a quantum mechanical Hamiltonian.

15.2.2 Quantum Angular Momentum in 2D

To extend the notion of angular momentum to quantum mechanical operators, it is most natural to start from the classical definition

$$\mathbf{L} = \mathbf{r} \times \mathbf{p} \tag{15.48}$$

so that the relevant component for 2D motion is the angular momentum about the z-axis, namely

$$L_z = xp_y - yp_x \tag{15.49}$$

Motivated by the position representation of operators, we replace the classical momentum components by their operator analogs and define

$$\begin{array}{ccc} \text{classical} & & \text{quantum mechanical} \\ L_z \longrightarrow & \hat{L}_z \equiv x\hat{p}_y - y\hat{p}_x = & \frac{\hbar}{i}\left(x\frac{\partial}{\partial y} - y\frac{\partial}{\partial x}\right) \end{array} \tag{15.50}$$

It is easy to show that $\hat{L}_z$ is Hermitian (P4.21), and that it is also the infinitesimal generator of rotations around the z-axis (P15.13), just as $\hat{p}_x$ is responsible for translations along the x-axis.

To express this more naturally in polar coordinates, we again use the defining relations

$$x = r\cos(\theta) \quad \text{and} \quad y = r\sin(\theta) \tag{15.51}$$

or their inverses

$$r = \sqrt{x^2 + y^2} \quad \text{and} \quad \tan(\theta) = \frac{y}{x} \tag{15.52}$$

and the chain rule to find

$$\frac{\partial}{\partial x} = \cos(\theta)\frac{\partial}{\partial r} - \frac{\sin(\theta)}{r}\frac{\partial}{\partial \theta} \tag{15.53}$$

$$\frac{\partial}{\partial y} = \sin(\theta)\frac{\partial}{\partial r} + \frac{\cos(\theta)}{r}\frac{\partial}{\partial \theta} \tag{15.54}$$

This then gives

$$\hat{L}_z = x\hat{p}_x - y\hat{p}_y = \frac{\hbar}{i}\frac{\partial}{\partial \theta} \tag{15.55}$$

and $\theta, \hat{L}_z$ can be seen (P15.13) to have many similarities with the conjugate pair $x, \hat{p}_x$.

The *eigenfunctions of angular momentum* (in 2D), labeled $\Theta_m(\theta)$, are then determined by the equation

$$\frac{\hbar}{i}\frac{d\Theta_m(\theta)}{d\theta} = \hat{L}_z\Theta_m(\theta) = L_z\Theta_m(\theta) = m\hbar\Theta_m(\theta) \tag{15.56}$$

$$\text{(operator } \hat{L}_z) \quad \text{(eigenvalue } L_z)$$

which yields

$$\Theta_m(\theta) = \frac{e^{im\theta}}{\sqrt{2\pi}} \tag{15.57}$$

where we have chosen to write the dimensionful angular momentum eigenvalue, L_z, in terms of the natural unit of $\hbar$. The normalization constant is chosen so as to satisfy

$$1 = \int_0^{2\pi} d\theta \, |\Theta_m(\theta)|^2 \tag{15.58}$$

which is natural for an angular wavefunction.

One difference between these solutions and momentum eigenstates arises because of the periodic nature of the variable θ; this presumably requires us to identify coordinates separated by $\theta = 2\pi$ as representing the same physical point,[4] that is,

$$\Theta_m(\theta + 2\pi) = \Theta_m(\theta) \quad \Longrightarrow \quad e^{i2\pi m} = 1 \tag{15.59}$$

or $m = 0, \pm 1, \pm 2, \ldots$ and the angular momentum is quantized

$$L_z = 0, \pm \hbar, \pm 2\hbar, \ldots \tag{15.60}$$

This quantization once again arises because of the need to impose (appropriate) boundary conditions. It also guarantees that eigenfunctions corresponding to different eigenvalues are orthogonal, namely

$$\langle n|m\rangle = \int_0^{2\pi} \Theta_n^*(\theta)\,\Theta_m(\theta)\,d\theta = \frac{1}{2\pi} \int_0^{2\pi} \left(e^{in\theta}\right)^* e^{im\theta}\,d\theta = \delta_{n,m} \tag{15.61}$$

These complex angular wavefunctions correspond most closely to the plane wave solutions (traveling waves) for momentum; linear combinations can yield $\Theta(\theta) = \sin(m\theta)$ or $\cos(m\theta)$, which are more like standing waves, and which may be more appropriate for some bound state problems, or for visualization purposes.

The appropriate Hamiltonian operator in Cartesian coordinates,

$$\frac{\hat{p}_x^2 + \hat{p}_y^2}{2m} + V(x,y) = -\frac{\hbar^2}{2m}\left(\frac{\partial^2}{\partial x^2} + \frac{\partial^2}{\partial y^2}\right) + V(x,y) = -\frac{\hbar^2}{2m}\nabla^2 + V(x,y) \tag{15.62}$$

can be written in polar coordinates by expressing the 2D gradient squared in terms of (r,θ) by extending the chain rule arguments used above. One finds (P15.14) that

$$\frac{\partial^2\psi(x,y)}{\partial x^2} + \frac{\partial^2\psi(x,y)}{\partial y^2} = \frac{1}{r}\frac{\partial}{\partial r}\left(r\frac{\partial\psi(r,\theta)}{\partial r}\right) + \frac{1}{r^2}\frac{\partial^2\psi(r,\theta)}{\partial\theta^2} \tag{15.63}$$

The Hamiltonian operator in polar coordinates can thus be written as

$$\hat{H}_{(r,\theta)} = -\frac{\hbar^2}{2\mu}\left(\frac{\partial^2}{\partial r^2} + \frac{1}{r}\frac{\partial}{\partial r} + \frac{1}{r^2}\frac{\partial^2}{\partial\theta^2}\right) + V(r,\theta) \tag{15.64}$$

where we have expanded the radial derivative $(\partial/\partial r)$ operators. We will also henceforward write the mass as μ to avoid confusion with the angular momentum quantum number m; this will also be appropriate for two-body

---

[4] This is true for the angular momentum associated with the orbital motion of particles; for the case of intrinsic angular momentum (spin), see Section 16.4.

problems where the use of the reduced mass μ is natural. The angular derivative term can be written in the form

$$\hat{T}_\theta = -\frac{\hbar^2}{2\mu r^2}\frac{\partial^2}{\partial \theta^2} = \frac{1}{2\mu r^2}\hat{L}_z^2 \tag{15.65}$$

which is indeed the obvious quantum operator analog of the classical energy of rotation in Eqn. (15.47). The appropriate Schrödinger equation is then

$$\hat{H}_{(r,\theta)}\psi(r,\theta) = E\psi(r,\theta) \tag{15.66}$$

along with the normalization condition for the corresponding probability density,

$$1 = \int_0^\infty r\,dr \int_0^{2\pi} d\theta\,|\psi(r,\theta)|^2 \tag{15.67}$$

This condition is associated with the fact that the probability of finding the particle simultaneously in the small coordinate intervals $(r, r + dr)$ and $(\theta, \theta + d\theta)$ is

$$d\text{Prob}(r,\theta) = |\psi(r,\theta)|^2\,r\,dr\,d\theta \tag{15.68}$$

and we will see that the additional factor of r in the "measure" is important.

The case of central force motion for which the potential has no angular dependence, that is, $V(\mathbf{r}) = V(r)$, is the most important and we note that:

- In this case, $[\hat{H}, \hat{L}_z] = 0$, so that the energy eigenfunctions will also be eigen-functions of the (planar) angular momentum; this fact also implies that the angular momentum will be a conserved quantity.

- The Schrödinger equation is separable in this case, so we can assume solutions of the form $\psi(r,\theta) = R(r)\,\Theta_m(\theta)$.

Performing the separation of variables in the Schrödinger equation, we find that

$$\frac{r^2}{R(r)}\left\{-\left(\frac{d^2 R(r)}{dr^2} + \frac{1}{r}\frac{dR(r)}{dr}\right) + \frac{2\mu}{\hbar^2}(V(r) - E)R(r)\right\} = \frac{1}{\Theta_m(\theta)}\frac{d^2\Theta_m(\theta)}{d\theta^2}$$
$$= -m^2 \tag{15.69}$$

so that the Schrödinger equation for the radial wavefunction (the quantum analog of Eqn. (15.43)) is

$$-\frac{\hbar^2}{2\mu}\left(\frac{d^2 R(r)}{dr^2} + \frac{1}{r}\frac{dR(r)}{dr}\right) + \left(V(r) + \frac{\hbar^2 m^2}{2\mu r^2}\right)R(r) = ER(r) \tag{15.70}$$

Given the already defined normalization properties of the $\Theta_m(\theta)$, the radial wavefunction must satisfy

$$1 = \int_0^\infty r\, dr\, |R(r)|^2 \tag{15.71}$$

A simple example of 2D motion for which this formulation is useful is the case of a mass connected to a light, but rigid, rod of length r_0, free to rotate around the origin; such a system is sometimes called a *rigid rotator* or *rotor*. In this case, the Hamiltonian is simply $\hat{H} = \hat{L}_z^2/2\mu r_0^2$ with eigenfunctions given by the $\Theta_m(\theta)$ and quantized energies given by $E_m = \hbar^2 m^2/2\mu r_0^2$. Solutions corresponding to $\pm m$ have the same quantized energies corresponding, in turn, to the equivalence of clockwise versus counterclockwise motion. The same result can be inferred from the complete radial Schrödinger equation Eqn. (15.70) if we assume that there is no potential $(V(r,\theta) = 0)$ and that the radius is fixed (so that spatial derivatives of $R(r)$ vanish). We turn to less trivial examples of rotational motion in the next section.

15.3 Quantum Systems with Circular Symmetry

15.3.1 Free Particle

The Schrödinger equation for a free particle in polar coordinates reads

$$-\frac{\hbar^2}{2\mu}\left(\frac{d^2 R(r)}{dr^2} + \frac{1}{r}\frac{dR(r)}{dr}\right) + \frac{\hbar^2 m^2}{2\mu r^2}R(r) = ER(r) \tag{15.72}$$

which can be written in terms of the dimensionless variable $z = kr$ (where $k = \sqrt{2\mu E/\hbar^2}$) as

$$\frac{d^2 R(z)}{dz^2} + \frac{1}{z}\frac{dR(z)}{dz} + \left(1 - \frac{m^2}{z^2}\right)R(z) = 0 \tag{15.73}$$

which can be recognized from the mathematical literature as *Bessel's equation* (see Appendix E.4.) Similarly to the case of a free particle in one dimension, it has two linearly independent solutions for each value of m^2, the so-called regular solution, $J_{|m|}(z)$, standardly labeled *cylindrical Bessel functions* of order $|m|$ (or Bessel functions of the first kind), and the irregular solutions, $Y_{|m|}(z)$, (Neumann functions or Bessel functions of the second kind); we will explore the mathematical properties and physical meaning of these solutions in this section.

We can exhibit the behavior of the solutions for large z by noting that the equation in this limit becomes approximately

$$\frac{d^2 R(z)}{dz^2} = -R(z) \tag{15.74}$$

so that the behavior is oscillatory, that is, $R(z) \to \sin(z)$, $\cos(z)$ or $\exp(\pm iz)$. We can do better by assuming a solution of the form

$$R(z) \longrightarrow z^\alpha \cos(z) \tag{15.75}$$

where we assume that $\alpha < 0$, and substitution into Eqn. (15.73) implies that (P15.15) the next order term $(z^{\alpha-1})$ also vanishes when $\alpha = -1/2$. These results help justify the well-known asymptotic expansions

$$J_{|m|}(z) \longrightarrow \sqrt{\frac{2}{\pi z}} \cos\left(z - \frac{|m|\pi}{2} - \frac{\pi}{4}\right) \left[1 + O(1/z^2)\right] \tag{15.76}$$

$$Y_{|m|}(z) \longrightarrow \sqrt{\frac{2}{\pi z}} \sin\left(z - \frac{|m|\pi}{2} - \frac{\pi}{4}\right) \left[1 + O(1/z^2)\right] \tag{15.77}$$

which are valid for $z \gg 0$. This behavior has an immediate physical interpretation as the *probability density times measure* gives

$$d\text{Prob}(r,\theta) = |R(r)\Theta_{|m|}(\theta)|^2 r \, dr \, d\theta$$

$$\propto \cos\left(kr - \frac{|m|\pi}{2} - \frac{\pi}{4}\right)^2 dr \, d\theta \tag{15.78}$$

implying that, in a spatially averaged sense, there is a uniform distribution of probability corresponding to constant speed motion everywhere in the plane; compare this to the case of the 2D free particle in Cartesian coordinates (P15.1) and the corresponding probability distribution. This can also be contrasted with the 1D case of the unstable harmonic oscillator (Section 9.5) where $\psi(x) \propto 1/\sqrt{x}$ as well, but which there corresponded to an (exponentially) accelerating particle. This is a reminder of the importance of the coordinate measure in the implementation of a probability interpretation. In Fig. 15.6, we plot $|J_0(kr)|^2$, both with and without the extra factor of r, to show the effect.

A similar analysis can be used to examine the $r \to 0$ behavior and we assume a series solution of the form

$$R(z) \to \sum_{s=\beta}^{\infty} a_s z^s = a_\beta z^\beta + a_{\beta+1} z^{\beta+1} + \cdots \tag{15.79}$$

with β to be determined. Once again, substitution into the differential equation Eqn. (15.73) yields the condition

$$\beta^2 = m^2 \quad \text{or} \quad \beta = \pm|m| \tag{15.80}$$

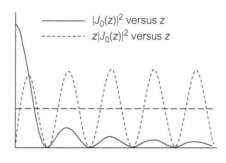

Figure 15.6. Plots of $|J_0(z)|^2$ (solid) and $z\,|J_0(z)|^2$ (dotted) versus z showing the effect of the "measure" for the free-particle wavefunction in two dimensions in polar coordinates. The horizontal dashed line corresponds to a probability distribution for constant speed motion in the plane.

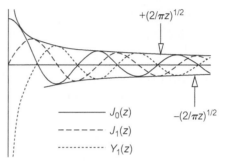

Figure 15.7. Plots of the regular $J_{0,1}(z)$ and irregular $Y_1(z)$ solutions of Bessel's equation, showing the small and large z behavior. The behavior for large z is consistent with Eqns (15.76) and (15.77).

The regular, that is, well-behaved at the origin, Bessel functions are convention-ally chosen to have $\beta = +|m|$, while the ill-behaved Neumann functions are described by $\beta = -|m|$ near the origin.[5] (The behavior of the two functions is somewhat similar to the exponentially growing and decaying solutions found in tunneling problems, the rotational kinetic term $\hbar^2 m^2/2\mu r^2$ playing the role of an "angular momentum barrier" in this case; see P15.16.)

Because of its divergence at the origin, it is often necessarily to exclude this solution by hand, that is, use the freedom to pick its coefficient in the most general solution to vanish. We plot $J_{0,1}(z)$ and $Y_1(z)$ in Fig. 15.7 for illustration.

The small r behavior of the Bessel functions solution is also intuitively physical. The probability of being 'near' the origin when the particle is in a state of angular momentum $\pm m\hbar$ is given by

$$d\text{Prob} \quad \propto \quad (kr)^{2|m|+1} \quad \text{when } r \to 0 \tag{15.81}$$

where the additional factor of r comes from the measure. This suppression can be understood as arising from the centrifugal barrier term in Eqn. (15.65), which demands a large cost in energy to be near the origin for rotating particles. The behavior of the first few Bessel functions, $J_m(x)$, for $m = 0, 1, 2, 3$ for small

[5] The behavior of $Y_0(z)$ near the origin requires special treatment as $m = 0$ in that case corresponds to a logarithmic behavior; specifically $Y_0(z) \to 2J_0(z)\log(\gamma z/2)/\pi$.

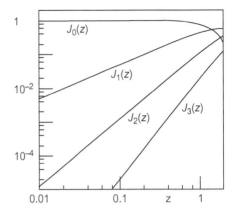

Figure 15.8. Plots of $J_m(z)$ versus z, on logarithmic scales, showing the power-law behavior of $J_m(z) \propto z^m$ for small z, arising from the centrifugal barrier.

argument is shown in Fig. 15.8 for illustration and the plot on semilog paper demonstrates the increasingly large power law behavior near the origin.

15.3.2 Circular Infinite Well

A simple use of the free-particle wavefunctions arises in the study of the infinite circular well,[6] defined by the potential

$$
V(r) = \begin{cases} 0 & \text{for } r < R \\ \infty & \text{for } r > R \end{cases} \tag{15.82}
$$

Inside the well, where the particle is free, the solutions are

$$
\psi(r,\theta) = J_{|m|}(kr)\Theta_m(\theta) \tag{15.83}
$$

or, perhaps more appropriately for bound states,

$$
J_{|m|}(kr)\sin(m\theta) \quad \text{and} \quad J_{|m|}(kr)\cos(m\theta) \tag{15.84}
$$

where we have excluded the irregular $Y_{|m|}(kr)$ solutions for the reasons discussed above. The boundary conditions at the edge of the well are satisfied for all values of θ provided that $J_m(kR) = 0$. If we label the nth zero of the mth Bessel function by $a_{(n,m)}$, we can see that the corresponding radial wavefunction will have $n_r = n - 1$ radial nodes; we note yet again that imposition of the the the boundary conditions has determined the quantized energies, in this case giving

$$
E_{n,m} = \frac{\hbar^2 k^2_{(n_r,m)}}{2\mu} = \frac{\hbar^2 a^2_{(n+1,h)}}{2\mu R^2} \tag{15.85}
$$

[6] For a discussion of the visualization of the solutions for this problem in both quantum and classical mechanics, see Robinett (1996a).

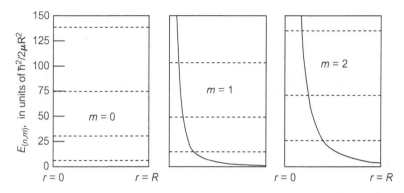

Figure 15.9. Part of the energy spectrum for the infinite circular well obtained using the Bessel function zeros in Eqn. (15.86). The solid curves correspond to the rotational kinetic energy (or centrifugal barrier) term, $\hbar^2 m^2 / 2\mu r^2$, in the Schrödinger equation. Note the resulting increase in energy of corresponding levels as m is increased.

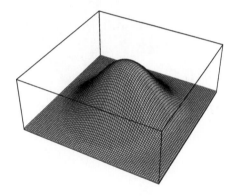

Figure 15.10. Plot of $|\psi(r,\theta)|^2$ for the ground state of the circular infinite well with $(n_r, m) = (0, 0)$.

The notation n_r is motivated by the fact that it counts the number of nodes in the radial wavefunction. Some of the lowest-lying zeroes are given by

$$m = 0: 2.40483 \quad 5.52008 \quad 8.65373 \quad \cdots$$
$$m = 1: 3.83171 \quad 7.01559 \quad 10.1735 \quad \cdots$$
$$m = 2: 5.13562 \quad 8.41724 \quad \cdots \qquad\qquad (15.86)$$
$$m = 3: 6.38016 \quad 9.76102 \quad \cdots$$

and part of the resulting energy spectrum is shown in Fig. 15.9. Each state with $m \neq 0$ state is doubly degenerate because the two values of $\pm|m|$, corresponding to rotations in opposite senses, give the same energy.

To see connections to both wave physics and classical particle motion, we plot, in Figs 15.10–15.12, $|\psi(r,\theta)|^2$ for several cases:

- In Fig. 15.10, we show the ground state corresponding to $m = 0$ and the first radial zero, that is $a_{(0,0)} = 2.404$, with no rotational kinetic energy and the

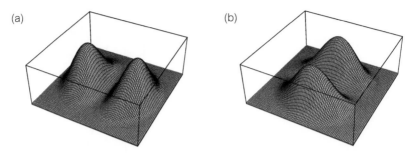

Figure 15.11. Plot of $|\psi(r,\theta)|^2$ for the lowest-lying $m = 1$ states; both the $\sin(\theta)$ (a) and $\cos(\theta)$ (b) cases are plotted to help visualize the double degeneracy.

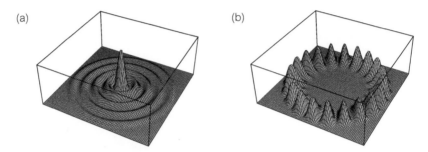

Figure 15.12. Plot of $|\psi(r,\theta)|^2$ for "radial" and "angular" states: Case (a) corresponds to a radially excited state with $(n_r, m) = (4, 0)$, with no angular momentum, while (b) is for $(n_r, m) = (0, 10)$ with large angular momentum (and hence rotational kinetic energy) and little radial kinetic energy (no radial nodes).

least amount of radial kinetic energy. The similarity to the shape of a circular drumhead is obvious.

- Figures 15.11(a) and (b) show $|\psi(r,\theta)|^2$ for the lowest lying states with $|m| = 1$ ($a_{(0,1)} = 3.8318...$), with both the $\cos(\theta)$ and the $\sin(\theta)$ solutions plotted to illustrate their similarity and to help visualize their degeneracy.

- Figure 15.12(a) shows a radially excited state (large n_r) but still with no angular momentum, $(n_r, m) = (4, 0)$, with $a_{(4,0)} = 14.43$. This corresponds to a classical particle bouncing back and forth through the origin in a particle interpretation or a spherically symmetric wave reflecting from the walls. We note that "corrals" of heavy atoms can be constructed which approximate infinite circular potential wells on surfaces and the measured electron densities in such a configuration closely matches these predictions; an example was shown in Fig. 1.3.

- Finally, Fig. 15.12(b) shows a state with large angular momentum ($m = 10$) but with the smallest radial quantum number possible, specifically $a_{(0,10)}$ with

$a_{(0,10)} = 14.48$. This corresponds most closely to a particle in such a well undergoing uniform circular motion and the expected peaking of the quantum wavefunction near the walls (classically the particle would, after all, roll around the inner edge) is apparent. This might be the quantum equivalent of a "roulette wheel." The total energies of the $(4, 0)$ and $(0, 10)$ states differ by less than 1%, but the distribution between radial and rotational kinetic energy is clearly very different.

It is, in principle, possible to construct localized wave packets and track the quasi-classical ballistic motion of particles bouncing in the circular well,[7] but we will not consider that here due to the technical complexity.

15.3.3 Isotropic Harmonic Oscillator

We return to the isotropic harmonic oscillator in two-dimensions, defined by the potential

$$V(r) = \frac{1}{2}Kr^2 = \frac{1}{2}\mu\omega^2 r^2 \qquad (15.87)$$

with the corresponding Schrödinger equation

$$-\frac{\hbar^2}{2\mu}\left(\frac{d^2 R(r)}{dr^2} + \frac{1}{r}\frac{dR(r)}{dr}\right) + \left(\frac{\hbar^2 m^2}{2\mu r^2} + \frac{\mu\omega^2}{2}r^2\right) R(r) = ER(r) \qquad (15.88)$$

A standard change of variables, $r = \rho y$ with $\rho^2 = \hbar/\mu\omega$, reduces this to

$$\frac{d^2 R(y)}{dy^2} + \frac{1}{y}\frac{dR(y)}{dy} + \left(\epsilon - y^2 - \frac{m^2}{y^2}\right) R(y) = 0 \qquad (15.89)$$

with the dimensionless energy eigenvalue $\epsilon = 2E/\hbar\omega$. The large y-dependence can be extracted as in Section 9.2.1, while the behavior near the origin is guaranteed to be of the form $y^{|m|}$ from Eqn. (15.80). We are thus led to write

$$R(y) = y^{|m|} e^{-y^2/2} G(y) \qquad (15.90)$$

leading to

$$\frac{d^2 G(y)}{dy^2} + \left(\frac{2|m| + 1}{y} - 2y\right)\frac{dG(y)}{dy} + (\epsilon - 2 - 2|m|)) G(y) = 0 \qquad (15.91)$$

A somewhat less obvious change of variables to $z = y^2$ then yields the differential equation

$$\frac{d^2 G(z)}{dz^2} + \frac{dG(z)}{dz}\left(\frac{|m| + 1}{z} - 1\right) + G(z)\left(\frac{\epsilon - 2(|m| + 1)}{4z}\right) = 0 \qquad (15.92)$$

---

[7] See Doncheski *et al.*. (2003) for details.

which, to the mathematically sophisticated can be recognized as *Laguerre's equation* as discussed in Appendix E.7. A series solution of the form $G(z) = \sum_{s=0}^{\infty} b_s z^s$ yields the recursion relation

$$\frac{b_{s+1}}{b_s} = \frac{(s + (|m| + 1)/2 - \epsilon/4)}{(s+1)(s+|m|+1)} \longrightarrow \frac{1}{s} \qquad (15.93)$$

in the limit of large s implying (again, as in Section 9.2.1) that $G(z) \sim e^z \sim e^{y^2}$, which would be inconsistent with the desired behavior in Eqn. (15.90). Once again we find the series must terminate, yielding a polynomial of finite degree in z. The quantized energies are given in terms of the maximum power of this polynomial, $s_{\max} = n_r$, by the condition

$$\epsilon = 2|m| + 2 + 4s_{\max} \equiv 2|m| + 2 + 4n_r \qquad (15.94)$$

and we note that n_r also counts the number of radial nodes of the resulting polynomial. This leads to the quantized energies

$$E_{n_r,m} = \hbar\omega \left(|m| + 2n_r + 1\right) \qquad (15.95)$$

and the corresponding $G(z)$ are *generalized Laguerre polynomials*, denoted as $L_{n_r}^{(|m|)}(z)$. The first few of these are given here for later use:

$$L_0^{(k)}(z) = 1$$
$$L_1^{(k)}(z) = 1 + k - z \qquad (15.96)$$
$$L_2^{(k)}(z) = \frac{1}{2}\left(2 + 3k + k^2 - 2z(k + 2) + z^2\right)$$

The resulting constant polynomials in the case of $n_r = 0$ are especially important for the classical limit.

The complete (but unnormalized) solutions in polar coordinates are then given by

$$\psi_{n_r,m}(r,\theta) \propto r^{|m|} e^{-r^2/2\rho^2} L_{n_r}^{(|m|)}(r^2/\rho^2) e^{im\theta}. \qquad (15.97)$$

The energy spectrum and degeneracies thusly derived from polar coordinates are shown in Fig. 15.13 and we see that the degeneracies agree with those found using Cartesian coordinates. The wavefunctions for a given energy level in the two different schemes are necessarily linear combinations of each other, which can be shown explicitly in simple cases (P15.24).

A particularly easy classical limit to exhibit in this case is that corresponding to uniform circular motion in which case one would use $|m| \gg 1$ and $n_r = 0$ corresponding to the minimum possible radial kinetic energy. In this case, the

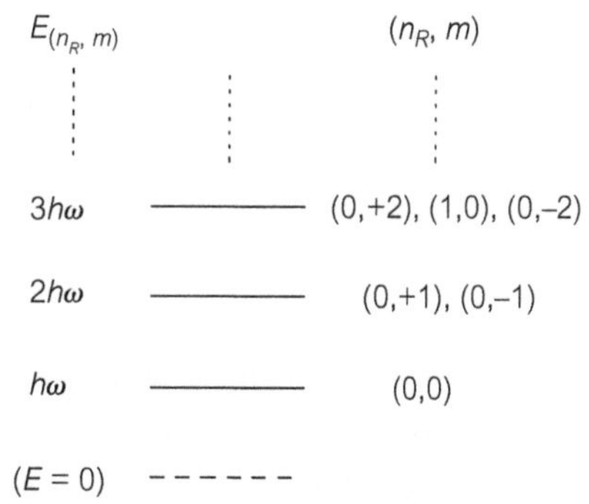

$E_{(n_R, m)}$ (n_R, m)

$3\hbar\omega$ ————   (0,+2), (1,0), (0,–2)

$2\hbar\omega$ ————   (0,+1), (0,–1)

Figure 15.13. Energy spectrum for the 2D harmonic oscillator problem as obtained in polar coordinates showing the same energy level degeneracy as in Fig. 15.5; values of (n_r, m) are shown for each level.

$\hbar\omega$ ————   (0,0)

$(E = 0)$ ------

Laguerre polynomials, $L_0^{|m|}(z)$, are constants so the radial probability density is proportional to

$$P(r) = |\psi_{0,m}(r,\theta)|^2 \sim r^{2|m|} e^{-r^2/\rho^2} \tag{15.98}$$

which has a maximum value when

$$0 = \frac{dP(r)}{dr} = \left(2|m|r^{2|m|-1} - \frac{2r^{2|m|+1}}{\rho^2}\right) e^{-r^2/\rho^2} \quad \text{or} \quad r_{\max}^2 = |m|\rho^2. \tag{15.99}$$

Recalling the definition ρ, we find that

$$r_{\max}^2 = \frac{|m|\hbar}{\mu\omega} = \frac{L_z}{\mu\omega} \tag{15.100}$$

The classical circular orbit, of constant radius r_0, is determined by Eqn. (15.43) where we take $\ddot{r}(t) = 0$ implying that

$$-\mu\omega^2 r_0 = -Kr_0 = F(r_0) = -\frac{L_z^2}{\mu r_0^3} \quad \text{or} \quad r_0^2 = \frac{L_z}{\sqrt{\mu K}} = \frac{L_z}{\mu\omega} \tag{15.101}$$

in agreement with Eqn. (15.100), and with the correspondence principle. Somewhat surprisingly, the form of this Schrödinger equation for a 2D simple harmonic oscillator (SHO) and its solutions are very similar to that for a charged particle in a uniform magnetic field (Section 18.5), which partly motivates our detailed study of it here.

15.4 **Questions and Problems**

Q15.1. Estimate the zero-point energy and spread in position of an electron bound to a (horizontal) surface because of gravity. Assume for simplicity that the potential is given by

$$V(z) = \begin{cases} \infty & \text{for } z < 0 \\ m_e g z & \text{for } z > 0 \end{cases} \tag{15.102}$$

What does this potential look like superimposed on Fig. 15.1?Repeat for the case of a neutron in the earth's gravitational potential. Can you imagine how you might see experimental evidence of *Quantum states of neutrons in Earth's gravitational field*[8]?

Q15.2. What would a plot of $|\phi(p_x, p_y)|^2$ versus (p_x, p_y) look like for the $(n, m) = (1, 1)$ state in the 2D infinite square well? How about for $(n, m) = (10, 10)$ or $(15, 30)$?

Q15.3. What would $|\phi(p_x, p_y; t)|^2$ look like for the "bouncing 2D wave packet" in Fig. 15.4 as a function of time?

Q15.4. Recall the wavefunctions in the circular well shown in Fig. 15.12 (a) and (b) corresponding to "radial" and "angular" motion, respectively. For the "radial" case in Fig. 15.12(a), it seems that there is a much larger probability of finding the particle near the origin than elsewhere. Is this consistent with what you know about the corresponding classical "motion"? The momentum space distributions for these two cases can be evaluated (numerically) and the resulting distributions are plotted in Fig. 15.14(a) and (b) Explain why they have the form they do; no numerical calculations are required, simply use your physical intuition, and think about what the momentum vectors in 2D would look like for the two cases of "radial" and "angular" motion. The small "bump" at $(p_x, p_y) = (0, 0)$ for the "radial" case seems to imply that there will be a reasonable chance of finding the particle with vanishing total momentum; is that true?

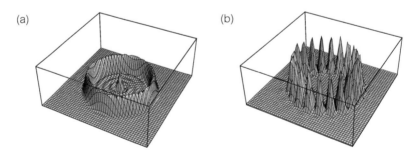

(a) (b)

Figure 15.14. Momentum-space probability distributions, $|\phi(p_x, p_y)|^2$ versus p_x, p_y, corresponding to the "radial" (a) and "angular" (b) wavefunctions in the circular infinite well of Fig. 15.12(a) and (b), respectively. Why do they look rather similar?

---

[8] See the paper of the same name by Nesvizhevsky *et al.*. (2002).

Q15.5. How would you solve the 2D harmonic oscillator problem using raising and lowering operators? Go as far as you can in generalizing the arguments of Section 13.2.

Q15.6. Are there any uncertainty relations in two-dimensions for pairs of variables like $x, \hat{p}_y$? Can you find an example of a state which has $\Delta x \, \Delta p_y = 0$ or arbitrarily small?

Q15.7. What are the appropriate commutation relations for θ and $\hat{L}_z$? Is there an associated uncertainty principle? Is there any problem[9] due to the fact that θ is only defined up to multiples of 2π, so that $\Delta\theta$ cannot be arbitrarily large?

Q15.8. "*Can you hear the shape of a drum*[10]?" The solution of the 2D Schrödinger equation for infinite wall boundaries of various shapes ("footprints") has many similarities with finding the allowed frequencies of vibration of 2D drumheads of the same shape, namely, solving the wave equation, with vanishing amplitudes on the variously shaped edges. To what extent do you think that knowing the "spectrum" of allowed energy eigenvalues (or allowed vibratory modes) allows you to determine the shape of the boundary?

P15.1. (a) Find the plane wave solutions of the time-independent free-particle Schrödinger equation in two-dimensions using Cartesian components. Show that they can be written in the form

$$\psi(\mathbf{r}; t) = e^{i(k_x x + k_y y - \omega t)} = e^{i(\mathbf{k}\cdot\mathbf{r} - \omega t)} \tag{15.103}$$

and find the dispersion relation relating $\mathbf{k}$ and ω.

(b) Explicitly construct a localized Gaussian wave packet with central momentum value $\mathbf{p}_0 = (p_0^x, p_0^y)$ with initial central position $\mathbf{r}_0 = (x_0, y_0)$. Calculate $\langle x \rangle_t$ and $\langle y \rangle_t$ for this state.

P15.2. Consider a 2D potential given by

$$V(x, y) = \begin{cases} 0 & \text{for } y < 0 \\ V_0 > 0 & \text{for } y > 0 \end{cases} \tag{15.104}$$

which is a 2D step-up potential.

(a) To examine plane wave scattering from such a step, consider a solution of the form

$$\psi(\mathbf{r}; t) = \begin{cases} Ie^{i(\mathbf{k}_1\cdot\mathbf{r} - \omega t)} + Re^{i(\mathbf{k}_1'\cdot\mathbf{r} - \omega t)} & \text{for } y < 0 \\ Te^{i(\mathbf{k}_2\cdot\mathbf{r} - \omega t)} & \text{for } y > 0 \end{cases} \tag{15.105}$$

where the wavevectors $\mathbf{k}_1, \mathbf{k}_1', \mathbf{k}_2$ are defined in Fig. 15.15. Match the wavefunction along the $z = 0$ boundary to find a relation between θ_1, θ_1', and θ_2 and compare to Snell's law of refraction.

---

[9] See, for example, Roy and Sannigrahi (1979).
[10] Which is the appropriately speculative title of a famous article by M. Kac (1966).

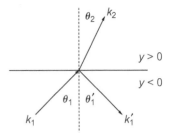

Figure 15.15. Wave numbers and reflection–refraction angles for 2D step-up potential.

(b) Match the derivatives of the wavefunction at the boundary (which direction?) to determine the reflection and transmission probabilities, namely, $|R/I|^2$ and $|T/I|^2$. Show that your results reduce to those in Section 11.2 for normal incidence, namely, $\theta_1 = 0$.

(c) For a given angle of incidence, θ, what is the minimum incident energy below which all of the incident particles will be reflected.

(d) Discuss how the notion of probability flux in Eqn. (4.32) can be generalized to two dimensions and how conservation of flux is realized in this problem.

P15.3. (a) Find the plane wave solutions to the Schrödinger equation for the 2D potential

$$V(x, y) = \begin{cases} 0 & \text{for } x > 0 \text{ and } y > 0 \\ +\infty & \text{otherwise} \end{cases} \qquad (15.106)$$

which is like a "corner (90°) reflector." What would you expect for the behavior of a wave packet incident on such a potential from various angles and for various incident energies?

(b) Can you construct "mirror" or "image" type solutions, by analogy with those discussed in Section 3.3 for one dimension? How many "images" do you need? Hint: Use your intuition from optics.

(c) Are there angles besides 90° between the two infinite wall barriers for which "mirror" or "image" solutions are easily obtained? For example, how about 45° or 60°?

P15.4. **Wave packet for projectile motion:**

(a) Write down the Schrödinger equation describing a particle moving in a vertical plane subject to a constant downward gravitational force, and show that it is separable.

(b) Use previously obtained results for the free particle (Section 3.2.2) and uniformly accelerating (Section 4.7.2) wave packets to write down a wave packet solution, $\psi_{WP}(x, y; t)$, for this problem.

(c) Evaluate $\langle x \rangle_t$ and $\langle y \rangle_t$ for the wave packet.

P15.5. Consider a wavefunction in the 2D infinite square box of Section 15.1.1 given at $t = 0$ by

$$\psi(x, y; 0) = Nx(L - x)y(L - y) \tag{15.107}$$

(a) Find N such that $\psi(x, y, 0)$ is normalized.

(b) What is the probability that a measurement of the energy of a particle described by this state would yield the ground state energy of the system at time $t = 0$? What is this probability at later times?

P15.6. Show that the exchange operator $\hat{E}_{x,y}$ in Eqn. (15.22) is Hermitian and that it's eigenvalues are ± 1.

P15.7. Accidental degeneracies?

(a) Find the energy eigenvalues for a particle in an infinite *rectangular* well with sides of lengths $L_1 \neq L_2$. Show that, in general, that no degenerate energy levels.

(b) Show that if L_1 and L_2 are commensurate, that is, if $L_1/L_2 = p/q$ is a ratio of integers, that two different levels characterized by pairs of integers (n_1, n_2) and $(pn_2/q, qn_1/p)$ can be degenerate. Show that an example is when $L_1 = 2L_2$ and the pairs $(4, 1)$ and $(2, 2)$ give rise to degenerate energy states. This phenomenon is often called "accidental degeneracy" as it is not due to any obvious symmetry. (Exchange symmetry is not an obviously useful idea for this asymmetric box.)

(c) For the special cases discussed in (b), consider a "bigger" square infinite well of size $L = qL_1 = pL_2$ on a side and show that the original box "fits into" the lower left-hand corner. Show that the degenerate wavefunctions in the original box, when extended to the larger box, are simply the standard degenerate pairs discussed in the text. This is illustrated in Fig. 15.16 for the explicit example in part (b).

P15.8. Isosceles triangle infinite well. Imagine the 2D infinite well defined by Eqn. (15.15) cut in half diagonally by the inclusion of another infinite wall along the $x = y$ direction, as in Fig. 15.17. The lower half of the potential is now an infinite well, but with an isosceles triangle ($45° - 45° - 90°$) footprint.

(a) Show that one of the linear combinations solutions in Eqn. (15.26) not only satisfies the Schrödinger equation for the new well, but also satisfies all of the relevant boundary conditions.

(b) What is the energy eigenvalue spectrum for this shape? Evaluate the 30 lowest-lying energy levels. Do you find any degeneracies?

P15.9. Show that there are N different states of the 2D SHO which have energy $(N + 1)\hbar\omega$, that is, calculate the degeneracy of each level. Use the solution in Cartesian components.

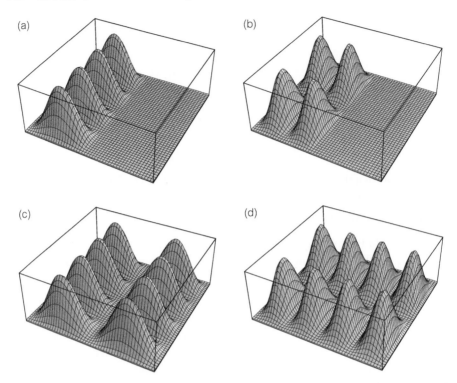

(a)   (b)

(c)   (d)

Figure 15.16. Accidental degeneracies in 2D rectangular boxes. Cases (a) and (b) correspond to the degenerate $(n_x, n_y) = (4, 1)$ and $(2, 2)$ levels in a box with $L_x = 2L_y$; (c) and (d) then correspond to the same levels in the related "extended" square box with $L = L_x = 2L_y$ where the levels are "naturally" degenerate.

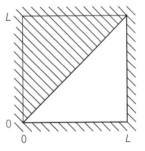

Figure 15.17. Isosceles ($45° − 45° − 90°$) infinite well footprint made by cutting a 2D square well in half along a diagonal.

P15.10. Evaluate the expectation values of x, y, $\hat{p}_x$, and $\hat{p}_y$ in any energy eigenstate of the 2D SHO of the form in Eqn. (15.28). Show that the expectation value of $\hat{L}_z$ also vanishes in any such state. How then can we have states of the 2D SHO with definite nonzero values of quantized angular momentum?

P15.11. Investigate possible accidental degeneracies in the energy spectrum of the non-isotropic 2D harmonic oscillator. What happens, for example, when $k_x = 4k_y$?

If you find any degeneracies, can you find any similarities with the discussion of P15.7?

P15.12. 2D harmonic oscillator wave packets.

(a) Using the explicit 1D SHO wave packets in Section 12.6.2, write down the probability distribution for a wave packet undergoing uniform circular motion in the isotropic 2D harmonic oscillator potential. Hint: Choose a wave packet for the x coordinate with appropriate initial position, but vanishing initial momentum and oppositely for the y coordinate packet.

(b) Calculate $\langle x \rangle_t$, $\langle y \rangle_t$, $\langle \hat{p}_x \rangle_t$, and $\langle \hat{p}_y \rangle_t$ for this state and show that they behave as expected.

(c) Calculate the expectation value of the angular momentum $\langle \hat{L}_z \rangle$ and show explicitly that it is conserved.

(d) Show that you can write the probability density for a wave packet representing counterclockwise motion in the form

$$|\psi(r,\theta;t)|^2 = \frac{1}{\pi L^2(t)} \exp\left(-(r^2 - 2rx_0 \cos(\theta - \omega t) + x_0^2\right) \quad (15.108)$$

(e) Repeat parts (a)–(c) for a wave packet representing a more general elliptical classical path.

(f) Repeat parts (a)–(c) for a wave packet under the influence of a "non-isotropic" spring of the form Eqn. (15.31).

P15.13. Angular momentum operator.

(a) Show that

$$e^{i\alpha \hat{L}_z} f(\theta) = f(\theta + \alpha) \quad (15.109)$$

(b) Calculate $[\hat{L}_z, \theta]$.

P15.14. Using the defining relations Eqn. (15.51) or (15.52), derive Eqns (15.53), (15.54), and (15.55) and show that

$$\frac{\partial^2}{\partial x^2} + \frac{\partial^2}{\partial y^2} = \frac{\partial^2}{\partial r^2} + \frac{1}{r}\frac{\partial}{\partial r} + \frac{1}{r^2}\frac{\partial^2}{\partial \theta^2} \quad (15.110)$$

P15.15. Substitute the trial solution Eqn. (15.75) into Bessel's equation and show that $\alpha = -1/2$ gives the next to leading behavior for large z.

P15.16. The *long-distance* (typically exponentially suppressed) behavior of quantum wavefunctions was discussed in Section 8.2.2. Use the same ideas, but with the rotational kinetic energy term, to derive the *short-distance* behavior of the solutions of the free-particle Schrödinger equation with circular symmetry. Specifically, assume that

$$\psi(r) \sim \exp\left(\pm\sqrt{\frac{2\mu}{\hbar^2}} \int^r \sqrt{V(r)}\, dr\right) \quad \text{where} \quad V(r) = \frac{m^2\hbar^2}{2\mu r^2} \quad (15.111)$$

and show that the suppression due to "tunneling" into the angular momentum barrier becomes a power law behavior.

P15.17. Classical probability distributions for the circular infinite well.

(a) Use the methods outlined in Section 5.1 to derive the classical radial probability distribution for a particle in the circular infinite well. Hint: use the energy relation in the form

$$E = \frac{1}{2}\mu\dot{r}^2 + \frac{L^2}{2\mu r^2} \tag{15.112}$$

to find an expression relating dr and dt, giving $P_{cl}(r)\, dr \propto dt$.

(b) Show that your result can be written in the form

$$P_{CL}(r) = \frac{r}{\sqrt{R^2 - R_{min}^2}\sqrt{r^2 - R_{min}^2}} \tag{15.113}$$

where $R_{min} = \sqrt{L^2/2\mu E}$ is the distance of closest approach. Discuss the limiting cases of purely radial and purely angular motion.

(c) Calculate $\langle r \rangle$ and Δr for discuss their behavior in the same limiting cases.

P15.18. Consider the wavefunction in the infinite circular well of radius a given by

$$\psi(r,\theta) = N(a - r)\sin^2(\theta) \tag{15.114}$$

(a) Find N such that ψ is properly normalized.

(b) What is the probability that a measurement of L_z in this state will yield $0\hbar$?; $\pm 1\hbar$?; $\pm 2\hbar$?; any other value? You should be able to obtain definite numerical answers for this part. Hint: What is the appropriate expansion theorem?

(c) What is the probability that a measurement of the position of the particle will find it in the inner half of the circle, that is, with $r < a/2$?

(d) What is the probability that a measurement of the energy finds this particle to be in the ground state of the well? Your answer will be in terms of integrals with Bessel functions which you do not need to evaluate numerically.

P15.19. For the infinite circular well of radius R, the angular wavefunctions, $\Theta_m(\theta)$, ensure that eigenstates with different values of m will be orthogonal. For a given value of m, show that one must have

$$\int_0^R dr\, r\, J_m(k_{(n_1,m)}r)\, J_m(k_{(n_2,m)}r) = 0 \quad \text{if} \quad n_1 \neq n_2 \tag{15.115}$$

If you have access to and expertize with an all-purpose computer mathematics package such as *Mathematica*®, confirm this by numerical integration for several cases if you can.

P15.20. Centrifugal force in the circular infinite well. Classically, a particle undergoing uniform circular motion would require a force given by

$$F_c(R) = \frac{\mu v^2}{R} = \mu \omega^2 R = \frac{\mu R^2 \omega^2}{R} = \frac{2T_{\text{rot}}}{R} \qquad (15.116)$$

to keep it in motion. Calculate the force exerted by the rapidly spinning ball in the infinite well problem by considering the change in energy when the wall is slowly moved outward a small amount dR, that is,

$$dW = \frac{\hbar^2 a_{(n_r,m)}^2}{2\mu R^2} - \frac{\hbar^2 a_{(n_r,m)}^2}{2\mu (R + dR)^2} = F \cdot dR \qquad (15.117)$$

Show that the force exerted by the wall on the particle is consistent with the classical result.

P15.21. Variational calculation for the circular infinite well.

(a) Make a variational estimate of the ground state energy of the circular infinite well with radius a. Since this state will necessarily have $m = 0$, this amounts to evaluating the energy functional

$$E[\psi] = \frac{\langle \psi | \hat{H}_r | \psi \rangle}{\langle \psi | \psi \rangle} \qquad (15.118)$$

where

$$\hat{H}_r = -\frac{\hbar^2}{2\mu} \left(\frac{d^2}{dr^2} + \frac{1}{r}\frac{d}{dr} \right) \qquad (15.119)$$

Assume a trial wavefunction of the form $\psi(r,\theta) = R(r) = a^\lambda - r^\lambda$ where λ is used as the variational parameter. Compare your answer to the exact ground state energy obtained from Eqn. (15.85) and the table of Bessel function zeroes in Eqn. (15.87).

(b) Because the angular wavefunctions $\Theta_m(\theta)$ form an orthogonal set, one can actually make a rigorous variational estimate for the lowest energy state for every value of m (Why?). Recalling the required behavior of the wavefunction near the origin, use a trial wavefunction of the form

$$\psi_{0,m}(r,\theta) = R_{(0,m)}(r) \, e^{im\theta} \quad \text{where} \quad R_{(0,m)}(r) = (a^\lambda - r^\lambda) r^m \qquad (15.120)$$

Show that the variational energy is given by

$$E_{\text{var}}(\lambda; m) = \frac{\hbar^2}{2\mu a^2} \frac{(1+m)(1+\lambda+m)(2+\lambda+2m)}{(\lambda+m)} \qquad (15.121)$$

with a minimum value

$$E_{\text{var}}^{\min}(m) = \frac{\hbar^2}{2\mu a^2} \left[\frac{(1+m)(4+2m+(3+m)\sqrt{2+m})}{\sqrt{2+m}} \right] \qquad (15.122)$$

Compare this to your answer in part (a) and to the exact answers for $m = 1, 2, 3, 4$ in Eqn. (15.87)

(c) Sketch the variational radial wavefunctions versus r for increasing values of m and note that the probability is increasingly peaked near the boundary. Using these approximate wavefunctions, find the value of r at which the probability, that is, $r|R_{0,m}(r)|^2$ is peaked, and show that it approaches a as $m \to \infty$.

P15.22. Semicircular infinite circular well. Consider a particle of mass μ in a "half" infinite circular potential well defined by

$$V(r, \theta) = \begin{cases} 0 & \text{for } 0 < \theta < \pi \text{ and } r < R \\ \infty & \text{otherwise} \end{cases} \qquad (15.123)$$

(a) Find the allowed energies and wavefunctions in terms of those of the "full" infinite circular well. Discuss the degeneracy of each level (if any).

(b) Show that the angular momentum operator, $\hat{L}_z$, is still Hermitian, and discuss why.

(c) Show that $\hat{L}_z$ no longer commutes with the Hamiltonian, so that $\langle \hat{L}_z \rangle_t$ need no longer be constant in time.

(d) Consider the wavefunction

$$\psi(r, \theta; 0) = AJ_0(k_{(0,1)}r) \sin(\theta) + BJ_0(k_{(0,2)}r) \sin(2\theta) \qquad (15.124)$$

Find the wavefunction for later times and evaluate $\langle \hat{L}_z \rangle_t$ and show that it is not constant. Show that any initial wavefunction with only even m or only odd m components will have a constant (and vanishing) angular momentum.

P15.23. 2D annular infinite well. Consider a 2D potential with circular symmetry corresponding to two infinite walls at $r = b, a$, defined via

$$V(r) = \begin{cases} \infty & \text{for } r < b \\ 0 & \text{for } b < r < a \\ \infty & \text{for } r > a \end{cases} \qquad (15.125)$$

(a) Find the allowed solutions and derive the condition which determines the energy eigenvalues for each value of m. Hint: The origin is excluded from the region where solutions are to be considered.

(b) Estimate the ground state energy when $\delta = a - b \ll a$, that is, the two walls are very close together. Is there any similarity in this limit to a long rectangular potential of dimensions $2\pi a \times \delta$?

P15.24. 2D harmonic oscillator solutions in polar coordinates.

(a) Confirm that the $L_{n_r}^{(k)}(z)$ given in Eqn. (15.97) actually solve Eqn. (15.92); for the case of $n_r = 0$, show that the solutions are trivial.

(b) Show that the degeneracy of the level with energy $E = N_s \hbar \omega$ is N_s.

(c) Compare the wavefunctions for the states with $E = (N+1)\hbar\omega$ in Cartesian and polar coordinates and show explicitly that they can be written as linear combinations of each other for the cases $N = 0, 1, 2$.

P15.25. Consider a particle of mass m moving in the 2D "one-quarter" isotropic harmonic oscillator potential

$$V(r,\theta) = \begin{cases} Kr^2/2 & \text{for } 0 < \theta < \pi/2 \\ \infty & \text{otherwise} \end{cases} \qquad (15.126)$$

This corresponds (very roughly) to a particle in two-dimensions, tied to a spring, which is tethered at a square corner.

(a) What are the allowed energy levels and wavefunctions in such a potential? How do they compare to the "full" harmonic oscillator?

P15.26. Consider a particle of mass m moving in a modified 2D harmonic potential given by

$$V(x,y) = \frac{K}{2}\left(x^2 + y^2\right) + \lambda xy \qquad (15.127)$$

(a) Show that the energy eigenvalue spectrum for this potential can be obtained explicitly. Hint: Make a change of variables from x, y to $\bar{x}, \bar{y}$, using a simple rotation, in order to eliminate the cross-term. Are there conditions on the allowed values of K, λ for this problem to be well-posed?

(b) Assuming that $\lambda \ll K$, write down the three lowest-energy energy eigenvalues.

(c) Now, treating the λxy term as a perturbation, and using the standard oscillator problem as the unperturbed system, calculate the first-, second-, and third-order corrections to the *ground state* energy using Eqns (10.112), (10.115), and (10.117), respectively. Hint: Use the matrix element results in Section 9.2.2. Compare your answers to the exact result as a series in λ.

(d) Finally, treating the next two lowest-lying states using degenerate perturbation theory (why?), find how they split in energy, and compare again to the exact result.

SIXTEEN

The Schrödinger Equation in Three Dimensions

Most "real-world" applications of quantum mechanics arise in three-dimensional systems where the Schrödinger equation for one particle takes the form

$$\hat{H}\psi(\mathbf{r}) = E\psi(\mathbf{r}) \tag{16.1}$$

where the Hamiltonian

$$\hat{H} = \frac{\hat{\mathbf{p}}^2}{2m} + V(\mathbf{r}) \tag{16.2}$$

is written in terms of the gradient in three dimensions, $\hat{\mathbf{p}} \equiv (\hbar/i)\nabla$. Two-body problems with a Hamiltonian of the form

$$\hat{H} = \frac{\hat{\mathbf{p}_1}^2}{2m_1} + \frac{\hat{\mathbf{p}_2}^2}{2m_2} + V(\mathbf{r}_1 - \mathbf{r}_2) \tag{16.3}$$

are intrinsically no more difficult as the use of relative and center-of-mass coordinates,

$$\mathbf{r}_{\text{rel}} = \mathbf{r}_1 - \mathbf{r}_2 \tag{16.4}$$

$$\mathbf{R}_{\text{cm}} = \frac{m_1\mathbf{r}_1 + m_2\mathbf{r}_2}{m_1 + m_2} \tag{16.5}$$

just as in one dimension (Section 14.3), leads to an effective one-particle Schrödinger equation

$$\left(\frac{\hat{\mathbf{p}}^2}{2\mu} + V(\mathbf{r})\right)\psi(\mathbf{r}) = E\psi(\mathbf{r}) \tag{16.6}$$

for the relative coordinate $\mathbf{r} \equiv \mathbf{r}_{rel}$ with the reduced mass μ as the mass parameter.

Explicit solutions of Eqn. (16.6) require the use of a specific coordinate system, the choice of which, in turn, usually depends on the form of the potential $V(\mathbf{r})$. Solutions in Cartesian coordinates, $\mathbf{r} = (x, y, z)$, which generalize the results of Chapter 5 and Section 15.1 are possible, and were implicitly used in the discussion in Section 7.3. For problems with a symmetry about a particular rotation axis, cylindrical coordinates given by $\mathbf{r} = (r, \theta, z)$ are often useful (see Section 18.5 where a uniform magnetic field singles out specific axis) and the two-dimensional polar coordinate results of Section 15.3 can be trivially extended. Solutions using less familiar systems such as parabolic coordinates (see P17.14) are even sometimes employed. For many three-dimensional problems for which there is no preferred axis, a separation of variables using spherical coordinates is most natural, and that is the subject of the next section.

16.1 Spherical Coordinates and Angular Momentum

For many important cases, the potential is spherically symmetric, that is, $V(\mathbf{r}) = V(r)$, and it is natural to attempt solutions using spherical coordinates, namely

$$x = r \sin(\theta) \cos(\phi)$$
$$y = r \sin(\theta) \sin(\phi)$$
$$z = r \cos(\theta) \tag{16.7}$$

or

$$r = \sqrt{x^2 + y^2 + z^2}$$
$$\tan(\phi) = y/x$$
$$\cos(\theta) = z/\sqrt{x^2 + y^2 + z^2} \tag{16.8}$$

as illustrated in Fig. 16.1.

The kinetic energy operator can always be written as

$$\hat{T} = \frac{\hat{p}^2}{2m} = -\frac{\hbar^2}{2m} \nabla^2 = -\frac{\hbar^2}{2m} \left(\frac{\partial^2}{\partial x^2} + \frac{\partial^2}{\partial y^2} + \frac{\partial^2}{\partial z^2} \right) \tag{16.9}$$

Using the defining relations for spherical coordinates, Eqns (16.7) and (16.8), one can (somewhat tediously) show that $\hat{T}$ can be written in the form

$$\hat{T} = -\frac{\hbar^2}{2m} \left[\frac{\partial^2}{\partial r^2} + \frac{2}{r} \frac{\partial}{\partial r} + \frac{1}{r^2} \left(\frac{\partial^2}{\partial \theta^2} + \cot(\theta) \frac{\partial}{\partial \theta} + \frac{1}{\sin^2(\theta)} \frac{\partial^2}{\partial \phi^2} \right) \right] \tag{16.10}$$

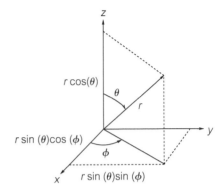

Figure 16.1. Definitions of spherical coordinates.

which, not surprisingly, is seen to measure the "wiggliness" of the wavefunction in the r, θ, and ϕ directions. Motivated by the results of Chapter 15, we expect an intimate connection between the θ and ϕ derivative terms, the rotational kinetic energy, and the angular momentum, now in three dimensions.

The vector angular momentum operator,

$$\hat{\mathbf{L}} = \mathbf{r} \times \hat{\mathbf{p}} \tag{16.11}$$

has the Cartesian components

$$\hat{L}_x = y\hat{p}_z - z\hat{p}_y \tag{16.12}$$

$$\hat{L}_y = z\hat{p}_x - x\hat{p}_z \tag{16.13}$$

$$\hat{L}_z = x\hat{p}_y - y\hat{p}_x \tag{16.14}$$

which are all obviously individually Hermitian. A change to spherical coordinates shows that

$$\hat{L}_z = \frac{\hbar}{i}\frac{\partial}{\partial\phi} \tag{16.15}$$

which is similar to Eqn. (15.55) in two-dimensions; one should note, however, the change in notation as the azimuthal angle in three dimensions is conventionally labeled ϕ. The other components have a slightly more complicated form, namely

$$\hat{L}_x = \frac{\hbar}{i}\left(-\sin(\phi)\frac{\partial}{\partial\theta} - \cot(\theta)\cos(\phi)\frac{\partial}{\partial\phi}\right) \tag{16.16}$$

and

$$\hat{L}_y = \frac{\hbar}{i}\left(\cos(\phi)\frac{\partial}{\partial\theta} - \cot(\theta)\sin(\phi)\frac{\partial}{\partial\phi}\right) \tag{16.17}$$

which together give information on the polar angle θ.

For an object with different moments of inertia about each of the three axes (I_x, I_y, I_z), the classical rotational kinetic energy is

$$T_{\text{rot}} = \frac{L_x^2}{2I_x} + \frac{L_y^2}{2I_y} + \frac{L_z^2}{2I_z} \tag{16.18}$$

which simplifies to

$$T_{\text{rot}} = \frac{L^2}{2I} \tag{16.19}$$

if $I_x = I_y = I_z = I$. For a single point particle in a spherically symmetric potential, one has $I = mr^2$ while for a pair of point masses the moment of inertia about the center of mass is $I = \mu r^2$. Motivated by this, we can easily show (P16.2) that

$$\hat{L}^2 = \hat{L}_x^2 + \hat{L}_y^2 + \hat{L}_z^2 = -\hbar^2 \left(\frac{\partial^2}{\partial \theta^2} + \cot(\theta) \frac{\partial}{\partial \theta} + \frac{1}{\sin^2(\theta)} \frac{\partial}{\partial \phi^2} \right) \tag{16.20}$$

which only acts on the angular degrees of freedom. This shows that the kinetic energy operator Eqn. (16.9) can be written in the shorthand form

$$\hat{T} = \hat{T}_r + \frac{\hat{L}^2}{2\mu r^2} \tag{16.21}$$

where the *radial kinetic energy operator* is

$$\hat{T}_r = -\frac{\hbar^2}{2\mu} \left(\frac{\partial^2}{\partial r^2} + \frac{2}{r} \frac{\partial}{\partial r} \right) \tag{16.22}$$

For a spherically symmetric potential, $V(r)$, the Hamiltonian is then clearly separable, and if we write a solution of the form $\psi(r, \theta, \phi) = R(r)Y(\theta, \phi)$ we obtain

$$-\frac{2\mu}{\hbar^2} \frac{r^2}{R(r)} \left(\hat{T}_r + V(r) - E \right) R(r) = \frac{1}{\hbar^2} \frac{1}{Y(\theta, \phi)} \left[\hat{L}^2 Y(\theta, \phi) \right] \tag{16.23}$$

or

$$F(r) = G(\theta, \phi) = \text{constant} = l(l+1) \tag{16.24}$$

We have written the separation constant as $l(l + 1)$ in anticipation of the result that the $Y(\theta, \phi)$ are the eigenfunctions of the square of the angular momentum operator given by

$$\hat{L}^2 Y(\theta, \phi) = l(l+1)\hbar^2 Y(\theta, \phi) \tag{16.25}$$

These eigenfunctions can be studied once and for all for any problem with spherical symmetry, and their properties will be discussed extensively in the next section.

For a given value of $l(l + 1)$, the corresponding dynamical equation which determines the energy eigenvalues is the *radial Schrödinger equation*,

$$-\frac{\hbar^2}{2\mu}\left(\frac{d^2 R(r)}{dr^2} + \frac{2}{r}\frac{dR(r)}{dr}\right) + \left(V(r) + \frac{l(l+1)\hbar^2}{2\mu r^2}\right)R(r) = ER(r) \quad (16.26)$$

The normalization condition

$$1 = \int d\mathbf{r}|\psi(\mathbf{r})|^2 = \int_0^\infty r^2\,dr\int d\Omega\,|\psi(r,\theta,\phi)|^2 \quad (16.27)$$

can then be enforced by demanding that

$$1 = \int_0^{2\pi} d\phi\int_0^\pi \sin(\theta)\,d\theta\,|Y(\theta,\phi)|^2 \quad (16.28)$$

and

$$1 = \int_0^\infty r^2\,dr\,|R(r)|^2 \quad (16.29)$$

separately. In three dimensions, an interesting simplification occurs if we write

$$R(r) = \frac{u(r)}{r} \quad (16.30)$$

in that the radial equation, Eqn. (16.26), becomes

$$-\frac{\hbar^2}{2\mu}\frac{d^2 u(r)}{dr^2} + \left(V(r) + \frac{l(l+1)\hbar^2}{2\mu r^2}\right)u(r) = Eu(r) \quad (16.31)$$

which is of the form of a standard one-dimensional Schrödinger equation with the inclusion of the centrifugal term $l(l + 1)\hbar^2/2\mu r^2$, similar to the results in two-dimensions in Section 15.2. The normalization condition is also similar as we now require that

$$\int_0^\infty |u(r)|^2\,dr = 1 \quad (16.32)$$

Because $R(r)$ should be well behaved at the origin, we must assume that

$$\lim_{r\to 0} u(r) = 0 \quad (16.33)$$

Taken together, this form is reminiscent of the "half" potential well problem considered in P9.10 for the harmonic oscillator. In such problems, which can be described by some potential $V(x)$ for $x > 0$, but with an infinite wall at the origin, we can make use of the odd solutions of the "full well" problem, as they satisfy the Schrödinger equation for $x > 0$ as well as the boundary condition at the wall. (See also the discussion in Section 8.3.2.)

In a central potential, the classical angular momentum vector, all of its components as well as its magnitude, will be conserved quantities and can, in principle, be known with arbitrary precision; we wish to understand to what extent this holds true in quantum systems. We first note that $\hat{L}^2$ commutes with the Hamiltonian since

$$[\hat{H}, \hat{L}^2] = [\hat{T}_r + V(r) + \frac{1}{2\mu r^2}\hat{L}^2, \hat{L}^2]$$

$$= [\hat{T}_r + V(r), \hat{L}^2] + \frac{1}{2\mu r^2}[\hat{L}^2, \hat{L}^2] = 0 \qquad (16.34)$$

since $\hat{L}^2$ acts only on angular variables and certainly commutes with itself. This implies that:

- The solutions of the Schrödinger equation for a rotationally invariant potential can have definite values of both the energy E and total angular momentum $l(l+1)\hbar^2$.

- The total angular momentum will be a constant in time.

While we may thus be able to know the "length" of the observable corresponding to $\hat{L}^2$ precisely, simultaneous exact measurements of the individual components are not possible as we have

$$[\hat{L}_x, \hat{L}_y] = [y\hat{p}_z - z\hat{p}_y, z\hat{p}_x - x\hat{p}_z]$$

$$= [y\hat{p}_z, z\hat{p}_x] - [y\hat{p}_z, x\hat{p}_z] - [z\hat{p}_y, z\hat{p}_x] + [z\hat{p}_y, x\hat{p}_z]$$

$$= y[\hat{p}_z, z]\hat{p}_x + x[z, \hat{p}_z]\hat{p}_y$$

$$= \frac{\hbar}{i}\left(y\hat{p}_x - x\hat{p}_y\right)$$

$$= i\hbar\hat{L}_z \neq 0 \qquad (16.35)$$

One similarly finds that

$$[\hat{L}_y, \hat{L}_z] = i\hbar\hat{L}_x \quad \text{and} \quad [\hat{L}_z, \hat{L}_x] = i\hbar\hat{L}_y \qquad (16.36)$$

both of which have the same form as Eqn. (16.35) with cyclic permutations of x, y, z.

On the other hand, one does find that

$$[\hat{L}_x, \hat{L}^2] = [\hat{L}_x, \hat{L}_x^2 + \hat{L}_y^2 + \hat{L}_z^2]$$

$$= \hat{L}_y[\hat{L}_x, \hat{L}_y] + [\hat{L}_x, \hat{L}_y]\hat{L}_y + \hat{L}_z[\hat{L}_x, \hat{L}_z] + [\hat{L}_x, \hat{L}_z]\hat{L}_z$$

$$= i\hbar\left(\hat{L}_y\hat{L}_z + \hat{L}_x\hat{L}_y - \hat{L}_z\hat{L}_y - \hat{L}_y\hat{L}_x\right)$$

$$= 0 \qquad (16.37)$$

as well as

$$[\hat{L}_y, \hat{L}^2] = [\hat{L}_z, \hat{L}^2] = 0 \tag{16.38}$$

so that one can measure the pairs (L_x, L^2) or (L_y, L^2) or (L_z, L^2) with arbitrary precision. Since each component operator also commutes with the Hamiltonian, we find that:

- The maximally large set of commuting operators for a general spherically symmetric potential will correspond to precisely determined values of the energy, the total angular momentum squared, and *one* component of L; the conventional choice is to look for simultaneous eigenfunctions of $\hat{H}, \hat{L}^2$, and $\hat{L}_z$. These will then be the maximally large set of conserved quantities.

We note that for certain potentials, such as the Coulomb potential considered in Chapter 17, there can be other conserved quantities because of additional special symmetries. Because the eigenfunctions of $\hat{L}^2$ and $\hat{L}_z$ are important for any spherically symmetric problem, we devote the next section to the examination of their properties.

16.2 Eigenfunctions of Angular Momentum

16.2.1 Methods of Derivation

The derivation of the eigenfunctions of the angular momentum operators is a standard one in many books on quantum mechanics. We will briefly discuss several of the usual methods of analysis of the properties of the $Y(\theta, \phi)$ before turning to attempts to visualize them, discussing their classical limit, and their application in important physical systems. We begin with a differential equation based approach, and then discuss the usefulness of operator methods to this problem.

The eigenvalue problem for $\hat{L}^2$ can be written as

$$\hat{L}^2 Y(\theta, \phi) = -\hbar^2 \left(\frac{\partial^2 Y}{\partial \theta^2} + \cot(\theta) \frac{\partial Y}{\partial \theta} + \frac{1}{\sin^2(\theta)} \frac{\partial^2 Y}{\partial \phi^2} \right)$$
$$= l(l+1)\hbar^2 Y(\theta, \phi) \tag{16.39}$$

The corresponding problem for $\hat{L}_z$ has already been discussed in the last chapter, in a different notation, where we found that

$$\Phi_m(\phi) = \frac{1}{\sqrt{2\pi}} e^{im\phi} \tag{16.40}$$

satisfies

$$\hat{L}_z \Phi_m(\phi) = \frac{\hbar}{i} \frac{\partial}{\partial \phi} \Phi_m(\phi) = m\hbar \Phi_m(\phi) \tag{16.41}$$

for integral values of m. We are thus led to write

$$Y_{l,m}(\theta, \phi) = \Theta_{l,m}(\theta) \Phi_m(\phi) \tag{16.42}$$

so that Eqn. (16.39) can be seen to be separable as

$$-\frac{\sin^2(\theta)}{\Theta_{l,m}(\theta)} \left[\frac{d^2 \Theta_{l,m}}{d\theta^2} + \cot(\theta) \frac{d\Theta_{l,m}}{d\theta} + l(l+1)\Theta_{l,m} \right] = \frac{1}{\Phi_m(\phi)} \frac{d^2 \Phi_m}{d\phi^2} = -m^2 \tag{16.43}$$

The new information on the total angular momentum eigenvalues, $l(l+1)$, is contained in the equation for $\Theta_{l,m}(\theta)$, namely

$$\sin^2(\theta) \frac{d^2 \Theta_{l,m}(\theta)}{d\theta^2} + \sin(\theta) \cos(\theta) \frac{d\Theta_{l,m}(\theta)}{d\theta}$$
$$+ \left(l(l+1) \sin^2(\theta) - m^2 \right) \Theta_{l,m}(\theta) = 0 \tag{16.44}$$

This equation can be turned into one of the "handbook" variety by noting that with the substitution, $z = \cos(\theta)$, Eqn. (16.44) becomes

$$(1 - z^2) \frac{d^2 \Theta_{l,m}(z)}{dz^2} - 2z \frac{d\Theta_{l,m}(z)}{dz} + \left(l(l+1) - \frac{m^2}{(1-z^2)} \right) \Theta_{l,m}(z) = 0 \tag{16.45}$$

which was studied by Legendre. We will not proceed further along these lines except to quote the result that the well-known solutions to this equation are polynomials in z, labeled $P_l^m(z)$, called the *associated Legendre polynomials*[1] which require l to be a nonnegative integer. We will determine and tabulate more of their properties using other methods below.

A possibly more profitable and interesting approach is to use ladder operators, factorization techniques, and a more formal notation, as with the harmonic oscillator in Section 13.2. For example, we can denote the angular wavefunctions in quantum state notation via

$$Y_{l,m}(\theta, \phi) \implies |Y_{l,m}\rangle \tag{16.46}$$

so that, for example,

$$\hat{L}^2 |Y_{l,m}\rangle = l(l+1)\hbar^2 |Y_{l,m}\rangle \tag{16.47}$$

---

[1] For details, see the mathematical handbook by Abramowitz and Stegun (1964), Arfken (1985), or Appendix E.6.

and the normalization condition on the angular wavefunctions reads

$$1 = \int d\Omega \, |Y_{l,m}(\theta,\phi)|^2 = \langle Y_{l,m} | Y_{l,m} \rangle \tag{16.48}$$

In this language, we can argue that

$$\langle Y_{l,m} | \hat{L}^2 | Y_{l,m} \rangle = \langle Y_{l,m} | \hat{L}_x^2 + \hat{L}_y^2 + \hat{L}_z^2 | Y_{l,m} \rangle \tag{16.49}$$

$$= \langle \hat{L}_x Y_{l,m} | \hat{L}_x Y_{l,m} \rangle + \langle \hat{L}_y Y_{l,m} | \hat{L}_y Y_{l,m} \rangle + \langle \hat{L}_z Y_{l,m} | \hat{L}_z Y_{l,m} \rangle$$

$$\langle Y_{l,m} | \hat{L}^2 | Y_{l,m} \rangle \geq 0$$

where have used the fact that the $\hat{L}_{x,y,z}$ are Hermitian, and that the norm of any vector is nonnegative. This immediately implies of course that $l(l + 1) \geq 0$, which we take to mean that $l \geq 0$ as well.

A pair of operators which will have the effect *raising* and *lowering* operators can be defined via

$$\hat{L}_+ = \hat{L}_x + i\hat{L}_y \quad \text{and} \quad \hat{L}_- = \hat{L}_x - i\hat{L}_y \tag{16.50}$$

which are not Hermitian, but which do satisfy $\hat{L}_+^\dagger = \hat{L}_-$ and $\hat{L}_-^\dagger = \hat{L}_+$. A useful product relation is

$$\hat{L}_+ \hat{L}_- = \left(\hat{L}_x + i\hat{L}_y \right) \left(\hat{L}_x - i\hat{L}_y \right)$$

$$= \hat{L}_x^2 + \hat{L}_y^2 - i[\hat{L}_x, \hat{L}_y]$$

$$\hat{L}_+ \hat{L}_- = \hat{L}_x^2 + \hat{L}_y^2 + \hbar \hat{L}_z \tag{16.51}$$

which gives

$$\hat{L}^2 = \hat{L}_+ \hat{L}_- + \hat{L}_z^2 - \hbar \hat{L}_z \tag{16.52}$$

with a similar derivation yielding

$$\hat{L}^2 = \hat{L}_- \hat{L}_+ + \hat{L}_z^2 + \hbar \hat{L}_z \tag{16.53}$$

Equally simple manipulations imply that

$$[\hat{L}_+, \hat{L}_-] = 2\hbar \hat{L}_z \tag{16.54}$$

$$[\hat{L}_+, \hat{L}_z] = -\hbar \hat{L}_+ \tag{16.55}$$

$$[\hat{L}_-, \hat{L}_z] = +\hbar \hat{L}_- \tag{16.56}$$

and

$$[\hat{L}_+, \hat{L}^2] = [\hat{L}_-, \hat{L}^2] = 0 \tag{16.57}$$

The ladder operators, $\hat{L}_{\pm}$, have no effect on the l quantum number when acting on a $|Y_{l,m}\rangle$ state since, because of Eqn. (16.57),

$$\hat{L}^2\left(\hat{L}_+|Y_{l,m}\rangle\right) = \hat{L}_+\left(\hat{L}^2|Y_{l,m}\rangle\right) = l(l+1)\hbar^2\left(\hat{L}_+|Y_{l,m}\rangle\right) \qquad (16.58)$$

and similarly for $\hat{L}_-$. On the other hand, using Eqn. (16.55), we have

$$\hat{L}_z\left(\hat{L}_+|Y_{l,m}\rangle\right) = \left(\hat{L}_+\hat{L}_z + \hbar\hat{L}_+\right)|Y_{l,m}\rangle = (m+1)\hbar\left(\hat{L}_+|Y_{l,m}\rangle\right) \qquad (16.59)$$

so that the application of the raising operator $\hat{L}_+$ to the state $|Y_{l,m}\rangle$ leaves the l value unchanged, but increases the m value by one unit of $\hbar$. This implies that

$$\hat{L}_+|Y_{l,m}\rangle \propto |Y_{l,m+1}\rangle \qquad (16.60)$$

with the similar relation for the lowering operator

$$\hat{L}_-|Y_{l,m}\rangle \propto |Y_{l,m-1}\rangle \qquad (16.61)$$

For a fixed value of l, the raising and lowering operators move one up and down the ladder in m in unit steps of $\hbar$ as shown in Fig. 16.2. The normalized versions these last two relations can be shown (P16.3) to be

$$\hat{L}_+|Y_{l,m}\rangle = \sqrt{l(l+1) - m(m+1)}\hbar|Y_{l,m+1}\rangle \qquad (16.62)$$

$$\hat{L}_-|Y_{l,m}\rangle = \sqrt{l(l+1) - m(m-1)}\hbar|Y_{l,m-1}\rangle \qquad (16.63)$$

More constraints can be obtained by using the fact that

$$\langle Y_{l,m}|\hat{L}_+\hat{L}_-|Y_{l,m}\rangle = \langle \hat{L}_-Y_{l,m}|\hat{L}_-Y_{l,m}\rangle \geq 0 \qquad (16.64)$$

which gives

$$\langle Y_{l,m}|\hat{L}^2 - \hat{L}_z^2 - \hbar\hat{L}_z|Y_{l,m}\rangle = \hbar^2\left(l(l+1) - m^2 + m\right) \geq 0 \qquad (16.65)$$

so that $l(l+1) \geq m(m-1)$; a similar restriction, namely that $l(l+1) \geq m(m+1)$ can be derived by considering the expectation value of $\hat{L}_-\hat{L}_+$. These bounds are illustrated in Fig. 16.3 from which it is clear that one obtains $-l \leq m \leq +l$

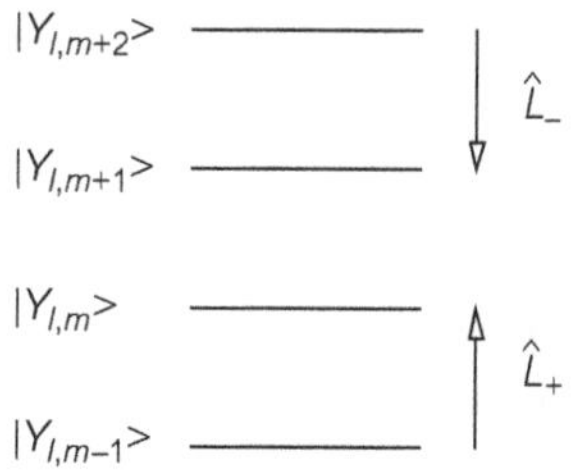

Figure 16.2. Effect of raising and lowering operators for the spherical harmonics, $|Y_{l,m}\rangle$.

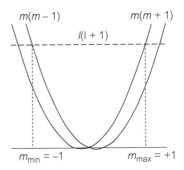

Figure 16.3. Bounds on maximum and minimum values of m for a given l.

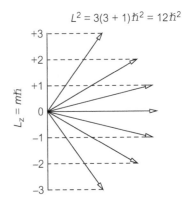

Figure 16.4. Vector diagram for quantized angular momentum.

for a given l value. We thus know that there is a state with a maximal value of m for each l, say m_+, and clearly the raising operator must annihilate that state, namely

$$\hat{L}_+ | Y_{l,m_+} \rangle = 0 \tag{16.66}$$

In this case, the inequality of Eqn. (16.65) is saturated and we have

$$l(l+1) = m_+(m_+ + 1) \quad \text{or} \quad m_+ = +l \tag{16.67}$$

and $\hat{L}_-$ acting on the state of minimal m implies that $m_- = -l$. We are finding that

• For a fixed value of (integral) l that the allowed values of m are given by

$$m = -l, -(l-1), \ldots, +(l-1), +l \tag{16.68}$$

• For a given value of l, there are $(2l+1)$ values of m given by Eqn. (16.68).

The quantized values of the magnitude of L^2 and L_z are often presented in a vector diagram as shown in Fig. 16.4, which will be useful in discussing the classical limit.

The raising and lowering operators are also helpful in constructing explicit solutions for the angular momentum eigenfunctions. Given a value of l, for the maximal value of $m_+ = l$, we have

$$Y_{l,l}(\theta, \phi) \propto \Theta_{l,l}(\theta)\, e^{il\phi} \tag{16.69}$$

which, by Eqn. (16.66), must satisfy

$$0 = \hat{L}_+ Y_{l,l}(\theta, \phi) = \hbar e^{i\phi} \left(\frac{\partial}{\partial \theta} + i \cot(\theta) \frac{\partial}{\partial \phi} \right) \left(\Theta_{l,l}(\theta)\, e^{il\phi} \right)$$

$$0 = \hbar\, e^{i\phi}\, e^{il\phi} \left(\frac{d\Theta_{l,l}}{d\theta} - l \cot(\theta) \Theta_{l,l}(\theta) \right) \tag{16.70}$$

This can be immediately integrated to give

$$\Theta_{l,l}(\theta) \propto \sin^l(\theta) \tag{16.71}$$

In the notation of the associated Legendre polynomials, this implies that $P_l^l(z) \propto (1 - z^2)^{l/2}$. This special case is useful for two reasons, one of which is that this form will be the appropriate one for the classical description of planar orbits. Perhaps more importantly, the other members of the "family" for a given l value which have lower values of m are easily obtained by repeated applications of the lowering operator, $\hat{L}_-$, for example,

$$Y_{l,l-1}(\theta, \phi) \propto \hat{L}_- Y_{l,l}(\theta, \phi)$$

$$= \hbar e^{-i\phi} \left(\frac{\partial}{\partial \theta} - i \cot(\theta) \frac{\partial}{\partial \phi} \right) \left[\sin^l(\theta)\, e^{il\phi} \right] \tag{16.72}$$

and so forth. By repeated applications of these methods, explicit representations for the $Y_{l,m}(\theta, \phi)$ can be constructed for any values of l, m. We will restrict ourselves to simply listing many of their most useful properties.

- The $Y_{l,m}(\theta, \phi)$ are collectively called *spherical harmonics* (by analogy with the eigenfunctions of the infinite well which are, in turn, similar to the various "harmonics" of a vibrating string)

- The properly normalized spherical harmonics are given by

$$Y_{l,m}(\theta, \phi) = (-1)^m \left[\frac{2l + 1}{4\pi} \frac{(l - m)!}{(l + m)!} \right]^{1/2} P_l^m(\cos(\theta))\, e^{im\phi} \tag{16.73}$$

at least for values of $m \geq 0$. For negative values of m, these results are extended via the relation

$$Y_{l,-m}(\theta, \phi) = (-1)^m Y_{l,m}^*(\theta, \phi) \tag{16.74}$$

Recall that the phase of an eigenfunction is arbitrary, so that the factors of -1 here are a convention.

- The associated Legendre polynomials appearing in Eqn. (16.73) can be constructed using the relation

$$P_l^m(z) = (-1)^{l+m} \frac{(l+m)!}{(l-m)!} \frac{1}{2^l l!} (1-z^2)^{-m/2} \left(\frac{d}{dz}\right)^{l-m} (1-z^2)^l \quad (16.75)$$

- For use in various problems, it is useful to have the simplest examples for $l = 0, 1, 2$:

$$Y_{0,0} = \frac{1}{\sqrt{4\pi}} \quad (16.76)$$

$$Y_{1,1} = -\sqrt{\frac{3}{8\pi}} e^{i\phi} \sin(\theta) \quad (16.77)$$

$$Y_{1,0} = \sqrt{\frac{3}{4\pi}} \cos(\theta) \quad (16.78)$$

$$Y_{2,2} = \sqrt{\frac{15}{32\pi}} e^{i2\phi} \sin^2(\theta) \quad (16.79)$$

$$Y_{2,1} = -\sqrt{\frac{15}{8\pi}} e^{i\phi} \sin(\theta) \cos(\theta) \quad (16.80)$$

$$Y_{2,0} = \sqrt{\frac{5}{16\pi}} (3\cos^2(\theta) - 1) \quad (16.81)$$

- From general principles, we know that the (properly normalized) $Y_{l,m}$ form an orthonormal set, namely that

$$\langle Y_{l',m'} | Y_{l,m} \rangle = \int d\Omega \, Y_{l',m'}^*(\theta, \phi) \, Y_{l,m}(\theta, \phi) = \delta_{l,l'} \, \delta_{m,m'} \quad (16.82)$$

- The following average values are sometimes useful, namely

$$\langle \sin^2(\theta) \rangle_{l,m} = \frac{2(l^2 + l - 1 + m^2)}{(2l-1)(2l+3)} \quad (16.83)$$

$$\langle \cos^2(\theta) \rangle_{l,m} = \frac{2(l^2 + l - 1/2 - m^2)}{(2l-1)(2l+3)} \quad (16.84)$$

where

$$\langle f(\theta, \phi) \rangle_{l,m} = \int d\Omega \, f(\theta, \phi) |Y_{l,m}(\theta, \phi)|^2 \quad (16.85)$$

- The spherical harmonics, that is, the set of all $Y_{l,m}(\theta, \phi)$, comprise a complete set of functions over the two-dimensional angular space described by θ, ϕ.

If one has a well-behaved angular function, $f(\theta, \phi)$, which itself is properly normalized, that is, $\int d\Omega \, |f(\theta, \phi)|^2 = 1$, then one can write

$$f(\theta, \phi) = \sum_{l=0}^{\infty} \sum_{m=-l}^{m=+l} a_{l,m} Y_{l,m}(\theta, \phi) \tag{16.86}$$

via the expansion theorem. Using Eqn. (16.82), the expansion coefficients can be obtained via

$$a_{l,m} = \int d\Omega \, Y_{l,m}^*(\theta, \phi) f(\theta, \phi) \tag{16.87}$$

and will satisfy

$$1 = \sum_{l=0}^{\infty} \sum_{m=-l}^{m=+l} |a_{l,m}|^2 \tag{16.88}$$

The $a_{l,m}$ have the usual probabilistic interpretation, specifically,

— The $|a_{l,m}|^2$ is the probability that a measurement of L^2 *and* L_z will yield the values $l(l+1)\hbar^2$ *and* $m\hbar$, respectively.

- In one-dimensional problems, the notion of parity was an important one; in that case, this was accomplished by the parity operator whose effect was to let $x \longrightarrow -x$. In three dimensions, the parity operator acts via

$$\hat{P}f(\mathbf{r}) = \hat{P}f(x, y, z) = f(-x, -y, -z) = f(-\mathbf{r}) \tag{16.89}$$

at least in Cartesian coordinates. From Fig. 16.1, we can see that in spherical coordinates the corresponding effect is

$$\psi(\mathbf{r}) \longrightarrow \psi(-\mathbf{r}) \implies \psi(r, \theta, \phi) \longrightarrow \psi(r, \pi - \theta, \pi + \phi) \tag{16.90}$$

From direct examination of the defining relations of the $Y_{l,m}$, we find that

$$Y_{l,m}(\theta + \pi, \phi) = (-1)^{l+2|m|} Y_{l,m}(\theta, \phi) = (-1)^l Y_{l,m}(\theta, \phi) \tag{16.91}$$

(since $2|m|$ is integral), so that the parity is given by $(-1)^l$; this is consistent with the fact that the eigenvalues of $\hat{P}$ are ± 1. For two-particle systems where $\mathbf{r} = \mathbf{r}_{rel} = \mathbf{r}_1 - \mathbf{r}_2$, the effect of particle interchange is to give $\mathbf{r}_1 \leftrightarrow \mathbf{r}_2$ so that $\mathbf{r} \longrightarrow -\mathbf{r}$. Knowledge of the parity of a two-particle system is then necessary for understanding the properties of the system under exchange, and is of fundamental importance for systems of indistinguishable particles.

Example 16.1.

Let us consider a rigid rotator whose Hamiltonian is given by $\hat{H} = \hat{\mathbf{L}}^2/2I$ and whose (angular) wavefunction is given by

$$\psi(\theta, \phi) = N\left[Y_{0,0}(\theta, \phi) + (1 + 3i)Y_{1,-1}(\theta, \phi) + 2Y_{2,-1}(\theta, \phi) + Y_{2,0}(\theta, \phi)\right] \tag{16.92}$$

The normalization constant N is obtained by invoking Eqn. (16.88) to find

$$1 = \sum_{l=0}^{\infty} \sum_{m=-l}^{m=+l} |a_{l,m}|^2 = N^2(1 + (1 + 9) + 4 + 1) \quad \longrightarrow \quad N = \frac{1}{4} \tag{16.93}$$

We then find the following probabilities,

$$\text{Prob}(l = 0) = \frac{1}{16}$$

$$\text{Prob}(m = 0) = \frac{1 + 1}{16} = \frac{1}{8}$$

$$\text{Prob}(L_z = -\hbar) = \frac{10 + 4}{16} = \frac{7}{8}$$

$$\text{Prob}(\mathbf{L}^2 = 6\hbar^2) = \frac{4 + 1}{16} = \frac{5}{16}$$

$$\text{Prob}(E = 2\hbar^2/2I) = \frac{10}{16} = \frac{5}{8} \tag{16.94}$$

The wavefunction at later times is given by

$$\psi(\theta, \phi; t) = N\left[Y_{0,0}(\theta, \phi)\,e^{-iE_0 t/\hbar} + (1 + 3i)\,Y_{1,-1}(\theta, \phi)\,e^{-iE_1 t/\hbar}\right.$$

$$\left. + [2Y_{2,-1}(\theta, \phi) + Y_{2,0}(\theta, \phi)]\,e^{-iE_2 t/\hbar}\right] \tag{16.95}$$

where $E_l = l(l + 1)\hbar^2/2I$. We then have the expectation values

$$\langle \hat{E} \rangle_t = \left(\frac{1}{16}\right)\frac{0(0 + 1)\hbar^2}{2I} + \left(\frac{10}{16}\right)\frac{1(1 + 1)\hbar^2}{2I} + \left(\frac{4 + 1}{16}\right)\frac{2(2 + 1)\hbar^2}{2I}$$

$$= \frac{50}{32}\frac{\hbar^2}{I}$$

$$\langle \hat{P} \rangle_t = \frac{1}{16}(1 - 10 + 4 + 1) = -\frac{2}{5} \tag{16.96}$$

for the energy and parity, respectively.

16.2.2 **Visualization and Applications**

The angular dependence of quantum wavefunctions, determined by the spherical harmonics, can be visualized in three-dimensional plots where the "radius" vector at a given value of (θ, ϕ) is given by the magnitude of $|Y_{l,m}(\theta, \phi)|^2$. For example, the spherically symmetric $|Y_{0,0}(\theta, \phi)|^2$ is shown in Fig. 16.5(a). For the case of $l = 1$, we plot $Y_{1,0}(\theta, \phi)$ in Fig. 16.5(b) in the same way, while for the $(l, m) = (1, \pm 1)$ states, we can take real linear combinations (as done in Fig. 15.12) proportional to $\sin(\phi)$ and $\cos(\phi)$, as shown in Figs 16.5 (c) and (d). The $l = 1$ plots may appear familiar from modern physics or chemistry as being related to p-orbitals in molecular bonding.

For larger values of l, we choose instead to plot

$$\left[Re(Y_{l,m}(\theta, \phi))\right]^2 \propto \left[P_l^m(\cos(\theta))\right]^2 \cos^2(m\phi) \qquad (16.97)$$

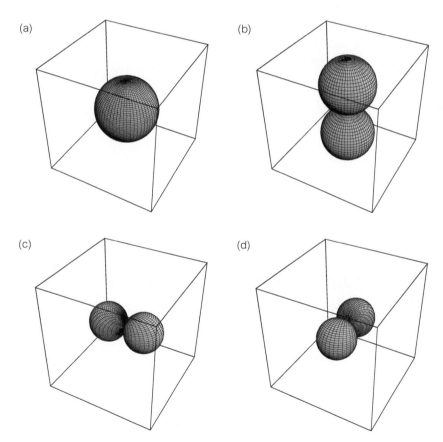

Figure 16.5. Visualization of the spherical harmonics, (a) $Y_{0,0}(\theta, \phi)$, (b) $Y_{1,0}(\theta, \phi)$, (c) $Re[Y_{1,1}(\theta, \phi)]$, and (d) $Im[Y_{1,1}(\theta, \phi)]$.

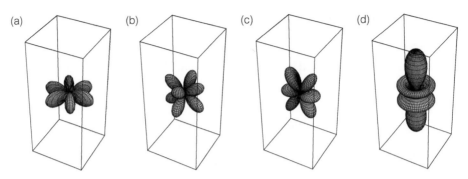

Figure 16.6. Visualization of the spherical harmonics, $Re[Y_{3,m}(\theta, \phi)]$, for (a) $m = 3$, (b) $m = 2$, (c) $m = 1$, and (d) $m = 0$.

For $m_{max} = +l$, the ϕ term has the maximum "wiggliness" in the azimuthal direction. As m is decreased from this value in unit increments, one node is "removed" from the ϕ direction and added to the θ direction as the Legendre polynomials acquire more and more nodes. This behavior is illustrated in Fig. 16.6 for the case of $l = 3$ and $m = 3, 2, 1, 0$. It is also consistent with the "sharing" of rotational kinetic energy given by

$$T_\phi = \left\langle \frac{\hat{L}_z^2}{2I} \right\rangle_{l,m} = \frac{\hbar^2}{2I} m^2 \tag{16.98}$$

and

$$T_\theta = \left\langle \frac{\hat{L}_x^2 + \hat{L}_y^2}{2I} \right\rangle_{l,m} = \frac{\hbar^2}{2I} \left(l(l+1) - m^2 \right) \tag{16.99}$$

The connection of the spherical harmonics to the description of "shapes" is familiar in *multipole expansions* in classical physics. Many textbooks on electromagnetism[2] show that the electric potential due to an arbitrary charge distribution at a large distance, R, from the origin has a systematic expansion of the form[3]

$$\phi(R) = \frac{1}{4\pi\epsilon_0} \left[\frac{1}{R} \int d\mathbf{r}\, \rho(\mathbf{r}) + \frac{1}{R^2} \int d\mathbf{r}\, z\, \rho(\mathbf{r}) \right.$$
$$\left. + \frac{1}{R^3} \int d\mathbf{r}\, (3z^2 - r^2)\, \rho(\mathbf{r}) + \cdots \right] \tag{16.100}$$

where $\rho(\mathbf{r})$ is the charge density; similar expressions are useful for the long-range gravitational potential of a mass distribution as well. The integral in the first term

---

[2] See, for example, Reitz, Milford, and Christy (1993).
[3] This form assumes a specific set of axes for simplicity.

simply counts the net charge and gives the *monopole term*, the second term gives the *electric dipole moment*, while the third describes the *quadrupole moment*. For a quantum mechanical distribution of charge described by a wavefunction of the form $\psi(\mathbf{r}) = R(r)Y_{l,m}(\theta, \phi)$, the charge density is $\rho(\mathbf{r}) = q|\psi(\mathbf{r})|^2$, and the quadrupole moment can be written in the form

$$Q = \int d\mathbf{r} \, (3z^2 - r^2) \, \rho(\mathbf{r}) \tag{16.101}$$

$$= q\langle R(r)|r^2|R(r)\rangle \, \langle Y_{l,m}|(3\cos^2(\theta) - 1)|Y_{l,m}\rangle \tag{16.102}$$

which is related to an average of $Y_{2,0}(\theta, \phi) \propto P_2^0(\cos(\theta))$. The correlations are

$$Q > 0 \longleftrightarrow \langle z^2 \rangle > \langle x^2 \rangle, \langle y^2 \rangle \longleftrightarrow \text{"prolate"}$$

$$Q < 0 \longleftrightarrow \langle z^2 \rangle > \langle x^2 \rangle, \langle y^2 \rangle \longleftrightarrow \text{"oblate"}$$

which is illustrated in Fig. 16.7 for two spheroidal shapes.

16.2.3 Classical Limit of Rotational Motion

One of the consequences of the conservation of angular momentum for the classical motion of point particles in a spherically symmetry potential is that their orbits must be planar. Because $\mathbf{L} = \mathbf{r}(t) \times \mathbf{p}(t)$, the plane containing $\mathbf{r}(t)$ must always be perpendicular to the vector $\mathbf{L}$ which is fixed in time. If we arbitrarily call this the z-direction, we also have $\mathbf{L} = L_z\hat{z}$.

We know that quantum mechanically one can have precisely known eigenvalues for both the operators $\hat{L}^2$ and $\hat{L}_z$ simultaneously. If, in addition, we had

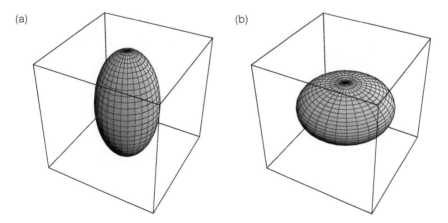

(a)   (b)

Figure 16.7. (a) Prolate and (b) oblate spheroids with quadrupole moments that are positive $(Q > 0)$ and negative $(Q < 0)$, respectively.

$L^2 = L_z^2$ for the eigenvalues, we then know that the values corresponding to both $\hat{L}_x$ and $\hat{L}_y$ would have to vanish identically which is inconsistent with Eqn. (16.36). We see this effect in Fig. 16.4 where the maximal projection of L along the z-axis will make an angle given by

$$\cos(\alpha) \equiv \frac{L_z}{|\mathbf{L}|} = \frac{l}{\sqrt{l(l+1)}} = \frac{1}{\sqrt{1+1/l}} \qquad (16.103)$$

since $m_{max} = +l$. The angle α, which measures the degree to which the total angular momentum vector can be made perpendicular to the classical plane of rotation, can thus be seen to become arbitrarily small for a macroscopic system since $\alpha \to 0$ as $l \to \infty$. An estimate of the rate at which this happens can be made by expanding the terms in Eqn. (16.103) in this limit to obtain

$$\cos(\alpha) \sim 1 - \frac{1}{2}\alpha^2 + \cdots = 1 - \frac{1}{2l} + \cdots \sim \frac{1}{\sqrt{1+1/l}} \qquad (16.104)$$

or $\alpha \propto 1/\sqrt{l}$.

This effect is also reflected in the behavior of the probability density for the angular variables given by the $|Y_{l,l}(\theta,\phi)|^2$ in this limit. The square of these spherical harmonics are plotted in Fig. 16.8 versus θ for increasingly large values of l: the tendency for the probability to be more and more concentrated in the plane corresponding to the classical orbit, that is, for $\theta = \pi/2$, is clear. One can show (P16.6) that the quantum probability distribution in this limit is given by

$$P_{ll}(\theta) = |Y_{l,l}(\theta,\phi)|^2 \propto \sqrt{l} \sin^{2l}(\theta) \qquad (16.105)$$

which exhibits the behavior shown in Fig. 16.8. The width of this distribution decreases with increasingly l, and one can estimate (in a somewhat "rough-and-ready" fashion) the spread in θ over which the probability of finding the particle

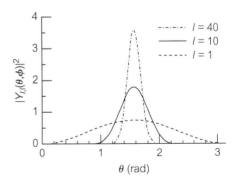

Figure 16.8. $|Y_{l,l}(\theta,\phi)|^2$ versus θ for increasingly large values of l, illustrating the approach to the classical limit. To see that the orbits become increasingly planar in this limit, just turn your head 90°.

is spread; calling this spread δ, we write

$$\sin^{2l}\left(\frac{\pi}{2} \pm \delta\right) = \cos^{2l}(\pm\delta)$$

$$= \left(1 - \frac{1}{2}\delta^2 + \cdots\right)^{2l}$$

$$= 1 - \delta^2 l + \cdots \qquad (16.106)$$

so that $\delta \propto 1\sqrt{l}$, which is similar to the result for the angle α. This is another nice example of the correspondence principle, and shows the classical limit of planar orbits in a striking way.

16.3 **Diatomic Molecules**

16.3.1 **Rigid Rotators**

Even though two atoms are known to bind via an interaction of the approximate form in Fig. 9.2, as a first approximation in our discussion of rotational states of molecules, we can consider a diatomic molecule to consist simply of two atoms with reduced mass μ with a constant separation r_0, that is, joined by the famous "massless, inextensible rod" of introductory mechanics problems. In this approximation, the Hamiltonian is just that of a *rigid rotator*

$$\hat{H} = \frac{\hat{L}^2}{2I} \qquad (16.107)$$

where $I = \mu r_0^2$ and the rotational energies are then given by

$$E_l = \frac{\hbar^2 l(l+1)}{2\mu r_0^2} = l(l+1)E_0 \qquad (16.108)$$

Transitions between states due to the emission or absorption of a photon are possible provided they satisfy certain *selection rules* (to be discussed below). In this case, the initial and final values of the angular momentum quantum number must change by one, that is, $\Delta l = \pm 1$; this implies, for example, that the photon energies measured in an absorption experiment will satisfy

$$E_\gamma^{(l)} = E_l - E_{l-1} = 2lE_0 \qquad (16.109)$$

for a purely rigid rotator. For diatomic molecules with $r_0 \sim 1 - 2\,\text{Å}, \mu \sim 1 - 20\,\text{amu}$, and $l \sim 1 - 10$, these energies are in range $E_\gamma \sim 10^{-3} - 10^{-4}$ eV; this corresponds to wavelengths of the order $\lambda \sim 1\,\text{mm} - 1$ cm, that is, in the far infrared to microwave regions.

This simple model implies that the ratio $E_\gamma^{(l)}/2l$ should be constant with increasing l for a given series of rotational lines. Data for H–Cl are plotted in Fig. 16.9 in this format (with the vertical scale greatly magnified.) to show that there is a systematic deviation (downward) for larger values of l. This effect can be understood semiquantitatively as being due to the "stretch" of the diatomic molecule in response to the increasing angular momentum. If we approximate the interatomic potential near equilibrium (r_0) by a harmonic oscillator of the form $V(r) = K(r - r_0)^2/2$, the classical force equation implies that

$$ma_c = \frac{L^2}{\mu r^3} = K(r - r_0) = F_c \tag{16.110}$$

where we have written the centripetal acceleration in terms of L. If we call the deviation from equilibrium (the "stretch") δ, we find

$$r - r_0 = \delta \approx \frac{L^2}{\mu r_0^3 K} \tag{16.111}$$

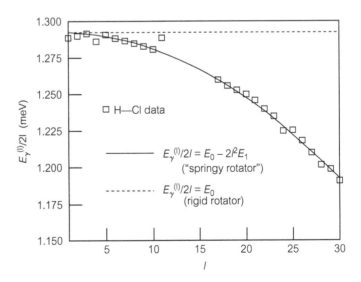

Figure 16.9. Photon energies for $\Delta l = \pm 1$ transitions for a diatomic molecule ($H - Cl$). The data are plotted to indicate the pattern of energy levels for a "springy rotator."

The energy now is given by

$$E = \frac{L^2}{2\mu r^2} + \frac{1}{2}K(r - r_0)^2$$

$$= \frac{L^2}{2\mu(r_0 + \delta)^2} + \frac{1}{2}K\delta^2$$

$$\approx \frac{L^2}{2\mu r_0^2} - \frac{L^4}{2\mu^2 r_0^6 K} + \mathcal{O}(L^6) + \cdots \tag{16.112}$$

Using the quantized values of L^2, this gives the expression

$$E_l = l(l+1)E_0 - [l(l+1)]^2 E_1 \quad \text{where} \quad E_1 \equiv \frac{\hbar^4}{2\mu^2 r_0^6 K} \tag{16.113}$$

for the spectrum of a "springy rotator"; systems with "stiffer springs," that is, ones with larger values of K will have their spectrum changed less. The corresponding photon energies satisfying the $|\Delta l| = 1$ selection rule would then have energies

$$E_\gamma^{(l)} = 2lE_0 - 4l^3 E_1 \quad \text{or} \quad \frac{E_\gamma^{(l)}}{2l} = E_0 - 2l^2 E_1 \tag{16.114}$$

and the solid line in Fig. 16.9 indicates a fit to the data of this form; the fitted values are

$$E_0 \sim 1.927 \times 10^{-3} \, \text{eV} \quad \text{and} \quad E_1 \sim 5.6 \times 10^{-5} \, \text{eV} \tag{16.115}$$

16.3.2 Molecular Energy Levels

To understand the rich structure of quantized energy levels available to diatomic molecules, we consider how one might calculate the energy spectrum of such a system from first principles, indicating what role the electronic and nuclear motions play, and how vibrational and rotational states are connected to electronic excitations.

For a diatomic molecule consisting of two atoms with nuclear charges Z_1, Z_2, there will be $Z_1 + Z_2$ electrons, and the complete multibody Hamiltonian can be formally written as

$$\hat{H} = \frac{\hat{P}_1^2}{2M_1} + \frac{\hat{P}_2^2}{2M_2} + \sum_{i=1}^{Z_1+Z_2} \frac{\hat{p}_i^2}{2m_e}$$

$$- \sum_{i=1}^{Z_1+Z_2} \frac{KZ_1 e^2}{|\mathbf{R}_1 - \mathbf{r}_i|} - \sum_{i=1}^{Z_1+Z_2} \frac{KZ_2 e^2}{|\mathbf{R}_2 - \mathbf{r}_i|} + \frac{KZ_1 Z_2 e^2}{|\mathbf{R}_1 - \mathbf{R}_2|}$$

$$+ \sum_{i\neq j=1}^{Z_1+Z_2} \frac{Ke^2}{|\mathbf{r}_i - \mathbf{r}_j|} \tag{16.116}$$

where $\mathbf{R}_{1,2}$ and $\mathbf{r}_i$ ($M_{1,2}$ and m_e) are the nuclear and electronic coordinates (masses), respectively; the mutual interactions are simply the Coulomb attractions or repulsions between all combinations of nuclei and electrons.

Because the nuclei are so much heavier than the electrons, the electrons can "respond" to changes in the nuclear positions rapidly, and it makes sense to initially approximate the nuclei as being fixed with some arbitrary separation $R = |\mathbf{R}_1 - \mathbf{R}_2|$. The electronic configuration, for the ground state or any excited state, can then (in principle, this is after all, an imagined calculation) be determined by solving the multiparticle Schrödinger equation. The "effective" Hamiltonian for the two nuclei can now be written in the form

$$\hat{H}_N = \frac{\hat{P}_1^2}{2M_1} + \frac{\hat{P}_1^2}{2M_1} + V(R) \tag{16.117}$$

where $V(R)$ now includes not only the Coulomb repulsion of the nuclei, but also the effective potential due to the electron configuration; the terms in Eqn. (16.116) involving the electron coordinates are averaged over.

This procedure can be repeated for different values of R, and can be used to "map out" the interatomic potential through which the nuclei interact. The schematic potential is shown in Fig. 16.10, both for a ground state electronic configuration, and for an electronic excited state; the two potentials are separated by roughly $1-10$ eV, that is, an atomic energy difference.

The vibrational energy levels considered in Section 9.3 are then due to the nuclear motion in each attractive well, and have energy splittings typically in the range $10^{-1}-10^{-2}$ eV range and are superimposed on each $V(R)$ curve as shown. The "spring constants" inferred from such data are of the order $K_{eff} \sim$ eV/Å^2 and are consistent with this picture as Å-size changes in the nuclear separation will change the energy of the electronic configuration by eVs. Finally, the ~~vibrational~~ *rotational* states are much more closely spaced with splittings of the order $10^{-3}-10^{-4}$ eV.

A useful mnemonic device for understanding this hierarchy of energy splittings is as follows: Typical electronic energies are determined by e, $\hbar$, m_e (as in

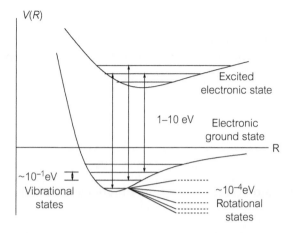

Figure 16.10. Schematic plot of the effective interatomic potential in the electronic ground state and in an excited electronic state. Levels indicating vibrational and rotational states are also indicated, along with their typical energy splittings.

Section 1.4) via

$$E_e \sim \frac{e^2}{a_0} \sim \frac{m_e e^4}{\hbar^2} \quad \text{where} \quad a_0 \sim \frac{\hbar^2}{m_e e^2} \tag{16.118}$$

The spring constants for molecular vibrational motion then scale as

$$K_{eff} \sim \frac{E_e}{a_0^2} \sim \frac{m_e^3 e^8}{\hbar^6} \tag{16.119}$$

so that the quantized energies go as

$$E_{vib} \sim \hbar \sqrt{\frac{K_{eff}}{M}} \sim \left(\frac{m_e e^4}{\hbar^2} \right) \sqrt{\frac{m_e}{M}} \tag{16.120}$$

where M is a nuclear mass. Finally, rotational states vary roughly as

$$E_{rot} \sim \frac{\mathbf{L}^2}{2\mu r_0^2} \sim \frac{\hbar^2}{M a_0^2} \sim \left(\frac{m_e e^4}{\hbar^2} \right) \frac{m_e}{M} \tag{16.121}$$

so that the electronic, rotational, and vibrational energies are roughly in the ratio

$$E_e / E_{vib} / E_{rot} : \ 1 \left/ \sqrt{\frac{m_e}{M}} \right/ \frac{m_e}{M} \tag{16.122}$$

Transitions involving changes in the electronic configuration will have typical energies in the $1-10$ eV region corresponding to visible wavelengths; because of the many rotational and vibrational states available in both the initial and final states, each electronic transition will actually correspond to many possible lines and *band spectra* are seen instead of sharp lines as in atomic spectra. Changes in the vibrational state within a given electronic configuration will still allow for many different rotational state, so that fine structure can be resolved. The energy levels of a "pure" rotator–vibrator system are given by

$$E_{n,l} = \hbar\omega(n + 1/2) + l(l + 1)E_0 \tag{16.123}$$

so that

$$E_\gamma^{(n,l)} = \hbar\omega + 2lE_0 \tag{16.124}$$

if we use the selection rule $\Delta n = \pm 1$ for vibrational state transitions, which we discuss next.

16.3.3 Selection Rules

Because of their importance in determining the observed patterns of emission and absorption in atomic and molecular spectroscopy (and beyond), we present here a "bare-bones" discussion of the physical principles underlying the selection rules[4], which we have freely used in our discussions of electromagnetic transitions between vibrational and rotational energy levels.

The description of the emission or absorption of a photon as a charged particle (or distribution of charges) changes its quantum state requires knowledge of the coupling of a charge to an external electromagnetic field as discussed in Section 18.2.2. The interaction term which is relevant for our discussion can be written in the form

$$\frac{1}{m}\hat{\mathbf{p}} \cdot \mathbf{A} = \frac{1}{m}\hat{\mathbf{p}} \cdot \boldsymbol{\epsilon}\, e^{-i\mathbf{k}\cdot\mathbf{r}} \tag{16.125}$$

where $\mathbf{A}$ is the so-called vector potential. In this case, it is written in terms of the polarization vector of the photon, $\boldsymbol{\epsilon}$, and its plane wavefunction where the wave number satisfies $k = 2\pi/\lambda = E_\gamma/\hbar c$.

The amplitude which, when squared, describes the probability for a radiative transition between initial and final states, ψ_i and ψ_f, is given by

$$\mathrm{Amp}(\psi_i \to \psi_f + \gamma) \propto \frac{1}{m}\left\langle \psi_f \left| \hat{\mathbf{p}} \cdot \boldsymbol{\epsilon}\, e^{-i\mathbf{k}\cdot\mathbf{r}} \right| \psi_i \right\rangle \tag{16.126}$$

If the bound state wavefunctions are localized on some length scale R, the argument of the exponential function will be at most of order kR over the region of integration where the overlap integral "gets its support," that is, is nonvanishing. For many atomic, molecular, and nuclear systems, this factor satisfies $kR \ll 1$

---

[4] For a more complete discussion, see, for example, Gasiorowicz (1996).

and the exponential can be approximated by unity; for example, in radiative transitions between atomic energy levels one has

$$kR \sim \frac{E_\gamma a_0}{\hbar c} \sim \frac{(2\,\text{eV})\,(1\,\text{Å})}{1973\,\text{eV}\,\text{Å}} \sim 10^{-3} \ll 1 \qquad (16.127)$$

We will see that this approximation corresponds to considering only electric dipole radiation (and not higher multipoles).

In this limit, we can write the required matrix element in the form

$$\frac{1}{m}\boldsymbol{\epsilon}\cdot\langle\psi_f|\hat{\mathbf{p}}|\psi_i\rangle = \boldsymbol{\epsilon}\cdot\frac{d}{dt}\langle\psi_f|\mathbf{r}|\psi_i\rangle \qquad (16.128)$$

which is seen to be essentially the time rate of change of the *dipole matrix element* between initial and final states. This expression can be simplified by evaluating the time-derivative using the relation

$$\frac{d}{dt}\langle\psi_f|\mathbf{r}|\psi_i\rangle = \frac{i}{\hbar}\langle\psi_f|[\hat{H},\mathbf{r}]|\psi_i\rangle = \frac{i(E_f - E_i)}{\hbar}\langle\psi_f|\mathbf{r}|\psi_i\rangle \qquad (16.129)$$

This gives the important result that the amplitude governing electric dipole radiation between two quantum states is proportional to the matrix element

$$\text{Amp}(\psi_i \to \psi_f + \gamma) \propto \boldsymbol{\epsilon}\cdot\langle\psi_f|\mathbf{r}|\psi_i\rangle \qquad (16.130)$$

and this dependence is the basis of many selection rules.

For example, for vibrational states described by eigenfunctions of the simple harmonic oscillator (considered here in one-dimensional, for simplicity), we find that the matrix elements satisfy

$$\langle\psi_{n_f}|x|\psi_{n_i}\rangle = 0 \quad \text{unless} \quad \Delta n = n_f - n_i = \pm 1 \qquad (16.131)$$

using the results of Eqn. (9.50).

In the case of the rigid rotator, the components of the vector $\mathbf{r} = (x, y, z)$ can be written in terms of the spherical harmonics $Y_{l=1,m}(\theta, \phi)$ with $m = 0, \pm 1$, so that we need to examine the structure of the matrix elements

$$\langle Y_{l_f,m_f}|Y_{1,m}|Y_{l_i,m_i}\rangle \qquad (16.132)$$

The azimuthal integrations over ϕ will have the general form

$$\int_0^{2\pi} d\phi \; e^{i(m_i+m-m_f)} \qquad (16.133)$$

which will vanish unless

$$\Delta m = m_f - m_i = m = +1, 0, -1. \qquad (16.134)$$

The selection rules for l are determined by the integration over the polar angle θ. Arguments similar to those leading to the addition of angular momentum rules in Section 16.5 imply that the product wavefunction $Y_{1,m}Y_{l_i,m_i}$ can be written in terms of individual spherical harmonics given by the addition rule $l_i+1 \longrightarrow l_i-1, l_i, l_i+1$. Because of the orthogonality of the spherical harmonics, the only overlap integrals which are then nonvanishing are those for which $l_f - l_i = +1, 0, -1$. However, when $l_f = l_i$ it is easy to see that Eqn. (16.132) vanishes by parity arguments since

$$\langle Y_{l,m}|x, y, z|Y_{l,m'}\rangle = 0 \tag{16.135}$$

This implies the selection rule

$$\Delta l = l_f - l_i = \pm 1 \tag{16.136}$$

and it is sometimes said that:

- "Quantum numbers change by one unit in dipole transitions."

Several other comments complete our discussion:

- The spectra of molecules consisting of identical atoms form a special case which can be examined on both physical and more formal grounds.

  — *Physical argument*: For diatomic molecules consisting of unlike atoms, such as C–O or H–Cl, there can be a permanent dipole moment which can then radiate when "shaken" (in transitions between vibrational states) or "spun" (in transitions between rotational states). For molecules of identical atoms, such as C_2 or O_2, the symmetry between the two atoms implies that no dipole moment can exist (after all, which way would it point?), and so electric dipole radiation is forbidden.

  — *Formal argument*: Such systems of indistinguishable particles must have wavefunctions which have the correct symmetry under exchange, that is

$$\psi(\mathbf{r}_1, \mathbf{r}_2) = \pm\psi(\mathbf{r}_2, \mathbf{r}_1) \tag{16.137}$$

  which implies that the wavefunction for the relative coordinate, $\mathbf{r} = \mathbf{r}_1 - \mathbf{r}_2$ must satisfy

$$\psi(\mathbf{r}) = \pm\psi(-\mathbf{r}) \tag{16.138}$$

  that is, have a definite parity. The eigenfunctions of the harmonic oscillator, for example, have parity given by $P = (-1)^n$, so that the allowed wavefunctions for indistinguishable particles must have either n odd or even and allowed states are separated by two units of n; since the selection rule Eqn. (16.131) implies that $\Delta n = \pm 1$, electric dipole transitions are forbidden.

- Under circumstances where electric dipole transitions are not allowed, the next term in the expansion of $e^{-i\mathbf{k}\cdot\mathbf{r}}$ is required; one must then consider the matrix elements of higher powers of $\mathbf{r}$ which can then connect states with $|\Delta n|$ or $|\Delta l|$ greater than one. Such transitions, corresponding to quadrupole, octopole, and the like terms will have much smaller amplitudes due to the extra powers of kR; they are not 'forbidden', but may well have unobservably small intensities in a given experiment.

16.4 Spin and Angular Momentum

To derive a quantum description of rotational motion, we have so far followed a path which is quite similar to that for quantizing classical oscillatory motion. Certain classical variables were generalized to quantum operators which yielded eigenvalue problems; the spherical harmonic functions, $Y_{l,m}(\theta,\phi)$, were obtained as the solutions corresponding to the quantized values of orbital angular momentum and its z component.

If we wish to associate the spin degree of freedom (for electrons, protons, quarks, etc.) with a half-integral value of angular momentum, a representation in terms of spherical harmonics is clearly inappropriate. To see this, if we formally attempt to use $l = 1/2$ and $m = \pm 1/2$, we find that

$$Y_{l,m} = Y_{1/2,\pm 1/2}(\theta,\phi) \propto \sqrt{\sin(\theta)}\, e^{\pm i\phi/2} \tag{16.139}$$

This identification does not satisfy any of the functional relations for the spherical harmonics since, for example,

$$\hat{L}_- Y_{1/2,+1/2} \propto \sqrt{\sin(\theta)} e^{\pm i\phi/2} \propto \frac{\cos(\theta)}{\sqrt{\sin(\theta)}} e^{-i\phi/2} \tag{16.140}$$

which is not, in turn, proportional to $Y_{1/2,-1/2}$ or even well-behaved at $\theta = 0$.

It is perhaps not surprising that this formulation for quantized values of angular momentum does not extend to intrinsic angular momentum or spin as the derivation leading to the $Y_{l,m}$ was based on generalizing classical orbital motion; clearly another representation of the angular momentum operators and their eigenfunctions is required which is more abstract.

As an example, consider a 3×3 matrix representation of the eigenfunctions corresponding to $l = 1$; this formulation can be easily generalized to any value of l using $(2l + 1) \times (2l + 1)$ matrices. We can explicitly write

$$L_x = \frac{\hbar}{2} \begin{pmatrix} 0 & \sqrt{2} & 0 \\ \sqrt{2} & 0 & \sqrt{2} \\ 0 & \sqrt{2} & 0 \end{pmatrix} \tag{16.141}$$

$$L_y = \frac{\hbar}{2i} \begin{pmatrix} 0 & \sqrt{2} & 0 \\ -\sqrt{2} & 0 & \sqrt{2} \\ 0 & -\sqrt{2} & 0 \end{pmatrix} \tag{16.142}$$

$$L_z = \hbar \begin{pmatrix} 1 & 0 & 0 \\ 0 & 0 & 0 \\ 0 & 0 & -1 \end{pmatrix} \tag{16.143}$$

The $2l+1$ values of L_z are located along the diagonal while the $\sqrt{2} = \sqrt{l(l+1)}$ is appropriate for $l = 1$. One can easily check (P16.11) that the usual commutation relations are obeyed as

$$[L_x, L_y] = i\hbar L_z \tag{16.144}$$

and that

$$L^2 = L_x^2 + L_y^2 + L_z^2 = \hbar^2 \begin{pmatrix} 2 & 0 & 0 \\ 0 & 2 & 0 \\ 0 & 0 & 2 \end{pmatrix} = 2\hbar^2 \mathbf{1} \tag{16.145}$$

The simultaneous eigenfunctions of L^2 and L_z with eigenvalues $l(l+1)\hbar^2 = 2\hbar^2$ and $L_z = +1, 0, -1$, respectively, are given by

$$v^{(+1)} = \begin{pmatrix} 1 \\ 0 \\ 0 \end{pmatrix}, \quad v^{(0)} = \begin{pmatrix} 0 \\ 1 \\ 0 \end{pmatrix}, \quad \text{and} \quad v^{(-1)} = \begin{pmatrix} 0 \\ 0 \\ 1 \end{pmatrix} \tag{16.146}$$

A general $l = 1$ "wavefunction" is then written in the form

$$\begin{pmatrix} \alpha \\ \beta \\ \gamma \end{pmatrix} = \alpha v^{(+1)} + \beta v^{(0)} + \gamma v^{(-1)} \tag{16.147}$$

where $|\alpha|^2 + |\beta|^2 + |\gamma|^2 = 1$ is the normalization condition. Raising and lowering operators are easily generalized as well and one has, for example,

$$L_+ = L_x + iL_y = \sqrt{2}\hbar \begin{pmatrix} 0 & 1 & 0 \\ 0 & 0 & 1 \\ 0 & 0 & 0 \end{pmatrix} \tag{16.148}$$

One can then check explicit relations such as

$$L_+ v^{(-1)} = \sqrt{2}\hbar \begin{pmatrix} 0 \\ 1 \\ 0 \end{pmatrix} = \sqrt{2}\hbar v^{(0)} \tag{16.149}$$

so that the normalizations of Eqn. (16.62) and (16.63) are maintained.

Example 16.2.

It is an instructive exercise to find the eigenvalues and eigenfunctions of the operator

$$L_\phi = \cos(\phi)L_x + \sin(\phi)L_y = \frac{\hbar}{\sqrt{2}} \begin{pmatrix} 0 & e^{-i\phi} & 0 \\ e^{+i\phi} & 0 & e^{-i\phi} \\ 0 & e^{+i\phi} & 0 \end{pmatrix} \tag{16.150}$$

Writing the eigenvalue problem as

$$L_\phi \begin{pmatrix} \alpha \\ \beta \\ \gamma \end{pmatrix} = L_\phi \begin{pmatrix} \alpha \\ \beta \\ \gamma \end{pmatrix} \tag{16.151}$$

we find the determinant condition on the eigenvalues L_ϕ,

$$\det \begin{pmatrix} L_\phi & -\hbar e^{-i\phi}/\sqrt{2} & 0 \\ -\hbar e^{+i\phi}/\sqrt{2} & L_\phi & -\hbar e^{-i\phi}/\sqrt{2} \\ 0 & -\hbar e^{+i\phi}/\sqrt{2} & L_\phi \end{pmatrix} = 0 \tag{16.152}$$

This implies that $L_\phi^3 - \hbar^2 L_\phi = 0$ or $L_\phi = +\hbar, 0, -\hbar$ as expected; the eigenfunction corresponding to $L_\phi = +\hbar$ is easily found to be

$$v_\phi^{(+1)} = \begin{pmatrix} e^{-i\phi}/\sqrt{2} \\ 1 \\ e^{+i\phi}/\sqrt{2} \end{pmatrix} \tag{16.153}$$

with similar results for $v_\phi^{(0)}$, $v_\phi^{(-1)}$.

A relabeling $\phi \to \phi + 2\pi$ should not have any effect on the results as the two labels for the same angle are physically equivalent. It is easy to check that $|v_{\phi+2\pi}|^2 = |v_\phi|^2$ so that the probability densities are invariant as they should; we note that the stronger condition, $v_{\phi+2\pi} = v_\phi$, also holds so that the wavefunction returns to the same phase upon one rotation. While such expectations might seem obvious for any classical system, the corresponding results for the nonclassical spin degree of freedom will be different.

This matrix representation of quantized angular momenta can now be generalized to treat spin-1/2 particles. Specifically, we define

$$S_x = \frac{\hbar}{2} \begin{pmatrix} 0 & 1 \\ 1 & 0 \end{pmatrix} \tag{16.154}$$

$$S_y = \frac{\hbar}{2i} \begin{pmatrix} 0 & 1 \\ -1 & 0 \end{pmatrix} \tag{16.155}$$

$$S_z = \frac{\hbar}{2} \begin{pmatrix} 1 & 0 \\ 0 & -1 \end{pmatrix} \tag{16.156}$$

which satisfy the standard commutation relations

$$[S_x, S_y] = i\hbar S_z \tag{16.157}$$

and cyclic permutations, and give

$$S^2 = \hbar^2 \begin{pmatrix} 3/4 & 0 \\ 0 & 3/4 \end{pmatrix} = \left[\frac{1}{2} \left(\frac{1}{2} + 1 \right) \hbar^2 \right] 1 \tag{16.158}$$

which is appropriate for $S = 1/2$. The eigenvectors of $\hat{S}^2$ and S_z are just the spinors introduced in Section 14.4,

$$\chi_+ = \begin{pmatrix} 1 \\ 0 \end{pmatrix} \quad \text{and} \quad \chi_- = \begin{pmatrix} 0 \\ 1 \end{pmatrix} \tag{16.159}$$

with eigenvalues $S_z = +1/2$ and $-1/2$, respectively; a general spinor is written as

$$\chi = \alpha^{(+)} \chi_+ + \alpha^{(-)} \chi_- = \begin{pmatrix} \alpha^{(+)} \\ \alpha^{(-)} \end{pmatrix} \tag{16.160}$$

We can repeat the exercise in Example 16.2 to examine the eigenvalues and eigenfunctions of the "rotated" spin matrix operator

$$S_\phi = \cos(\phi) S_x + \sin(\phi) S_y = \frac{\hbar}{2} \begin{pmatrix} 0 & e^{-i\phi} \\ e^{+i\phi} & 0 \end{pmatrix} \tag{16.161}$$

The eigenvalues of S_ϕ are found to be $+\hbar/2, -\hbar/2$ as expected with corresponding eigenvectors

$$v^{(+1/2)} = \frac{1}{\sqrt{2}} \begin{pmatrix} e^{-i\phi/2} \\ e^{+i\phi/2} \end{pmatrix} \quad \text{and} \quad v^{(-1/2)} = \frac{1}{\sqrt{2}} \begin{pmatrix} e^{-i\phi/2} \\ -e^{+i\phi/2} \end{pmatrix} \tag{16.162}$$

Both spinors satisfy the relation

$$\left| v_{\phi+2\pi}^{(\pm 1/2)} \right|^2 = \left| v_\phi^{(\pm 1/2)} \right|^2 \tag{16.163}$$

so the probability density is unchanged by the relabeling of angle as it should. The spinor wavefunctions or amplitudes, however, do change sign since

$$v_{\phi+2\pi}^{(\pm 1/2)} = -v_{\phi}^{(\pm 1/2)} \tag{16.164}$$

and a rotation by 4π is required to return them to their original phase. This phase change under rotations is typical of fermionic spin-1/2 wavefunctions and has no classical analog; as with any phase behavior, it is best tested in interference experiments, and this possibility will be discussed in Section 18.7.1.

Before proceeding, we note that it is often useful to write the spin matrices in dimensionless form by defining

$$S = \frac{\hbar}{2}\sigma \tag{16.165}$$

where

$$\sigma_x = \begin{pmatrix} 0 & 1 \\ 1 & 0 \end{pmatrix}, \quad \sigma_y = \begin{pmatrix} 0 & -i \\ i & 0 \end{pmatrix}, \quad \text{and} \quad \sigma_z = \begin{pmatrix} 1 & 0 \\ 0 & -1 \end{pmatrix} \tag{16.166}$$

which satisfy

$$[\sigma_i, \sigma_j] = 2i\epsilon_{ijk}\sigma_k \tag{16.167}$$

where $i, j, k = x, y, z$ in any permutation and $\epsilon_{i,j,k}$ is the totally antisymmetric symbol (Appendix F.1). The σ are called the *Pauli matrices* and they are Hermitian (that is $M^\dagger = (M^*)^T = M$.) If the behavior of the spinor wavefunction under rotations is decidedly "unclassical," the dynamics of a charged particle with spin-1/2 does have some classical analogs, namely, their interactions with an external magnetic field.

We will see in Section 18.5 that the Hamiltonian for a charged particle in a uniform magnetic field will have a term of the form

$$\hat{H} = -\left(\frac{q}{2m}\right)\hat{L} \cdot B = -\hat{M} \cdot B \tag{16.168}$$

where we have defined a magnetic moment operator via

$$\hat{M} = \frac{q}{2m}\hat{L} \tag{16.169}$$

Equation (16.168) is just the analog of the classical energy of a magnetic dipole in an external **B** field. The relation Eqn. (16.169) between the magnetic moment and angular momentum is consistent with that of a classically rotating point particle.

Particles with intrinsic spin often have microscopic magnetic moments given by the similar relation

$$M = g\frac{q}{2m}S \tag{16.170}$$

where q, m are the charge and mass, respectively, and g is a dimensionless number called the *gyromagnetic ratio*. As the name implies, g measures the ratio of the magnetic moment to the (intrinsic) angular momentum vector. For a classical rotating particle or a classical spinning object for which the mass and charge densities are proportional, one can show that $g = 1$ (P16.12). In contrast, the g value for a point-like, charged spin-1/2 particle can be derived using the machinery of relativistic quantum mechanics[5] (via the so-called *Dirac equation*) and one finds $g = 2$. Small, calculable corrections to this value arise due to effects from quantum field theory; for the electron these have the form

$$g_e = 2 + \frac{\alpha}{\pi} - 0.65696 \left(\frac{\alpha}{\pi}\right)^2 + \cdots \approx 2.0023193048 \qquad (16.171)$$

where α is the electromagnetic fine-structure constant; we will take $g_e = 2$ for simplicity. The magnetic moment vector for such a point-like spin-1/2 particle can then be written in the form

$$\mathbf{M} = \frac{g}{2}\left(\frac{q\hbar}{2m}\right)\boldsymbol{\sigma} = \mu\boldsymbol{\sigma} \qquad (16.172)$$

where the *magneton* is defined via $\mu = q\hbar/2m$.

Based on the prediction of the Dirac theory, for the neutron and proton we would expect to have

$$\mu_p = \frac{e\hbar}{2m_p} \equiv \mu_N \qquad \mu_n = 0 \qquad (16.173)$$

since the charge of the neutron vanishes. Instead, one finds

$$\mu_p \approx 2.79\mu_N \qquad \mu_n \approx -1.91\mu_N \qquad (16.174)$$

or more precisely

$$\frac{g_p}{2} = 2.79284739(6) \quad \text{and} \quad \frac{g_n}{2} = -1.9130427(5) \qquad (16.175)$$

These large deviations from point-like structure can be understood if the neutron and proton are bound states of more fundamental constituents, and are best explained in terms of the *quark model* (P16.18).

The magnetic moments of such particles provide a "handle" with which to manipulate their spin orientation via their interactions with externally applied magnetic fields. The interaction Hamiltonian is given by

$$\hat{H} = -\mathbf{M} \cdot \mathbf{B} = -\frac{g\mu}{2}\boldsymbol{\sigma} \cdot \mathbf{B} \qquad (16.176)$$

---

[5] See, for example, Griffiths (1987).

For a uniform field along the z direction, the dynamical equation of motion for the spinor wavefunction is simply the Schrödinger equation, which in this case has the form

$$i\hbar \frac{\partial}{\partial t} \chi(t) = \hat{H}\chi(t) = -\frac{g\mu}{2} B\sigma_z \chi(t) \tag{16.177}$$

or

$$\begin{pmatrix} \dot{\alpha}^{(+)}(t) \\ \dot{\alpha}^{(-)}(t) \end{pmatrix} = i \left(\frac{g\mu B}{2\hbar} \right) \begin{pmatrix} \alpha^{(+)}(t) \\ -\alpha^{(-)}(t) \end{pmatrix} \tag{16.178}$$

which has the solutions

$$\alpha^{(\pm)}(t) = \alpha^{(\pm)}(0) \, e^{\pm i\omega t} \tag{16.179}$$

where $\omega = g\mu B/2\hbar$. These solutions correspond to energies

$$E_\pm = \hbar\omega_\pm = \pm\frac{g\mu B}{2} \tag{16.180}$$

where the spin is parallel $(-)$ or antiparallel $(+)$ to the field direction.

Example 16.3. Spin precession in a uniform field

Suppose that initially one has a spinor wavefunction which is an eigenvector of S_x with eigenvalue $+\hbar/2$, that is, one knows that the spin is "pointing" along the $+x$ direction; this corresponds to

$$\chi(0) = \frac{1}{\sqrt{2}} \begin{pmatrix} 1 \\ 1 \end{pmatrix} \tag{16.181}$$

If a uniform magnetic field in the z direction is applied, the time-development of this spinor is given by Eqn. (16.179) so that

$$\chi(t) = \frac{1}{\sqrt{2}} \begin{pmatrix} e^{+i\omega t} \\ e^{-i\omega t} \end{pmatrix} \tag{16.182}$$

At later times, the expectation value of the spin along the x axis is given by

$$
\begin{aligned}
\langle S_x \rangle_t &= \langle \chi(t)|S_x|\chi(t) \rangle \\
&= \left[\frac{1}{\sqrt{2}} (e^{-i\omega t}, e^{+i\omega t}) \right] \left[\frac{\hbar}{2} \begin{pmatrix} 0 & 1 \\ 1 & 0 \end{pmatrix} \right] \left[\frac{1}{\sqrt{2}} \begin{pmatrix} e^{+i\omega t} \\ e^{-i\omega t} \end{pmatrix} \right] \\
&= \frac{\hbar}{2} \cos(2\omega t)
\end{aligned}
\tag{16.183}
$$

(Continued)

while $\langle S_y \rangle_t = -\hbar \sin(2\omega t)/2$. We see that:

- The magnetic moment precesses around the z axis at an angular velocity (2ω) which is twice the rate at which the phase of the spinor wavefunction changes (ω).

This precession can also be derived (P16.13) in a more formal way which makes its classical analog more obvious. This phenomenon is the basis for one-experimental test of the spinor nature of the spin-1/2 fermion wavefunction (Section 18.7.1).

16.5 **Addition of Angular Momentum**

For classical particles which do not interact with each other, the total value of some observable quantities, such as energy, momentum, and the like, is often obtained by simply summing the values for each particle. This can occur in quantum mechanical systems as well where "noninteracting" often implies that the multiparticle wavefunction can be written in a factorized, product form.

For example, for two noninteracting energy eigenstates, one might have

$$\psi(x_1, x_2; t) = \left[\psi_{E_1}(x_1) e^{-iE_1 t/\hbar} \right] \left[\psi_{E_2}(x_2) e^{-iE_2 t/\hbar} \right] \tag{16.184}$$

which is also an energy eigenstate since

$$\hat{E}\psi(x_1, x_2; t) = i\hbar \frac{\partial}{\partial t} \psi(x_1, x_2; t) = (E_1 + E_2)\psi(x_1, x_2; t) \tag{16.185}$$

Free particle momentum eigenstates also behave in this manner since

$$\hat{P}_{\text{tot}} \left[e^{i\mathbf{p}_1 \cdot \mathbf{r}_1/\hbar} e^{i\mathbf{p}_2 \cdot \mathbf{r}_2/\hbar} \right] = (\mathbf{p}_1 + \mathbf{p}_2) \left[e^{i\mathbf{p}_1 \cdot \mathbf{r}_1/\hbar} e^{i\mathbf{p}_2 \cdot \mathbf{r}_2/\hbar} \right] \tag{16.186}$$

where $\hat{P}_{\text{tot}} = (\hbar/i)(\nabla_1 + \nabla_2)$.

For two particles which are described by eigenstates of $\hat{L}_1^2, \hat{L}_2^2$ and $\hat{L}_{1,z}, \hat{L}_{2,z}$, respectively, the situation is more complex. The operators corresponding to the total angular momentum squared and its corresponding z component are given by

$$\hat{J}^2 = \left(\hat{L}_1 + \hat{L}_2 \right)^2 = \hat{L}_1^2 + 2\hat{L}_1 \cdot \hat{L}_2 + \hat{L}_2^2 \tag{16.187}$$

$$\hat{J}_z = \hat{L}_{1,z} + \hat{L}_{2,z} \tag{16.188}$$

and we wish to find the corresponding eigenstates, $\psi_{(J,J_z)}$, which satisfy

$$\hat{J}^2 \psi_{(J,J_z)} = J(J+1)\hbar^2 \, \psi_{(J,J_z)} \quad \text{and} \quad \hat{J}_z \psi_{(J,J_z)} = J_z \hbar \, \psi_{(J,J_z)} \tag{16.189}$$

We especially wish to know if simple products of the respective spherical harmonics will also be eigenstates of these operators. We will thus consider fixed values of l_1 and l_2 with all of their respective values of m_1, m_2.

A product wavefunction of the form $Y_{(l_1, m_1)}(\theta_1, \phi_1) Y_{(l_2, m_2)}(\theta_2, \phi_2)$ is clearly an eigenstate of $\hat{J}_z$ since

$$\hat{J}_z Y_{l_1, m_1} Y_{l_2, m_2} = \left(\hat{L}_{1,z} + \hat{L}_{2,z} \right) Y_{l_1, m_1} Y_{l_2, m_2}$$

$$= (m_1 + m_2) \hbar Y_{l_1, m_1} Y_{l_2, m_2} \tag{16.190}$$

or $J_z = (m_1 + m_2)\hbar$. The same is *not* true for $\hat{J}^2$ because of the "offending cross term" in $\hat{J}^2 = (\hat{L}_1 + \hat{L}_2)^2$, namely

$$\hat{L}_1 \cdot \hat{L}_2 = \hat{L}_{1,x} \hat{L}_{2,x} + \hat{L}_{1,y} \hat{L}_{2,y} + \hat{L}_{1,z} \hat{L}_{2,z} \tag{16.191}$$

since individual spherical harmonics cannot be simultaneous eigenfunctions of all three components of $\hat{L}$. More concretely, the operator $\hat{L}_1 \cdot \hat{L}_2$ acting on a specific $Y_{l_1, m_1} Y_{l_2, m_2}$ will introduce new values of m_1, m_2. This can be seen most easily by noting that

$$\hat{J}^2 = \hat{L}_1^2 + 2\hat{L}_1 \cdot \hat{L}_2 + \hat{L}_2^2$$

$$= \hat{L}_1^2 + \hat{L}_2^2 + 2\hat{L}_{1,z} \hat{L}_{2,z} + \hat{L}_{1,+} \hat{L}_{2,-} + \hat{L}_{1,-} \hat{L}_{2,+} \tag{16.192}$$

and we recall that the action of the raising and lowering operators for each label can be obtained via

$$\hat{L}_+ Y_{l,m} = \sqrt{l(l+1) - m(m+1)} \, \hbar Y_{l,m+1}$$

$$= \sqrt{(l+m+1)(l-m)} \, \hbar Y_{l,m+1} \tag{16.193}$$

and

$$\hat{L}_- Y_{l,m} = \sqrt{(l-m+1)(l+m)} \, \hbar Y_{l,m-1} \tag{16.194}$$

We can, however, hope to find linear combinations of spherical harmonics which do satisfy Eqn. (16.189), namely

$$\psi_{(J, J_z)} = \sum_{m_1 = -l_1}^{m_1 = +l_1} \sum_{m_2 = -l_2}^{m_2 = +l_2} C(J, J_z; l_1, m_1, l_2, m_2) Y_{l_1, m_1} Y_{l_2, m_2} \tag{16.195}$$

Such an expansion is called a *Clebsch–Gordan series* and the $C(J, J_z, l_1, m_1, l_2, m_2)$ are referred to as the *Clebsch–Gordan coefficients*. The allowed values of J and J_z can then be determined by looking for eigenstates of this form.

We will proceed by simply stating the result for the possible values of J, which are allowed in such an expansion, providing some *prima facie* evidence to justify the answer, and then discuss the general procedure for its proof.

One finds that for a fixed value of l_1 and l_2:

- The allowed values for the total angular momentum quantum number, J, are in the range

$$J = |l_1 - l_2|, \ |l_1 + l_2 - 1|, \ \ldots, \ l_1 + l_2 - 1, \ l_1 + l_2 \tag{16.196}$$

or equivalently, that J occurs in integral steps in the range

$$J_{min} = |l_1 - l_2| \leq J \leq (l_1 + l_2) = J_{max}. \tag{16.197}$$

These limiting values are consistent with a classical picture of vector addition as when L_1 and L_2 are aligned or parallel (anti-aligned or antiparallel) their magnitude would be $L_1 + L_2$ ($|L_1 - L_2|$). In addition, the enumeration of distinct quantum states is consistent in the two pictures. For fixed values of l_1 and l_2 there are a total of $(2l_1 + 1)(2l_2 + 1)$ product wavefunctions of the form $Y_{l_1,m_1} Y_{l_2,m_2}$; for the eigenfunctions of total angular momentum, $\psi_{(J,J_z)}$, the equivalent counting is given by (P16.15)

$$\sum_{J=|l_1-l_2|}^{J=l_1+l_2} (2J + 1) = (2l_1 + 1)(2l_2 + 1) \tag{16.198}$$

as well. The values of m_1, m_2, and J_z are then obviously determined via $J_z = m_1 + m_2$ which implies that

$$C(J, J_z; l_1, m_1, l_2, m_2) = 0 \quad \text{if} \quad m_1 + m_2 \neq J_z \tag{16.199}$$

This observation also suggests that we analyze the Clebsch–Gordan series by looking systematically at $\psi_{(J,J_z)}$ for specific values of J_z.

Consider, for example, the unique product wavefunction with the maximal value of J_z, namely, $J_{max} = l_1 + l_2$; it is given by

$$\psi_{(J,J_z)} = Y_{l_1,l_1}(\theta_1, \phi_1) Y_{l_2,l_2}(\theta_2, \phi_2) \tag{16.200}$$

which is obviously an eigenfunction of $\hat{J}_z = \hat{L}_{1,z} + \hat{L}_{2,z}$ with $J_z = (l_1 + l_2)\hbar$. It is also an eigenfunction of $\hat{J}^2$ since, using Eqn. (16.192) and the fact that the $\hat{L}_{1,+}$ and $\hat{L}_{2,+}$ operators annihilate both spherical harmonics, we find

$$\hat{J}^2 \psi_{(J,J_z)} = [l_1(l_1 + 1) + l_2(l_2 + 1) + 2l_1 l_2] \hbar^2 \ \psi_{(J,J_z)}$$
$$= [(l_1 + l_2)(l_1 + l_2 + 1)]\hbar^2 \ \psi_{(J,J_z)}$$
$$= J(J + 1)\hbar^2 \ \psi_{(J,J_z)} \tag{16.201}$$

or $J = l_1 + l_2$. This is, of course, the only possible value for $J_z = l_1 + l_2$.

For the next lowest value of J_z, there are two possible terms which can contribute to an eigenfunction of $\hat{J}_z$ with $J_z = J_{max} - 1 = (l_1 + l_2 - 1)$, so we write

$$\psi_{(J,J_z)} = \alpha \left[Y_{l_1,l_1-1}(\theta_1,\phi_1)\, Y_{l_2,l_2}(\theta_2,\phi_2) \right] + \beta \left[Y_{l_1,l_1}(\theta_1,\phi_1)\, Y_{l_2,l_2-1}(\theta_2,\phi_2) \right]$$

(16.202)

In this case we have

$$\hat{J}^2 \psi_{(J,J_z)} = \left[\alpha(l_1(l_1+1) + l_2(l_2+1) + 2(l_1-1)l_2 + \beta(\sqrt{4l_1l_2})) \right] \hbar^2 Y_{l_1,l_1-1} Y_{l_2,l_2}$$

$$+ \left[\alpha(\sqrt{4l_1l_2}) + \beta(l_1(l_1+1) + l_2(l_2+1) \right.$$

$$+ 2(l_2-1)l_1) \left] \hbar^2 Y_{l_1,l_1} Y_{l_2,l_2-1} \right.$$

$$= J(J+1)\hbar^2 \left[\alpha Y_{l_1,l_1-1} Y_{l_2,l_2} + \beta Y_{l_1,l_1} Y_{l_2,l_2-1} \right]$$

$$= J(J+1)\hbar^2 \psi_{(J,J_z)}.$$

(16.203)

Equating coefficients, we find a set of coupled equations for α and β which can be written in the form

$$\mathbf{M} \begin{pmatrix} \alpha \\ \beta \end{pmatrix} = J(J+1) \begin{pmatrix} \alpha \\ \beta \end{pmatrix}$$

(16.204)

where

$$\mathbf{M} = \begin{pmatrix} l_1(l_1+l_2+1) + l_2(l_1+l_2-1) & 2\sqrt{l_1l_2} \\ 2\sqrt{l_1l_2} & l_1(l_1+l_2-1) + l_2(l_1+l_2+1) \end{pmatrix}$$

(16.205)

This matrix eigenvalue problem Eqn. (16.204) will have a solution provided that

$$\det |\mathbf{M} - J(J+1)\mathbf{1}| = 0$$

(16.206)

or equivalently

$$[l_1(l_1+l_2+1)l_2(l_1+l_2-1)-J(J+1)] \cdot [l_1(l_1+l_2-1) + l_2(l_1+l_2+1)-J(J+1)]$$

$$= 4l_1l_2$$

(16.207)

$$= (2l_1)(2l_2) \quad \text{Choice 1}$$

$$= (-2l_2)(-2l_1) \quad \text{Choice 2}$$

where we have written the right-hand side in two (hopefully) suggestive ways, as the two solutions to Eqn. (16.207) correspond to the two choices. The second of these yields

$$l_1(l_1 + l_2 + 1) + l_2(l_1 + l_2 - 1) + 2l_2 = J(J+1)$$

and

$$l_1(l_1 + l_2 - 1) + l_2(l_1 + l_2 + 1) + 2l_1 = J(J + 1) \qquad (16.208)$$

which both give

$$(l_1 + l_2)(l_1 + l_2 + 1) = J(J + 1) \implies J = l_1 + l_2 \qquad (16.209)$$

The first choice in Eqn. (16.207) corresponds to

$$l_1(l_1 + l_2 + 1) + l_2(l_1 + l_2 - 1) - 2l_1 = J(J + 1)$$
$$l_1(l_1 + l_2 - 1) + l_2(l_1 + l_2 + 1) - 2l_2 = J(J + 1) \qquad (16.210)$$

which implies

$$(l_1 + l_2)(l_1 + l_2 - 1) = J(J + 1) \implies J = l_1 + l_2 - 1 \qquad (16.211)$$

These two possibilities for J are the only ones consistent with $J_z = J_{max} - 1 = l_1 + l_2 - 1$. The values of α and β can be obtained in either case, and the normalized wavefunctions can be written as

$$\psi_{(l_1+l_2,l_1+l_2-1)} = \sqrt{\frac{l_1}{l_1 + l_2}} Y_{l_1,l_1-1} Y_{l_2,l_2} + \sqrt{\frac{l_2}{l_1 + l_2}} Y_{l_1,l_1} Y_{l_2,l_2-1} \qquad (16.212)$$

$$\psi_{(l_1+l_2-1,l_1+l_2-1)} = \sqrt{\frac{l_2}{l_1 + l_2}} Y_{l_1,l_1-1} Y_{l_2,l_2} - \sqrt{\frac{l_1}{l_1 + l_2}} Y_{l_1,l_1} Y_{l_2,l_2-1} \qquad (16.213)$$

which are obviously mutually orthogonal.

One can proceed in a similar fashion and construct all of the (J,J_z) eigenstates in this manner, but it is also possible to derive the same states more "mechanically" by repeated use of the "total" lowering operator $\hat{J}_- = \hat{L}_{1,-} + \hat{L}_{2,-}$.

To accomplish this, we note first that one must formally have

$$\hat{J}_- \psi_{(l_1+l_2,l_1+l_2)} = \sqrt{2(l_1 + l_2)} \hbar \, \psi_{(l_1+l_2,l_1+l_2-1)} \qquad (16.214)$$

since $\hat{J}$ satisfies all of the usual properties of an angular momentum operator. We then also have, more explicitly,

$$\hat{J}_- \psi_{(l_1+l_2,l_1+l_2)} = \left(\hat{L}_{1,-} + \hat{L}_{2,-} \right) Y_{l_1,l_1} Y_{l_2,l_2}$$
$$= \sqrt{2l_1} Y_{l_1,l_1-1} Y_{l_2,l_2} + \sqrt{2l_2} Y_{l_1,l_1} Y_{l_2,l_2-1} \qquad (16.215)$$

which together with (16.214) give the form in Eqn. (16.212). The remaining combination Eqn. (16.213) can be constructed by finding the orthogonal linear combination of states.

This procedure is the basis for a systematic method of constructing the entire set of eigenfunctions of both $\hat{J}^2$ and $\hat{J}_z$, namely:

• Start with the (unique) state of largest J_z as in Eqn. (16.200), which gives $\psi_{(J_{max},J_{max})}$.

• Use $\hat{J}_-$ on this wavefunction as in Eqns (16.214) and (16.215) to obtain $\psi_{J_{max},J_{max}-1}$; the state orthogonal to this is necessarily $\psi_{(J_{max}-1,J_{max}-1)}$.

• Operate on   both these states with $\hat{J}_-$ to obtain $\psi_{(J_{max},J_{max}-2)}$ and $\psi_{(J_{max}-1,J_{max}-2)}$, respectively; the state orthogonal to both of these must, in turn, be $\psi_{(J_{max}-2,J_{max}-2)}$.

• Repeat until all the $\psi_{(J,J_z)}$ states are constructed; along the way, $\hat{J}_-$ will eventually begin to annihilate states it acts on until only the state $\psi_{(J_{max},-J_{max})}$ is left.

This procedure is illustrated below.

$\psi_{(J_{max},J_{max})}$

$\downarrow$ via $\hat{J}_-$

$\psi_{(J_{max},J_{max}-1)}$ $\xrightarrow{\text{orthogonality}}$ $\psi_{(J_{max}-1,J_{max}-1)}$

$\downarrow$ via $\hat{J}_-$ $\downarrow$ via $\hat{J}_-$

$\psi_{(J_{max},J_{max}-2)}$ $\psi_{(J_{max}-1,J_{max}-2)}$ $\xrightarrow{\text{orthogonality}}$ $\psi_{(J_{max}-2,J_{max}-2)}$

and so forth.

Example 16.4. Two-electron spin wavefunctions

We illustrate the utility of this procedure by deriving the spin wavefunctions of a two electron system. If we label the spinors of the two spin-1/2 particles as $\chi_\pm^{(1)}$ and $\chi_\pm^{(2)}$, respectively, we are interested in eigenfunctions of the total spin angular momentum operator squared, $\hat{S}^2 = (\hat{S}_1 + \hat{S}_2)^2$. Using the spin addition rules, we know that the total spin quantum number, S, will be given by

$$S_1 + S_2 = 1/2 + 1/2 \quad \longrightarrow \quad S = 0, 1 \tag{16.216}$$

with a total of four states; these will be labeled by (S, S_z) as $(1, +1)$, $(1, 0)$, $(1, -1)$, and $(0, 0)$.
We obviously have

$$\chi(1,+1) = \chi_+^{(1)} \chi_+^{(2)} \tag{16.217}$$

(Continued)

as the state of highest S_z. Acting on this with the "total" lowering operator $\hat{S}_- = \hat{S}_{1,-} + \hat{S}_{2,-}$ requires the effect of the individual raising and lowering operators on single-particle spinors

$$\hat{S}_- \chi_+ = \hbar \chi_- \quad \text{and} \quad \hat{S}_+ \chi_- = \hbar \chi_+ \tag{16.218}$$

which are consistent with Eqns (16.193) and (16.194). We then have

$$\sqrt{2}\hbar \, \chi_{(1,0)} = \hat{S}_- \chi_{(1,+1)}$$
$$= \left(\hat{S}_{1,-} + \hat{S}_{2,-} \right) \chi_+^{(1)} \chi_+^{(2)}$$
$$= \chi_+^{(1)} \chi_-^{(2} + \chi_-^{(1)} \chi_+^{(2} \tag{16.219}$$

or

$$\chi_{(1,0)} = \frac{1}{\sqrt{2}} \left(\chi_+^{(1)} \chi_-^{(2} + \chi_-^{(1)} \chi_+^{(2)} \right) \tag{16.220}$$

The orthogonal combination is

$$\chi_{(0,0)} = \frac{1}{\sqrt{2}} \left(\chi_+^{(1)} \chi_-^{(2} - \chi_-^{(1)} \chi_+^{(2)} \right) \tag{16.221}$$

Acting on $\psi_{(0,0)}$ with $\hat{S}_-$ once gives

$$\hat{S}_- \psi_{(0,0)} = \left(\hat{S}_{1,-} + \hat{S}_{2,-} \right) \frac{1}{\sqrt{2}} \left(\chi_+^{(1)} \chi_-^{(2)} - \chi_-^{(1)} \chi_+^{(2)} \right) = 0 \tag{16.222}$$

as expected, since $\hat{S}_- \chi_{(0,0)} \propto \chi_{(0,-1)}$, which does not exist. The last state is obtained using

$$\sqrt{2}\hbar \chi_{(1,-1)} = \hat{S}_- \chi_{(1,0)}$$
$$= \left(\hat{S}_{1,-} + \hat{S}_{2,-} \right) \frac{1}{\sqrt{2}} \left(\chi_+^{(1)} \chi_-^{(2} + \chi_-^{(1)} \chi_+^{(2)} \right)$$
$$= \frac{1}{\sqrt{2}} 2 \, \chi_-^{(1)} \chi_-^{(2)} \tag{16.223}$$

or $\chi_{(1,-1)} = \chi_-^{(1)} \chi_-^{(2)}$, also as expected.

We note that the Clebsch–Gordan coefficients in these spinor wavefunctions have exactly the same form as given in Eqns (16.212) and (16.213) with $l_1 = l_2 = 1/2$. This formulation in terms of raising and lowering operators treats integral and half-integral angular momenta in the same manner. Similar results can be obtained for the addition of orbital and intrinsic angular momenta, $\mathbf{L} + \mathbf{S}$, (P16.16), which are useful in the study of single-electron atoms. The addition of more than two spins or orbital angular momenta requires only repeated use of the Clebsch–Gordan series for two angular momenta. The allowed values of J

and the corresponding wavefunctions corresponding to a sum such as $L_1 + L_2 + L_3 + \cdots + L_N$ can be obtained by "adding" the first two, then "adding" the third to all possible results of the first step, and so on. This is illustrated in P16.17, and in the next example.

Example 16.5. Deuteron wavefunction

The deuteron is a bound state of two spin-1/2 particles, the neutron and proton. The ground state is known to have total angular momentum $J = 1$ and even parity, and we wish to construct the most general wavefunction which satisfies these requirements; for definiteness, we will consider $\psi_{(J,J_z)} = \psi_{(1,+1)}$. This problem requires the angular momentum or spinor wavefunctions corresponding to the angular momentum addition problem

$$S_1 + S_2 + L = \frac{1}{2} + \frac{1}{2} + l \tag{16.224}$$

The spin degrees of freedom can first be combined, as in Example 16.4, to give $S_{tot} = 0, 1$ with corresponding wavefunctions $\chi_{(0,0)}$ and $\chi_{(1,\pm1)}$, $\chi_{(1,0)}$.

For $l = 0$ (s-state), the total angular momentum J comes from the spin degrees of freedom alone and we have

$$\psi^S_{(1,+1)} = R_S(r)Y_{(0,0)}(\theta, \phi)\chi_{(1,+1)} = R_S(r)Y_{(0,0)}(\theta, \phi)\chi_+^{(p)}\chi_+^{(n)} \tag{16.225}$$

where $R_S(r)$ is a normalized $l = 0$ radial wavefunction.

When $l = 1$ (p-states), there are two possible combinations which give $J = 1$. If $S_{tot} = 0$, then we have

$$\psi^{p,A}_{(1,+1)} = R_p(r)Y_{1,+1}(\theta, \phi)\frac{1}{\sqrt{2}}\left(\chi_+^{(p)}\chi_-^{(n)} - \chi_-^{(p)}\chi_+^{(n)}\right) \tag{16.226}$$

For the case of $S_{tot} + l = 1 + 1$ we can use the Clebsch–Gordan coefficients derived in Eqn. (16.213) to show that

$$\psi^{p,B}_{(1,+1)} = R_p(r)\left[\frac{1}{\sqrt{2}}Y_{1,+1}(\theta, \phi)\chi_{(1,0)} - \frac{1}{\sqrt{2}}Y_{1,0}(\theta, \phi)\chi_{(1,+1)}\right] \tag{16.227}$$

$$= R_p(r)\left[\frac{1}{2}Y_{1,+1}(\theta, \phi)(\chi_+^{(p)}\chi_-^{(n)} + \chi_-^{(p)}\chi_+^{(n)}) - \frac{1}{\sqrt{2}}Y_{1,0}(\theta, \phi)\chi_+^{(p)}\chi_+^{(n)}\right]$$

We recall, however, that the parity of the spherical harmonics is given by $(-1)^l$, so that these states have odd parity and thus cannot contribute to the even parity deuteron ground state wavefunction.

The final possibility occurs when $l = 2$ (d-state); in this case $S_{tot} = 0$ cannot combine with $l = 2$ to give $J = 1$, so we require the Clebsch–Gordan series for the $(1, +1)$ state occurring in the angular momentum sum $1 + 2$. The necessary expressions are given in P16.19, and we

(Continued)
find that

$$\psi_{(1,+1)}^d = R_d(r)\left[\sqrt{\frac{3}{5}}Y_{2,+2}(\theta,\phi)\chi_-^{(p)}\chi_-^{(n)} - \sqrt{\frac{3}{10}}Y_{2,+1}(\theta,\phi)\right.$$

$$\left. \times \frac{1}{\sqrt{2}}(\chi_+^{(p)}\chi_-^{(n)} + \chi_-^{(p)}\chi_+^{(n)}) + \sqrt{\frac{1}{10}}Y_{2,0}(\theta,\phi)\chi_+^{(p)}\chi_+^{(n)}\right] \quad (16.228)$$

For $l \geq 3$ it is not possible to construct a $J = 1$ wavefunction.

The observed magnetic moment of the deuteron system provides information on the structure of the ground state wavefunction. The total magnetic moment of this composite system is due to the intrinsic magnetic moments of the neutron and proton as well as the orbital angular momentum of the (charged) proton, but not the (neutral) neutron. The magnetic moment operator (along the z axis) can then be written as

$$M_z = \mu_N\left(\frac{1}{2}\frac{L_z}{\hbar} + \frac{g_p}{2}\sigma_z^{(p)} + \frac{g_n}{2}\sigma_z^{(n)}\right) \quad (16.229)$$

In the center-of-mass system, the angular momentum of the proton is half that of the total system; this accounts for the factor of $1/2$ multiplying L_z.

In a pure S-wave state, the magnetic moment is given by

$$\langle\psi_{(1,+1)}|M_z|\psi_{1,+1}\rangle = \left[\frac{g_p + g_n}{2}\right]\mu_N = (0.8798047(7))\mu_N \quad (16.230)$$

which is only slightly different that the observed value of

$$M_z^{(\text{exp})} = 0.8574376(4)\mu_N \quad (16.231)$$

The small discrepancy is still much larger than the measured errors and can be partly explained by assuming that the ground state wavefunction has a small admixture of the d-state in Eqn. (16.228), that is,

$$\psi_{(1,+1)}^{\text{mix}} = a_s\psi_{(1,+1)}^s + a_d\psi_{(1,+1)}^d \quad (16.232)$$

where $a_s^2 + a_d^2 = 1$. Using the explicit form of Eqn. (16.228), one can show (P16.20) that

$$\langle\psi_{(1,+1)}^{\text{mix}}|M_z|\psi_{1,+1}^{\text{mix}}\rangle/\mu_N = a_s^2\left[\frac{g_p + g_n}{2}\right] + a_d^2\left[\frac{3 - (g_p + g_n)}{4}\right] \quad (16.233)$$

The experimental value can then be explained by assuming $a_d^2 \approx 0.04$ or a roughly 4% mixture of d-wave in the ground state. This calculation ignores possibly important spin–orbit interactions, relativistic effects, and meson exchanges which can also affect the magnetic moment. Information from other deuteron observables, however, including its quadrupole moment (P16.21), and from scattering experiments confirm the approximately 4% d-state admixture[6].

---

[6] The understanding of magnetic moments of nuclei in terms of nuclear models is discussed in many texts on nuclear physics, for example, Krane (1988).

16.6 **Free Particle in Spherical Coordinates**

We close this chapter with a brief discussion of the solutions of the free-particle Schrödinger equation in three dimensions in the language of spherical coordinates. They will be used extensively in our discussion of quantum scattering in Chapter 20.

The radial equation for a free particle simply reads

$$-\frac{\hbar^2}{2\mu}\left(\frac{d^2R(r)}{dr^2} + \frac{2}{r}\frac{dR(r)}{dr}\right) + \frac{l(l+1)\hbar^2}{2\mu r^2}R(r) = ER(r) \tag{16.234}$$

and the obvious change of coordinates $z = kr$, where $\hbar k = \sqrt{2mE}$, gives the dimensionless equation

$$\frac{d^2R(z)}{dz^2} + \frac{2}{z}\frac{dR(z)}{dz} + \left(1 - \frac{l(l+1)}{z^2}\right)R(z) \tag{16.235}$$

There is an obvious similarity to the corresponding radial equation in two dimensions in Eqn. (15.72), and we use our experience there to simplify Eqn. (16.235). In the two-dimensional case, the solutions for large z were of the form $\sin(z)/\sqrt{z}, \cos(z)/\sqrt{z}$; the factors of $\sqrt{z}$ (when squared) were understood to exactly compensate the 'measure' factor $(r\,dr)$ to give a constant amplitude wavefunction, consistent with a free particle. In the three-dimensional case, we would then expect a $1/z$ dependence at large z since $d\mathbf{r} \propto r^2\,dr$. Motivated by this connection, we look for solutions of the form $R(z) = F(z)/\sqrt{z}$ and find that the radial equation has the form

$$\frac{d^2F(z)}{dz^2} + \frac{1}{z}\frac{dF(z)}{dz} + \left(1 - \frac{(l+1/2)^2}{z^2}\right)F(z) \tag{16.236}$$

This is exactly of the form of the standard (or cylindrical) Bessel differential equation (Appendix E.4), but for half-integral values of $m = l + 1/2$. The corresponding solutions are called *spherical Bessel functions*, and conventionally written as

$$j_l(z) = \sqrt{\frac{\pi}{2z}}J_{l+1/2}(z) \quad \text{and} \quad n_l(z) = \sqrt{\frac{\pi}{2z}}N_{l+1/2}(z) \tag{16.237}$$

where the $j_l(z)$ $(n_l(z))$ are called the regular (irregular) solutions since they are well behaved (diverge) near the origin.

The spherical Bessel functions are actually simpler in form than their two-dimensional relatives as they can be written in closed form, namely

$$j_l(z) = (-z)^l\left(\frac{1}{z}\frac{d}{dz}\right)^l\left(\frac{\sin(z)}{z}\right) \tag{16.238}$$

and

$$n_l(z) = -(-z)^l \left(\frac{1}{z} \frac{d}{dz} \right)^l \left(\frac{\cos(z)}{z} \right) \tag{16.239}$$

The first few functions have the explicit form given by:

$$j_0(z) = \frac{\sin(z)}{z} \quad n_0(z) = -\frac{\cos(z)}{z} \tag{16.240}$$

$$j_1(z) = \frac{\sin(z)}{z^2} - \frac{\cos(z)}{z} \quad n_1(z) = -\frac{\cos(z)}{z^2} - \frac{\sin(z)}{z} \tag{16.241}$$

and

$$j_2(z) = \left(\frac{3}{z^3} - \frac{1}{z} \right) \sin(z) - \frac{3}{z^2} \cos(z) \tag{16.242}$$

$$n_2(z) = -\left(\frac{3}{z^2} - \frac{1}{z} \right) \cos(z) - \frac{3}{z^2} \sin(z) \tag{16.243}$$

From general arguments (P16.22), we know that the wavefunctions must go as z^{+l} or $z^{-(l+1)}$ near the origin, and with the standard normalization one has

$$j_l(z) \sim \frac{z^l}{1 \cdot 3 \cdot 5 \cdots (2l + 1)} = \frac{z^l}{(2l + 1)!!} \tag{16.244}$$

$$n_l(z) \sim -\frac{1 \cdot 3 \cdot 5 \cdots (2l - 1)}{z^{l+1}} = \frac{(2l - 1)!!}{z^{l+1}} \tag{16.245}$$

for $z \ll l$, where $(2l + 1)!! \equiv (2l + 1) \cdot (2l - 1) \cdots 5 \cdot 3 \cdot 1$. The asymptotic behavior for $z \gg l$ is given by

$$j_l(z) \sim \frac{1}{z} \sin \left(z - \frac{l\pi}{2} \right) \quad \text{and} \quad n_l(z) \sim -\frac{1}{z} \cos \left(z - \frac{l\pi}{2} \right) \tag{16.246}$$

16.7 **Questions and Problems**

Q16.1. A classical point particle at the end of a proverbial "massless stick pivoted at one end without friction", will rotate with constant angular velocity with a well-defined angular coordinate given by $\phi(t) = \omega t + \phi_0$. How would you construct a quantum wave packet which gives this in the classical limit?

Q16.2. In geography, the position on the surface of the earth which would correspond to a parity transformation of your current location would be at an *antipode*. Can you find two cities which are (approximately) related by parity, as in Eqn. (16.90)?

Q16.3. Products of momentum eigenfunctions, as in Eqn. (16.186), are also eigen-functions of $\hat{P}_{tot}^2 = (\hat{p}_1 + \hat{p}_2)^2$. Explain why the "cross-terms" in $\hat{P}^2$ do not "mess things up" in the way that the cross-terms in $\hat{J}^2 = (\hat{L}_1 + \hat{L}_2)^2$ do.

Q16.4. The $H - Cl$ molecule has a vibrational frequency of roughly 9×10^{13} Hz. Can you use that information, the reduced mass of the molecule, and a value of $r_0 \sim 1$ Å to estimate the value of E_l for the "springy rotator" description of its rotational spectra in Eqn. (16.113), and then compare to the experimental fit?

Q16.5. In the early days of nuclear physics, the decay of the neutron was a puzzle as the only particles directly observed in the process were consistent with the process $n \rightarrow pe^-$ where the electron was observed with a continuous spec-trum of energies. Why were people worried about the lack of conservation of angular momentum and energy? On the basis of these issues, Wolfgang Pauli predicted the existence of the massless (but difficult to observe) neutrino so that the process is really $n \rightarrow pe^-\bar{\nu}_e$. How did his idea "rescue" the fundamental conservations laws?

P16.1. Show that the angular momentum operators $\hat{L}_x$ and $\hat{L}_y$ in spherical coordinates (Eqns (16.16) and (16.17) are Hermitian by confirming that, for example,

$$\langle \hat{L}_x \rangle = \int_0^\pi \sin(\theta) \, d\theta \int_0^{2\pi} d\phi \, f^*(\theta,\phi) \, \hat{L}_x f(\theta,\phi) \tag{16.247}$$

is real. What assumptions do you have to make about the arbitrary $f(\theta,\phi)$ you used?

P16.2. (a) Starting with Eqn. (16.9) and using the chain rule, show that the kinetic energy operator in spherical coordinates is given by Eqn. (16.10).

(b) Using the expressions for $\hat{L}_{x,y,z}$ in spherical coordinates, show that the total angular momentum squared in spherical coordinates is given by Eqn. (16.20).

P16.3. Derive Eqns (16.62) and (16.63) by defining

$$\hat{L}_\pm|Y_{l,m}\rangle = C_{l,m}^{(\pm)}|Y_{l,m\pm1}\rangle \tag{16.248}$$

using relations such as

$$\langle Y_{l,m}|\hat{L}_+\hat{L}_-|Y_{l,m}\rangle = \langle \hat{L}_- Y_{l,m}|\hat{L}_- Y_{l,m}\rangle \tag{16.249}$$

and Eqns (16.52) and (16.53).

P16.4. (a) Evaluate the expectation value of $\hat{L}_{x,y}$ and $\hat{L}_{x,y}^2$ in a state described by a spherical harmonic, for example, calculate

$$\langle Y_{l,m}|\hat{L}_{x,y}|Y_{l,m}\rangle \tag{16.250}$$

and so forth.

(b) Assume that you have a general angular wavefunction of the form $\psi = \sum_{l,m} a_{l,m}|Y_{l,m}\rangle$. Express the expectation values of $\hat{L}_x$ and $\hat{L}_y$ in terms of the $a_{l,m}$. Hint: Use the definitions of raising and lowering operators and Eqns (16.62) and (16.63)

P16.5. (a) Calculate the following commutators

$$[\hat{L}_y, x], \quad [\hat{L}_y, y], \quad [\hat{L}_y, z] \tag{16.251}$$

and

$$[\hat{L}_y, \hat{p}_x], \quad [\hat{L}_y, \hat{p}_y], \quad [\hat{L}_y, \hat{p}_z] \tag{16.252}$$

(b) Using these results, show that if $|\psi\rangle$ is an eigenstate of $\hat{L}_y$, then $\langle\psi|\hat{p}_x|\psi\rangle$ and $\langle\psi|\hat{p}_z|\psi\rangle$ both vanish. Similarly show that $\langle\psi|x|\psi\rangle$ and $\langle\psi|z|\psi\rangle$ vanish.

P16.6. Using Eqn. (16.73), show that the angular probability density corresponding to $P_{l,l}(\theta) = |Y_{l,l}(\theta,\phi)|^2$ satisfies

$$P_{l,l}(\theta) \propto \sqrt{l}\sin^{2l}(\theta) \tag{16.253}$$

Hint: Use Stirling's approximate formula (Appendix E.9) for the factorial function, namely

$$x! \sim e^{-x}x^x\sqrt{2\pi x}\left(1 + \mathcal{O}\left(\frac{1}{x}\right)\right) \tag{16.254}$$

which is useful for $x \gg 1$.

P16.7. An angular wavefunction is given by

$$f(\theta,\phi) = N\left[1 + \sin(\phi)\sin(\theta)\right] \tag{16.255}$$

(a) Find N such that $f(\theta,\phi)$ is properly normalized.

(b) Find the expansion of $f(\theta,\phi)$ in spherical harmonics.

(c) For a particle described by $f(\theta,\phi)$, what are $\langle\hat{L}_z\rangle$, ΔL_z, $\langle L^2\rangle$?

(d) What are the probabilities if finding $L_z = -\hbar$?, $L_z = +2\hbar$?

P16.8. Show that the expectation value of θ in the state $Y_{l,l}(\theta,\phi)$ is identically equal to $\pi/2$, independent of l. If you have the ability to perform numerical integrations (e.g., by using a multipurpose mathematical language such as Mathematica), evaluate

$$(\Delta\theta)^2 = \left\langle\left(\theta - \frac{\pi}{2}\right)^2\right\rangle = \int d\Omega \, |Y_{l,l}(\theta,\phi)|^2 \left(\theta - \frac{\pi}{2}\right)^2 \tag{16.256}$$

and show that for large l it approaches $\Delta\theta = 1/\sqrt{2l}$.

P16.9. (a) Show that the quadrupole moment of a particle of charge e in a classical planar circular orbit of radius R is given by $Q = -eR^2$.

(b) For the same particle described by the spherical harmonic $Y_{l,l}(\theta, \phi)$, show that

$$Q = -e\left(\frac{2l}{2l+3}\right)\langle r^2\rangle \tag{16.257}$$

and comment on the approach to the classical limit.

P16.10. Nuclear rotational levels. Quantized rotational energy levels are observed in heavy nuclei. For nuclei with even numbers of both neutrons and protons (so-called even–even nuclei), the nuclear shape is symmetric with respect to a reflection in the origin, and this restricts the allowed quantum numbers for rotation just as for molecules with identical atoms.

(a) Since the nuclear orientation, described by (θ, ϕ), is only defined up to a rotation by $180°$, show that the fact that the rotational wavefunction should be single-valued implies that only *even* values of the angular momentum are allowed, that is, $J = 0, 2, 4, \ldots$, and so forth.

(b) Data[2] for one such system (^{160}Dy) is given here:

J	E(keV)
0	0
2	86.8
4	283.8
6	581.2
8	967.2

Show that the data are approximately described by the formula

$$E_J = \frac{\hbar^2}{2I}J(J+1) \tag{16.258}$$

(c) Estimate the the numerical value of the nuclear moment of inertia, I, and compare to the classical value for a rigid, rotating sphere of uniform density, namely

$$I = \frac{2}{5}MR^2 \tag{16.259}$$

where $M = 160$ amu and $R(A) \approx (1.2-1.4)A^{1/3}$ F. Comment on your result.

(d) What would the classical angular frequency, ω, of the $J = 8$ state be?

(e) Try to fit the data to the form of a "spring rotator" in Eqn. (16.113) by plotting $E/J(J+1)$ versus $J(J+1)$.

[2] From Johnson, Ryde, and Hjorthh (1972).

P16.11. (a) Verify that the 3×3 matrix representation of the angular momentum operators in Eqns (16.141)–(16.143) satisfy the commutation relations $[L_x, L_y] = i\hbar L_z$, $[L_y, L_z] = i\hbar L_x$, and $[L_z, L_x] = i\hbar L_y$.

(b) What is the matrix form of the lowering operator, L_-, corresponding to Eqn. (16.148).

(c) Verify the commutation relations in Eqns (16.55)–(16.56) for the matrix representations of L_+, L_-, and L_z.

(d) Show that the eigenvalues of L_ϕ in Example 16.2 are $L_\phi = +\hbar$, 0, $-\hbar$. Find the three corresponding eigenvectors.

P16.12. (a) The magnetic moment of a planar current loop is given by $|M| = IA$ where I is the current and A is the area enclosed. For a charged particle undergoing classical uniform circular motion, show that $|M| = qrv/2$. Since the orbital angular momentum is $|L| = mvr$, show that the classical gyromagnetic ratio is given by Eqn. (16.169) with $g = 1$.

(b) A particle moving in an inverse square law force field can also have elliptical orbits. Calculate the gyromagnetic ratio for any such orbit and show that $g = 1$. Hint: Use the classical relations discussed in Section 18.2. The area of an ellipse is $A = \pi ab$ where a, b are the semimajor and semiminor axes.

(c) Show that you obtain the same result for g for a charged disk with uniform charge and mass density rotating about its center, as well as for a uniform charged sphere. Hint: Consider the disk as lots of little loops, and the sphere as lots of little disks.

P16.13. Precession of a magnetic dipole in a uniform field.

(a) Consider a magnetic moment given by

$$\mathbf{M} = \frac{q}{2m}\mathbf{L} \tag{16.260}$$

described by the Hamiltonian $\hat{H} = -\mathbf{M} \cdot \mathbf{B}$ with $\mathbf{B}$ a uniform field. Use the quantum mechanical evolution equation for expectation values

$$\frac{d}{dt}\langle \hat{L} \rangle_t = \frac{i}{\hbar}\langle [\hat{H}, \hat{L}] \rangle_t \tag{16.261}$$

to show that

$$\frac{d}{dt}\langle \hat{L} \rangle = \left(\frac{q}{2m}\right)\langle \hat{L} \rangle_t \times \mathbf{B} = \langle \mathbf{M} \rangle_t \times \mathbf{B} \tag{16.262}$$

and show that this reproduces the result of Example 16.3. Discuss the connection with the classical equation of motion for the precession of a magnetic dipole.

(b) Consider the time-development of $\langle \hat{L} \cdot \hat{L} \rangle_t$ in a similar manner and show that it is a constant. Hint: Generalize the result in P12.15 for the expectation values of products of operators.

P16.14. Conservation of probability for precessing spinors.

(a) Calculate the Hamiltonian, $\hat{H}$, for the interaction of a spin-1/2 particle in a (possibly time-dependent) magnetic field with arbitrary direction,

$$\mathbf{B}(t) = (B_x(t), B_y(t), B_z(t)) \tag{16.263}$$

and show that $\hat{H}$ is Hermitian.

(b) Use the Schrödinger equation,

$$i\hbar \frac{\partial}{\partial t} \chi(t) = \hat{H}\chi(t) \tag{16.264}$$

and its complex conjugate equation to show that $|\chi(t)|^2 = |\alpha_+(t)|^2 + |\alpha_-(t)|^2$ is constant in time so that probability is conserved.

P16.15. Explicitly confirm the summation in Eqn. (16.198) using the results of Appendix D.2.

P16.16. Adding spin and orbital angular momenta. Consider the angular momentum "addition" problem of a spin $S = 1/2$ particle combining with orbital angular momentum L. Using the methods in Section 16.5, show explicitly that the allowed values of J are $l + 1/2$ and $l - 1/2$. Do this by writing

$$\psi_{(J,J_z)} = \alpha Y_{l,m}(\theta,\phi)\chi_+ + \beta Y_{l,m+1}(\theta,\phi)\chi_- \tag{16.265}$$

and finding the values for J which give

$$\hat{J}^2 \psi_{(J,J_z)} = \left(\hat{L} + \hat{S}\right)^2 \psi_{(J,J_z)} = J(J+1)\hbar^2 \, \psi_{(J,J_z)} \tag{16.266}$$

Also find the Clebsch–Gordan coefficients for both solutions and show, for example, that

$$\psi_{(l+1/2,m+1/2)} = \sqrt{\frac{l+m+1}{2l+1}} Y_{(l,m)}\chi_+ + \sqrt{\frac{l-m}{2l+1}} Y_{(l,m+1)}\chi_- \tag{16.267}$$

and find an expression for $\psi_{(l-1/2,m+1/2)}$.

P16.17. Addition of three spins. Consider the product wavefunction of three spin-1/2 particles, that is,

$$\chi_{(S,S_z)} = \sum_{\pm} C\left[S, S_z; \frac{1}{2}, \pm\frac{1}{2}; \frac{1}{2}, \pm\frac{1}{2}; \frac{1}{2}, \pm\frac{1}{2}\right] \chi_{\pm}^{(1)} \chi_{\pm}^{(2)} \chi_{\pm}^{(3)} \tag{16.268}$$

where the summation is over all of the $2 \times 2 \times 2 = 8$ possible values of S_z; this corresponds to the problem of "adding" three spins. Use the spin addition rules to show that

$$\frac{1}{2} + \frac{1}{2} + \frac{1}{2} \longrightarrow (0,1) + \frac{1}{2} \longrightarrow \left.\frac{1}{2}\right|_A \quad \text{or} \quad \left.\frac{1}{2}\right|_B \quad \text{or} \quad \frac{3}{2} \tag{16.269}$$

where $1/2|_{A,B}$ are two distinct spin 1/2 states. Find the product wavefunctions corresponding to the possible values of S. Do this in two steps by "adding" the first two spins to find the the possible $S = 0, 1$ wavefunctions, and then adding the third spin-1/2. Verify that the wavefunctions are mutually orthogonal and that the action of the raising and lowering operators $S_{\pm,tot} = S_{\pm,1} + S_{\pm,2} + S_{\pm,3}$ moves one up and down in S_z appropriately in a state of given S.

P16.18. Neutron and proton spin in the quark model: An exercise with spinors. The nonpoint-like values of the gyromagnetic ratios for the neutron (n) and proton (p) is one piece of evidence that they are composite objects. A simple version of the *quark model* states that they can each be thought of as bound states of more fundamental, point-like, spin-1/2 objects called *quarks*; the *up-quark* with $Q_u = +2/3$ (in terms of the basic charge e) and the *down-quark* with $Q_d = -1/3$ with a common mass m_q. The proton is taken to be a $p = (u_1 u_2 d)$ bound state where $u_{1,2}$ are two different u-type quarks while $N = (u d_1 d_2)$. This model can be used to understand the values of the n, p magnetic moments as follows: The magnetic moment operator for the composite system can be written as

$$M_{z,tot} = \sum_{i=u,d} \left(\frac{Q_i e\hbar}{2m_q} \right) \sigma_z^{(i)} \tag{16.270}$$

The spin-dependent part of the proton wavefunction can be written as

$$\chi_+^{(p)} = \frac{1}{\sqrt{6}} \left(\chi_+^{(u_1)} \chi_-^{(u_2)} \chi_+^{(d)} + \chi_-^{(u_1)} \chi_+^{(u_2)} \chi_+^{(d)} - 2\chi_+^{(u_1)} \chi_+^{(u_2)} \chi_-^{(d)} \right) \tag{16.271}$$

Evaluate the expectation value of the magnetic moment of the proton by showing that

$$\langle \chi_+^{(p)} | M_{z,tot} | \chi_+^{(p)} \rangle = \left(\frac{e\hbar}{2m_q} \right) \langle \chi_+^{(p)} | Q_u \sigma_z^{(u_1)} + Q_u \sigma_z^{(u_2)} + Q_d \sigma_z^{(d)} | \chi_+^{(p)} \rangle$$

$$= \frac{e\hbar}{2m_q} \frac{1}{3} (4Q_u - Q_d)$$

The corresponding result for the neutron is obtained by letting $u \leftrightarrow d$. Show that this implies that

$$\frac{\mu_n}{\mu_p} = -\frac{2}{3} \tag{16.272}$$

and compare this to the experimental values

$$\frac{\mu_n}{\mu_p} = \frac{-1.913}{2.793} = -0.685 \tag{16.273}$$

P16.19. Calculating Clebsch–Gordan coefficients. Using the methods outlined in Sec. 16.5, calculate the Clebsch–Gordan coefficients for $J_z = J_{max} - 2 = l_1 + l_2 - 2$.

(a) For $J_z = l_1 + l_2$ and $J_z = l_1 + l_2 - 1$, act on Eqns (16.212) and (16.213) with the lowering operator.

(b) For $J = l_1 + l_2 - 2$, use orthogonality and show that your result can be written in the form

$$\psi_{(l_1+l_2-2,l_1+l_2-2)} = \alpha\, Y_{l_1,l_1-2}\, Y_{l_2,l_2} + \beta\, Y_{l_1,l_1-1}\, Y_{l_2,l_2-1} + \gamma\, Y_{l_1,l_1}\, Y_{l_2,l_2-2} \tag{16.274}$$

where

$$\alpha = \sqrt{\frac{l_2(2l_2+1)}{(l_1+l_2-1)(2l_1+2l_2-1)}} \tag{16.275}$$

$$\beta = -\sqrt{\frac{(2l_1-1)(2l_2-1)}{(l_1+l_2-1)(2l_1+2l_2-1)}} \tag{16.276}$$

$$\gamma = \sqrt{\frac{l_1(2l_1+1)}{(l_1+l_2-1)(2l_1+2l_2-1)}} \tag{16.277}$$

P16.20. (a) Using the wavefunction in Eqn. (16.232), show that the matrix element of the magnetic moment operator, Eqn. (16.229), gives the result in Eqn. (16.233).

(b) Show that $a_d^2 \approx 0.04$ reproduces the observed magnetic moment.

P16.21. Quadrupole moment of the deuteron. The quadrupole moment of the deuteron can be defined via

$$eQ = \frac{1}{4}e \int d\mathbf{r}\, |\psi(\mathbf{r})|^2(3z^2 - r^2) \tag{16.278}$$

where the "extra" factor of 1/4 arises from the fact that only the proton contributes to the charge distribution. Write this as

$$Q = \frac{1}{4}\int d\mathbf{r}\, r^2\, |\psi^{(mix)})(\mathbf{r}|^2\, (3\cos^2(\theta) - 1) \tag{16.279}$$

and use the explicit deuteron wavefunction in Eqn. (16.232) to show that

$$Q = a_s a_d \frac{\sqrt{2}}{10}\langle r^2 \rangle_{sd} - a_d^2 \frac{1}{20}\langle r^2 \rangle_{dd} \tag{16.280}$$

where

$$\langle r^2 \rangle_{sd} = \langle R_s|r^2|R_d \rangle \quad \text{and} \quad \langle r^2 \rangle_{dd} = \langle R_d|r^2|R_d \rangle \tag{16.281}$$

Explain how this result could be used to check the magnitude and *sign* of a_d.

P16.22. Find the short-distance behavior of the three-dimensional radial equation, Eqn. (16.234), by assuming a solution of the form $R(r) = r^\alpha G(r)$, and show that $\alpha = +l$ and $-(l+1)$.

P16.23. Variational estimate for the spherical infinite well. Obtain a variational estimate of the ground state energy of of a particle of mass m in a spherical infinite well of radius R and compare it to the exact answer. Hint: Use a trial solution of the form $R(r; a) = N(R^a - r^a)$.

P16.24. Solve for the energy eigenvalues and wavefunctions for a particle of mass m in a three-dimensional square box of length L on each side. Generalize to a rectangular parallelepiped of sides L_1, L_2, L_3. Discuss the degeneracies, accidental or otherwise for various cases.

P16.25. The harmonic oscillator in three dimensions. The isotropic harmonic oscillator in three dimensions is defined by the potential

$$V(\mathbf{r}) = \frac{1}{2}m\omega^2\mathbf{r}^2 = \frac{1}{2}m\omega^2(x^2 + y^2 + z^2) \tag{16.282}$$

(a) Solve the energy eigenvalues and eigenfunctions using Cartesian coordinates. Show that the energies can be written as

$$E_{n_x,n_y,n_z} = \left(n_x + n_y + n_z + \frac{3}{2}\right)\hbar\omega = \left(N + \frac{3}{2}\right)\hbar\omega = E_N \tag{16.283}$$

Show that the number of states with energy E_N (i.e. the degeneracy) is given by $(N + 1)(N + 2)/2$.

(b) Solve this problem using spherical coordinates making use of the same kind of variable changes used in the 1D and 2D oscillators. Show that the same energy spectrum is obtained which can be written in the form

$$E_{n_r,l} = \left(2n_r + l + \frac{3}{2}\right)\hbar\omega \tag{16.284}$$

where n_r counts the number of radial nodes. Compare the degeneracy of each level in this notation to that found in part (a). Show that the wavefunctions can be written in the form

$$\psi(r, \theta, \phi) \propto y^l \, e^{-y^2/2} \, L_{n_r}^{l+1/2}(y^2) \, Y_{l,m}(\theta, \phi) \tag{16.285}$$

where $r = \rho y$ and $\rho^2 = \hbar/\mu\omega$. Show that the wavefunctions for the first few lowest lying levels in the two solution schemes (Cartesian versus spherical) are linear combinations of each other.

SEVENTEEN
The Hydrogen Atom

One of the first great triumphs of classical mechanics was the analysis of the motion of two bodies (earth-sun, moon-earth, or, perhaps, even earth-apple) moving under the influence of the inverse square law of gravitation using the, then-new, tool of calculus. The use of sophisticated computational methods later allowed for the precise calculation of the trajectories of a collection of particles, the solar system. The predictions obtained became so accurate that observed deviations from theory could confidently be used as a signal of new physics phenomena; examples include the prediction of new planets,[1] or the need for general relativistic effects as inferred by Einstein.[2]

The situation is somewhat similar in quantum mechanics. The study of the hydrogen atom system, consisting of an electron and proton interacting (to first approximation) only via their Coulomb attraction, was one of the first successes of both the early quantum ideas of Bohr, and later of the complete machinery of wave and matrix mechanics. Extensions to include other effects, such as relativity and spin, as well as the description of multielectron atoms and the resulting understanding of the periodic table, constitute one of the triumphs of theoretical physics of the twentieth century.

In this chapter, we focus on some of the simplest aspects of hydrogen and hydrogen-like atoms, with a brief discussion of some of the complications arising from more complex atomic systems, using helium and lithium as examples.

17.1 Hydrogen Atom Wavefunctions and Energies

We begin by considering hydrogen-like atoms consisting of a single electron and a point-like, positively charged nucleus of mass M and charge $+Ze$; this includes not only hydrogen itself (where $Z = 1$) but also other atomic states such as

---

[1] Adams and Leverrier independently deduced the mass and orbital parameters of Neptune using Newtonian celestial mechanics.

[2] See, for example, Weinberg (1972).

singly ionized helium, He^+ ($Z = 2$), doubly ionized lithium, Li^{++} ($Z = 3$), and so forth. The Hamiltonian for the two-particle system is

$$\hat{H} = \frac{\hat{p}_e^2}{2m_e} + \frac{\hat{p}_N^2}{2M} - \frac{KZe^2}{|\mathbf{r}_e - \mathbf{r}_N|} \tag{17.1}$$

The usual change to center-of-mass and relative coordinates produces the Hamiltonian

$$\hat{H} = \frac{\hat{p}^2}{2\mu} - \frac{KZe^2}{r} \tag{17.2}$$

where $\mathbf{r} = \mathbf{r}_e - \mathbf{r}_N$ is the relative coordinate, $\mathbf{p}$ is its corresponding momentum operator, and $\mu = m_e M/(m_e + M)$ is the reduced mass.

Just as with the harmonic oscillator, the energy spectrum and wavefunctions for the hydrogen atom problem can be derived using operator methods,[3] but we will follow a straightforward differential equation approach. A solution in spherical coordinates of the standard form, $\psi(\mathbf{r}) = R(r)Y_{l,m}(\theta, \phi)$, then gives the radial equation

$$\frac{d^2 R(r)}{dr^2} + \frac{2}{r}\frac{dR(r)}{dr} + \frac{2\mu}{\hbar^2}\left(E + \frac{KZe^2}{r} - \frac{l(l+1)\hbar^2}{2\mu r^2}\right) R(r) = 0 \tag{17.3}$$

where we write $E = -|E|$ when looking for negative energy bound states. The method of solution is then a standard one for which we only present the "highlights." A suitable change of variables,

$$\rho = \left(\frac{8\mu|E|}{\hbar^2}\right)^{1/2} r \tag{17.4}$$

reduces Eqn. (17.3) to dimensionless form

$$\frac{d^2 R}{d\rho^2} + \frac{2}{r}\frac{dR}{d\rho} + \left(\frac{\sigma}{\rho} - \frac{l(l+1)}{\rho^2} - \frac{1}{4}\right) R = 0 \tag{17.5}$$

The parameter σ is given by

$$\sigma = \frac{KZe^2}{\hbar^2}\left(\frac{\mu}{2|E|}\right)^{1/2} = Z\alpha\left(\frac{\mu c^2}{2|E|}\right)^{1/2} \tag{17.6}$$

where, for convenience, we again use the *fine structure constant*, $\alpha \equiv Ke^2/\hbar c$, as introduced in Section 1.4. We note that this change of variable depends on the value of the quantized energy, $|E_n|$, and so will vary with quantum number. The

---

[3] See, for example, Ohanian (1990) for such a derivation at the level of this text.

behavior of $R(\rho)$ for both large and small ρ is easily extracted (P17.1) giving $R(\rho) \longrightarrow e^{-\rho/2}$ and $R(\rho) \longrightarrow \rho^l$ in those two limits. These can be incorporated by writing

$$R(\rho) = G(\rho)\, \rho^l\, e^{-\rho/2} \tag{17.7}$$

to obtain

$$\frac{d^2 G(\rho)}{d\rho^2} + \frac{dG(\rho)}{d\rho}\left(\frac{2l+2}{\rho} - 1\right) + G(\rho)\left(\frac{\sigma - l - 1}{\rho}\right) = 0 \tag{17.8}$$

A series solution to Eqn. (17.8) of the form $G(\rho) = \sum_{s=0}^{\infty} a^s \rho_s$ yields the recursion relation

$$\frac{a_{s+1}}{a_s} = \frac{s - (\sigma - l - 1)}{(s+1)(s+2l+1)} \longrightarrow \frac{1}{s} \implies G(\rho) \longrightarrow e^{\rho} \tag{17.9}$$

unless the series terminates. The integral value of s for which the solutions of Eqn. (17.8) are then well behaved, $s_{\max}$, counts the number of radial nodes, so we find

$$s_{\max} \equiv n_r = \sigma - l - 1 \tag{17.10}$$

or

$$\sigma = n_r + l + 1 \equiv n \tag{17.11}$$

where n is called the *principal quantum number*; it satisfies $n = 1, 2, \ldots$ and $n \geq l + 1$.

From Eqn. (17.6), we see that the energies are quantized in terms of n since

$$E_n = -\frac{1}{2}\mu c^2 (Z\alpha)^2 \frac{1}{n^2} \approx -Z^2 \frac{13.6\,\text{eV}}{n^2} \tag{17.12}$$

when $M \gg m_e$ This form is useful because:

- It shows how the Balmer formula for the spectral lines for hydrogen-like atoms arises. The wavelengths of the photons emitted in transitions from one state to another are given by

$$\hbar c \frac{2\pi}{\lambda} = \hbar\omega = E_\gamma = E_m - E_n \tag{17.13}$$

or

$$\frac{1}{\lambda} = R_\infty \left(1 + \frac{m_e}{M}\right)^{-1} Z^2 \left(\frac{1}{n^2} - \frac{1}{m^2}\right) \tag{17.14}$$

This expression gives the so-called *Rydberg constant*

$$R_\infty \equiv \frac{m_e c \alpha^2}{4\pi \hbar} \approx 1.097 \times 10^7\,\text{m}^{-1} \approx \frac{1}{911}\,\text{Å}^{-1} \tag{17.15}$$

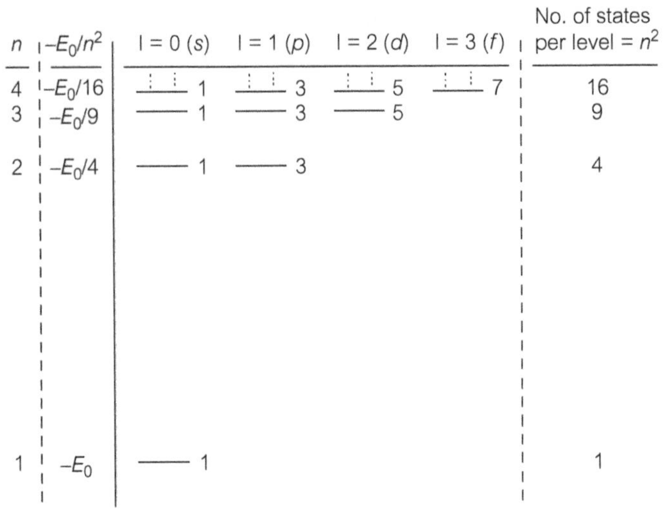

n	$-E_0/n^2$	$l = 0$ (s)	$l = 1$ (p)	$l = 2$ (d)	$l = 3$ (f)	No. of states per level $= n^2$
4	$-E_0/16$	┈ 1	┈ 3	┈ 5	┈ 7	16
3	$-E_0/9$	── 1	── 3	── 5		9
2	$-E_0/4$	── 1	── 3			4
1	$-E_0$	── 1				1

Figure 17.1. Part of the energy spectrum for hydrogen-like atoms; the energies correspond to $-E_0 = -\mu c^2 (Z\alpha)^2/2$. Standard spectroscopic notation is used for the various angular momentum states, and the degeneracies of each level are also shown.

in terms of fundamental constants of nature; Eqn. (17.14) also explicitly exhibits the dependence on the nuclear charge and mass.

- When written as

$$E_n = -\frac{1}{2}\mu \left(\frac{Z\alpha c}{n}\right)^2 = -\frac{1}{2}\mu v^2 \tag{17.16}$$

it shows that the magnitude of the typical velocity is given by $v \sim c(Z\alpha/n)$, which, in turn, sets the scale for when relativistic effects become important.

The energy spectrum is illustrated in Fig. 17.1 in such a way as to emphasize that the energy levels do not depend on the angular momentum quantum number l (as, for example, in Fig. 15.9 for the two-dimensional circular well) but only on n.

The radial wavefunctions are written in terms of the variable ρ, which we can now express in the form

$$\rho = \left(\frac{8\mu|E|}{\hbar^2}\right)^{1/2} r = \left(\frac{2Z}{na_0}\right) r \tag{17.17}$$

where we have defined the *Bohr radius* via

$$a_0 = \frac{\hbar^2}{m_e Ke^2} = \frac{\hbar}{m_e c\alpha} \approx 0.529 \text{ Å} \tag{17.18}$$

which is defined for an infinitely heavy nucleus ($\mu \to m_e$); this length sets the scale for many atomic physics problems. Finally, this implies that the quantized energies can also be written in the form

$$E_n = -\frac{1}{2}\left(\frac{Ke^2}{a_0}\right)\frac{Z^2}{n^2} \tag{17.19}$$

which is sometimes useful as it emphasizes the electrostatic origin of the binding energy. Note that the expressions for the energies in Eqns (17.12) and (17.19), and the radial variable (17.17) depend only on the combination Z/n.

Comparison of Eqn. (17.8) with standard forms (as in Appendix E.7) shows that the solutions $G(\rho)$ can be expressed in terms of Laguerre polynomials; this gives for the radial wavefunction

$$R(r) = R_{n,l}(\rho) \quad \propto \quad \rho^l \, L_{n-l-1}^{2l+1}(\rho) \, e^{-\rho/2} \tag{17.20}$$

The properly normalized radial wavefunctions for the lowest-lying states (along with the traditional spectroscopic label) are:

$$1s : R_{1,0}(r) = 2\left(\frac{Z}{a_0}\right)^{3/2} e^{-Zr/a_0} \tag{17.21}$$

$$2s : R_{2,0}(r) = 2\left(\frac{Z}{2a_0}\right)^{3/2} \left[1 - \left(\frac{Zr}{2a_0}\right)\right] e^{-Zr/2a_0} \tag{17.22}$$

$$2p : R_{2,1}(r) = \frac{1}{\sqrt{3}}\left(\frac{Z}{2a_0}\right)^{3/2} \frac{Zr}{a_0} e^{-Zr/2a_0} \tag{17.23}$$

$$3s : R_{3,0}(r) = 2\left(\frac{Z}{3a_0}\right)^{3/2} \left[1 - 2\left(\frac{Zr}{3a_0}\right) + \frac{2}{3}\left(\frac{Zr}{3a_0}\right)^2\right] e^{-Zr/3a_0} \tag{17.24}$$

$$3p : R_{3,1}(r) = \frac{4\sqrt{2}}{9}\left(\frac{Z}{3a_0}\right)^{3/2} \left(\frac{Zr}{a_0}\right)\left[1 - \frac{1}{2}\left(\frac{Zr}{3a_0}\right)\right] e^{-Zr/3a_0} \tag{17.25}$$

$$3d : R_{3,2}(r) = \frac{2\sqrt{2}}{27\sqrt{5}}\left(\frac{Z}{3a_0}\right)^{3/2} \left(\frac{Zr}{a_0}\right)^2 e^{-Zr/3a_0} \tag{17.26}$$

We plot the corresponding radial probability densities (including the measure) as $P^{(n,l)}(r) = r^2|R_{n,l}(r)|^2$ versus r in Fig. 17.2 for the lowest-lying states. The general form for the special case $R_{n,n-1}(r)$ is easy to derive (P17.3) and is given by

$$R_{n,n-1}(r) = \left[\left(\frac{2Z}{na_0}\right)^{2n+1} \frac{1}{(2n)!}\right]^{1/2} r^{n-1} e^{-Zr/na_0} \tag{17.27}$$

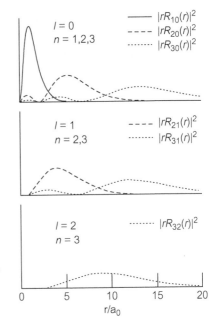

Figure 17.2. Plots of the radial probability densities for the hydrogen atom, $P(r) = r^2 |R_{n,l}(r)|^2$ versus r/a_0; the $n = 1, 2, 3$ states are shown by solid, dashed, and dotted lines, respectively.

It will also be useful to have at hand expressions for various expectation values for the radial variable,

$$\langle r^k \rangle = \int d\mathbf{r}\, r^k\, |\psi(\mathbf{r})|^2 = \int_0^\infty r^{k+2}\, |R_{n,l}(r)|^2\, dr \tag{17.28}$$

which we collect here without proof:

$$\langle r \rangle = \frac{1}{2}\left(\frac{a_0}{Z}\right)\left[3n^2 - l(l+1)\right] \tag{17.29}$$

$$\langle r^2 \rangle = \frac{1}{2}\left(\frac{a_0 n}{Z}\right)^2\left[5n^2 + 1 - 3l(l+1)\right] \tag{17.30}$$

$$\left\langle \frac{1}{r} \right\rangle = \frac{1}{n^2}\left(\frac{Z}{a_0}\right) \tag{17.31}$$

$$\left\langle \frac{1}{r^2} \right\rangle = \left(\frac{Z}{a_0}\right)^2\left[\frac{1}{n^3(l+1/2)}\right] \tag{17.32}$$

$$\left\langle \frac{1}{r^3} \right\rangle = \left(\frac{Z}{a_0}\right)^3\left[\frac{1}{n^3(l(l+1)(l+1/2)l)}\right] \tag{17.33}$$

$$\left\langle \frac{1}{r^4} \right\rangle = \left(\frac{Z}{a_0}\right)^4\left[\frac{3n^2 - l(l+1)}{2n^5(l-1/2)l(l+1/2)(l+1)(l+3/2)}\right] \tag{17.34}$$

These solutions form a complete and orthonormal set of states in terms of which any (properly behaved) 3D wavefunction (in r, θ, ϕ) can be expanded, namely

$$\psi(r,\theta,\phi) = \sum_{n=1}^{\infty} \sum_{l=0}^{n-1} \sum_{m=-l}^{+l} a_{n,l,m}\, R_{n,l}(r)\, Y_{l,m}(\theta,\phi) \tag{17.35}$$

as well as satisfying

$$\langle \psi_{n,l,m} | \psi_{n',l',m'} \rangle = \left(\int_0^{\infty} r^2\, R_{n,l}^*(r)\, R_{n',l'}(r)\, dr \right) \left(\int d\Omega\, Y_{l,m}^*(\theta,\phi) Y_{l',m'}(\theta,\phi) \right)$$

$$= \delta_{n,n'}\, \delta_{l,l'}\, \delta_{m,m'} \tag{17.36}$$

17.2 The Classical Limit of the Quantum Kepler Problem

The motion of a particle under the influence of an inverse square force law is one of the best studied of all classical mechanics problems.[4] We briefly review here some aspects of the solutions in the context of the hydrogen atom problem, using $V(r) = -Ke^2/r$; the scaling with Z for ions is easily obtained.

The trajectories for bound state motion (i.e. for $E < 0$) consist of elliptical paths as in Fig. 17.3 with the center-of-mass at one focus (labeled F); circular orbits are a special case. (The case of unbound motion where $E > 0$, which is of relevance to scattering problems, gives rise to hyperbolic orbits which can be studied in a similar fashion.) The semimajor and semiminor axes, a and b, are given in terms of the energy and angular momentum via

$$|E| = \frac{Ke^2}{2a} = \frac{L^2}{2\mu b^2} \tag{17.37}$$

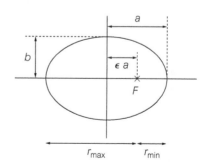

Figure 17.3. Definition of parameters for elliptical orbits.

[4] For a concise treatment, see Barger and Olsson (1995).

where μ is the reduced mass. They are also related by

$$b = a\sqrt{1 - \epsilon^2} \quad \text{or} \quad \epsilon^2 = 1 - \frac{b^2}{a^2} = 1 - \frac{2L^2|E|}{\mu(Ke^2)^2} \tag{17.38}$$

where ϵ is the eccentricity of the elliptical orbit; $\epsilon = 0$ corresponds to circular orbits, while $\epsilon \to 1$ gives purely radial motion with no angular momentum.

The parametric equation determining the elliptical orbit can be written in the form

$$r(\theta) = \frac{\alpha}{1 + \epsilon \cos(\theta)} \quad \text{where} \quad \alpha = \frac{L^2}{\mu Ke^2} = a(1 - \epsilon^2) \tag{17.39}$$

Using this expression, one finds that the distances of closest and furthest approach[5] are given by

$$r_{min} = a(1 - \epsilon) \quad \text{and} \quad r_{max} = a(1 + \epsilon) \tag{17.40}$$

The period of the orbit, τ, is given in terms of the semimajor axis via *Kepler's third law*, namely

$$\tau^2 = \left(\frac{4\pi^2\mu}{Ke^2}\right) a^3 \tag{17.41}$$

We can now use the corresponding quantum results, namely

$$E = E_n = -\frac{Ke^2}{2a_0}\frac{1}{n^2} \quad \text{and} \quad L^2 = l(l+1)\hbar^2 \tag{17.42}$$

to see that $a = a_0 n^2$ is the semimajor axis; the facts that (i) the classical energy for the Kepler problem depends only on the value of the semimajor axis a (via Eqn. (17.37)) while (ii) the quantum energy depends only on n, are obviously related.

The "effective potential" in which the electron moves is then

$$V_{\text{eff}}(r) = -\frac{Ke^2}{r} + \frac{l(l+1)\hbar^2}{2\mu r^2} \tag{17.43}$$

and is shown in Fig.17.4 for (n, l) equal to $(20, 19), (20, 15)$, and $(20, 0)$ to illustrate the classical turning points of the motion.

We can use Eqns (17.37) and (17.38) to show that

$$\epsilon^2 = 1 - \frac{l(l+1)}{n^2} \tag{17.44}$$

[5] In this context, these classical radii might be called "peri-nucleus" and "apo-nucleus" respectively, by analogy with "perigee/perihelion" and "apogee/aphelion" for the corresponding quantities for earth/solar orbit.

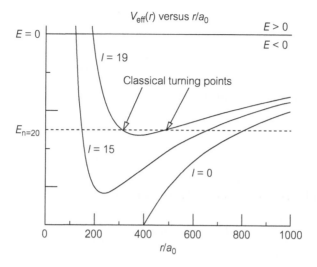

Figure 17.4. Effective potential for $n = 20$ and various l values; the dashed horizontal line shows the $n = 20$ energy level, and the intersections with $V_{eff}(r)$ give the classical turning points.

Classical circular motion corresponds to having no nodes in the radial wavefunction ($n_r = 0$) which gives $l = n - 1$ or

$$\epsilon^2 = \frac{1}{n} \tag{17.45}$$

so that truly circular orbits are achieved only for $n \to \infty$. The corresponding quantum wavefunctions are given by Eqn. (17.27) as $R_{n,n-1} \propto r^{n-1} \, e^{-r/na_0}$ with corresponding probability density for r given by

$$P(r) \quad \propto \quad r^2 \, |R_{n,n-1}(r)|^2 = r^{2n} \, e^{-2r/na_0} \tag{17.46}$$

The maximum of this distribution is given by

$$\frac{dP(r)}{dr} = 2n \, r^{2n-1} e^{-r/na_0} - \frac{2}{na_0} r^{2n} e^{-r/na_0} = 0 \quad \text{or} \quad r = a_0 n^2 \tag{17.47}$$

This is just the classical radius for circular motion for this case, in which $\epsilon \to 0$ and $r_{max} = r_{min}$.

The classical probability distributions can also be derived by standard methods, and compared with the quantum results for any value of n and l. To do this, we first write

$$E = \frac{1}{2}\mu v^2 - \frac{Ke^2}{r} + \frac{L^2}{2\mu r^2} \tag{17.48}$$

in the form

$$v = \frac{dr}{dt} = \sqrt{\frac{2}{\mu}\left(E + \frac{Ke^2}{r} - \frac{L^2}{2\mu r^2}\right)} \tag{17.49}$$

or substituting the quantum values

$$dt = \sqrt{\frac{a_0 n^2 \mu}{Ke^2}} \left[\frac{r \, dr}{\sqrt{2ra_0 n^2 - l(l+1)n^2 a_0^2 - r^2}} \right] \tag{17.50}$$

With the period is given by Eqn. (17.41), we use

$$d\text{Prob} \equiv P_{\text{CL}}(r)dr = \frac{dt}{\tau/2} \tag{17.51}$$

to find

$$P_{\text{CL}}^{(n,l)}(r) = \frac{1}{\pi a_0 n^2} \frac{r}{\sqrt{2ra_0 n^2 - l(l+1)n^2 a_0 - r^2}} \tag{17.52}$$

which is defined only in the interval $(r_{\text{min}}, r_{\text{max}})$ with

$$r_{\text{min,max}} = a_0 n^2 \left(1 \mp \sqrt{1 - \frac{l(l+1)}{n^2}} \right) \tag{17.53}$$

These classical distributions can be compared with the quantum wavefunctions[6] (including the measure.)

$$P_{\text{QM}}^{(n,l)} = r^2 |R_{n,l}(r)|^2 \tag{17.54}$$

and we show results in Fig. 17.5 for the same set of (n, l) values in Fig. 17.4.

Finally, we note that wave packet solutions for the hydrogen atom in circular[7] or elliptical[8] orbits can be constructed which spread in time, as expected from earlier one- and two-dimensional examples. Just as with the infinite well, however, Coulomb wave packets can also exhibit so-called *revivals* or *reformations* in which the wave packet returns to a well-defined localized state. Localized electronic wave packets can be excited in Rydberg atoms using picosecond laser pulses,[9] and revival phenomena have been observed[10]; such experiments provide a unique laboratory for the study of the connection between quantum and classical dynamics.

[6] See Rowe (1987) for similar images.
[7] See, for example, Brown (1973) for an early example.
[8] See, for example, Nauenberg (1989).
[9] See the review by Alber and Zoller (1991).
[10] For a review of *Quantum Wave Packet Revivals*, see Robinett (2004).

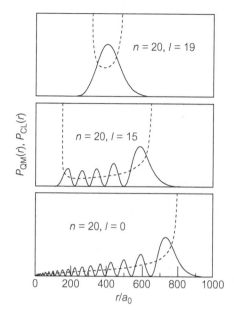

Figure 17.5. The classical (dashed) and quantum (solid) probability distributions for the radial coordinate; $P_{CL}(r)$ and $P_{QM}(r)$ versus r/a_0 are shown for the same (n, l) values, and over the same range, as in Fig. 17.4.

Figure 17.6. Precessing elliptical orbits in a non-$1/r^2$ force field and the corresponding behavior of the Runge–Lenz vector.

Many attractive potentials admit approximately elliptical orbits, but the inverse square force law is known to be special in that the orientation of the ellipse (as measured, say, by the direction of the semimajor axis) stays fixed with time; that is, the ellipse does not precess, as shown in Fig. 17.6. This is related to the fact that while for *any* central potential the angular momentum vector **L** is conserved, there is an additional conserved vector, unique to the $1/r$ potential. This quantity is called the **Lenz–Runge vector** and can be written (classically) as

$$\mathbf{R} = \frac{\mathbf{r}}{r} - \left(\frac{1}{mKe^2}\right)\mathbf{p} \times \mathbf{L} \qquad (17.55)$$

It can be shown (**P17.17**) that classically

- The vector **R** points along the semimajor axis of the ellipse, as in Fig. 17.7,
- It is fixed in time (i.e. conserved), and

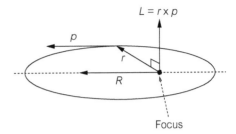

Figure 17.7. Relationship of the Runge–Lenz and angular momentum vectors for a classical orbit in a $1/r$ potential.

- It has magnitude equal to the eccentricity, that is

$$\mathbf{R}^2 = 1 - \frac{2L^2|E|}{\mu(Ke^2)^2} \tag{17.56}$$

This last relation implies an extra degeneracy in the problem due to this simple relation between three conserved quantities. Even relatively small deviations from the $1/r^2$ force law of classical gravity, as due, for example, to the corrections from general relativity,[11] will "break the symmetry," allow $\mathbf{R}$ to vary in time, and cause a precession of the elliptical orbit; for example, the precession of the orbit of Mercury is due to non-$1/r^2$ effects arising from corrections to the Newtonian potential given by general relativity.

The corresponding quantum vector operator, given by appropriately replacing $\mathbf{p}$ and $\mathbf{L}$ by their operator counterparts, is,

$$\hat{\mathbf{R}} = \frac{\mathbf{r}}{r} - \left(\frac{1}{2mKe^2}\right)\left[\hat{\mathbf{p}} \times \hat{\mathbf{L}} - \hat{\mathbf{L}} \times \hat{\mathbf{p}}\right] \tag{17.57}$$

which can be shown (P17.18) to commute with the Hamiltonian and therefore to be conserved in the quantum version of the Kepler problem. (Note the "anti-symmetrized" quantum version of the classical expression involving the vector cross-product.)

This has profound consequences as the remarkable simplicity of the hydrogen atom energy spectrum is, in part, due to this additional conserved quantity. The conservation of angular momentum only guarantees that the energy eigenvalues of all of the $2l + 1$ different m values for a given value of l will be degenerate in any purely central potential. The conservation of $\mathbf{R}$ can be shown to imply that the states with $l = 0, 1, \ldots, n - 1$ for a given value of the principal quantum number n will also be degenerate; this gives the total of

$$\sum_{l=0}^{n-1}(2l + 1) = n^2 \tag{17.58}$$

---

[11] See Marion and Thornton (2004).

states with same energy. Any additional physical effects which alter the $1/r$ potential, such as relativistic effects (this time special relativity), will then cause splittings of the energy levels and give rise to a less "symmetric" spectrum.

Example 17.1. Level splittings in non-Coulombic potentials

As a pedagogical example[12], consider the slightly modified Coulomb potential given by

$$V(r) = -\frac{ZKe^2}{r}\left(1 + b\frac{a_0}{r}\right) \tag{17.59}$$

where b is dimensionless. The Schrödinger equation for this potential can be solved using the same techniques as above (P17.13) and one finds that the energy levels are now

$$E_{n,l} = -\frac{1}{2}\mu c^2 (Z\alpha)^2 \frac{1}{[n + \sqrt{(l+1/2)^2 - 2b} - (l+1/2)]^2} \tag{17.60}$$

The energy levels now depend on both n and l with the larger angular momentum states affected less by the non-Coulombic interaction. (Can you provide an explanation of this last fact?)

17.3 **Other "Hydrogenic" Atoms**

There are a number of two-particle, atomic bound state systems, many of whose properties can be understood rather simply on the basis of straightforward extensions of the hydrogen atom.

17.3.1 **Rydberg Atoms**

An atom in which one electron is highly excited (but not ionized), and hence far from the remaining "ionic core," exhibits hydrogenic-like behavior for sufficiently large values of the principal quantum number, n. This is plausible since for sufficiently large distances from the inner electrons, the excited electron will be influenced only by the net charge of the core and not by its detailed structure; the total charge (nuclear charge plus core electrons) will be $+e$, just as for hydrogen. Such highly excited states are often called *Rydberg atoms.*[13]

---

[12] More realistic effects which give rise to level splittings will be considered in P17.11 and P17.12.
[13] For reviews, see Stebbings (1983) or Gallagher (1994).

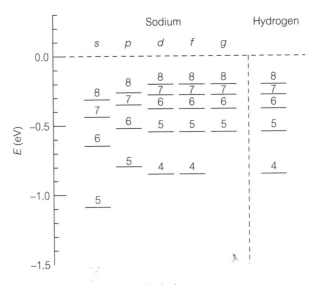

Figure 17.8. Energy levels for sodium compared to hydrogen.

The alkali metals (the elements in Column IA of the periodic table, lithium (Li), sodium (Na), potassium (K), cesium (Cs), etc.) are often used for studies of Rydberg states as they consist of a single unpaired electron outside a closed (noble gas atom) electron shell and therefore most readily exhibit hydrogen-like behavior; they are the "work horses" of atomic physics research and applications.[14] The incredible frequency resolution of modern tunable lasers makes it possible to populate individual Rydberg states with values of principal quantum number $n > 200$.

The energy levels of such an alkali metal are illustrated in Fig. 17.8, where the spectrum of sodium is compared with that for hydrogen. The states with the highest orbital angular momentum l show the closest agreement for a given n as they represent circular orbits which have the largest angular momentum barrier near the origin. Such orbits are kept furthest away from the core region where they would experience the largest deviations from the $1/r$ potential and are called *nonpenetrating*. States with lower values of l correspond classically to elliptical orbits of large eccentricity which can make excursions into the core region giving rise to nonhydrogenic behavior; they are called *penetrating orbits*. The classical elliptical orbits for the angular momentum states allowed for $n = 6$ are shown in Fig. 17.9 where the size of the "ionic core" is (very) roughly indicated by the star.

---

[14] Recall that the fundamental unit of time, the second, is now *defined* as being "*equal to* 9,192,631,770 *periods of the radiation corresponding to the transition between the two hyperfine levels of the ground state of* [133]*Cs.*"

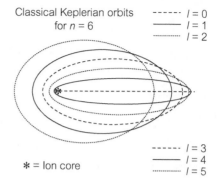

Classical Keplerian orbits for $n = 6$

------ $l = 0$
——— $l = 1$
.... $l = 2$

------ $l = 3$
——— $l = 4$
.... $l = 5$

* = Ion core

Figure 17.9. Classical elliptical orbits corresponding to the $n = 6$ levels of sodium for allowed values of $l = 0, 1, 2, 3, 4, 5$. Orbits with large values of l are less "penetrating," that is, have less of an overlap with the ion core.

The spectrum of such Rydberg states is characterized by an *effective quantum number*, n^*, or the related notion of a *quantum defect*, δ_l, defined via

$$E_{n,l} = -\frac{E_0}{(n^*)^2} = -\frac{E_0}{(n - \delta_l)^2} \tag{17.61}$$

where $E_0 = \mu c^2 \alpha^2 / 2 \approx 13.6\,\text{eV}$. The degeneracy of the spectrum for the pure Coulomb potential is lifted by the additional effects of the ionic core, and δ_l measures the splitting due to these interactions. In the simple model of Example 17.1, for example, the quantum defect due to the addition of a small b/r^2 potential is

$$\delta_l = -\left[\sqrt{(l + 1/2)^2 - 2b} - (l + 1/2)\right] \approx \frac{b}{l} \quad \text{for} \quad l \gg 1 \tag{17.62}$$

For small values of the quantum defect, one can write

$$E_{n,l} = -\frac{E_0}{n^2(1 - \delta_l/n)^2} \approx -\frac{E_0}{n^2} - \frac{2E_0\delta_l}{n^3} + \cdots \tag{17.63}$$

and the fact that the energy shift goes as $1/n^3$ can be understood as being due to the fact that the probability of the valence electron being near the core scales as n^{-3}.

Rydberg atoms are extremely loosely bound, and are characterized by their large sizes. Because $\langle r \rangle \sim a_0 n^2$, such atoms can approach almost macroscopic sizes; for $n = 100$, the diameter of a Rydberg atom is $\sim 1\mu\text{m}$, roughly the size of a biological cell, and classical arguments are often more appropriate than quantum ones, as mentioned in Example 1.3.

17.3.2 **Muonic Atoms**

Hydrogen-like atoms can be formed in which a *muon* is bound to a nucleus of charge Ze; such states are called *muonic atoms.*[15] Muons are "heavy electrons"

---

[15] For a review, see Hüfner, Scheck, and Wu (1977).

in that they have the same electric charge $(-e)$ and spin $(1/2)$, but are roughly 207 times heavier ($m_\mu c^2 = 105.6\,\text{MeV}$ instead of $m_e c^2 = 0.511\,\text{MeV}$). Beams of muons can be slowed in matter, captured by charged nuclei in a highly excited state (i.e. large values of n, l), and undergo successive radiative decays to the ground state with a characteristic time $\tau_{\text{capture}} \sim 10^{-13} - 10^{-14}$ s; the classical period of rotation once in the ground state is roughly $\tau_{\text{orbit}} \sim 10^{-18}/Zs$ (P17.21). Even though muons are unstable (decaying because of the weak interaction via $\mu \rightarrow e + \nu_\mu + \bar{\nu}_e$), their lifetime, $\tau_\mu \approx 2 \times 10^{-6}$ s, is long compared to the other characteristic timescales in the problem.[16] Because it is not identical to the other electrons present in the atom, there is no exclusion principle constraint on the muon, and it can exist in any energy level. The short lifetime and the lack of suitably intense muon beams guarantees that atoms with two or more muons, for which the Pauli principle would be important, cannot be realized in the lab.

The dominating feature of such *muonic atoms* is the extremely small size of the Bohr orbits, especially the ground state, which can be obtained from a simple scaling of the results for hydrogen in Eqn. (17.18); the muonic Bohr radius is roughly

$$a_0^{(\mu)} = a_0^{(e)} \frac{m_e}{m_\mu} \approx 0.0025\,\text{Å} = 250\,F \tag{17.64}$$

which is roughly halfway between atomic (Å) and nuclear (F) length scales when measured logarithmically; atoms with all their electrons replaced by muons would be truly Lilliputian,[17] and would have a very different chemistry (P17.22). Such scaling arguments clearly fail for muonic atoms formed with nuclei of sufficiently large Z; for example, for Bi^{209} which has $Z = 83$ ($A = 209$) the ground state radius obtained from scaling H-atom results would be roughly $\langle r \rangle \sim 250\,F/83 \sim 3\,F$ while the nuclear radius is approximately $R_N \sim 7\,F$; the simplifying assumption of a point-like charged nucleus is strongly violated in that limit.

Such arguments do imply that the muon spends much more time "inside" the nucleus than does the electron; the hydrogen wavefunctions in Eqns (17.21)–(17.26), along with the appropriate "measure", imply that the probability of finding either particle inside a radius $R \ll a_0^{(e,\mu)}/Z$ scales roughly as

$$\text{Prob}_{n,l}(r < R) \quad \propto \quad \left(\frac{ZR}{n a_0^{(e,\mu)}} \right)^{2l+3} \tag{17.65}$$

---

[16] If one scales the muon lifetime to the average human lifespan, the capture time to the ground state (orbital period) would be roughly $10-100$ s (1 ms).

[17] Recall that the size of the inhabitants (and other flora and fauna) described by Jonathan Swift (1726) in Gulliver's Travels was consistently 12 times smaller than "normal."

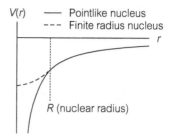

Figure 17.10. Electrostatic potential for point-like nucleus, that is, Coulomb's law (solid) and finite radius nucleus (dashed) in Eqn. (17.66).

The effect of the finite size of the nucleus on the observable energy levels arises from the fact that the nuclear charge is "spread out" and thus leads to a modification of the Coulomb potential at short distances. If, for example, the nuclear charge is modeled as a uniform, spherically symmetric charge density of radius R, the electrostatic potential energy takes the form (P17.23)

$$V(r) = \begin{cases} -(ZKe^2/2R^3)(3R^2 - r^2) & \text{for } r < R \\ -ZKe^2/r & \text{for } r > R \end{cases} \tag{17.66}$$

as shown in Fig. 17.10. If ZR/na_0 is not too large, the effects of the modification can be estimated by perturbation theory (P17.24) using

$$V(r) = -\frac{ZKe^2}{r} + \Delta V(r) \tag{17.67}$$

where

$$\Delta V(r) = -\frac{ZKE^2}{r} + \frac{ZKe^2}{2R^3}(3R^2 - r^2) \quad \text{for } r < R \tag{17.68}$$

and vanishing elsewhere. If the energy level shifts are too large, one can solve the radial equation using Eqn. (17.66) directly (P17.25). For this simple model, the nuclear radius R is the only variable, and some of the first systematic measurements of the variation of nuclear size with atomic number were carried out using data from muonic transitions.

17.4 **Multielectron Atoms**

The continued study of the hydrogen atom can provide a model system in which one can most simply discuss relativistic effects (P17.12), the interactions of the orbital motion and spin degree of freedom of the electron (Section 18.6.2), and the effects of external electric (Section 18.4.2) and magnetic (Section 18.6.1) fields.

Table 17.1. The ionization potentials (in eV) for neutral and partially ionized atoms

Z	Element	I	II	III	IV	V
1	H	13.59844				
2	He	24.58741	54.41778			
3	Li	5.39172	75.64018	122.45429		
4	Be	9.32263	17.21116	153.89661	217.71865	
5	B	8.29803	25.15484	37.93064	259.3721	340.22580

If, however, we wish to gain more insight into the structure of matter, we should discuss the structure of multielectron atoms; in the next two sections we focus on the ground state properties of the simplest two-electron (helium-like) and three-electron (lithium-like) atoms (and ions) as they elucidate many of the general properties of atomic structure, which, in turn, results in the periodic table of the elements.

As we focus on the ground states of such systems, we will find it useful to collect in Table 17.1 a compilation of the ionization potentials of neutral and partially ionized atoms; we show values corresponding to the first five elements (i.e. values of the nuclear charge Z ranging from 1 to 5) from which we extract many of the experimental values quoted below.

The traditional notation of spectroscopy is used, so that I indicates the neutral atom, II the singly ionized atom, and so forth. For example, the energy required to extract a single electron from neutral lithium is $E_I(Li) = 5.39172$ eV; the second electron removed from singly ionized Li^+ "costs" $E_{II}(Li) = 75.64108$ eV while it requires $E_{III}(Li) = 122.4529$ eV to remove the last electron from the hydrogen-like Li^{++}. Taken together, these imply that the total electronic binding energy of lithium is

$$-(E_I(Li) + E_{II}(Li) + E_{III}(Li)) = -203.4862 \text{ eV} \qquad (17.69)$$

We can already make good use of Table 17.1 by noting that the measured ionization potentials of the hydrogen-like atoms H, He^+, Li^{++}, and so forth are given by the uppermost value in each column; they are seen to be in the ratio

$$
\begin{array}{ccccc}
E_I(H) & : E_{II}(He) & : E_{III}(Li) & : E_{IV}(B) & : E_V(Be) \\
13.59844 & : 54.41778 & : 122.4529 & : 217.71865 & : 340.22580 \\
1 & : 4.001766 & : 9.004923 & : 16.01056 & : 25.01947
\end{array}
\qquad (17.70)
$$

and to compare well with the Z^2 scaling predicted by Eqn. (17.12).

17.4.1 **Helium-Like Atoms**

In contrast to the elegant, nonprecessing, elliptical orbits found in the classical two-body problem evolving under an inverse square law force, the classical three-body problem corresponding to the helium atom exhibits extremely complex motion, even displaying chaotic behavior. Many of the semiclassical methods (such as Bohr–Sommerfeld quantization) used in the early days of quantum mechanics, are built on knowledge of the classical trajectories, and such chaotic solutions prevented the broad application of these techniques to the helium atom and other systems[18] The situation circa 1922 is described[19] by the following comments:

Numerous attempts have been made to construct quantum theory models of the normal helium. In Bohr's own model...the two electrons revolve about the nucleus at extremities of a diameter... In no case is the agreement satisfactory, so that apparently none of these models can be correct if the Sommerfeld quantization conditions are accepted.... we must bear in mind that the extreme chemical stability of helium indicates that the arrangement of the two electrons is particularly simple and symmetrical, for an electron revolving in an orbit outside that of its mate would presumably be a valence electron.

This difficulty is also apparent in the lack of closed-form solutions of the Schrödinger equation. The most accurate studies of the helium atom have utilized variational methods instead, and we will follow that approach in what follows.

The spatial coordinates of the two electrons in a helium-like atom are shown in Fig. 17.11 as measured from the nucleus; because of the large nuclear mass, we will make the approximation that the nucleus is at rest. The Hamiltonian for the two-electron system is then

$$\hat{H} = \frac{\hat{\mathbf{p}}_1^2}{2m_e} + \frac{\hat{\mathbf{p}}_2^2}{2m_e} - \frac{ZKe^2}{r_1} - \frac{ZKe^2}{r_2} + \frac{Ke^2}{|\mathbf{r}_1 - \mathbf{r}_2|} \tag{17.71}$$

where Z is the nuclear charge. The neutral helium atom corresponds to $Z = 2$, but one can also consider the family of singly ionized atoms or ions, namely, Li^+, B^{++}, Be^{+++}; we will only present numerical values explicitly for helium.

The binding energy of helium can be read from Table 17.1 to be

$$-(E_I(H) + E_{II}(H)) = -(24.58741 + 54.41778) \text{ eV} \approx -79.0 \text{ eV} = E_{\text{exp}} \tag{17.72}$$

---

[18] Einstein evidently understood the connection between what is now called chaos and the problems with the quantum theory of the helium atom. The semiclassical quantization of the helium atom using periodic orbit trajectories has, in fact, only been accomplished relatively recently; for a discussion, see Heller and Tomsovic (1993) and Gutzwiller (1990).

[19] Comments by Van Vleck (1922) in an article entitled *"The Dilemma of the Helium Atom."*

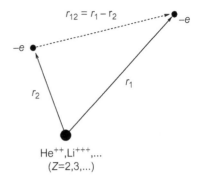

Figure 17.11. Coordinates used for two-electron atoms. For neutral helium, one has $He = He^{++} + 2e^-$, while for singly ionized lithium, one has $Li^+ = Li^{+++} + 2e^-$, and so forth.

If, for the moment, we ignore the effect of the electron–electron repulsion (the last term in Eqn. (17.71), the Schrödinger equation is separable. The ground-state position-space wavefunction is simply the product of hydrogen-like ground state solutions, but we also require an antisymmetric spin wavefunction to satisfy the spin-statistics theorem for the two indistinguishable electrons; the result is

$$\psi(\mathbf{r}_1, \mathbf{r}_2) = \psi_{100}(\mathbf{r}_1)\psi_{100}(\mathbf{r}_2)\chi_{(0,0)} \tag{17.73}$$

where

$$\chi_{(0,0)} = \frac{1}{\sqrt{2}}\left(\chi_\uparrow^{(1)}\chi_\downarrow^{(2)} - \chi_\downarrow^{(1)}\chi_\uparrow^{(2)}\right) \tag{17.74}$$

The corresponding energy

$$E_0 = 2E_H = 2\left(-\frac{Ke^2}{2a_0}\right)Z^2 \tag{17.75}$$

or

$$E_0 = 2(-13.6 \text{ eV})4 = -108.8 \text{ eV} \tag{17.76}$$

for helium, which is obviously too low.

We can estimate the effect of the electron–electron repulsion, using perturbation theory, by evaluating

$$\Delta E_{ee} = \left\langle \psi(\mathbf{r}_1, \mathbf{r}_2)\left|\frac{Ke^2}{|\mathbf{r}_1 - \mathbf{r}_2|}\right|\psi(\mathbf{r}_1, \mathbf{r}_2)\right\rangle \tag{17.77}$$

$$= \int r_1^2 \, dr_1 \, d\Omega_1 \int r_2^2 \, dr_2 \, d\Omega_2 \, |\psi_{100}(\mathbf{r}_1) \, \psi_{100}(\mathbf{r}_2)|^2 \frac{Ke^2}{|\mathbf{r}_1 - \mathbf{r}_2|}$$

The physical meaning of this expression can be made clearer by recalling that the local charge density of each electron is given by

$$\rho(r_{1,2}) = -e|\psi(\mathbf{r}_{1,2})|^2 \tag{17.78}$$

so that Eqn. (17.77) reduces to

$$\Delta E_{ee} = K \int dr_1 dr_2 \frac{\rho(\mathbf{r_1})\rho(\mathbf{r_2})}{|\mathbf{r_1} - \mathbf{r_2}|} = \int \frac{Kdq_1 dq_2}{|\mathbf{r_1} - \mathbf{r_2}|} \qquad (17.79)$$

which is just the total Coulomb energy due to the two electron charge distributions.

To evaluate this expression, we first recall that

$$|\mathbf{r_1} - \mathbf{r_2}| = \sqrt{r_1^2 + r_2^2 - 2r_1 r_2 \cos(\theta_{1,2})} \qquad (17.80)$$

where $\theta_{1,2}$ is the angle between $\mathbf{r}_{12}$. We can use the arbitrariness in the definition of the coordinate system for this (or any other) problem to choose to define the z-axis such that it "points along" one of the vectors, say $\mathbf{r_2}$, so that $\theta_{12} = \theta_2$. In this case we have

$$\int d\Omega_2 \frac{1}{|\mathbf{r_1} - \mathbf{r_2}|} = \int_0^{2\pi} d\phi_2 \int_0^\pi \sin(\theta_2) \, d\theta_2 \, \left(r_1^2 + r_2^2 - 2r_1 r_2 \cos(\theta_2)\right)^{-1/2}$$

$$= 2\pi \frac{\left(r_1^2 + r_2^2 - 2r_1 r_2 \cos(\theta_2)\right)^{1/2}}{r_1 r_2} \Bigg|_0^\pi$$

$$= \frac{2\pi}{r_1 r_2} \left(\sqrt{(r_1 + r_2)^2} - \sqrt{(r_1 - r_2)^2}\right)$$

$$= \frac{2\pi}{r_1 r_2} [(r_1 + r_2) - |r_1 - r_2|]$$

$$= \begin{cases} 4\pi/r_1 & \text{for } r_1 > r_2 \\ 4\pi/r_2 & \text{for } r_1 < r_2 \end{cases} \qquad (17.81)$$

The radial integrals can then be performed giving

$$\Delta E_{ee} = \frac{16Ke^2 Z^6}{a_0^6} \int_0^\infty r_1^2 \, dr_2 \, e^{-Zr_1/a_0}$$

$$\times \left(\int_0^{r_1} dr_2 \, r_2^2 \frac{1}{r_1} e^{-Zr_2/a_0} + \int_{r_1}^\infty dr_2 \, r_2^2 \frac{1}{r_2} e^{-Zr_2/a_0}\right)$$

$$= \frac{5}{8} \frac{Ke^2}{a_0} Z \qquad (17.82)$$

and the estimate

$$E_{\text{pert}} = E_0 + \Delta E_{ee} = -\frac{Ke^2}{a_0} Z^2 + \frac{5}{8} \frac{Ke^2}{a_0} Z \qquad (17.83)$$

For helium, this yields the perturbation theory estimate

$$E_{\text{pert}} = -108.8 + 34.0 = -74.8 \, \text{eV} \qquad (17.84)$$

which is much closer to the observed value.

While this value takes into account the interaction of the electrons, it makes no allowance for the fact that one electron will effectively "shield" the other from the nuclear charge some fraction of the time. This effect can be taken into account in a variational trial wavefunction of the same form as Eqn. (17.73), but with the physical nuclear charge, Z, replaced by an "effective charge," Z^*, which is taken as the variational parameter; we thus take

$$\psi_T(\mathbf{r}_1, \mathbf{r}_2) = \psi_{100}(\mathbf{r}_1; Z^*)\psi_{100}(\mathbf{r}_2; Z^*) \qquad (17.85)$$

where we have suppressed the spinor wavefunction which plays no role in the energy minimization. The energy functional,

$$E[\psi_T] = \langle \psi_T | \hat{H} | \psi_T \rangle \qquad (17.86)$$

is most easily evaluated by rewriting the Hamiltonian in a clever way, namely

$$\hat{H} = \left(\frac{\hat{\mathbf{p}}_1^2}{2m_e} - \frac{Z^* K e^2}{r_1} \right) + \left(\frac{\hat{\mathbf{p}}_2^2}{2m_e} - \frac{Z^* K e^2}{r_2} \right)$$

$$+ \frac{(Z^* - Z)K e^2}{r_1} + \frac{(Z^* - Z)K e^2}{r_2}$$

$$+ \frac{K e^2}{|\mathbf{r}_1 - \mathbf{r}_2|} \qquad (17.87)$$

The terms in Eqn. (17.86) can be evaluated in turn. For example, one has

$$\left\langle \psi_T \left| \frac{\hat{\mathbf{p}}_{1,2}^2}{2m_e} - \frac{Z^* K e^2}{r_{1,2}} \right| \psi_T \right\rangle = -\frac{K e^2}{2a_0}(Z^*)^2 \qquad (17.88)$$

since $\psi_{100}(\mathbf{r}; Z^*)$ is the ground state wavefunction corresponding to the effective charge Z^*. The expectation value of $\langle 1/r \rangle$ in Eqn. (17.31) implies that

$$\left\langle \psi_T \left| \frac{(Z^* - Z)K e^2}{r_{1,2}} \right| \psi_T \right\rangle = (Z^* - Z)\frac{K e^2}{a_0}Z^* \qquad (17.89)$$

while Eqn. (17.82) generalizes to

$$\Delta E_{ee}(Z^*) = \left\langle \psi_T \left| \frac{K e^2}{|\mathbf{r}_1 - \mathbf{r}_2|} \right| \psi_T \right\rangle = \frac{5}{8}\frac{K e^2}{a_0}Z^* \qquad (17.90)$$

The variational estimate of the energy is then

$$E[\psi_T] = E_{var}(Z^*) = -\frac{Ke^2}{a_0}\left(4Z^*(Z - 5/16) - 2(Z^*)^2\right) \qquad (17.91)$$

which can be minimized to yield

$$\frac{dE_{var}(Z^*)}{dZ^*} = 0 \implies Z^* = Z - \frac{5}{16} \qquad (17.92)$$

In this picture, each electron is "inside the orbit" of the other (and hence shields the nuclear charge) something like $1/3$ of the time. The variational minimum energy can then be written in the form akin to Eqn. (17.75)

$$E_{var}^{(min)} = -2\left(\frac{Ke^2}{2a_0}\right)\left(Z - \frac{5}{16}\right)^2 \qquad (17.93)$$

which gives

$$E_{var} = -77.45 \text{ eV} \qquad (17.94)$$

for helium, which is lower than the perturbation theory result and hence rigorously closer to the experimental value. We collect the various estimates for *He* in Table 17.2; the reader is encouraged to fill in the remaining entries using the data in Table 17.1 and Eqns (17.75), (17.83), and (17.93) and compare with the experimental values for other states.

One can systematically improve on this simple variational estimate[20] by including more variational parameters to model angular and radial correlations, and by including the effects of nuclear motion and relativity.[21]

For excited states in which the two electrons can be in different energy levels, and hence have different spatial wavefunctions, the role of spin becomes more important. For the ground state, only the antisymmetric spin state corresponding to $S = 0$ is allowed but for other states, the total wavefunction can have the proper

Table 17.2. Experimental values and theoretical estimates of the ground state energies of various helium-like atoms

Z	Element	$E_{exp}(eV)$	$E_0 = 2E_H(eV)$	$\Delta E_{ee}(eV)$	$E_{pert}(eV)$	$E_{var}(eV)$
2	He	-79.0	-108.8	$+34.0$	-74.79	-77.45
3	Li$^+$					
4	Be^{++}					
5	B^{+++}					

[20] This approach was first used by Hylleraas (1928).
[21] A nice discussion at the level of this text appears in Park (1992).

symmetry under exchange with an $S=1$ spin wavefunction; such states are called *para-helium* ($S = 0$ or spin-singlet) and *ortho-helium* ($S = 1$ or spin triplet), respectively. For example, if one electron is excited to the (nlm) hydrogen-like state, we can write the combinations

$$S = 0: \frac{1}{\sqrt{2}} \left[\psi_{100}(\mathbf{r}_1)\psi_{nlm}(\mathbf{r}_2) + \psi_{100}(\mathbf{r}_2)\psi_{nlm}(\mathbf{r}_1) \right] \chi_{(0,0)} \qquad (17.95)$$

$$S = 1: \frac{1}{\sqrt{2}} \left[\psi_{100}(\mathbf{r}_1)\psi_{nlm}(\mathbf{r}_2) - \psi_{100}(\mathbf{r}_2)\psi_{nlm}(\mathbf{r}_1) \right] \chi_{(1,S_z)} \qquad (17.96)$$

where the χ_{1,S_z} are given in Example 16.4.

The perturbation theory estimate of the energies of these two states can be seen to be of the form

$$E_{\text{pert}} = (E_1 + E_n) + J_{nl} \pm K_{nl} \qquad (17.97)$$

where the $+, -$ corresponds to $S = 0, S = 1$, respectively. The Coulomb energy, J_{nl} has the same form as Eqn. (17.77) with one ψ_{100} replaced by ψ_{nlm}, but with the same physical interpretation. The so-called *exchange energy* is given by

$$K_{nl} = \int d\mathbf{r}_1 \int d\mathbf{r}_2 \psi_{100}(\mathbf{r}_1) \; \psi_{nlm}(\mathbf{r}_2) \frac{Ke^2}{|\mathbf{r}_1 - \mathbf{r}_2|} \psi_{100}(\mathbf{r}_2) \; \psi_{nlm}(\mathbf{r}_1) \qquad (17.98)$$

and has no classical analog as it arises from the symmetry constraints on the quantum wavefunction due to indistinguishability. While the integral defining K_{nl} can be evaluated quite generally, we only require the fact that it is positive.

This implies that the spin-singlet ($S = 0$) states are shifted up in energy relative to the spin-triplet ($S = 1$) levels. The shift arises because the spatially symmetric (antisymmetric) wavefunction of the $S = 0$ ($S = 1$) state imply that the electrons are closer together (further apart) yielding more (less) Coulomb repulsion. The level structure of helium is shown in Fig. 17.12 illustrating the effect. The dotted lines indicate the value of hydrogen energy levels as measured from the first ionization energy, $E_{II}(He)$.

17.4.2 Lithium-Like Atoms

The next most complicated systems are three-electron atoms such as Li^+, Be^+, B^{++}, and so forth, and we can fairly easily extend a variational analysis to such states. The Hamiltonian is the generalization of Eqn. (17.71), namely

$$\hat{H} = \sum_{i=1}^{3} \left(\frac{\hat{p}_i^2}{2m_e} + \frac{ZKe^2}{r_i} \right) + \sum_{i>j=1}^{3} \frac{Ke^2}{|\mathbf{r}_i - \mathbf{r}_j|} \qquad (17.99)$$

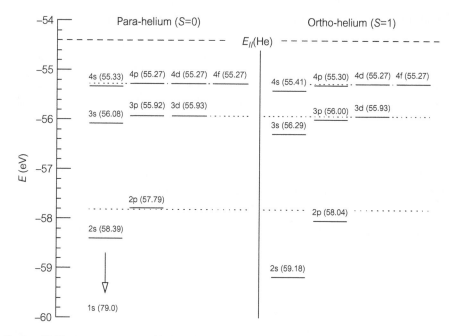

Figure 17.12. Energy spectrum of helium; both $S = 0$ (ortho-helium) and $S = 1$ (para-helium) levels are shown (solid). The energy of singly ionized He corresponding to $E_{//}(He) = 2^2 E_I(H) \approx -54.4\,eV$ is shown for reference. The states are compared to the energy levels for hydrogen as measured from $E_{//}(He)$ (dotted lines).

The ground state configuration will consist of two electrons in the $1s$ state with opposite spins and one electron (either spin-up or spin-down) in one of the first excited states; for simplicity, we consider it in the $2s$ state. A properly normalized and antisymmetrized variational wavefunction which describes this configuration is given by the Slater determinant

$$\psi_T(\mathbf{r}_1, \mathbf{r}_2, \mathbf{r}_3; Z_1^*, Z_2^*) = \frac{1}{\sqrt{3!}} \begin{vmatrix} \psi_1(\mathbf{r}_1; Z_1^*)\chi_\uparrow^1 & \psi_1(\mathbf{r}_1; Z_1^*)\chi_\downarrow^1 & \psi_2(\mathbf{r}_1; Z_2^*)\chi_\uparrow^1 \\ \psi_1(\mathbf{r}_2; Z_1^*)\chi_\uparrow^2 & \psi_1(\mathbf{r}_2; Z_1^*)\chi_\downarrow^2 & \psi_2(\mathbf{r}_2; Z_2^*)\chi_\uparrow^2 \\ \psi_1(\mathbf{r}_3; Z_1^*)\chi_\uparrow^3 & \psi_1(\mathbf{r}_3; Z_1^*)\chi_\downarrow^3 & \psi_2(\mathbf{r}_3; Z_2^*)\chi_\uparrow^3 \end{vmatrix}$$

(17.100)

$$= \frac{1}{\sqrt{6}} \left(\psi_1(\mathbf{r}_1; Z_1^*)\psi_1(\mathbf{r}_2; Z_1^*)\psi_2(\mathbf{r}_3; Z_2^*)\chi_\uparrow^1\chi_\downarrow^2\chi_\uparrow^3 + \cdots \right)$$

(17.101)

where $\psi_{1,2}(\mathbf{r}; Z_{1,2}^*)$ are the $n = 1, 2s$-state hydrogen wavefunctions in Eqns (17.21) and (17.22). We allow for the possibility of different effective charges (Z_1^*, Z_2^*) as we expect the electron in the outer orbital will be shielded more effectively from the nucleus by the inner-shell electrons.

Table 17.3. One- and two-parameter variational estimates of the ground state energies of lithium-like atoms

Z	Element	E_{exp} (eV)	$Z_1 = Z_2$	E_{var} (eV)	Z_1	Z_2	E_{var} (eV)
3	Li	−203.5	2.537	−196.5	2.67	1.37	−200.8
4	Be$^+$	−389.9	3.537	−382.0	3.67	2.44	−386.2
5	B^{++}	−637.2	4.537	−628.9	4.67	3.45	−632.8

The variational energy can be evaluated in a straightforward (if tedious) manner using the same tricks described above to obtain

$$E[\psi] = E(Z_1, Z_2) = \frac{Ke^2}{a_0} \left(-Z_1^2 - \frac{Z_2^2}{8} + 2(Z_1 - Z) + \frac{(Z_2 - Z)Z_2}{4} \right.$$

$$\left. + \frac{5}{8}Z_1 + f(Z_1, Z_2) \right) \tag{17.102}$$

where

$$f(Z_1, Z_2) = \frac{2Z_1 Z_2}{(2Z_1 + Z_2)^5} \left(8Z_1^4 + 20Z_1^3 Z_2 + 12Z_1^2 Z_2^2 + 10Z_1 Z_2^3 + Z_2^4 \right) \tag{17.103}$$

(We will henceforth write Z_1^*, Z_2^* as Z_1, Z_2 for typographical convenience.) We can easily minimize Eqn (17.102) analytically for the case when $Z_1 = Z_2 = Z^*$ (P17.30) or numerically when both Z_1, Z_2 are allowed to "float," and we show the results in Table 17.3 for several lithium-like atoms for both cases; the experimental values are evaluated using the data in Table 17.1.

Several comments can be made:

- The two-parameter variational energy estimate is lower, and hence closer to the ground state energy, than the one-parameter fit; this is guaranteed since the latter is simply a special case of the former.

- The effective charges in the 1s and 2s states are different in just the way expected above. The values of Z_1 for the two inner-shell electrons are almost identical to those preferred by the helium-like atoms; namely, $Z^* = Z - 5/16$. This effect can be understood on the basis of a Gauss' law argument; the inner electrons should "feel" no effect of a spherically symmetric distribution of charge "outside" themselves.

- The values of the effective charges of the outermost electron are roughly consistent with the observed "first" ionization potentials in Table 17.1. The energy required to ionize the first electron from Li, Be, and B are given by

$E_I = 5.39$ eV, 17.21 eV, and 37.93 eV, respectively. If we estimate the effective charge seen by the $2s$ electron via

$$E_I = -\frac{13.6 \text{ eV}}{2^2}(Z^*)^2 \tag{17.104}$$

we find $Z^* = 1.26$, 2.31, and 3.34; these values are to be compared to values of Z_2 from the variational estimate in the second-to-last row of Table 17.3.

Finally, we can obtain some intuition about the spatial distribution of charge in the simplest atoms considered so far by plotting the radial probability density, $P(r) = r^2|\psi(r)|^2$, for finding a single electron in each system. We use the exact result for hydrogen and the optimized variational wavefunctions for helium and lithium. In the latter cases, we ask for the probability of finding a specific electron, say 1, at some radius; these distributions are obtained by integrating Eqns (17.85) and (17.101) over the radial coordinates of the unobserved electrons in each case. The results are shown in Fig. 17.13 where the value of the RMS radius, $\sqrt{\langle r^2 \rangle}$, is also indicated as a measure of the size of the atomic system.

17.4.3 **The Periodic Table**

These results serve to motivate some final comments on the structure of the periodic table arising from the continued "filling" of energy levels with more electrons.

- The inner core electrons in a heavier atom gives rise to a non-Coulombic potential which lifts the degeneracy among l values for a given n; as discussed in the context of Rydberg atoms, the ($l=0$) s-states are most "penetrating," see the largest effective charge, and hence are lower in energy that the corresponding ($l=1$) p-levels, and are "filled in" first. The electron configuration of the first

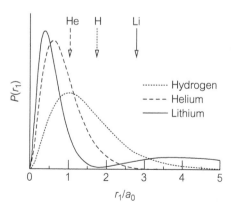

Figure 17.13. Single-electron radial probability density ($P(r_1)$) for hydrogen (dotted), helium (dashed), and lithium (solid). The exact solution in Eqn. (17.21) is used for hydrogen, while the variational wavefunctions of Eqns (17.85) and (17.100) are used for helium and lithium. The arrows indicate the location of the RMS radius.

10 elements (in standard spectroscopic notation) reflects this fact with the assignments

Z	Element	Configuration
1	H	$1s^1$
2	He	$1s^2$
3	Li	$1s^2 2s^1$
4	Be	$1s^2 2s^2$
5	B	$1s^2 2s^2 2p^1$
6	C	$1s^2 2s^2 2p^2$
7	N	$1s^2 2s^2 2p^3$
8	O	$1s^2 2s^2 2p^4$
9	F	$1s^2 2s^2 2p^5$
10	Ne	$1s^2 2s^2 2p^6$

where the first (numerical) label corresponds to the n value, the second the angular momentum, and the exponent counts the number of electrons in each l state.

- The elements corresponding to completely filled levels are said to have "closed electron shells." The large energy gap to the next excited state implies that the elements with $Z=2$ (helium) and $Z=10$ (neon) will be chemically inert, and these correspond to the first two of the *noble gases.*[22]

- There is a relatively large energy difference between the $3p$ and $3d$ levels so that $Z=18$ also corresponds to an inert gas; the configurations up to the next noble gas are

11	Na	$1s^1 2s^2 2p^6 3s^1$
12	Mg	$1s^2 2s^2 2p^6 3s^2$
13	Al	$1s^2 2s^2 2p^6 3s^2 3p^1$
14	Si	$1s^2 2s^2 2p^6 3s^2 3p^2$
15	P	$1s^2 2s^2 2p^6 3s^2 3p^3$
16	S	$1s^2 2s^2 2p^6 3s^2 3p^4$
17	Cl	$1s^2 2s^2 2p^6 3s^2 3p^5$
18	Ar	$1s^2 2s^2 2p^6 3s^2 3p^6$

---

[22] As its name implies, helium was first discovered, not via its "terrestrial" chemical properties, but rather by spectroscopic measurements of the sun's chromosphere. Evidence for argon was obtained by comparing the density of atmospheric nitrogen with that prepared by other means.

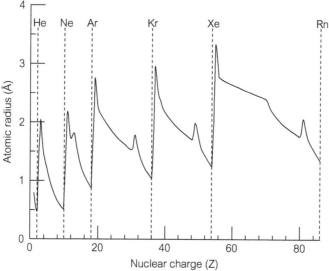

Figure 17.14. (Calculated) atomic radii for elements in the periodic table, by atomic number; the correlation with the closed electron shells corresponding to noble gases is clear.

- The level splittings becomes so great for the $n = 3$ levels that the $4s$ level actually lies lower than the $3d$ level so that potassium ($Z = 19$) has the structure $1s^2 2s^2 2p^6 3s^2 3p^6 4s^1$.

- The pattern of ionization energies in Fig. 1.5 reflects the partial and complete filling of energy levels; more nearly closed shells are harder to ionize. (See Fig. 17.14). The *polarizability*[23] of each element is a measure of its tendency to interact with an external electric field, distorting in order to minimize the field energy, and shows the opposite correlation with nuclear charge; more nearly closed shells have a decreased polarizability. To see the strong anticorrelation between these two atomic properties, we plot their values for each element in Fig. 17.15, partly for future reference.

17.5 **Questions and Problems**

Q17.1. If the volume occupied by a hydrogen atom in its ground state were scaled to be the size of your fist, how big a volume would the Rydberg state of hydrogen with the largest value of n observed (roughly $n \approx 350$) be in these scaled units? How big would a muonic atom with $Z = 50$ in its ground state be in these units?

---

[23] See Section 18.4.1.

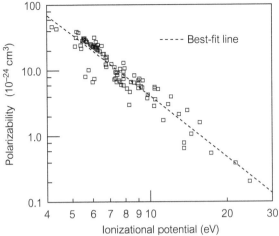

Figure 17.15. Atomic polarizability plotted against ionization potential (eV) for elements in the periodic table; the straight line corresponds to a "best straight-line fit" through the data points.

Q17.2. In Bohr's original paper deriving the Balmer law for the spectrum of hydrogen, Bohr speculates on why emission from states with large values of n had not been observed in laboratory experiments while values up to $n \approx 30$ (at that time) had been detected in astrophysical spectra. Discuss why this might be so by considering two additional facts: (i) the hydrogen gas in astrophysical situations is far less dense than in laboratory experiments, but there are large regions of it, (ii) the radius of the Bohr orbits increases like n^2, so that the geometrical cross section scales like n^4.

Q17.3. What would the energy spectrum of hydrogen-like atoms look like in two dimensions? What would be the structure of the periodic table in two dimensions? (For one attempt at this "what-if" question, see Asturias and Aragón (1985).)

Q17.4. We have seen that when two operators do not commute, the quantum mechanical equivalent of their product is often given by a *symmetric* combination, such as $xp \rightarrow (x\hat{p} + \hat{p}x)/2$. (Recall, for example, the quantum covariance in P4.18.) Why then, in Eqn. (17.57), is the *antisymmetric* combination of $\hat{p}$ and $\hat{L}$ used?

Q17.5. What property of the muon is responsible for the interest shown in *muon-catalyzed fusion* (μCF)[24]? What one property of the muon would you change if you wanted to make μCF more probable? Hint: Recall the discussion in Section 11.4.4 on the role of the Coulomb barrier in inhibiting fusion reactions.

P17.1. Show that the small and large ρ behavior of the hydrogen atom wavefunctions are given by Eqn. (17.7).

---

[24] See Jones, Rafelski, and Monkhorst (1989).

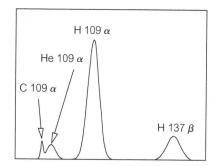

Figure 17.16. Schematic picture of spectral lines for highly excited Rydberg atoms as seen by radio astronomers.

P17.2. Radio astronomers searching for radiation generated when electrons and protons recombine have obtained data[25] showing Balmer lines such as in Fig. 17.16; the units on both axes have been suppressed (since they're "strange" astronomy units), so this is a qualitative question. The $n\alpha$ ($n\beta$) labels correspond to transitions between hydrogenic energy levels of the type $(n+1) \to n$ $((n+2) \to n)$ for the various elements listed. Show that the three 109 lines and the one 137 line are consistent with the Balmer formula in Eqn. (17.14), when reduced mass effects are included. Hint: What is being plotted on the horizontal axis? Evaluate the wavelengths of the lines being described and discuss why astronomers detected them.

P17.3. Show explicitly that the "quasi-circular" radial wavefunctions $R_{n,n-1}(r)$ in Eqn. (17.27) satisfy the radial equation and are properly normalized.

P17.4. For the "quasi-circular" wavefunctions with $l = n - 1$, evaluate $\Delta r^2 = \langle r^2 \rangle - \langle r \rangle^2$. Show that $(\Delta r)^2 / \langle r^2 \rangle \approx 1/2n$.

P17.5. Prove the relation, valid only for S-states,

$$|\psi_{n,0,0}(0)|^2 = \frac{\mu}{2\pi \hbar^2} \left\langle \psi_{n,0,0} \left| \frac{dV(r)}{dr} \right| \psi_{n,0,0} \right\rangle \qquad (17.105)$$

for any spherically symmetric potential. Use it to then evaluate the S-state wavefunction at the origin squared for any value of n and compare to the explicit cases listed in Eqns (17.21), (17.22), and (17.24).

P17.6. Variational estimate for the hydrogen atom. (a) Use the variational method to estimate the ground state energy of the hydrogen atom by using the trial wavefunction

$$\psi(r; a) = e^{-r/a} \qquad (17.106)$$

where a is the variational parameter. You should, of course, get the correct answer since this is the correct functional form.
(b) Repeat, but use a trial wavefunction of the form

$$\psi(r; b) = e^{-r^2/2b^2} \qquad (17.107)$$

---

[25] Adapted from Kleppner (1986).

and find the fractional error made in the energy. Show (numerically) that the overlap of the true ground state wavefunction and the variational estimate is roughly

$$\langle \psi_{(1,0,0)}(r) | \psi(r, b_{min}) \rangle \approx 0.9782. \tag{17.108}$$

(c) Repeat part (b), but use the trial wavefunction

$$\psi(r; c) = \frac{1}{(r^2 + c^2)^2} \tag{17.109}$$

P17.7. Virial theorem.

(a) Generalize the proof of the virial theorem in P12.12 to three dimensions and confirm it for the case of the Coulomb potential by showing that

$$\langle \hat{T} \rangle_{n,l} = \left\langle \frac{\hat{p}^2}{2m} \right\rangle_{n,l} = -\frac{1}{2} \left\langle \frac{Ke^2}{r} \right\rangle_{n,l} = \frac{1}{2} \langle \mathbf{r} \cdot \nabla V(r) \rangle_{n,l} \tag{17.110}$$

(b) Use these results to determine the fraction of energy stored in radial kinetic, rotational kinetic, and potential energy in a hydrogen atom energy eigenstate characterized by (n, l).

P17.8. Tritium decays. The nucleus consisting of two neutrons and a single proton (hence with $Z = +1$) is called the *triton*, t, (by analogy with the deuteron); the system consisting of a single electron and a triton is called a *tritium atom*. The triton (or ^3H) is unstable against radioactive decays via the process $t \rightarrow {}^3He + e^- + \bar{\nu}_e$ where ^3He consists of one neutron and two protons (i.e. $Z = +2$). On an atomic timescale, the nuclear decay and the ejection of the $e^- + \bar{\nu}_e$ happen almost instantaneously. After such a decay, the bound electron suddenly finds itself in Coulomb field with twice the nuclear charge.

(a) What is the probability that an electron which was originally in the ground state will remain in the ground state of the new system? in the $n = 2$ s-state?

(b) What is the probability of being in an $l = 1$ state after the decay?
Hint: This is just an "expansion in eigenstates problem" similar to Example 6.3 or the "sudden approximation" of time-dependent perturbation theory in Section 10.5.3, but in the context of a real physical system.

P17.9. Momentum-space wavefunctions for hydrogen. Calculate the momentum space wavefunctions for the three lowest-lying s-wave states of hydrogen, $\psi_{n,l,m}(r)$, that is, for $n = 1, 2, 3$ and $l = m = 0$. The appropriate three-dimensional Fourier transform is given by

$$\phi_{n,l,m}(\mathbf{p}) = \frac{1}{\sqrt{(2\pi\hbar)^3}} \int d\mathbf{r} \, e^{-i\mathbf{p}\cdot\mathbf{r}/\hbar} \, \psi_{n,l,m}(\mathbf{r}) \tag{17.111}$$

Hint: Since there is no preferred direction for s-states, write $\mathbf{p} \cdot \mathbf{r} = pr \cos(\theta)$ and let θ be measured along the z-axis so that

$$\phi(\mathbf{p}) = \frac{1}{\sqrt{(2\pi\hbar)^2}} \int_0^\infty r^2 \, dr \, \psi(r) \int_0^{2\pi} d\phi \int_0^\pi \sin(\theta) \, e^{-ipr\cos(\theta)/\hbar}$$

(17.112)

For example, for the ground state, show that

$$|\phi_{1,0,0}(\mathbf{p})|^2 = \frac{8}{\pi^2} \frac{p_0^5}{(p^2 + p_0^2)^4}$$

(17.113)

where $p_0 \equiv \hbar/a_0$. Compare this expression to the experimental data of Lohmann and Weigold (1981) in Fig. 1.4.

P17.10. For Compton scattering from unbound electrons, the relation between the wavelength of the scattering X-rays and the incident one is

$$\lambda' - \lambda = \frac{2\pi\hbar}{m_e c} (1 - \cos(\theta))$$

(17.114)

where θ is the angle of the scattered photon from the incident direction. If the scattering instead takes place from electrons bound in a hydrogen atom, show that there will be a spread in the scattered wavelengths of the order $\delta\lambda'/\lambda \sim \alpha$ due to the momentum spread of the target electrons. Hint: Use the result of the previous problem.

P17.11. Relativistic wave equation. The Klein–Gordon equation (first discussed in Section 3.1) can be used to solve the hydrogen atom problem in a relativistically correct way; it is only appropriate for spin-0 particles, and thus does not apply directly to electrons and the "real" hydrogen atom. It is useful for pionic (π) and kaonic (K) atoms, and does give some idea of the form of the relativistic corrections.

(a) Assuming a solution of the form $\phi(\mathbf{r}) = R(r) Y_{l,m}(\theta, \phi)$, show that the Klein–Gordan wave equation for the Coulomb potential, namely

$$\left(E + \frac{ZKe^2}{r} \right)^2 \phi(\mathbf{r}) = \left(\hat{p}^2 c^2 + (mc^2)^2 \right)^2 \phi(\mathbf{r})$$

(17.115)

gives the radial equation

$$\frac{d^2 R(r)}{dr^2} + \frac{2}{r} \frac{dR(r)}{dr} \left(\frac{B}{r} - \frac{C}{r^2} - |A| \right) R(r) = 0$$

(17.116)

where

$$|A| = \frac{|E^2 - (mc^2)^2|}{(\hbar c)^2}, \quad B = \frac{2EZKe^2}{(\hbar c)^2}, \quad \text{and} \quad C = l(l+1) - (Z\alpha)^2$$

(17.117)

(b) Attempt a solution of the form $R(r) = r^\gamma e^{-r/\beta} G(r)$ and show that $\beta = 1/|A|$ and

$$\gamma = \sqrt{(l+1/2)^2 - (Z\alpha)^2} - 1/2 \tag{17.118}$$

and obtain the differential equation for $G(r)$.

(c) Show that the condition for the power series solution for $G(r)$ to terminate is

$$\frac{B}{2\sqrt{|A|}} = (n_r + l + 1) + \gamma - l = n + \gamma - l \tag{17.119}$$

where n_r and n are the usual quantum numbers.

(d) Show that the quantized energies now depend on *both n and l* (in contrast to the nonrelativistic case), and are given by

$$E_{n,l} = mc^2 \left(1 + \frac{(Z\alpha)^2}{(n + \sqrt{(l+1/2)^2 - (Z\alpha)^2} - (l+1/2))^2}\right)^{-1/2} \tag{17.120}$$

(e) Expand this result in powers of $Z\alpha$ and show that

$$E_{n,l} = mc^2 \left[1 - \frac{(Z\alpha)^2}{2n^2}\left(1 + \frac{(Z\alpha)^2}{n}\left(\frac{1}{l+1/2} - \frac{3}{4n}\right) + \cdots\right)\right] \tag{17.121}$$

(f) Compare the $\mathcal{O}(Z\alpha)^2$ terms to the energy levels in Eqn. (17.12), and the $\mathcal{O}(Z\alpha)^4$ terms to the relativistic corrections derived in the next problem.

P17.12. Relativistic corrections in hydrogen. In Section 1.1, we showed that the expansion of the relativistically correct kinetic energy is given by

$$E = \sqrt{\mathbf{p}^2 c^2 + (mc^2)^2} \approx mc^2 + \frac{\mathbf{p}^2}{2m} - \frac{1}{8}\frac{(\mathbf{p}^2)^2}{m^3 c^2} + \cdots \tag{17.122}$$

Evaluate the effect of the $\mathcal{O}((\mathbf{p}^2)^2)$ term (now expressed as an operator) on the hydrogen atom wavefunctions using perturbation theory. Show that the first-order energy shift for the state $\psi_{n,l,m}$ is given by

$$\Delta E_{n,l}^{REL} = -\frac{1}{2}mc^2(Z\alpha)^4\frac{1}{n^3}\left(\frac{2}{2l+1} - \frac{3}{4n}\right) \tag{17.123}$$

Hint: Write $\hat{p}^2$ in terms of the unperturbed Hamiltonian using

$$\hat{p}^2 = 2m\left(\hat{H} + \frac{ZKe^2}{r}\right) \tag{17.124}$$

and use the average values of $\langle r^n \rangle$.

P17.13. Modified Coulomb potential. Solve the Schrödinger equation for the potential

$$V(r) = -\frac{ZKe^2}{r}\left(1 + b\frac{a_0}{r}\right) \tag{17.125}$$

and show that the energy levels are given by

$$E_{n,l} = -\frac{1}{2}\mu c^2 (Z\alpha)^2 \frac{1}{[n + \sqrt{(l+1/2)^2 - 2b} - (l+1/2)]^2} \tag{17.126}$$

where $n = n_r + l + 1$ has its usual meaning. Hint: Use the method outlined in P17.11.

P17.14. Hydrogen atom in parabolic coordinates.

(a) Show that the Hamiltonian for the hydrogen atom problem is also separable in *parabolic coordinates,*[26] namely

$$q_1 = r - z = r(1 - \cos(\theta)) \tag{17.127}$$

$$q_2 = r + z = r(1 + \cos(\theta)) \tag{17.128}$$

$$\phi = \phi \tag{17.129}$$

where (r, θ, ϕ) are the usual spherical coordinates. Hint: If you use the expression

$$\nabla^2 = \frac{4}{(q_1 + q_2)}\left[\frac{\partial}{\partial q_1}\left(q_1\frac{\partial}{\partial q_1}\right) + \frac{\partial}{\partial q_2}\left(q_2\frac{\partial}{\partial q_2}\right)\right] + \frac{1}{q_1 q_2}\frac{\partial^2}{\partial \phi^2} \tag{17.130}$$

you should prove it first.

(b) Try a solution of the form

$$\psi(q_1, q_2, \phi) = Q_1(q_1)\, Q_2(q_2)\, e^{im\phi} \tag{17.131}$$

obtain the equations for $Q_{1,2}$ and show how the quantized eigenvalues arise. Note: This form of solution is useful when the z-axis is singled out for some reason, as with the Stark effect in Section 18.4.2, or in scattering problems as in Chapter 19; it also provides more evidence for the remarkable symmetry properties of this problem.

P17.15. Classical probability distributions. Show explicitly that the form of the classical probability distribution in Eqn. (17.52) is properly normalized, that is, show that

$$1 = \int_{r_{\min}}^{r_{\max}} P_{\text{CL}}^{(n,l)}(r)\, dr = \frac{1}{\pi a_0 n^2}\int_{r_{\min}}^{r_{\max}} dr \frac{r\, dr}{\sqrt{2ra_0 n^2 - l(l+1)n^2 a_0^2 - r^2}} \tag{17.132}$$

---

[26] See, for example Schiff (1955).

Evaluate the *classical* expectation values $\langle r \rangle$, $\langle r^2 \rangle$, and $\langle 1/r \rangle$ and compare to the quantum results in Eqns (17.29), (17.30), and (17.31).

P17.16. Bohr–Sommerfeld quantization of the hydrogen atom. A semiclassical picture of the hydrogen atom considers only planar orbits with variables r, θ and their corresponding momenta.

(a) Show that the WKB-type quantization condition on the angular variable gives

$$\int_{\theta_{min}}^{\theta_{max}} L \, d\theta = n_\theta h = 2\pi n_\theta \hbar \quad \text{or} \quad L = n_\theta \hbar \qquad (17.133)$$

(b) The corresponding condition for the radial coordinate is

$$\int_{r_{min}}^{r_{max}} p_r \, dr = \int_{r_{min}}^{r_{max}} \sqrt{2\mu} \sqrt{E - \frac{L^2}{2\mu r^2} + \frac{ZKe^2}{r}} \, dr = \pi n_r \hbar \quad (17.134)$$

Use the results of Appendix D.1 to evaluate the integrals and show that the resulting energy quantization condition is

$$E = -\frac{1}{2}\mu c^2 \frac{(Z\alpha)^2}{(n_r + n_\theta)^2} \qquad (17.135)$$

P17.17. Lenz–Runge vector I: Classical case.

(a) Using the definition of Eqn. (17.55) and the equations of motion, show that **R** is constant in time, namely that $\dot{\mathbf{R}} = 0$.

(b) Derive Eqn. (17.56) by squaring Eqn. (17.55)

(c) Show that the expression for $\mathbf{r} \cdot \mathbf{R} = r|\mathbf{R}| \cos(\theta)$ gives the equation for an ellipse in Eqn. (17.39).

P17.18. Lenz–Runge vector II: Quantum case.

(a) Using the definition in Eqn. (17.57), show that $\hat{\mathbf{R}}$ is conserved by proving that it commutes with the Hamiltonian

$$\hat{H} = \frac{\hat{p}^2}{2\mu} - \frac{Ke^2}{r} \qquad (17.136)$$

(b) Show that $\hat{\mathbf{R}}$ is Hermitian.

(c) Show that the vector operator defined via

$$\hat{\mathbf{S}} = \sqrt{\frac{\mu (Ke^2)^2}{2|E|}} \hat{\mathbf{R}} \qquad (17.137)$$

satisfies

$$[\hat{S}_x, \hat{L}_y] = i\hbar \hat{S}_z \quad \text{and} \quad [\hat{S}_x, \hat{S}_y] = i\hbar \hat{L}_z \qquad (17.138)$$

and all the usual permutations. (This requires a reasonable amount of algebra.)

(d) Define the generalized angular momentum operators

$$\hat{M}_{i,j} = \left(\frac{\hbar}{i}\right)\left(x_i\frac{\partial}{\partial x_j} - x_j\frac{\partial}{\partial x_i}\right) \tag{17.139}$$

for $i, j = 1, 2, 3$. Show that they satisfy the commutation relation

$$\left[\hat{M}_{ij}, \hat{M}_{kl}\right] = i\hbar\left(\hat{M}_{ik}\delta_{jl} + \hat{M}_{jl}\delta_{ik} - \hat{M}_{il}\delta_{jk} - \hat{M}_{jk}\delta_{il}\right) \tag{17.140}$$

Show that the $\hat{M}_{ij}$ are equivalent to the familiar $\hat{L}$ with the identifications

$$\hat{M}_{12} = \hat{L}_z, \qquad \hat{M}_{23} = \hat{L}_x, \quad \text{and} \quad \hat{M}_{31} = \hat{L}_y \tag{17.141}$$

Finally, show that $\hat{S}$ can be included by extending the index label to 4 and identifying

$$\hat{M}_{14} = \hat{S}_x, \qquad \hat{M}_{24} = \hat{S}_y, \quad \text{and} \quad \hat{M}_{34} = \hat{S}_z \tag{17.142}$$

Note: If one thinks of the $i, j = 1, 2, 3$ sector as representing angular momentum operators in three dimensions, this exercise shows that there is actually a kind of four-dimensional rotational invariance to the inverse square law problem, giving it its enhanced symmetry properties.

P17.19. Model for the quantum defect in alkali atoms. For values of the angular momentum l of the excited electron in a Rydberg atom satisfying $l > l_{core}$ (where l_{core} is the maximum possible angular momentum of the inner shell electrons) there is little actual penetration of the ionic core. One of the most important interactions which is then responsible for the quantum defect is the so-called *polarization energy* of the outer electron with the dipole field of the core (which is induced by the electron itself). This potential has the form

$$V_{POL}(r) = -\frac{1}{2}\frac{\alpha_D(Ke^2)^2}{r^4} \sim -\frac{b}{2}\frac{Ke^2a_0^3}{r^4} \tag{17.143}$$

where the *dipolar polarizability*, $\alpha_D = ba_0^3/K$, is written in terms of a dimensionless constant b as suggested by results from Section 18.4.2. Evaluate the effect of this potential on hydrogen atom states of arbitrary n, l using perturbation theory, and show that the resulting quantum defect scales as $\delta_l \sim 1/l^5$.

P17.20. Scattering of Rydberg atoms.[27] The cross sections for collisions (of a certain type) of highly excited $l = 0$ sodium atoms are listed below along with the effective quantum number, n^*.

[27] The data, and the fit to data, are taken from Gallagher *et al.*(1982).

State	n^*	$\sigma \ (10^9 \ \text{Å}^2)$
16s	15.2	0.78 ± 0.18
18s	17.2	1.25 ± 0.13
20s	19.2	2.16 ± 0.66
23s	22.2	3.8 ± 1.0
25s	24.2	3.8 ± 1.0
27s	26.2	5.8 ± 1.8

and the data can be "fit" via the formula

$$\sigma = (3.3 \pm 0.6) \times 10^4 (n^*)^{3.7 \pm 0.5} \qquad (17.144)$$

Plot these data along with the "fit" on log–log paper to see if this is true. Can you explain why the observed dependence on n^* is consistent with $\sigma \propto (n^*)^4$?

P17.21. Estimate the classical period of motion in a hydrogen-like state with quantum numbers n, l for a given reduced mass and nuclear charge Z. Confirm the estimate for muonic atoms in Section 17.3.2.

P17.22. Scaling laws for muon-chemistry. Suppose that the muon (assumed for the moment to be stable) replaced each electron in all atoms. If we call the ratio of the muon-to-electron mass $R = m_\mu / m_e \approx 200$, by what power of R (if any) would the following quantities scale ?

(a) Size of typical atom.

(b) Binding energies.

(c) Density of matter.

(d) Effective "spring constant" in a diatomic molecule.

(e) Typical vibrational energy level in diatomic molecule.

(f) Typical rotational energy level in diatomic molecule.

P17.23. Non-Coulombic charge densities. Use Gauss's law to show that the electric field for a uniform spherical distribution of charge of radius R and total charge Ze is given by

$$|E(r)| = \begin{cases} ZKer/R^3 & \text{for } r < R \\ ZKe/r^2 & \text{for } r > R \end{cases}$$

The use this result to derive Eqn. (17.66).

P17.24. Estimate the shift in the $1s, 2s$, and $2p$ energy levels due to the modified Coulomb potential in Eqn. (17.68) using perturbation theory. Assume that $ZR/a_0^{(e,\mu)} \ll 1$. For the ground state, for what value of Z does perturbation theory begin to become unreliable, that is when does $E^{(1)}/E^{(0)} \sim 0.1$?

P17.25. Numerical solutions of the Schrödinger equation. With the change of variables discussed in Section 16.1, namely, $u(r) = rR(r)$, rewrite the three-dimensional Schrödinger equation for s-states in the form shown in Eqn. (16.31).

You may have already written a short computer program to numerically integrate the Schrödinger equation in one dimension to analyze problems in Chapter 10. Use your program to numerically solve for the ground state energy of the muonic atom with $Z = 60$ and $R = 6.3\ F$ with the "smeared out" Coulomb potential in Eqn. (17.66). Compare your answer to the result obtained by simply "scaling" the hydrogen result using the muon mass and appropriate charge Z and also to the experimental value of roughly $E_0 \approx -6.88\ \text{MeV}$. How does your wavefunction (if your program can generate it) differ from the Bohr ground state? Hint: If the step size in your program is $\Delta x = \epsilon$, use the starting data $u(0) = 0$ and $u(\Delta x = \epsilon) = \Delta x$.

P17.26. Applications of the Feynman–Hellmann theorem. Recall that the Feynman–Hellman theorem (from P4.32) required that

$$\frac{\partial E(\lambda)}{\partial \lambda} = \left\langle \frac{\partial \hat{H}(\lambda)}{\partial \lambda} \right\rangle \qquad (17.145)$$

if the Hamiltonian (and hence any wavefunction solution) depends on some parameter λ in some well-defined way, $\hat{H} = \hat{H}(\lambda)$ so that

$$\langle E \rangle(\lambda) = \langle \psi(\lambda)|\hat{H}(\lambda)|\psi(\lambda) \rangle \qquad (17.146)$$

Confirm the Feynman–Hellmann theorem for hydrogen-atom wavefunctions explicitly by calculating both sides of Eqn. (17.145) for the Coulomb potential using the nuclear charge Z as the parameter. Repeat using the reduced mass, μ, as the parameter.

P17.27. Helium-like atoms. Calculate the values in the blank entries in Table 17.2. Use Table 17.1 to calculate the experimental values and Eqns (17.75), (17.83), and (17.93) to evaluate the zero-order, perturbation theory and variational estimates.

P17.28. Hydrogen ion. The hydrogen ion consists of a hydrogen atom with an additional electron; it is a helium-like atom but with $Z = 1$. Estimate its ground state energy using the variational formula in Eqn. (17.93), and compare your result to the minimum energy of a hydrogen atom and a free electron. Does your result imply that the hydrogen ion does not have a bound state?

P17.29. Evaluate the exchange integral in Eqn. (17.98) for the case of $(nlm) = (200)$, and use your result to compare to the ortho-para-helium splitting for the $2s$ state as read from Fig. 17.12.

P17.30. Derive the variational estimate in Eqn. (17.102) for the lithium ground state. Find the value of $Z = Z_1 = Z_2$, which minimizes Eqn. (17.102).

EIGHTEEN

Gravity and Electromagnetism in Quantum Mechanics

In introductory physics courses, we are introduced to (usually at some length) two of the four fundamental forces of nature,[1] namely, gravity and electromagnetism. In this chapter, we discuss some examples of the description of these forces and their effects in quantum mechanical terms.

18.1 Classical Gravity and Quantum Mechanics

The observation that gravity is the most obvious force in the macroscopic world[2] while electromagnetism is the dominant interaction at the microscopic level of atoms and molecules is easily explained by two facts:

1. The gravitational force between an electron and a proton is 40 orders of magnitude smaller than the electrostatic attraction between them (P18.1), independent of their separation.

2. The electric charges of the electron and proton, while of opposite sign, are experimentally remarkably close in magnitude; the best experimental limits[3] imply that $|Q_p + Q_e|/e \leq 10^{-21}$. This implies that normal matter, in bulk, is electrically neutral allowing gravitational interactions to dominate since its interaction strength (i.e. its mass) increases with the size of the system, while the net charge is roughly zero.

---

[1] The other important interactions, the so-called weak and strong interactions, are of most relevance in the subatomic domain, and are discussed in any good text on subatomic (i.e. nuclear or elementary particle) physics; see, for example, Perkins (1996).
[2] Everyone, after all, has fallen down.
[3] Taken from the Review of Particle Properties (2002).

These facts suggest that it might not be possible to find a terrestrial system in which gravity and quantum mechanics simultaneously play an important and fundamental role.[4]

Gravitational effects are occasionally used to novel effect in laboratory atomic systems; one notable example[5] is the "atomic fountain" which "launches" cesium atoms upward in an apparatus used to make precise frequency measurements, and uses the weak pull of gravity to increase the interaction time. This type of geometry is currently being used as one of the primary atomic frequency standards in the United States; this is, in turn, an integral part of the global position system (SPS). Other electrically neutral particles, most notably neutrons, have been used in a variety of experiments which test both the "classical mechanics" of subatomic particles and the quantum effects of local gravity on the quantum mechanical wavefunction, and that is the main topic of this section.

Beams of reactor neutrons have been measured[6] to "fall" in the earth's gravitational field, following parabolic trajectories just like textbook projectiles.[7] This is purely a classical effect and would be a "wave packet" limit of a particle in a linear potential (as in P15.4).

A much more interesting quantum interference effect was observed using *two* beams of neutrons which were split and then allowed to recombine, as in Fig. 18.1. All such interference experiments rely on the wave phenomenon of the addition of different wave amplitudes, differing by a phase; the electron interference patterns in Fig. 1.2 are an example where the phase difference arises from a difference in path length, and we review this effect in Section 18.8.

In one version of this experiment,[8] the paths followed by the two neutron beams differ in that they find themselves in regions of different gravitational potential energy. The resulting phase difference is most easily estimated by using

Figure 18.1. Geometry for experiment showing interference effects as a result of the quantum phase of the neutron wavefunction due to the earth's gravitational potential. The neutrons follow the two paths *PQR* and *PSR*, which can be rotated about the horizontal axis *PR*.

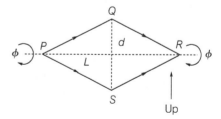

---

[4] Recall, however, the compact astrophysical objects discussed in Section 7.4.3.
[5] See Gibble and Chu (1993).
[6] See Dabbs *et al.* (1965) who even measure the local acceleration of gravity using neutron fall to be $g_{exp} = 975.4 \pm 3.1$ cm/s^2, compared to the known local value of $g = 979.74$ cm/s^2 at the site of the experiment.
[7] The usual approximation of neglecting air resistance is, in this case, presumably an excellent one!
[8] See Colella and Overhauser (1980) for details and the original references.

WKB-type wavefunctions (as in Section 10.3.1), that is, we use

$$\psi(x_1) \sim \psi(x_2) \exp\left(i \int_{x_1}^{x_2} p(x)\, dx / \hbar \right) \tag{18.1}$$

for the phase of the neutron wavefunction. If we consider the gravitational potential, $V(y) = mgy$ (where $y = y(x) = (d/L)x$ along the path) as a small perturbation, we have

$$p(x) = \sqrt{2m(E - V(x))} \approx \sqrt{2mE}\left(1 - \frac{1}{2}\frac{V(x)}{E} + \cdots \right) \tag{18.2}$$

since we assume that $V(x) \ll E$. The first term will be the same for both upper and lower arms (provided the path lengths are identical), so that the interference comes from the second term and depends on the path followed through the potential. We then have the resulting total phase increase in going along the upper arm PQR given by

$$\phi_{\text{PQR}} = -\frac{1}{\hbar}\sqrt{\frac{m}{2E}}\, 2 \int_0^L mgy(x)\, dx = -\frac{m^2 g}{\hbar p}\frac{dL}{2} \tag{18.3}$$

where we use $E = p^2/2m$. The total difference in phase between the upper (PQR) and lower (PSR) paths can then be written as

$$\Delta\phi = -\frac{m^2 g A \lambda}{2\pi\hbar^2} \tag{18.4}$$

where $A = 2dL$ is the area of the PQRS surface, and we have used the deBroglie relation. Using values which are relevant to the real apparatus, namely, $A \approx (2-3 \text{ cm})^2$, and thermal neutrons ($E \approx k_B T \approx 0.03$ eV), we find that $\Delta\phi \approx 28$; this would correspond to $\Delta\phi/2\pi \approx 5$ fringe shifts in a standard interference experiment. To eliminate systematic effects and make the interference pattern more obvious, the neutron counting rate was measured at point R (as this measured the total recombined amplitude squared) as the device was rotated around the axis PR. As the angle is changed, the effective path difference and relative phase is altered; this resulted in the pattern shown in Fig. 18.2, which illustrates the effect. More sophisticated versions of this experimental technique[9] have even proved sensitive to the earth's rotation, and succeeded in measuring the resulting Coriolis effect on the neutron's path.

---

[9] See Staudenmann et al. (1980).

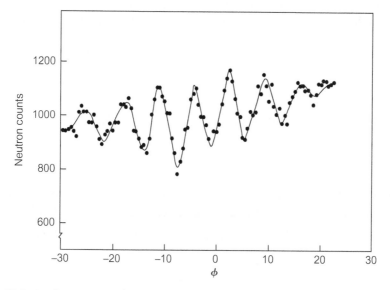

Figure 18.2. Interference pattern from neutron interferometer showing the effect of gravity on the phase of the neutron wavefunction; data are from Colella, Overhauser, and Werner *et al.* (1975).

18.2 **Electromagnetic Fields**

The classical equations governing the motion of a charged particle moving under the influence of external electric and magnetic fields are given by Newton's law of motion with the Lorentz force law, namely,

$$\underset{\text{Newton's law}}{m\ddot{\mathbf{a}}(t) =} \quad \underset{\text{Lorentz force}}{\mathbf{F}} \quad = q\left[\mathbf{E}(\mathbf{r}, t) + \mathbf{v}(t) \times \mathbf{B}(\mathbf{r}, t)\right] \tag{18.5}$$

We will learn below how to incorporate the interactions of electromagnetic (hereafter EM) fields with charged particles into a quantum description of matter, but we first review the classical description of the EM fields themselves.

18.2.1 **Classical Electric and Magnetic Fields**

One of the major intellectual triumphs of classical physics[10] was the coherent presentation and extension of the then-known laws of electricity and magnetism

---

[10] Feynman (1963) has said "*there can be little doubt that the most significant event of the 19th century will be judged as Maxwell's discovery of the laws of electrodynamics. The American Civil War will pale into provincial insignificance in comparison with this important scientific event of the same decade.*"

by Maxwell. In modern language, these can be written in the following form:

$$\nabla \cdot \mathbf{E}(\mathbf{r}, t) = \frac{1}{\epsilon}\rho(\mathbf{r}, t) \qquad\qquad \text{Gauss' law} \qquad (18.6)$$

$$\nabla \cdot \mathbf{B}(\mathbf{r}, t) = 0 \qquad\qquad \text{``no-name'' law} \qquad (18.7)$$

$$\nabla \times \mathbf{E}(\mathbf{r}, t) = -\frac{\partial}{\partial t}\mathbf{B}(\mathbf{r}, t) \qquad\qquad \text{Faraday's law} \qquad (18.8)$$

$$\nabla \times \mathbf{B}(\mathbf{r}, t) = \mu \mathbf{J}(\mathbf{r}, t) + \mu\epsilon\frac{\partial}{\partial t}\mathbf{E}(\mathbf{r}, t) \qquad \text{Ampere's law} \qquad (18.9)$$

The electric ($\mathbf{E}$) and magnetic ($\mathbf{B}$) fields are determined by the local charge (ρ) and current ($\mathbf{J}$) densities; as a simple example, recall that Gauss' law implies that the electric field arising from a point electric charge in vacuum is given by

$$\mathbf{E}(\mathbf{r}) = \frac{q}{4\pi\epsilon_0}\frac{\hat{\mathbf{r}}}{r^2} = \frac{Kq}{r^2}\hat{\mathbf{r}} \qquad (18.10)$$

giving the familiar inverse-square law for the Coulomb force. The corresponding equation for the magnetic field, Eqn. (18.7), implies that there are no point magnetic charges. We also note that Maxwell's equations are often written in an even more compact form using the auxiliary fields $\mathbf{D} = \epsilon\mathbf{E}$ and $\mathbf{B} = \mu\mathbf{H}$ using the fundamental constants of electricity (the electric permittivity, ϵ) and magnetism (the magnetic permeability, μ).

The important concept of the conservation of electric charge is implicitly contained in Maxwell's equations, and can be recovered by taking $\nabla \cdot$ (Ampere's law) and using Gauss' law to find

$$\frac{\partial\rho(\mathbf{r}, t)}{\partial t} + \nabla \cdot \mathbf{J}(\mathbf{r}, t) = 0 \qquad (18.11)$$

This so-called *equation of continuity* has the identical form as that arising from the conservation of probability in Section 4.2.

One of the most important consequences of Maxwell's equations is the connection with the EM wave equation. One can, for example, take $\nabla \times$ (Ampere's law), and use Eqns (18.7) and (18.8), and the vector identity

$$\nabla \times (\nabla \times \mathbf{B}) = \nabla (\nabla \cdot \mathbf{B}) - \nabla^2\mathbf{B} \qquad (18.12)$$

to show that

$$\frac{\partial^2\mathbf{B}(\mathbf{r}, t)}{\partial t^2} = \frac{1}{\epsilon_0\mu_0}\nabla^2\mathbf{B}(\mathbf{r}, t) = c^2\nabla^2\mathbf{B}(\mathbf{r}, t) \qquad (18.13)$$

in vacuum and in the absence of charges; the electric field $\mathbf{E}(\mathbf{r}, t)$ satisfies an identical wave equation. The derivation of the connection between the three fundamental constants $c = 1/\sqrt{\epsilon_0\mu_0}$ was one of the main results of Maxwell's contribution. Plane wave solutions of Eqn. (18.13) of the form

$$B(r, t) = B_0 e^{i(k\cdot r - \omega t)} \quad \text{and} \quad E(r, t) = E_0 e^{i(k\cdot r - \omega t)} \tag{18.14}$$

must satisfy $\omega^2 = |k|^2 c^2$. They must also, however, be consistent with each of individual Maxwell equations; Faraday's law, for example, implies that

$$k \times E_0 = -\omega B_0 \tag{18.15}$$

while Ampere's law requires that

$$k \times B_0 = c^2 \omega E_0 \tag{18.16}$$

These imply that the electric and magnetic fields in an EM wave must be *transverse*, that is, perpendicular to the direction of propagation determined by k, and must have magnitudes related by $|B_0| = |E_0|/c$.

A final important result describing the flow of energy can be obtained by taking $B \cdot$ (Faraday's law) $-E \cdot$ (Ampere's law) (for simplicity, in the case where there is no current density) where one finds

$$\frac{\partial u(r, t)}{\partial t} + \nabla S(r, t) = 0 \tag{18.17}$$

The quantity $u(r, t)$ is given by

$$u(r, t) = u_E(r, t) + u_B(r, t) = \frac{1}{2}\epsilon|E(r, t)|^2 + \frac{1}{2\mu}|B(r, t)|^2 \tag{18.18}$$

and can be shown to correspond to the local energy density stored in the electric and magnetic fields. The so-called *Poynting vector*, given by

$$S(r, t) = \frac{1}{\mu}E(r, t) \times B(r, t) \tag{18.19}$$

then describes the rate of energy flow per unit time per unit area; S also has the units of *intensity*, namely, power per unit area. Equation (18.17) then expresses the manner in which energy "flows" through a system. Using the relativistic connection between energy and momentum, we also note that S/c has the units of (momentum per unit area per unit time) or (force per area) or pressure and can be used to describe *radiation pressure*. The related quantity, S/c^2, has the units of *momentum density* or momentum per unit volume; we can then write

$$P_{rad} = \frac{S}{c} \quad \text{and} \quad \frac{dp}{dV} = \frac{S}{c^2} \tag{18.20}$$

Example 18.1. Laser fields

Modern high-power lasers used in atomic physics research can achieve intensities of the order $I = 10^{22}$ W/m^2. The strength of the electric field component corresponding to this intensity is given by

$$|\mathbf{S}| = \frac{1}{\mu_0}|\mathbf{E}||\mathbf{B}| = \frac{1}{\mu_0 c}|\mathbf{E}|^2 = I \qquad (18.21)$$

so that $|\mathbf{E}| \sim 2 \times 10^{12}$ N/C; this can also be expressed in somewhat different units as $|\mathbf{E}| \sim 200$ V/Å. A standard value to which to compare this field strength is the magnitude of the electric field felt by the electron in the hydrogen atom, namely, the field arising from a charge e at a distance of roughly $a_0 \sim 0.53$ Å. This "typical" atomic field strength is approximately

$$E_c \equiv \frac{Ke}{a_0^2} = \left(\frac{Ke^2}{\hbar c}\right)\left(\frac{\hbar c}{a_0^2}\right)\frac{1}{e} \approx \left(\frac{1}{137}\right)\left(\frac{1973\,eV \cdot \text{Å}}{(0.53\,\text{Å})^2}\right)\frac{1}{e} \approx 51\,\frac{V}{\text{Å}} \quad (18.22)$$

Another useful relation arises when one thinks of the laser pulse as an ensemble of photons, each carrying quantized energy $\hbar\omega$ at the speed of light. The Poynting vector can then be written in the form

$$|\mathbf{S}| = n_\gamma \hbar\omega c \qquad (18.23)$$

where n_γ is the number density (number per unit volume) of photons. For a laser of the intensity above, operating at $\lambda = 1000\,nm = 10^4$ Å, corresponding to a photon energy $\hbar\omega \approx 1$ eV, the relation Eqn. (18.23) implies that there are roughly 2×10^{22} γ/m^3. An atom in such a laser field can find itself immersed in a relatively dense "photon gas."

Example 18.2. Uncertainty principle constraints on E and B fields

It is clear from the discussions above that the fact that EM radiation can carry momentum is not a consequence of any particle-like (i.e. photon) interpretation, but arises naturally from classical considerations. We can, however, combine the expression for the momentum density in EM fields,

$$\frac{d\mathbf{p}}{dV} = \frac{1}{\mu_0 c^2}\mathbf{E} \times \mathbf{B} \qquad (18.24)$$

with the standard Heisenberg uncertainty principle to "motivate" (not prove) an interesting limit to the measurability of EM field strengths. If we take one component of Eqn. (18.24) and consider the momentum in some small volume, δV, we can write

$$\frac{p_x}{\delta V} \sim (\mathbf{E} \times \mathbf{B})_x = \frac{1}{\mu_0 c^2}\left(E_y B_z - E_z B_y\right) \qquad (18.25)$$

(Continued)
We can argue that the *uncertainty* in this momentum component will be of the same order as the uncertainties in the values of the field components, so that

$$\left(\delta V \frac{1}{\mu_0 c^2} \Delta E_y \Delta B_z\right) \Delta x \quad \sim \quad \Delta p_x \Delta x \gtrsim \frac{\hbar}{2} \tag{18.26}$$

where the last bound comes from the quantum mechanical uncertainty principle. Taken together, these imply that

$$\Delta E_y \Delta B_z \gtrsim \frac{\hbar}{\mu_0 c^2} \frac{1}{\Delta x \delta V} \tag{18.27}$$

This interesting result can be derived in a somewhat more careful way by examining the ways in which one might attempt to simultaneously measure any two such components of the EM fields[11]

It suggests that the values of the EM fields become increasingly uncertain as one attempts to measure them on increasingly smaller distance scales; the values of the **E** and **B** fields, and hence the energy contained in the EM field, can be thought of as fluctuating wildly at short distances.

In classical mechanics, one often finds it useful to use the concept of a potential energy function $V(\mathbf{r})$ instead of the Newtonian force $\mathbf{F}(\mathbf{r})$; the connection between the two is $\mathbf{F}(\mathbf{r}) = -\nabla V(\mathbf{r})$. This relation makes it clear that different choices of the "zero of potential" can have no physical meaning as $V(\mathbf{r}) \rightarrow V(\mathbf{r}) + V_0$ yields the same measurable force.

A similar, but more subtle and deep situation arises in electrodynamics where one can express the (physical) electric and magnetic fields in terms of scalar $(\phi(\mathbf{r}, t))$ and vector $(\mathbf{A}(\mathbf{r},t))$ potentials via

$$\mathbf{B}(\mathbf{r}, t) = \nabla \times \mathbf{A}(\mathbf{r}, t) \tag{18.28}$$

$$\mathbf{E}(\mathbf{r}, t) = -\nabla \phi(\mathbf{r}, t) - \frac{\partial}{\partial t} \mathbf{A}(\mathbf{r}, t) \tag{18.29}$$

It is easy to show that the fields produced by the potentials $\phi, \mathbf{A}$ will be identical to those produced by the new potentials $\phi', \mathbf{A}'$ provided they are related by the transformations

$$\mathbf{A}'(\mathbf{r}, t) = \mathbf{A}(\mathbf{r}, t) + \nabla f(\mathbf{r}, t) \tag{18.30}$$

$$\phi'(\mathbf{r}, t) = \phi(\mathbf{r}, t) - \frac{\partial}{\partial t} f(\mathbf{r}, t) \tag{18.31}$$

[11] See Landé (1951).

where $f(\mathbf{r}, t)$ is an arbitrary scalar function. There are thus an infinite number of different EM potentials which correspond to a given configuration of measureable fields. Such a change in potentials is called a *gauge transformation*, and will be seen to play an important role in the quantum mechanical treatment of charged particle interactions.

Example 18.3. Scalar and vector potentials

We will consider below the familiar cases of charged particles acted on by uniform electric and magnetic fields, and we consider the EM potentials, ϕ and $\mathbf{A}$, which can describe these cases.

A uniform electric field, given by $\mathbf{E}(\mathbf{r}, t) = \mathbf{E}_0$, can be obtained from potentials

$$\phi(\mathbf{r}, t) = -\mathbf{E}_0 \cdot \mathbf{r} \quad \text{and} \quad \mathbf{A}(\mathbf{r}, t) = 0 \qquad (18.32)$$

which is the standard choice. However, the same field can be described by the potentials

$$\phi'(\mathbf{r}, t) = 0 \quad \text{and} \quad \mathbf{A}'(\mathbf{r}, t) = -\mathbf{E}_0 t \qquad (18.33)$$

It is easy to see that the transformation relating these two sets of potentials via Eqns (18.30) and (18.31) is generated by $f(\mathbf{r}, t) = -\mathbf{E}_0 \cdot \mathbf{r}t$.

A uniform magnetic field in the z direction, $\mathbf{B} = B_0\hat{\mathbf{z}} = (0, 0, B_0)$ can be obtained from the potentials

$$\phi(\mathbf{r}, t) = 0 \quad \text{and} \quad \mathbf{A}(\mathbf{r}, t) = \frac{B_0}{2}(-y, x, 0) \qquad (18.34)$$

One of the many other possible choices is

$$\phi'(\mathbf{r}, t) = 0 \quad \text{and} \quad \mathbf{A}'(\mathbf{r}, t) = B_0(-y, 0, 0) \qquad (18.35)$$

and these two sets can be seen to be gauge equivalent with $f(\mathbf{r}, t) = -B_0 xy/2$.

18.2.2 E and B Fields in Quantum Mechanics

The standard procedure we have adopted to extend the equations of motion of classical mechanics to the quantum Schrödinger equation (in position space) has been through a generalization of the Hamiltonian formulation of classical mechanics. For the free particle in one dimension this consisted of simply making the identification

$$H_{\text{classical}} = \frac{p^2}{2m} \quad \Longrightarrow \quad \hat{H} = \frac{\hat{p}^2}{2m} \qquad (18.36)$$

We show in Appendix G that the classical Hamiltonian function appropriate for a charged particle (of charge q) acted on by external electric and magnetic fields (now in three dimensions) is given by

$$H_{\text{classical}} = \frac{1}{2m} \left(\mathbf{p} - q\mathbf{A}(\mathbf{r}, t) \right)^2 + q\phi(\mathbf{r}, t). \tag{18.37}$$

This choice for $H_{\text{classical}}$ reproduces the correct equations of motion in Eqn. (18.5).

The corresponding quantum mechanical Hamiltonian is obtained by replacing the momentum variable by its operator counterpart thereby giving the Schrödinger equation

$$\hat{H}\psi(\mathbf{r}, t) = \hat{E}\psi(\mathbf{r}, t) = i\hbar \frac{\partial}{\partial t}\psi(\mathbf{r}, t) \tag{18.38}$$

where

$$\hat{H} = \frac{1}{2m} \left(\hat{\mathbf{p}} - q\mathbf{A}(\mathbf{r}, t) \right)^2 + q\phi(\mathbf{r}, t) \tag{18.39}$$

One must, of course, be careful of the ordering of any differential operators, so we find

$$\left[\hat{\mathbf{p}} - q\mathbf{A}(\mathbf{r}, t) \right]^2 \psi(\mathbf{r}) = -\hbar^2 \nabla^2 \psi(\mathbf{r}, t) + iq\hbar \nabla \cdot [\mathbf{A}(\mathbf{r}, t)\psi(\mathbf{r}, t)]$$
$$+ iq\hbar \mathbf{A}(\mathbf{r}, t) \cdot [\nabla \psi(\mathbf{r}, t)] + q^2 [\mathbf{A}(\mathbf{r}, t) \cdot \mathbf{A}(\mathbf{r}, t)]\psi(\mathbf{r}, t) \tag{18.40}$$

The classical equations of motion depend on the E and B fields themselves, and are obviously invariant under any gauge transformation. The Hamiltonian which now appears in the Schrödinger equation, however, depends on the EM potentials explicitly, and does change its form under such a transformation. Specifically, under the change in potentials given by Eqns (18.30) and (18.31), the original Hamiltonian, Eqn. (18.39), is replaced by

$$\hat{H}' = \frac{1}{2m} \left(\hat{\mathbf{p}} - q\mathbf{A}'(\mathbf{r}, t) \right)^2 + q\phi'(\mathbf{r}, t)$$
$$= \frac{1}{2m} \left(\hat{\mathbf{p}} - q\mathbf{A}(\mathbf{r}, t) - q\nabla f(\mathbf{r}, t) \right)^2$$
$$+ q\phi'(\mathbf{r}, t) - q\frac{\partial}{\partial t}f(\mathbf{r}, t) \tag{18.41}$$

which does not manifestly imply the same physical solutions. It is not hard to show, however, that if one also simultaneously changes the original wavefunction $\psi(\mathbf{r}, t)$ by a (possibly time- and space-dependent) phase factor, namely

$$\psi'(\mathbf{r}, t) = \psi(\mathbf{r}, t)\, e^{iqf(\mathbf{r},t)/\hbar} \tag{18.42}$$

then

$$\hat{H}\psi(\mathbf{r}, t) = \hat{E}\psi(\mathbf{r}, t) = i\hbar\frac{\partial}{\partial t}\psi(\mathbf{r}, t)$$

$$\Downarrow \qquad\qquad (18.43)$$

$$\hat{H}'\psi'(\mathbf{r}, t) = \hat{E}\psi'(\mathbf{r}, t) = i\hbar\frac{\partial}{\partial t}\psi'(\mathbf{r}, t)$$

The probability densities corresponding to ψ' and ψ are identical because

$$|\psi'(\mathbf{r}, t)|^2 = |\psi(\mathbf{r}, t)|^2 \qquad\qquad (18.44)$$

and the gauge transformation makes no change in the observable physics of the system. (Recall P6.4 where a change in the "zero of potential" is discussed, and a similar result is found, namely, a simple change of phase.) The solutions obtained in different gauges may well look quite different, but must correspond to the same physical energy eigenvalues and be related via Eqn. (18.42); it is often useful to use this freedom of gauge to choose the ϕ and $\mathbf{A}$ which make the problem most tractable.

In the next two sections we deal with several special cases corresponding to uniform electric and magnetic fields where a problem can be solved explicitly in separate gauges, and where the connections between the solutions are easily confirmed; these examples also show that different properties of the solution of a given physical problem may be more apparent in one gauge or another.

18.3 Constant Electric Fields

We first consider the action of a uniform electric field, for simplicity in the $+x$ direction, that is, $\mathbf{E}_0 = E_0\hat{\mathbf{x}}$, on an otherwise free particle of charge q. The classical solutions correspond to free-particle motion in the y, z directions and uniform acceleration in the x direction, namely

$$x(t) = \frac{qE_0}{2m}t^2 + v_{0x}t + x_0 \qquad\qquad (18.45)$$

$$y(t) = v_{0y}t + y_0 \qquad\qquad (18.46)$$

$$z(t) = v_{0z}t + z_0 \qquad\qquad (18.47)$$

The corresponding quantum problem can be defined using the standard gauge choice in Eqn. (18.32), by the Hamiltonian operator

$$\hat{H} = \frac{\hat{\mathbf{p}}^2}{2m} - qE_0x = \left(\frac{\hat{p}_x^2}{2m} - eE_0x\right) + \frac{\hat{p}_y^2}{2m} + \frac{\hat{p}_z^2}{2m} \qquad\qquad (18.48)$$

The system is clearly separable and the wavefunction can be written in the form

$$\psi(\mathbf{r}, t) = \psi(x, t)\, e^{i(p_y y - p_y^2 t/2m)/\hbar}\, e^{i(p_z z - p_z^2 t/2m)/\hbar} \tag{18.49}$$

The problem of uniform acceleration in one dimension was discussed in Section 4.7.2 where a Gaussian wave packet solution was found with the form

$$\psi(x, t) = \frac{1}{\sqrt{\beta\sqrt{\pi}(1 + it/t_0)}}\, e^{iFt(x - Ft^2/6m)/\hbar}$$

$$\times \exp\left(-(x - Ft^2/2m)^2/2\beta^2(1 + it/t_0)\right) \tag{18.50}$$

In Eqn. (18.50), $F = qE_0$, and the initial Gaussian wave packet is obviously

$$\psi(x, 0) = \frac{1}{\sqrt{\beta\sqrt{\pi}}}\, e^{-x^2/2\beta^2} \tag{18.51}$$

In solving this problem, we have explicitly used the most familiar gauge in which

$$\phi(\mathbf{r}, t) = -Ex \quad \text{and} \quad \mathbf{A}(\mathbf{r}, t) = 0 \tag{18.52}$$

but we can just as well use the gauge equivalent set

$$\phi(\mathbf{r}, t) = 0 \quad \text{and} \quad \hat{\mathbf{A}}(\mathbf{r}, t) = -E_0 \hat{\mathbf{x}} t \tag{18.53}$$

These two are related by the gauge function $f(x, t) = -E_0 x t$.

In the new gauge, the y and z dependence is unchanged, but the 1D Schrödinger equation for the x behavior now reads

$$\hat{H}\psi'(x) = \left(\hat{H}_0 + t\hat{H}_1 + t^2 \hat{H}_2\right)\psi'(x, t)$$

$$= \left(\frac{\hat{p}_x^2}{2m} + \frac{qE_0 t}{m}\hat{p}_x + \frac{1}{2m}(qE_0 t)^2\right)\psi'(x, t)$$

$$= i\hbar\frac{\partial}{\partial t}\psi'(x, t) \tag{18.54}$$

where the terms linear in $\hat{p}_x$ simplify because the vector potential does not depend on position. As in Section 12.5, we can formally solve this problem by integration of the initial value problem and find

$$\psi'(x, t) = e^{-i(\hat{H}_0 t + \hat{H}_1 t^2/2 + \hat{H}_2 t^3/3)/\hbar}\,\psi'(x, 0) \tag{18.55}$$

In this special case, the three component pieces, $\hat{H}_{0,1,2}$ commute with each other so that their order in the exponential is unimportant and the time-development operator can be written in the simpler form

$$\psi'(x, t) = e^{-i(\hat{H}_2 t^3/3)/\hbar}\, e^{-i(\hat{H}_1 t^2/2)/\hbar}\, e^{-i(\hat{H}_t)/\hbar}\psi'(x, 0)$$

$$= e^{-iF^2 t^3/6m\hbar}\, e^{-iFt^2\hat{p}_x/2m\hbar}\, e^{-i\hat{p}_x^2 t/2m\hbar}\psi'(x, 0) \tag{18.56}$$

and we can consider the effect of each operator in turn. The time-development operator for the free particle, namely, $\hat{H}_0 = \hat{p}_x^2/2m$ has a simple effect when acting on a Gaussian initial state (Example 12.4), that is,

$$\psi'(x, t) = e^{-i\hat{H}_0 t/\hbar} \left(\frac{1}{\sqrt{\beta\sqrt{\pi}}} e^{-x^2/2\beta^2} \right) = \frac{1}{\sqrt{\beta(1 + it/t_0)}\sqrt{\pi}} e^{-x^2/2\beta^2(1+it/t_0)}$$

(18.57)

corresponding to spreading, but no translation of the central value. The effect of the second operator can be made clear by recalling (P12.6) that the momentum operator generates translations in space via

$$e^{ia\hat{p}/\hbar}\psi(x, t) = \psi(x+a, t)$$

(18.58)

so that the spreading wave packet of Eqn. (18.57) is translated via

$$x \longrightarrow x - \frac{qE_0 t^2}{2m} = x - \frac{Ft^2}{2m}$$

(18.59)

The final factor, $e^{-i\hat{H}_2 t^3/3\hbar}$, is a simple phase. Taken together, these three operators acting on the initial wavefunction give

$$\psi'(x, t) = \frac{1}{\sqrt{\beta\sqrt{\pi}(1 + it/t_0)}} e^{-i(F^2 t^3/6m)/\hbar} e^{-(x - Ft^2/2m)^2/2\beta^2(1+it/t_0)}$$

(18.60)

where $F = qE_0$. This explicitly satisfies

$$\psi'(x, t) = e^{-iqf(x,t)/\hbar} \psi(x, t) = e^{-iFxt/\hbar} \psi(x, t)$$

(18.61)

as expected from the gauge transformation.

18.4 **Atoms in Electric Fields: The Stark Effect**

One of the simplest ways to probe and manipulate the energy level structure and wavefunction of an atom is to place it in a constant external electric field. We will consider the quantum version of this problem for hydrogen where the resulting pattern of energy level shifts is called the *Stark effect*; we begin, however, by considering a simplified classical model to gain some intuition.

18.4.1 **Classical Case**

As a simple model of a (classical) hydrogen atom, consider a positive point charge $+e$, representing the proton, embedded in a uniformly charged spherical cloud of radius a_0 (the Bohr radius) with total charge $-e$. The resulting charge density is

$$\rho(\mathbf{r}) = \begin{cases} -3e/4\pi a_0^3 & \text{for } r < a_0 \\ 0 & \text{for } r > a_0 \end{cases} \tag{18.62}$$

Gauss's law can then be used to derive (as in P17.23) the corresponding electric field giving

$$\mathbf{E}(\mathbf{r}) = \begin{cases} -Ke\mathbf{r}/a_0^3 & \text{for } r < a_0 \\ -Ke\hat{\mathbf{r}}/r^2 & \text{for } r > a_0 \end{cases} \tag{18.63}$$

with corresponding potential energy

$$V(\mathbf{r}) = \begin{cases} -Ke^2/2a_0(3 - r^2/a_0^2) & \text{for } r < a_0 \\ -Ke^2/r & \text{for } r > a_0 \end{cases} \tag{18.64}$$

If an external electric field $\mathbf{E}_0$ (say in the z direction) is applied, the electron cloud and "nucleus" will shift in opposite directions by a net amount r_0 until a new equilibrium situation is achieved as in Fig. 18.3. The new system has the separation of charge characteristic of an electric dipole and the "atom" has been *polarized* by the external field. If we assume that the spherical shape of the electron cloud is unchanged for sufficiently small external fields, the net displacement can be determined by balancing the forces on the proton giving

$$0 = F_{\text{external}} + F_{\text{cloud}} = eE_0 - \frac{Ke^2 r_0}{a_0^3} \quad \text{or} \quad r_0 = E_0 a_0^3/Ke \tag{18.65}$$

The resulting induced *dipole moment* of the system is $p \equiv er_0 = E_0 a_0^3/K$. The new situation can also be visualized in terms of the potential energy function in

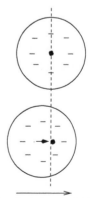

Figure 18.3. Classical picture of polarization of a charge distribution by an external electric field.

Applied electric field induces dipole moment

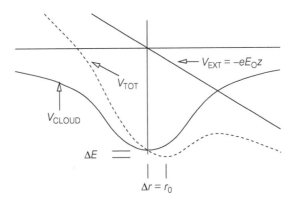

Figure 18.4. Energy shift due to applied electric field.

Fig. 18.4 where the new minimum is apparent. It is easy to show that the new system has a potential energy, which is lower by an amount

$$\Delta V = \Delta E = -\frac{E_0^2 a_0^3}{2K} \tag{18.66}$$

The interaction energy of an electric dipole in an external field is given by $-\mathbf{p} \cdot \mathbf{E}_0$ so an atom with a permanent dipole moment would have an energy shift linear in the applied field. For the situation above, there is only an induced dipole moment so that energy shift is necessarily quadratic in E_0; the *electric dipole polarizability* (labeled α) is often defined from the observed energy shift ΔE via

$$\Delta E = -\frac{1}{2}\alpha E_0^2 \tag{18.67}$$

The simple model above then implies that $\alpha \propto a_0^3/K$ with a numerical coefficient which varies with the assumed charge distribution. We note that this energy shift has the dimensions given by product of the energy density in the external field ($u_E = \epsilon |E|^2/2$) times a typical atomic or molecular volume $V \sim a_0^3$. This result also gives the form for the interaction potential of a charged particle with a neutral atom or molecule, the so-called *polarization potential,* via

$$V_{POL}(r) \sim -\frac{1}{2}\alpha \left(\frac{Ke}{r^2}\right)^2 \propto (\alpha K)\frac{Ke^2}{r^4} \propto \frac{Ke^2 a_0^3}{r^4} \tag{18.68}$$

Note that the introduction of the external (linear) potential $V_{ext} = -eE_0 z$ makes the potential "turn over" for sufficiently large values of $|z|$. The existence of a potential minimum even for arbitrarily large values of E_0 guarantees that there will always be a classical bound state; quantum effects, such as tunneling and zero-point energy, will make the problem in real atoms more interesting.

18.4.2 **Quantum Stark Effect**

We now turn to the quantum description of hydrogen atom in an external field. Being somewhat careful initially, we note that the potential felt by the two charged particles (proton and electron) in an external field in the $+z$ direction is

$$V(\mathbf{r}_e, \mathbf{r}_p) = -eE_0 z_p + eE_0 z_e = eE_0(z_e - z_p) = eE_0 z \qquad (18.69)$$

where $z = z_{\text{rel}} = z_e - z_p$ is the relative coordinate. The complete Hamiltonian is now

$$\hat{H} = \frac{\hat{\mathbf{p}}^2}{2\mu} - \frac{Ke^2}{r} + eE_0 z \qquad (18.70)$$

For sufficiently weak fields, we can use perturbation theory[12] to treat the additional term representing the external field; we will consider its effects on both the ground state and first excited states as examples of second-order and degenerate state perturbation theory, respectively.

For the ground state, $\psi_{100}(\mathbf{r})$, the first-order shift in energy is given by

$$E_1^{(1)} = \langle \psi_{100} | eE_0 z | \psi_{100} \rangle = eE_0 \int d\mathbf{r} \, |\psi_{100}(\mathbf{r})|^2 \, z = 0 \qquad (18.71)$$

from familiar parity arguments. This is also consistent with the classical arguments in which we expect no energy shift linear in the applied field. We must then consider the second-order shift given by

$$E_1^{(2)} = \sum_{n=2}^{\infty} \sum_{l=0}^{n-1} \sum_{m=-l}^{+l} \frac{|\langle \psi_{nlm} | eE_0 z | \psi_{100} \rangle|^2}{E_1^{(0)} - E_n^{(0)}} \qquad (18.72)$$

Writing $z = r \cos(\theta)$, we require the matrix elements

$$\langle \psi_{nlm} | z | \psi_{100} \rangle = \int d\mathbf{r} \, [R_{n,l}(r) Y_{l,m}^*(\theta, \phi)] \, [r \cos(\theta)] \, [R_{1,0}(r) Y_{0,0}(\theta, \phi)] \qquad (18.73)$$

The angular integration can be easily performed since

$$Y_{0,0} = \frac{1}{\sqrt{4\pi}} \quad \text{while} \quad \cos(\theta) = \sqrt{\frac{4\pi}{3}} Y_{1,0}(\theta, \phi) \qquad (18.74)$$

which contributes a factor of

$$\int d\Omega \, Y_{l,m}^*(\theta, \phi) \frac{1}{\sqrt{3}} Y_{1,0}(\theta, \phi) = \frac{1}{\sqrt{3}} \delta_{l,1} \delta_{m,0} \qquad (18.75)$$

[12] It has been said (Condon and Shortley, 1951) that the treatment of the Stark effect in hydrogen was the first application of perturbation theory in quantum mechanics.

when we use the orthonormality properties of the spherical harmonics. This result is in the form of a selection rule. It also simplifies the overlap integrals required for the radial piece of Eqn. (18.73) as we now only require

$$\int_0^\infty dr\, r^2\, R_{n,l}(r)\, r\, R_{1,0}(r)\, \delta_{1,0}\, \delta_{m,0} = \int_0^\infty dr\, r^3 R_{n,1}(r)\, R_{1,0}(r) \qquad (18.76)$$

However, the complete evaluation of the second-order shift requires not only the bound states in Eqn. (18.72), but also the continuum ($E_k > 0$) states as well, giving a contribution in schematic form of

$$E_1^{(2)} = \int_{E_k>0} \frac{|\langle \psi_k | eE_0 z | \psi_{100}\rangle|^2}{E_1^{(0)} - E_k}\, dk \qquad (18.77)$$

where $\psi_k(\mathbf{r})$ denotes the unbound (scattering) states.[13] This complicates the explicit evaluation of the second-order shift, but luckily other methods, including the use of parabolic coordinates[14] (in which the the problem is still separable) or subtle tricks[15] exist which give the complete result in closed form as

$$E_1^{(2)} = -\frac{9}{4}\frac{E_0^2\, a_0^3}{K} \qquad (18.78)$$

consistent with our classical arguments.

For the first excited state of hydrogen, corresponding to $n = 2$, we have four states which are degenerate in energy in the absence of an external field,

$$n = 2, l = 0 \qquad \psi_{200} = \psi_A \qquad (18.79)$$

and

$$n = 2, l = 1, m = \left\{ \begin{array}{ll} +1 & \psi_{2,1,+1} \equiv \psi_B \\ 0 & \psi_{2,1,0} \equiv \psi_C \\ -1 & \psi_{2,1,-1} \equiv \psi_D \end{array} \right\} \qquad (18.80)$$

Using the formalism of degenerate state perturbation theory (Section 10.5.2), we then are required, in principle, to diagonalize the 4×4 matrix corresponding

---

[13] See Ruffa (1973) for an explicit calculation of just this contribution.
[14] See, for example, Bethe and Salpeter (1957).
[15] A number of textbooks discuss the method due to Dalgarno and Lewis (1955).

to the equation

$$\begin{pmatrix} E_2^{(0)} + \langle V \rangle_{AA} & \langle V \rangle_{AB} & \langle V \rangle_{AC} & \langle V \rangle_{AD} \\ \langle V \rangle_{BA} & E_2^{(0)} + \langle V \rangle_{BB} & \langle V \rangle_{BC} & \langle V \rangle_{BD} \\ \langle V \rangle_{CA} & \langle V \rangle_{CB} & E_2^{(0)} + \langle V \rangle_{CC} & \langle V \rangle_{CD} \\ \langle V \rangle_{DA} & \langle V \rangle_{DB} & \langle V \rangle_{DC} & E_2^{(0)} + \langle V \rangle_{DD} \end{pmatrix} \begin{pmatrix} \psi_A \\ \psi_B \\ \psi_C \\ \psi_D \end{pmatrix}$$

$$= E_2^{(1)} \begin{pmatrix} \psi_A \\ \psi_B \\ \psi_C \\ \psi_D \end{pmatrix} \tag{18.81}$$

where $V = eE_0 z$. The problem simplifies considerably since the matrix elements connecting $\psi_B = \psi_{2,1,+1}$ to the other states through this interaction vanish since

$$\langle \psi_B | z | \psi_{A,C,D} \rangle = 0 \text{ because they have different } m \text{ values} \tag{18.82}$$

while

$$\langle \psi_B | z | \psi_B \rangle = 0 \text{ because of parity} \tag{18.83}$$

The same is true of the matrix elements for $\psi_D = \psi_{2,1,-1}$; these two states effectively decouple from the problem, and their energy levels are unchanged. That leaves the 2×2 subspace corresponding to $\psi_{A,C}$. The matrix elements connecting these two also simplify as

$$\langle \psi_A | z | \psi_A \rangle = 0 = \langle \psi_C | z | \psi_C \rangle \tag{18.84}$$

by parity, while

$$\langle \psi_A | z | \psi_C \rangle = -3a_0 \tag{18.85}$$

by direct calculation (P18.8). So, in the $\psi_{A,C}$ subspace, we have the determinant

$$\det \begin{pmatrix} E_2^{(0)} - E_2^{(1)} & -3eE_0 a_0 \\ -3eE_0 a_0 & E_2^{(0)} - E_2^{(1)} \end{pmatrix} = 0 \tag{18.86}$$

or

$$E_2^{(1)} = E_2^{(0)} \pm 3eE_0 a_0 \tag{18.87}$$

and the energies of the $\psi_{2,0,0}$ and $\psi_{2,1,0}$ states are split as shown in Fig. 18.5. The two solutions labeled by $\pm$ correspond to the (normalized) wavefunctions

$$\psi^{(+)} = \frac{1}{\sqrt{2}} (\psi_{2,0,0} - \psi_{2,1,0}) \quad \text{and} \quad \psi^{(-)} = \frac{1}{\sqrt{2}} (\psi_{2,0,0} + \psi_{2,1,0}) \tag{18.88}$$

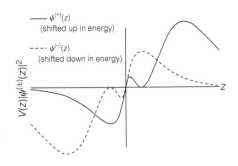

Figure 18.5. $V(z) = eE_0z$ times $|\psi^{\pm}(0, 0, z)|^2$ for the linear combination states shifted up, (+), and down, (−), in energy in the linear Stark effect.

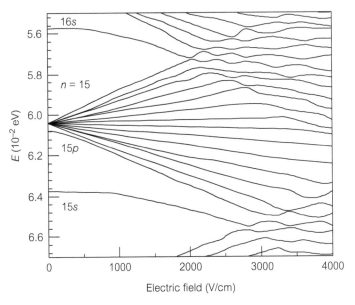

Figure 18.6. Energy versus applied electric field for highly excited states of lithium ($m = 0$ states) illustrating the first-order (linear in E) and second-order (quadratic in E) Stark effects. Data taken from Zimmerman *et al.* (1979).

To visualize this result, in Fig. 18.5 we plot $V(z)|\psi^{(\pm)}(0, 0, z)|^2$ versus z to show the distribution of potential energy in the two cases; the + (−) combinations are obviously shifted up (down) in energy as expected.

It is important to stress that the presence of an energy shift which is linear or first order in E_0, depends critically on the presence of degenerate energy levels; linear combinations of the $\psi_{2,1,0}$ and $\psi_{2,0,0}$ states are required to produce the dipole moments which can interact via $E = -\mathbf{p} \cdot \mathcal{E}$, and these states must be degenerate (or nearly so) for this mixing to occur.

To exemplify these remarks further, in Fig. 18.6 we show part of the energy spectrum of lithium Rydberg atoms as a function of the strength of the applied field. As discussed in Section 17.3.1, the states with the lowest values of l are shifted down in energy due to their interactions with the inner electron core; in this case the $15s$ ($l=0$) state is lowered in energy much more than the $15p$ ($l=1$) state while the remaining ones with $l = 2, 3, \ldots, 14$ are still almost degenerate. We note the following distinctive features:

- For small field strengths (less than roughly 2500 V/cm), the energy shift due to the external field for the $15s$ state (as well as the $16s$ state) is quadratic instead of linear due to its "isolation" from any nearby degenerate states. The polarizability of these states can be estimated from the data (P18.14)

- For quite small fields, the $15p$ state also has a quadratic energy shift but as the Stark level shifts of the $n = 15$ states become comparable to the initial splitting of the $15p$ state from the rest, it "joins in" and contributes to the linear Stark effect pattern. States do not have to be exactly degenerate to require use of degenerate perturbation theory[16] as discussed in Section 10.5.2.

- The remaining $n = 15$ states show the standard Stark effect of a linear energy shift with a large number of splittings due to the large degeneracy.

- For large enough fields, the perturbation theory predictions become unreliable and the pattern of energy levels becomes highly complex. Note however, the many avoided level crossings which are characteristic of "level repulsion," as mentioned in Section 10.5.

We briefly mention two other effects which arise in the quantum description of atomic energy levels in an external electric field:

- The combined Coulomb plus external field potential felt by an electron (illustrated in Fig. 18.7) implies that there are no absolutely stable bound states in such a potential as there is always a possibility of quantum tunneling. One

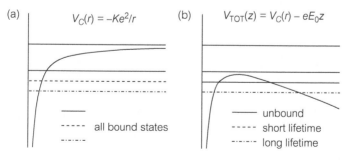

Figure 18.7. Potential energy due to Coulomb plus external electric fields showing the possibility of field ionization and tunneling.

[16] For a nice discussion of degeneracy effects and dipole moments, see Gasiorowicz (1996).

can derive estimates (P18.15) of the tunneling probability which confirm that the lifetimes in modest fields are much longer than the age of the universe, consistent with intuition. If, however, the external field is increased enough then the potential "turns over" allowing states to become unbound, leading to so-called *field ionization*. Simple estimates of the field necessary to unbind the nth state seem to be confirmed by experiment.

- The emission of photons via radiative transitions from one state to another forms the basis of atomic and molecular spectroscopy, which, in turn, provides the evidence for the quantized energy levels calculated in quantum mechanics. The related (inverse) process of single-photon absorption is conceptually the same; electrons absorb photons of energy $\hbar\omega = |E_n - E_{n'}|$ in transitions from one level to another or are ionized from the n-th level if $E_\gamma = \hbar\omega > |E_n|$. Sufficiently intense beams of photons of energy *less than* $|E_n|$ can still ionize atoms[17] via the process of *multiphoton ionization (MPI)* which is illustrated schematically in Fig. 18.8. The absorption of the "first" photon would not normally allow the electron to be freed, but if the number density of photons is high enough so that there is a chance of a second (or third and so forth) interaction, the ionization can occur as a multistep process. A similar process is called *above threshold ionization (ATI)* where the final state energy of the electron is measured to be

$$E_e = -|E_{\text{binding}}| + n\hbar\omega \qquad (18.89)$$

with $n > 1$ implying that more than one photon absorption has been used to "kick" the electron into the continuum of free-particle states. Clearly the probability for a k-step ionization process scales with the intensity, I, as I^k. The effect is illustrated schematically in Fig. 18.9 where the same laser beam

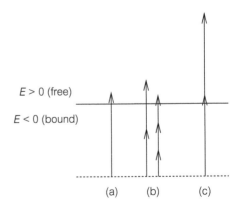

Figure 18.8. Schematic diagram showing photon absorption leading to (a) single photon ionization, (b) multiphoton ionization (MPI), and (c) above threshold ionization (ATI).

[17] For a nice review of multiphoton processes in atoms, see Delone and Krainov (1994).

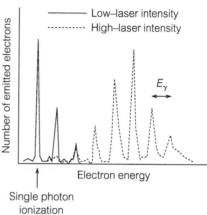

Figure 18.9. Number of emitted photoelectrons versus their kinetic energy for two laser intensities; for low laser intensities (solid), single photon ionization dominates, while for larger laser intensities (dashed), the absorption of more than one photon becomes appreciable. The distance between peaks is $E_\gamma = \hbar\omega$ indicating that the kinetic energies are given by $mv^2/2 = nE_\gamma - E_0$.

is used in both (a) and (b) (meaning that the frequency ω or energy of each photon E_γ is unchanged), but the intensity is increased near the critical value for the onset of the multiphoton process.

18.5 **Constant Magnetic Fields**

We next consider the problem of a charged particle (charge q) in a uniform magnetic field in the $+z$ direction. The classical solutions consist of helical trajectories, that is, uniform translational motion in the direction parallel to the field $\mathbf{B}_0$ and uniform rotational motion in the plane perpendicular to $\mathbf{B}_0$; for a positive charge q, the circular motion is in the clockwise direction when viewed from above ($z > 0$), as in Fig. 18.10.

To discuss the quantum version, we choose the gauge where

$$\mathbf{A}(\mathbf{r}, t) = \frac{B_0}{2}(-y, x, 0) \tag{18.90}$$

and consider other choices in the problems. The Hamiltonian from Eqn. (18.39) can be written as

$$\hat{H} = \frac{\hat{\mathbf{p}}^2}{2\mu} - \frac{q}{2\mu}(\hat{\mathbf{p}} \cdot \mathbf{A} + \mathbf{A} \cdot \hat{\mathbf{p}}) + \frac{1}{2\mu}\left(\frac{qB_0}{2}\right)^2 (x^2 + y^2) \tag{18.91}$$

The middle terms can be combined and written in the form

$$-\frac{q}{\mu}\frac{\hbar}{i}\frac{B_0}{2}\left(-\frac{\partial}{\partial y}x + \frac{\partial}{\partial x}y\right) = -\frac{qB_0}{2\mu}\hat{L}_z \tag{18.92}$$

Figure 18.10. Classical helical motion of a positively charged particle in a uniform magnetic field.

This form is clearly a generalization of the classical result

$$\hat{H} = -\mathbf{M} \cdot \mathbf{B} \tag{18.93}$$

where the magnetic moment is given by

$$\mathbf{M} = g \left(\frac{q}{2\mu} \right) \mathbf{L} \tag{18.94}$$

with the classical value of $g = 1$, as found in P16.12 for a rotating particle.

We can define the so-called *Larmor frequency*

$$\omega_L = \frac{qB_0}{2\mu} = \frac{\omega_c}{2} \tag{18.95}$$

which is half the *cyclotron frequency* which corresponds to the classical frequency of circular motion determined via the Lorentz force law, namely

$$F = \mu a_c \quad \longrightarrow \quad qvB_0 = \mu \frac{v^2}{r} = \mu \omega_c^2 r \quad \text{or} \quad \omega_c = \frac{qB_0}{\mu} \tag{18.96}$$

We then write the Hamiltonian in the form

$$\hat{H} = \frac{\hat{\mathbf{p}}^2}{2\mu} - \omega_L \hat{L}_z + \frac{1}{2} \mu \omega_L^2 (x^2 + y^2) \tag{18.97}$$

The corresponding Schrödinger equation is most conveniently solved in cylindrical coordinates where we assume a solution of the form

$$\psi(\mathbf{r}) = \psi(r, \theta, z) = R(r) \, e^{im\theta} \, e^{ik_z z} \tag{18.98}$$

This solution is an eigenfunction of both the $\hat{L}_z$ and z kinetic energy terms which together have contributions to the total energy given by $-m\hbar\omega_L$ and $\hbar^2 k_z^2/2\mu$, respectively. The remaining two-dimensional problem is essentially the planar harmonic oscillator (in polar coordinates) discussed extensively in Section 15.3.3. The resulting solutions are given by

$$\psi(r, \theta, z) \propto r^{|m|} \, e^{-r^2/2\rho^2} \, L_{n_r}^{|m|}(r^2/\rho^2) \, e^{im\theta} \, e^{ik_z z} \tag{18.99}$$

where $\rho^2 = \hbar/\mu\omega_L$ with energy eigenvalues

$$E_{n_r,m} = \hbar\omega_L\,(2n_r + |m| - m + 1) + \frac{\hbar^2 k_z^2}{2\mu} \qquad (18.100)$$

The equally spaced harmonic-oscillator-like energy states are often called *Landau levels*.

As discussed for the two-dimensional oscillator, the classical limit of uniform circular motion in the plane corresponds to the minimum number of radial nodes ($n_r = 0$). In that case, large values of $|m|$ of either sign could give macroscopic energies corresponding to motion in either clockwise or counterclockwise directions. Here, Eqn. (18.100) implies that only $m = -|m|$, that is, $L_z = -|m|\hbar$ or clockwise motion gives that limit, as expected for the motion of a positive charge.

Using the wavefunction solutions in Eqn. (18.99) for $n_r = 0$ and large $|m|$, one can show that the probability density is strongly peaked at a (planar) radius coordinate given by

$$r^2 = \frac{|m|\hbar}{\mu\omega_L} = \frac{|L_z|}{\mu\omega_L} \qquad (18.101)$$

To compare this to the classical case, we can use arguments similar to those in Section 1.4 and Section 15.3.3, and the Lorentz force law, to find

$$\frac{\mu v^2}{r} = \mu a_c = F = qvB_0 \qquad (18.102)$$

The relationship between velocity and momentum in the presence of a magnetic vector potential is generalized (P18.7) to be

$$\mu\mathbf{v} = \mathbf{p} - q\mathbf{A} \qquad (18.103)$$

so that

$$\mu\,(\mathbf{r} \times \mathbf{v})_z = (\mathbf{r} \times \mathbf{p})_z - q\,(\mathbf{r} \times \mathbf{A})_z \qquad (18.104)$$

or

$$-\mu r v = -|L_z| - \frac{1}{2}qB_0 r^2 \qquad (18.105)$$

Equations (18.102) and (18.105) can then be combined to give

$$r^2 = \frac{2|L_z|}{qB_0} = \frac{|L_z|}{\mu\omega_L} \qquad (18.106)$$

which is consistent with the large quantum number limit.

The energies of the system now form a continuous spectrum because of the motion in the z direction. Even for a fixed value of k_z, however, there is still an infinite degeneracy in the quantized energy levels corresponding to the fact that $|m| - m = 0$ for all values of positive m. This degeneracy has its origin in the arbitrariness in the initial conditions for the planar orbits; the x and y coordinates of the center of the circular orbit are not specified even if the total energy and k_z are. This is strictly true only for a region of uniform field which is of infinite extent; if the field is confined in the x and y directions by a box with sides L, not all circular orbits will "fit into" the box. In this more realistic case, the degeneracy, N_d, of each level[18] can be shown to be

$$N_d = \frac{qB_0L^2}{\pi\hbar} \tag{18.107}$$

which scales as the area as expected.

18.6 Atoms in Magnetic Fields

It has been said that "*Magnetism is inseparable from quantum mechanics...*"[19] since it can be shown that systems interacting via purely classical mechanics in statistical equilibrium can exhibit no magnetic moments, even in response to externally applied magnetic fields. Of the many possible manifestations of magnetic effects in quantum mechanical systems, in this section, we consider only three:

1. The problem of a one-electron atom subject to an external B field (the Zeeman effect).

2. The interactions of the electron spin in an atom due to the "internal" magnetic field caused by the orbital motion of the atomic constituents themselves (the so-called spin-orbit coupling).

3. The magnetic couplings of two spin magnetic moments in an atom, giving rise to so-called hyperfine splittings in atoms and contributing to spin–spin level shifts in other systems.

18.6.1 The Zeeman Effect: External B Fields

To study the effect of a constant external magnetic field (oriented in the z-direction for definiteness) on a one-electron atom, we consider the

---

[18] For a careful discussion of the boundary conditions which give rise to this estimate, see Peierls (1955).

[19] See Kittel (1971); He goes on to say "and were the value of $\hbar$ to go to zero, the loss to the science of magnetism is one of the catastrophes that would overwhelm the universe."

Hamiltonian for such a system (in a standard gauge) given by

$$\hat{H} = \left(\frac{\hat{p}^2}{2m_e} - \frac{KZe^2}{r} \right) + \frac{eB}{2m_e} \hat{L}_z + \frac{e^2 B^2}{8m_e}(x^2 + y^2)$$

$$= \hat{H}_{Coul} + \hat{H}_1 + \hat{H}_2 \tag{18.108}$$

where $q = -e$ and we assume $\mu \approx m_e$. This is strictly valid only for a spinless electron as we have ignored the coupling of its intrinsic angular momentum to the external field; this complication is discussed in P18.22.

For sufficiently small applied fields, we can initially neglect the term quadratic in B, the so-called *diamagnetic term*; in this case we see that the eigenfunctions of the standard Coulomb problem remain solutions since the spherical harmonics are also eigenfunctions of $\hat{L}_z$. The shift in energy due to the external field is then given by

$$E_B^{(1)} Y_{l,m} = \hat{H}_1 Y_{l,m} = \left[\left(\frac{e\hbar}{2m_e} B \right) m \right] Y_{l,m} \tag{18.109}$$

or

$$E_B^{(1)} = \left[\frac{e\hbar}{2m_e} B \right] m = (\mu_e B) \, m = m\hbar\omega_L \tag{18.110}$$

We have written this in two complementary forms:

- One form implicitly use the *Bohr magneton*, $\mu_e = e\hbar/2m_e$, whose numerical value[20] is $\mu_e = 5.788 \times 10^{-5}$ eV/T; this emphasizes its identification as being the interaction energy of the magnetic moment due to orbital motion with the external field.

- The second form is written in terms which are reminiscent of the quantized Landau energy level spacing for a charged particle discussed in Section 18.5.

The introduction of $\hat{H}_1$ reduces the (unexpectedly large) symmetry of the pure Coulomb problem by picking out a specific axis, and the energy spectrum is correspondingly "disrupted'; L_z, however, remains a constant of the motion and the problem is still exactly soluble. The inclusion of $\hat{H}_2$ completely destroys the symmetry, but for small fields the effect of $\hat{H}_2$ can then be estimated via perturbation theory using the standard solutions for hydrogen-like atoms; using the angular and radial overlap integrals in Eqn. (16.83) and Eqn. (17.30), and the fact that

$$x^2 + y^2 = r^2 \sin^2(\theta) \tag{18.111}$$

---

[20] In this section, as elsewhere, the appropriate MKSA unit of magnetic field is the Tesla (T) which is related to the Gauss (G) via $1 \, T = 10^4 \, G$.

we find that

$$E_B^{(2)} \approx \frac{e^2 B^2 a_0^2}{8m_e} F(n, l, m) \tag{18.112}$$

where

$$F(n, l, m) = \left[n^2 (5n^2 + 1 - 3l(l+1)) \frac{(l^2 + l - 1 - m^2)}{(2l-1)(2l+3)} \right] \tag{18.113}$$

when evaluated in the state $R_{n,l,m} = R_{n,l}(r) Y_{l,m}(\theta, \phi)$ and where a_0 is the Bohr radius. The prefactor in Eqn. (18.112) can be written in the form

$$\frac{e^2 B^2 a_0^2}{8m_e} = \frac{1}{4} \frac{(\hbar \omega_L)^2}{|E_1|} \tag{18.114}$$

where the E_n are the Bohr energy levels for the Coulomb problem. To see how this scales with n, we can choose for definiteness $l = n-1$ and $m=0$ and find that

$$E_B^{(2)} \approx \frac{e^2 B^2 a_0^2}{16m_e} n^4 \tag{18.115}$$

In the limit when the effects of $\hat{H}_2$ can no longer be considered as a small perturbation, other calculational tools (such as variational methods or diagonalization of matrices) must be employed to estimate the energy eigenvalues using the full Hamiltonian of Eqn. (18.108).

Considering only $E_B^{(1)}$, we see that for a given value of l, the $2l + 1$ degenerate levels corresponding to different values of m are split by equal amounts and the pattern of energy level shifts is shown in Fig. 18.11. Despite the seemingly large number of possible new transitions, the selection rule for the magnetic quantum

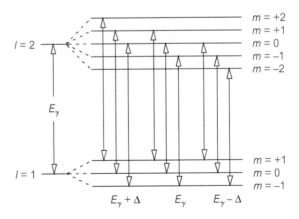

Figure 18.11. Energy level splittings for P ($l=1$) and D ($l=2$) states illustrating the linear Zeeman effect.

number for dipole radiation (see Section 16.3.3), namely, $\Delta m = +1, 0, -1$, and the uniform splitting imply that the absorption or emission line corresponding to the transition in the $B = 0$ case (E_γ) is only split into three distinct lines $(E_\gamma + \mu_e B \Delta m)$. This pattern of shifts in spectral lines is called the (*linear*) *ordinary Zeeman effect*, and it is observed for small field strengths in atoms in which the total electronic spin is zero; the more usual case where one must also consider the electron spin exhibits more structure, and is called the *anomalous Zeeman effect*.

To better understand the relative magnitudes of the energy splittings induced by $\hat{H}_{1,2}$, we first note that the difference between adjacent energy levels in hydrogen goes as

$$\Delta E_n \sim 13.6 \, \text{eV} \left(\frac{1}{(n-1)^2} - \frac{1}{n^2} \right) \sim 27 \, \text{eV} \frac{1}{n^3} \tag{18.116}$$

The maximum splitting due to $\hat{H}_1$ between the uppermost $(m = +l)$ and lowermost $(m = -l)$ scales roughly as

$$\Delta E_B^{(1)} \sim (5.8 \times 10^{-5} \frac{\text{eV}}{\text{T}}) B(2n) \tag{18.117}$$

since $n \geq l$. This implies that

$$\frac{\Delta E_B^{(1)}}{\Delta E_n} \sim \left(\frac{B}{2.3 \times 10^5 \, \text{T}} \right) n^4 \tag{18.118}$$

Typical values of the magnetic field range from $B \sim 0.5 \, \text{G} = 5 \times 10^{-5} \, \text{T}$ (corresponding to the earth's intrinsic field) or less (if shielding is provided) to $B \sim 5 - 10 \, \text{T}$, which is now standardly available in laboratories. For low-lying states $(n = 1, 2, ...)$, we see that the (linear) Zeeman splittings are always much smaller than the differences between energy levels; however, Eqn. (18.118) implies that the two become comparable in a 1 Tesla field for $n \sim 22$. We can also write the ratio in Eqn. (18.118) in a more symbolic fashion in the form

$$\frac{\Delta E_B^{(1)}}{\Delta E_n} = \left(\frac{B}{B_c} \right) n^4 \quad \text{with} \quad B_c \equiv \frac{\hbar}{ea_0^2} = \frac{\Phi_B}{\pi a_0^2} = 2.3 \times 10^5 \, \text{T} \tag{18.119}$$

where $\Phi_B = \pi \hbar / e$ is written in terms of fundamental constants, and has the units of a magnetic flux.

For such large fields (B) or principal quantum numbers (n), we must consider the effect of $\hat{H}_2$, and we can use the perturbation theory estimate of Eqn. (18.114) to write

$$\frac{\Delta E_B^{(2)}}{\Delta E_n} \approx \left(\frac{B}{4B_c} \right)^2 n^7 \tag{18.120}$$

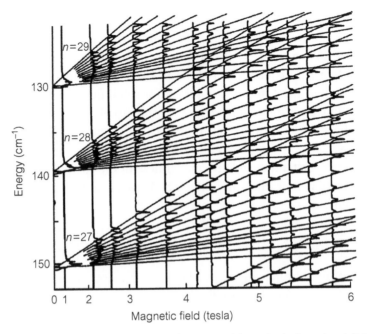

Figure 18.12. Energy versus applied magnetic field squared (note the horizontal scale!) illustrating the effect of diamagnetic $\hat{H}_2$ term in the Hamiltonian; data taken from Zimmerman *et al.* (1978) from measurements of lithium atoms.

This condition implies that the *quadratic Zeeman effect* (with energy shifts varying as B^2) will dominate over the more familiar linear Zeeman effect for highly excited Rydberg states (n large) or for sufficiently large magnetic field strengths[21]

As an example of the effect of the diamagnetic term in the Hamiltonian, $\hat{H}_2$, in the large field limit, we show in Fig. 18.12 the spectra of highly excited ($n = 27 - 29$) sodium Rydberg states in magnetic fields up to $B = 6\ T$. The experiment was performed by selectively choosing specific m states so that the linear Zeeman effect is absent. The theoretical predictions (obtained by matrix diagonalization of the Hamiltonian) are shown in Fig. 18.12 (solid lines) along with the experimental excitation curves (turn your head 90°) at various values of the external field. The energy (note the units) is plotted versus B^2 (note the horizontal scale also) to show the quadratic (parabolic) dependence on b; note also the avoided level crossings.

---

[21] It is estimated that when $B > 100\ T$ one has $E_B^{(2)} > E_B^{(1)}$. Such large values of the magnetic field can be found at the surfaces of collapsed astrophysical objects such as white dwarf or neutron stars, and quadratic Zeeman effects due to them have likely been seen in spectral lines from such objects.

18.6.2 **Spin-Orbit Splittings: Internal B Fields**

Even in the absence of externally applied magnetic fields, the charged particles (and their associated magnetic moments) in an atom are still subject to magnetic forces due to the "motion" of the other charges. To understand the physical origin of this effect, we consider first the classical problem a charged particle in uniform circular motion, as in Fig. 18.13; the circulating charge is equivalent to a current loop and the magnetic field at the center of the loop can be calculated classically using the Biot–Savart law. The small contribution from one "part of the circuit" can be integrated to obtain the familiar result

$$dB = \frac{\mu_0}{4\pi} \frac{Idl \times r}{r^3} = \frac{\mu_0}{4\pi} \frac{dqv \times r}{r^3} \Longrightarrow B_{cent} = \frac{\mu_0}{4\pi} \frac{qv}{r^2} \hat{k} \qquad (18.121)$$

If we

- Associate the circulating charge $q = Ze$ with the nuclear charge (remember that the electron and nucleus are in orbit around each other so the electron "sees" a positively charged current loop),

- Recall that the fundamental constant of magnetism can be rewritten using $c^2 = 1/\mu_0\epsilon_0$, and

- Use the classical relation for the angular momentum $|L| = rp = rmv$,

we can estimate that an atomic electron "sees" an effective magnetic field of the order

$$B_{cent} = \frac{\mu_0}{4\pi} \frac{Ze}{mr^3} L = \frac{Ze^2}{4\pi\epsilon_0} \frac{1}{mc^2 r^3} L \qquad (18.122)$$

We can make an initial estimate of the magnitude of this field by letting $|L| \sim \hbar$ and $r \sim a_0$ and find that $B_{cent} \sim 12\ T$; this is much larger than typical external fields, and can cause sizable level splittings.

The magnetic moment associated with the electron, namely

$$M = \frac{gq}{2m} S = -\frac{e}{m_e} S \qquad (18.123)$$

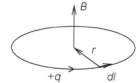

Figure 18.13. Classical picture of the magnetic field at center of current loop.

will then have an interaction energy in the atomic field given by

$$E = -\mathbf{M} \cdot \mathbf{B} = \frac{ZKe^2}{m^2c^2}\frac{1}{r^3}\mathbf{L} \cdot \mathbf{S} \tag{18.124}$$

where $K = 1/4\pi\epsilon_0$ as usual.

This classical estimate of the *spin-orbit (SO) interaction* (so-called because it exhibits the coupling between the electron spin and the orbital motion) neglects other important relativistic effects (note the explicit factors of c which appear), and the corresponding quantum mechanical expression is most convincingly derived by a nonrelativistic reduction of the Dirac equation for the electron. When this is done, the precise expression for the spin–orbit coupling term differs by only a factor of $1/2$ from our simple estimate, giving the spin–orbit coupling term

$$\hat{H}_{SO} = \frac{ZKe^2}{2m^2c^2}\frac{1}{r^3}\hat{\mathbf{L}} \cdot \hat{\mathbf{S}} \tag{18.125}$$

To evaluate the effect of this interaction on the spectrum of a one-electron atom, we first need to calculate the effect of the $\mathbf{L} \cdot \mathbf{S}$ term. The total angular momentum due to orbital and spin contributions will be given by

$$l + 1/2 \implies j = l + 1/2 , \ l - 1/2 \tag{18.126}$$

with wavefunctions discussed in P16.16. The more formal relation among the corresponding operators

$$\mathbf{J} = \mathbf{L} + \mathbf{S} \tag{18.127}$$

can be manipulated to yield

$$\mathbf{J}^2 = (\mathbf{L} + \mathbf{S})^2 \implies \mathbf{L} \cdot \mathbf{S} = \frac{1}{2}\left(\mathbf{J}^2 - \mathbf{L}^2 - \mathbf{S}^2\right) \tag{18.128}$$

Recalling that S is fixed, we then have

$$\mathbf{L} \cdot \mathbf{S} = \frac{1}{2}\left(j(j+1) - l(l+1) - \frac{3}{4}\right)\hbar^2 \tag{18.129}$$

or

$$\mathbf{L} \cdot \mathbf{S} = \begin{cases} l/2 & \text{for } j = l + 1/2 \\ -(l+1)/2 & \text{for } j = l - 1/2 \end{cases} \tag{18.130}$$

which implies that states with $j = l+1/2$ ($l-1/2$) are shifted up (down) in energy. For states with $l > 0$, the original $2(2l+1)$ degenerate levels are split into two distinct sets of states with $2(l+1/2)+1 = 2l+2$ and $2(l-1/2)+1 = 2l$ levels, respectively. For $l=0$, only $j=1/2$ is allowed and there no splitting. This pattern of energy level shifts is sometimes called *Lande's interval rule.*

The fact that $\hat{H}_{SO}$ is relativistic in origin suggests that its effects may be small for low Z (recall Eqn. (17.16)), and allows us to use perturbation theory to estimate its size. We require the expression

$$\left\langle \frac{1}{r^3} \right\rangle_{n,l} = \left(\frac{Z^3}{a_0^3} \right) \frac{2}{n^3 l(2l+1)(l+1)} \tag{18.131}$$

The splittings due to this spin–orbit coupling can then be written as

$$\Delta E_{n,l}^{SO} = \frac{Ke^2}{2m_e^2 c^2} \frac{Z^4 \hbar^2}{a_0^3} \frac{1}{n^3 l(2l+1)(l+1)} \left\{ \begin{array}{ll} +l & \text{for } j = l+1/2 \\ -(l+1) & \text{for } j = l-1/2 \end{array} \right\} \tag{18.132}$$

We note that this expression is not obviously well-defined for $l = 0$, being a ratio of the form $0/0$. A careful derivation using the Dirac equation shows that the s-state energy levels are, in fact, shifted by just the amount predicted by Eqn. (18.132) by simply canceling the factors of l for the $j = l+1/2$ case.[22]

Using the expression for the unperturbed energy levels, we can rewrite Eqn. (18.132) in the more compact form

$$\frac{\Delta E_{n,l}^{SO}}{|E_n|} = (Z\alpha)^2 \frac{1}{nl(2l+1)(l+1)} \left\{ \begin{array}{ll} +l & \text{for } j = l+1/2 \\ -(l+1) & \text{for } j = l-1/2 \end{array} \right\} \tag{18.133}$$

Several comments can be made:

- The factor of $(Z\alpha)^2$ in Eqn. (18.133) implies that spin–orbit splittings are typically $Z^2 \times 10^{-5}$ smaller than the unperturbed energy levels; this justifies our use of perturbation theory.

- At this same level of precision, we must also include relativistic corrections to the non-relativistic approximation for the kinetic energy (as in Section 1.4), which we standardly use. We can use the results of P17.12 in the form

$$\frac{\Delta E_{n,l}^{rel}}{|E_n|} = -\frac{(Z\alpha)^2}{n} \left[\frac{2}{2l+1} - \frac{3}{4n} \right] \tag{18.134}$$

---

[22] This result may seem counter-intuitive as we would expect no spin-orbit contribution for s-states for which there is no angular momentum; in fact, a more careful evaluation of the effect of Eqn. (18.125) for the case of $l=0$, not based on perturbation theory, shows that $<1/r^3>$ is indeed finite, so that it gives no effect on s-states. There is, however, another term which arises in the nonrelativistic reduction of the Dirac equation (which has no simple classical correspondence) which is only nonvanishing for s-states, and which gives exactly the effect as Eqn. (18.132) which we then take as *operationally* correct for all values of l; for a thorough discussion, see Condon and Shortley (1951).

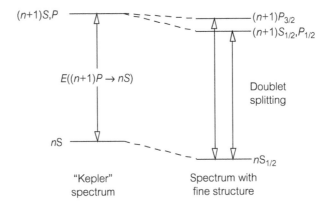

Figure 18.14. Splittings of energy levels due to fine structure (spin-orbit plus relativistic kinetic energy effects) on S and P states; the "doublet" splitting relevant for $(n + 1)P \to nS$ transitions is due to the spin-orbit coupling only.

and taken together with Eqn. (18.133), these two relativistic corrections successfully reproduce the results of the Dirac equation analysis up to order $(Z\alpha)^4$, namely

$$\frac{\Delta E_{n,l}^{rel} + \Delta E_{n,l}^{SO}}{|E_n|} = -\frac{(Z\alpha)^2}{n} \left[\frac{2}{2j+1} - \frac{3}{4n} \right] \tag{18.135}$$

for both $j = l \pm 1/2$. The spin splittings due to these combined relativistic effects are shown schematically in Fig. 18.14 for adjacent S and P states. The spectroscopic notation nL_j is used to distinguish states of different principle quantum number (n), orbital angular momentum $(L = S, P, D, F, \ldots$ for $l = 0, 1, 2, 3, \ldots)$ and total angular momentum $j = l \pm 1/2$.

- The splitting of the $(n+1)P \to nS$ transition line into two distinct lines is due to the energy difference between the $P_{3/2,1/2}$ energy levels, which, in turn arises solely from the spin–orbit coupling since the relativistic correction is the same for fixed l, n. This splitting is responsible for the distinctive pattern of *doublets* in the line spectra of alkali atoms; the pair of closely spaced lines called the *sodium doublet*[23] is perhaps the most famous example, and corresponds to the transition of the outermost electron from the $4P \to 3S$ level.

Spin–orbit coupling terms are also present in nuclear systems where they play an important role in understanding the observed pattern of the so-called magic numbers implied by nuclear shell structure. Just as with atomic systems, large

---

[23] This fact is the basis for the only suggestion for a quantum mechanics experiment in this book, namely, looking at the emission spectrum of sodium (using a diffraction grating) by appropriately igniting some table salt, that is, sodium chloride; see Crawford (1968) for more concrete suggestions.

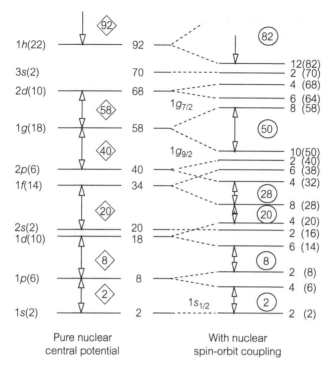

Figure 18.15. Energy level scheme for typical central nuclear potential (left) and including spin–orbit interactions (right). The inclusion of the spin–orbit couplings reproduces the observed "magic numbers".

energy gaps between relatively closely spaced sets of quantized energy levels in nuclei can give rise to especially stable configurations as the available states are "filled" with neutrons and protons; many observable properties are correlated with nuclei whose Z or N is equal to 2, 8, 20, 28, 50, 82, 126.

Calculations for various models described by purely central potentials, $V(\mathbf{r}) = V(r)$, as in Fig. 18.15, give levels consistent with the first few such "magic numbers," but fail to reproduce the observed pattern for heavier nuclei. If one includes a strong spin–orbit coupling[24] of the form $V_{SO}(r)\,\mathbf{L}\cdot\mathbf{S}$ (with the appropriate sign for $V_{SO}(r)$), the level structure is changed dramatically for larger angular momentum states and nicely reproduces observed nuclear shell structure. For example, the lowest-lying $l = 4$ state, here labeled $1g(18)$, can accommodate $2(2\cdot4+1) = 18$ spin-1/2 neutrons or protons; the total angular momentum of a single nucleon in such a state is $J = L+S = 4+1/2 = 7/2$ (8 states) and $J = 9/2$ (10 states), and the fine-structure interaction splits these two combinations into different shells.

---

[24] This observation was made independently by Maria Goeppart Mayer and H. D. Jensen for which they shared the Nobel prize in 1963.

18.6.3 **Hyperfine Splittings: Magnetic Dipole–Dipole Interactions**

The magnetic moment of an atomic electron can also interact with the magnetic dipole field associated with the nuclear magnetic moment. In the context of hydrogen (and other atomic systems) the resulting spin–spin interactions give rise to small level shifts called *hyperfine splittings* (hereafter *h.f.s*), and are also present in other "hydrogenic" bound state systems.

The magnetic field from a *point* dipole, **M**, is given by

$$\mathbf{B}(\mathbf{r}) = \frac{\mu_0}{4\pi} \left(\frac{3(\mathbf{M} \cdot \mathbf{r})\,\mathbf{r}}{r^5} - \frac{\mathbf{M}}{r^3} + \frac{8\pi}{3} \mathbf{M}\,\delta(\mathbf{r}) \right) \tag{18.136}$$

The first two terms are the familiar result derived in most standard texts on electricity and magnetism as the dipole field of a distant current loop; the third term is more subtle,[25] and arises when one considers point dipoles. The corresponding classical energy of one dipole in the field of another is then simply

$$\begin{aligned}
E &= -\mathbf{M} \cdot \mathbf{B} = \hat{H}_{\text{dip}} \\
&= -\frac{\mu_0}{4\pi} \left(\frac{3(\mathbf{M}_N \cdot \mathbf{r})(\mathbf{M}_e \cdot \mathbf{r})}{r^5} - \frac{\mathbf{M}_N \cdot \mathbf{M}_e}{r^3} + \frac{8\pi}{3} \mathbf{M}_N \cdot \mathbf{M}_e\,\delta(\mathbf{r}) \right)
\end{aligned} \tag{18.137}$$

where we have specialized to nuclear (N) and electronic (e) moments given by

$$\mathbf{M}_N = g_N \frac{Ze}{2M}\,\mathbf{S}_N \quad \text{and} \quad \mathbf{M}_e = -g_e \frac{e}{2m_e}\,\mathbf{S}_e \approx -\frac{e}{m_e}\,\mathbf{S}_e \tag{18.138}$$

where M, Z, and g_N are the nuclear mass, charge, and gyromagnetic ratio, respectively. The Hamiltonian describing this dipole–dipole interaction is given by this form where $\mathbf{S}_{e,N}$ are associated with spin operators.

The effect of this spin–spin interaction on the ground state of a hydrogen-like atom can be estimated using perturbation theory via $\Delta E_{1,0}^{(h.f.s)} = \langle \psi_{1,0,0} | \hat{H}_{\text{dip}} | \psi_{1,0,0} \rangle$, which we can evaluate using several observations:

1. The expectation value of the "standard" dipole term in Eqn. (18.137) vanishes when evaluated in the spherically symmetric ground state.

2. The "point dipole" term then contributes a factor of

$$\langle \psi_{1,0,0} | \delta(\mathbf{r}) | \psi_{1,0,0} \rangle = |\psi_{1,0,0}(0)|^2 = \frac{1}{\pi} \left(\frac{Z}{a_0} \right)^3 \tag{18.139}$$

3. The magnetic permeability can be replaced in favor or $\mu_0 = 1/(c^2 \epsilon_0)$.

[25] For a nice discussion, see Griffiths (1999).

4. The product of spin operators can be performed using the standard trick, namely

$$S_e \cdot S_N = \frac{1}{2}\left(S^2 - S_e^2 - S_N^2\right) \qquad (18.140)$$

where $S = S_e + S_N$ in general. For the case of hydrogen when the total spin can be $S = 1/2 + 1/2 = 0, 1$, we have

$$S_e \cdot S_p = \hbar^2 \begin{cases} 1(1+1) - 3/4 - 3/4 = 1/2 & \text{for } S = 1 \\ 0 - 3/4 - 3/4 = -3/2 & \text{for } S = 0 \end{cases} \qquad (18.141)$$

These can be combined to find the spin-dependent shifts to the ground state levels of hydrogen, namely

$$\Delta E_{1,0}^{(h.f.s)} = \frac{Ke^2}{2m_e^2 c^2} \frac{Z^4 \hbar^2}{a_0^3} \left[g_p \frac{4}{3} \frac{m_e}{m_p} \right] \begin{cases} +1 & \text{for } S = 1 \\ -3 & \text{for } S = 0 \end{cases} \qquad (18.142)$$

which we have put in a form which can be compared more readily to the expression for fine-structure splitting in Eqn. (18.132). We note that:

- The $S = 0$ or $^1S_0$ state is split down relative to the $S = 1$ or $^3S_1$ state (where the notation ^SL_J is used).

- This hyperfine splitting is suppressed relative to the fine-structure effects by the factor in square brackets, namely, $m_e/M \lesssim 1/1800$ and motivates its name.

- The photon energy corresponding to transitions from the $S = 1$ to $S = 0$ states is roughly $E_\gamma = 5.9 \times 10^{-6}$ eV which corresponds to a frequency and wavelength of roughly $f \approx 1420$ Hz and $\lambda \approx 21$ cm, respectively. Interstellar hydrogen atoms can undergo collisional excitations which populate the (slightly) excited $S = 1$ state from the ground state ($S = 0$), and the resulting 21 cm radio emission line is extensively used by astronomers to map the concentrations of hydrogen gas; for example, Doppler-shifted line profiles of the 21 cm line spectrum have been used to map out the spiral-arm structure of our galaxy. This physical feature of the most basic atom in nature has been described[26] as "*a unique, objective standard frequency, which must be known to every observer in the universe*"; and has given rise to suggestions[27] that it be used (or at least monitored) for interstellar communications.

---

[26] See Cocconi and Morrison (1959).
[27] For a guide to the scientific literature on search for extraterrestrial civilizations, see Kuiper (1989).

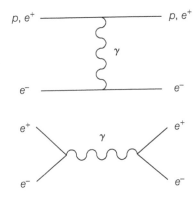

Figure 18.16. Feynman diagrams leading to hyperfine splittings for e^+e^- bound states.

The spin-spin level splittings due to magnetic dipole–dipole interactions are present in other systems as well. In the e^+e^- positronium system, for example, the effect is present with two important changes:

- Because both particles have the same mass, the hyperfine splitting is not suppressed relative to spin-orbit or fine-structure effects as the m_e/M_p term in Eqn. (18.142) is now of order unity.

- An additional physical mechanism contributes to the effective interaction in Eqn. (18.137) due to the possibility of "annihilation"-type interactions as, shown in Fig. 18.16.

18.7 Spins in Magnetic Fields

18.7.1 Measuring the Spinor Nature of the Neutron Wavefunction

We have focused on the interaction of charged particles with $\mathbf{E}$ and $\mathbf{B}$ fields. Electrically neutral particles can have nontrivial EM interactions via the coupling of a magnetic dipole moment with external fields. This was discussed in Section 16.4 in connection with the spinor wavefunction of spin-1/2 particles where it was shown that such an interaction can provide a "handle" on the precession of the spin vector. This fact was the basis for the observation[28] that the phase behavior of the neutron spinor wavefunction, with its predicted phase change of -1 on rotation by 2π, could be tested in such systems.

[28] First made by Bernstein (1967) and Aharonov and Susskind (1967).

The angle through which a magnetic moment precesses in a field in a small time dt is given by Eqns (16.179) and (16.180) as

$$d\theta = \omega_L t = \frac{\mu B}{\hbar} dt \qquad (18.143)$$

This can be integrated over any path over which the neutron moves with constant speed v to yield

$$\theta = \int d\theta = \frac{\mu_n}{\hbar} \int B \, dt = \frac{g\mu}{v\hbar} \int_{\text{path}} \mathbf{B} \cdot d\mathbf{l} \qquad (18.144)$$

and v is the classical speed. Recall that the neutron's magnetic moment is roughly $\mu_n = -1.93\,\mu_N$ where $\mu_N = e\hbar/2m_n$.

After having its magnetic moment precess through the angle θ, the spinor wavefunction becomes

$$\begin{pmatrix} \alpha^+(\theta) \\ \alpha^-(\theta) \end{pmatrix} = \begin{pmatrix} e^{+i\theta/2} & 0 \\ 0 & e^{-i\theta/2} \end{pmatrix} \begin{pmatrix} \alpha^+(0) \\ \alpha^-(0) \end{pmatrix} = \begin{pmatrix} \alpha^+(0)e^{+i\theta/2} \\ \alpha^-(0)e^{-i\theta/2} \end{pmatrix}$$

$$(18.145)$$

Now consider the experiment described schematically by Fig. 18.17 in which a beam of neutrons is split between two paths, one which traverses a magnetic field, and one in a field-free region. If the initial neutron beam is unpolarized (equal amounts of "up" and "down"), then the neutrons along the path ABD experience no change in phase and

$$\psi_{ABD} = \psi_A = \frac{1}{\sqrt{2}} \begin{pmatrix} 1 \\ 1 \end{pmatrix} \qquad (18.146)$$

The pieces of the neutron spin wavefunction which do traverse the magnetic field pick up phases given by

$$\psi_{ACD}(\theta) = \begin{pmatrix} e^{+i\theta/2} & 0 \\ 0 & e^{-i\theta/2} \end{pmatrix} \psi_A = \frac{1}{\sqrt{2}} \begin{pmatrix} e^{+i\theta/2} \\ e^{-i\theta/2} \end{pmatrix} \qquad (18.147)$$

When the beams are recombined, the total wavefunction is given by

$$\psi_{TOT}(\theta) = \psi_{ABD}(\theta) + \psi_{ACD}(\theta) = \frac{1}{\sqrt{2}} \begin{pmatrix} 1 + e^{+i\theta/2} \\ 1 + e^{-i\theta/2} \end{pmatrix} \qquad (18.148)$$

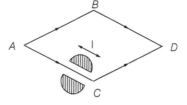

Figure 18.17. Schematic representation of experiment designed to test the phase change on rotation of the neutron's spinor wavefunction. The neutrons following the path ACD traverse a region of magnetic field, while those which follow ABD do not.

so that the probability density is proportional to

$$|\psi_{TOT}(\theta)|^2 = \frac{1}{2}\left(\left|1 + e^{+i\theta/2}\right|^2 + \left|1 + e^{+i\theta/2}\right|^2\right) \quad \propto \quad 1 + \cos\left(\frac{\theta}{2}\right)$$
(18.149)

The counting rate at a detector at point D measures the beam intensity and hence this probability; if one scales the intensity obtained with various amounts of "magnetic path length" $\int_{\text{path}} \mathbf{B} \cdot d\mathbf{l}$ to that with the field turned off, one should then have

$$\frac{I(\theta)}{I(0)} = \frac{|\psi_{TOT}(\theta)|^2}{|\psi_{TOT}(0)|^2} = \frac{1 + \cos(\theta/2)}{2}$$
(18.150)

This formula exhibits the typical spin-1/2 phase change upon rotation by 2π; constructive interference corresponding to a return to the same phase is only seen after a rotation of the magnetic moment by an angle of $\theta = 4\pi$.

Two sets of experiments using neutron interferometers (as in Section 18.1) were performed[29] soon after the predictions were made and results from one of them are shown in Fig. 18.18. Numerical values for this data set are analyzed in P18.27.

18.7.2 Spin Resonance

The steady precession of a magnetic moment around the direction of a static field can exhibit striking resonance effects if a time-varying magnetic field is applied

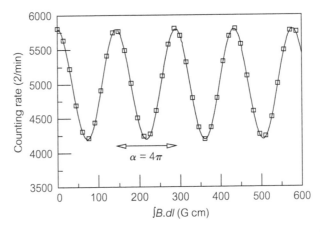

Figure 18.18. Measured neutron intensity versus magnetic path length illustrating the "sign flip" for a spin-1/2 wavefunction rotated through 2π. Data from Rauch *et al.* (1975).

---

[29] See Werner *et al.* (1975) and Rauch *et al.* (1975).

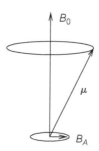

Figure 18.19. Magnetic moment, μ, precessing around static field **B**$_0$ plus additional, rotating magnetic field **B**$_A$.

at right angles to the original field. This can be understood in two relatively simple ways:

- We illustrate in Fig. 18.19 a large, uniform field in the z direction and the precession of the magnetic moment **M** (and hence spin direction) around it. The presence of a small, rotating **B**$_A$ field will, in general, have little effect as the torque it induces on **M** will average out to zero if it rotates at a different rate (or even different direction) than does **M**; if, however, the magnetic field "stays in phase" with the moment by rotating at a rate $\omega = \omega_{precess} = g\mu B/\hbar$, its torque can act continuously and induce dramatic changes in the rotational motion.

- We have seen (Section 16.4) that in a uniform field, the energy levels of the spin-1/2 particle are given by

$$E_\pm = \pm\hbar\omega_{phase} = \pm\frac{g\mu B}{2} \qquad (18.151)$$

which implies a splitting between the parallel and antiparallel configurations of spins given by $\Delta E = g\mu B$; this implies that EM radiation of energy $E_\gamma = \hbar\omega = \Delta E$ or angular frequencies $\omega = g\mu B/\hbar$ can preferentially be absorbed.

In either viewpoint, the spin system will exhibit resonance behavior when subjected to external magnetic fields when $\omega = \omega_{precess}$, and we will analyze the dynamical equations of motion for the spinor wavefunction in a somewhat formal way in order to derive the exact form of the resonance response.

We assume a large, static field (**B**$_0$) in the z direction and a rotating field of arbitrary magnitude (**B**$_A$) at right angles to it, given by

$$\mathbf{B}_0 = (0, 0, B_0) \qquad (18.152)$$

$$\mathbf{B}_A = (B_A \cos(\omega t), -B_A \sin(\omega t), 0) \qquad (18.153)$$

The spin Hamiltonian is then given by

$$\hat{H} = -\frac{g\mu}{2} \mathbf{B} \cdot \sigma = -\frac{g\mu}{2} \left(B_A^{(x)} \sigma_x + B_A^{(y)} \sigma_y + B_A^{(z)} \sigma_z \right)$$

$$= -\frac{g\mu}{2} \left(\begin{array}{cc} B_0 & B_A e^{+i\omega t} \\ B_A e^{-i\omega t} & -B_0 \end{array} \right) \qquad (18.154)$$

where the spin matrices are defined in Eqn. (16.166). The corresponding Schrödinger equation for the spinor coordinates is

$$i\hbar \frac{\partial}{\partial t} \left(\begin{array}{c} \alpha_+(t) \\ \alpha_-(t) \end{array} \right) = \hat{H} \left(\begin{array}{c} \alpha_+(t) \\ \alpha_-(t) \end{array} \right) \qquad (18.155)$$

which can be written in the form

$$\left(\begin{array}{c} \dot{\alpha}_+(t) \\ \dot{\alpha}_-(t) \end{array} \right) = \frac{i}{2} \left(\begin{array}{cc} \omega_0 & \omega_A e^{+i\omega t} \\ \omega_A e^{-i\omega t} & -\omega_0 \end{array} \right) \left(\begin{array}{c} \alpha_+(t) \\ \alpha_-(t) \end{array} \right) \qquad (18.156)$$

where we have defined two new frequencies,

$$\omega_0 = \frac{g\mu B_0}{\hbar} \quad \text{and} \quad \omega_A = \frac{g\mu B_A}{\hbar} \qquad (18.157)$$

in addition to the frequency ω describing the time rate of change of the external field. These then give the coupled differential equations

$$2\dot{\alpha}_+(t) = i \left(\omega_0 \alpha_+(t) + \omega_A e^{+i\omega t} \alpha_-(t) \right) \qquad (18.158)$$

$$2\dot{\alpha}_-(t) = i \left(\omega_A e^{-i\omega t} \alpha_+(t) - \omega_0 \alpha_-(t) \right) \qquad (18.159)$$

We assume that the spinor is initially in the "up" direction so that the

$$\left(\begin{array}{c} \alpha_+(0) \\ \alpha_-(0) \end{array} \right) = \left(\begin{array}{c} 1 \\ 0 \end{array} \right) \qquad (18.160)$$

and attempt to find the solutions of Eqn. (18.156) subject to these initial conditions.

Motivated by the time-dependence of the spinor wavefunction when $B_A = 0$, we attempt a solution of the forms

$$\alpha_+(t) = A_+ e^{i\omega_+ t} \quad \text{and} \quad \alpha_-(t) = A_- e^{i\omega_- t} \qquad (18.161)$$

so that Eqns (18.158) and (18.159) become

$$2\omega_+ A_+ = \omega_0 A_+ + \omega_A A_- e^{i(\omega + \omega_- - \omega_+)t} \qquad (18.162)$$

$$2\omega_- A_- = \omega_A A_+ e^{-i(\omega + \omega_+ - \omega_-)t} - \omega_0 A_- \qquad (18.163)$$

The time-dependence can be eliminated if we choose

$$\omega_+ = \omega_- + \omega \qquad (18.164)$$

which gives now the coupled *algebraic* equations in matrix form

$$\begin{pmatrix} (2\omega_+ - \omega_0) & -\omega_A \\ -\omega_A & (2\omega_- + \omega_0) \end{pmatrix} \begin{pmatrix} A_+ \\ A_- \end{pmatrix} = 0 \qquad (18.165)$$

For a consistent solution, the determinant of coefficients must vanish and if we combine this condition with Eqn. (18.164) we find that

$$\omega_- = -\frac{\omega}{2} \pm \frac{1}{2}\sqrt{(\omega - \omega_0)^2 + \omega_A^2} = -\frac{\omega}{2} \pm \Delta\omega \qquad (18.166)$$

so that $\omega_+ = +\omega/2 \pm \Delta\omega$.

As with any set of coupled second-order differential equations, the result for $\alpha_-(t)$ will consist of a linear combination of the two independent solutions, namely

$$\alpha_-(t) = Ae^{-i\omega t/2}e^{+i\Delta\omega t} + Be^{-i\omega t/2}e^{-i\Delta\omega t} \qquad (18.167)$$

The initial condition $\alpha_-(0) = 0$ implies that $B = -A$ while its derivative is given by $\dot{\alpha}_-(0) = 2iA\Delta\omega$. Substituting these values into Eqn. (18.159) and using the other initial condition on the spinor upper component, namely, $\alpha_+(0) = 1$, implies that

$$2iA = \frac{i\omega_A}{2\Delta\omega} \qquad (18.168)$$

Our "big result" is then that the total time-dependence is given by

$$\alpha_-(t) = \frac{i\omega_A}{2\Delta\omega} \sin(\Delta\omega t) e^{-i\omega t/2} \qquad (18.169)$$

This spinor amplitude gives information on the probability that the spin will have "flipped" to a state anti-parallel to $\mathbf{B}_0$ at later times since

$$\text{Prob(spin down)}(t) = |\alpha_-(t)|^2 = \left[\frac{\omega_A^2}{\omega_A^2 + (\omega - \omega_0)^2}\right] \sin^2(\Delta\omega t) \qquad (18.170)$$

Observations on this result include:

• The prefactor in Eqn. (18.170), namely,

$$\frac{\omega_A^2}{\omega_A^2 + (\omega - \omega_0)^2} \qquad (18.171)$$

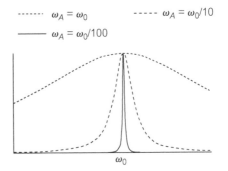

Figure 18.20. Probability of a spin-flip versus frequency of applied field ω showing Lorentzian line shape for spin-resonance; different values of ω_A are shown indicating the effect of the magnitude of B_A on the "sharpness" of the resonance peak.

exhibits the *Lorentzian line shape* discussed in P4.9 which is typical of resonance phenomena, and we will illustrate its form for various values of ω_A in Fig. 18.20. The spin-flip amplitude has its maximum value when

$$\omega = \omega_0 = \frac{g\mu B_0}{\hbar} = \omega_{\text{precess}} \qquad (18.172)$$

as expected, but we also note that the "sharpness" of the resonance is proportional to ω_A. In order to measure the resonant frequency as precisely as possible (which is the hallmark of the method), we want to make B_A small which often means that the sample must be shielded from stray fields not under the control of the experimenter (the earth's field, for example, or stray radio-frequency (RF) signals).

- When applied to an unpaired electron spin, this method is called *electron spin resonance* or *ESR* and for typical external field strengths of $B_0 \approx 0.3\ T$, the resonant frequencies, wavelengths, and photon energies are

$$f \approx 10\ \text{GHz}, \qquad \lambda \approx 3\ \text{cm}, \quad \text{and} \quad E_\gamma = \Delta E \approx 4 \times 10^{-5}\ \text{eV} \quad (18.173)$$

so that ESR usually requires microwave techniques. When applied to the nuclear magnetic moments of free or unpaired protons (or other nuclei with nonvanishing spins) the technique is called *nuclear magnetic resonance or NMR*; the corresponding resonance parameters for protons in a $1\ T$ field are

$$f \approx 40\ \text{MHz}, \quad \lambda \approx 7\ \text{m}, \quad \text{and} \quad E_\gamma = \Delta E \approx 2 \times 10^{-7}\ \text{eV} \quad (18.174)$$

which shows that RFs are required for the time-dependent fields.

- Even with large field strengths B_0, the energy "gain" from being parallel to the field is much smaller than ordinary thermal fluctuations so that there is only a small excess of "spin-up" (N_+) versus "spin-down" (N_-) states in a typical sample. For example, the population ratio in the two states is given by their respective Boltzmann factors, namely

$$\frac{N_+}{N_-} = \frac{e^{-E_+/kT}}{e^{-E_-/kT}} = e^{-\Delta E/kT} \tag{18.175}$$

where k is Boltzmann's constant and T is the temperature. At room temperatures, where $kT \approx 1/40$ eV, this implies that

$$\left(\frac{N_+ - N_-}{N_+ + N_-}\right) \sim \frac{1}{2}\left(1 - e^{-\Delta E/kT}\right) \sim \frac{\Delta E}{2kT} \approx 4 \times 10^{-6} \tag{18.176}$$

for the proton values above.

- Electron spin resonance and NMR have been extensively applied in the fields of nuclear and solid state physics as well as in chemistry. The precision with which frequency measurements can be made can then be translated into very accurate determinations of nuclear magnetic moments. Nuclei with known moments can then be used to probe the electronic environment in solids or determine the structure of complex molecules. The technique is perhaps best known, however, when it is applied to probing the protons in the human body where it is known as *magnetic resonance imaging* or *MRI*[30]

18.8 The Aharonov–Bohm Effect

We have so far considered the EM interactions of charged particles and magnetic dipoles in regions where there is a nonvanishing **B** field. We now describe a strikingly "quantum" phenomenon in which the phase of the wavefunction of a charged particle is changed even in a field-free region; the phase can be calculated in terms of the vector potential **A** and shows that the potentials themselves can play an important role.

We have stressed that a given configuration of electric and magnetic fields can be derived from an arbitrarily large number of different EM potentials, ϕ and **A**. This fact even carries over to the case where there are no magnetic fields present. The restriction

$$\nabla \times \mathbf{A}(\mathbf{r}, t) = \mathbf{B}(\mathbf{r}, t) = 0 \tag{18.177}$$

in a field-free region can still be satisfied by the gradient of any scalar function, namely, $\mathbf{A}(\mathbf{r}, t) = \nabla f(\mathbf{r}, t)$; this fact is also consistent with the gauge transformation Eqns (18.30) and (18.31) starting with a vanishing **A**. This relation can be

---

[30] For a review of the physical principles and clinical applications of MRI, see Partain *et al.* (1988).

inverted to give

$$f(\mathbf{r}, t) = \int^{\mathbf{r}} d\mathbf{r}' \cdot \mathbf{A}(\mathbf{r}', t) \qquad (18.178)$$

so that the phase factor connecting a free particle wavefunction to its gauge equivalent partner can be written as

$$\psi'(\mathbf{r}, t) = e^{iqf(\mathbf{r},t)/\hbar} [\psi_{\text{free}}(\mathbf{r}, t)] = \exp\left(iq\int^{\mathbf{r}} d\mathbf{r}' \cdot \mathbf{A}(\mathbf{r}', t)/\hbar\right) \psi_{\text{free}}(\mathbf{r}, t)$$

$$(18.179)$$

The phase factor is seemingly arbitrary as it depends on the choice of gauge, and is unmeasurable by itself as $|\psi'|^2 = |\psi|^2$.

We have already been reminded in Sections 18.1 and 18.7.1 that interference experiments are necessary to probe the relative phases of wavefunctions so, before proceeding further, we recall the wave physics behind electron (or any wave) interference experiments as illustrated in Fig. 18.21. The wavefunctions from the source, S, travel different path lengths and acquire different phase factors. When the electron beams are recombined the wave amplitude can be expressed as

$$\psi^O_{tot} = \psi^O_1 + \psi^O_2 = \psi^S_1 e^{ikl_1} + \psi^S_2 e^{ikl_2} = \left(\psi^S_1 + \psi^S_2 e^{i\theta}\right) e^{ikl_1} \qquad (18.180)$$

where

$$\theta = k(l_2 - l_1) = k\Delta l = \frac{2\pi}{\lambda}\Delta l \qquad (18.181)$$

This implies the familiar result that a difference in path length equal to an integral number of wavelengths will give constructive interference.

Aharonov and Bohm[31] suggested a conceptually simple version of such an interference experiment to probe the quantum phase induced by the vector potential which is illustrated in Fig. 18.22. A small region of magnetic field is present (in the circular area shown), but the particles move essentially classically

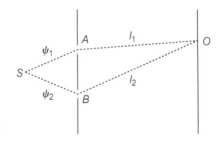

Figure 18.21. Path length geometry for classical wave interference.

[31] See Aharonov and Bohm (1957).

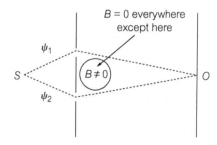

Figure 18.22. Schematic representation of Aharonov–Bohm experiments.

in field-free regions along the paths SAO and SBO. The geometrical path length along SAO and SBO are identical, but there can still be a phase difference in the two paths due to a variation in the vector potential traversed, despite the fact that neither particle is in a magnetic field. Similarly to Eqn. (18.180), we have

$$\psi_{tot}^O = \psi_1^O + \psi_2^O = \psi_1^S e^{i\theta_1} + \psi_2^S e^{i\theta_2} = \left(\psi_1^S + \psi_2^S e^{i(\theta_2-\theta_1)}\right) e^{i\theta_1} \quad (18.182)$$

where

$$\theta_1 = \frac{q}{\hbar}\int_{SAO}^{\mathbf{r}} d\mathbf{r}' \cdot \mathbf{A}(\mathbf{r}', t) \quad \text{and} \quad \theta_2 = \frac{q}{\hbar}\int_{SBO}^{\mathbf{r}} d\mathbf{r}' \cdot \mathbf{A}(\mathbf{r}', t) \quad (18.183)$$

The phase difference which can give rise to interference effects can be written as

$$\Delta\theta = \theta_2 - \theta_1$$

$$= \frac{q}{\hbar}\left(\int_{SAO}^{\mathbf{r}} d\mathbf{r}' \cdot \mathbf{A}(\mathbf{r}', t) - \int_{SBO}^{\mathbf{r}} d\mathbf{r}' \cdot \mathbf{A}(\mathbf{r}', t)\right)$$

$$= \frac{q}{\hbar}\oint_{SAOBS} d\mathbf{r}' \cdot \mathbf{A}(\mathbf{r}', t) \quad (18.184)$$

The line integral of $\mathbf{A}$ around the closed path $SAOBS$ can be rewritten using Stoke's theorem in the form of an area integral

$$\Delta\theta = \frac{q}{\hbar}\int_{area} \nabla' \times \mathbf{A}(\mathbf{r}', t) \cdot d\mathbf{S} = \frac{q}{\hbar}\int_{area} \mathbf{B} \cdot d\mathbf{S} = \frac{q}{\hbar}\Phi_B \quad (18.185)$$

where Φ_B is the magnetic flux enclosed by the path. This result is striking as it says that

- The phase of the wavefunction does depend on the (gauge-dependent) vector potential $\mathbf{A}$ in a nontrivial way, even in a region where the physical magnetic field vanishes.

- That phase dependence gives rise to observable interference effects.

- The quantity which actually determines the interference pattern, however, is the enclosed magnetic flux, which is a perfectly gauge invariant quantity.

The experimental verification of this prediction was first performed[32] by the use of a micron size iron "whisker" containing the field.

18.9 **Questions and Problems**

Q18.1. The word "gauge" is most often associated with a standard or scale of measurement, especially of length. Examples include the thickness of wire, the distance between rails of a railway, and the like. How do you think that such a word came to be applied to the transformations of EM fields in Eqns (18.30) and (18.31)? Any good encyclopedia of science (or the web) may provide the historical background.

Q18.2. In the discussion of the Stark effect, we found that only states with $m = 0$ contributed to the second-order perturbation theory result. Discuss why this should be so, concentrating on the relationship between the perturbation eE_0z and $\hat{L}_z$.

Q18.3. Suppose one wished to calculate the Stark shifts for a hydrogen-like ion. How would form of the perturbation change, and what additional approximations, if any, would one have to make? By what factor would the polarizability for the ground state of such ions change, that is, how would Eqn. (18.78) scale with Z? How would you attempt to evaluate the polarizability of the helium atom?

Q18.4. Evaluate the "generalized slope" of the best fit line in log–log plot in Fig. 17.15, and discuss the relationship between polarizability and ionization potential that it implies.

Q18.5. Develop a sports metaphor for multiple photon ionization by considering how a single individual might not be able to throw (kick, hit) a ball through a given long distance, but that a team of individuals might, and how the likelihood would scale with team size.

Q18.6. One is often told that a microwave oven works along these lines:

- A time-dependent external electric field interacts with the permanent dipole moment of the water molecules in the food causing them to rotate.

- The resulting rotational kinetic energy is transferred to neighboring molecules, resulting in the desired increase in overall temperature.

For an oven operating at $f \approx 2.5$ GHz, evaluate the energy of a single microwave photon, and compare it to the minimum energy necessary to excite a typical rotational state as discussed in Section 16.2.2. Based on your answer, decide whether the microwave oven is a classical or a quantum device, and then explain briefly how you think it works.

[32] See Chambers (1960).

Q18.7. The energy levels of the Rydberg states of lithium shown in Fig. 18.12 are labeled on the vertical axis as "*Energy* (cm^{-1})". Show that the numerical values given there corresponding to the $n = 29, 28, 27$ states (for vanishing magnetic field) are consistent with what you know about the Balmer formula.

Q18.8. What kind of spectroscopic evidence would show that sunspots are correlated with regions of high magnetic field activity?

Q18.9. It was mentioned that the 21 cm hyperfine line of hydrogen might be useful as a "universal" standard of frequency. The Pioneer-10 and -11 spacecraft carried engraved plaques which were designed to be the first material artifacts of mankind designed to escape the solar system carrying a message (as opposed, for example, to EM transmissions). These "cosmic greeting cards" specified all distances and times in terms of this frequency.[33] What do you think the message consisted of? What would you have included in such a message? What fundamental concepts in physics would you think are important and possible to communicate to another intelligent civilization?

Q18.10. How would one use NMR phenomena as the basis for an imaging technique as in MRI? How does one know where in the body the RF photons from the external field were absorbed?

P18.1. **Gravitational versus electrostatic forces.** (a) Calculate the ratio of the gravitational force to the electrostatic force between a proton and an electron.

(b) A rough limit on any possible difference between the magnitudes of the electron and proton charges can be inferred from cosmology; we will use the notation $|Q_p + Q_e| = \Delta e$ for any such difference. The electrostatic repulsion between two (supposedly neutral) hydrogen atoms could, in principle, overwhelm their gravitational attraction if Δe were too large; such a repulsion could make the universe expand at a rate which is much larger than observed. Show that demanding that any such repulsive force be less than 10% of the mutual gravitational attraction of two hydrogen atoms implies that $\Delta e/e \lesssim 10^{-18}$.

P18.2. **Neutron wave optics.**[34] (a) What is the wavelength λ and wave number k for thermal neutrons, that is, ones for which $E = p^2/2m = k_B T/2$ with $T = 300$ K?

(b) The behavior of neutrons in matter can be characterized by an effective *index of refraction* given by

$$n = \sqrt{1 - \frac{4\pi Nb}{k^2}} \qquad (18.186)$$

where N is the number of nuclei per unit volume and b is the so-called *scattering length*. Estimate the quantity $1 - n$ for thermal neutrons if $b \approx 5\ F$ and

---

[33] For a brief review, see Robinett (2001).

[34] For a review of applications of neutron scattering to solid state physics, see Dobbrzynski and Blinowski (1994).

$N \approx 10^{29} \text{ m}^{-3}$.

(c) Using (b), estimate the "glancing angle" at which thermal neutrons incident on matter will experience total internal reflection. Can you imagine how this effect is used to provide a "clean" neutron beam far from a "dirty" reactor? Hint: How do fiber optic cables work?

(d) Neutrons are projected horizontally from a height H with velocity v and follow parabolic trajectories. Show that the critical height, H_c, at which all of the neutrons "skid" along the surface because of internal reflection is given by

$$H_c = \frac{2\pi m_n^2}{g\hbar^2} Nb \tag{18.187}$$

This effect is the basis for the *neutron-gravity refractometer*.

P18.3. Photon properties.

(a) How many photons per second enter your eye from a 100 W lightbulb one meter away; estimate the size of your iris in such a situation.

(b) What is the number density of photons (n_γ/m^3) one kilometer away from a 50,000 W AM radio station operating at $f = 1000$ kHz; assume that the antenna radiates its energy uniformly (which is not a very good approximation.)

(c) What is the radiation pressure (in N/m^2) due to sunlight at the earth's surface; the solar luminosity is roughly 4×10^{26} W. Compare it to atmospheric pressure. Why do you "feel" the sunlight's energy but not its momentum?

(d) Solar radiation pressure is responsible for "sweeping" the solar system clean of dust particles below a certain size; estimate the radius of dust particles for which the radiation force overcomes the gravitational attraction of the sun. Is this pressure large enough to "push" anything else around, even in space? (See Clarke (1972) and Wright (1992) for some ideas on this subject.)

P18.4. Show that the vector potential for a uniform field in an arbitrary direction, $\mathbf{B}_0$, can be written as $\mathbf{A} = -\mathbf{r} \times \mathbf{B}_0/2$.

P18.5. Show that the combined effects of a gauge transformation on the Hamiltonian via Eqn. (18.41) and on the wavefunction via Eqn. (18.42) reproduces the original Schrödinger equation.

P18.6. For a system described by the Hamiltonian

$$\hat{H} = \frac{1}{2\mu} \left(\mathbf{p} - q\mathbf{A} \right)^2 \tag{18.188}$$

show that the probability current which satisfies the equation of continuity,

$$\frac{\partial}{\partial t} \left(\psi^*(\mathbf{r}, t)\psi(\mathbf{r}, t) \right) + \nabla \cdot \mathbf{J}(\mathbf{r}, t) = 0 \tag{18.189}$$

is given by

$$\mathbf{J} = \frac{\hbar}{2\mu i}\left(\psi^*\nabla\psi - \psi\nabla\psi^* - \frac{2iq}{\hbar}\mathbf{A}|\psi|^2\right) \tag{18.190}$$

P18.7. EM potentials, quantum mechanics, and the Lorentz force.

(a) Using the quantum mechanical Hamiltonian in Eqn. (18.39), show that the time-dependence of the expectation value of position is given by

$$m\frac{d}{dt}\langle x\rangle_t = \langle\hat{p}_x - qA_x\rangle_t \equiv \langle\hat{\Pi}_x\rangle_t \tag{18.191}$$

Hint: Use the general result in Eqn. (12.88), but note that since x does not depend explicitly on time, only the commutator term is relevant. The combination $\hat{\Pi} = \hat{p} - \mathbf{A}$ is the appropriate mechanical momentum operator for problems with EM potentials.

(b) This last result seems to suggest that $d\langle x\rangle_t/dt$ might depend on the particular vector potential, $\mathbf{A}$, used to evaluate the expectation value. First show that $\langle x\rangle_t$ is invariant under gauge transformations given by Eqns (18.30), (18.31), and especially (18.42), because $|\psi'|^2 = |\psi|^2$. Then show that $d\langle x\rangle_t/dt$ is also invariant, but in a more interesting way.

(c) Show that the Hamiltonian can then be written in the form

$$\hat{H} = \frac{1}{2m}(\hat{p} - q\mathbf{A})^2 + q\phi \equiv \frac{\hat{\Pi}^2}{2m} + q\phi \tag{18.192}$$

but examine the commutation relations among the $\hat{\Pi}_{x,y,z}$. For example, show that $[\hat{\Pi}_x, \hat{\Pi}_y] = iq\hbar B_z$.

(d) Finally, show that

$$m\frac{d^2}{dt^2}\langle x\rangle_t = \frac{d}{dt}\langle\hat{p}_x - qA_x\rangle_t = q\left\langle E_x + \frac{1}{2m}\left\{\left(\hat{\Pi}\times\mathbf{B}\right)_x - \left(\mathbf{B}\times\hat{\Pi}\right)_x\right\}\right\rangle_t \tag{18.193}$$

using the definitions of the **E** and **B** fields in terms of the potentials in Eqns (18.28) and (18.29). Hint: In this case, the vector potential **A** *does* depend on time, so include both terms in Eqn. (12.88). Discuss why this is the appropriate generalization of the classical Lorentz force equation. Repeat part (b) to show that $d^2\langle x\rangle_t/dt^2$ is invariant under gauge transformations.

P18.8. Second-order Stark shift.

(a) Evaluate the contribution to the second-order Stark shift for hydrogen in Eqn. (18.72) from the $n = 2$ and $n = 3$ states. How much of the *total* second-order shift in Eqn. (18.78) is due to these two states?

(b) One can show that the contribution from each (bound) $\psi_{n,1,0}$ state to the ground state Stark shift in Eqn. 18.72) is given by

$$-\left[\frac{(2n)^9(n-1)^{2n-6}}{3(n+1)^{2n+6}}\right]\frac{E_0^2 a_0^3}{K} \tag{18.194}$$

Compare your answers to part (a) for $n = 2, 3$ with this general expression. Write a short computer program to sum these contributions until you feel you have converged to a stable answer. What fraction of the *total* shift is due to the bound states alone?

(c) Evaluate the matrix element in Eqn. (18.85).

P18.9. Stark effect for electric field in arbitrary direction. In Section 18.4.2 we considered a uniform electric field in a specific direction, $V(\mathbf{r}) = e\mathcal{E}z$. Suppose the field is in an arbitrary direction, namely, $V(\mathbf{r}) = e\mathcal{E}\cdot\mathbf{r}$ where $\mathcal{E} = (\mathcal{E}_x, \mathcal{E}_y, \mathcal{E}_z)$.

(a) Why must the results for the first- and second-order shifts for the ground state be the same as in Eqns (18.71) and (18.78)?

(b) Show explicitly that this is so. Use the relations

$$x = -\sqrt{\frac{2\pi}{3}} r \left(Y_{1,1}(\theta,\phi) - Y_{1,-1}(\theta,\phi) \right) \tag{18.195}$$

$$y = i\sqrt{\frac{2\pi}{3}} r \left(Y_{1,1}(\theta,\phi) + Y_{1,-1}(\theta,\phi) \right) \tag{18.196}$$

(c) What happens to the shifts in the first excited states? Is the pattern of splitting the same? Are the linear combinations which split the same?

(d) Explicitly work out the splitting of the first excited state for a uniform field of the form $V(\mathbf{r}) = e\mathcal{E}x$

P18.10. Variational method for Stark effect. Use the variational method to estimate the ground state energy for the Stark Hamiltonian $\hat{H} = \hat{H}_{\text{Coul}} + e\mathcal{E}z$ using the trial wavefunction

$$\psi(r) = \cos(\theta)\,\psi_{1,0,0}(\mathbf{r}) + \sin(\theta)\,\psi_{2,1,0}(\mathbf{r}) \tag{18.197}$$

with θ as the variational parameter. Evaluate the energy shift due to the electric field by calculating $\Delta E = E(\mathcal{E} \neq 0) - E(\mathcal{E} = 0)$, show that it is proportional to $\mathcal{E}^2$, and evaluate α using Eqn. (18.67).

P18.11. Work out the pattern of level splittings for the Stark effect for the $n = 3$ case of hydrogen. You will have to consider the nine possible states corresponding to s, p, d states.

P18.12. Stark effect for a three-dimensional harmonic oscillator.

(a) Consider a three-dimensional isotropic oscillator with potential

$$V(x, y, z) = \frac{1}{2}K(x^2 + y^2 + z^2) \tag{18.198}$$

with an external field given by $V(z) = e\mathcal{E}z$. Show that the energy eigenvalues and wavefunctions can be obtained exactly by generalizing P9.9.

(b) Use your results to obtain the polarization, α, for the ground state.

(c) What happens to the first and second excited state? Is there a linear Stark effect for any set of levels? Give a reason for your answer.

(d) If the oscillator is nonisotropic, so that

$$V(x, y, z) = \frac{1}{2}(K_x x^2 + K_y y^2 + K_z z^2) \qquad (18.199)$$

show that the problem is still soluble exactly. Show that the energy shift due to the external field in an arbitrary direction, $V(\mathbf{r}) = e\mathbf{E} \cdot \mathbf{r}$, can be written in the form

$$\Delta E = -\frac{1}{2}\mathcal{E} \cdot \alpha \cdot \mathcal{E} \qquad (18.200)$$

where α is now a *polarization tensor*.

P18.13. Consider a particle of mass m and charge e in a three-dimensional cubical box of side L. Estimate the dipole polarizability of the particle in the ground state. Is there a first-order Stark effect for the first excited state (which is triply degenerate)?

P18.14. Use the data in Fig. 18.6 to estimate the polarizability of the (nondegenerate) $15s$ state. Compare the value you obtain with the ground state value given by $\alpha = 9a_0^3/2K$ and explain any differences.

P18.15. Use the tunneling formula for field emission in Section 11.4.1 as a very rough guide to estimate the lifetime, due to tunneling, of the electron in a hydrogen atom placed in an external electric field $\mathcal{E}$. (This amounts to neglecting the effect of the Coulomb attraction.) Show that this gives

$$\tau_{tunnel} \sim \tau_H \exp\left(\frac{4\sqrt{2}}{3}\sqrt{\frac{m_e W}{\hbar^2}}\frac{W}{e\mathcal{E}}\right) \qquad (18.201)$$

where τ_H is a characteristic atomic timescale. Obtain a numerical estimate of the lifetime by using $W \sim E_1 \sim 13.6 \text{ eV}, \mathcal{E} = 2000 \text{ V/m}$, and τ_H as the classical orbital period for hydrogen.

P18.16. An electron in an external electric field has the potential energy function

$$V(\mathbf{r}) = -\frac{Ke^2}{r} - e\mathcal{E}z \qquad (18.202)$$

Use this to argue that a Rydberg atom in a state of effective quantum number n^* will be ionized by an external field of strength

$$\mathcal{E} = \frac{E_c}{16(n^*)^4} \qquad (18.203)$$

where E_c is the "typical" atomic field in Eqn. (18.22).

P18.17. Multiphoton ionization.

(a) Imagine that the values below have been obtained in an experiment looking for multiphoton ionization of hydrogen atoms from their ground state; they give the yield of emitted electrons versus the laser intensity, I.

I (TW/cm^2)	Electrons (10^{10})
4	3
7	18
20	300
60	8000

From the "data" (which is not from any particular experiment), estimate the value of k in a power-law fit to the data of the form

$$\text{electron yield} \propto (\text{laser power})^k \tag{18.204}$$

(b) Estimate how many photons are required for multiphoton ionization of hydrogen if the photon wavelength used was $\lambda = 2480\text{Å}$ and compare to your answer in part (a).

P18.18. Uniform magnetic field in different gauges. The consider the problem of a charged particle in a uniform field in the $+z$ direction described by the vector potential $\mathbf{A} = (0, xB_0, 0)$.

(a) Show that the Hamiltonian can be written in the form

$$\hat{H} = \frac{\hat{p}_x^2}{2\mu} + \frac{(qB_0)^2}{2\mu}x^2 + \frac{\hat{p}_y^2}{2\mu} - \frac{qB_0}{\mu}\hat{p}_y + \frac{\hat{p}_z^2}{2\mu} \tag{18.205}$$

(b) Try a solution of the form

$$\psi(x, y, z) = X(x)\, e^{i(k_y y + k_z z)} \tag{18.206}$$

and show that the equation for $X(x)$ reduces to that for a shifted harmonic oscillator in as P9.9.

(c) Find the quantized energy eigenvalues, and show that they give the same spectrum as Eqn. (18.100) with the same degeneracy.

(d) The wavefunctions corresponding to this solution, and those for the case where $\mathbf{A} = B_0/2(-y, x, 0)$ should be related by a simple phase, as in Eqn. (18.42). Is it easy to see this relationship?

P18.19. Two-dimensional harmonic oscillator in a magnetic field. A particle of mass μ and charge $q > 0$ moves in a plane subject to a two-dimensional isotropic harmonic oscillator potential, $V(x, y) = K(x^2 + y^2)/2$. A uniform magnetic field in the $+z$ direction is also applied. Using the methods of Section 18.5, find the allowed energy eigenvalues.

P18.20. Velocity selectors. The motion of a charged particle in combined (but constant) electric E_0 and magnetic B_0 fields can be quite complicated in general. For a particle with initial velocity given by

$$v_{cross} = \frac{E_0 \times B_0}{|B_0|^2} \qquad (18.207)$$

the electric and magnetic forces cancel, and the particle moves in a straight line trajectory. This is the basis for a classical "velocity selector." Solve for the energy eigenvalue spectrum and wavefunctions for the quantum version of this problem for uniform electric and magnetic fields in the $-x$ and $+z$ directions, respectively; use the potentials

$$\phi(r) = E_0 x \quad \text{and} \quad A(r) = (0, xB_0, 0) \qquad (18.208)$$

Discuss to what extent any "velocity selection effect" is still present in the quantum system.

P18.21. Hydrogen atom in a different gauge. One almost always solves the hydrogen atom problem by (implicitly) using the gauge

$$\phi(r, t) = \frac{Ke}{r} \quad \text{and} \quad A(r, t) = 0 \qquad (18.209)$$

to describe the Coulomb field of the proton.

(a) Show that this configuration can also be derived by the potentials

$$\phi(r, t) = 0 \quad \text{and} \quad A(r, t) = -\frac{Ke\,r}{r^2}t \qquad (18.210)$$

and find the gauge function $f(r, t)$ which connects it to the choice in (a).

(b) Set up the corresponding Hamiltonian, $\hat{H}'$, and show explicitly that

$$\psi'_{nlm} = \exp(iqf(r, t)/\hbar)\psi_{nlm} \qquad (18.211)$$

solves the Schrödinger equation in the new gauge.

(c) Discuss how the semi-classical arguments in Section 1.4 are changed in this case; recall that when there is a vector potential, one has $m\,v = p - q\,A$.

P18.22. The "anomalous" Zeeman effect. If we include the effect of electron spin and its coupling to the external magnetic field in the Zeeman effect, we need to consider the Hamiltonian

$$\hat{H}_1 = \frac{eB}{2m_e}(L_z + 2S_z) \qquad (18.212)$$

if we assume that $g_e \approx 2.0$ for the electron. Evaluate the expectation value of this term in the appropriate coupled states in P16.16 for both $j = l + 1/2$ and $j = l - 1/2$ and show that your result can be expressed in the form

$$\Delta E_B^{(1)} = \frac{e\hbar B}{m_e}\frac{m_j}{(2l+1)}\begin{cases} l+1 & \text{for } j = l + 1/2 \\ l & \text{for } j = l - 1/2 \end{cases} \qquad (18.213)$$

P18.23. Diamagnetism. The quadratic term in the Hamiltonian for a uniform magnetic field in the standard gauge is given by

$$\frac{e^2 B^2}{8m}(x^2 + y^2) \tag{18.214}$$

(a) Using perturbation theory, show that the corresponding change in energy is

$$\Delta E = \frac{e^2 B^2}{12m}\langle r^2 \rangle \tag{18.215}$$

for spherically symmetric distributions.

(c) Using the definition, $\Delta E = -\boldsymbol{\mu} \cdot \mathbf{B}$ to show that the induced magnetic dipole is

$$\mu = -\frac{e^2 \langle r^2 \rangle}{6m} B \tag{18.216}$$

An induced magnetic moment opposite to the applied field is termed *diamagnetic*.

(d) Can you derive this result classically by looking at the magnetic moment induced by the charges rotating in the uniform field?

P18.24. Islands of isomerism. Using the nuclear shell energy level diagram in Fig. 18.15, explain why there are many long-lived excited states (so-called *isomers*) for odd-A nuclei for values of either N or Z *just below* the observed magic numbers. Hint: The rate for radiative decays between two energy levels is proportional to $(kR)^{2l+1}$ where l is the change in angular momentum between the initial and final state; $E_i - E_f = \Delta E = E_\gamma = \hbar kc$ is the photon energy in the transition.

P18.25. Use the results of P16.25 to show that the "magic numbers" for the three-dimensional harmonic oscillator potential are $2, 8, 20, 40, 70, 112, 168, \ldots$, and so forth. Hint: Use the degeneracy of the spectrum to find the *total* number of spin-1/2 particles which can be accommodated upon the closing of each shell. Since the harmonic oscillator has such a simple spectrum, the shell structure is obvious.

P18.26. (a) Use the expression in Eqn. (18.142) to find the numerical value of $\Delta E, f$, and λ for the energy, frequency, and wavelength of the photon emitted in the $^3S_1 \rightarrow {}^3S_0$ hyperfine transition in hydrogen.

(b) Generalize your result to find the corresponding quantity for deuterium. You will need to know that (i) the deuteron magnetic moment is $\mu_D = 0.8798\mu_N$, (ii) $M_D \approx 2m_p$, and (iii) the deuteron spin is $S = 1$.

P18.27. Use the data in Fig. 18.18 to show that the angle θ through which the neutron magnetic moment must precess to change its phase by -1 is $352 \pm 19°$. Use the following numerical values:

1. The observed "period" in the magnetic path length is $\int_{\text{path}} \mathbf{B} \cdot d\mathbf{l} = 1.44 \pm 0.08 \ T \cdot m$.

2. The gyromagnetic ratio of the neutron is $g_n = -1.93$.

3. The neutron magnetic moment is given by $g_n e\hbar/2m_n$.

4. The wavelength of the neutrons used was 1.82 ± 0.01 Å. This combined with the mass of the neutron 1.67×10^{-27} kg gives the neutron velocity v.

P18.28. Flux quantization. Consider an electron moving in the magnetic field geometry of the Aharonov–Bohm effect in Section 18.8. Show that upon making one complete circuit around the localized magnetic field region that the electron wavefunction acquires a phase given by $\exp(ie\Phi/\hbar)$. Use the fact that the electron wavefunction must be single valued at any point to show that this implies that magnetic flux must be quantized, that is,

$$\Phi_n = \left(\frac{2\pi \hbar}{e}\right) n \quad \text{where } n = 0, 1, 2, \ldots \tag{18.217}$$

NINETEEN

Scattering in Three Dimensions

While much of our knowledge of the microscopic world has been gleaned from bound state problems (e.g. the spectroscopy of atoms, molecules, nuclei, etc.), scattering experiments have also provided a wealth of information on the nature of the constituents of matter and their interactions on atomic scales and below.[1] Because much of the formalism of scattering is similar in the classical and quantum formulations, we begin by discussing some generalities shared by the two descriptions.

A typical scattering experiment has the layout already shown schematically in Fig. 11.1. A beam of incident scatterers with a given flux or intensity (number of particles per unit area per unit time) impinges on some target, described here by a scattering potential; this flux can be written as

$$J_{\text{inc}} = \frac{dN_{\text{inc}}}{dA\,dt} \tag{19.1}$$

Classically, individual scatterers are deflected along well-defined trajectories, while in the quantum case the "scattered" wavefunction gives probabilistic information on the likelihood of finding a particle scattered through a given angle. In either case, the number of particles per unit time which are detected in a small region of solid angle, $d\Omega$, located at a given angular deflection specified by (θ, ϕ), can be counted; this can be written as

$$\frac{dN_{\text{sc}}}{d\Omega\,dt} \tag{19.2}$$

The *differential cross-section* for scattering is defined by the ratio of these two quantities via

$$\frac{d\sigma}{d\Omega}(\theta, \phi) = \left(\frac{dN_{\text{sc}}}{d\Omega\,dt}\right) \Big/ \left(\frac{dN_{\text{inc}}}{dA\,dt}\right) \tag{19.3}$$

[1] The sentence you just read, for example, was "detected" by performing a scattering experiment involving optical photons and a system of biological analog detectors.

From this definition, it is clear that $d\sigma/d\Omega$ has the dimensions of an area. The *total cross-section* corresponds to scatterings through any angle and is given by

$$\sigma = \int d\Omega \, \frac{d\sigma}{d\Omega} = \int_0^{2\pi} d\phi \int_0^\pi \sin(\theta) \, d\theta \, \frac{d\sigma}{d\Omega}(\theta, \phi) \qquad (19.4)$$

19.1 Classical Trajectories and Cross-Sections

For a particle obeying classical mechanics, the trajectory for the unbound motion corresponding to a scattering event is deterministically predictable, given the interaction potential and the initial conditions; the path of any scatterer in the incident beam can be followed, in principle, and its angular deflection determined as precisely as required. (We ignore complications arising from the possible sensitive dependence on initial conditions seen in classically chaotic systems.)

For a central force, any given trajectory will be planar (why?) and we illustrate two paths corresponding to slightly different initial conditions in Fig. 19.1; as the z-axis is often defined to be the "beam line," we show events in the $z - y$ plane, with the center of the (here repulsive) scattering potential at the origin. The equations of motion for any incident particle can be integrated (perhaps numerically, if necessary) to find the trajectory via

$$m\ddot{\mathbf{r}}(t) = \mathbf{F}(r) \qquad (19.5)$$

where we use the initial conditions

$$y(t = -\infty) = b \quad \dot{y}(t = -\infty) = 0$$

$$z(t = -\infty) = -\infty \quad v_\infty \equiv \dot{z}(t = -\infty) = \sqrt{2mE} \qquad (19.6)$$

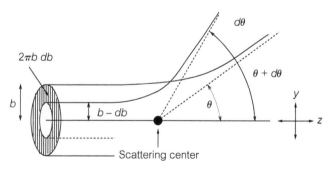

Figure 19.1. Scattering trajectories corresponding to different impact parameters, b, give different scattering angles, θ. All of the particles in the beam in the hatched region of area $d\sigma = 2\pi \, b \, db$ are scattered into the angular region $(\theta, \theta + d\theta)$.

where we have used the fact that the initial kinetic energy is given by $E = mv_\infty^2/2$. In this language, b is called the *impact parameter*, and would be the distance of closest approach to the scattering center in the absence of any force.

In Fig. 19.1, two trajectories, differing by a small amount db, and scattering into angles separated by $d\theta$ are shown. The number of particles scattered per unit time into the angular region $(\theta, \theta + d\theta)$ with any value of ϕ can be written as

$$\frac{dN_{sc}}{dt} = d\sigma \, \frac{dN_{inc}}{dA \, dt} \quad \text{where} \quad d\sigma = 2\pi \, b \, db \tag{19.7}$$

which we can also write as

$$\frac{d\sigma}{d\theta} = 2\pi \, b(\theta) \left| \frac{db(\theta)}{d\theta} \right| \tag{19.8}$$

The absolute value is required since θ increases (larger angle scattering) as $b(\theta)$ decreases.

Since we have "integrated" over all ϕ values, in this case $d\Omega = 2\pi \sin(\theta) d\theta$, and using the definition of Eqn. (19.3), we find that

$$\frac{d\sigma}{d\Omega} = \left[\frac{1}{2\pi \sin(\theta)} \right] 2\pi \, b(\theta) \left| \frac{db(\theta)}{d\theta} \right| = \frac{b(\theta)}{\sin(\theta)} \left| \frac{db(\theta)}{d\theta} \right| \tag{19.9}$$

Knowledge of $b(\theta)$, obtained directly from Newton's laws or other methods, is then sufficient to calculate the scattering cross-section. At this stage, the relation of one initial parameter, namely, $b = y(t = -\infty)$, on the scattering process, that is, $\theta(b)$, is evident; the influence of $v_\infty = \dot{z}(t = -\infty)$ or, equivalently, the energy E is less explicit, but arises through the equations of motion to give $\theta(b)$ as well.

Example 19.1. "Specular" scattering from a hard sphere

A simple case which can be treated using only geometrical methods involves the scattering of small, light particles from a heavy, impenetrable sphere of radius R. A classical trajectory is shown in Fig. 19.2 where the incident particles experience equal angle ($\alpha = \alpha_{inc} = \alpha_{ref}$) scattering relative to the tangent plane at the intersection point; hence the name "specular" or mirror-like. Instead of using an exclusively geometrical approach, however, we will determine $\theta(b)$ using a more 'functional' approach which can be generalized to other problems (P19.1–P19.2).

From the diagram it is clear that the scattering angle satisfies $\theta = 2\phi$ while $\tan(\phi) = |dy/dz|$, evaluated at $y = b$ where b is the impact parameter. The spherical scattering

(Continued)

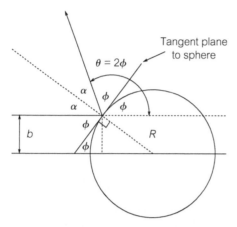

Figure 19.2. Geometry for "specular" scattering from a hard sphere.

surface is determined by $x^2 + y^2 + z^2 = R^2$ so that in the $y - z$ plane we have

$$\tan\left(\frac{\theta}{2}\right) = \tan(\phi) = \left|\frac{dy}{dz}\right| = \left|\frac{z}{y}\right| = \frac{\sqrt{R^2 - b^2}}{b} \tag{19.10}$$

Solving for $b(\theta)$ gives

$$b(\theta) = R\cos\left(\frac{\theta}{2}\right) \quad \text{so that} \quad \left|\frac{db}{d\theta}\right| = \frac{R}{2}\sin\left(\frac{\theta}{2}\right) \tag{19.11}$$

Taken together, and using the fact that $\sin(\theta) = 2\sin(\theta/2)\cos(\theta/2)$, these imply that

$$\frac{d\sigma}{d\theta} = \pi R^2 \sin\left(\frac{\theta}{2}\right)\cos\left(\frac{\theta}{2}\right) = \frac{\pi R^2}{2}\sin(\theta) \tag{19.12}$$

or equivalently

$$\frac{d\sigma}{d\Omega} = \frac{R^2}{4} \tag{19.13}$$

The total cross-section is

$$\sigma = \int d\Omega \frac{d\sigma}{d\Omega} = \frac{R^2}{4}\int d\Omega = \pi R^2 \tag{19.14}$$

which is just the effective cross-sectional area presented to the scatters by the sphere; this is just the so-called *geometrical cross-section* of the object. A cross-section which is independent of polar angle θ, as in Eqn. (19.13), is called *isotropic*, and is consistent with the simple spherical shape of the scatterer in this example.

(Continued)

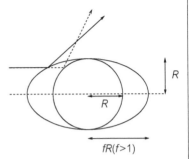

Figure 19.3. Ellipsoidal hard scatters.

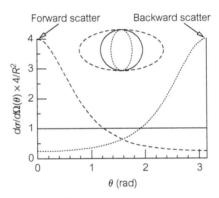

Figure 19.4. Angular dependence of differential cross-sections for various ellipsoidal shapes. Spherical or $f = 1$ (solid), and ellipsoidal shapes with $f > 1$ (dashed) and $f < 1$ (dotted) are illustrated. Values of $f = 1, f > 1$, and $f < 1$ give isotropic, forward-peaked, and backward-peaked scattering, respectively.

A less trivial case is evident when one considers scattering from ellipsoids as illustrated in Fig. 19.3. The length of the axis along the z direction is multiplied by a dimensionless factor f while the transverse (x, y) dimensions are unchanged (so as to keep the geometrical cross-section presented to the incident beam the same). The surface of the ellipsoid is now determined by the relation $x^2 + y^2 + z^2/f^2 = R^2$ and one can show (P19.1) that

$$b(\theta) = R\frac{1}{\sqrt{1 + f^2 \tan^2(\theta/2)}} \tag{19.15}$$

corresponding to the differential cross-section

$$\frac{d\sigma}{d\Omega} = \frac{R^2}{4}\left[\frac{2f}{1 + f^2 + (1 - f^2)\cos(\theta)}\right]^2 \tag{19.16}$$

which still satisfies $\sigma = \int d\Omega \, (d\sigma/d\Omega) = \pi R^2$ as intended. The resulting angular dependence is shown in Fig. 19.4 for various values of f; we note that $f > 1$ ($f < 1$) corresponds to more forward (backward) scattering consistent with more glancing (large angle) collisions than with a sphere.

While this is a very artificial example (there is no energy dependence, for example), it does illustrate how the angular dependence of the scattering cross-section can give information on the form of the scattering potential, in this case the "shape" of the scatterers.

For scattering from nontrivial central forces, the trajectory can be obtained from the equations of motion or by using energy methods; the latter case is often more direct as we do not require detailed knowledge of the time-dependence of the coordinates, that is $r(t), \theta(t)$, but rather the path in space, that is $r(\theta)$. For example, one can rewrite

$$E = \frac{1}{2}m\dot{r}^2 + \frac{L^2}{2mr^2} + V(r) \tag{19.17}$$

in the form

$$\sqrt{\frac{2}{m}\left(E - \frac{L^2}{2mr^2} - V(r)\right)} = \frac{dr}{dt} = \frac{dr}{d\theta}\frac{d\theta}{dt} = \frac{dr}{d\theta}\dot{\theta} \tag{19.18}$$

The angular momentum in the process can be written as $L = mr^2\dot{\theta}$, and one can write

$$d\theta = \left(\frac{L}{r^2\sqrt{2m(E - L^2/2mr^2 - V(r))}}\right) dr \tag{19.19}$$

so that the angle through which the particle moves as it travels between two radial distances is

$$\Delta\theta = \int_{r_1}^{r_2} dr \frac{L}{r^2\sqrt{2m(E - L^2/2mr^2 - V(r))}} \tag{19.20}$$

For a scattering process, we have the situation pictured in Fig. 19.5. The angle Θ is the angular deflection experienced by the particle as it moves from infinity to the distance of closest approach, r_{min}; since the scattering event is symmetrical about this point, the net scattering angle is easily seen to satisfy

$$2\Theta + \theta = \pi \quad \text{or} \quad \Theta = \frac{\pi}{2} - \frac{\theta}{2} \tag{19.21}$$

The angular momentum can be determined by the impact parameter from the relation

$$L = mv_\infty b \tag{19.22}$$

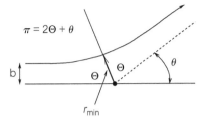

Figure 19.5. Scattering trajectories in a central potential showing the distance of closest approach, r_{min}.

which can be written in terms of the total energy since $E = mv_\infty^2/2$ giving

$$L = b\sqrt{2mE} \tag{19.23}$$

Using Eqn. (19.19), one has

$$\Theta = \int_{r_{min}}^{\infty} dr \, \frac{b}{r^2\sqrt{1 - b^2/r^2 - V(r)/E}} \tag{19.24}$$

This makes it clear that the scattering trajectories are determined by the energy E, the angular momentum L (via b), and the form of the potential. If this can be evaluated explicitly, one obtains $b(\Theta) = b((\pi-\theta)/2)$, and one can then evaluate the differential cross-section using Eqn. (19.9). The integral can be evaluated in closed form for only a few special cases, and it is lucky that the inverse square law force is one of them.

Example 19.2. Classical Coulomb scattering

The potential corresponding to the scattering of charged particles via Coulomb's law can be written in the general form

$$V_C(r) = \frac{A}{r} \tag{19.25}$$

where the constant $A = \pm Z_1 Z_2 K e^2$ allows for the electrostatic interaction of like sign and opposite sign point charges of arbitrary magnitudes. The necessary integral in Eqn. (19.24) has the form

$$\Theta = \int_{r_{min}}^{\infty} dr \, \frac{b}{r\sqrt{r^2 - (A/E)r - b^2}} \tag{19.26}$$

where r_{min} is determined by the vanishing of the denominator, namely

$$r_{min} = \sqrt{b^2 + (A/2E)^2} + (A/2E) \tag{19.27}$$

(Continued)

We note that for large energies ($E \rightarrow \infty$) or vanishing charges ($A \rightarrow 0$), one has $r_{min} \rightarrow b$ as expected. The integral can be evaluated using the results of Appendix D.1 and using trigonometric identities, one finds

$$\cos(\Theta) = \frac{a}{\sqrt{a^2 + b^2}} \tag{19.28}$$

where $a \equiv A/2E$. This corresponds to

$$b(\theta) = a \cot\left(\frac{\theta}{2}\right) \quad \text{and} \quad \frac{db(\theta)}{d\theta} = -\frac{a}{2\sin^2(\theta/2)} \tag{19.29}$$

One then finds that

$$\frac{d\sigma}{d\theta} = 2\pi b(\theta)\left|\frac{db(\theta)}{d\theta}\right| = \pi a^2 \frac{\cos(\theta/2)}{\sin^3(\theta/2)} \tag{19.30}$$

or

$$\frac{d\sigma}{d\Omega} = \frac{a^2}{4}\frac{1}{\sin^4(\theta/2)} = \left[\frac{KZ_1Z_2e^2}{4E}\right]^2 \frac{1}{\sin^4(\theta/2)} \tag{19.31}$$

This result is called the *Rutherford formula*, and exhibits the famous $1/\sin^4(\theta/2)$ dependence which is characteristic of Coulomb scattering from a point-like charge.

If one attempts to evaluate the total cross-section using Eqn. (19.4), one obtains an infinite result. This divergence is due to the essentially infinite range of the $1/r$ potential, and the "infinity" in the integral comes from scatterings as $\theta \rightarrow 0$; from Eqn. (19.29) we can see that this corresponds to arbitrarily large impact parameters where the Coulomb potential can just "tickle" the incoming particle, causing a scatter through an arbitrarily small angle, but still contribute to the total cross-section. This infinity is not, however, realized in nature due to the overall electrical neutrality of matter. If, for example, we scatter electrons from charged nuclei then for sufficiently large values of b, the beam electrons are outside the atomic electron cloud, the system appears electrically neutral, and Coulomb scattering from the nucleus is shielded.

Given the special nature of the $1/r$ potential, it may not be surprising that the classical formula can be evaluated exactly; what is more unexpected is that the Rutherford formula also emerges from the quantum treatment as well. (See Section 19.3.)

19.2 **Quantum Scattering**

19.2.1 **Cross-Section and Flux**

While it is possible to describe quantum scattering in three dimensions using wave packets, we will instead use an approach similar to that discussed extensively in 1D, namely, the scattering of plane wave states. In three dimensions, the incident plane wave impinges on the scattering potential as in Fig. 11.1, giving rise to a spherical wavefront. Plane wave solutions representing incident particles of momentum **p** will have the form

$$\psi_{inc}(\mathbf{r}) = \frac{1}{L^{3/2}} e^{i\mathbf{p}\cdot\mathbf{r}/\hbar} = \frac{1}{L^{3/2}} e^{i\mathbf{k}\cdot\mathbf{r}} \tag{19.32}$$

where we include the (arbitrary) factor $1/L^{3/2}$ to denote the dimensions of a three-dimensional wavefunction. We will often assume that the beam of scatterers is incident along the z direction, so that $\mathbf{k}\cdot\mathbf{r} = kz = kr\cos(\theta)$.

The scattered wave will have the form

$$\psi_{sc}(\mathbf{r}) = \frac{1}{L^{3/2}}\left[f(\hat{\mathbf{n}} = \hat{\mathbf{r}})\frac{e^{ikr}}{r}\right] = \frac{1}{L^{3/2}}f(\theta,\phi)\frac{e^{ikr}}{r} \tag{19.33}$$

where $\hat{\mathbf{n}} = \hat{\mathbf{r}}$ denotes a unit vector in the direction of the detector located at angular position (θ,ϕ). The quantity $f(\theta,\phi)$ has the dimensions of length and is called the *scattering amplitude*; we will see that it determines the scattering cross-section. The total wavefunction has the form

$$\psi(\mathbf{r}) = \psi_{inc}(\mathbf{r}) + \psi_{sc}(\mathbf{r}) = \frac{1}{L^{3/2}}\left(e^{i\mathbf{k}\cdot\mathbf{r}} + f(\theta,\phi)\frac{e^{ikr}}{r}\right) \tag{19.34}$$

The flux of incident and scattered particles can be related to the probability flux for this wavefunction, namely

$$\mathbf{j}(\mathbf{r}) = \frac{\hbar}{2mi}\left(\psi^*(\mathbf{r})\nabla\psi(\mathbf{r}) - \psi(\mathbf{r})\nabla\psi^*(\mathbf{r})\right) \tag{19.35}$$

where we can evaluate the two terms in

$$\nabla\psi(\mathbf{r}) = \nabla\psi_{inc}(\mathbf{r}) + \nabla\psi_{sc}(\mathbf{r}) \tag{19.36}$$

separately. We clearly have

$$\nabla\psi_{inc}(\mathbf{r}) = \frac{1}{L^{3/2}}\left[i\mathbf{k}e^{i\mathbf{k}\cdot\mathbf{r}}\right] \tag{19.37}$$

while using the expression for the gradient operator in spherical coordinates we find

$$\nabla\psi_{sc}(\mathbf{r}) = \frac{1}{L^{3/2}}\left[\hat{\mathbf{r}}f(\theta,\phi)\left\{ik\frac{e^{ikr}}{r} - \frac{e^{ikr}}{r^2}\right\} + \hat{\theta}\frac{1}{r}\frac{\partial f(\theta,\phi)}{\partial\theta}\frac{e^{ikr}}{r}\right.$$
$$\left. + \hat{\phi}\frac{1}{r}\frac{\partial f(\theta,\phi)}{\partial\phi}\frac{e^{ikr}}{r}\right] \tag{19.38}$$

where we have used a standard expression for the gradient operator in terms of spherical coordinates when acting on $\psi_{sc}(\mathbf{r})$.

That part of the flux from Eqn. (19.35) which contains only $\psi_{inc}(\mathbf{r})$ has the familiar form

$$\mathbf{j}_{inc} = \frac{1}{L^3}\frac{\hbar k}{m} = \frac{1}{L^3}\mathbf{v} \tag{19.39}$$

along the initial direction specified by $\mathbf{k}$ or the velocity $\mathbf{v}$. This quantity has the dimensions of number of particles per unit area per unit time, and we identify it with Eqn. (19.1). The flux corresponding to $\psi_{sc}(\mathbf{r})$ has the form

$$\mathbf{j}_{sc} = \hat{\mathbf{r}}\frac{1}{L^3}\left(\frac{p}{m}\right)\frac{|f(\theta,\phi)|}{r^2} + \mathcal{O}\left(\frac{1}{r^3}\right) \tag{19.40}$$

where we will be able to eventually neglect the terms of order $1/r^3$ or higher. The number of particles which are scattered into a small area of detector per unit time is given by

$$\mathbf{j}_{sc} \cdot d\mathbf{A} = \left[\hat{\mathbf{r}}\frac{1}{L^3}\left(\frac{p}{m}\right)\frac{|f(\theta,\phi)|}{r^2}\right] \cdot [\hat{\mathbf{r}}\,r^2\,d\Omega]$$

$$= \frac{1}{L^3}\left(\frac{p}{m}\right)|f(\theta,\phi)|^2\,d\Omega \tag{19.41}$$

where r and $d\Omega$ are the distance to and the angle subtended by the detector, respectively. This implies that

$$\frac{dN_{sc}}{d\Omega\,dt} = \frac{1}{L^3}\frac{p}{m}|f(\theta,\phi)|^2 \tag{19.42}$$

and we see that any terms which drop off as $1/r^3$ or faster will not contribute to a measurement at macroscopic distances r; this justifies the neglect of such terms in Eqn. (19.40). There are also cross-terms in Eqn. (19.35) which arise due to the "interference" between ψ_{inc} and ψ_{sc}. They all contain factors of the form

$$e^{\pm ikr(1-\cos(\theta))} \tag{19.43}$$

and we note that for a detector far away from the scattering center, the factor in the exponential satisfies $kr >>> 1$. This implies that the phase of these contributions oscillates very rapidly, and the positive and negative, real and imaginary parts average to zero.[2]

Finally, using the definition of cross-section in Eqn. (19.3) and Eqns (19.39) and (19.42), we find that the differential cross-section is simply given by the

---

[2] This is not true, of course, for $\theta \to 0$, but then this contribution is not counted as having been scattered anyway.

scattering amplitude via

$$\frac{d\sigma(\theta,\phi)}{d\Omega} = |f(\theta,\phi)|^2 \tag{19.44}$$

The problem of determining the cross-section in quantum mechanics then reduces to finding $f(\theta,\phi)$ for a given incident energy and scattering potential.

19.2.2 Wave Equation for Scattering and the Born Approximation

The information on scattering probabilities encoded in $f(\theta,\phi)$ is presumably contained in the large r behavior of the Schrödinger wavefunction for unbound states. One approach to the evaluation of the scattering amplitude involves rewriting the Schrödinger equation in a form which "builds in" as much of the information on the initial conditions of a scattering problem as possible. We can trivially rewrite

$$\left(-\frac{\hbar^2}{2m}\nabla^2 + V(\mathbf{r})\right)\psi(\mathbf{r}) = E\psi(\mathbf{r}) \tag{19.45}$$

in the form

$$\left(\nabla^2 + k^2\right)\psi(\mathbf{r}) = \left(\frac{2m}{\hbar^2}\right)V(\mathbf{r})\psi(\mathbf{r}) \tag{19.46}$$

where $k^2 = 2mE$, and which is more reminiscent of the classical wave equation. Using standard techniques for the study of such differential equations, we can rewrite Eqn. (19.46) as an *integral equation* in the form

$$\psi(\mathbf{r}) = e^{i\mathbf{k}\cdot\mathbf{r}} - \frac{m}{2\pi\hbar^2}\int \frac{e^{ik|\mathbf{r}-\mathbf{r}'|}}{|\mathbf{r}-\mathbf{r}'|}V(\mathbf{r}')\,\psi(\mathbf{r}')\,d\mathbf{r}' \tag{19.47}$$

which is called the *Lippman–Schwinger equation*.

We will not describe the manipulations by which Eqn. (19.47) can be derived, but will rather content ourselves in showing that it is equivalent to Eqn. (19.46); this can be accomplished by acting on both sides of Eqn. (19.47) with the "wave operator" $\nabla^2 + k^2$ and reproducing Eqn. (19.46). The first term in Eqn. (19.47) clearly satisfies

$$\left(\nabla^2 + k^2\right)e^{i\mathbf{k}\cdot\mathbf{r}} = (-k^2 + k^2)e^{i\mathbf{k}\cdot\mathbf{r}} = 0 \tag{19.48}$$

The "wave operator" acts on the second term inside the integral sign and requires evaluation of

$$\left(\nabla^2 + k^2\right)\left(\frac{e^{ik|\mathbf{r}-\mathbf{r}'|}}{|\mathbf{r}-\mathbf{r}'|}\right) \tag{19.49}$$

This, in turn, means we need to calculate

$$\nabla^2 \left(\frac{e^{ikr}}{r} \right) = -k^2 \left(\frac{e^{ikr}}{r} \right) + e^{ikr} \nabla^2 \left(\frac{1}{r} \right) \tag{19.50}$$

A standard way to "derive" the action of the gradient squared on the function $1/r$ is to appeal to a familiar result from electrostatics. The electric field of a point charge (located at the origin for definiteness) will certainly satisfy Gauss's law in the form

$$\nabla \cdot \mathbf{E}(\mathbf{r}) = -\frac{1}{\epsilon_0} \rho(\mathbf{r}) = -\frac{1}{\epsilon_0} q \delta(\mathbf{r}) \tag{19.51}$$

where the three-dimensional delta function can be understood to mean

$$\delta(\mathbf{r}) = \delta(x) \, \delta(y) \, \delta(z) \tag{19.52}$$

and this form is consistent with the charge density of truly point-like charge. The Coulomb field can be written in the form

$$\frac{q}{4\pi\epsilon_0} \frac{\hat{\mathbf{r}}}{r^2} = \mathbf{E}(\mathbf{r}) = -\nabla\phi(\mathbf{r}) = -\nabla\left(\frac{q}{4\pi\epsilon_0} \frac{1}{r} \right) \tag{19.53}$$

which when combined with Eqn. (19.51) yields

$$\nabla^2 \left(\frac{1}{r} \right) = -4\pi \, \delta(\mathbf{r}) \tag{19.54}$$

We then find that

$$\left(\nabla^2 + k^2 \right) \left(\frac{e^{ik|\mathbf{r}-\mathbf{r}'|}}{|\mathbf{r}-\mathbf{r}'|} \right) = -4\pi \, e^{ik|\mathbf{r}-\mathbf{r}'|} \delta(\mathbf{r}-\mathbf{r}') = -4\pi \, \delta(\mathbf{r}-\mathbf{r}') \tag{19.55}$$

This implies that

$$\left(\nabla^2 + k^2 \right) \psi(\mathbf{r}) = -\frac{m}{2\pi\hbar^2} (-4\pi) \int \delta(\mathbf{r}-\mathbf{r}') \, V(\mathbf{r}') \, \psi(\mathbf{r}') \, d\mathbf{r}'$$

$$= \left(\frac{2m}{\hbar^2} \right) V(\mathbf{r}) \, \psi(\mathbf{r}) \tag{19.56}$$

as hoped.

The usefulness of Eqn. (19.47) in extracting the scattering amplitude can be seen by expanding the exponential in the limit $|\mathbf{r}| \gg |\mathbf{r}'|$, that is, at an observation point far from the region where the scattering potential is acting. One has

$$\frac{e^{ik|\mathbf{r}-\mathbf{r}'|}}{|\mathbf{r}-\mathbf{r}'|} \approx \frac{e^{ikr}}{r} e^{-ik\hat{\mathbf{r}}\cdot\mathbf{r}'} = \frac{e^{ikr}}{r} e^{-i\mathbf{k}'\cdot\mathbf{r}'} \tag{19.57}$$

where $\mathbf{k}' \equiv k\hat{\mathbf{r}}$ is the wave vector in the direction of the detected scattered particle. We thus have

$$\psi(\mathbf{r}) \stackrel{r\,\text{large}}{\longrightarrow} e^{i\mathbf{k}\cdot\mathbf{r}} - \left[\frac{m}{2\pi\hbar^2}\int e^{-i\mathbf{k}'\cdot\mathbf{r}'}\,V(\mathbf{r}')\,\psi(\mathbf{r}')\,d\mathbf{r}'\right]\frac{e^{ikr}}{r} \qquad (19.58)$$

which is exactly of the form Eqn. (19.34), with the scattering amplitude given by

$$f(\theta,\phi) = -\frac{m}{2\pi\hbar^2}\int e^{-i\mathbf{k}'\cdot\mathbf{r}}\,V(\mathbf{r})\,\psi(\mathbf{r})\,d\mathbf{r} \qquad (19.59)$$

where we have dropped the prime label on the integration variable for convenience; the dependence on the scattering angles, (θ,ϕ), is contained in the scattered wave vector, $\mathbf{k}' = k\hat{\mathbf{r}}$.

The evaluation of the scattering amplitude via Eqn. (19.59) still requires knowledge of the *exact* unbound wavefunction, $\psi(\mathbf{r})$, for a given energy in the field of the potential. A formal solution of Eqn. (19.47) which lends itself to a systematic approximation method can be obtained by iteration as follows:

- The incident plane wave will be the solution in the absence of the scattering potential, so label

$$\psi^{(0)}(\mathbf{r}) = e^{i\mathbf{k}\cdot\mathbf{r}} \qquad (19.60)$$

as the zero-th order solution.

- Substitute this into the integral on the right-hand side of Eqn. (19.47); writing

$$G(\mathbf{r},\mathbf{r}') = -\frac{m}{2\pi\hbar^2}\frac{e^{ik|\mathbf{r}-\mathbf{r}'|}}{|\mathbf{r}-\mathbf{r}'|} \qquad (19.61)$$

for simplicity, this yields the "next best guess"

$$\psi^{(1)}(\mathbf{r}) = \psi^{(0)}(\mathbf{r}) + \int G(\mathbf{r},\mathbf{r}')\,\psi^{(0)}(\mathbf{r}')\,V(\mathbf{r}')\,d\mathbf{r}' \qquad (19.62)$$

- Iterate by using $\psi^{(1)}$ to obtain $\psi^{(2)}$ and so forth; the solution can then be written as an infinite series in the form

$$\psi(\mathbf{r}) = \psi^{(0)}(\mathbf{r}) + \int d\mathbf{r}'\,G(\mathbf{r},\mathbf{r}')\,V(\mathbf{r}')\,\psi^{(0)}(\mathbf{r}')$$

$$+ \int d\mathbf{r}'\int d\mathbf{r}''\,G(\mathbf{r},\mathbf{r}')\,V(\mathbf{r}')\,G(\mathbf{r}',\mathbf{r}'')\,V(\mathbf{r}'')\,\psi^{(0)}(\mathbf{r}'')$$

$$+ \cdots \qquad (19.63)$$

If the scattering potential is sufficiently weak,[3] then the first nontrivial iteration is used; this is called the *Born approximation* and gives the scattering

---

[3] See Saxon (1968) for a discussion of the conditions under which this is a valid assumption.

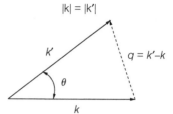

Figure 19.6. Incident (k) and scattered (k') wave numbers showing momentum transfer, $\Delta p = \hbar q = \hbar(k' - k)$.

amplitude as

$$f_B(\theta, \phi) = -\frac{m}{2\pi \hbar^2} \int e^{-i\mathbf{k}' \cdot \mathbf{r}} V(\mathbf{r}) \, e^{i\mathbf{k} \cdot \mathbf{r}} \, d\mathbf{r}$$

$$= -\frac{m}{2\pi \hbar^2} \int e^{-i\mathbf{q} \cdot \mathbf{r}} V(\mathbf{r}) \, d\mathbf{r} \tag{19.64}$$

The vector $\mathbf{q} \equiv \mathbf{k}' - \mathbf{k}$ is related to the momentum change of the particle as a result of the collision since $\mathbf{q} = (\mathbf{p}' - \mathbf{p})/\hbar$; to this order of approximation, the scattering amplitude is simply the (three-dimensional) Fourier transform of the potential, $V(\mathbf{r})$, with respect to $\mathbf{q}$. For central potentials for which there is no interesting ϕ dependence, it is useful to write $q = |\mathbf{q}|$ in terms of the scattering angle using the geometry in Fig. 19.6. We find that

$$q^2 = (\mathbf{k}' - \mathbf{k})^2 = 2k^2 - 2\mathbf{k}' \cdot \mathbf{k}$$

$$= 2k^2(1 - \cos(\theta))$$

$$= 4k^2 \sin^2(\theta/2) \tag{19.65}$$

or

$$q = 2k \sin(\theta/2) = q_{max} \sin(\theta/2) \tag{19.66}$$

with $q_{max} = 2k = 2\sqrt{2mE}/\hbar$; the angular dependence of the scattering angle depends only on q, so we often write $f(\theta, \phi) = f(q)$. In this case, we can choose the polar axis to lie along the z direction so that $\mathbf{q} \cdot \mathbf{r} = qr \cos(\theta)$ and

$$\int e^{-i\mathbf{q} \cdot \mathbf{r}} V(r) \, d\mathbf{r} = \int_0^\infty dr \, r^2 \, V(r) \int_0^{2\pi} d\phi \int_0^\pi d\theta \, \sin(\theta) e^{-iqr \cos(\theta)}$$

$$= \frac{2\pi}{iq} \int_0^\infty dr \, r \, V(r) \left(e^{iqr} - e^{-iqr} \right)$$

$$= \frac{4\pi}{q} \int_0^\infty dr \, r \, \sin(qr) \, V(r) \tag{19.67}$$

or

$$f_B(q) = -\frac{2m}{\hbar^2 q} \int_0^\infty dr \, r \, \sin(qr) \, V(r) \tag{19.68}$$

which is sometimes useful.

Example 19.3. Scattering from a finite well

As an example, let us calculate the scattering cross-section for the finite attractive well in three dimensions, defined via

$$V(r) = \begin{cases} -V_0 & \text{for } r < a \\ 0 & \text{for } r > a \end{cases} \tag{19.69}$$

using the Born approximation; besides illustrating several rather general aspects of scattering theory, this potential has some relevance to the (short range) nucleon–nucleon potential as noted in Section 8.3.2.

The scattering amplitude from Eqn. (19.68) is given by

$$\begin{aligned}
f_B(q) &= \frac{2mV_0}{\hbar^2 q} \int_0^a r \sin(qr)\, dr \\
&= \frac{2mV_0 a^3}{\hbar^2} \left(\frac{\sin(qa)}{(qa)^3} - \frac{\cos(qa)}{(qa)^2} \right) \\
&= \frac{2mV_0 a^3}{3\hbar^2} \left(3 \frac{j_1(z)}{z} \right) \\
&= \left(\frac{2mV_0 a^3}{3\hbar^2} \right) G(z)
\end{aligned} \tag{19.70}$$

where $z = qa$ and $f_B(q)$ can be seen to have the appropriate dimensions. We have made use of the fact that the result can be written in terms of the spherical Bessel function since

$$j_1(z) = \frac{\sin(z)}{z^2} - \frac{\cos(z)}{z} \tag{19.71}$$

and written $G(z)$ in such a way that $G(0) = 1$ for convenience.

The differential cross-section is simply given by

$$\frac{d\sigma}{d\Omega}(\theta) = |f_B(q)|^2 = \left(\frac{2mV_0 a^3}{3\hbar^2} \right)^2 |G(z)|^2 \tag{19.72}$$

where $z = aq_{max} \sin(\theta/2)$. We exhibit the angular dependence of the scattering in Fig. 19.7 for several values of $z_{max} = aq_{max} = 2a\sqrt{2mE}/\hbar$ corresponding to low and high energies; z_{max} is an appropriate figure of merit as its definition implies that

$$E = \frac{z_{max}^2 \hbar^2}{8ma^2} \tag{19.73}$$

and it measures E relative to a typical bound state energy scale. The plot in Fig. 19.7 shows some typical behavior of the differential cross-section for scattering from a finite range potential, and we make some general comments based on this one example.

(Continued)

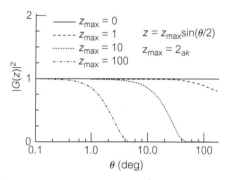

Figure 19.7. Plot of differential cross-section (via $|G(z)|^2$) versus angle θ for scattering from finite well of radius a, for various values of ka, showing forward diffractive peak.

- **Low energies:** For low energies, the scattering is uniform in angle implying an isotropic cross-section. This result can be seen more generally from Eqn. (19.64) when we note that low energies implies that $\mathbf{q} \cdot \mathbf{r} \to 0$ so that

$$f_B(q) \approx -\frac{m}{2\pi\hbar^2} \int V(r)\, d\mathbf{r} \approx -\frac{m}{2\pi\hbar^2} V(0)\, \Delta^3 r \qquad (19.74)$$

where $\Delta^3 r$ is roughly the volume over which the potential is nonvanishing; in our example this gives the exact result, namely

$$f_B(q) = \frac{m}{2\pi\hbar^2} V_0 \left(\frac{4\pi}{3} a^3 \right) = \frac{2mV_0 a^3}{3\hbar^2} \qquad (19.75)$$

More generally, $f_b(q)$ is independent of angle for low energies.

- **High energies:** For high energies, the cross-section obviously becomes more and more peaked in the forward ($\theta \approx 0$) direction, and the total cross-section (given, recall, by the "area" under the $d\sigma/d\Omega$ versus θ curve weighted by $2\pi \sin(\theta)$) becomes smaller. Both of these effects can also be understood rather generally from Eqn. (19.64).

  — When the argument of the exponential phase, $2ka \sin(\theta/2)$, becomes large, the phase factor oscillates rapidly and the contribution to the total integral from that region becomes small due to cancellations; a rough cutoff for when this happens is when

$$2ka \sin\left(\frac{\theta}{2}\right) \longrightarrow ka\theta \approx 1 \qquad (19.76)$$

  As the energy increases, so that $ka \gg 1$, this occurs for smaller and smaller angles, and the differential cross-section becomes nonnegligible only for angles satisfying

$$\theta_{max} \lesssim \frac{1}{ka} \qquad (19.77)$$

(Continued)

This effect is reminiscent of classical wave diffraction, and the forward peak is often called the main diffractive peak for this reason. The location of the first zero of $G(z)$, determined by $j_1(z) = 0$ can be used to specify the first diffraction minimum; one finds $2ka \sin(\theta/2) \approx 4.49$. Not coincidentally, the condition for the diffraction pattern for circular apertures in optics (familiar from problems involving the design of optical instruments) is quite similar to this relation (P19.5).

— The decrease of the total cross-section with energy can now be understood by integrating $d\sigma/d\Omega$; one has roughly

$$\sigma = \int d\Omega \, \frac{d\sigma}{d\Omega} \approx 2\pi \int_0^{1/ka} |f(\theta)|^2 \, \sin(\theta) \, d\theta$$

$$\approx 2\pi |f(0)|^2 \int_0^{1/ka} \theta \, d\theta$$

$$\approx \frac{\pi |f(0)|^2}{a^2 k^2}$$

$$\sigma \propto \frac{1}{E}. \tag{19.78}$$

19.3 **Electromagnetic Scattering**

Because of its importance, we consider the scattering of a charged particle via the Coulomb potential separately. Much of the symmetry of the $1/r^2$ force carries over from bound state problems to scattering, yielding simple, closed form solutions for the cross-section for an inverse square law force; deviations from that result can then signal the presence of new physics as did the splitting of degenerate energy levels in the hydrogen atom spectrum.

The long range of the $1/r$ potential means that many of the standard formulations of scattering theory are not directly applicable.[4] We will be content with quoting some standard results using the Born approximation.

In order to calculate the scattering amplitude for the Coulomb potential, it is necessary to introduce a convergence factor and write

$$V_C(r) = \frac{Ke^2}{r} \quad \longrightarrow \quad V_C(r; \mu) = \frac{Ke^2}{r} e^{-\mu r} \tag{19.79}$$

for the interaction of two like sign charges $\pm e$. This form allows us to evaluate the Fourier transform explicitly, and then let $\mu \to 0$ at the end of the calculation.

[4] For a very thorough discussion and further references, see Schiff (1968) and Sakurai (1994).

The Born approximation then gives

$$f_B(\theta) = f_B(q) = -\left(\frac{2m}{\hbar^2}\right)\frac{Ke^2}{q}\int_0^\infty dr\, e^{-\mu r}\sin(qr)$$

$$= -\left(\frac{2mKe^2}{\hbar^2}\right)\frac{1}{\mu^2 + q^2}$$

$$\longrightarrow -\left(\frac{2mKe^2}{\hbar^2}\right)\frac{1}{q^2} \quad \text{as } \mu \to 0 \tag{19.80}$$

The differential cross-section can then be written as

$$\frac{d\sigma}{d\Omega} = \left(\frac{2mKe^2}{\hbar^2 q^2}\right)^2 = \left(\frac{Ke^2}{4E}\right)^2 \frac{1}{\sin^4(\theta/2)} \tag{19.81}$$

using the definitions of $q = 2k\sin(\theta/2)$ and $\hbar k = \sqrt{2mE}$. Somewhat amazingly, this is the same *Rutherford cross-section* result obtained using purely classical techniques in Example 19.2, and some comments can be made:

- The fact that this cross-section does not contain any explicit factors of $\hbar$ when expressed in terms of the observable energy E is special to the case of Coulomb scattering, and is also suggestive of the unique correspondence to the classical result.

- The result is easily generalized to the scattering of charges $Z_1 e$ by a potential of charge $Z_2 e$ by simply letting $e^2 \to Z_1 Z_2 e^2$, and has been verified for many systems; data for scattering of oxygen nuclei ($Z_1 = 8$) on gold nuclei ($Z_2 = 79$) is shown in Fig. 19.8 confirming the characteristic $1/\sin^4(\theta/2)$ dependence.

- The exact expression for the Coulomb scattering amplitude can be derived by solving the Schrödinger equation in parabolic coordinates[5] with the result

$$f_C(\theta) = f_B(\theta)\, e^{-i\eta\ln(\sin^2(\theta/2))+2i\beta} \tag{19.82}$$

where

$$\eta \equiv \frac{mZ_1 Z_2 Ke^2}{\hbar^2 k} = \frac{Z_1 Z_2 Ke^2}{\hbar v} \tag{19.83}$$

where v is the velocity of the particle. The additional phase factor is given in terms of η via

$$e^{2i\beta} = \frac{\Gamma(1 + i\eta)}{\Gamma(1 - i\eta)} \tag{19.84}$$

---

[5] See, for example, Schiff (1968).

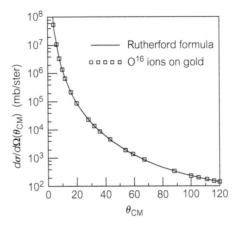

Figure 19.8. Differential cross-section for Coulomb scattering of oxygen on gold nuclei illustrating the Rutherford formula; data taken from Bromley, Kuehner, and Almqvist (1961).

where $\Gamma(x)$ is the generalized factorial function discussed in Appendix E.9. The η factor *does* depend explicitly on $\hbar$, and so is a true quantum effect, but it does not seem to have any observable effects as $|f_C(\theta)|^2 = |f_B(\theta)|^2$. Interference effects, such as occur in the scattering of identical particles, can show the effects of such phases, and we will discuss this in Section 19.5.2.

- It is interesting to note that the same (exact) result can be used to give the Coulomb wavefunction in the other limit, namely, $r \approx 0$. One obtains

$$|\psi_C(r \to 0)|^2 \propto \frac{2\eta\pi}{v(e^{2\eta\pi} - 1)} \tag{19.85}$$

For very slow particles, corresponding to $\eta \gg 1$, we find different behavior for the attractive ($\eta < 0$) and repulsive ($\eta > 0$) case, namely

$$|\psi_C(0)|^2 \approx \begin{cases} \frac{2\pi|\eta|}{v} & \text{for } \eta < 0 \\ \frac{2\pi\eta}{v}e^{-2\pi\eta} & \text{for } \eta > 0 \end{cases} \tag{19.86}$$

The exponential suppression for like sign charges is simply the Gamow tunneling factor derived in Section 11.4.3.

Another case of interest is the scattering of a point charge (here of magnitude e) from a distribution of charge which is not point-like (but with total charge still equal to e); a classic example is the scattering of electrons from charged nuclei where the fact that the nuclear charge is "spread out" over a finite size becomes apparent. This situation can be handled by recalling that the potential of a point charge in the field of a nontrivial charge density can be written in the form

$$V(r) = Ke^2 \int \frac{\rho(r')\,dr'}{|r - r'|} \tag{19.87}$$

where we have written the scattering charge density as $e\rho(r)$; $\rho(r)$ is normalized such that

$$\int dr'\rho(r') = 1 \tag{19.88}$$

or Z for an object with larger charge ($Q = Ze$); a point charge corresponds to $\rho(r) = \delta(r)$, and Eqn. (19.87) reproduces Coulomb's law in that limit.

If we substitute this into the Born amplitude we require the Fourier transform

$$\int dr\, e^{-iq\cdot r} \left[\int dr' \frac{\rho(r')}{|r - r'|} \right] = \left[\int dr'\, \rho(r')\, e^{-iq\cdot r'} \right] \left[\int dr \frac{e^{-iq\cdot(r-r')}}{|r - r'|} \right]$$

$$\equiv F(q) \left[\int ds\, \frac{1}{s}\, e^{-iq\cdot s} \right] \tag{19.89}$$

where we have written $s = r - r'$ for convenience. The last term in Eqn. (19.89) is simply the Fourier transform of the Coulomb potential which, using regularization tricks, we have evaluated above. The first term, defined via

$$F(q) = \int dr\, \rho(r)\, e^{-iq\cdot r} \tag{19.90}$$

is the Fourier transform of the (dimensionless) charge density, and it describes the modification to the Rutherford cross-section due to the "smearing out" of the charge; $F(q)$ is called the *form factor*. The two terms are seem to "factorize" from each other, so that one can write

$$\frac{d\sigma}{d\Omega} = \left(\frac{d\sigma}{d\Omega} \right)_R |F(q)|^2 \tag{19.91}$$

so that $|F(q)|$ can be measured as the ratio of the observed cross-section to the Rutherford prediction, $(d\sigma/d\Omega)_R$, as one varies q. Several comments can be made:

- The charge density is normalized such that

$$F(0) = \int dr\, \rho(r) = 1 \tag{19.92}$$

The physical interpretation of this result is that small values of q correspond to very long wavelengths, and such probes "see" only the total charge of the scatterer and cannot resolve its structure; small q can also arise if $\theta \approx 0$ and no scattering occurs.

- We can expand $F(\mathbf{q})$ for small q and find

$$
F(\mathbf{q}) = \int d\mathbf{r}\, \rho(\mathbf{r})\, e^{-i\mathbf{q}\cdot\mathbf{r}}
$$

$$
= \int \rho(\mathbf{r})\, d\mathbf{r} - i \sum_j q_j \int x_j \rho(\mathbf{r})
$$

$$
- \frac{1}{2} \sum_{j,k} q_j q_k \int x_j\, x_k\, \rho(\mathbf{r})\, d\mathbf{r} + \cdots
$$

$$
= 1 - i \sum_j q_j \langle x_j \rangle - \frac{1}{2} \sum_{j,k} q_j\, q_k \langle x_j x_k \rangle + \cdots \tag{19.93}
$$

where

$$
\langle f(\mathbf{r}) \rangle = \int f(\mathbf{r})\, \rho(\mathbf{r})\, d\mathbf{r} \tag{19.94}
$$

For a spherically symmetric charge distribution with $\rho(\mathbf{r}) = \rho(r)$, the average values satisfy

$$
\langle x_j \rangle = 0 \quad \text{and} \quad \langle x_j x_k \rangle = \frac{1}{3}\langle r^2 \rangle \delta_{j,k} \tag{19.95}
$$

since $\langle x^2 \rangle = \langle y^2 \rangle = \langle z^2 \rangle$. This implies that

$$
F(\mathbf{q}) = F(q) = 1 - \frac{1}{6} q^2 \langle r^2 \rangle + \cdots \tag{19.96}
$$

and knowledge of the form factor for small but finite q values gives information on the spatial extent of the charge distribution, it's "charge radius," $\langle r^2 \rangle$.

Example 19.4. Nuclear sizes

Much of the information on the size of nuclei has been obtained using the experimental determination of form factors from electron–nucleus scattering experiments. As a simple model of the charge distribution of a nucleus, consider the density given by

$$
\rho(\mathbf{r}) = \rho(r) = \begin{cases} 3/4\pi R^3 & \text{for } r < R \\ 0 & \text{for } r > R \end{cases} \tag{19.97}
$$

that is, a constant nuclear charge density up to sharply defined nuclear radius, R. The form factor for this distribution is given by the same Fourier transform considered in Example 19.3 and we find that

$$
F(q) = \frac{3 j_1(kR)}{qR} = 3\left(\frac{\sin(qR)}{(qR)^3} - \frac{\cos(qR)}{(qR)^2} \right) \tag{19.98}
$$

(Continued)

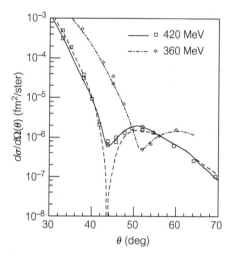

Figure 19.9. Differential cross-section for electron scattering on oxygen nuclei. Note that the first "diffraction dip" moves outward in angle as the energy is decreased. The dashed curve corresponds to a calculation with a nuclear charge density with "sharp edges," while the solid curves are for a more realistic parameterization which is smoother. Adapted from Ehrenberg *et al.* (1959).

The differential cross-section for electron scattering on oxygen (O^{16}) nuclei for two different incident energies is shown in Fig. 19.9 where a pronounced dip is seen at roughly 44° for the 420 MeV data. The dashed curve is the Rutherford cross-section multiplied by the form factor in Eqn. (19.98) with an appropriately chosen value of R. The simple model accounts for many of the gross features of the actual data; the assumption of a "sharp" nuclear edge is responsible for the bad fit near the first diffraction minimum; more realistic nuclear models with a smoother nuclear surface work better and give a less deep diffraction "dip".

To make quantitative contact with the data, we note that electrons of these energies are ultrarelativistic implying that their energies satisfy $E = pc = \hbar k c$. Using the value of the first zero of $j_1(z)$ mentioned above, we find that

$$qR = 4.49 \implies R = \frac{4.49\,\hbar c}{2E\sin(\theta/2)} \approx 2.8\,F \qquad (19.99)$$

A common parameterization of nuclear sizes which gives a good fit to many nuclei is given by $R(A) = (1.2\,F)A^{1/3}$ where A is the atomic mass of the nucleus; here, $A = 16$, which gives $R \approx 3\,F$. The data for lower energy exhibits the same diffraction minimum, but at larger angles, consistent with Eqn. (19.99).

We can also use form factor ideas to understand the shielding of the nuclear charge by its atomic electron cloud, and how this effect gives a total cross-section for scattering via neutral atoms which does not diverge, as the simple Rutherford formula suggests. For example, we might write the charge density for a neutral hydrogen atom in its ground state as

$$e\rho_H(r) = +e\delta(\mathbf{r}) - e|\psi_{1,0,0}(\mathbf{r})|^2 \tag{19.100}$$

where $\psi_{1,0,0}(\mathbf{r})$ is the electron wavefunction; this corresponds to a positive point-like nucleus and "spread-out" negative electron cloud. Using

$$|\psi_{1,0,0}(\mathbf{r})|^2 = \frac{1}{\pi a_0^3} e^{-2r/a_0} \tag{19.101}$$

we find that the form factor is

$$F(q) = 1 - \frac{1}{(1 + (qa_0/2)^2)^2} \tag{19.102}$$

so that

$$|F(q \to 0)|^2 \approx \left| 1 - \left(1 - 2\frac{q^2 a_0^2}{4} + \cdots \right) \right|^2 \approx \frac{q^4 a_0^4}{4} \tag{19.103}$$

When multiplied by the Rutherford cross-section in Eqn. (19.91), this eliminates the $1/\sin^4(\theta/2)$ divergence in the cross-section.

One final comment can be made concerning the scattering of point charges (such as electrons) by other subatomic particles via electromagnetic forces. Besides its charge which gives rise to Rutherford scattering via the Coulomb interaction, a proton, for example, has an intrinsic magnetic moment. An incident electron can then also experience magnetic scattering, calculable classically, for example, via the Lorentz force $\mathbf{F} = -e\mathbf{v} \times \mathbf{B}$. The form of the magnetic potential for a point magnetic dipole,

$$\mathbf{A}(\mathbf{r}) = \frac{\mu_0}{4\pi} \frac{\mathbf{m} \times \mathbf{r}}{r^3} \sim \frac{1}{r^2} \tag{19.104}$$

shows that it is of shorter range ($1/r^2$) than the Coulomb interaction ($1/r$) so that electrostatic interactions will dominate at long distances; on the other hand, this also implies that the magnetic force can come to dominate the scattering if r is small enough. Using Eqn. (19.27), we note that the classical distance of closest approach for an inverse law force can be written

$$r_{min} = \left(\frac{A}{2E} \right) \left[\frac{1}{\sin(\theta/2)} + 1 \right] \tag{19.105}$$

which implies that magnetic scattering can become important for sufficiently large energies and/or scattering angles. This effect is clearly observed in elastic electron–proton scattering where the magnetic contribution comes to dominate the cross-section for $q^2 \gtrsim 2\,\text{GeV}^2$; it is even more important in electron–neutron scattering (at all energies) as it is the dominant effect due to the vanishing charge of the neutron.

19.4 **Partial Wave Expansions**

The methods used so far to calculate scattering amplitudes have the flavor of perturbation theory, and in this section we present an alternative method of evaluating $f(\theta)$ which is based more directly on matching of solutions of the Schrödinger equation with the form of the incident and scattered wave of Eqn. (19.34); it also makes connection somewhat more directly to the notion of classical scattering trajectories being determined by energy and angular momentum values.

We first recall that the solutions of the free particle radial in spherical coordinates,

$$-\frac{\hbar^2}{2m}\left(\frac{d^2R(r)}{dr^2} + \frac{2}{r}\frac{dR(r)}{dr}\right) + \frac{l(l+1)}{2mr^2}R(r) = ER(r) \qquad (19.106)$$

which are well behaved at the origin have the form

$$R_{E,l}(r) = j_l(kr) \quad \longrightarrow \quad \frac{\sin(kr - l\pi/2)}{kr} \qquad (19.107)$$

where $k = \sqrt{2mE}/\hbar$. Far from the interaction point, where the potential is negligible ($V(r) \approx 0$), the scattered wavefunction must also satisfy Eqn. (19.106), but with the more general solution

$$R_{sc}(r) = \alpha_l j_l(kr) + \beta_l n_l(kr) \qquad (19.108)$$

since one is far from the origin, which is the only place where the $n_l(kr)$ are poorly behaved. We can then write the α_l, β_l in a suggestive form and examine the large r limit and find

$$R_{sc}(r) = a_l\left(\cos(\delta_l)j_l(lr) - \sin(\delta_l)n_l(kr)\right)$$

$$\longrightarrow \frac{a_l\left(\cos(\delta_l)\sin(kr - l\pi/2) + \sin(\delta_l)\cos(kr - l\pi/2)\right)}{kr}$$

$$= a_l\left[\frac{\sin(kr - l\pi/2 + \delta_l)}{kr}\right] \qquad (19.109)$$

This result is similar to the one-dimensional scattering examples considered in Section 11.3 where the reflected and transmitted waves differed in amplitude and phase from the incident plane wave. For this reason, δ_l is called the *phase shift of the lth partial wave*. We note that

- Attractive (repulsive) potentials imply that $\delta_l > 0$ ($\delta_l < 0$) corresponding to the wave being "pulled in" ("pushed out") by the scattering center resulting in a phase delay (advance).

The phase shifts can, in principle, be extracted from the exact radial wavefunctions, if known, for each l value by examining their large r behavior which is guaranteed to be of the form in Eqn. (19.108); from that form, one has

$$-\frac{\beta_l}{\alpha_l} = \frac{\sin(\delta_l)}{\cos(\delta_l)} = \tan(\delta_l) \tag{19.110}$$

The complete solution of the scattering wavefunction can then be written for large r in the form

$$\psi(\mathbf{r}) \longrightarrow \sum_{l=0}^{\infty} \sum_{m=-l}^{+l} a_l \left(\frac{\sin(kr - l\pi/2 + \delta_l)}{kr} \right) Y_{l,m}(\theta, \phi)$$

$$\longrightarrow \sum_{l=0}^{\infty} a_l \left(\frac{\sin(kr - l\pi/2 + \delta_l)}{kr} \right) P_l(\cos(\theta)) \tag{19.111}$$

for a central potential which implies no ϕ and hence no m dependence. (Recall that the $P_l(y)$ are the (ordinary or $m = 0$) Legendre polynomials.) We then wish to match this form onto the standard scattering solution for a plane wave incident along the z-axis

$$\psi = e^{ikz} + f(\theta)\frac{e^{ikr}}{r} \tag{19.112}$$

to determine the scattering amplitude.

To accomplish this, we use the relation

$$e^{ikz} = e^{ikr\cos(\theta)} = \sum_{l=0}^{\infty} (2l + 1)\, i^l j_l(kr)\, P_l(\cos(\theta)) \tag{19.113}$$

which we quote, but do not prove. The strategy is then to:

- Equate Eqn. (19.111) with Eqn. (19.112), using the identity Eqn. (19.113).

- Use the trigonometric relation

$$\sin(z) = \frac{e^{iz} - e^{-iz}}{2i} \tag{19.114}$$

to rewrite the sine functions in terms of complex exponentials, and

- Equate the coefficients of the linearly independent terms proportional to e^{ikr} and e^{-ikr}.

The coefficients of the e^{-ikr} terms can be seen to imply that

$$a_l = (2l + 1)i^l e^{i\delta_l} \tag{19.115}$$

while the e^{ikr} terms impose a relation on the scattering amplitude, namely

$$f(\theta) = \frac{2i}{k} \sum_{l=0}^{\infty} \left[a_l e^{i\delta_l} - (2l + 1)i^l \right] e^{-il\pi/2} P_l(\cos(\theta)) \tag{19.116}$$

Using Eqns (19.115) and (19.116) and the fact that $e^{-il\pi/2} = (1/i)^l$ we find the very useful result

$$f(\theta) = \frac{1}{2ik} \sum_{l=0}^{\infty} e^{-il\pi/2} (2l + 1) \, i^l \, [e^{2i\delta_l} - 1] \, P_l(\cos(\theta))$$

$$= \frac{1}{k} \sum_{l=0}^{\infty} (2l + 1) \, e^{i\delta_l} \, \sin(\delta_l) \, P_l(\cos(\theta)) \tag{19.117}$$

The total cross-section is then given by

$$\sigma = \int d\Omega \, \frac{d\sigma}{d\Omega} = \int d\Omega \, |f(\theta)|^2 = \frac{4\pi}{k^2} \sum_{l=0}^{\infty} (2l + 1) \, \sin^2(\delta_l) \tag{19.118}$$

where we use the orthonormality properties of the Legendre polynomials, namely

$$\int d\Omega \, P_l(\cos(\theta)) \, P_{l'}(\cos(\theta)) = \frac{4\pi}{(2l + 1)} \delta_{l,l'} \tag{19.119}$$

Example 19.5. Hard sphere scattering

One of the simplest scattering problems in classical mechanics is that of scattering from an impenetrable sphere considered in Example 19.1. We consider here the same problem in quantum mechanics using the method of partial waves; the appropriate central potential is given by

$$V(r) = \begin{cases} 0 & \text{for } r > a \\ +\infty & \text{for } r < a \end{cases} \tag{19.120}$$

The solution for $r \leq a$ must vanish, and it must also match onto the most general free-particle solution for a given partial wave for $r > a$, namely

$$R_l(r) = \alpha_l j_l(kr) + \beta_l n_l(kr) \tag{19.121}$$

(Continued)

which implies that

$$\frac{\beta_l}{\alpha_l} = -\frac{j_l(ka)}{n_l(ka)} \tag{19.122}$$

Using Eqn. (19.110), we find that

$$\tan(\delta_l) = \frac{j_l(ka)}{n_l(ka)} \quad \text{or} \quad \sin^2(\delta_l) = \frac{j_l^2(ka)}{j_l^2(ka) + n_l^2(ka)} \tag{19.123}$$

Using the expansion in Section 16.6 for the spherical Bessel functions, we find that for $ka \ll 1$ that

$$\tan(\delta_l) \approx \frac{(ka)^{(2l+1)}}{(2l+1)(1 \cdot 3 \cdots (2l-1))^2} \tag{19.124}$$

so that for low enough energies, only the $l = 0$ partial wave will contribute appreciably to the scattering; the angular momentum barrier effectively keeps the scatterers apart and only s-wave scattering is important. This is also consistent with the conclusions drawn from the Born approximation, namely that low-energy scattering from finite potentials has an isotropic angular distribution. Conversely, if the effective range of some interaction is roughly R, then the contribution to the effective potential due to rotational motion will equal the incident energy when

$$\frac{l(l+1)\hbar^2}{2mR^2} \approx E \quad \text{or} \quad l_{max} \approx \frac{\sqrt{2mE}}{\hbar} R \approx kR \tag{19.125}$$

implying angular momentum values up to of the order l_{max} will contribute to the scattering amplitude, consistent with Eqn. (19.124).

In this example, when $ka \ll 1$, we have $\sin(\delta_0) \approx \delta_0 \approx ka$ which implies that the total cross-section is

$$\sigma = \frac{4\pi}{k^2} \sin^2(\delta_0) \approx \frac{4\pi k^2 a^2}{k^2} = 4\pi a^2 \tag{19.126}$$

Based on our experience with the Rutherford cross-section, we might have expected that any cross-section which does not contain an explicit factor of $\hbar$ would necessarily correspond to the classical result, but we find here a cross-section which is four times larger than the geometrical area presented to the scatterers. To understand better the approach to the classical limit, we consider now the limit where $ka \gg 1$ as well; in that case, one finds

$$\tan(\delta_l) \longrightarrow -\frac{\sin(kr - l\pi/2)}{\cos(kr - l\pi/2)} \tag{19.127}$$

(Continued)

so that $\delta_l \longrightarrow -(kr - l\pi/2) >> 1$ in magnitude so that many partial waves contribute to the total cross-section. We can estimate the summation in Eqn. (19.118) in the following heuristic manner:

- Each contributing factor of $\sin^2(\delta_l)$ with a large argument will average to $1/2$ as that is the value of $\sin^2(x)$ when averaged over many cycles.

- The summation will be cut off at a value of $l_{max} \approx ka$ and we approximate

$$\sum_{l=0}^{l_{max}} (2l + 1) \approx \int_0^{l_{max}} (2l + 1)\, dl \approx l_{max}^2 \approx (ka)^2 \qquad (19.128)$$

- The total cross-section would then be given by

$$\sigma = \frac{4\pi}{k^2} \left(\frac{1}{2}\right) (ka)^2 = 2\pi a^2 \qquad (19.129)$$

While this is not a proof, a careful numerical evaluation of the exact cross-section given by

$$\sigma = \frac{4\pi}{k} \sum_{l=0}^{\infty} (2l + 1) \frac{j_l^2(ka)}{(j_l^2(ka) + n_l^2(ka))} \qquad (19.130)$$

confirms the semiquantitative argument and the results are shown in Fig. 19.10; even at high energy, the total cross-section is still twice the classical value. This result is perhaps more surprising than Eqn. (19.126), since when $ka >> 1$ we have $2\pi a >> \lambda$ and we could, in principle, make wave packets of dimensions much smaller than the scattering object, which should follow classical trajectories as in Section 19.1.

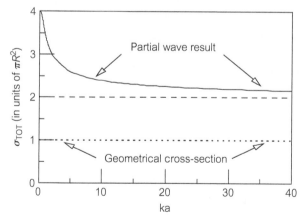

Figure 19.10. Energy dependence (via ka) of the total cross section for scattering from a hard sphere using partial wave expansions. Note that even for large energies, the total cross-section is given by $\sigma_{tot} = 2\pi R^2$ instead of the expected classical geometrical cross-section, $\sigma_{cl} = \pi R^2$.

(Continued)

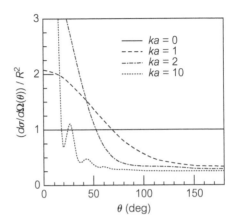

Figure 19.11. Differential cross-section for scattering from a hard sphere using a partial wave expansion. For increasing energies (or ka), the "extra" πR^2 (compared to the classical result) in the total cross-section is all in the very forward direction and unobservable.

The answer to where the "extra" πa^2 comes from can be seen much more clearly by examination of the differential cross-section, $d\sigma(\theta)/d\Omega$ versus θ, for increasingly large values of ka as shown in Fig. 19.11. In units of a^2, the classical differential cross-section is given by Eqn. (19.13) yielding the isotropic distribution

$$\left[\frac{d\sigma}{d\Omega}(\theta)\right]_{CL}\frac{1}{a^2} = \frac{1}{4} \tag{19.131}$$

and for $ka \ll 1$ we obtain four times that result as noted above. On the other hand, for $ka \gg 1$, the cross-section shows an obvious forward (diffractive) peaking consistent with $\theta_{max} \sim 1/ka$ as mentioned above with a long "tail" which approaches the "flat" value given by Eqn. (19.131). Figure 19.14 shows that the "extra" cross-section is all contained in the forward diffraction peak; for macroscopic systems, the value of θ_{max} is so small as to make this contribution unobservable (P19.9), and the remaining cross-section is consistent with the purely classical result.

19.5 **Scattering of Particles**

Thus far, we have considered only so-called *potential scattering* where particles interact with fixed external sources of potential, $V(\mathbf{r})$. This can only be an idealization as such processes do not conserve momentum, as illustrated in Fig. 19.6;

the momentum transfer $\Delta \mathbf{p} = \hbar \mathbf{q}$ to the particle comes from "nowhere." It is more realistic to consider the collisions of two bodies which scatter under the influence of their mutual interaction potential, $V(\mathbf{r}_1 - \mathbf{r}_2)$. The formalism developed in Section 14.3.2 for separating two-body problems using relative and center-of-mass coordinates is again directly applicable in this situation. The equivalent one-particle Schrödinger equation for the relative coordinate, now with reduced mass μ, will give the information on the scattering process. All of the results we have derived for scattering can then be taken over directly, provided the center-of-mass coordinate is fixed; many of the complications of analyzing scattering experiments will arise for simple geometrical reasons due to the motion of the center-of-mass of the scattering system.

In the next two sections, we describe (i) how the description of the scattering process depends on the motion of the center-of-mass of the two-body system and (ii) how the spin-statistics theorem affects the scattering of indistinguishable particles.

19.5.1 **Frames of Reference**

For a two-body problem, the center-of-mass coordinate is given by

$$\mathbf{R}_{cm} = \frac{m_1 \mathbf{r}_1 + m_2 \mathbf{r}_2}{m_1 + m_2} \tag{19.132}$$

which also gives a relation involving the momenta, namely

$$\mathbf{P}_{cm} = M \mathbf{V}_{cm} = m_1 \mathbf{v}_1 + m_2 \mathbf{v}_2 = \mathbf{p}_1 + \mathbf{p}_2 \tag{19.133}$$

Let us first analyze scattering in a frame of reference in which $\mathbf{R}_{cm}$ is fixed so that the total momentum vanishes, as in Fig. 19.12; in this frame, the particles approach each other with equal and opposite momenta (but not necessarily velocities). This is called the *center-of-mass frame*, hereafter abbreviated as CM, and is the one in which our scattering formalism (Born approximation, partial wave analysis, etc.) applies directly. We will carefully distinguish between the center-of-mass variables (cm) defined above and the center-of-mass frame of reference (CM).

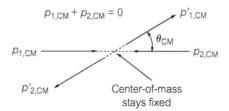

Figure 19.12. Center-of-mass frame of reference for scattering.

Using conservation of energy for an elastic collision and conservation of momentum in the special case where $\mathbf{P}_{cm} = 0$, one can show that that the *magnitude* of $\mathbf{p}_{1,CM}$ (and hence $\mathbf{p}_{2,CM}$) cannot change, but only be scattered through an angle θ_{CM} as shown. This also implies that the speed of each particle is unchanged in the collision, for example

$$|\mathbf{v}'_{1,CM}| = |\mathbf{v}_{1,CM}| \tag{19.134}$$

where we use the subscript CM to remind ourselves that this is generally only true in this special frame of reference. This effect can be easily visualized in one dimension where conservation of kinetic energy and momentum read

$$\frac{p_1^2}{2m_1} + \frac{p_2^2}{2m_2} = E_{CM} \quad \text{and} \quad p_1 + p_2 = 0 \tag{19.135}$$

which corresponds to the intersection of a line and an ellipse, as shown in Fig. 19.13; the two solutions correspond to a change in direction of $p_{1,CM}$ and $p_{2,CM}$, but a constant magnitude for each.

Actual scattering experiments are more often realized by colliding beams of particles with stationary targets, as shown in Fig. 19.14; this arrangement is conventionally called the *laboratory frame of reference*. This notation is something of a misnomer (or at least an anachronism) as many modern experiments do utilize colliding beams of particles of equal and opposite momenta[6] where the center-of-mass system is realized directly in the laboratory; nevertheless, we will use the standard notation. In this arrangement, the total momentum of the system is nonvanishing and given by

$$\mathbf{P}_L = \mathbf{P}_{cm} = m_1\mathbf{v}_{1,L} + m_2\mathbf{v}_{2,L} = m_1\mathbf{v}_{1,L} \neq 0 \tag{19.136}$$

where we use the subscript L to indicate quantities measured in the lab frame of reference.

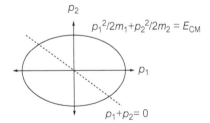

Figure 19.13. Conservation of energy and momentum for a two-body collision in one dimension.

---

[6] This can occur, for example, at facilities which make use of the same arrangements of magnetic fields to accelerate both particles and and their antiparticles in opposite directions simultaneously; since the particle–antiparticle pair have the same mass but opposite charge, the acceleration process gives them equal but opposite momenta; examples include electron–positron and proton–antiproton colliders.

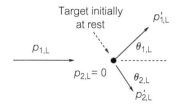

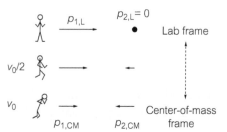

Figure 19.14. Laboratory frame of reference with target at rest.

Figure 19.15. Transformation from lab frame to center-of-mass frame.

The velocities, angles, and energies measured by an observer in the lab frame will not correspond directly to those necessary to analyze the experiment in the CM frame, and our task is to be able to "translate" from one frame to the other. To "view" or analyze this system in the center-of-mass, we would have to "run along side" the lab system (as in Fig. 19.15) at a velocity v_0 such that $v_{1,2}$ are seen to have their values changed so as to satisfy

$$\mathbf{P}_{cm} = m_1(\mathbf{v}_{1,L} - \mathbf{v}_0) + m_2(-\mathbf{v}_0) = 0 \tag{19.137}$$

which implies that

$$\mathbf{v}_0 = \left(\frac{m_1}{m_1 + m_2}\right)\mathbf{v}_{1,L} \tag{19.138}$$

The velocities and momenta of each particle before the collision in the CM frame are then given by

$$\mathbf{v}_{1,CM} = \mathbf{v}_{1,L} - \mathbf{v}_0 = \left(\frac{m_2}{m_1 + m_2}\right)\mathbf{v}_{1,L} \tag{19.139}$$

$$\mathbf{v}_{2,CM} = -\mathbf{v}_0 = -\left(\frac{m_1}{m_1 + m_2}\right)\mathbf{v}_{1,L} \tag{19.140}$$

and

$$\mathbf{P}_{1,CM} = m_1\mathbf{v}_{1,CM} = \mu\mathbf{v}_{1,L} = -\mathbf{P}_{2,CM} \tag{19.141}$$

The kinetic energy available in the CM system for the collision can be written in the form

$$E_{CM} = \frac{1}{2} m_1 \left(v_{1,CM}\right)^2 + \frac{1}{2} m_2 \left(v_{2,CM}\right)^2$$

$$= \frac{1}{2} \left(\frac{m_1 m_2}{m_1 + m_2}\right) \left(v_{1,L}\right)^2$$

$$= \left(\frac{\mu}{m_1}\right) \left[\frac{1}{2} m_1 \left(v_{1,L}\right)^2\right]$$

$$E_{CM} = \frac{\mu}{m_1} E_{1,L} \tag{19.142}$$

This reminds us that the kinetic energy, among many other observables, is a reference frame-dependent quantity. We note the two limits

$$E_{CM} = \left(\frac{m_2}{m_1 + m_2}\right) E_{1,L} \approx \begin{cases} E_{1,L} & \text{for } m_2 \gg m_1 \\ E_{1,L}/2 & \text{for } m_1 = m_2 \end{cases} \tag{19.143}$$

To relate the angles measured in the lab frame to those required for the analysis in the CM frame, we note that the velocity vectors of the scattered particle (1) in the two frames are related by

$$\mathbf{v}'_{1,CM} = \mathbf{v}'_{1,L} - \mathbf{v}_0 \quad \text{or} \quad \mathbf{v}'_{1,CM} + \mathbf{v}_0 = \mathbf{v}'_{1,L} \tag{19.144}$$

The corresponding relations amongst the components parallel and perpendicular to the initial direction can be written as

$$v_{1,CM} \cos(\theta_{CM}) + v_0 = v'_{1,L} \cos(\theta_L) \tag{19.145}$$

$$v_{1,CM} \sin(\theta_{CM}) = v'_{1,L} \sin(\theta_L) \tag{19.146}$$

where we have used Eqn. (19.134).

Taking the ratio of Eqns (19.146) to (19.145) we find that

$$\tan(\theta_L) = \frac{\sin(\theta_{CM})}{\cos(\theta_{CM}) + \gamma} \tag{19.147}$$

where

$$\frac{v_0}{v_{1,CM}} = \frac{m_1}{m_2} \equiv \gamma \tag{19.148}$$

This relation between the angles can also be usefully written in the forms

$$\cos(\theta_L) = \frac{\cos(\theta_{CM}) + \gamma}{\sqrt{1 + \gamma^2 + 2\gamma \cos(\theta_{CM})}} \tag{19.149}$$

$$\sin(\theta_L) = \frac{\sin(\theta_{CM})}{\sqrt{1 + \gamma^2 + 2\gamma \cos(\theta_{CM})}} \tag{19.150}$$

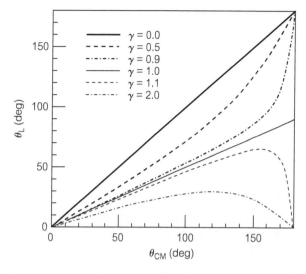

Figure 19.16. Lab (θ_L) versus center-of-mass (θ_{CM}) for various values of $\gamma = m_1/m_2$.

We illustrate the relation between the lab and CM angles in Fig. 19.16 for several values of γ. Several limits of this relation should be noted:

• When the target mass is much heavier than the projectile, we have $m_2 \gg m_1$ which implies that $\gamma \ll 1$ and $\theta_L \approx \theta_{CM}$; this limit corresponds to potential scattering where the heavy target can act to conserve momentum.

• The special case of $m_2 = m_1$ implies that

$$\tan(\theta_L) = \frac{\sin(\theta_{CM})}{\cos(\theta_{CM}) + 1} = \tan\left(\frac{\theta_{CM}}{2}\right) \qquad (19.151)$$

so that $\theta_{CM} = 2\theta_L$; this is important for the case of scattering of identical particles.

• For any value of $m_1 > m_2$, the incident projectile can no longer scatter backward in the lab frame; even for head-on collisions, the heavier incident particle now proceeds forward. Even more interestingly, there are now *two* center-of-mass angles which correspond to a given lab angle, as shown in Fig. 19.17; it is interesting to ponder how one could disentangle the two contributions.

• When the projectile is much heavier than the target, that is $m_2 \ll m_1$, Eqns (19.147) and (19.149) imply that there is a maximum angle through which m_1 can be scattered in the lab frame which goes as $\theta_L \to 1/\gamma$ for $\gamma \gg 1$; there is not much scattering of a moving truck from a stationary pebble.

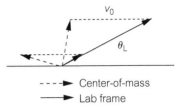

Figure 19.17. Two velocity vectors (dashed arrows) with different θ_{CM} which give velocity vectors in the lab frame (solid arrows) corresponding to the same lab angle, θ_L; this is only possible when $m_a > m_2$.

The *total scattering cross-sections* in the two frames must be equal, since σ measures the total probability of particles being scattered in any direction which should be independent of the frame of reference. The differential cross-sections, however, will be frame-dependent because they are defined using different angles; we can only say that the total number of particles scattered into solid angle $d\Omega_L$ at angle (θ_L, ϕ_L) (call it dN_L) must be the same as scattered into $d\Omega_{CM}$ at (θ_{CM}, ϕ_{CM}) (labeled dN_{CM}). We thus have

$$dN_L \propto \left(\frac{d\sigma}{d\Omega}\right)_L \sin(\theta_L)\, d\theta_L\, d\phi_L = \left(\frac{d\sigma}{d\Omega}\right)_{CM} \sin(\theta_{CM})\, d\theta_{CM}\, d\phi_{CM} \propto dN_{CM}$$

(19.152)

so that

$$\left(\frac{d\sigma}{d\Omega}\right)_L = \left(\frac{d\sigma}{d\Omega}\right)_{CM} \left[\frac{\sin(\theta_{CM})}{\sin(\theta_L)}\right]\left(\frac{d\theta_{CM}}{d\theta_L}\right)$$

(19.153)

The differential relation between $d\theta_L$ and $d\theta_{CM}$ can be derived by differentiating both sides of Eqn. (19.147) giving

$$\frac{1}{\cos^2(\theta_L)}\, d\theta_L = \frac{1 + \gamma\cos(\theta_{CM})}{(\cos(\theta_{CM}) + \gamma)^2}\, d\theta_{CM}$$

(19.154)

Then using Eqn. (19.149) we find that

$$\left(\frac{d\sigma}{d\Omega}\right)_L = \frac{\left(1 + \gamma^2 + 2\gamma\cos(\theta_{CM})\right)^{3/2}}{(1 + \gamma\cos(\theta_{CM}))}\left(\frac{d\sigma}{d\Omega}\right)_{CM}$$

(19.155)

For the special case of equal mass particles ($\gamma = 1$), we find the useful relation

$$\left(\frac{d\sigma}{d\Omega}\right)_{CM} = \frac{1}{4\cos(\theta_L)}\left(\frac{d\sigma}{d\Omega}\right)_L$$

(19.156)

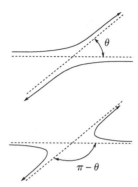

Figure 19.18. Indistinguishable scattering geometries.

19.5.2 **Identical Particle Effects**

We have seen that the bound state wavefunctions of systems of indistinguishable particles have to satisfy strong symmetry constraints imposed by the spin-statistics theorem. The cross-sections describing the scattering of identical particles will have to satisfy similar constraints as they are derived from the large distance behavior of similar unbound wavefunctions.

To see how indistinguishability effects can play a role in the measurement of a scattering cross-section, we show in Fig. 19.18 two classical trajectories describing the scattering of identical mass particles in the center-of-mass; the two configurations for which the scattering angles are related via $(\theta, \phi) \to (\pi - \theta, \pi + \phi)$ give rise to the same "signal" in the detectors. In a classical description of scattering where we could, in principle, follow the individual trajectories, we would simply add the two contributing scattering probabilities, that is, cross-sections, to obtain

$$\left(\frac{d\sigma}{d\Omega}\right)_{\text{CL}} = \frac{d\sigma(\theta)}{d\Omega} + \frac{d\sigma(\pi - \theta)}{d\Omega} \tag{19.157}$$

In the quantum mechanical case, just as with any wave phenomena, we must *first* combine the scattering amplitudes and *then* square to obtain the differential cross-section. The relevant amplitudes, namely, $f(\theta)$ and $f(\pi - \theta)$, correspond to the position-space wavefunctions of the two-particle system, and information on the spin degrees of freedom must also be included so as to obtain a total wavefunction with the appropriate symmetry under exchange, just as with the bound state case; systems of bosons (fermions) must have a total scattering amplitude which is symmetric (antisymmetric) under the interchange of the two-particle labels.

The simplest case corresponds to spinless bosons with $J = 0$. Since there are no spin-wavefunctions, the scattering amplitude is simply the symmetric combination which gives

$$\left(\frac{d\sigma}{d\Omega}\right)_{J=0} = |f(\theta) + f(\pi - \theta)|^2$$

$$= \frac{d\sigma(\theta)}{d\Omega} + \frac{d\sigma(\pi - \theta)}{d\Omega} + 2Re[f(\theta)f^*(\pi - \theta)] \qquad (19.158)$$

This form already imposes interesting constraints on the scattering of identical spinless particles as it implies that only *even* partial wave amplitudes will contribute to such scattering; this arises since $\cos(\pi - \theta) = -\cos(\theta)$ and the Legendre polynomials satisfy $P_l(-y) = (-1)^l P_l(y)$.

For spin-1/2 particles, the spinor wavefunctions (Example 16.4) describe a total spin of $S = 0$ or 1 which are respectively antisymmetric and symmetric under exchange. The corresponding "spatial" wavefunctions must then be the combinations which are symmetric and antisymmetric under $\theta \to \pi - \theta$ in order for the total wavefunction to satisfy the spin-statistics for fermions. In addition, we recall two other facts about the "spin-counting"; (i) there are a total of $2S + 1 = 3$ of the $S = 1$ states which can contribute and only a single $S = 0$ state and (ii) in an experiment involving unpolarized particles, all $3 + 1 = 4$ combinations of spins are equally likely, and we should average these possibilities. All of these effects taken together give the cross-section

$$\left(\frac{d\sigma}{d\Omega}\right)_{J=1/2} = \frac{3}{4}|f(\theta) - f(\pi - \theta)|^2 + \frac{1}{4}|f(\theta) + f(\pi - \theta)|^2$$

$$= \frac{d\sigma(\theta)}{d\Omega} + \frac{d\sigma(\pi - \theta)}{d\Omega} - Re[f(\theta)f^*(\pi - \theta)] \qquad (19.159)$$

The Clebsch–Gordan coefficients for the spin sum $1 + 1$ can be used to show that the spinor wavefunctions for the net spin $J = 2, 1, 0$ are symmetric, antisymmetric, and symmetric, respectively, so that performing the "spin counting" as above, we find for the scattering of identical $J = 1$ particles,

$$\left(\frac{d\sigma}{d\Omega}\right)_{J=1} = \frac{5+1}{9}|f(\theta) + f(\pi - \theta)|^2 + \frac{3}{9}|f(\theta) - f(\pi - \theta)|^2$$

$$= \frac{d\sigma(\theta)}{d\Omega} + \frac{d\sigma(\pi - \theta)}{d\Omega} - \frac{2}{3}Re[f(\theta)f^*(\pi - \theta)] \qquad (19.160)$$

These examples can be further generalized (P19.10) to show that the cross-section for the scattering of identical spin J particles (whether bosons or fermions) can be written in the form

$$\left(\frac{d\sigma}{d\Omega}\right)_J = \frac{d\sigma(\theta)}{d\Omega} + \frac{d\sigma(\pi - \theta)}{d\Omega} + 2\frac{(-1)^{2J}}{(2J + 1)}Re[f(\theta)f^*(\pi - \theta)] \qquad (19.161)$$

For scattering via the Coulomb potential, the form of the cross-terms can be calculated exactly using Eqn. (19.82) giving

$$Re[f_C(\theta)f_C^*(\pi - \theta)] \propto \frac{1}{\sin^2(\theta/2)\cos^2(\theta/2)} Re\left[e^{-i\eta \ln(\sin^2(\theta/2))}e^{+i\eta \ln(\cos^2(\theta/2))}\right]$$

$$= \frac{1}{\sin^2(\theta/2)\cos^2(\theta/2)} Re[e^{-i\eta \ln(\tan^2(\theta/2))}]$$

$$= \frac{1}{\sin^2(\theta/2)\cos^2(\theta/2)} \cos(\eta \ln(\tan^2(\theta/2))) \qquad (19.162)$$

where

$$\eta \equiv \frac{Z_1 Z_2 K e^2 \mu}{\hbar^2 k} = Z_1 Z_2 \frac{K e^2}{\hbar c}\sqrt{\frac{\mu c^2}{2 E_{cm}}} \qquad (19.163)$$

This form explicitly shows how the classical limit of Eqn. (19.157) is reached; if we formally allow $\hbar \to 0$, then $\eta \to \infty$, and the interference term oscillates so rapidly that any measurement which accepts particles in a finite range of θ (as any real detector must do) will effectively sample the cross-term over many cycles of its argument, and hence average to zero. We also see that the Born approximation for the scattering amplitude misses this essential physics, since in that case we have

$$Re[f_B(\theta)f_B^*(\pi - \theta)] \propto \frac{1}{\sin^2(\theta/2)\cos^2(\theta/2)} \qquad (19.164)$$

corresponding to $\eta = 0$

Example 19.6. Coulomb scattering of identical spin-zero particles

We examine to what extent the effects of indistinguishability can be observed by discussing an experiment involving the scattering of spinless carbon nuclei; specifically, we consider C^{12} for which $Z = 6$ and $\mu = M/2 = 6$ amu. The cross-section for electromagnetic scattering of such particles is given by Eqn. (19.158) with the the cross-term in Eqn. (19.162), giving the so-called *Mott cross-section*

$$\frac{d\sigma}{d\Omega} = \left(\frac{Z^2 K e^2}{4 E_{CM}}\right)^2 \left[\frac{1}{\sin^4(\theta/2)} + \frac{1}{\cos^4(\theta/2)} + \frac{2\cos(\eta \ln(\tan^2(\theta/2)))}{\sin^2(\theta/2)\cos^2(\theta)}\right] \qquad (19.165)$$

where

$$\eta = 0.158 Z^2 \sqrt{\frac{\mu(\text{amu})}{E_{CM}(\text{MeV})}} \qquad (19.166)$$

(Continued)

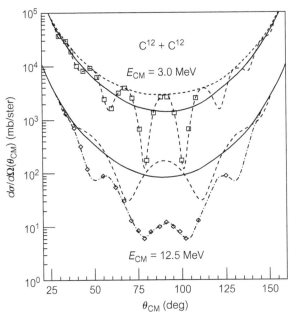

Figure 19.19. Differential cross-section for the elastic scattering of spinless $C^{12} + C^{12}$ nuclei versus center-of-mass angle θ_{CM} for two values of E_{CM}. The upper set of curves correspond to $E_{CM} = 3.0$ *MeV* while the lower ones are for $E_{CM} = 12.5$ *MeV*. In each case, the solid lines are the Rutherford prediction (Eqn. (19.81)) while the dashed line is the Mott prediction (Eqn. (19.165)) using the exact Coulomb scattering amplitude. For the top set, the dotted line shows the Mott prediction but using the approximate Born amplitudes which miss the η dependence. For energies below the Coulomb barrier, pure electromagnetic scattering dominates and the Mott prediction agrees very well with experiment; for higher energies, nuclear interactions also contribute and the dot-dash line for the lower set shows a fit including both Coulomb and nuclear scattering.

We show in Fig. 19.19 the experimental data for two different energies; the upper curves correspond to $E_{CM} = 3.0$ MeV, while the lower ones are for $E_{CM} = 12.5$ MeV. The solid curves in both cases show the prediction for the "classical" result of Eqn. (19.157), while the dashed lines show the Mott prediction of Eqn. (19.165) which includes the interference effects; the dotted curve for the lower energy data indicates the Mott result, but with $\eta = 0$ corresponding to the Born amplitude.

The 3.0 MeV data clearly agrees with the Mott cross-section, and gives clear evidence for the interference effects due to quantum mechanics *and* the exact Coulomb scattering amplitudes. The higher energy data (fitted with the dot-dash curve) does not agree with purely electromagnetic scattering as described by quantum mechanics because of the need to include the strong nuclear force; we briefly discuss these effects below.

(Continued)

For two charged objects of finite radii R_1 and R_2, the electrostatic potential energy when they "just touch" is given

$$E_{cm}^{top} \approx \frac{Z_1 Z_2 K e^2}{R_{sep}} \qquad (19.167)$$

where $R_{sep} \approx (R_1 + R_2)$ and E_{cm}^{top} refers to the "top" of the Coulomb barrier. Using $R = (1.2\,F)A^{1/3}$ for each nucleus, this implies that $E_{cm}^{top} \approx 9.4$ MeV for this system. At the lower energy, the two nuclei are kept sufficiently far apart by their Coulomb repulsion that they never "feel" the finite range nuclear force; for higher energies, nuclear interactions come into play, and scattering amplitudes due to this force must be incorporated in addition to the purely electromagnetic scattering. We can even roughly estimate the number of partial waves which must be taken into account to fit the data by using energy conservation to write

$$E_{cm} = E_{Coulomb} + E_{rot}$$

$$12.5\,\text{MeV} = 9.4\,\text{MeV} + \frac{l(l+1)\hbar^2}{2\mu R_{sep}^2} \qquad (19.168)$$

which implies that $l(l+1) \approx 31.6$ or $l_{max} \approx 5.1$; since only even partial waves contribute in this case (why?), the experimental analysis actually required $l_{max} = 6$ to obtain agreement with the data.

In Chapter 18, we noted that deviations from the purely Coulombic predictions for the detailed structure of energy levels for the hydrogen atom signaled the need for new physics input, such as relativity, spin–orbit coupling, and finite nuclear size effects. We now see that deviations from pure electromagnetic scattering (when all other quantum effects are properly included) can reveal the presence of new interactions, in this case the strong nuclear force. The systematic study of the electromagnetic interactions of electron–positron pairs (in center-of-mass $e^+ - e^-$ colliders) is another example where purely (quantum) electrodynamic interactions can dominate at low energies, with additional "annihilation"-type contributions required due to the particle–antiparticle symmetry of the system, as in Fig. 18.16. At higher energies, however, evidence for the *weak interactions* becomes obvious, and even comes to dominate the reactions.

19.6 **Questions and Problems**

Q19.1. What would the picture corresponding to Fig. 17.7 for the angular momentum and Lenz–Runge vectors look like for the case of unbound, hyperbolic orbits found in the Coulomb scattering problem.

Q19.2. If we think of a total cross-section as an effective area presented to scatterers, discuss why relativistic "boosts," that is, changes in coordinate systems moving at constant speed relative to one another, should have no effect on σ_{TOT}. Hint: How do length contractions work in relativity?

Q19.3. If the incident particles are heavier than the target particles (initially at rest) in a scattering experiment, scatterings from two distinct center-of-mass angles give contributions to the same lab angle. How do you tell one scatter from the other (or can you?)

Q19.4. Can you use the methods outlined in this chapter (e.g. the Born approxima-tion or partial wave expansions) to discuss quantum scattering in two or one dimensions? (For references, see Eberly (1965) (1D) and Lapidus (1982a) or Adhikari (1986) (2D).)

P19.1. (a) Use the method outlined in Example 19.1 to find the hard scattering cross-section for the ellipsoidal shape discussed there. Specifically, show that

$$\tan\left(\frac{\theta}{2}\right) = \left|\frac{dy}{dx}\right| = \frac{\sqrt{R^2 - b^2}}{fb} \tag{19.169}$$

and thereby derive Eqn. (19.15).

(b) Show that this gives the cross-section in Eqn. (19.16), and integrate it explicitly to see that $\sigma = \pi R^2$.

P19.2. Find the differential cross-section for classical hard-scattering from a parabol-oid shape; in this case, the surface is given by the relation

$$x = a\left(\frac{y^2 + z^2}{R^2} - 1\right) \tag{19.170}$$

for $y^2 + z^2 \le R^2$.

(a) Show that there is a minimum angle below which there is no scattering, given by

$$\tan\left(\frac{\theta_{min}}{2}\right) = \frac{R}{2a} \tag{19.171}$$

(b) Show that the cross-section can be written in the form

$$\frac{d\sigma}{d\Omega} = \frac{1}{16}\frac{R^4}{a^2}\frac{1}{\sin^4(\theta/2)} \tag{19.172}$$

(c) Confirm that the total cross-section is still the geometrical one given by $\sigma = \pi R^2$.

(d) Sketch $(d\sigma(\theta)/d\Omega)/\sigma$ versus θ for $a = 0.5R, R, 5R$ and discuss your results.

(e) Does the $1/\sin^4(\theta/2)$ dependence have anything to do with Rutherford scattering?

P19.3. Calculate the classical scattering cross-section from an impenetrable sphere of radius a by using the "integrating the potential" approach in Eqn. (19.24) and show that your answer agrees with the geometrically derived result in Example 19.1. The appropriate potential for this case is

$$V(r) = \begin{cases} +\infty & \text{for } r < a \\ 0 & \text{for } r > a \end{cases} \tag{19.173}$$

P19.4. Classical two-dimensional scattering. Use the methods and ideas outlined in Section 19.1 to discuss "specular" scattering in two dimensions. First define and then calculate the analogs of the differential and total cross-sections for scattering from a "hard circle" and then a "hard ellipse" of arbitrary parameter f and discuss your results.

P19.5. The condition for the first diffraction minimum for a circular aperture is quoted in many introductory physics books and is given by $\sin(\theta) \approx 1.22\lambda/D$. Compare this to the corresponding angle for scattering from a finite well discussed in Example 19.3.

P19.6. Using the Born approximation, calculate the differential and total cross-sections for the two potentials

$$V_1(r) = V_0 \left(\frac{a}{r}\right) e^{-r/a} \quad \text{and} \quad V_2(r) = V_0 \, e^{-r^2/a^2} \tag{19.174}$$

Show that your answers can be written in the form

$$\sigma_1 = 4\pi \left(\frac{2mV_0a^3}{\hbar^2}\right)^2 \frac{1}{(1 + 4a^2k^2)} \tag{19.175}$$

$$\sigma_2 = 2 \left(\frac{mV_0\pi \, a^3}{2\hbar^2}\right)^2 \frac{(1 - e^{-2a^2k^2})}{(ka)^2} \tag{19.176}$$

Discuss whether the general conclusions about scattering in the low- and high-energy limits discussed in Example 19.3 are satisfied. Do the differential cross-section show any interesting structure (dips or zeroes) as a function of θ because of diffraction effects? Why or why not?

P19.7. The form factor of the proton obtained from electron scattering is fit rather well over the region $(\hbar q)^2 \approx 0-2 \text{ GeV}^2$ with a *dipole formula*, namely

$$F(q^2) = \frac{1}{(1 + (\hbar q/Mc)^2)^2} \tag{19.177}$$

where $Mc^2 \approx 0.84 \text{ GeV} = 840 \text{ MeV}$.

(a) Show that the charge radius of the proton is roughly $\sqrt{\langle r^2 \rangle} \approx 0.88$ F.

(b) To what distribution of charge, $\rho(\mathbf{r})$, does this correspond?

Note: The form factors for the corresponding magnetic scattering for both proton and neutron have similar forms[7] implying that the charge and "magnetism" for both particles is distributed in the same way.

---

[7] See, for example, Perkins (2000).

P19.8. Partial waves for the finite well. Consider particles of mass m scattered form a three-dimensional finite well given by

$$V(r) = \begin{cases} -V_0 & \text{for } r < a \\ 0 & \text{for } r > a \end{cases} \qquad (19.178)$$

and shown in Fig. 19.20. For low enough energies, we need only consider s-wave scattering so, using the results of Section 16.6, write a solution of the form

$$u(r) = rR(r) = \begin{cases} A\sin(\kappa r) & \text{for } r < a \\ B\sin(kr) + C\cos(kr) & \text{for } r > a \end{cases} \qquad (19.179)$$

where $\hbar^2 k^2 = 2mE$ and $\hbar^2 \kappa^2 = 2m(E + V_0)$.

(a) Impose the appropriate boundary conditions at $r = a$, find the ratio C/B and show that that the $l = 0$ phase shift is given by

$$\delta_0 \approx (ka)\left(\frac{\tan(\kappa a)}{\kappa a} - 1\right) \qquad (19.180)$$

when $ka \ll 1$.

(b) Find the total cross-section and show that there are energies for which this approximation to σ becomes large, in fact, formally infinite in this approximation. Return to the derivation and show that at these energies a better approximation is $\sin(\delta_0) \approx 1$ so that

$$\sigma = \frac{4\pi}{k^2} \qquad (19.181)$$

Such energies correspond to *resonant states*

(c) Show that there are energies for which the cross-section (in this approximation) vanishes. Show that these energies correspond to the *transmission* resonances considered in Section 11.3.1.

(d) Consider the case of a repulsive well with $-V_0 \longrightarrow +V_0$. Show that the expression above for the total cross-section can be written as

$$\sigma = 4\pi a^2 \left(\frac{\tanh(\kappa' a)}{\kappa' a} - 1\right)^2 \qquad (19.182)$$

and find an expression for κ'. Show that when $V_0 \to +\infty$ that one obtains the $l = 0$ partial wave result for the total cross-section.

P19.9. Classical versus quantum scattering. If we scatter one gram BB's with a speed of one cm/s from a bowling ball, inside what angles is the "other half" of the quantum cross-section, that is, the forward diffractive peak, contained?

P19.10. Justify the formula in Eqn. (19.161) for the cross-section for scattering of indistinguishable spin J particles.

Table 19.1. Differential cross-sections (in $mb/ster$) in the lab frame versus lab angle for $\alpha - \alpha$ scattering at $E_L = 0.6$ MeV

θ_L	θ_{cm}	Experimental $(d\sigma/d\Omega)_L$	Experimental $(d\sigma/d\Omega)_{cm}$	Theory $(d\sigma/d\Omega)_{J=0}$	Theory $(d\sigma/d\Omega)_{J=1/2}$	Theory $(d\sigma/d\Omega)_{CL}$
15	30	187.4				
20	40	47.6				
25	50	18.2				
30	60	12.2				
35	70	11.4				
40	80	11.4				
45	90	10.8				
50	100	9.35				
55	110	7.90				
60	120	7.01				
65	130	8.26				
70	140	16.6				
75	150	48.6				

P19.11. Mott scattering for α particles. The data[8] in Table 19.1 for the differential cross-section for $\alpha - \alpha$ scattering were obtained in the lab frame of reference with $E_L = 0.6$ MeV corresponding to (via Eqn. (19.143)) $E_{cm} = 0.3$ MeV. The lab scattering angles, θ_L have been transformed into θ_{cm} using Eqn. (19.151). All cross-section values are in mb/ster (where 1 mb $= 10^{-3}$ b) and have a statistical error of roughly 1%. The α particles have $J = 0$.

(a) Use the transformation from lab to center-of-mass differential cross-sections to fill in the fourth column.

(b) Use the Mott scattering formula in Eqn. (19.165) to fill in the fifth column, and compare your results to the experimental data; use $Z = 2$ and $\mu = M_\alpha/2 = 2$ amu.

(c) Also calculate the classical and $J = 1/2$ cross-sections in Eqns (19.157) and (19.159) and complete the last two columns for comparison.

(d) Repeat the Coulomb barrier argument in Example 19.6 to see if you would expect purely electromagnetic scattering to dominate for these data.

---

[8] Taken from Heydenburg and Temmer (1956).

APPENDIX A

Dimensions and MKS-type Units for Mechanics, Electricity and Magnetism, and Thermal Physics

Many introductory physics textbooks begin with a discussion of the dimensions and the small number of dimensionful units necessary to fully describe physical quantities. In this Appendix we briefly review and collect some of the dimensional equivalences which readers may find useful in solving problems or checking derivations. We then collect the values of important physical constants, in these units, in Appendix B.

For purely mechanical systems, the standard MKS set of units, using the meter (m), kilogram (kg), and second (s) as basic dimensionful quantities, is presumably familiar. We collect below some of the more important connections.

The extension to include a fundamental dimension related to electromagnetism (advocated by Giorgi, and included in the MKSA system) uses the Ampere (A) as the basic quantity, but for purposes of simplicity, we will here use the basic unit of charge, the Coulomb (C), noting that the substitution $C = A \cdot s$ allows one to translate. For thermal physics problems, the degree Kelvin (or kelvin, here written simply as K) is used. We will not require the two additional basic units (candela and mole) used in the Systéme Internationale (SI) d'Unités.

For completeness, we note that many of the standard electromagnetic relationships, in these units, are given through the textbook, namely

- the Coulomb force and potential are given by Eqns (1.28) and (1.29), respectively the Lorentz force law is given by Eqn. (18.5),

- Maxwell's equations are given by Eqns (18.6)–(18.9),

- the Poynting vector is given by Eqn. (18.19),

- the scalar (ϕ) and vector $(\mathbf{A})$ potentials are given by Eqns (18.28) and (18.29)

Table A.1. Dimensions for quantities in mechanics

Quantity	Symbol (Name)	Units
Mass		kg
Length		m
Time		s
Speed	v	m/s
Acceleration	a	m/s^2
Force	F (Newton)	kg $\cdot$ m/s^2
Energy	E (Joule)	N $\cdot$ m or J
Power	P (Watt)	J/s or N $\cdot$ m/s
Momentum	p	kg $\cdot$ m/s or N/s
Moment of inertia	I	kg $\cdot$ m^2
Angular momentum	L	kg $\cdot$ m^2/s
Wavelength	λ	m
Wave number	k, κ	1/m
Frequency	f (Hertz)	1/s
Angular frequency	$\omega = 2\pi f$	rad/s

Table A.2. Dimensions of quantities in electromagnetism and thermal physics

Quantity	Symbol (Name)	Units
Charge	q (Coulomb)	C
Current	I (Ampere)	A $=$ C/s
Electric potential	V (Volt)	J/C or V
Electric field	E	N/C or V/m
Electric dipole moment	p	C $\cdot$ m
Electric quadrupole moment	Q	C $\cdot$ m^2
Electric permittivity	ϵ	C^2/(N $\cdot$ m^2)
Capacitance	F (farad)	C/V or C^2/J
Resistance	Ω (Ohm)	V/A or J $\cdot$ s/C^2
Electrical resistivity	ρ	$\Omega \cdot$ m or J $\cdot$ s $\cdot$ m/C^2
Electrical conductivity	σ	A/(m $\cdot$ V) or C^2/(J $\cdot$ m $\cdot$ s)
Magnetic field (induction)	T (Tesla)	N $\cdot$ s/(C $\cdot$ m)
Magnetic permeability	μ	N $\cdot$ s^2/C^2
Magnetic flux	Φ_B	Tesla $\cdot$m^2 or N $\cdot$ m $\cdot$/C^2
Magnetic dipole moment	m	A $\cdot$m^2 or C $\cdot$ m^2/s
Magnetic permeability	μ	N $\cdot$ s^2/C^2
Auxiliary magnetic field	H	A/m or C/(m $\cdot$ s)
Poynting vector	$E \times H$	J/(m^2 $\cdot$ s)
Degree (absolute scale)	Kelvin	K

A.1 **Problems**

PA.1 Dimensions for familiar quantities I. Add entries to Table A.1 for (a) pressure, (b) density, and (c) torque.

PA.2 Dimensions for familiar quantities II. Add entries to Table A.2 for (a) thermal conductivity, (b) specific heat, and (c) entropy. (You may have to look up some definitions for these.)

APPENDIX B

Physical Constants, Gaussian Integrals, and the Greek Alphabet

B.1 **Physical Constants**

We collect values of various physical constants used in the text. For dimensionful quantities involved in electricity and magnetism, we consistently use the same MKSA or SI ("Système International") units, as in Appendix A, unless specifically noted. In addition to the usual units of mass (kg), length (m), and time (s), for simplicity, we often express less familiar dimensionful quantities in terms of force (Newton, N), energy (Joule, J), and charge (Coulomb, C).

Planck's constant	$\hbar = 1.055 \times 10^{-34} \text{ J} \cdot \text{s}$
	$= 6.582 \times 10^{-16} \text{ eV} \cdot \text{s}$
	$h = 2\pi\hbar = 6.626 \times 10^{-34} \text{ J} \cdot \text{s}$
speed of light (in vacuum)	$c = 2.9979 \times 10^8 \text{ m/s}$
	$\hbar c = 1973 \text{ eV} \cdot \text{Å}$
	$= 197.3 \text{ MeV} \cdot \text{F}$
electron mass	$m_e = 9.11 \times 10^{-31} \text{ kg}$
	$m_e c^2 = 0.511 \text{ MeV}$
proton mass	$m_p = 1.67 \times 10^{-27} \text{ kg}$
	$m_p c^2 = 938.3 \text{ MeV}$
neutron mass	$m_n c^2 = 939.6 \text{ MeV}$
muon rest mass	$m_\mu c^2 = 105.7 \text{ MeV}$
fundamental charge	$e = 1.60 \times 10^{-19} \text{ C}$
electric permittivity	$\epsilon_0 = 8.85 \times 10^{-12} \text{ C}^2/(\text{N} \cdot \text{m}^2)$
	$K = 1/4\pi\epsilon_0 = 8.98 \times 10^9 \text{ N} \cdot \text{m}^2/\text{C}^2$

(*Continued*)

$1 \ amu = 1.66 \times 10^{-27} \ kg$

	$Ke^2 = 2.30 \times 10^{-28} \text{ J} \cdot \text{m}$
	$= 14.4 \ eV \cdot \text{Å}$
	$= 1.44 \text{ MeV} \cdot \text{F}$
fine structure constant	$\alpha = Ke^2 \hbar c = 1/137.0$
permeability constant	$\mu_0 = 4\pi \times 10^{-7} \text{ N} \cdot \text{s}^2/\text{C}^2$
flux quantum	$\Phi_B = 4.14 \times 10^{-15} \text{ T} \cdot \text{m}^2$
Boltzmann constant	$k_B = 1.38 \times 10^{-23} \text{ J}/(\text{molecule} \cdot \text{K})$
	$= 8.617 \times 10^{-5} \text{ MeV/K}$
thermal energy at	$k_B T = 1/39 \text{ eV}$
$T = 300 \ K$	
Avogadro constant	$N_0 = 6.02 \times 10^{23} \text{ molecules/mole}$
Gas constant	$R = N_0 k_B = 8.31 \text{ J}/(\text{mole} \cdot \text{K})$
Bohr radius	$a_0 = 0.53 \text{ Å}$
Rydberg constant	$R_\infty = 1.10 \times 10^7 \text{ m}^{-1}$
Rydberg energy	$E_0 = m_e c^2 \alpha^2/2 = 13.6 \text{ eV}$
Gravitational constant	$G = 6.67 \times 10^{-11} \text{ N} \cdot \text{m}^2/\text{kg}^2$
solar mass	$M_\odot = 1.99 \times 10^{30} \text{ kg}$
earth mass	$M_e = 5.98 \times 10^{24} \text{ kg}$
moon mass	$M_m = 7.36 \times 10^{22} \text{ kg}$
mean earth-sun distance	$AU = 1.50 \times 10^{11} \text{ m}$

Some useful conversion factors are:

$$1 \text{ Å} = 10^{-10} \text{ m}$$
$$1 \text{ F} = 10^{-15} \text{ m}$$
$$1 \text{ ly (lightyear)} = 9.46 \times 10^{15} \text{ m}$$
$$1 \text{ pc (parsec)} = 3.09 \times 10^{16} \text{ m}$$
$$1 \text{ eV} = 1.69 \times 10^{-19} \text{ J}$$
$$1 \text{ barn} = 10^{-28} \text{ m}^2$$
$$1 \text{ G (Gauss)} = 10^{-4} \text{ T (Tesla)}$$

Some often used prefixes for powers of ten are:

P	peta	10^{15}		m	milli	10^{-3}
T	tera	10^{12}		μ	micro	10^{-6}
G	giga	10^{9}		n	nano	10^{-9}
M	mega	10^{6}		p	pico	10^{-12}
k	kilo	10^{3}		f	femto	10^{-15}

B.2 **The Greek Alphabet**

Alpha	A	α	Nu	N	ν
Beta	B	β	Xi	Ξ	ξ
Gamma	Γ	γ	Omicron	O	o
Delta	Δ	δ	Pi	Π	π
Epsilon	E	ϵ	Rho	R	ρ
Zeta	Z	ζ	Sigma	Σ	σ
Eta	H	η	Tau	T	τ
Theta	Θ	θ	Upsilon	Υ	υ
Iota	I	ι	Phi	Φ	ϕ
Kappa	K	κ	Chi	X	χ
Lambda	Λ	λ	Psi	Ψ	ψ
Mu	M	μ	Omega	Ω	ω

B.3 **Gaussian Probability Distribution**

Finding the probability that a variable represented by a Gaussian probability density with mean value μ and standard deviation σ will have a value in some finite region (a, b) requires the evaluation of the "area under the curve" given by

$$\text{Prob}[x \in (a, b)] = \int_a^b dx\, P(x; \mu, \sigma) \tag{B.1}$$

where

$$P(x; \mu, \sigma) = \frac{1}{\sigma\sqrt{2\pi}} e^{-(x-\mu)^2/2\sigma^2} \tag{B.2}$$

Any such problem can be "standardized" in terms of a dimensionless variable by writing

$$z = \frac{x - \mu}{\sigma} \tag{B.3}$$

where z measures the "distance" of x away from the mean μ, in units of σ. All of the information required to evaluate such probabilities can be tabulated once and for all in the form of a cumulative probability distribution using this standardized normal random variable by calculating

$$F(z) = \frac{1}{\sqrt{2\pi}} \int_{-\infty}^z e^{-t^2/2}\, dt \tag{B.4}$$

This corresponds to the probability of finding the variable t anywhere in the interval $(-\infty, z)$. It is defined such that $F(0) = 0.5$, corresponding to half the probability being on either side of μ. The integral defining $F(z)$ can be evaluated numerically and values are shown in Table B.1. They can be extended to negative values of z by using $F(-z) = 1 - F(z)$. Finally, the probability of finding the standardized variable in the interval (z_{min}, z_{max}) is given by

$$\text{Prob}[z \in (z_{min}, z_{max})] = F(z_{max}) - F(z_{min}) \tag{B.5}$$

Example B.1. Normal distributions

As an example of the use of Table B1, we can calculate the probability that a measurement of a variable corresponding to a Gaussian distribution with $\mu = 7$ and $\sigma = 2$ will find it in the interval $(5.8, 9.4)$. The corresponding range in the standardized variables are

$$z_{min} = \frac{5.8 - 7}{2} = -0.6 \quad \text{and} \quad z_{max} = \frac{9.4 - 7}{2} = 1.2 \tag{B.6}$$

The probability in this interval is

$$\text{Prob}[z \in (-0.6, 1.2)] = F(1.2) - F(-0.6)$$
$$= 0.8849 - (1.0000 - 0.7257) = 0.6016 \tag{B.7}$$

or about 60% of the total.

Table B.1. Values of the Cumulative Gaussian Probability Distribution Defined by the Integral in Eqn. (B.4)

z	$F(z)$	z	$F(z)$	z	$F(z)$
0.0	0.5000	1.0	0.8413	2.0	0.9722
0.1	0.5398	1.1	0.8643	2.1	0.9821
0.2	0.5793	1.2	0.8849	2.2	0.9861
0.3	0.6179	1.3	0.9032	2.3	0.9893
0.4	0.6554	1.4	0.9192	2.4	0.9918
0.5	0.6915	1.5	0.9332	2.5	0.9938
0.6	0.7257	1.6	0.9452	2.6	0.9953
0.7	0.7580	1.7	0.9554	2.7	0.9965
0.8	0.7881	1.8	0.9641	2.8	0.9974
0.9	0.8159	1.9	0.9713	2.9	0.9981
1.0	0.8413	2.0	0.9772	3.0	0.9987

B.4 **Problems**

PB.1. Verify the probabilities of measuring a Gaussian distribution in the intervals $(\mu-\sigma, \mu+\sigma)$, $(\mu-2\sigma, \mu+2\sigma)$, and $(\mu-3\sigma, \mu+3\sigma)$ as discussed in Example 4.2. How far away from μ (in terms of σ) should one go (symmetrically) on either side to have half of the probability contained under the Gaussian integral?

APPENDIX C

Complex Numbers and Functions

Because the formalism of quantum mechanics requires the manipulation of complex variables, we review here some of the basic definitions and formulae governing their properties. The imaginary unit, i, is defined[1] via $i = \sqrt{-1}$, and a general *complex number* is given by

$$z = a + ib \tag{C.1}$$

where a, b themselves have purely real values. The values a and b are called the *real* and *imaginary* parts of z, respectively; these are often written in the form

$$a = Re(z) \quad \text{and} \quad b = Im(z) \tag{C.2}$$

Complex numbers obey standard algebraic relations. For example, if $z_{1,2} = a_{1,2} + ib_{1,2}$, we have addition and subtraction given by

$$z_1 \pm z_2 = (a_1 \pm a_2) + i(b_1 \pm b_2) \tag{C.3}$$

while multiplication is carried out by using the distributive law, giving

$$z_1 z_2 = (a_1 + ib_1)(a_2 + ib_2) = (a_1 a_2 - b_1 b_2) + i(a_1 b_2 + a_2 b_1) \tag{C.4}$$

More complicated functions can often be evaluated using series expansions. For example, for θ real we can write

$$e^{i\theta} = 1 + (i\theta) + \frac{(i\theta)^2}{2!} + \frac{(i\theta)^3}{3!} + \cdots$$

$$= \left(1 - \frac{\theta^2}{2} + \cdots\right) + i\left(\theta - \frac{\theta^3}{3!} + \cdots\right)$$

$$e^{i\theta} = \cos(\theta) + i\sin(\theta) \tag{C.5}$$

using the series expansions in Appendix D.2.

---

[1] Engineers often use the notation $j = \sqrt{-1}$.

The *complex conjugate* of a complex number is defined via

$$z^* \equiv a - ib \tag{C.6}$$

that is, by letting $i \to -i$. A useful relation is

$$|z|^2 \equiv zz^* = (a + ib)(a - ib) = a^2 + b^2 \tag{C.7}$$

which defines the *modulus*, $|z|$, of a complex number via

$$|z| = \sqrt{a^2 + b^2} \tag{C.8}$$

This quantity is the analog of the absolute value of a real number. We can make use of the identity (C.5) to also write

$$a + ib = z \equiv |z|e^{i\theta} = |z|\cos(\theta) + i|z|\sin(\theta) \tag{C.9}$$

where θ is called the *phase* or *argument* of the complex number z; it is given by

$$\tan(\theta) = \frac{b}{a} = \frac{Im(z)}{Re(z)} \tag{C.10}$$

This form for complex numbers is useful as it shows that

$$|z_1 z_2| = |z_1||z_2| \tag{C.11}$$

We obviously have $z^* = |z|e^{-i\theta}$, so complex conjugation "flips the phase" of z, but keeps its modulus fixed. A complex number with $|z| = 1$, that is, of the form $z = e^{i\theta}$, is often said to be "just a phase". A general complex number can be represented as a point (or vector) in the complex plane, as shown in Fig. C.1, and addition and subtraction can be given a vector interpretation.

Some useful formulae (for θ real) are

$$\cos(\theta) = \frac{e^{i\theta} + e^{-i\theta}}{2} \quad \text{and} \quad \sin(\theta) = \frac{e^{i\theta} - e^{-i\theta}}{2i} \tag{C.12}$$

which are easily derived by combining Eqn. (C.5) and its complex conjugate. One also has

$$\cos(i\theta) = \cosh(\theta) \quad \text{and} \quad \sin(i\theta) = i\sinh(\theta) \tag{C.13}$$

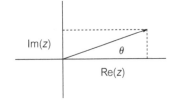

Figure C.1. Representation of a complex number, z, as a vector in the complex plane.

Other familiar trig identities are easily proved using complex notation. For example, one has

$$2 \sin \left(\frac{\alpha + \beta}{2} \right) \cos \left(\frac{\alpha - \beta}{2} \right)$$

$$= 2 \left(\frac{e^{i(\alpha+\beta)/2} - e^{-i(\alpha+\beta)/2}}{2i} \right) \left(\frac{e^{i(\alpha-\beta)/2} + e^{i(\alpha-\beta)/2}}{2} \right)$$

$$= \sin(\alpha) + \sin(\beta) \tag{C.14}$$

One also has

$$\sin(\alpha \pm \beta) = \sin(\alpha) \cos(\beta) \pm \cos(\alpha) \sin(\beta) \tag{C.15}$$

$$\cos(\alpha \pm \beta) = \cos(\alpha) \cos(\beta) \mp \sin(\alpha) \sin(\beta) \tag{C.16}$$

and the related special cases for double-angle formulae,

$$\sin(2\alpha) = 2 \sin(\alpha) \cos(\alpha) \tag{C.17}$$

$$\cos(2\alpha) = \cos^2(\alpha) - \sin^2(\alpha) \tag{C.18}$$

or

$$\sin^2(\alpha) = \frac{1}{2}[1 - \cos(2\alpha)] \tag{C.19}$$

$$\cos^2(\alpha) = \frac{1}{2}[1 + \cos(2\alpha)] \tag{C.20}$$

C.1 **Problems**

PC.1. Calculate the result of dividing two complex numbers; specifically, if

$$Re(w) + i Im(w) = w = \frac{z_1}{z_2} = \frac{a_1 + i b_1}{a_2 + i b_2} \tag{C.21}$$

find explicit expressions for $Re(w), Im(w)$.

PC.2. Find a general expression for the modulus of $|w|$ if

$$w = \exp \left(\frac{1}{a + ib} \right) \tag{C.22}$$

where a, b are real; find a numerical value if $a = 2$ and $b = 1$.

PC.3. Verify the identity

$$|e^{i\alpha} + e^{i\beta}| = \left| 2 \cos \left(\frac{\alpha + \beta}{2} \right) \right| \tag{C.23}$$

if α, β are real. Derive a similar identity for $|e^{i\alpha} - e^{i\beta}|$.

PC.4. If $z = 2 - 3i$, find $\sqrt{z}$.

PC.5. Show that the following identities hold for any positive integer n,

$$e^{-\pi i n^2/2} = \begin{cases} +1 & \text{for } n \text{ even} \\ -i & \text{for } n \text{ odd} \end{cases} = \left(\frac{1-i}{2}\right) + \left(\frac{1+i}{2}\right)(-1)^n$$

$$= \frac{1}{\sqrt{2}}\left[e^{-i\pi/4} + e^{+i\pi/4}(-1)^n\right] \qquad (\text{C}.24)$$

APPENDIX D

Integrals, Summations, and Calculus Results

D.1 Integrals

In this section we collect many of the nontrivial indefinite and definite integrals which may be needed for most of the derivations or exercises in the text. Some of them are evaluated using sophisticated methods (such as contour integration, discussed briefly in Section D.4), but we are only interested in using this collection as a reference. The reader is urged to consult other mathematical handbooks or especially to make use of symbolic manipulation programs such as *Mathematica*® or *Maple*®.

We begin by recalling that the simple rule for the differentiation of product functions

$$\frac{d}{dx}[f(x)\,g(x)] = \frac{df(x)}{dx}g(x) + f(x)\frac{dg(x)}{dx} \tag{D.1}$$

is the basis for the integration by parts (or IBP) method which we use frequently, namely

$$\int_a^b dx\,\frac{df(x)}{dx}g(x) = -\int_a^b dx f(x)\frac{dg(x)}{dx} + \left[f(x)\,g(x)\right]_a^b \tag{D.2}$$

Some standard indefinite integrals:

$$\int \frac{dx}{x^2 + a^2} = \frac{1}{a}\tan^{-1}\left(\frac{x}{a}\right) \tag{D.3}$$

$$\int \frac{dx}{a^2 - x^2} = \frac{1}{2a}\log\left(\frac{a + x}{a - x}\right) \quad (a^2 > x^2) \tag{D.4}$$

$$\int (\sin(ax))\,(\sin(bx))\,dx = \frac{\sin(a - b)x}{2(a - b)} - \frac{\sin(a + b)x}{2(a + b)} \quad (a^2 \neq b^2) \tag{D.5}$$

$$\int (\cos(ax))\,(\cos(bx))\,dx = \frac{\sin(a - b)x}{2(a - b)} + \frac{\sin(a + b)x}{2(a + b)} \quad (a^2 \neq b^2) \tag{D.6}$$

$$\int (\sin(ax))\,(\cos(bx))\ dx = -\frac{\cos(a-b)x}{2(a-b)} - \frac{\cos(a+b)x}{2(a+b)} \qquad (a^2 \neq b^2)$$
(D.7)

$$\int dx\, x \sin(ax) = \frac{1}{a^2}\sin(ax) - \frac{x}{a}\cos(ax)$$
(D.8)

$$\int dx\, x \cos(ax) = \frac{1}{a^2}\cos(ax) + \frac{x}{a}\sin(ax)$$
(D.9)

$$\int dx\, x^2 \sin(ax) = \frac{2x}{a^2}\sin(ax) - \frac{(a^2x^2-2)}{a^3}\cos(ax)$$
(D.10)

$$\int dx\, x^2 \cos(ax) = \frac{2x}{a^2}\cos(ax) + \frac{(a^2x^2-2)}{a^3}\sin(ax)$$
(D.11)

$$\int dx\, x^4 \sin(ax) = \frac{4x(a^2x^2-6)}{a^4}\sin(ax)$$
$$- \frac{(a^4x^4 - 12a^2x^2 + 24)}{a^5}\cos(ax)$$
(D.12)

$$\int dx\, x^4 \cos(ax) = \frac{4x(a^2x^2-6)}{a^4}\cos(ax)$$
$$+ \frac{(a^4x^4 - 12a^2x^2 + 24)}{a^5}\sin(ax)$$
(D.13)

$$\int dx\, x \sin^2(ax) = \frac{x^2}{4} - \frac{x\sin(2ax)}{4a} - \frac{\cos(2ax)}{8a^2}$$
(D.14)

$$\int dx\, x \cos^2(ax) = \frac{x^2}{4} + \frac{x\sin(2ax)}{4a} + \frac{\cos(2ax)}{8a^2}$$
(D.15)

$$\int dx\, x^2 \sin^2(ax) = \frac{x^3}{6} - \left(\frac{x^2}{4a} - \frac{1}{8a^3}\right)\sin(2ax) - \frac{x\cos(2ax)}{4a^2}$$
(D.16)

$$\int dx\, x^2 \cos^2(ax) = \frac{x^3}{6} + \left(\frac{x^2}{4a} - \frac{1}{8a^3}\right)\sin(2ax) + \frac{x\cos(2ax)}{4a^2}$$
(D.17)

$$\int e^{ax}\ dx = \frac{1}{a}e^{ax}$$
(D.18)

$$\int x\, e^{ax}\ dx = \frac{1}{a^2}(ax-1)e^{ax}$$
(D.19)

$$\int x^2\, e^{ax}\ dx = \frac{1}{a^3}(a^2x^2 - 2ax + 2)e^{ax}$$
(D.20)

Some definite integrals:

$$\int_{-\infty}^{+\infty} \frac{\sin(x)}{x} \, dx = \pi \tag{D.21}$$

$$\int_{-\infty}^{+\infty} \frac{\sin^2(x)}{x^2} \, dx = \pi \tag{D.22}$$

$$\int_{-\infty}^{+\infty} \frac{\sin^4(x)}{x^2} \, dx = \frac{\pi}{2} \tag{D.23}$$

$$\int_{-\infty}^{+\infty} \frac{(1 - \cos(x))}{x^2} \, dx = \pi \tag{D.24}$$

$$\int_{-\infty}^{+\infty} \frac{(1 - \cos(x))^2}{x^2} \, dx = \pi \tag{D.25}$$

$$\int_{-\infty}^{+\infty} \frac{\sin(x) \cos(x)}{x} \, dx = \frac{\pi}{2} \tag{D.26}$$

$$\int_{-\infty}^{+\infty} \frac{\sin(x) \cos(mx)}{x} \, dx = \begin{cases} 0 & \text{for } |m| > 1 \\ \pi/2 & \text{for } |m| = 1 \\ \pi & \text{for } |m| < 1 \end{cases} \tag{D.27}$$

$$\int_{-\infty}^{+\infty} \frac{\sin(x_1 - x) \sin(x_2 - x)}{(x - x_1)(x - x_2)} \, dx = \pi \, \frac{\sin(x_1 - x_2)}{(x_1 - x_2)} \tag{D.28}$$

$$\int_0^\infty \frac{\cos(mx)}{x^2 + a^2} \, dx = \frac{\pi}{2|a|} e^{-|ma|} \tag{D.29}$$

$$\int_0^\infty \frac{\cos(mx) \cos(nx)}{x^2 + a^2} \, dx = \frac{\pi}{a} \left(e^{-|(m-n)a|} + e^{-|(m+n)a|} \right) \tag{D.30}$$

$$\int_0^\infty \frac{\sin(mx) \sin(nx)}{x^2 + a^2} \, dx = \frac{\pi}{a} \left(e^{-|(m-n)a|} - e^{-|(m+n)a|} \right) \tag{D.31}$$

$$\int_0^\infty \cos(mx) \, e^{-ax} \, dx = \frac{a}{a^2 + m^2} \quad (a > 0) \tag{D.32}$$

$$\int_0^\infty \sin(mx) \, e^{-ax} \, dx = \frac{m}{a^2 + m^2} \quad (a > 0) \tag{D.33}$$

The following integrals make use of the Euler Gamma function ($\Gamma(z)$), the generalized factorial function, as discussed in Appendix **E**.9.

$$\int_0^\infty x^n \, e^{-x} \, dx = n! = \Gamma(n + 1) \tag{D.34}$$

$$\int_0^\infty dx \, x^n \, e^{-(ax)^m} = \frac{1}{ma^{n+1}} \Gamma\left(\frac{n+1}{m}\right) \tag{D.35}$$

$$\int_0^1 \frac{x^m \, dx}{\sqrt{1-x^n}} = \frac{\Gamma(1/2)\Gamma((m+1)/n)}{n\Gamma(1/2+(m+1)/n)} \tag{D.36}$$

$$\int_0^1 x^{m-1} (1-x)^{n-1} \, dx = \frac{\Gamma(n)\Gamma(n)}{\Gamma(m+n)} \tag{D.37}$$

$$\int_0^\infty \frac{x^a \, dx}{(m+x^b)^c} = \frac{m^{(a+1-bc)/b}}{b} \left[\frac{\Gamma((a+1)/b))\,\Gamma(c-(a+1)/b)}{\Gamma(c)}\right] \tag{D.38}$$

Integrals containing Gaussian terms of the form $\exp(-ax^2)$ are of special importance and we discuss their evaluation in slightly more detail. The standard trick for the evaluation of the basic integral

$$I \equiv \int_{-\infty}^{+\infty} dx \, \exp(-x^2) = \sqrt{\pi} \tag{D.39}$$

is to consider

$$I^2 = I \cdot I = \left(\int_{-\infty}^{+\infty} dx \, \exp(-x^2)\right) \cdot \left(\int_{-\infty}^{+\infty} dy \, \exp(-y^2)\right)$$

$$= \int_{-\infty}^{+\infty} \int_{-\infty}^{+\infty} dx \, dy \, \exp(-x^2 - y^2)$$

$$= \int_0^\infty \int_0^{2\pi} r \, dr \, d\theta \, \exp(-r^2)$$

$$= 2\pi \int_0^\infty dr \, r \, \exp(-r^2)$$

$$I^2 = \pi \tag{D.40}$$

so that $I = \sqrt{\pi}$. The more general basic integral is

$$I(a) = \int_{-\infty}^{+\infty} dx \, \exp(-ax^2) = \sqrt{\frac{\pi}{a}} \tag{D.41}$$

and a related one is

$$I(a, b) \equiv \int_{-\infty}^{+\infty} dx \, \exp(-ax^2 - bx)$$

$$= \int_{-\infty}^{+\infty} dx \, \exp(-a(x^2 + bx/a + b^2/4a^2 - b^2/4a)$$

$$= \exp(b^2/4a) \int_{-\infty}^{+\infty} dx \, \exp(-a(x + b/a)^2)$$

$$I(a, b) = \exp(b^2/4a)\sqrt{\frac{\pi}{a}} \tag{D.42}$$

where we have used a standard method of completing the square and shifting variables. One can generalize these expressions further by "differentiating under the integral sign" to obtain

$$J(a, b; n) = \int_{-\infty}^{+\infty} dx \, x^n \exp(-ax^2 - bx)$$

$$= \left(-\frac{\partial}{\partial b}\right)^n I(a, b) = \left(-\frac{\partial}{\partial b}\right)^n \left[\exp(b^2/4a)\sqrt{\frac{\pi}{a}}\right] \tag{D.43}$$

For example, one has

$$J(a, b; 1) \equiv \int_{-\infty}^{+\infty} x \exp(-ax^2 + bx) \, dx = \left(\frac{b}{2a}\right) \exp(b^2/4a)\sqrt{\frac{\pi}{a}} \tag{D.44}$$

$$J(a, b; 2) \equiv \int_{-\infty}^{+\infty} x^2 \exp(-ax^2 + bx) \, dx = \left(\frac{b^2 + 2a}{4a^2}\right) \exp(b^2/4a)\sqrt{\frac{\pi}{a}} \tag{D.45}$$

and so forth. For even values of $n = 2k$ we can also evaluate $J(a, b; n = 2k)$ by using

$$J(a, b; 2k) = \int_{-\infty}^{+\infty} x^{2k} e^{-ax^2 - bx} \, dx = \left(-\frac{\partial}{\partial a}\right)^k I(a, b) \tag{D.46}$$

Integrals involving $\cos(kx)$ and $\sin(kx)$ terms and Gaussian integrands can also be done by using identities such as $\cos(kx) = [\exp(+ikx) + \exp(-ikx)]/2$ to obtain

$$\int_{-\infty}^{+\infty} e^{-ax^2 - bx} \cos(kx) \, dx = +\sqrt{\frac{\pi}{a}} \, e^{(b^2 - k^2)/4a} \cos(kb/2a) \tag{D.47}$$

$$\int_{-\infty}^{+\infty} e^{-ax^2 - bx} \sin(kx) \, dx = -\sqrt{\frac{\pi}{a}} \, e^{((b^2 - k^2)/4a} \sin(kb/2a) \tag{D.48}$$

Integrals containing $ax^2 + bx + c$ arise in the study of the classical limit of the hydrogen atom and elsewhere. If we define $X = ax^2 + bx + c$ and $q = 4ac - b^2$, one has

$$\int \frac{dx}{\sqrt{X}} = -\frac{1}{\sqrt{-a}} \sin^{-1}\left(\frac{2ax + b}{\sqrt{-q}}\right) \qquad (a < 0) \tag{D.49}$$

$$\int \frac{x \, dx}{\sqrt{X}} = \frac{\sqrt{X}}{a} - \frac{b}{2a} \int \frac{dx}{\sqrt{X}} \tag{D.50}$$

$$\int \frac{x^2\, dx}{\sqrt{X}} = \left(\frac{x}{2a} - \frac{3b}{4a^2}\right)\sqrt{X} + \frac{3b^2 - 4ac}{8a^2}\int \frac{dx}{\sqrt{X}} \tag{D.51}$$

$$\int \frac{x^3\, dx}{\sqrt{X}} = \left(\frac{x^2}{3a} - \frac{5bx}{12a^2} + \frac{5b^2}{8a^3} - \frac{2c}{3a^2}\right)\sqrt{X} + \left(\frac{3bc}{4a^2} - \frac{5b^3}{16a^3}\right)\int \frac{dx}{\sqrt{X}} \tag{D.52}$$

$$\int \frac{dx}{x\sqrt{X}} = \frac{1}{\sqrt{-c}}\sin^{-1}\left(\frac{bx + 2c}{|x|\sqrt{-q}}\right) \qquad (c < 0) \tag{D.53}$$

D.2 Summations and Series Expansions

We collect here some useful results which evaluate the summations of certain finite and infinite series.

$$\sum_{k=1}^{k=N} x^k = \frac{1 - x^{N+1}}{1 - x} \tag{D.54}$$

$$\sum_{k=1}^{k=N} k = \frac{N(N+1)}{2} \tag{D.55}$$

$$\sum_{k=1}^{k=N} k^2 = \frac{N(N+1)(2N+1)}{6} \tag{D.56}$$

The *Riemann zeta function* is defined via

$$\zeta(s) = 1 + \frac{1}{2^s} + \frac{1}{3^s} + \cdots = \sum_{n=1}^{\infty} \frac{1}{n^s} \tag{D.57}$$

Some special cases are

$$\zeta(2) = \frac{\pi^2}{6} \qquad \zeta(4) = \frac{\pi^4}{90} \qquad \zeta(6) = \frac{\pi^6}{945} \qquad \zeta(8) = \frac{\pi^8}{9450} \tag{D.58}$$

One can also show that

$$\zeta_{odd}(s) \equiv \frac{1}{1} + \frac{1}{3^s} + \frac{1}{5^s} + \cdots = \sum_{n=1}^{\infty} \frac{1}{(2n-1)^s} = \left(1 - \frac{1}{2^s}\right)\zeta(s) \tag{D.59}$$

so that

$$\zeta_{odd}(2) = \frac{\pi^2}{8} \qquad \text{and} \qquad \zeta_{odd}(4) = \frac{\pi^4}{96} \tag{D.60}$$

$$\zeta_{odd}(6) = \frac{63}{64} \cdot \frac{\pi^6}{945} = \frac{\pi^6}{960}$$

One also has:

$$S(x) = \sum_{k=1}^{\infty} \frac{1}{((2k-1)^2 - x^2)} = \frac{\pi}{2x} \tan\left(\frac{\pi x}{2}\right) \tag{D.61}$$

$$T(x) = \sum_{k=1}^{\infty} \frac{1}{((2k-1)^2 - x^2)^2} = \frac{\pi}{16x^3}\left[\pi x \sec^2\left(\frac{\pi x}{2}\right) - 2\tan\left(\frac{\pi x}{2}\right)\right] \tag{D.62}$$

The *Taylor series expansion* of a (well-behaved) function $f(x)$ about the point $x = a$ is given by

$$f(x) = f(a) + f'(a)(x - a) + \frac{1}{2!}f''(a)(x - a)^2 + \cdots$$

$$= \sum_{n=0}^{\infty} \frac{f^{(n)}(a)}{n!}(x - a)^2 \tag{D.63}$$

where

$$f^{(n)}(a) = \frac{d^n f(x)}{dx^m}\bigg|_{x=a} \tag{D.64}$$

is the nth derivative of $f(x)$ evaluated at $z = a$. Familiar examples include:

$$(1 \pm x)^n = 1 \pm nx + \frac{n(n-1)}{2!}x^2 \pm \frac{n(n-1)(n-2)}{3!}x^3 + \cdots$$

$$= \sum_{k=1}^{\infty} (\pm 1)^k \frac{n!}{(n-k)!k!}x^k \quad \text{for } |x| < 1 \tag{D.65}$$

$$\sqrt{1 \pm x} = 1 \pm \frac{x}{2} - \frac{x^2}{8} \pm \frac{x^3}{16} + \cdots \quad \text{for } |x| < 1 \tag{D.66}$$

$$e^x = 1 + x + \frac{x^2}{2!} + \frac{x^3}{3!} + \cdots = \sum_{k=0}^{\infty} \frac{x^k}{k!} \quad \text{for all real } x \tag{D.67}$$

$$\ln(1 + x) = x - \frac{x}{2} + \frac{x^3}{3} + \cdots = \sum_{k=1}^{\infty} (-1)^k \frac{x^k}{k} \quad \text{for } -1 < x \leq +1 \tag{D.68}$$

$$\sin(x) = x - \frac{x^3}{3!} + \frac{x^5}{5!} + \cdots = \sum_{k=1}^{\infty} (-1)^k \frac{x^{2k-1}}{(2k-1)!} \quad \text{for all real } x \tag{D.69}$$

$$\cos(x) = 1 - \frac{x^2}{2!} + \frac{x^4}{4!} + \cdots = \sum_{k=0}^{\infty} (-1)^k \frac{x^{2k}}{(2k)!} \quad \text{for all real } x \tag{D.70}$$

$$\tan(x) = x - \frac{x^3}{3!} + \frac{2x^5}{15} + \cdots \quad \text{for } |x| < \pi/2 \tag{D.71}$$

A useful tool to investigate the convergence of a series expansion is the *ratio test*. If an infinite summation is defined via

$$S = \sum_{n=0}^{\infty} \rho_n \qquad \text{(D.72)}$$

the limit of successive ratios is defined via

$$\rho = \lim_{n \to \infty} \frac{\rho_{n+1}}{\rho_n} \qquad \text{(D.73)}$$

One then knows that

- The series *converges* (S is finite) if $\rho < 1$,
- The series *diverges* (S is infinite) if $\rho > 1$,
- The test is inconclusive (the series may either converge or diverge) if $\rho = 1$.

If the terms in the series (i) alternate in sign, (ii) decrease in magnitude (each one smaller than the one before it), and the terms approach zero, then the series is known to converge by *Liebniz's theorem*.

It is often useful to recall the definition of the (one-dimensional) integral as the "area under the curve." The trapezoidal approximation to the area under $f(x)$ in the interval (a, b) is obtained by splitting the interval into N equal pieces of size $h = (b - a)/N$ which gives

$$\int_a^b dx\, f(x) \approx F_N(a, b) \equiv h \left(\frac{1}{2} f(a) + \sum_{n=1}^{N-1} f(a + nh) + \frac{1}{2} f(b) \right) \qquad \text{(D.74)}$$

This expression can form the basis for the simplest of numerical integration programs if necessary. The *Euler–Maclaurin formula* describes the difference between these two approximations via

$$\left[\int_a^b dx\, f(x) \right] - F_N(a, b) = -\left. \frac{B_2}{2!} h^2 f'(x) \right|_a^b - \left. \frac{B_4}{4!} h^4 f'''(x) \right|_a^b + \cdots \qquad \text{(D.75)}$$

where the B_n are the Bernoulli numbers the first few of which are

$$B_0 = 1 \qquad B_2 = \frac{1}{6} \qquad B_4 = -\frac{1}{30} \qquad B_6 = \frac{1}{42} \qquad \text{(D.76)}$$

D.3 **Assorted Calculus Results**

The gradient-squared operator or *Laplacian operator* in rectangular (Cartesian), cylindrical (polar), or spherical coordinates is given by

$$\nabla^2 f(x, y, z) = \frac{\partial^2 f}{\partial x^2} + \frac{\partial^2 f}{\partial y^2} + \frac{\partial^2 f}{\partial z^2} \tag{D.77}$$

$$\nabla^2 f(r, \theta, z) = \frac{1}{r}\frac{\partial}{\partial r}\left(r\frac{\partial f}{\partial r}\right) + \frac{1}{r^2}\frac{\partial^2 f}{\partial \theta^2} + \frac{\partial^2 f}{\partial z^2} \tag{D.78}$$

$$\nabla^2 f(r, \theta, \phi) = \frac{1}{r^2}\frac{\partial}{\partial r}\left(r^2\frac{\partial f}{\partial r}\right) + \frac{1}{r^2 \sin(\theta)}\frac{\partial}{\partial \theta}\left(\sin(\theta)\frac{\partial f}{\partial \theta}\right)$$
$$+ \frac{1}{r^2 \sin^2(\theta)}\frac{\partial^2 f}{\partial \phi^2} \tag{D.79}$$

If one changes variables in a multidimensional integral, one must also apply the appropriate transformation in the "infinitesimal measure" as well. For example, if one changes variables via

$$x, y \implies u(x, y), v(x, y) \tag{D.80}$$

then one has the relation

$$du\, dv = J(w, v; x, y)\, dx\, dy = \det\begin{pmatrix} \partial u/\partial x & \partial u/\partial y \\ \partial v/\partial x & \partial v/\partial y \end{pmatrix} dx\, dy \tag{D.81}$$

with similar extensions to more dimensions; the function $J(u, v; x, y)$ is called the *Jacobian* of the transformation. You should be able to use this approach to derive the familiar result that $dx\, dy\, dz = d\mathbf{r} = r^2\, dr \sin(\theta)\, d\theta\, d\phi$.

D.4 **Real Integrals by Contour Integration**

Large numbers of useful real integrals, especially ones involving integration over the entire real line, can be done using simple contour integration techniques, making use of complex variables. We *very* briefly review the rudimentary complex analysis and "tricks of the trade" needed to implement many such integrals, leaving detailed discussions to undergraduate texts on mathematical methods.

The basic result from complex analysis which is required is the *residue theorem* which simplifies the evaluation of integrals of complex functions about a closed curve C in the complex plane, using only knowledge of the structure of the poles (and essential singularities) that are enclosed by C. The appropriate connection is given by

$$\oint_C f(z) \, dz = 2\pi i \sum_i \mathcal{R}_i \tag{D.82}$$

where for a function $f(z)$ which has a pole of order n at $z = z_0$, the residue, $\mathcal{R}_i$, is given by

$$\mathcal{R}_i = \frac{1}{(n-1)!} \left\{ \left(\frac{d}{dz}\right)^{n-1} \left[(z - z_0)^n f(z)\right] \right\}_{z \to z_0} \tag{D.83}$$

The closed contour C is assumed to have a counterclockwise orientation, while if it is completed in a clockwise direction, an overall minus sign is added to the right-hand side of Eqn. (D.82). A judicious choice of an appropriate contour is often all that is needed to use the residue theorem to aid in the evaluation of integrals along the real-axis, and we present two exemplary cases.

Simple poles: Consider the real integral

$$I_1 = \int_{-\infty}^{+\infty} \frac{dx}{(1 + x^2)^3} \tag{D.84}$$

which is clearly square-integrable and convergent. Generalize this to the complex line integral given by

$$\mathcal{I}_1 = \oint_C \frac{dz}{(1 + z^2)^3} \tag{D.85}$$

over the contour shown in Fig. C.2(a), considering the limit that $R \to \infty$ so that the semicircle C_1 eventually extends to infinity, while the part of C along the real line reproduces I_1. For those values on the semicircle C_1, we can write the complex variable z in the form

$$z = Re^{i\theta} \quad \text{so that} \quad dz = iRe^{i\theta} \, d\theta \quad \text{giving} \quad \frac{dz}{(1 + z^2)^3} \to \frac{ie^{-5i\theta}}{R^5} \tag{D.86}$$

which becomes arbitrarily small as $R \to \infty$.

On the one hand, the complex integral over C is the sum of the desired real integral and that over the semicircular curve C_1 in the form

$$\mathcal{I}_1 = \int_{-R}^{+R} \frac{dx}{(1 + x^2)^3} + \int_{C_1 \, (R \to \infty)} \frac{dz}{(1 + z^2)^2} \to I_1 \tag{D.87}$$

in the limit that $R \to \infty$, since the contribution from C_1 vanishes. We can, however, also evaluate $\mathcal{I}_1$ using the residue theorem with $f(z) = 1/(1 + z^2)^3$,

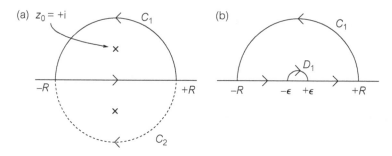

Figure D.1. Contours used in the evaluation of real integrals.

which has simple poles of order 3 at $z_0 = \pm i$. Using the contour C (semicircle C_1 plus real line) in Fig. D.1(a), which encloses the pole at $z_0 = +i$, we find

$$I_1 = \mathcal{I}_1 = 2\pi i \frac{1}{2!} \left\{ \left(\frac{d}{dz}\right)^2 \left[(z-i)^3 \frac{1}{(z-i)^3(z+i)^3} \right] \right\}_{z \to +i}$$

$$= \pi i \left(\frac{(-3)(-4)}{(2i)^5} \right) = \frac{3\pi}{8} \tag{D.88}$$

If we consider the related contour consisting of the semicircle C_2 (dashed contour) and the real line (enclosing the pole at $z_0 = -i$), we obtain the same result (recall the additional minus sign if the contour orientation is clockwise.)

Deformed contours: Consider the integral

$$I_2 = \int_{-\infty}^{+\infty} \frac{\sin(x)}{x} \, dx \tag{D.89}$$

which despite appearances is everywhere finite and convergent. The integrand is well behaved at $x = 0$ (since $\lim_{x\to 0} \sin(x)/x = 1$) and while the large $|x|$ dependence of $1/x$ would yield a logarithmic divergence by itself, the alternation of signs due to the oscillatory $\sin(x)$ gives convergence; think of the integral as an infinite sum of terms (the positive and negative areas defining the area under the integrand) with alternating signs, and of decreasing magnitude, which guarantees so-called *conditional convergence*.

In this case we choose to write I_2 in terms of the imaginary part (Im) of an already complex integral in the form

$$I_2 = Im \left[\int_{-\infty}^{+\infty} \frac{e^{ix}}{x} \, dx \right] \equiv Im[\tilde{I}] \tag{D.90}$$

and use contour integration over the deformed semicircle shown in Fig. C.2(b), where the region near $z = 0$ is treated more carefully, with a smaller semicircle of radius ϵ. The contour integral over C_1, D_1 and the integrals over the ranges

$(-R, -\epsilon)$ and $(+\epsilon, +R)$ constitute the desired contour C, but since it encloses no poles we have

$$\mathcal{I}_2 \equiv \int_C \frac{e^{iz}}{z} \, dz = 2\pi i \sum_i \mathcal{R}_i = 0 \tag{D.91}$$

via the residue theorem. The integral $\mathcal{I}_2$ can also be split up into the four contributions

$$\mathcal{I}_2 = \left[\int_{-R}^{-\epsilon} + \int_{+\epsilon}^{+R} \frac{e^{ix}}{x} \, dx \right] + \int_{C_1} \frac{e^{iz}}{z} \, dz + \int_{D_1} \frac{e^{iz}}{z} \, dz \tag{D.92}$$

eventually taking the twin limits $\epsilon \to 0$ and $R \to \infty$. The first two terms give $\tilde{I}$ in this limit, while for C_1 we use the same parameterization as in Eqn. (D.86) and it is easy to show that the contribution is exponentially suppressed as $\exp(-R\sin(\theta))$ as $R \to \infty$. Finally, for the contribution around D_1, we use $z = \epsilon \exp(i\theta)$ which gives

$$\int_{D_1} \frac{e^{iz}}{z} \, dz = \int_\pi^0 \frac{e^{i\epsilon e^{i\theta}}}{\epsilon e^{i\theta}} \left(i\epsilon e^{i\theta} \, d\theta \right) \longrightarrow i \int_\pi^0 d\theta = -i\pi \tag{D.93}$$

as $\epsilon \to 0$. From Eqns (D.91) and (D.92) we then find that $0 = \mathcal{I}_2 = \tilde{I} + 0 - i\pi$ so that

$$\int_{-\infty}^{+\infty} \frac{e^{ix}}{x} \, dx = \tilde{I} = i\pi \quad \text{and} \quad I_2 = \text{Im}[\tilde{I}] = \text{Im}[i\pi] = \pi \tag{D.94}$$

A large number of other similar integrals can be done using the deformed contour shown in Fig. C.2(b). For example, you should be able to show that

$$\int_{-\infty}^{+\infty} \frac{[1 - \cos(y)]}{y^2} \, dy = \pi \tag{D.95}$$

by taking the real part of a related integral, and extending it to the contour used above.

D.5 **Plotting**

The functional relationship between two variables is often best exemplified or analyzed (or even discovered in the first place) by plotting the "data" in an appropriate manner. In this section, we briefly recall some of the basics of plotting techniques; because linear relations are easiest to visualize, many standard tricks rely on graphing data in such a way as to yield a straight line.

For variables which are connected by an exponential relation one has

$$\text{exponential:} \quad y = ae^{bx} \implies \ln(y) = \ln(a) + bx \qquad \text{(D.96)}$$

which suggests that one plot $\ln(y)$ versus x; this gives a so-called *semilog plot*. A straight-line fit on such a plot implies an exponential relation, and the "generalized slope" is given by

$$b = \frac{(\ln(y_2) - \ln(y_1))}{(x_2 - x_1)} = \frac{\ln(y_2/y_1)}{(x_2 - x_1)} \qquad \text{(D.97)}$$

The value of a (or $\ln(a)$) plays the role of an "intercept" and can be extracted from any point on the line once b is known; if the point with $x = 0$ is included, then $y(0) = a$ is the most obvious choice.

For power-law relations of the form

$$\text{power-law:} \quad y = cx^d \implies \ln(y) = \ln(c) + d\ln(x) \qquad \text{(D.98)}$$

it is best to graph $\ln(y)$ versus $\ln(x)$ giving a *log–log plot* where the "generalized slope" is now

$$d = \frac{(\ln(y_2) - \ln(y_1))}{(\ln(x_2) - \ln(x_1))} = \frac{\ln(y_2/y_1)}{\ln(x_2/x_1)} \qquad \text{(D.99)}$$

D.6 **Problems**

PD.1. Derive any of the integrals in Eqns (D.5)–(D.7) by using complex exponentials.

PD.2. Derive any of the integrals in Eqns (D.8)–(D.11) using IBP techniques.

PD.3. Derive the integral in Eqn. (D.8) by differentiating both sides of the relation

$$\int \cos(ax)\, dx = \frac{1}{a}\sin(ax) \qquad \text{(D.100)}$$

with respect to a.

PD.4. Evaluate $J(a, b; n)$ in Eqn. (D.43) for $n = 2, 4$ by differentiating with respect to b. Then evaluate those two cases using Eqn. (D.46) by differentiating with respect to a and confirm you get the same answers.

PD.5. Derive Eqn. (D.59).

PD.6. Evaluate the integral in Eqn. (D.24) using contour integration.

PD.7. At very low temperatures, the heat capacity (at constant volume) of metals is expected to given by an expression of the form

$$C_V = \gamma T + AT^3 \qquad \text{(D.101)}$$

Given experimental values for T and $C_V(T)$, what would be the best way to plot the data to confirm such a relation and to most easily extract γ and A?

APPENDIX E

Special Functions

In this section, we collect some of the well-known properties of many of the special functions considered in this text. Most are quoted here without detailed proofs or derivations, while some have been discussed in a more physical context throughout the book.

E.1 Trigonometric and Exponential Functions

Although they are presumably familiar to all students, we briefly discuss the properties of the trigonometric and exponential functions. Since many of the special functions found in mathematical physics arise as solutions to similar differential equations, it is useful to recall here that:

- The differential equation

$$\frac{d^2 f(x)}{dx^2} = -k^2 f(x) \tag{E.1}$$

  has the (conventionally normalized) trig function solutions $f(x) = \sin(kx), \cos(kx)$, while

- The differential equation

$$\frac{d^2 f(x)}{dx^2} = +\kappa^2 f(x) \tag{E.2}$$

  has exponential solutions $f(x) = e^{\kappa x}, e^{-\kappa x}$.

The intuitive physical connections of these solutions with the oscillatory motion of a particle near a potential energy minimum (case (E.1)) versus the "runaway" (or damped) motion of a particle moved away from an unstable potential maximum (case (E.2)) can often be generalized to other differential equations to help understand the physical origin of the behavior of the solutions.

In each case above, as with any second-order ordinary differential equation, we obtain two, linearly independent solutions, $f_1(x)$, $f_2(x)$. The most general solution is then obtained by taking a linear combination $a_1 f_1(x) + a_2 f_2(x)$ and using the boundary conditions (in quantum mechanics) or initial conditions (in classical mechanics) to determine the arbitrary coefficients.

E.2 **Airy Functions**

The Airy differential equation is written in the form

$$\frac{d^2 f(x)}{dx^2} = xf(x) \tag{E.3}$$

Here we note the following:

- This problem is related to the quantum version of a particle moving under the influence of a uniform force.

- It also appears in the context of matching WKB-type (Chapter 10) solutions near classical turning points, where the potential energy function can be approximated (locally) as a linear potential.

- Using Eqn. (E.2) as a model, for $x > 0$, we expect exponentially damped or growing solutions; these should be consistent with the tunneling wavefunctions of Section 8.2.2.

- For $x < 0$ we expect oscillatory solutions with the period of oscillation *decreasing* for increasing $|x|$ as the "effective wave number" grows like $k^2 \sim |y|$.

The two linearly independent solutions are labeled $Ai(x)$ and $Bi(x)$ and are shown in Fig. E1. If we introduce the natural variable, $\zeta = 2x^{3/2}/3$, these solutions can be expanded for large values of $|x|$ as follows:

$$Ai(x) \longrightarrow \frac{1}{2} \frac{1}{\sqrt{\pi}\sqrt{x}} e^{-\zeta} \left[1 - \frac{c_1}{\zeta} + \cdots \right] \tag{E.4}$$

$$Ai(-x) \longrightarrow \frac{1}{\sqrt{\pi}\sqrt{x}} \left[\sin\left(\zeta + \frac{\pi}{4}\right) - \cos\left(\zeta + \frac{\pi}{4}\right) \frac{c_1}{\zeta} + \cdots \right] \tag{E.5}$$

$$Bi(x) \longrightarrow \frac{1}{\sqrt{\pi}\sqrt{x}} e^{\zeta} \left[1 + \frac{c_1}{\zeta} + \cdots \right] \tag{E.6}$$

$$Bi(-x) \longrightarrow \frac{1}{\sqrt{\pi}\sqrt{x}} \left[\cos\left(\zeta + \frac{\pi}{4}\right) + \sin\left(\zeta + \frac{\pi}{4}\right) \frac{c_1}{\zeta} + \cdots \right] \tag{E.7}$$

where $c_1 = 5/72$.

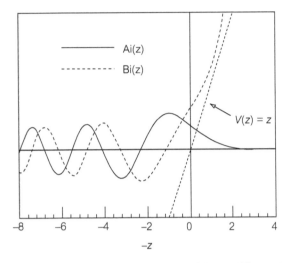

Figure E.1. Linearly independent solutions, $Ai(z)$ and $Bi(z)$, of the Airy differential equation.

E.3 Hermite Polynomials

The differential equation

$$\frac{d^2 h_n(z)}{dz^2} - 2z\frac{dh_n(z)}{dz} + 2nh_n(z) = 0 \tag{E.8}$$

is called *Hermite's equation* and has the solutions given by *Rodriges' formula*

$$h_n(z) = (-1)^n e^{z^2} \frac{d^n}{dz^n}\left(e^{-z^2}\right) \tag{E.9}$$

which are polynomials of degree n. The solutions are defined over the interval $(-\infty, +\infty)$ and satisfy the normalization condition

$$\int_{-\infty}^{+\infty} [h_n(z)]^2\, e^{-z^2}\, dz = 2^n n!\sqrt{\pi} \tag{E.10}$$

The first few Hermite polynomials are given by

$$h_0(z) = 1 \qquad h_1(z) = 2z \qquad h_2(z) = 4z^2 - 2 \qquad h_3(z) = 8z^3 - 12z \tag{E.11}$$

E.4 **Cylindrical Bessel Functions**

The free-particle Schrödinger and wave equation in two dimensions (three dimensions), when written in polar (cylindrical) coordinates, leads to the equation

$$\frac{d^2 R_m(z)}{dz^2} + \frac{1}{z}\frac{dR_m(z)}{dz} + \left(1 - \frac{m^2}{z^2}\right) R_m(z) = 0 \qquad (\text{E.12})$$

where we consider integral values of m. The solutions are generically called *cylindrical Bessel functions*, and for each $|m|$, the two linearly independent solutions are labeled $J_{|m|}(z)$ (Bessel functions of the first kind) or $Y_{|m|}(z)$ (Neumann functions or Bessel functions of the second kind). Their limiting behavior and properties are discussed and displayed graphically in Section 15.3.1.

E.5 **Spherical Bessel Functions**

The free-particle Schrödinger equation in three-dimensions written in spherical coordinates yields another version of Bessel's equation, namely

$$\frac{d^2 R_l(z)}{dz^2} + \frac{2}{z}\frac{dR_l(z)}{dz} + \left(1 - \frac{l(l+1)}{z^2}\right) R_l(z) = 0 \qquad (\text{E.13})$$

with l an integer. Its solutions are the *spherical Bessel functions*, $j_l(z)$ and $n_l(z)$, which can be written in a standard form, in terms of the cylindrical Bessel functions

$$j_l(z) = \sqrt{\frac{\pi}{2z}}J_{l+1/2}(z) \quad \text{and} \quad n_l(z) = \sqrt{\frac{\pi}{2z}}Y_{l+1/2}(z) \qquad (\text{E.14})$$

and are discussed in Section 16.6.

E.6 **Legendre Polynomials**

The (associated) Legendre's differential equation is written in the form

$$(1 - z^2)\frac{d^2 \Theta_{l,m}(z)}{dz^2} - 2z\frac{d\Theta_{l,m}(z)}{dz} + \left(l(l+1) - \frac{m^2}{(1 - z^2)}\right)\Theta_{l,m}(z) = 0$$

$$(\text{E.15})$$

The solutions are the *associated Legendre polynomials* given by

$$P_l^m(z) = (-1)^m \frac{(1-z^2)^{m/2}}{2^l \, l!} \left(\frac{d}{dz} \right)^{l+m} (z^2 - 1)^l \qquad \text{(E.16)}$$

for $m > 0$ and extended to negative m via

$$P_l^{-m}(z) = (-1)^m \frac{(l-m)!}{(l+m)!} P_l^m(z) \qquad \text{(E.17)}$$

They are defined over the interval $(-1, 1)$ and the normalization is such that

$$\int_{-1}^{+1} dz \, P_l^m(z) \, P_{l'}^m(z) = \int_0^\pi \sin(\theta) \, d\theta \, P_l^m(\cos(\theta)) \, P_{l'}^m(\cos(\theta))$$

$$= \frac{2}{2l+1} \frac{(l+m)!}{(l-m)!} \delta_{l,l'} \qquad \text{(E.18)}$$

The special case of $m = 0$ gives the *Legendre polynomials* which are defined via

$$P_l(z) \equiv P_l^{m=0}(z) \qquad \text{(E.19)}$$

which satisfy the differential equation

$$(1-z^2)\frac{d^2 P_l(z)}{dz^2} - 2z\frac{dP_l(z)}{dz} + l(l+1) P_l(z) = 0 \qquad \text{(E.20)}$$

E.7 Generalized Laguerre Polynomials

The differential equation

$$\frac{d^2 G(z)}{dz^2} + \left(\frac{\alpha - 1}{z} + 1 \right) \frac{dG(z)}{dz} + nG(z) = 0 \qquad \text{(E.21)}$$

is called *Laguerre's equation* and has polynomial solutions labeled as

$$G(z) = L_n^{(\alpha)}(z) \qquad \text{(E.22)}$$

which can be generated using *Rodrigues' formula*

$$L_n^{(\alpha)}(z) = \frac{e^z}{n! \, z^\alpha} \left(\frac{d}{dz} \right)^n \left[z^{n+\alpha} e^{-z} \right] \qquad \text{(E.23)}$$

The solutions are defined over the interval $(0, +\infty)$ and satisfy the normalization condition

$$\int_0^{+\infty} dz \, z^\alpha e^{-z} \left[L_n^{(\alpha)}(z) \right]^2 = \frac{\Gamma(n+\alpha+1)}{n!} \qquad \text{(E.24)}$$

E.8 **The Dirac δ-Function**

The Dirac δ-function was introduced and discussed extensively in Section 2.4 and here we only list some additional properties and identities. We recall that

$$\int_a^b dx\, f(x)\, \delta(x - c) = \begin{cases} f(c) & \text{if } a < c < b \\ 0 & \text{otherwise} \end{cases} \tag{E.25}$$

that is, the value of $f(x=c)$ is picked out from the integrand, or not, depending on whether the singularity is contained in the region of integration, or not. One can also derive (or justify) the following results.

$$\delta(ax) = \frac{1}{|a|}\, \delta(x) \tag{E.26}$$

$$\delta(x^2 - a^2) = \delta[(x - a)(x + a)]$$

$$= \frac{1}{|x + a|}\delta(x - a) + \frac{1}{|x - a|}\delta(x + a)$$

$$\delta(x^2 - a^2) = \frac{1}{2|a|}\left(\delta(x - a) + \delta(x + a)\right) \tag{E.27}$$

which is a special case of the more general relation

$$\delta[f(x)] = \sum_i \frac{\delta(x - x_i)}{|df/dx|_{x=x_i}} \tag{E.28}$$

where the sum is over all possible roots of $f(x_i) = 0$. Finally, recall that the *step-* or *Heaviside-function* is defined via

$$\Theta(x - a) = \begin{cases} 0 & \text{for } x < a \\ 1 & \text{for } x > a \end{cases} \tag{E.29}$$

and is given by

$$\Theta'(x - a) = \delta(x - a) \tag{E.30}$$

One can show that $\delta(x)$ can be obtained by taking the limit of the family of functions

$$\delta_\epsilon(x) = \frac{\epsilon}{\pi}\, \frac{\sin^2(x/\epsilon)}{x^2} \quad \text{as } \epsilon \to 0 \tag{E.31}$$

E.9 The Euler Gamma Function

Using integration by parts techniques, it is easy to derive Eqn. (D.34), namely

$$\int_0^\infty dx\, x^n\, e^{-x} = n(n-1)(n-2)\cdots 3\cdot 2\cdot 1 \equiv n! \qquad (E.32)$$

where $n!$ is read as "n-factorial." The integral can be generalized to noninteger values of n, and the *Gamma function* is defined in this way via

$$\int_0^\infty dx\, x^{n-1}\, e^{-x} \equiv \Gamma(n) \quad \text{for } n \neq 0, -1, -2, -3, \ldots \qquad (E.33)$$

For positive integers, it reduces to the factorial function

$$\Gamma(n) = (n-1)! \quad \text{for integral } n > 0 \qquad (E.34)$$

and also satisfies

$$\Gamma(n+1) = n\Gamma(n) \qquad (E.35)$$

$$\Gamma(n)\Gamma(1-n) = \frac{\pi}{\sin(n\pi)} \qquad (E.36)$$

Other special values are

$$\Gamma\left(\frac{1}{2}\right) = 2\int_0^\infty e^{-t^2}\, dt = \sqrt{\pi} \qquad (E.37)$$

which can be evaluated since this is now a Gaussian integral. This can be combined with Eqn. (E.36) (for nonnegative integral n) to give

$$\Gamma\left(n+\frac{1}{2}\right) = \frac{1\cdot 3\cdot 5\cdots(2n-1)}{2^n} = \frac{(2n-1)!!}{2^n}\sqrt{\pi} \qquad (E.38)$$

which implicitly defines the double-factorial function. Finally, we note that *Stirling's formula* can be used to estimate the value of the factorial function for large argument, namely

$$\Gamma(n+1) = n! \sim \sqrt{2\pi n}\left(\frac{n}{e}\right)^n\left(1 + \frac{1}{12n} + \frac{1}{288n^2} + \cdots\right) \qquad (E.39)$$

E.10 Problems

PE.1. Show that the solutions to the Airy differential equation can be written in terms of cylindrical Bessel functions (satisfying Eqn. (E.12) of fractional ($n = \pm 1/3$)

order. For example, a standard result is that

$$Ai(-x) = \frac{1}{3}\sqrt{x}\left[J_{1/3}(y) + J_{-1/3}(y)\right] \tag{E.40}$$

where $y = 2x^{3/2}/3$.

PE.2. Estimate the value of 20! using Stirling's formula and compare to the exact value.

APPENDIX F

Vectors, Matrices, and Group Theory

We collect here some of the most basic definitions and properties of real, finite-dimensional vectors and matrices. We intentionally ignore all of the subtleties regarding the precise definitions of vectors and tensors and supply only the "bare necessities". Many comments on the generalization of these ideas to complex vectors and infinite dimensional spaces are given in the text. We then briefly describe some of the rudiments of group theory.

F.1 Vectors and Matrices

We will take vectors to be ordered N-tuples of numbers, for example,

$$\mathbf{x} = (x_1, x_2, \ldots, x_N) \quad \text{and} \quad \mathbf{y} = (y_1, y_2, \ldots, y_N) \tag{F.1}$$

along with a *dot-* or *inner-product* of the form

$$\mathbf{x} \cdot \mathbf{y} = \sum_{i=1}^{N} x_i y_i = x_1 y_1 + x_2 y_2 + \cdots + x_N y_N \tag{F.2}$$

The *norm* (or generalized length) of the vector is taken to be

$$|x| = \sqrt{\mathbf{x} \cdot \mathbf{x}} \tag{F.3}$$

A *matrix* will be defined to be a square $N \times N$ array of the form

$$\mathbf{M} = \begin{pmatrix} M_{11} & M_{12} & \cdots & M_{1N} \\ M_{21} & M_{22} & \cdots & M_{2N} \\ \vdots & \vdots & \ddots & \vdots \\ M_{N1} & M_{N2} & \cdots & M_{NN} \end{pmatrix} \tag{F.4}$$

The *unit matrix* is given by

$$
1 = \begin{pmatrix} 1 & 0 & \cdots & 0 \\ 0 & 1 & \cdots & 0 \\ \vdots & \vdots & \ddots & 0 \\ 0 & 0 & \cdots & 1 \end{pmatrix}
\tag{F.5}
$$

Multiplication of a vector by a matrix *on the left* (as with operators) gives a vector, that is,

$$
\mathbf{x}' = \mathbf{M} \cdot \mathbf{x}
\tag{F.6}
$$

In component form one can write

$$
(x')_i = (\mathbf{M} \cdot \mathbf{x})_i = \sum_{j=1}^{N} M_{ij} x_j
\tag{F.7}
$$

or more explicitly

$$
\mathbf{M} \cdot \mathbf{x} = \begin{pmatrix} M_{11} & M_{12} & \cdots & M_{1N} \\ M_{21} & M_{22} & \cdots & M_{2N} \\ \vdots & \vdots & \ddots & \vdots \\ M_{N1} & M_{N2} & \cdots & M_{NN} \end{pmatrix} \cdot \begin{pmatrix} x_1 \\ x_2 \\ \vdots \\ x_N \end{pmatrix}
$$

$$
= \begin{pmatrix} M_{11}x_1 + M_{12}x_2 + \cdots + M_{1N}x_N \\ M_{21}x_1 + M_{22}x_2 + \cdots + M_{2N}x_N \\ \vdots \\ M_{N1}x_1 + M_{N2}x_2 + \cdots + M_{NN}x_N \end{pmatrix}
\tag{F.8}
$$

The product of two matrices is again a matrix with the component definition

$$
(\mathbf{M} \cdot \mathbf{N})_{ik} = \sum_{j=1}^{N} M_{ij} N_{jk}
\tag{F.9}
$$

or

the (ik)-th element of $\mathbf{M} \cdot \mathbf{N}$

$\Updownarrow$

(the ith row of $\mathbf{M}$) dotted into (the kth column of $\mathbf{N}$)

The *transpose* of the matrix $\mathbf{M}$, labeled $\mathbf{M}^{\mathrm{T}}$, is obtained by "reflecting" all of its elements along the diagonal (the $i = j$ components staying fixed), that is,

$$
(\mathbf{M}^{\mathrm{T}})_{ij} = (\mathbf{M})_{ji} = M_{ji}
\tag{F.10}
$$

so that

$$
\begin{pmatrix} a & b & c \\ d & e & f \\ g & h & i \end{pmatrix}^{\mathrm{T}} = \begin{pmatrix} a & d & g \\ b & e & h \\ c & f & i \end{pmatrix}
\tag{F.11}
$$

The generalization of this to complex matrices is the *adjoint* or *Hermitian conjugate* defined via

$$
\mathbf{M}^{\dagger} = (\mathbf{M}^{\mathrm{T}})^* = (\mathbf{M}^*)^{\mathrm{T}}
\tag{F.12}
$$

which "flips" the matrix elements *and* takes their complex conjugate.

The equivalent of the expectation value of an operator in a quantum state is given by

$$
\langle x|M|x\rangle \sim \mathbf{x} \cdot \mathbf{M} \cdot \mathbf{x} = \sum_{j=1}^{N} \sum_{k=1}^{N} x_j M_{jk} x_k
\tag{F.13}
$$

A matrix transformation of the form Eqn. (F.6) generally changes the norm of the vector since

$$
\begin{aligned}
\mathbf{x}' \cdot \mathbf{x}' = \sum_{i} x_i' x_i' &= \sum_{i=1}^{N} \left(\sum_{j=1}^{N} M_{ij} x_j \right) \left(\sum_{k=1}^{N} M_{ik} x_k \right) \\
&= \sum_{j,k=1}^{N} x_j \left[\sum_{i} (M^T)_{jk} M_{ik} \right] x_k \\
&= \sum_{j,k=1}^{N} x_j P_{jk} x_k \\
&\neq \sum_{j} x_j x_j = \mathbf{x} \cdot \mathbf{x}
\end{aligned}
\tag{F.14}
$$

unless one has

$$
\sum_{i} (M^T)_{ji} M_{ik} = P_{jk} = \delta_{j,k} \quad \text{or} \quad \mathbf{M}^{\mathrm{T}} \cdot \mathbf{M} = \mathbf{P} = \mathbf{1}
\tag{F.15}
$$

Matrices satisfying Eqn. (F.15) are said to be *orthogonal*.

Finally, the *determinant* of a matrix is a number formed from the elements of the matrix via

$$
\det(\mathbf{M}) = \sum_{i_1, i_2, \dots, i_N = 1}^{N} \epsilon_{(i_1, i_2, \dots, i_N)} M_{1, i_1} M_{2, i_2} \cdots M_{N, i_N}
\tag{F.16}
$$

The *totally antisymmetric symbol*,[1] $\epsilon_{(i_1,i_2,\ldots,i_n)}$, is defined via

$$\epsilon_{(i_1,i_2,\ldots,i_N)} = \begin{cases} +1 & \text{if } (i_1, i_2, \ldots, i_N) \text{ is an} \\ & \quad \text{even permutation of}(1, 2, \ldots, N) \\ -1 & \text{if it is an odd permutation} \\ 0 & \text{otherwise} \end{cases} \tag{F.17}$$

It vanishes if any two of its indices are the same and is antisymmetric under the interchange of any pair of indices. Each term in the determinant then consists of a product of one element from each row, with appropriate signs. For example,

$$\det(\mathbf{A}) = \det \begin{pmatrix} a_{11} & a_{12} \\ a_{21} & a_{22} \end{pmatrix} = a_{11}a_{22} - a_{12}a_{21} \tag{F.18}$$

and

$$\det(\mathbf{B}) = \det \begin{pmatrix} b_{11} & b_{12} & b_{13} \\ b_{21} & b_{22} & b_{23} \\ b_{31} & b_{32} & b_{33} \end{pmatrix}$$

$$= b_{11}b_{22}b_{33} + b_{12}b_{23}b_{31} + b_{13}b_{32}b_{21} - b_{13}b_{31}b_{22}$$
$$- b_{11}b_{23}b_{32} - b_{12}b_{21}b_{33} \tag{F.19}$$

One important property of determinants is that the interchange of any two rows (or columns) gives the same value, but with an additional factor of (-1); this follows from the definition in Eqn. (F.16) and the antisymmetry of the ϵ symbol.
Equations of the form

$$\mathbf{M} \cdot \mathbf{v}_\lambda = \lambda \mathbf{v}_\lambda \tag{F.20}$$

are called *eigenvalue problems* and λ is the *eigenvalue* and $\mathbf{v}_\lambda$ the corresponding *eigenvector*. We can also write this as

$$(\mathbf{M} - \lambda \mathbf{1}) \cdot \mathbf{v} = 0 \tag{F.21}$$

or in matrix form as

$$\begin{pmatrix} M_{11} - \lambda & M_{12} & \cdots & M_{1N} \\ M_{21} & M_{22} - \lambda & \cdots & M_{2N} \\ \vdots & \vdots & \ddots & \vdots \\ M_{N1} & M_{N2} & \cdots & M_{NN} - \lambda \end{pmatrix} = 0 \tag{F.22}$$

This is equivalent to a set of N linear equations in N unknowns and the condition for a solution to exist is that

$$\det(\mathbf{M} - \lambda \mathbf{1}) = 0 \tag{F.23}$$

---

[1] It is also called the Levi-Civita symbol.

and this condition determines the allowed eigenvalues λ. Real matrices for which

$$\mathsf{M}^\mathsf{T} = \mathsf{M} \tag{F.24}$$

are called *symmetric*, while complex matrices for which

$$\mathsf{M}^\dagger = \mathsf{M} \tag{F.25}$$

are called *Hermitian* and both have the properties:

- The eigenvalues of M are real.
- The eigenvectors of M corresponding to different eigenvalues are orthogonal.

Example F.1. Eigenvalues and eigenvectors of a simple matrix

The eigenvalues of the matrix

$$\mathsf{M} = \begin{pmatrix} 23 & -36 \\ -36 & 2 \end{pmatrix} \tag{F.26}$$

are determined by the condition

$$\det \begin{pmatrix} 23 - \lambda & -36 \\ -36 & 2 - \lambda \end{pmatrix} = \lambda^2 - 25\lambda - 1250 = 0 \tag{F.27}$$

which has solutions $\lambda_1 = 50$ and $\lambda_2 = -25$. The eigenvector corresponding to λ_1 can be found by insisting that

$$\begin{pmatrix} 23 - 50 & -36 \\ -36 & 2 - 50 \end{pmatrix} \begin{pmatrix} a \\ b \end{pmatrix} = 0 \tag{F.28}$$

or

$$\mathbf{v}_1 = \begin{pmatrix} 4/5 \\ -3/5 \end{pmatrix} \tag{F.29}$$

when normalized so that $\mathbf{v}_1 \cdot \mathbf{v}_1 = 1$. One similarly finds that

$$\mathbf{v}_2 = \begin{pmatrix} 3/5 \\ 4/5 \end{pmatrix} \tag{F.30}$$

(with $\mathbf{v}_2 \cdot \mathbf{v}_2 = 1$ by construction) and we confirm that $\mathbf{v}_1 \cdot \mathbf{v}_2 = 0$.

Finally, some useful identities involving the scalar and cross-products of vectors are

$$\mathbf{A} \times (\mathbf{B} \times \mathbf{C}) = \mathbf{B}(\mathbf{C} \cdot \mathbf{A}) - \mathbf{C}(\mathbf{A} \cdot \mathbf{B}) \tag{F.31}$$

$$\mathbf{A} \cdot (\mathbf{B} \times \mathbf{C}) = \mathbf{B} \cdot (\mathbf{C} \times \mathbf{A}) = \mathbf{C} \cdot (\mathbf{A} \times \mathbf{B}) \tag{F.32}$$

$$(\mathbf{A} \times \mathbf{B}) \cdot (\mathbf{C} \times \mathbf{D}) = (\mathbf{A} \cdot \mathbf{C})(\mathbf{B} \cdot \mathbf{D}) - (\mathbf{A} \cdot \mathbf{D})(\mathbf{B} \cdot \mathbf{C}) \tag{F.33}$$

F.2 **Group Theory**

We conclude with a brief definition of a mathematical *group*. A set of elements $G = \{g_1, g_2, \ldots\}$ along with a binary operation (often called "group multiplication") denoted by $g_1 \cdot g_2$ constitutes a group if it satisfies four conditions:

1. The product of any two group elements is also a group element, that is, $g_3 = g_1 \cdot g_2$ is from G if g_1, g_2 are also; the group is closed under multiplication.
2. The group multiplication is associative, namely

$$(g_1 \cdot g_2) \cdot g_3 = g_1 \cdot (g_2 \cdot g_3) \tag{F.34}$$

3. There is a unique group element, labeled I or the identity element, which satisfies

$$I \cdot g_i = g_i \cdot I = g_i \tag{F.35}$$

for all $g_i \in G$.

4. Every group element, g_i, has a unique inverse element, labeled g_i^{-1} which satisfies

$$g_i \cdot g_i^{-1} = g_i^{-1} \cdot g_i = I \tag{F.36}$$

The set of group elements can be finite or infinite. Groups for which the multiplication gives the same answer in either order, that is, for which $g_i \cdot g_j = g_j \cdot g_i$ for every pair of group elements is called a *commutative* or *Abelian group*.

F.3 **Problems**

P.F.1. Show that $(A \cdot B \cdots Y \cdot Z)^T = Z^T \cdot Y^T \cdots B^T \cdot A^T$.

P.F.2. Show that the cross-product of two vectors can be written in the form

$$A \times B = \det \begin{pmatrix} \hat{i} & \hat{j} & \hat{k} \\ A_x & A_y & A_z \\ B_x & B_y & B_z \end{pmatrix} \tag{F.37}$$

P.F.3. Find the eigenvalues and eigenvectors of the Hermitian matrix

$$M = \begin{pmatrix} 4 & 3+2i \\ 3-2i & -5 \end{pmatrix} \tag{F.38}$$

and show explicitly that the two eigenvectors are orthogonal. Note that the dot-product of two *complex* vectors is defined via $v_1^* \cdot v_2$.

APPENDIX G

Hamiltonian Formulation of Classical Mechanics

In this appendix, we briefly review some aspects of the Hamiltonian formulation of classical mechanics. The Hamiltonian function for a single particle described by a single coordinate, call it x, is a function of x and the so-called "conjugate momentum," p_x. Examples of such coordinate pairs include

- x and $p_x = m\dot{x}$ (ordinary momentum) for translational motion and
- θ and $p_\theta = L_z = mr^2\dot{\theta}$ (angular momentum) for a rotational system.

The Hamiltonian is written as $H = H(x, p_x)$ and x and p_x are initially considered as independent variables.

The dynamical equations of motion for $x(t)$ and $p_x(t)$ are *Hamilton's equations*, namely

$$\frac{dx}{dt} = \dot{x} = \frac{\partial H}{\partial p_x} \tag{G.1}$$

$$-\frac{dp_x}{dt} = -\dot{p}_x = \frac{\partial H}{\partial x} \tag{G.2}$$

To see the equivalence to Newtonian mechanics, note that the Hamiltonian function

$$H(x, p_x) = \frac{p_x^2}{2m} + V(x) \tag{G.3}$$

gives the equations

$$\dot{x} = \frac{p_x}{m} \quad \text{and} \quad -\dot{p}_x = \frac{\partial V(x)}{\partial x} \equiv -F(x) \tag{G.4}$$

or

$$m\ddot{x} = \dot{p}_x = F(x) \tag{G.5}$$

The Hamiltonian function for the degrees of freedom of more than one particle (in one dimension) can be written

$$H = H(x_i, p_i) = \sum_i \frac{p_i^2}{2m_i} + \sum_i V_i(x_i) + \sum_{i>j} V_{ij}(x_i - x_j) \qquad \text{(G.6)}$$

with the corresponding equations

$$\frac{dx_i}{dt} = \dot{x}_i = \frac{\partial H}{\partial p_i} \qquad \text{(G.7)}$$

$$-\frac{dp_i}{dt} = -\dot{p}_i = \frac{\partial H}{\partial x_i} \qquad \text{(G.8)}$$

For two functions which depend on the coordinates of a multivariable problem (and possibly the time coordinate explicitly), $g = g(x_i, p_i; t)$ and $h = h(x_i, p_i; t)$, the *Poisson bracket* is defined via

$$[g, h] = \sum_k \left(\frac{\partial g}{\partial x_k} \frac{\partial h}{\partial p_k} - \frac{\partial g}{\partial p_k} \frac{\partial h}{\partial x_k} \right) \qquad \text{(G.9)}$$

Any such arbitrary function can depend on time either from an explicit t dependence or via the coordinates $x_i(t), p_i(t)$; a convenient way of exhibiting the time-development of a function is

$$\frac{dg}{dt} = \frac{\partial g}{\partial t} + \sum_k \left(\frac{\partial g}{\partial x_k} \frac{dx_k}{dt} + \frac{\partial g}{\partial p_k} \frac{dp_k}{dt} \right)$$

$$= \frac{\partial g}{\partial t} + \sum_k \left(\frac{\partial g}{\partial x_k} \frac{\partial H}{\partial p_i} - \frac{\partial g}{\partial p_k} \frac{\partial H}{\partial x_i} \right)$$

$$= \frac{\partial g}{\partial t} + [g, H] \qquad \text{(G.10)}$$

Note the similarity between this classical relation and Eqn. (12.88) for the time rate of change of expectation values of quantum operators.

Example G.1.

Using the Poisson bracket formalism, we can show that angular momentum is conserved for a central potential in three dimensions. The Hamiltonian function is given by

$$H = \frac{1}{2m} \left(p_x^2 + p_y^2 + p_z^2 \right) + V(r) \qquad \text{(G.11)}$$

where $r = \sqrt{x^2 + y^2 + z^2}$. We note that the force is given $\mathbf{F} = -\nabla V(r)$ so that

$$\frac{\partial V(r)}{\partial x} = (-F)\frac{x}{r} \qquad \text{(G.12)}$$

and so forth.

(Continued)

Considering, for definiteness, the z component of angular momentum given by $L_z = xp_y - yp_x$, one can show that

$$\frac{dL_z}{dt} = \frac{\partial L_z}{\partial t} + [L_z, H]$$

$$= \left[\frac{\partial L_z}{\partial x} \frac{\partial H}{\partial p_x} - \frac{\partial L_z}{\partial p_x} \frac{\partial H}{\partial x} \right] + \left[\frac{\partial L_z}{\partial y} \frac{\partial H}{\partial p_y} - \frac{\partial L_z}{\partial p_y} \frac{\partial H}{\partial y} \right] + \left[\frac{\partial L_z}{\partial z} \frac{\partial H}{\partial p_z} - \frac{\partial L_z}{\partial p_z} \frac{\partial H}{\partial z} \right]$$

$$= \left[(p_y) \left(\frac{p_x}{m} \right) - (-y) \left(-F\frac{x}{r} \right) \right] + \left[(-p_x) \left(\frac{p_y}{m} \right) - (x) \left(-F\frac{y}{r} \right) \right]$$

$$\frac{dL_z}{dt} = 0 \qquad\qquad\qquad\qquad\qquad\qquad\qquad (G.13)$$

This relation also suggests that the Poisson bracket of two functions is the classical quantity which can be generalized in quantum mechanics to the commutator of two operators

$$[\hat{g}, \hat{h}] \equiv \hat{g}\hat{h} - \hat{h}\hat{g} \qquad\qquad\qquad (G.14)$$

via

$$[g, h]_{\text{Poisson}} \longrightarrow [\hat{g}, \hat{h}] = i\hbar [g, h]_{\text{Poisson}} \qquad\qquad (G.15)$$

The Hamiltonian for a charged particle acted on by electromagnetic fields is written in terms of the potentials, $\phi(\mathbf{r}, t)$ and $\mathbf{A}(\mathbf{r}, t)$ via

$$H(\mathbf{r}, \mathbf{p}) = \frac{1}{2m} \left(\mathbf{p} - q\mathbf{A}(\mathbf{r}, t) \right)^2 + q\phi(\mathbf{r}, t) \qquad\qquad (G.16)$$

To prove this requires one to show that the corresponding Hamilton's equations reproduce Newton's laws with the Lorentz force. The classical Hamiltonian in Eqn. (G.16) can be written more explicitly as

$$H = \frac{1}{2m} \left(p_x^2 + p_y^2 + p_z^2 \right) - \frac{q}{m} \left(p_x A_x + p_y A_y + p_z A_z \right)$$

$$+ \frac{q^2}{2m} \left(A_x^2 + A_y^2 + A_z^2 \right) + q\phi \qquad\qquad (G.17)$$

Hamilton's equations for the x and p_x coordinates in this case become

$$\dot{x} = \frac{\partial H}{\partial p_x} = \frac{p_x}{m} - \frac{q}{m} A_x \quad \text{or} \quad m\dot{x} = p_x - qA_x \qquad (G.18)$$

and

$$
\begin{aligned}
-\dot{p}_x &= \frac{\partial H}{\partial x} \\
&= -\frac{q}{m}\left(p_x\frac{\partial A_x}{\partial x} + p_y\frac{\partial A_y}{\partial x} + p_z\frac{\partial A_z}{\partial x}\right) \\
&\quad + \frac{q^2}{2m}\left(A_x\frac{\partial A_x}{\partial x} + A_y\frac{\partial A_y}{\partial x} + A_z\frac{\partial A_x}{\partial x}\right) + q\frac{\partial \phi}{\partial x}
\end{aligned}
\tag{G.19}
$$

These can be combined by differentiating Eqn. (G.18) with respect to t provided one recalls that

$$
m\ddot{x} = \dot{p}_x - q\frac{dA_x}{dt} = \dot{p}_x - q\left(\frac{\partial A_x}{\partial t} + \dot{x}\frac{\partial A_x}{\partial x} + \dot{y}\frac{\partial A_x}{\partial y} + \dot{z}\frac{\partial A_x}{\partial z}\right)
\tag{G.20}
$$

since $A = A(x(t), y(t), z(t); t)$ depends on time explicitly (through the t) and implicitly (through the time-dependent positions). The resulting equation for the x variable is then

$$
m\ddot{x} = q\left(\left\{-\frac{\partial \phi}{\partial x} - \frac{\partial A_x}{\partial t}\right\} + \left\{\dot{y}\left[\frac{\partial A_y}{\partial x} - \frac{\partial A_x}{\partial y}\right] + \dot{z}\left[\frac{\partial A_z}{\partial x} - \frac{\partial A_x}{\partial z}\right]\right\}\right)
\tag{G.21}
$$

or

$$
m\ddot{x} = q\left(E_x + (\mathbf{v} \times \mathbf{B})_x\right)
\tag{G.22}
$$

since

$$
\mathbf{E} = -\nabla\phi(\mathbf{r}, r) - \frac{\partial}{\partial t}\mathbf{A}(\mathbf{r}, t) \quad \text{and} \quad \mathbf{B} = \nabla \times \mathbf{A}(\mathbf{r}, t)
\tag{G.23}
$$

The Hamiltonian formalism also provides a way to describe classical probability distributions for position and momentum which can be compared to their quantum counterparts (as discussed in Sections 5.1 and 9.4). One starts with the notion of the *classical phase space*. For one particle in one dimension, this is the space of possible values of x and p, as in Fig. G1; for N particles in three dimensions, it is a $6N$-dimensional space corresponding to the possible values of $\mathbf{r}_i$, $\mathbf{p}_i$. Given a set of initial conditions, the solutions obtained from Newton's (or Hamilton's) equations for $x(t)$ and $p(t)$ trace out a trajectory in the phase space. For example, for a harmonic oscillator with initial conditions $x(0) = A$ and $\dot{x}(0) = 0$, the solutions are obviously

$$
x(t) = A\cos(\omega t) \quad \text{and} \quad p(t) = -\frac{A\omega}{m}\sin(\omega t)
\tag{G.24}
$$

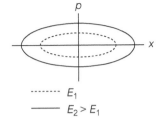

Figure G.1. Phase space diagram (plot of allowed values of p and x) for a single particle in a harmonic oscillator potential. The allowed "trajectories" in the parameter space in this case are determined by Eqn. (G.24).

which gives the elliptical path in phase space shown in Fig. G1. The form of this "trajectory" can be determined, even if we specify only the total energy, via the relation

$$E = H = \frac{p^2}{2m} + \frac{1}{2}m\omega^2 x^2 \qquad (G.25)$$

The *phase space distribution*, $\rho(x, p)$, for a given value of E can then be written in the form

$$\rho(x, p) = K\delta(E - H(p, x)) \qquad (G.26)$$

where the normalization constant is determined by the condition that

$$\int dx \int dp \, \rho(x, p) = 1 \qquad (G.27)$$

The classical probability densities for position (x) or momentum (p) can be derived by integrating over the variable which is not specified; for example,

$$P_{CL}(x) = \int dp \, \rho(x, p) \quad \text{and} \quad P_{CL}(p) = \int dx \, \rho(x, p) \qquad (G.28)$$

As an example, consider the Hamiltonian with a general potential energy function $V(x)$

$$H = \frac{p^2}{2m} + V(x) \qquad (G.29)$$

If we write $p_0(x) = \sqrt{E - V(x)}$, the corresponding classical distribution in x will be given by

$$P_{CL}(x) = K \int dp \, \delta\left(E - \left(\frac{p^2}{2m} - V(x)\right)\right)$$

$$\propto \int dp \, \delta\left(p^2 - 2m(E - V(x))\right)$$

$$\equiv \int dp \, \delta\left(p^2 - p_0^2(x)\right)$$

$$= \frac{1}{2|p_0(x)|} \int dp \left[\delta(p - p_0(x)) + \delta(p + p_0(x))\right]$$

$$\propto \frac{1}{\sqrt{E - V(x)}} \tag{G.30}$$

since $p_0^2(x) = 2m(E - V(x)$. This is the same result obtained in Sections 5.1 and 9.4.1, using more intuitive methods.

G.1 **Problems**

PG.1. Write down Hamilton's equations for the angular variable pair θ and $p_\theta = mr^2\theta$ where the Hamiltonian is

$$H = \frac{p_\theta^2}{2I} + V(\theta) \tag{G.31}$$

and show how the standard equations for rotational motion arise.

PG.2. Show that the Poisson bracket of the position and momentum coordinates of a multiparticle system satisfy

$$[x_i, x_j] = 0, \quad [p_i, p_j] = 0, \quad \text{and} \quad [x_i, p_j] = \delta_{i,j} \tag{G.32}$$

PG.3. Consider a harmonic oscillator for which the classical Hamiltonian is

$$H = \frac{p_x^2}{2m} + \frac{1}{2}m\omega^2 x^2 \tag{G.33}$$

Use Eqn. (G.10) to show that the function

$$\phi(x, p_x; t) = i\left(\log(A) - \log(x - ip/m\omega)\right) - \omega t \tag{G.34}$$

is actually independent of time for this system. What is the physical significance of this variable?

PG.4. For the classical Kepler problem, defined by the Hamiltonian

$$H = \frac{1}{2m}\mathbf{p}^2 - \frac{k}{r} \tag{G.35}$$

show that the *Lenz–Runge vector* defined by

$$\mathbf{R} = \frac{\hat{\mathbf{r}}}{r} - \left(\frac{1}{mk}\right)\mathbf{p} \times \mathbf{L} \tag{G.36}$$

is a constant of the motion, that is it is conserved. Do this by showing that $d\mathbf{R}/dt = 0$ using the Poisson bracket formalism.

PG.5. Using the Hamiltonian corresponding to a linear confining potential

$$H = \frac{p^2}{2m} + C|x| \tag{G.37}$$

and Eqn. (G.28), derive $P_{CL}(p)$.

PG.6. What does the classical phase space diagram look like for one particle in the one-dimensional infinite well, that is, what is the analog of Fig. G1? For the particle in the potential of PG.5? For an unbound free particle? For a particle subject to a constant force given by $V(x) = -Fx$? For a particle bouncing up and down on a table, with no energy loss, under the influence of gravity?

References

Physics, like the arts and humanities, is a human endeavor, and no text should fail to acknowledge the accomplishments of the men and women who discovered, codified, or have helped us understand its content. In most cases, the complete titles for journal articles are included in order to give a better "flavor" of the original paper.

[1] Abramowitz, M. and I. A. Stegun (1964). *Handbook of Mathematical Functions*, National Bureau of Standards.

[2] Adhikari, S. K. (1986). Quantum scattering in two dimensions, *Am. J. Phys.* **54**, 362.

[3] Aharonov, Y. and D. Bohm (1957). Significance of electromagnetic potentials in the quantum theory, *Phys. Rev.* **115**, 485.

[4] —— and L. Susskind (1967). Observability of the sign change of spinors under 2π rotations, *Phys. Rev.* **158**, 1237.

[5] Alber, G. and P. Zoller (1991). Laser excitation of electronic wave packets in Rydberg atoms, *Phys. Rep.* **199**, 231.

[6] Albiol, F., S. Navas, and M. V. Andres (1993). Microwave experiments on electromagnetic evanescent waves and tunneling effect, *Am. J. Phys.* **61**, 165.

[7] Andrews, M. (1998). Wave packets bouncing off walls, *Am. J. Phys.* **66**, 252.

[8] Andrews, M. R., C. G. Townsend, H. -J. Miesner, D. S. Durfee, D. M. Kurn, and W. Ketterle (1997). Observation of interference between two Bose condensates, Science **275**, 637.

[9] Arfken, G. (1985). *Mathematical methods for physicists*, Academic Press, Orlando.

[10] Arndt, M. O. Nairz, J. Vos-Andreae, C. Keller, V. van der Zouw, and A. Zeilinger (1999). Wave-particle duality of C_{60} molecules, Nature **401**, 680. See also O. Nariz, M. Arndt, and A. Zeilinger, Quantum interference experiments with large molecules, *Am. J. Phys.* **71**, 319 (2003).

[11] Balian, R. and C. Bloch (1970). Distribution of eigenfrequencies for the wave equation in a finite domain I. Three-dimensional problem with smooth boundary surface, *Ann. Phys.* **60**, 401–477.

[12] Barger, V. and M. Olsson (1995). *Classical Mechanics: A Modern Perspective*, Second Edition, McGraw-Hill, New York.

[13] Baym, G. (1976). *Lectures on quantum mechanics*, W. A. Benjamin, Reading.

[14] Belloni, M., M. A. Doncheski, and R. W. Robinett (2004). Wigner quasi-probability distribution for the infinite square well: Energy eigenstates and time-dependent wave packets, *Am. J. Phys.* **72**, 1183.

[15] —— (2005). Exact results for "bouncing" Gaussian wave packets, *Phys. Scripta* **71**, 136.

[16] Bernstein, H. J. (1967). Spin precession during interferometry of fermions and the phase factor associated with rotations through 2π radians, *Phys. Rev. Lett.* **18**, 1102.

[17] Bethe, H. A. and E. E. Salpeter (1957). *Quantum mechanics of one-and two-electron atoms*, Springer-Verlag, Berlin.

[18] —— and R. Jackiw (1986). *Intermediate quantum mechanics*, Third Edition, Benjamin Cummings, Menlo Park.

[19] Blinder, S. M. (1968). Evolution of a Gaussian wavepacket, *Am. J. Phys.* **36**, 525.

[20] Brown, L. S. (1973). Classical limit of the hydrogen atom, *Am. J. Phys.* **41**, 525.

[21] Burrows, A. (1990). Neutrinos from supernova explosions, *Ann. Rev. Nucl. Part. Sci.* **40**, 181.

[22] Butkov, E. (1968). *Mathematical physics*, Addison-Wesley, Menlo Park.

[23] Carnal, O., M. Sigel, T. Sleator, H. Takuma, and J. Mlynek (1991). Imaging and focusing of atoms by a Fresnel zone plate, *Phys. Rev. Lett.* **67**, 3231. See also O. Carnal and J. Mlynek, Young's double-slit experiment with atoms: A simple atom interferometer, *Phys. Rev. Lett.* **66**, 2689 (1991).

[24] Chambers, R. G. (1960). Shift of an electron interference pattern by enclosed magnetic flux, *Phys. Rev. Lett.* **5**, 3.

[25] Chapman, M. S., C. R. Ekstrom, T. D. Hammond, R. Rubenstein, J. Schmiedmayer, S. Wehinger, and D. E. Pritchard (1995). Optics and interferometry with Na_2 molecules, *Phys. Rev. Lett.* **74**, 4783.

[26] Chen, J. (1993). Introduction to scanning tunneling microscopy, Oxford University Press, New York.

[27] Cohen-Tannoudji, C. N. and W. D. Phillips (1990). New mechanisms for laser cooling, Phys. Today, September, 34.

[28] Clarke, A. (1972). *The wind from the sun*, Harcourt Brace Jovanovich, New York.

[29] Cocconi, Giuseppe and P. Morrison (1959). Searching for interstellar communications, Nature **184**, 844.

[30] Colella, R. and A. W. Overhauser (1980). Neutrons, gravity and quantum mechanics, *Am. Sci.* **68**, 70.

[31] ———A. W. Overhauser, and S. A. Werner (1975). Observation of gravitationally induced quantum interference, *Phys. Rev. Lett.* **34**, 1472.

[32] Condon, E. U. and G. H. Shortley (1951). *The theory of atomic spectra*, Cambridge University Press, Cambridge.

[33] Crawford, F. S. (1968). *Waves*, Volume 3 of the Berkeley Physics Series, McGraw-Hill, New York.

[34] Crawford, F. (1971). Culvert whistlers, *Am. J. Phys.* **39**, 610.

[35] ——— (1988). Culvert whistlers revisited, *Am. J. Phys.* **56**, 752.

[36] Crommie, M. F., C. P. Lutz, and D. M. Eigler (1993). Imaging standing waves in a 2-dimensional electron-gas, *Nature* **363**, 524.

[37] Dabbs, J., J. Harvey, D. Paya, and F. Horstmann (1965). Gravitational acceleration of free neutrons, *Phys. Rev.* **139**, B756.

[38] Davisson, C. and L. H. Germer (1927). Diffraction of electrons by a crystal of nickel, *Phys. Rev.* **30**, 705.

[39] Dalgarno, A. and J. T. Lewis (1955). The exact calculation of long-range forces between atoms by perturbation theory, *Proc. Roy. Soc.* **A233**, 70.

[40] DeLange, O. L. and R. E. Raab (1991). *Operator Methods in Quantum Mechanics*, Clarendon Press, Oxford.

[41] Delone, N. B. and V. P Krainov (1994). *Multiphoton Processes in Atoms*, Springer-Verlag, Berlin.

[42] Dobrzynski, L. and K. Blinowski (1994). *Neutrons and Solid State Physics*, Ellis Horwood (Simon and Schuster), New York.

[43] Doncheski, M. A. and R. W. Robinett (1999). Anatomy of a quantum "bounce," *Eur. J. Phys.* **20**, 29.

[44] —— (2000). Comparing classical and quantum probability distributions for an asymmetric infinite well, *Eur. J. Phys.* **21**, 217.

[45] —— S. Heppelmann, R. W. Robinett, and D. C. Tussey (2003). Wave packet construction in two-dimensional quantum billiards: Blueprints for the square, equilateral triangle, and circular cases, *Am. J. Phys.* **71**, 541.

[46] Eberly, J. H. (1965). Quantum scattering theory in one-dimension, *Am. J. Phys.* **33**, 771.

[47] Ehrenberg, H. F., R. Hofstadter, U. Meyer-Berkhour, D. G. Ravenhhall, and S. E. Sobotten (1959). High-energy electron scattering and the charge distribution of Carbon-12 and Oxygen-16, *Phys. Rev.* **113**, 666.

[48] E. Elizalde, and A. Romeo (1991). Essentials of the Casimir effect and its computation, *Am. J. Phys.* **59**, 711.

[49] Eisberg, R. and R. Resnick (assisted by David Caldwell and J. Richard Christman) (1985). *Quantum physics of atoms, molecules, solids, nuclei, and particles*, Second Edition, Wiley, New York.

[50] Epstein, S. (1960). Application of the Rayleigh–Schrödinger perturbation theory to the delta-function potential, *Am. J. Phys.* **28**, 495.

[51] Fabre, C., M. Gross, J. M. Raimond, and S. Haroche (1983). Measuring atomic dimensions by transmission of Rydberg atoms through micrometre size slits, *J. Phys. B: At. Mol. Phys* **16**, L671.

[52] Feynman, R. P. (1963). *The Feynman lectures on physics*, Addison-Wesley, Reading, MA.

[53] Fowles, G. R. and G. L. Cassiday (1999). *Analytical mechanics*, Harcourt Brace, Fort Worth.

[54] Freeman, R. R. and D. Kleppner (1974). Core polarization and quantum defects in high-angular momentum states of alkali atoms, *Phys. Rev.* **A14**, 1614.

[55] French, A. P. and E. F. Taylor (1971). Qualitative plots of bound state wavefunctions, *Am. J. Phys.* **39**, 961.

[56] Gallagher, T. F. (1994). *Rydberg atoms*, Cambridge University Press, Cambridge.

[57] —— K. A. Safinya, F. Gounand, J. F. Delpech, W. Sandner, and R. Kachru (1982). Resonant Rydberg-atom Rydberg atom collisions, *Phys. Rev.* **A25**, 1905.

[58] Gähler, R. and A. Zeilinger (1991). Wave-optical experiments with very cold neutrons, *Am. J. Phys.* **59**, 316.

[59] Gamow, G. (1946). *Mr. Tompkins in Wonderland; Stories of c, G, and h*, MacMillan, New York. Reprinted in *Mr. Tompkins in paperback*, Cambridge University Press, Cambridge in 1965.

[60] Gasiorowicz, S. (1996). *Quantum physics*, Second Edition, Wiley, New York.

[61] Gibble, K. and S. Chu (1993). Laser-cooled Cs frequency standard and a measurement of the frequency shift due to ultracold collisions, *Phys. Rev. Lett.* **70**, 1771.

[62] Gilbert, L., M. Belloni, M. A. Doncheski, and R. W. Robinett (2005). More on the asymmetric infinite square well: Energy eigenstates with zero curvature, *Eur. J. Phys.* **26**, 815.

[63] Griesen, K. (1966). End to the cosmic-ray spectrum? *Phys. Rev. Lett.* **16**, 748.

[64] Griffiths, D. (1987). *Introduction to elementary particles*, Harper and Row, Cambridge.

[65] —— (1998). *Introduction to electrodynamics*, Third Edition, Prentice Hall, Englewood Cliffs.

[66] Gutzwiller, M. C. (1990). *Chaos in classical and quantum mechanics*, Springer-Verlag, New York.

[67] Heller, E. J. and S. Tomsovic (1993). Postmodern quantum mechanics, Phys. Today, July, 38.

[68] Hess, G. B. and W. M. Fairbank (1967). Measurements of angular momentum in superfluid helium, *Phys. Rev. Lett.* **19**, 216.

[69] Heydenburg, N. P. and G. M. Temmer (1956). Alpha-alpha scattering at low energies, *Phys. Rev.* **104**, 123.

[70] Hüfner, J., F. Scheck, and C. S. Wu (1977). In *Muon physics*, V. Hughes and C. S. Wu, eds. Academic Press, New York.

[71] Hylleraas, E. A. (1928). The fundamental state of the helium atom, *Z. Phys.* **48**, 469; Calculation of the energy of helium, *Z. Phys.* **54**, 347 (1929); The ground term of the two-electron problem: H^-, H, Li^+, Be^{++}, etc., *Z. Phys.* **65**, 209 (1930).

[72] Johnson, A., H. Ryde, and S. A. Hjorth (1972). Nuclear moment of inertia at high rotational frequencies, *Nucl. Phys.* **A179**, 753.

[73] Jones, W. (1980). Earnshaw's theorem and the stability of matter, *Eur. J. Phys.* **1**, 85.

[74] Jones, S. E., J. Rafelski, and H. Monkhorst (1989). *Muon-catalyzed fusion*, American Institute of Physics, New York.

[75] Kac, M. (1966). Can you hear the shape of a drum?, *Am. Math. Monthly* **73**, 1.

[76] Keith, D. W., C. R. Ekstrom, Q. A. Turchette, and D. E. Pritchard (1991). An interferometer for atoms, *Phys. Rev. Lett.* **66**, 2693.

[77] Kittel, C. (1971). *Introduction to solid state physics*, 4th Edition, Wiley, New York.

[78] Kleber, M. (1994). Exact solutions for time-dependent phenomena in quantum mechanics, *Phys. Rep.* **236**, 331.

[79] Kleppner, D. (1986). An introduction to Rydberg atoms In *Atoms in unusual situations*, J. P. Briand, ed. Plenum Press, New York.

[80] Krane, K. (1988). *Introductory nuclear physics*, Wiley, New York.

[81] Kuiper, T. B. H. (1989). Resource Letter ETC-1: Extraterrestrial civilizations, *Am. J. Phys.* **57**, 12.

[82] Kronig, R. L. and W. B. Penney (1931). Quantum mechanics of electrons in crystal lattices, *Proc. Roy. Soc.* (London), **130A**, 499.

[83] Landau, L. D. and E. M. Lifschitz (1965). *Quantum mechanics: Non-relativistic theory*, Oxford, New York.

[84] Landé, A. (1951). *Quantum mechanics*, Pitman, London.

[85] Lapidus, I. R. (1970). One-dimensional model of a diatomic ion, *Am. J. Phys.* **38**, 905.

[86] —— (1982a). Quantum-mechanical scattering in two dimensions, *Am. J. Phys.* **50**, 45.

[87] —— (1982b). One-dimensional hydrogen atom in an infinite square well, *Am. J. Phys.* **50**, 563.

[88] —— (1982c). Resonance scattering from a double δ-function potential, *Am. J. Phys.* **50**, 663.

[89] —— (1987). Particle in a square well with a δ-function potential, *Am. J. Phys.* **55**, 172.

[90] Laughlin, R. B. (1983). Anomalous quantum Hall effect: An incompressible quantum fluid with fractionally charged excitations, *Phys. Rev. Lett.* **50**, 1395.

[91] Lieb, E. (1976). The stability of matter, *Rev. Mod. Phys.* **48**, 553.

[92] Lieber, M. (1975). Quantum mechanics in momentum space: an illustration, *Am. J. Phys.* **43**, 486.

[93] Lighthill, M. J. (1958). *Introduction to fourier analysis and generalised functions*, Cambridge University Press, Cambridge.

[94] Lohmann, B. and E. Weigold (1981). Direct measurement of the electron momentum probability distribution in atomic hydrogen, *Phys. Lett.* **86A**, 139.

[95] Marion, J. B. and S. T. Thornton (2004). *Classical dynamics of particles and systems*, 5th Edition, Brooks Cole, Belmont, CA.

[96] Mathews, J. and R. L. Walker (1970). *Mathematical methods of physics*, Second Edition, Benjamin, Menlo Park.

[97] Meekhof, D. M., C. Monroe, B. E. King, W. M, Itano, and D. J. Wineland (1996). Generation of nonclassical motional states of a trapped atom, *Phys. Rev. Lett.* **76**, 1796.

[98] Merli, P. G., G. F. Missiroli, and G. Pozzi (1976). On the statistical aspect of electron interference phenomena. *Am. J. Phys.* **44**, 306.

[99] Migdal, A. B. and V. P. Krainov (1969). *Approximation methods in quantum mechanics*, Benjamin, New York.

[100] Millikan, R. A. (1916). A direct photoelectric determination of Planck's "h," *Phys. Rev.* **7**, 355.

[101] Millikan, R. A. and C. F. Eyring (1926). Laws governing the pulling of electrons out of metals by intense electrical fields, *Phys. Rev.* **27**, 51.

[102] Milburn, R. H. (1963). Electron scattering by an intense polarized photon field, *Phys. Rev. Lett.* **10**, 75.

[103] Müller, A. T. T. Tsong (1969). *Field Ion microscopy: principles and applications*, American Elsevier, New York.

[104] Nauenberg, M. (1989). Quantum wave packets on Kepler elliptic orbits, *Phys. Rev.* **A40**, 1133.

[105] Nesvizhevsky, V. V. *et al.* (2002). Quantum states of neutrons in the Earth's gravitational field, *Nature* **415**, 297.

[106] Nussenzveig, H. M. (1992). *Diffraction effects in semiclassical scattering*, Cambridge University Press, New York.

[107] Ohanian, H. C. (1990). *Principles of quantum mechanics*, Prentice-Hall, Englewood Cliffs.

[108] Park, D. (1992). *Introduction to the quantum theory*, McGraw-Hill, New York.

[109] Partain, C. L., R. R. Price, J. A. Patton, M. V. Kulkarni, and A. E. James (1988). *Magnetic resonance imaging*, W. B. Saunders Company, Philadelphia.

[110] Review of Particle Properties (2004). Particle Data Group, *Phys. Lett.* **B 592**, 1.

[111] Perkins, D. H. (2000). *Introduction to high energy physics*, Fourth Edition, Cambridge University Press, Cambridge.

[112] Peierls, R. E. (1955). *The quantum theory of solids*, Clarendon Press, Oxford.

[113] Pippard, A. B. Pippard (1978). *The physics of vibration*, Vol. 1, Cambridge University Press, Cambridge.

[114] —— (1983). *The physics of vibration*, Vol. 2, Cambridge University Press.

[115] Press, W. H., B. P. Flannery, S. A Teukolsky, and W. T. Vetterling (2002). *Numerical recipe in C++: The art of scientific computing*, Second Edition, Cambridge University Press, Cambridge.

[116] Prodan, J., A. Migdall, W. D. Phillips, I. So, H. Metcalf, and J. Dalibard (1985). Stopping atoms with laser light, *Phys. Rev. Lett.* **54**, 992.

[117] Ramberg, E. and G. A. Snow (1990). Experimental limit on a small violation of the Pauli principle, *Phys. Lett.* **B238**, 438.

[118] Rauch, H., Z. Zeilinger, G. Badurek, A. Wilfing, W. Bauspiess, and U. Bonse (1975). Verification of coherent spinor rotation of fermions, *Phys. Lett.* **54A**, 425.

[119] Reif, F. (1965). *Fundamentals of statistical and thermal physics*, McGraw-Hill, New York.

[120] Reitz, J. R., F. J. Milford, and R. W. Christy (1993). *Foundations of electromagnetic theory*. Addison-Wesley, New York.

[121] Robinett, R. W. (1995). Quantum and classical probability distributions for position and momentum, *Am. J. Phys.* **63**, 823.

[122] —— (1996a). Visualizing the solutions for the circular infinite well in quantum and classical mechanics, *Am. J. Phys.* **64**, 440.

[123] —— (1996b). Quantum mechanical time development operator for the uniformly accelerated particle, *Am. J. Phys.* **64**, 803.

[124] —— (2001). Spacecraft artifacts as physics teaching resources, *Phys. Teach.* **39**, 476.

[125] —— (2004). Quantum wave packet revivals, *Phys. Rep.* **392**, 1.

[126] Rowe, E. G. P. (1987). The classical limit of quantum mechanical hydrogenic radial distributions, *Eur. J. Phys.* **8**, 81.

[127] Roy, C. L. and A. B. Sannigrahi (1979). Uncertainty relation between angular momentum and angle variable, *Am. J. Phys.* **47**, 965.

[128] Ruffa, A. R. (1973). Continuum wave functions in the calculation of sums involving off-diagonal matrix elements, *Am. J. Phys.* **41**, 234.

[129] Salis, G., B. Graf, K. Ensslin, K. Campman, K. Maranowski, and A. C. Gossard, (1997). Wave function spectroscopy in quantum wells with tunable electron density, *Phys. Rev. Lett.* **79**, 5106.

[130] Sakurai, J. J. (1994). *Modern quantum mechanics* (Revised Edition), Addison-Wesley, Reading.

[131] Saxon, D, S, (1968). *Introductory quantum mechanics*, McGraw-Hill, New York.

[132] Schiff, L. I. (1968). *Quantum mechanics*, McGraw-Hill, New York.

[133] Schleich, W. P. (2001). *Quantum optics in phase space*, Wiley-VCH, Berlin.

[134] Schollkopf, W. and J. Toennies (1994). Nondestructive mass selection of small van der Waals clusters, Science **266**, 1345.

[135] Scully, M. O. and M. S. Zubairy (1997). *Quantum optics*, Cambridge, Cambridge.

[136] Segre, C. U., and J. D. Sullivan (1976). Bound-state wave packets, *Am. J. Phys.* **44**, 729.

[137] Shapiro, S., and S. Teukolsky (1983). *Black holes, white dwarfs, and neutron stars*, Wiley, New York.

[138] Shayegan, M., T. Sajoto, M. Santos, and C. Silvestre (1988). Realization of a quasi-three-dimensional modulation-doped semiconductor structure, *Appl. Phys. Lett.* **53**, 791.

[139] Staelin, D. H. and E. C. Reifenstein III (1968). Pulsating radio sources near the Crab Nebula, Science **162**, 1481

[140] Staudenmann, J., S. Werner, R. Colella, and A. W. Overhauser (1980). Gravity and inertia in quantum mechanics, *Phys. Rev.* **A21**, 1419.

[141] Stebbings, R. F. and F. B. Dunning (1983). *Rydberg states of atoms and molecules*, Cambridge University Press, Cambridge.

[142] Stroscio, J. A. and D. M. Eigler (1991). Atomic and molecular manipulation with the scanning tunneling microscope, Science **254**, 1319.

[143] Stroscio, J. A. and W. J. Kaiser (editors) (1993). *Scanning tunneling microscopy*, Academic Press, Boston.

[144] Styer, D. F. (1990). The motion of wave packets through their expectation values and uncertainties, *Am. J. Phys.* **58**, 742.

[145] Styer, D. F. (2001). Quantum revivals versus classical periodicity in the infinite square well, *Am. J. Phys.* **69**, 56.

[146] Sundaram, M., A. C. Gossard, J. H. English, and R. M. Westervelt (1988). Remotely-doped graded potential well structures, *Superlattices and Microstructures* **4**, 683.

[147] Swift, J. (1726). *Travels into several remote nations of the world, in four parts, by Lemuel Gulliver*, Printed for Benj. Motte, London. For a modern copy, using a later edition, see *The annotated Gulliver's travels* (1980), edited by Isaac Asimov, Clarkson Potter, New York.

[148] Symon, K. (1971). *Mechanics*, Third Edition, Addison-Wesley, Reading.

[149] Tsong, T. T. (1990). *Atom probe field ion microscopy*, Cambridge University Press, Cambridge.

[150] Urey, H. C., F. G. Brickwedde, and G. Murphy (1932). A Hydrogen isotope of mass 2 and its concentration, Phys. Rev. **40**, 1.

[151] Van Vleck, J. H. (1922). The dilemma of the helium atom, *Phil. Mag.* **44**, 842.

[152] von Klitzing, K., G. Dorda, and M. Pepper (1980). New method for high accuracy determination of the fine structure constant based on quantized Hall Resistance, *Phys. Rev. Lett.* **45**, 494.

[153] von Klitzing, K. (1987). The quantum Hall effect, in *The physics of the two-dimensional electron gas*, J. T. Devreese and F. M. Peeters, eds. Plenum Press, New York.

[154] Walker, F. W., D. Miller, and F. Feiner (1983). Chart of the Nuclides, 13th Edition, Knolls Atomic Power Laboratory (for the Department of Energy), General Electric Company, San Jose.

[155] Wals, J. *et al.* (1994). Observation of Rydberg wave packet dynamics in a Coulombic and magnetic field, *Phys. Rev. Lett.* **72**, 3783.

[156] Weinberg, S. (1972). *Gravitation and cosmology*, Wiley, New York.

[157] Werner, S. A., R. Collela, A. W. Overhauser, and C. F. Eagen (1975). Observation of the phase shift of a neutron due to precession in a magnetic field, *Phys. Rev. Lett.* **35**, 1053.

[158] Wigner, E. (1932). On the quantum correction for thermodynamic equilibrium, *Phys. Rev.* **40**, 749.

[159] Wright, J. (1992). *Space sailing*, Gordon and Breach, Langhorn, Pennsylvania.

[160] Yeazell, J. A., M. Mallalieu, J. Parker, and C. R. Stroud, Jr. (1989). Classical periodic motion of atomic-electron wave packets, *Phys. Rev.* **A40**, 5040.

[161] Yeazell, J. A., M. Mallalieu, and C. R. Stroud, Jr. (1990). Observation of the collapse and revival of a Rydberg electronic wave packet, *Phys. Rev. Lett.* **64**, 2007.

[162] Ying, X., K. Karraï, H. D. Drew, M. Santos, and M. Shayegan (1992). Collective cyclotron resonance of an inhomogeneous electron gas, *Phys. Rev.* **B46**, 1823.

[163] Zeilinger, A., R. Gähler, C. G. Shull, W. Treimer, and W. Mampe (1988). Single- and double-slit diffraction of neutrons, *Rev. Mod. Phys.* **60**, 1067.

[164] Zimmerman, M. L., J. C. Castro, and D. Kleppner (1978). Diamagnetic structure of Na Rydberg states, *Phys. Rev. Lett.* **40**, 1083.

[165] —— M. G. Littman, M. M. Kash, and D. Kleppner (1979). Stark structure of the Rydberg states of alkali-metal atoms, *Phys. Rev.* **A20**, 2251.

Index

Math identities and other results

Gaussian integrals:

$$I^{(0)}(a, b) \equiv \int_{-\infty}^{+\infty} e^{-ax^2 - bx}\, dx = \sqrt{\pi/a}\,\exp(b^2/4a)$$

$$I^{(1)}(a, b) \equiv \int_{-\infty}^{+\infty} x e^{-ax^2 - bx}\, dx = (-b/2a)\sqrt{\pi/a}\,\exp(b^2/4a)$$

$$I^{(2)}(a, b) \equiv \int_{-\infty}^{+\infty} x^2 e^{-ax^2 - bx}\, dx = [(b^2 + 2a)/4a^2]\sqrt{\pi/a}\,\exp(b^2/4a)$$

Other integrals:

$$\int \sin^2(ax)\, dx = x/2 - \sin(2ax)/4a$$

$$\int \cos^2(ax)\, dx = x/2 + \sin(2ax)/4a$$

$$\int \sin(mx)\sin(nx)\, dx = +\sin[(m-n)x]/2(m-n) - \sin[(m+n)x]/2(m+n)$$

$$\int \cos(mx)\cos(nx)\, dx = +\sin[(m-n)x]/2(m-n) + \sin[(m+n)x]/2(m+n)$$

$$\int \sin(mx)\cos(nx)\, dx = -\cos[(m-n)x]/2(m-n) - \cos[(m+n)x]/2(m+n)$$

Trig identities:

$$\sin(a \pm b) = \sin(a)\cos(b) \pm \cos(a)\sin(b)$$
$$\cos(a \pm b) = \cos(a)\cos(b) \mp \sin(a)\sin(b)$$
$$\sin(a)\sin(b) = \cos(a-b)/2 - \cos(a+b)/2$$
$$\cos(a)\cos(b) = \cos(a-b)/2 + \cos(a+b)/2$$
$$\sin(a)\cos(b) = \sin(a-b)/2 + \sin(a+b)/2$$
$$\cos(a)\sin(b) = \sin(a+b)/2 - \sin(a-b)/2$$
$$\exp(\pm iz) = \cos(z) \pm i\sin(z)$$

Free-particle Gaussian wave packets (Chapters 3 and 4):

$$\phi_{(G)}(p, t) = \phi_0(p) e^{-ip^2 t/2m\hbar} = \left[\sqrt{\frac{\alpha}{\sqrt{\pi}}}\, e^{-\alpha^2(p-p_0)^2/2}\, e^{-ipx_0/\hbar}\right] e^{-ip^2 t/2m\hbar}$$

$$\psi_{(G)}(x, t) = \frac{1}{\sqrt{\sqrt{\pi}\alpha\hbar(1 + it/t_0)}}\, e^{ip_0(x-x_0)/\hbar}\, e^{-ip_0^2 t/2m\hbar}$$
$$e^{-(x-x_0-p_0 t/m)^2/2(\alpha\hbar)^2(1+it/t_0)}$$

$$|\psi_{(G)}(x, t)|^2 = \frac{1}{\sqrt{\pi}\beta_t}\, e^{-(x-x_0-p_0 t/m)^2/\beta_t^2}$$

$$\langle p \rangle_t = p_0, \quad \langle p^2 \rangle_t = p_0^2 + \frac{1}{2\alpha^2}, \quad \text{and} \quad \Delta p_t = \Delta p_0 = \frac{1}{\alpha\sqrt{2}}$$

$$\langle x \rangle_t = x(t) \equiv x_0 + p_0 t/m, \quad \langle x^2 \rangle_t = [x(t)]^2 + \frac{\beta_t^2}{2}, \quad \text{and} \quad \Delta x_t = \frac{\beta_t}{\sqrt{2}}$$

where

$$\beta_t \equiv \alpha\hbar\sqrt{1 + t^2/t_0^2} \quad \text{and} \quad t_0 \equiv m\hbar\alpha^2 = \frac{m\hbar}{2(\Delta p_0)^2} = \frac{2m(\Delta x_0)^2}{\hbar}$$

Printed in the USA/Agawam, MA
August 20, 2020

760109.007